T0180710

Lecture Notes in Electrical Engineering

Volume 459

About this Series

"Lecture Notes in Electrical Engineering (LNEE)" is a book series which reports the latest research and developments in Electrical Engineering, namely:

- Communication, Networks, and Information Theory
- Computer Engineering
- Signal, Image, Speech and Information Processing
- Circuits and Systems
- Bioengineering

LNEE publishes authored monographs and contributed volumes which present cutting edge research information as well as new perspectives on classical fields, while maintaining Springer's high standards of academic excellence. Also considered for publication are lecture materials, proceedings, and other related materials of exceptionally high quality and interest. The subject matter should be original and timely, reporting the latest research and developments in all areas of electrical engineering.

The audience for the books in LNEE consists of advanced level students, researchers, and industry professionals working at the forefront of their fields. Much like Springer's other Lecture Notes series, LNEE will be distributed through Springer's print and electronic publishing channels.

More information about this series at http://www.springer.com/series/7818

Yingmin Jia · Junping Du
Weicun Zhang
Editors

Proceedings of 2017 Chinese Intelligent Systems Conference

Volume I

Editors
Yingmin Jia
Beihang University
Beijing
China

Junping Du
Beijing University of Posts
and Telecommunications
Beijing
China

Weicun Zhang
University of Science and
Technology Beijing
Beijing
China

ISSN 1876-1100 ISSN 1876-1119 (electronic)
Lecture Notes in Electrical Engineering
ISBN 978-981-13-4891-4 ISBN 978-981-10-6496-8 (eBook)
https://doi.org/10.1007/978-981-10-6496-8

Printed on acid-free paper

This Springer imprint is published by Springer Nature
The registered company is Springer Nature Singapore Pte Ltd.
The registered company address is: 152 Beach Road, #21-01/04 Gateway East, Singapore 189721, Singapore

Contents

Recursive State Estimation for Discrete-Time Nonlinear Complex Networks

Jian Sun and Wenling Li

Abstract This paper studies the state estimation problem for a class of discrete-time nonlinear complex networks. The purpose is to design a recursive state estimator by using the variance-constrained approach such that the variance of the estimation error is not more than the prescribed upper bound. By adopting the structure of the extended Kalman filter (EKF), the gain matrix is determined by minimizing the trace of the prescribed upper bound matrix. It is shown that the estimator can be developed by solving two Riccati-like difference equations. A numerical example is provided to illustrate the effectiveness of the proposed estimator.

Keywords Extend Kalman filter · Complex network · Variance-constrained

1 Introduction

Complex networks have received considerable attention in the past years due to their potential applications in many fields such as brain structures, social interactions, and so on. In order to further understand the dynamical behaviors of complex networks, the state estimation problem has been a significant topic because it helps understand the intrinsic structure of the networks.

Because of the large scale of complex networks with complicated connections between nodes, the state estimation problem of complex networks becomes more challenge than isolated systems. In order to overcome the challenge, many strategies have been proposed to develop estimators for complex networks [1–9]. For example,

J. Sun · W. Li (✉)
The Seventh Research Division and the Center for Information and Control,
Beihang University (BUAA), Beijing 100191, China
e-mail: lwlmath@buaa.edu.cn

J. Sun
e-mail: sy1503104@buaa.edu.cn

Y. Jia et al. (eds.), *Proceedings of 2017 Chinese Intelligent Systems Conference*, Lecture Notes in Electrical Engineering 459,
https://doi.org/10.1007/978-981-10-6496-8_1

In [2], an H_∞ state estimator has been developed for a class of complex networks with uncertain coupling strength and incomplete measurements. In [8], the state estimation problem has been studied for a class of coupled outputs discrete-time networks with stochastic measurements. Recently, the variance-constrained approach has been used to develop a recursive state estimator for complex networks with missing measurements in [6]. As shown in [6], the specific idea of the variance-constrained approach is to guarantee an optimized upper bound on the state estimation error covariance and the gain matrix is determined with respect to the trace of the upper bound matrix. It should be pointed out that the drawback of the estimator is that all the states of the nodes are formulated into an augmented vector and all the gain matrices for the nodes are determined simultaneously according to the augmented vector. Thus, the computational cost might be very high for complex networks with a large number of nodes.

In this paper, we attempt to develop a recursive state estimator for discrete-time nonlinear complex networks. Unlike the augmented approach in [6], the non-augmented approach is adopted such that the variance of the estimation error for each node is not more than the prescribed upper bound and the gain matrix can be determined by minimizing the trace of the prescribed upper bound matrix. The advantage of the proposed estimator is that the computational cost is lower than the augmented estimator for all nodes. Finally, a numerical example involving localization of four interacting mobile robots is provided to verify the effectiveness of the proposed estimator.

The rest of this paper is organized as follow. The state estimation problem for nonlinear complex networks is formulated in Sect. 2. A state estimator is developed based on EKF for nonlinear complex networks is shown under certain conditions. In Sect. 4, simulations results are provided to illustrate the effectiveness of the proposed estimator. Conclusion is drawn in Sect. 5.

2 Problem Formulation and Preliminaries

We consider the following complex network with N nodes

$$x_{i,k+1} = f(x_{i,k}) + \sum_{j=1}^{N} a_{ij}\Gamma(x_{j,k} - x_{i,k}) + B_{i,k}w_{i,k} \tag{1}$$

$$y_{i,k} = h(x_{i,k}) + v_{i,k} \tag{2}$$

where k is the time instant and i is the index of node. $x_{i,k+1} \in \mathbb{R}^n$ is the state vector of the i-th node, $y_{i,k} \in \mathbb{R}^m$ is the measurement output, $f(x_{i,k})$ and $h(x_{i,k})$ are the nonlinear functions. $w_{i,k}$ and $v_{i,k}$ are independent process noise and measurement noise with covariances $Q_{i,k}$ and $R_{i,k}$, respectively. $B_{i,k}$ is bounded matrices. $\Gamma = \text{diag}\{\gamma_1, \gamma_2, ..., \gamma_n\}$ is a matrix linking the j-th node state, and $\mathrm{W} = (a_{ij})$ is the coupling configuration matrix of the network with $a_{ij} \geq 0 (i \neq j)$ but not all zero.

In this paper, the recursive estimator to be designed is of the following form:

$$\hat{x}_{i,k+1|k} = f(\hat{x}_{i,k|k}) + \sum_{j=1}^{N} a_{ij}\Gamma(\hat{x}_{j,k|k} - \hat{x}_{i,k|k}) \tag{3}$$

$$\hat{x}_{i,k+1|k+1} = \hat{x}_{i,k+1|k} + K_{i,k+1}\left[y_{i,k+1} - h(\hat{x}_{i,k+1|k})\right] \tag{4}$$

where $\hat{x}_{i,k|k}$ is the estimation of $x_{i,k}$ at time k, $\hat{x}_{i,k+1|k}$ is the one step prediction at time k, $K_{i,k+1}$ is the filter gain to be determined. As in the EKF, the one step prediction and the updated estimation errors are defined as

$$e_{i,k+1|k} = x_{i,k+1} - \hat{x}_{i,k+1|k}, \quad e_{i,k+1|k+1} = x_{i,k+1} - \hat{x}_{i,k+1|k+1} \tag{5}$$

According to the definition of the estimation errors, the corresponding estimation error covariance matrices can be defined as

$$P_{i,k+1|k} = \mathbb{E}\{e_{i,k+1|k}e_{i,k+1|k}^T\}, \quad P_{i,k+1|k+1} = \mathbb{E}\{e_{i,k+1|k+1}e_{i,k+1|k+1}^T\} \tag{6}$$

The aim of this paper is to design an recursive state estimator of form (3)–(4) and there exists a sequence of positive-definite matrices $\Psi_{i,k+1|k}$ and $\Psi_{i,k+1|k+1}$ satisfying

$$P_{i,k+1|k} \leq \Psi_{i,k+1|k}, \quad P_{i,k+1|k+1} \leq \Psi_{i,k+1|k+1} \tag{7}$$

The gain matrix $K_{i,k+1}$ can be determined by minimizing the trace of the upper bound matrix $\Psi_{i,k+1|k+1}$ for each node.

Remark 1 In [6], an augmented upper bound matrix is provided for all the updated estimation error covariance, i.e., $P_{k+1|k+1} \leq \Psi_{k+1|k+1}$ and $e_{k+1|k+1} = [e_{1,k+1|k+1}^T, \cdots, e_{N,k+1|k+1}^T]^T$. It can be seen that the size of the upper bound matrix is $Nn \times Nn$ which becomes larger as the dimension of the state vector and the number of nodes increases. Compared with [6], as shown in (7), the non-augmented vector approach is adopted to design the state estimator, which facilitates the development of distributed algorithms.

The following lemmas are important for deriving the gain matrix.

Lemma 1 *Given compatible dimensions matrices A, H, E, and F, $FF^T \leq I$, let X be a symmetric positive definite matrix, and let γ be an arbitrary constant such that $\gamma^{-1}I - EXE^T > 0$. Then, the following matrix inequality holds:*

$$(A + HFE)X(A + HFE)^T \leq A(X^{-1} - \gamma E^T E)^{-1}A^T + \gamma^{-1}HH^T$$

Lemma 2 *For $0 \leq k \leq n$, suppose that $X = X^T > 0$, $S_k(X) = S_k^T(X) \in \mathbb{R}^{L\times L}$, and $G_k(X) = G_k^T(X) \in \mathbb{R}^{L\times L}$, if there exists $V = V^T > X$ such that $S_k(X) \geq S_k(V)$ and $G_k(X) \geq S_k(X)$ then the solutions N_k and M_k to the following difference equations*

$$N_k = S_k(N_{k-1}), \quad M_k = G_k(M_{k-1}), \quad N_0 = M_0 > 0$$

satisfy $N_k \leq M_k$.

3 Main Results

In this section, the recursive filter is developed as follows. According to (1), (3) and (5), we have

$$e_{i,k+1|k} = f(x_{i,k}) - f(\hat{x}_{i,k|k}) + \sum_{j=1}^{N} a_{ij}\Gamma(e_{j,k|k} - e_{i,k|k}) + B_{i,k}w_{i,k} \tag{8}$$

By utilizing the Taylor series expansion around $\hat{x}_{i,k|k}$, the nonlinear function $f(x_{i,k})$ can be linearized as follows:

$$f(x_{i,k}) = f(\hat{x}_{i,k|k}) + F_{i,k}e_{i,k|k} + o(|e_{i,k|k}|) \tag{9}$$

where $F_{i,k} = \frac{\partial f(x_{i,k})}{\partial x_{i,k}}|_{x_{i,k}=\hat{x}_{i,k|k}}$ is the Jacobian matrix and $o(|e_{i,k|k}|) = U_{i,k}W_{i,k}e_{i,k|k}$ represents the high-order terms of the Taylor series expansion, where $U_{i,k}$ is the problem dependent scaling matrix and $W_{i,k}$ is the unknown time-varying matrix representing for the linearization errors satisfying $W_{i,k}W_{i,k}^T \leq I$ [10–19].

Substituting (9) into (8) yields

$$e_{i,k+1|k} = \sum_{j=1}^{N} \left(F_{i,k} + U_{i,k}W_{i,k} - a_{ij}\Gamma\right) e_{i,k|k} + \sum_{j=1}^{N} a_{ij}\Gamma e_{j,k|k} + B_{i,k}w_{i,k} \tag{10}$$

Similarly, the update estimation error can be rewritten as follows:

$$e_{i,k+1|k+1} = \left[I - K_{i,k+1}\left(H_{i,k+1} + V_{i,k}M_{i,k}\right)\right] e_{i,k+1|k} - K_{i,k+1}v_{i,k+1} \tag{11}$$

where the Taylor series expansion is adopted to address the nonlinear function

$$h(x_{i,k}) = h(\hat{x}_{i,k+1|k}) + H_{i,k+1}e_{i,k+1|k} + o(|e_{i,k+1|k}|) \tag{12}$$

with $H_{i,k+1} = \frac{\partial h(x_{i,k})}{\partial x_{i,k}}|_{x_{i,k}=\hat{x}_{i,k+1|k}}$ being the Jacobian matrix. The high-order terms can be represented by $o(|e_{i,k+1|k}|) = V_{i,k}M_{i,k}e_{i,k+1|k}$ where $V_{i,k}$ is the problem dependent scaling matrix and $M_{i,k}$ is the unknown time-varying matrix representing for the linearization errors satisfying $M_{i,k}M_{i,k}^T \leq I$. According to (6), the one step prediction error covariance can be derived as

$$P_{i,k+1|k} = \sum_{j=1}^{N}\sum_{l=1}^{N} \left(\mathrm{F}_{i,k} + U_{i,k}W_{i,k} - a_{ij}\Gamma\right) P_{i,k|k}\left(F_{i,k} + U_{i,k}W_{i,k} - a_{il}\Gamma\right)^T$$
$$+ \sum_{j=1}^{N}\sum_{l=1}^{N} \mathbb{E}\left[\left(\mathrm{F}_{i,k} + U_{i,k}W_{i,k} - a_{ij}\Gamma\right) e_{i,k|k}\left(a_{il}\Gamma e_{l,k|k}\right)^T\right]$$
$$+ \sum_{j=1}^{N}\sum_{l=1}^{N} \mathbb{E}\left[\left(a_{ij}\Gamma e_{j,k|k}\right) e_{i,k|k}^T\left(\mathrm{F}_{i,k} + U_{i,k}W_{i,k} - a_{il}\Gamma\right)^T\right]$$
$$+ \sum_{j=1}^{N}\sum_{l=1}^{N} \mathbb{E}\left[\left(a_{ij}\Gamma e_{j,k|k}\right)\left(a_{il}\Gamma e_{l,k|k}\right)^T\right] + B_{i,k}Q_{i,k}B_{i,k}^T \tag{13}$$

According to (8), the estimation error covariance can be derived as

$$P_{i,k+1|k+1} = \left[I - K_{i,k+1}\left(H_{i,k+1} + V_{i,k}M_{i,k}\right)\right] P_{i,k+1|k}$$
$$\times \left[I - K_{i,k+1}\left(H_{i,k+1} + V_{i,k}M_{i,k}\right)\right]^T + K_{i,k+1}R_{i,k+1}K_{i,k+1}^T \tag{14}$$

Theorem 1 *Consider the one step prediction error and the filtering estimation error covariance matrices $P_{i,k+1|k}$ and $P_{i,k+1|k+1}$. Let $\beta_i (i = 1, 2)$ be positive scalars. if the following two Riccati-like difference equations*

$$\Psi_{i,k+1|k} = 2\sum_{j=1}^{N}\left(F_{i,k} - a_{ij}\Gamma\right)\left[\Psi_{i,k|k}^{-1} - (1+r)\beta_1 I\right]^{-1}\left(F_{i,k} - a_{ij}\Gamma\right)^T$$
$$+ 2\beta_1^{-1} U_{i,k}U_{i,k}^T + 2\sum_{j=1}^{N} a_{ij}\bar{a}_{il}\Gamma\Psi_{j,k|k}\Gamma^T + B_{i,k}Q_{i,k}B_{i,k}^T \tag{15}$$

$$\Psi_{i,k+1|k+1} = \left(I - K_{i,k+1}H_{i,k+1}\right)\left(\Psi_{i,k+1|k}^{-1} - \beta_2 I\right)^{-1}\left(I - K_{i,k+1}H_{i,k+1}\right)^T$$
$$+ K_{i,k+1}\left(\beta_2^{-1}V_{i,k}V_{i,k}^T + R_{i,k+1}\right)K_{i,k+1}^T \tag{16}$$

where $\bar{a}_{il} = \sum_{l=1}^{N} a_{il}$ with initial condition $\Psi_{i,0|0} = P_{i,0|0} > 0$ have the positive definite solutions $\Psi_{i,k+1|k}$ and $\Psi_{i,k+1|k+1}$, for all $0 \le k \le N$, the constraints $\frac{1}{1+r}\beta_1^{-1}I - \Psi_{i,k|k} > 0$ and $\beta_2^{-1}I - \Psi_{i,k+1|k} > 0$ are satisfied, then the gain $K_{i,k+1}$ is given by

$$K_{i,k+1}=\left(\Psi_{i,k+1|k}^{-1}-\beta_2 I\right)^{-1}H_{i,k+1}^T \times\left[H_{i,k+1}\left(\Psi_{i,k+1|k}^{-1}-\beta_2 I\right)^{-1}H_{i,k+1}^T+\left(\beta_2^{-1}V_{i,k}V_{i,k}^T+R_{i,k+1}\right)\right]^{-1} \tag{17}$$

Proof Following the elementary inequality $xy^T+yx^T\le xx^T+yy^T$ and adding second items to third items, we can obtain

$$\begin{aligned}&\sum_{j=1}^{N}\sum_{l=1}^{N}\mathbb{E}\left[\left(\mathrm{F}_{i,k}+U_{i,k}W_{i,k}-a_{ij}\Gamma\right)e_{i,k|k}\left(a_{il}\Gamma e_{l,k|k}\right)^T\right]\\&+\sum_{j=1}^{N}\sum_{l=1}^{N}\mathbb{E}\left[\left(a_{ij}\Gamma e_{j,k|k}\right)e_{i,k|k}^T\left(\mathrm{F}_{i,k}+U_{i,k}W_{i,k}-a_{il}\Gamma\right)^T\right]\\&\le\sum_{j=1}^{N}\sum_{l=1}^{N}\mathbb{E}\left[\left(\mathrm{F}_{i,k}+U_{i,k}W_{i,k}-a_{ij}\Gamma\right)e_{i,k|k}e_{i,k|k}^T\left(\mathrm{F}_{i,k}+U_{i,k}W_{i,k}-a_{il}\Gamma\right)^T\right]\\&+\sum_{j=1}^{N}\sum_{l=1}^{N}\mathbb{E}\left[\left(a_{ij}\Gamma e_{j,k|k}\right)\left(a_{il}\Gamma e_{l,k|k}\right)^T\right]\end{aligned} \tag{18}$$

Thus, we have

$$\begin{aligned}P_{i,k+1|k}\le&\,2\sum_{j=1}^{N}\sum_{l=1}^{N}\mathbb{E}\left[\left(\mathrm{F}_{i,k}+U_{i,k}W_{i,k}-a_{ij}\Gamma\right)P_{i,k|k}\left(\mathrm{F}_{i,k}+U_{i,k}W_{i,k}-a_{il}\Gamma\right)^T\right]\\&+2\sum_{j=1}^{N}\sum_{l=1}^{N}\mathbb{E}\left[\left(a_{ij}\Gamma e_{j,k|k}\right)\left(a_{il}\Gamma e_{l,k|k}\right)^T\right]\end{aligned} \tag{19}$$

According to the elementary inequality and Lemma 1, we have

$$\begin{aligned}&2\sum_{j=1}^{N}\sum_{l=1}^{N}\mathbb{E}\left[\left(\mathrm{F}_{i,k}+U_{i,k}W_{i,k}-a_{ij}\Gamma\right)P_{i,k|k}\left(\mathrm{F}_{i,k}+U_{i,k}W_{i,k}-a_{il}\Gamma\right)^T\right]\\&\le 2\sum_{j=1}^{N}\mathbb{E}\left[\left(\mathrm{F}_{i,k}+U_{i,k}W_{i,k}-a_{ij}\Gamma\right)P_{i,k|k}\left(\mathrm{F}_{i,k}+U_{i,k}W_{i,k}-a_{ij}\Gamma\right)^T\right]\\&\le 2\sum_{j=1}^{N}\left(\mathrm{F}_{i,k}-a_{ij}\Gamma\right)\left[P_{i,k|k}^{-1}-(1+r)\beta_1 I\right]^{-1}\left(\mathrm{F}_{i,k}-a_{ij}\Gamma\right)^T+2\beta_1^{-1}U_{i,k}U_{i,k}^T\end{aligned} \tag{20}$$

For the fourth of inequality (14), we have

$$
\begin{aligned}
&2\sum_{j=1}^{N}\sum_{l=1}^{N}\mathbb{E}\left[\left(a_{ij}\Gamma e_{j,k|k}\right)\left(a_{il}\Gamma e_{l,k|k}\right)^{T}\right] \\
&= \mathbb{E}\left[\sum_{j=1}^{n}\sum_{l=1}^{n} a_{ij}a_{il}\Gamma(e_{j,k|k}e_{l,k|k}^{T} + e_{l,k|k}e_{j,k|k}^{T})\Gamma^{T}\right] \\
&\leq \mathbb{E}\left[\sum_{j=1}^{n}\sum_{l=1}^{n} a_{ij}a_{il}\Gamma(e_{j,k|k}e_{j,k|k}^{T} + e_{l,k|k}e_{l,k|k}^{T})\Gamma^{T}\right] = 2\sum_{j=1}^{n} a_{ij}\bar{a}_{il}\Gamma P_{j,k|k}\Gamma^{T}
\end{aligned} \tag{21}
$$

So we can obtain

$$
\begin{aligned}
P_{i,k+1|k} \leq{}& 2\sum_{j=1}^{N}\left(\mathrm{F}_{i,k} - a_{ij}\Gamma\right)\left[P_{i,k|k}^{-1} - (1+r)\,\beta_1 I\right]^{-1}\left(\mathrm{F}_{i,k} - a_{ij}\Gamma\right)^{T} \\
&+ 2\beta_1^{-1}\,U_{i,k}U_{i,k}^{T} + 2\sum_{j=1}^{N} a_{ij}\bar{a}_{il}\Gamma P_{j,k|k}\Gamma^{T} + B_{i,k}Q_{i,k}B_{i,k}^{T}
\end{aligned} \tag{22}
$$

According to Lemma 2, let $\Psi_{i,0|0} \geq P_{i,0|0}$

$$
\begin{aligned}
\Psi_{i,k+1|k} ={}& 2\sum_{j=1}^{N}\left(\mathrm{F}_{i,k} - a_{ij}\Gamma\right)\left[\Psi_{i,k|k}^{-1} - (1+r)\,\beta_1 I\right]^{-1}\left(\mathrm{F}_{i,k} - a_{ij}\Gamma\right)^{T} \\
&+ 2\beta_1^{-1}\,U_{i,k}U_{i,k}^{T} + 2\sum_{j=1}^{N} a_{ij}\bar{a}_{il}\Gamma \Psi_{j,k|k}\Gamma^{T} + B_{i,k}Q_{i,k}B_{i,k}^{T}
\end{aligned} \tag{23}
$$

We have $P_{i,k+1|k} \leq \Psi_{i,k+1|k}$

By applying Lemma 1 to (14), we can obtain

$$
\begin{aligned}
P_{i,k+1|k+1} ={}& \left[I - K_{i,k+1}\left(H_{i,k+1} + V_{i,k}M_{i,k}\right)\right]P_{i,k+1|k} \\
&\times\left[I - K_{i,k+1}\left(H_{i,k+1} + V_{i,k}M_{i,k}\right)\right]^{T} + K_{i,k+1}R_{i,k+1}K_{i,k+1}^{T} \\
\leq{}& \left(I - K_{i,k+1}H_{i,k+1}\right)\left(P_{i,k+1|k}^{-1} - \beta_2 I\right)^{-1}\left(I - K_{i,k+1}H_{i,k+1}\right)^{T} \\
&+ K_{i,k+1}\left(\beta_2^{-1}V_{i,k}V_{i,k}^{T} + R_{i,k+1}\right)K_{i,k+1}^{T}
\end{aligned} \tag{24}
$$

Following the same lines to $P_{i,k+1|k} \leq \Psi_{i,k+1|k}$, we have

$$
\begin{aligned}
\Psi_{i,k+1|k+1} ={}& \left(I - K_{i,k+1}H_{i,k+1}\right)\left(P_{i,k+1|k}^{-1} - \beta_2 I\right)^{-1}\left(I - K_{i,k+1}H_{i,k+1}\right)^{T} \\
&+ K_{i,k+1}\left(\beta_2^{-1}V_{i,k}V_{i,k}^{T} + R_{i,k+1}\right)K_{i,k+1}^{T}
\end{aligned} \tag{25}
$$

where $\Psi_{i,k+1|k+1}$ is the upper bound of $P_{i,k+1|k+1}$
The filtering gain $K_{i,k+1}$ is designed to minimize the upper bound $\Psi_{i,k+1|k+1}$, according to (25) we have

$$\frac{\partial tr(\Psi_{i,k+1|k+1})}{\partial K_{i,k+1}} = -2\left(\Psi_{i,k+1|k}^{-1} - \beta_2 I\right)^{-1} H_{i,k+1}^T + 2K_{i,k+1}\Lambda_i K_{i,k+1}^T \tag{26}$$

where $\Lambda_i = \left[H_{i,k+1}\left(\Psi_{i,k+1|k}^{-1} - \beta_2 I\right)^{-1} H_{i,k+1}^T + \left(\beta_2^{-1} V_{i,k} V_{i,k}^T + R_{i,k+1}\right)\right]$
let $\frac{\partial tr(\Psi_{i,k+1|k+1})}{\partial K_{i,k+1}} = 0$, the filtering gain $K_{i,k+1}$ can be obtained as

$$K_{i,k+1} = \left(\Psi_{i,k+1|k}^{-1} - \beta_2 I\right)^{-1} H_{i,k+1}^T \Lambda_i \tag{27}$$

This completes the proof.

4 Numerical Example

In this section, a numerical example that involving an RSS-based indoor localization of mobile robots is presented to illustrate the usefulness of the established state estimation scheme. The kinematic for the robot can be represented by [20]

$$x_{i,k+1} = \begin{bmatrix} \xi_{i,k} + \varrho_{i,k}\cos\theta_{i,k} \\ \zeta_{i,k} + \varrho_{i,k}\sin\theta_{i,k} \\ \theta_{i,k} + \eta_{i,k} \end{bmatrix} + c\sum_{j=1}^{4} \Gamma(x_{j,k} - x_{i,k}) + \begin{bmatrix} w_{i,k}^x \\ w_{i,k}^y \\ w_{i,k}^\theta \end{bmatrix}$$

where $x_{i,k} = [\xi_{i,k}, \zeta_{i,k}, \theta_{i,k}]^T$, $(\xi_{i,k}, \zeta_{i,k})$ and $\theta_{i,k}$ denote the position and the orientation of the i-th robot, respectively. $(\varrho_{i,k}, \eta_{i,k})$ denotes the velocity vector. $w_{i,k} = (w_{i,k}^x, w_{i,k}^y, w_{i,k}^\theta)$ is zero mean white Gaussian noise with covariance $Q_{i,k}$.

The RSS measurement is generated by [21]

$$y_{i,k} = K - 10\eta\log(d_{i,k}) + v_{i,k} \tag{28}$$

$$d_{i,k} = \sqrt{(\xi_{i,k} - r_{i,x})^2 + (\zeta_{i,k} - r_{y,i})^2}, i = 1, 2, ..., 4 \tag{29}$$

where $(r_{i,x}, r_{i,y})$ is the position of the i-th sensor. $K = 9$ and $\eta = 2.5$ denote the transmission power and the path loss exponent, respectively. In the simulations, four sensor are locked at (2.5, 1), (2.5, 5.5), (7.5, 1), (7.5, 5.5) in meters. $Q_{i,k} = 0.001I_3 (i = 1, 2, 3, 4)$, $R_{i,k} = I_4$. $c = 0.001$ $v_{i,k} = 0.15$ (i = 1, 2, 3, 4), $\Gamma = 0.001I_3$, $U = 0.01I_3$, $V = 0.02I_4$. The trajectories of four robots are shown in Fig. 1. To illustrate the effects of the coupling behaviors we make a comparison between the coupling behaviors

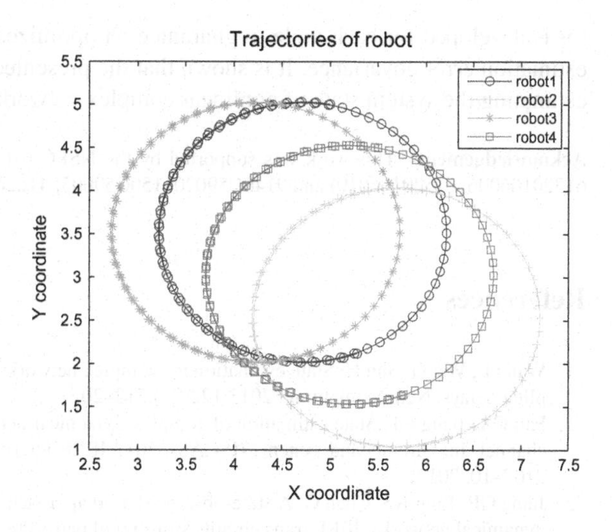

Fig. 1 Trajectories of robots

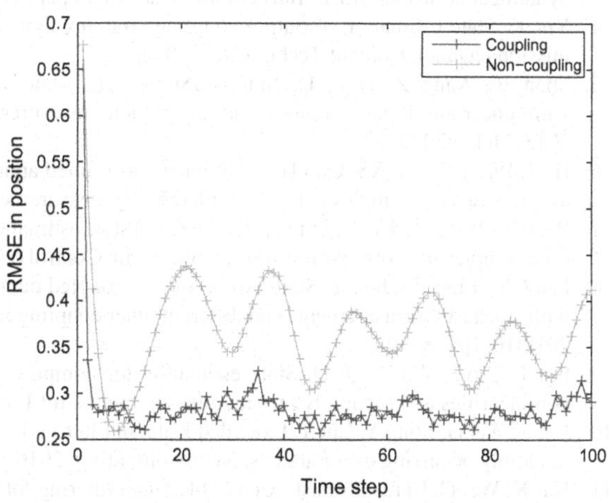

Fig. 2 RMSE in position versus time instants

and the noncoupling behaviors (the EKF without considering coupling behaviors) and the averaged root mean square errors (RMSE) over four robots are used. It can be see that the designed state estimators have a satisfactory tracking performance and the proposed coupling filter is superior to the state estimator, which indicates the usefulness of the established state estimation scheme (Fig. 2).

5 Conclusion

In this paper, we proposed a recursive state estimator by adopting a non-augmented vector approach for nonlinear complex networks, Compared with [6], the estimator is developed for each node to guarantee an optimized upper bound of the state estimation error covariance. It is shown that the presented scheme performs well in estimating the system state of nonlinear complex networks.

Acknowledgements This work was supported by the NSFC (61573031, 61327807, 61532006, 61320106006, 61520106010) and 9140C590201150C59005, 4162070.

References

1. Wang L, Wei G, Shu H. State estimation for complex networks with randomly occurring coupling delays. Neurocomputing. 2013;122(25):513–20.
2. Fan CX, Jiang GP. State estimation of complex dynamical network under noisy transmission channel. In: Circuits and systems (ISCAS), 2012 IEEE international symposium. vol. 57, p. 2107–10, 2012.
3. Jiang GP, Tang KS, Chen G. A state-observer-based approach for synchronization in complex dynamical networks. IEEE trans circuits syst I regul pap. 2006;53(12):2739–45.
4. Yin X. State estimation of output-coupling complex dynamical networks under noisy transmission channel. Comput Technol Dev. 2013.
5. Shen B, Wang Z, Ding D, Shu HuiSheng. H_∞ state estimation for complex networks with uncertain inner coupling and incomplete measurements. Neural Netw Learn Syst. 2013;24(12):2027–37.
6. Hu J, Wang Z, Liu XS, Gao H. A variance-constrained approach to recursive state estimation for time-varying complex networks with missing measurements. Automatica. 2016;64:155–62.
7. Wang L, Wang Z, Wei G, Song Y. Event-based state estimation for a class of nonlinear discrete-time complex networks with stochastic noises. In: Control conference. IEEE; 2015. p. 1804–09.
8. Fan CX, Yang F, Zhou Y. State estimation for coupled output discrete-time complex network with stochastic measurements and different inner coupling matrices. Int J Control Autom Syst. 2012;10(3):498–505.
9. Liu J, Cao J, Wu Z, Qi Q. State estimation for complex systems with randomly occurring nonlinearities and randomly missing measurements. Int J Syst Sci. 2005;24:1233–42.
10. Liu S, Wei G, Song Y, Liu Y. Extended Kalman filtering for stochastic nonlinear systems with randomly occurring cyber attacks. Neurocomputing. 2016.
11. Kai X, Wei C, Liu L. Robust extended Kalman filtering for nonlinear systems with stochastic uncertainties. IEEE Trans Syst Man Cybern Part A Syst Hum. 2010;40(2):399–405.
12. Hu J, Liu S, Ji D, Li S. On co-design of filter and fault estimator against randomly occurring nonlinearities and randomly occurring deception attacks. Int J Gen Syst. 2016;45(5):619–32.
13. Ding D, Shen Y, Song Y, Wang Y. Recursive state estimation for discrete time-varying stochastic nonlinear systems with randomly occurring deception attacks. Int J Gen Syst. 2016;45(5):548–60.
14. Hu J, Wang Z, Gao H, Stergioulas LK. Extended Kalman filtering with stochastic nonlinearities and multiple missing measurements. Automatica. 2012;48(8):2007–15.
15. Zheng X, Fang H. Recursive state estimation for discrete-time nonlinear systems with event-triggered data transmission, norm-bounded uncertainties and multiple missing measurements. Int J Robust Nonlinear Control. 2016; doi:10.1002/rnc.3527.
16. Kai X, LiangDong L, YiWu L. Robust extended Kalman filtering for nonlinear systems with multiplicative noises. Optim Control Appl Methods. 32: 47–63. doi:10.1002/oca.928.
17. Wang L, Wei G, Shu H. State estimation for complex networks with randomly occurring coupling delays. Neurocomputing. 2013;122(55):513–20.
18. Hu J, Wang Z, Shen B, Gao H. Quantised recursive filtering for a class of nonlinear systems with multiplicative noises and missing measurements. Int J Control. 2013;86(4):650–63.

19. Reif K, Unbehauen R. Stochastic stability of the discrete-time extended Kalman filter. IEEE Trans Autom Control. 1999;44:714–28.
20. Chen X, Jia Y. Indoor localization for mobile robots using lampshade corners as landmarks: visual system calibration, feature extraction and experiments. Int J Control Autom Syst. 2014;12(6):1313–22.
21. Li W, Jia Y, Du J. RSS-based joint detection and tracking in mixed LOS and NLOS environments. Digit Signal Process. 2015;43:38–46.

Adaptive Observer Design for Quasi-one-sided Lipschitz Nonlinear Systems

Jun Huang, Lei Yu and Minjie Shi

Abstract This paper deals with the adaptive observer design problem for quasi-one-sided Lipschitz nonlinear systems. First, some useful assumptions are presented for the observer design purpose. Then, under the assumptions, an adaptive observer is constructed for the nonlinear system. Finally, a numerical example is given to illustrate the effectiveness of the proposed method.

Keywords Adaptive observer · Quasi-one-sided Lipschitz · Nonlinear systems

1 Introduction

With the development of theory and application, state estimation or observer for nonlinear systems has been studied widely in the field of control [1–4]. The literature of observer design problem mainly focuses on two aspects. One is that different types of observers for nonlinear systems have been investigated, such as adaptive observers [5, 6], H_∞ observers [7, 8], interval observers [9, 10] and so on. The other is that different design methods have been proposed for observers, [11] presented a linear matrix inequality (LMI) approach to design observer for one-sided Lipschitz nonlinear systems. [12] used Riccati equations to dealt with full-order and reduced-order observers design problem for one-sided Lipschitz nonlinear systems. In view of different conditions on set-valued functions in Lur'e differential inclusion systems, [13] constructed the observer for the system under dissipative approach. [14, 15] designed the observer for the system by positive real method.

Recently, one-sided Lipschitz condition has been introduced for the control theory. [16, 17] firstly studied one-sided Lipschitz condition for observer design problem in nonlinear systems. The region of one-sided Lipschitz constant is much larger than that of traditional Lipschitz condition, because the one-sided Lipschitz constant can be zero or even negative while the Lipschitz constant must be positive.

J. Huang (✉) · L. Yu · M. Shi
School of Mechanical and Electrical Engineering, Soochow University, Suzhou 215021, China
e-mail: cauchyhot@163.com

Y. Jia et al. (eds.), *Proceedings of 2017 Chinese Intelligent Systems Conference*, Lecture Notes in Electrical Engineering 459,
https://doi.org/10.1007/978-981-10-6496-8_2

This means that one-sided Lipschitz condition is more general than Lipschitz condition. Following the line of [16–18] considered the observer design problem and presented the standard one-sided Lipschitz condition. [11, 12] gave different methods to design observers for one-sided Lipschitz nonlinear systems respectively. [19] improved the result of [16] and derived sufficient condition for the existence of observer of one-sided Lipschitz nonlinear systems.

It is worth noting that P-one-sided Lipschitz condition (modified one-sided Lipschitz condition) and quasi-one-sided Lipschitz condition have been introduced in [16, 17]. [16] illustrated that P-one-sided Lipschitz condition is less conservative than one-sided Lipschitz condition because a fixed symmetric definite matrix P broadens the range of nonlinear function $f(x)$ in the system. Meanwhile, [17] introduced a quasi-one-sided Lipschitz condition to estimate the influence of nonlinear function on the observer and verified that the quasi-one-sided Lipschitz condition is an extension of the P-one-sided Lipschitz condition. Under quasi-one-sided Lipschitz condition, [20] considered the robust stabilization problem of nonlinear networked control systems and established a less conservative sufficient condition. [21] addressed the problem of state feedback and output feedback control for a class of nonlinear systems, and the nonlinearity was assumed to satisfy global quasi-one-sided Lipschitz condition. For one-sided Lipschitz and quasi-one-sided Lipschitz systems, [22] presented fault detection algorithm based on sliding mode observer. [23] employed both Lyapunov function approach and linear matrix inequality technique to study finite-time H_∞ control problem of quasi-one-sided Lipschitz nonlinear systems with parameter uncertainties. It should be noted that the nonlinear function in [20–23] satisfies quasi-one-sided Lipschitz condition as well as quadratically inner-bounded condition. If the nonlinear function in the system is not quadratically inner-bounded, how can we design the observer for the system? There seems no available answer to it.

Motivated by above discussion, this paper investigates the adaptive observer design problem for quasi-one-sided Lipschitz nonlinear systems with uncertain parameters. The nonlinear function of the system does not satisfy the property of quadratic inner-boundedness. The contribution of this paper mainly lies in two aspects: One is that the adaptive observer design method is extended to quasi-one-sided Lipschitz nonlinear systems, and the quasi-one-sided Lipschitz term is more general than the standard one-sided Lipschitz term. The other is that a tractable method is provided to deal with the nonlinear function without quadratic inner-boundedness. The paper is organized as follows: Sect. 2 presents the problem formulation and some preliminaries. Section 3 designs the adaptive observer for quasi-one-sided Lipschitz nonlinear systems. Section 4 illustrates the effectiveness of the designed adaptive observer by numerical example.

The notations used in this paper are as follows: $\|x\|$ and $\|A\|$ mean the Euclidean norm of the vector x and the Euclidean norm of the matrix A respectively, x^T is the transposition of the vector x, and A^T denotes the transposition of the matrix A. I is the identity matrix. $P > (<)0$ represents the positive (negative) definite matrix P with $P = P^T$. $\langle \cdot, \cdot \rangle$ stands for the inner product, i.e., given $x, y \in R^n$, then $\langle x, y \rangle = x^T y$.

2 Problem Formulation and Preliminaries

Consider the following nonlinear system

$$\begin{cases} \dot{x} = Ax + Bf_1(Hx, u) + Df_2(x, u)\theta, \\ y = Cx, \end{cases} \tag{1}$$

where $x \in R^n$ is the state of the system, $u \in R^r$ is the control input, and $y \in R^q$ is the measurable output. $\theta \in R^l$ is the unknown constant vector, $f_1(\cdot,\cdot) : R^m \times R^r \to R^m$ and $f_2(\cdot,\cdot) : R^n \times R^r \to R^{k\times l}$ are given nonlinear functions. A, B, H, D, C are known real matrices with appropriate dimensions. Without loss of generality, it is assumed that C is of full row rank.

Firstly, let us recall some basics of one-sided Lipschitz function, more details can be referred to [16, 17].

Definition 1 *(Standard one-sided Lipschitz condition)* $f(v, u)$ is said to be one-sided Lipschitz with respect to v if there exists a constant $\rho \in R$ such that

$$\langle f(v, u) - f(w, u), v - w \rangle \leq \rho \|v - w\|^2, \tag{2}$$

holds for any $v, w \in R^m$, where ρ is called the one-sided Lipschitz constant. The real number ρ can be positive, zero, or even negative.

Definition 2 *(P-one-sided Lipschitz condition)* $f(v, u)$ is said to be P-one-sided Lipschitz with respect to v if there exists a matrix $P > 0$ such that

$$\langle \Phi(v, u) - \Phi(w, u), v - w \rangle \leq \nu_P \|v - w\|^2, \tag{3}$$

holds for any $v, w \in R^m$, where $\Phi(v, u) = Pf(v, u)$. The real number ν_P is called the P-one-sided Lipschitz constant, which can be positive, zero, or even negative.

Definition 3 *(Quasi-one-sided Lipschitz condition)* $f(v, u)$ is said to be quasi-one-sided Lipschitz with respect to v if there exist matrices $P > 0$ and M such that

$$\langle \Phi(v, u) - \Phi(w, u), v - w \rangle \leq (v - w)^T M (v - w), \tag{4}$$

holds for any $v, w \in R^m$, where $\Phi(v, u) = Pf(v, u)$. The real symmetric matrix M is called the quasi-one-sided Lipschitz constant matrix.

In order to obtain the main result, the following assumptions and lemmas are needed.

Assumption 1 $f_1(v, u)$ is quasi-one-sided Lipschitz function, i.e., there exist matrices $P > 0$ and M such that

$$\langle \Phi_1(v, u) - \Phi_1(w, u), v - w \rangle \leq (v - w)^T M (v - w), \tag{5}$$

holds for any $v, w \in R^m$, where $\Phi_1(v, u) = Pf_1(v, u)$.

Assumption 2 $f_2(x, u)$ is Lipschitz function, i.e., there exists a constant $\eta > 0$ such that

$$\|f_2(x, u) - f_2(\hat{x}, u)\| \le \eta \|x - \hat{x}\|, \tag{6}$$

holds for any $x, \hat{x} \in R^n$, and the constant η is Lipschitz constant.

Assumption 3 The unknown parameter vector θ is bounded with $\mu > 0$, i.e.,

$$\|\theta\| \le \mu. \tag{7}$$

Assumption 4 Let $\beta = \eta\mu\|D\|$, where η and μ are defined in (6) and (7). There exist constant $\varepsilon > 0$, matrices $Q > 0$, L, F, N such that

$$\begin{aligned} &Q(A - LC) + (A - LC)^T Q + \beta Q^2 \\ &+ 2(H - FC)^T M(H - FC) + (\beta + \varepsilon)I \le 0, \end{aligned} \tag{8}$$

$$B^T Q = P(H - FC), \tag{9}$$

$$D^T Q = NC. \tag{10}$$

Lemma 1 *[24] Let $\varphi : R^+ \to R$ be a known function. If φ is uniformly continuous and the integral $\int_0^\infty \varphi(s)ds$ exists, then $\lim_{t\to\infty} \varphi(t) = 0$.*

Lemma 2 *For a given matrix $S = \begin{bmatrix} S_{11} & S_{12} \\ S_{12}^T & S_{22} \end{bmatrix}$ with $S_{11}^T = S_{11}$ and $S_{22}^T = S_{22}$, the following conditions are equivalent:*

(1) $S < 0$,
(2) $S_{11} < 0$, $S_{22} - S_{12}^T S_{11}^{-1} S_{12} < 0$,
(3) $S_{22} < 0$, $S_{11} - S_{12} S_{22}^{-1} S_{12}^T < 0$.

3 Main Result

The adaptive observer for the system (1) is designed as

$$\begin{cases} \dot{\hat{x}} = A\hat{x} + Bf_1(H\hat{x} + F(y - C\hat{x}), u) \\ \qquad + Df_2(\hat{x}, u)\hat{\theta} + L(y - C\hat{x}), \\ \hat{y} = C\hat{x}, \end{cases} \tag{11}$$

with

$$\dot{\hat{\theta}} = f_2^T(\hat{x}, u)N(y - C\hat{x}). \tag{12}$$

Denote that $e = x - \hat{x}$, $\tilde{\theta} = \theta - \hat{\theta}$, $v = Hx$, $w = H\hat{x} + F(y - C\hat{x})$ and $\tilde{f}_1 = f_1(v, u) - f_1(w, u)$, $\tilde{f}_2 = f_2(x, u) - f_2(\hat{x}, u)$, $\hat{f}_2 = f_2(\hat{x}, u)$. Subtracting (11) from (1) results in the error system:

$$\dot{e} = (A - LC)e + B\tilde{f}_1 + D\tilde{f}_2\theta + D\hat{f}_2\tilde{\theta}. \tag{13}$$

We can now state the theorem as follows.

Theorem 1 *If Assumptions 1–4 hold, then* (11) *is an adaptive observer for the system* (1)*, i.e., $e(t)$ of the error system* (13) *satisfies* $\lim_{t\to\infty} e(t) = 0$.

Proof Let us consider the following Lyapunov function candidate

$$V = e^T Q e + \tilde{\theta}^T \tilde{\theta}. \tag{14}$$

Along the trajectories of the error system (13), the derivative of V can be calculated as

$$\begin{aligned} \dot{V} &= 2e^T Q\dot{e} + 2\tilde{\theta}^T \dot{\tilde{\theta}} \\ &= 2e^T Q(A - LC)e + 2e^T QB\tilde{f}_1 + 2e^T QD\tilde{f}_2\theta \\ &\quad + 2e^T QD\hat{f}_2\tilde{\theta} + 2\tilde{\theta}^T \dot{\tilde{\theta}}. \end{aligned} \tag{15}$$

By Assumption 1 and (9), we have

$$\begin{aligned} 2e^T QB\tilde{f}_1 &= 2e^T (H - FC)^T P\tilde{f}_1 \\ &= 2\langle P\tilde{f}_1, (H - FC)e\rangle \\ &\leq 2e^T (H - FC)^T M(H - FC)e. \end{aligned} \tag{16}$$

By Assumptions 2 and 3, the following holds

$$\begin{aligned} 2e^T QD\tilde{f}_2\theta &\leq 2\|e^T Q\|\|D\|\|\tilde{f}_2\|\|\theta\| \\ &\leq 2\eta\mu\|D\|\|e^T Q\|\|e\| \\ &\leq \beta(e^T Q^2 e + e^T e). \end{aligned} \tag{17}$$

From (10) and (12), we can obtain that

$$\begin{aligned} 2e^T QD\hat{f}_2\tilde{\theta} + 2\tilde{\theta}^T \dot{\tilde{\theta}} &= 2e^T QD\hat{f}_2\tilde{\theta} - 2\tilde{\theta}^T \dot{\hat{\theta}} \\ &= 2e^T QD\hat{f}_2\tilde{\theta} - 2\tilde{\theta}^T \hat{f}_2^T D^T Qe = 0. \end{aligned} \tag{18}$$

Substituting (16)–(18) into (15) yields

$$\begin{aligned}\dot{V} \le e^T[Q(A - LC) + (A - LC)^T Q + \beta(Q^2 + I) \\ + 2(H - FC)^T M(H - FC)]e.\end{aligned} \tag{19}$$

It follows from (8) that

$$\dot{V} \le -\varepsilon e^T e. \tag{20}$$

Integrating both sides from 0 to t yields

$$V(t) \le V(0) - \int_0^t \varepsilon e^T(s)e(s)ds. \tag{21}$$

Since $V(t) > 0$, (21) implies $\int_0^t \varepsilon e^T(s)e(s)ds \le V(0)$, which deduces to

$$\lim_{t\to\infty} \int_0^t \varepsilon e^T(s)e(s)ds \le V(0) < \infty. \tag{22}$$

By Lemma 1, we obtain

$$\lim_{t\to\infty} \varepsilon e^T(t)e(t) = 0, \tag{23}$$

which means that

$$\lim_{t\to\infty} e(t) = 0. \tag{24}$$

Thus, the proof is completed.

Remark 1 The function $f_1(v, u)$ does not satisfy the quadratically inner-bounded condition, thus the methods proposed in the former works such as [20–23] can not be applied in this paper.

Remark 2 Different from the system with P-one-sided Lipschitz term studied in [25], we consider the system with quasi-one-sided Lipschitz term and uncertain parameters in this paper. Besides, the pair (P, M) can be obtained by solving quasi-one-sided Lipschitz condition constraint (5).

Remark 3 By using elimination of matrix variables, the observer gain L can be given as $L = \frac{\sigma}{2}Q^{-1}C^T$ in Theorem 1.

Remark 4 As we know, how to find a suitable one-sided Lipschitz function which can include standard one-sided Lipschitz function, P-one-sided Lipschitz function and quasi-one-sided Lipschitz function is still an open problem, and the function that we use is affected by the method of converting nonlinear matrix inequality to linear matrix inequality, the pratical model, the generality of one-sided Lipschitz nonlinear system and so on.

4 Numerical Example

Consider the system (1) with

$$A=\begin{bmatrix}-1 & -2\\ 1 & -1.2\end{bmatrix},\quad C=\begin{bmatrix}1 & 0\end{bmatrix},\quad H=\begin{bmatrix}1 & 1\\ 0 & 1\end{bmatrix},$$

$$B=\begin{bmatrix}1 & 0\\ 0 & 1\end{bmatrix},\quad D=\begin{bmatrix}0.1\\ -0.1\end{bmatrix},\quad u=\sin t,$$

$$f_1(Hx,u)=\begin{bmatrix}0.5u\\ \sin(x_1+x_2)-x_2^{\frac{1}{3}}\end{bmatrix},\quad f_2(x,u)=\sin x_1.$$

It was verified that $f_1(Hx,u)$ is not quadratically inner-bounded in [25]. Meanwhile, according to the mean-value theorem, there exists non-zero $\xi_0\in(\min(x_2,\hat{x}_2),\max(x_2,\hat{x}_2))$ and $\xi=(\xi_1,\xi_2)\in Co(x,\hat{x})$ such that

$$\begin{aligned}
&\langle Pf_1(Hx,u)-Pf_1(H\hat{x},u),H(x-\hat{x})\rangle\\
&=\alpha[(\cos(H_1\xi),-\cos(H_1\xi))](x-\hat{x})(x_2-\hat{x}_2)\\
&\quad-\frac{\alpha}{3}\xi_0^{-\frac{2}{3}}(x_2-\hat{x}_2)^2\\
&\le\alpha|(\cos(H_1\xi),-\cos(H_1\xi))|\,||(x-\hat{x})(x_2-\hat{x}_2)||\\
&\le(H(x-\hat{x}))^T\begin{bmatrix}1.42\alpha & 0\\ 0 & 1.42\alpha\end{bmatrix}H(x-\hat{x}),
\end{aligned}\tag{25}$$

where $H_1=\begin{bmatrix}1 & 1\end{bmatrix}$, $f_1(Hx,u)$ is quasi-one-sided Lipschitz and $M=v_P$ $I=\begin{bmatrix}1.42\alpha & 0\\ 0 & 1.42\alpha\end{bmatrix}$ for any $P=diag\{\lambda,\alpha\}$.

In the meantime, $f_2(x,u)$ is globally Lipschitz with the Lipschitz constant $\eta=1$. Selecting the unknown parameter $\theta=0.5$, we can obtain that $\mu=0.5$. Let $P=I$, this means $v_P=1.42$. Based on the above results, we solve (8)–(10) by Scilab, then $Q=\begin{bmatrix}41.2694 & 1\\ 1 & 1\end{bmatrix}$, $L=\begin{bmatrix}171.6597\\ -111.1394\end{bmatrix}$, $F=\begin{bmatrix}-40.2694\\ -1\end{bmatrix}$, $N=4.0269$, $\varepsilon=0.0475$. We now use the Simulink in Matlab to complete the simulation. As shown in Figs. 1, 2, the estimated state trajectories of the adaptive observer (11) converge to the state trajectories of the system (1). Figure 3 shows the estimated parameter $\hat{\theta}$, and it does not converge to the nominal value $\theta=0.5$. From the simulation results, we can conclude the observer design method is valid in this paper.

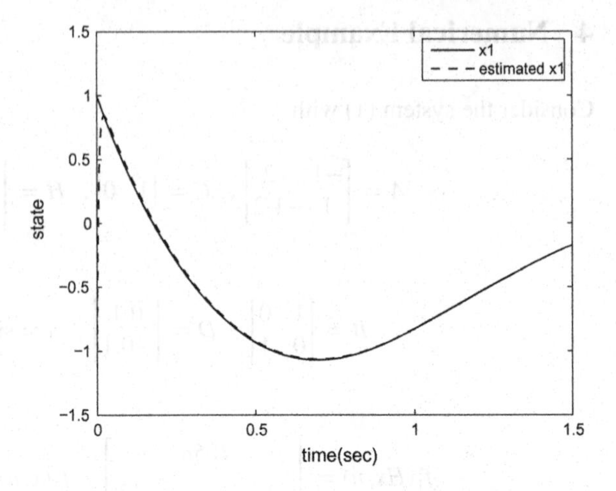

Fig. 1 The state x_1 of system (1) and the estimated state $\hat{x}_1$ of adaptive observer (11)

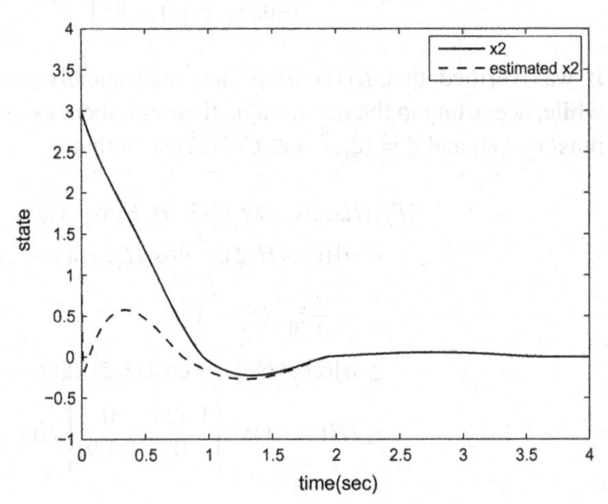

Fig. 2 The state x_2 of system (1) and the estimated state $\hat{x}_2$ of adaptive observer (11)

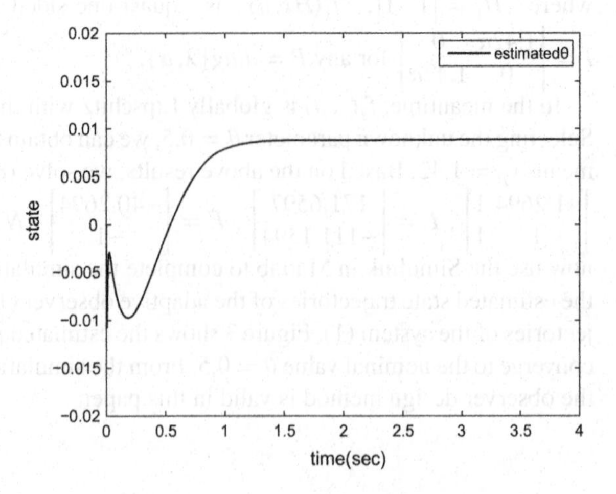

Fig. 3 The estimated parameter $\hat{\theta}$ of adaptive observer (11)

5 Conclusion

We consider the adaptive observer design problem for quasi-one-sided Lipschitz nonlinear systems in this paper. Based on some assumptions, we construct an adaptive observer for the system. Then, we verify that the error system is asymptotically stable by Lyapunov stability theory. Finally, we also simulate the numerical example to show the effectiveness of the proposed method.

Acknowledgements The authors are grateful for the National Natural Science Foundation of China (61403267, 61403268), Natural Science Foundation of Jiangsu Province of China (BK20130322), and China Postdoctoral Science Foundation (2017M611903).

References

1. Krener A, Respondek W. Nonlinear observer with linearizable error dynamics. SIAM J Control Optim. 1985;23(2):197–216.
2. Arcak M, Kokotovic P. Nonlinear observers: a circle criterion design and robustness analysis. Automatica. 2001;37(12):1923–30.
3. Ibrir S. Circle-criterion approach to discrete-time nonlinear observer design. Automatica. 2007;43(8):1432–41.
4. Starkov K, Coria L, Aguilar L. On synchronization of chaotic systems based on the Thau observer design. Commun Nonlinear Sci Numer Simul. 2012;17(1):17–25.
5. Huang J, Han Z, Cai X, Liu L. Adaptive full-order and reduced-order observers for the Lur'e differential inclusion system. Commun Nonlinear Sci Numer Simul. 2011;16(7):2869–79.
6. Huang J, Han Z. Adaptive non-fragile observer design for the uncertain Lur'e differential inclusion system. Appl Math Model. 2013;37(1–2):72–81.
7. Pertew A, Marquez H, Zhao Q. H_∞ observer design for Lipschitz nonlinear systems. IEEE Trans Autom Control. 2006;51(7):1211–6.
8. Zhang W, Su H, Su S, Wang D. Nonlinear H_∞ observer design for one-sided Lipschitz systems. Neurocomputing. 2014;145:505–11.
9. Zheng G, Efimov D, Bejarano F, Perruquetti W, Wang H. Interval observer for a class of uncertain nonlinear singular systems. Automatica. 2016;71:159–68.
10. He Z, Xie W. Control of non-linear switched systems with average dwell time: interval observer-based framework. IET Control Theory Appl. 2016;10(1):10–6.
11. Zhang W, Su H, Liang Y, Han Z. Non-linear observer design for one-sided Lipschitz systems: an linear matrix inequality approach. IET Control Theory Appl. 2012;6(9):1297–303.
12. Zhang W, Su H, Wang H, Han Z. Full-order and reduced-order observers for one-sided Lipschitz nonlinear systems using Riccati equations. Commun Nonlinear Sci Numer Simul. 2012;17(12):4968–77.
13. Osorio M, Moreno J. Dissipative design of observers for multivalued nonlinear systems. In: Proceedings of the 45th IEEE conference on decision and control. 2006.
14. Doris A, Juloski A, Mihajlovic N, Heemels W, Wouw N, Nijmeijer H. Observer designs for experimental non-smooth and discontinuous systems. IEEE Trans Control Syst Technol. 2008;16(6):1323–32.
15. Brogliato B, Heemels W. Observer design for Lur'e systems with multivalued mappings: a passivity approach. IEEE Trans Autom Control. 2009;54(8):1996–2001.
16. Hu G. Observers for one-sided Lipschitz non-linear systems. IMA J Math Control Inf. 2006;23(4):395–401.

17. Hu G. A note on observer for one-sided Lipschitz non-linear systems. IMA J Math Control Inf. 2008;25(3):297–303.
18. Abbaszadeh M, Marquez H. Nonlinear observer design for one-sided Lipschitz systems. In: Proceedings of the American control conference. 2010.
19. Zhao Y, Tao J, Shi N. A note on observer design for one-sided Lipschitz nonlinear systems. Syst Control Lett. 2010;59(1):66–71.
20. Chen Z, He Y, Wu M. Robust stabilization of nonlinear networked control systems with quasi-one-sided Lipschitz condition. In: Proceedings of the 2010 IEEE international conference on mechatronics and automation. 2010.
21. Fu F, Hou M, Duan G. Stabilization of quasi-one-sided Lipschitz nonlinear systems. IMA J Math Control Inf. 2013;30(2):169–84.
22. Li L, Yang Y, Zhang Y, Ding S. Fault estimation of one-sided Lipschitz and quasi-one-sided Lipschitz systems. In: Proceedings of the 33rd Chinese control conference. 2014.
23. Song J, He S. Finite-time H_∞ control for quasi-one-sided Lipschitz nonlinear systems. Neurocomputing. 2015;149(Part C):1433–9.
24. Kristic M, Modestino J, Deng H. Stabilization of nonlinear uncertain systems. New York: Springer; 1998.
25. Zhang W, Su H, Zhu F, Bhattacharyya S. Improved exponential observer design for one-sided Lipschitz nonlinear systems. Int J Robust Nonlinear Control. 2016;26(18):3958–73.

A Feature Selection Method Based on Information Gain and BP Neural Network

Xingyun Wang, Min Zuo and Lihui Song

Abstract Data mining and machine learning fields are facing with a great challenge of mass data with high dimensionality. Feature selection can contribute a lot to address this issue with the concept of reducing the number of features by eliminating the redundant and irrelevant ones while preserving the information of original features maximally. This paper analyzes and compares two common feature selection methods, then puts forward a novel method for feature selection based on information gain and BP neural network (IGBP). The experimental result shows that IGBP method can reduce the time cost and improve the accuracy of the model at the meantime. The scientificity and superiority of IGBP are demonstrated in this paper, making it an efficient approach to deal with high-dimensional data.

Keywords Feature selection · Data mining · BP neural network · Information gain · IGBP method

1 Introduction

In recent years, data mining theories and methods have been constantly applied to practical fields, revealing a great value and potential in all walks of life [1, 2]. However in internet era, data mining task becomes extremely difficult even impossible to accomplish due to the rapid increase in data size and dimensionality and the restriction of computing capability on the contrary. Under this circumstance, data mining technology confronts a test of both time cost and accuracy.

X. Wang · M. Zuo (✉)
School of Computer and Information Engineering, Beijing Technology and Business University, Beijing 100048, China
e-mail: zuomin1234@163.com

L. Song
College of Mechanical and Electrical Engineering, Yanching Institute of Technology, Sanhe 065201, Hebei, China
e-mail: 651265386@qq.com

Y. Jia et al. (eds.), *Proceedings of 2017 Chinese Intelligent Systems Conference*, Lecture Notes in Electrical Engineering 459,
https://doi.org/10.1007/978-981-10-6496-8_3

As a key step of the data mining technology, feature selection has crucial impacts on training time and training results [3]. In real-world applications, the dataset of high dimensionality contains numerous features which could be irrelevant to the output classification and could have correlative dependence on each other leading to a redundancy. If all these features are adopted to conduct the data mining process, the redundant part will weaken the model's interpreting ability while the irrelevant part will result in a high time consumption [4]. Therefore, the central research of this paper is to find the optimal subset of features by using a feature selection method which is effective and efficient. And once the subset is clarified, we can get the consequence as accurately as possible in limited time without reducing the size of dataset.

2 Related Work

Feature selection is used to eliminate some irrelevant and (or) redundant features from a large number of them, making the chosen subset reserves the characteristics of the original set as much as possible [5–7]. Feature selection is helpful to reduce the dimensionality of data and remain a high relevance between features and the output classification, shorten the training time and improve learning performance.

2.1 *Feature Selection Method*

Feature selection methods are used to assess the subset of features with some criterions, determining the optimal subset directly. And the feature selection method can be categorized into two types: filter and wrapper. The filter method is independent of the selected classifier as a pre-processing step which evaluates each feature based on its inner effect on the output classification. The wrapper method uses a single learner as a black box to evaluate the subset of features according to their predictive performance [8].

2.2 *Information Gain*

Entropy as a wildly used measure standard in information theory can be taken when apply the filter method to evaluate features [9]. A high level of entropy means the feature contains more unpredictability than a lower one. Whereas, a low level of entropy stands for more precision of information.

The information gain represents the decrease of expected entropy of a feature when partition the dataset with it. In practice, features are ranked according to the information gain. The higher the feature scores in the ranking list, the less information remains after partition, and the more suitable of the feature for classification.

2.3 *BP Neural Network*

The feature selection method based on BP neural network is a typical one among numerous wrapper methods. BP neural network is capable of mapping any n-dimension to m-dimension [10]. It can classify inputs that have different values into different categories accurately after being training well enough. Moreover, BP neural network is suitable for both continuous and discretized features and has strong robustness on null values and even errors [11]. In the application of feature selection, BP neural network trains and builds network model through using different feature sets as the input, measuring the feature sets by the evaluation indicator of the model after building.

3 IGBP Method

This paper analyzed and compared the methods of filter and wrapper adequately, we found these two methods do not exist oppositely. The filter method proceeds prior to the data mining procedure while the wrapper method is conducted in the middle of data mining procedure. Additional, the methods of filter and wrapper have some peculiarities respectively. Applying the filter method based on information gain is convenient and requires small computational complexities. But the chosen subset of features may not be the optimum for a certain classifier. On the other hand, the wrapper method based on BP neural network has the advantage of accuracy and a weakness as well [12]. Theoretically, to validate all combinations of the feature sets requires the method of exhaustion. And it is impossible to complete the task in limited time when the number of the features is too large considering that the training speed will decrease sharply in every single procedure.

Considering that the method of filter is simple while the wrapper method is accurate, we can keep the advantages of both methods and weaken their disadvantages by using the two in a combination way. Therefore this paper proposed a novel method which combines the filter and wrapper methods. The combination method is based on information gain and BP neural network (IGBP). In detail, IGBP applies the filter method based on information gain to evaluate each feature firstly. Then it choses different feature sets according to the result from the first step to build BP neural network models. At last, the final evaluation of a certain feature set is judged by the performance of the model. IGBP method makes up the lack of using a single wrapper method and gets rid of the inaccuracy problem by using a single filter method.

4 Experiment

In this section, we conducted a data mining experiment to demonstrate the IGBP method. The dataset for experiments is extracted from CFDA (China Food and Drug Administration) Sampling and Monitoring System. This dataset contains 14,517 instances, recording food features and detection conclusions. There are nearly 70 features in the dataset, however most of them are not suitable for the data mining task. 20 features are chosen manually for the follow-up experiment. The detection conclusions are 'qualified' and 'unqualified' which correspond to the positive and negative class as the output classification of the BP neural network. Within the dataset, there are 10,000 instances with positive values and 4517 instances with negative values.

4.1 Pre-evaluation

After a pre-processing work, we imported the well-organized dataset into WEKA (Waikato Environment for Knowledge Analysis) and apply the InfoGainAttributes Eval algorithm to evaluate these 20 features. Table 1, 2 and 3 shows the result.

As the result shows, the initial 20 features have different influence degrees on the output classification. And they can be divided into 3 groups. Group A contains 6 features of which the score is greater than 0.05, representing that these features are strongly indicative of the output classification. The features in group A are {Food Category, Sampling Date, Sampled Co. Province, Production Co. Province, Production Date, Shelf Life}. Group B are {Sampling Package, Sample Form, Quality Rank, Annual Sales, Unit Price, Sampling Site, Food Function}. These 7 features score greater than 0.01 and less than 0.05 representing a less importance to the output classification. The features in group C hardly have any influence on the output classification. They score less than 0.01, including {Storage State, Sample Origin, Sample Package, Use of Food, Export or Not, Sampling Approach, Area Type}. The follow-up experiments are performed on the basis of the score result.

Table 1 The score result of features in group A

No.	Feature	Average merit	Average rank
1	Food Category	0.383	1
2	Sampling Date	0.171	2
3	Sampled Co. Province	0.129	3
4	Production Co. Province	0.125	4
5	Production Date	0.069	5
6	Shelf Life	0.068	6

Table 2 The score result of features in group B

No.	Feature	Average merit	Average rank
7	Sampling Package	0.026	7
8	Sample Form	0.025	8
9	Quality Rank	0.023	9
10	Annual Sales	0.023	10
11	Unit Price	0.022	11
12	Sampling Site	0.02	12
13	Food Function	0.019	13

Table 3 The score result of features in group C

No.	Feature	Average merit	Average rank
14	Storage State	0.007	14
15	Sample Origin	0.006	15
16	Sample Package	0.004	16
17	Use of Food	0.003	17
18	Export or Not	0.001	18
19	Sampling Approach	0	19
20	Area Type	0	20

4.2 Evaluation Model 1

We adopted total 20 features (20-subset) as the input set to train the network model (model 1) with the Multilayer Perceptron algorithm. The training process lasts 38 min and the result shows that model 1 has an accuracy of 90.5352% to classify instances into correct classifications.

The TPR (True Positive Rate) of model 1 is 0.981 with a 0.262 FPR (False Positive Rate). And its ROC (receiver operating characteristic) area is 0.9172 which represents an average performance of the model. The ROC curve of model 1 is shown in Fig. 1.

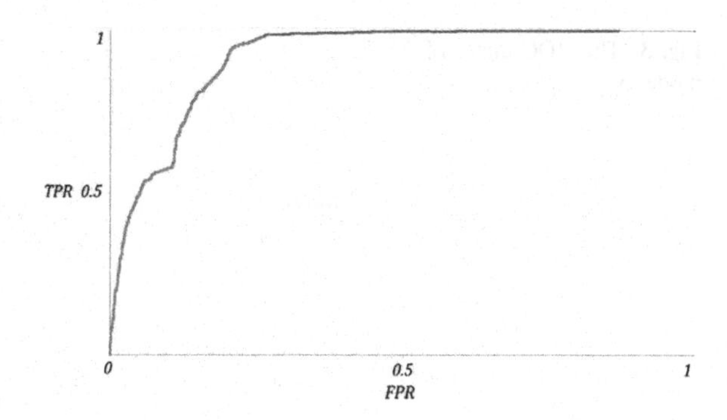

Fig. 1 The ROC curve of model 1

4.3 Evaluation Model 2

In the next stage, we removed the last 7 features in group C according to the score result and repeated the same training process above with the 13-subset of features to build the network model (model 2). The training process of model 2 lasts 30 min which is 21% time cost lower than model 1. Meanwhile, the accuracy of model 2 is 93.6006% which improves 3.1% compared with model 1.

Model 2 has a 0.964 TPR and a 0.126 FPR and its ROC area is 0.967 which is greater than model 1, namely the performance of model 2 improves significantly compared with model 1. Tues the chosen subset of features is more suitable for data mining tasks than the set contains the total 20 features. Figure 2 shows the ROC curve of model 2.

4.4 Evaluation Model 3

The last experiment chose 6 features (6-subset) in group A as the input to build the BP neural network (model 3). It takes 20 min and results in an accuracy of 89.2% which is the least in all these experiments.

Model 3 has the TPR of 0.921 and the FPR of 0.171 with the ROC area of 0.935. Figure 3 shows the ROC curve of model 3.

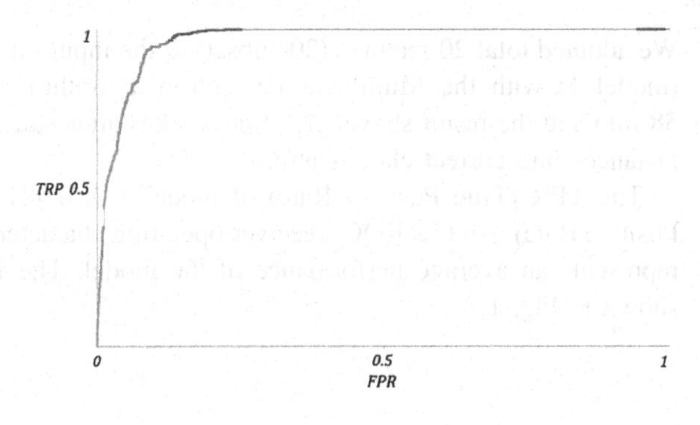

Fig. 2 The ROC curve of model 2

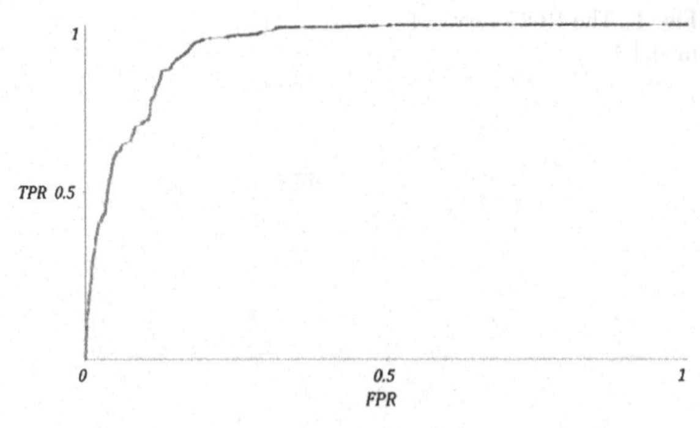

Fig. 3 The ROC curve of model 3

4.5 Discussion of Experimental Results

The summary of experimental results is given in Table 4.

According to the experimental results, the optimal subset of features is {Food Category, Sampling Date, Sampled Co. Province, Production Co. Province, Production Date, Shelf Life, Sampling Package, Sample Form, Quality Rank, Annual Sales, Unit Price, Sampling Site, Food Function}. When choosing the 13-subset of features as the input of BP neural network, it can result in a 93.6% accuracy and a satisfactory ROC area of 0.967. The model of 13-subset performs well enough to conduct subsequent data mining work.

As a typical case of the wrapper method, BP neural network requires a huge time cost in every training process due to its complicate structure. If we adopt BP neural network individually to verify every subset of features, the repeated training process will cost immeasurable time. Considering the 20-subset from the beginning, taking the method of exhaustion requires $2^{20} - 1$ times of training which is an impossible work in limited time. The IGBP method presented in this paper can find the optimum by eliminating features one after another according to the criterion of information gain. The worst case in the example above only takes 20 times of training, making the data mining task possible to accomplish.

On the other hand, IGBP method also avoids the issue of inaccuracy to adapt the filter method along. As the last experiment illustrates, if we chose the 6-subset in group A without any other evaluation to conduct the data mining process, the model will result in a low accuracy which means the 6-subset is not the optimum. Therefore IGBP method reserves the advantage of efficiency of the filter method and remedies its disadvantage of inaccuracy by combining the wrapper method.

Table 4 The summary of experimental results

Model	Indicator				
	Time cost (min)	Accuracy (%)	TPR	FPR	ROC
Model 1	38	90.5	0.981	0.262	0.917
Model 2	30	93.6	0.964	0.126	0.967
Model 3	20	89.2	9.921	0.171	0.935

5 Conclusion

Feature selection plays an important role in data mining and machine learning fields since it has great influences on both processes and results of the task. Choosing a proper subset of features to perform data mining could reduce the time cost and improve the performance of the model simultaneously. This paper proposed the IGBP method after making sufficient analyses on two types of feature selection methods. Through performing the evaluation processes repeatedly, the feasibility as well as the superiority of the IGBP method when facing the mass data have been proved, providing the subsequent data mining work with theoretical and practical basis.

Acknowledgements This work was supported by The National Key Technology R&D Program of China (2015BAK36B04).

References

1. Agarwal S. Data mining: data mining concepts and techniques. In: 2013 international conference on machine intelligence and research advancement (ICMIRA). IEEE; 2013. p. 203–7.
2. Almasoud AM, Al-Khalifa HS, Al-Salman A. Recent developments in data mining applications and techniques. In: 2015 tenth international conference on digital information management (ICDIM). IEEE; 2015. p. 36–42.
3. Liu H, Motoda H. Feature selection for knowledge discovery and data mining. Springer Science & Business Media; 2012.
4. Guyon I, Elisseeff A. An introduction to variable and feature selection. J Mach Learn Res. 2003;3(March):1157–82.
5. González A, Pérez R. Selection of relevant features in a fuzzy genetic learning algorithm. IEEE Trans Syst Man Cybern Part B (Cybern). 2001;31(3):417–25.
6. Yu L, Liu H. Efficient feature selection via analysis of relevance and redundancy. J Mach Learn Res. 2004;5(October):1205–24.
7. Sheikhpour R, Sarram MA, Gharaghani S, et al. A survey on semi-supervised feature selection methods. Pattern Recogn. 2017;64:141–58.
8. Kohavi R, John GH. Wrappers for feature subset selection. Artif Intell. 1997;97(1–2):273–324.
9. Wu G, Xu J. Optimized approach of feature selection based on information gain. In: 2015 international conference on computer science and mechanical automation (CSMA). IEEE; 2015. p. 157–61.
10. Najah A, El-Shafie A, Karim OA, et al. Application of artificial neural networks for water quality prediction. Neural Comput Appl. 2013;22(1):187–201.
11. Ennett CM, Frize M, Walker CR. Influence of missing values on artificial neural network performance. Stud Health Technol Inform. 2001;1:449–53.
12. Ding S, Li H, Su C, et al. Evolutionary artificial neural networks: a review. Artif Intell Rev. 2013;1–10.

Fault Diagnosis of Hoist Braking System Based on Improved Particle Swarm Optimization Algorithm

Lei Yao, Fu Zhong Wang and Su Min Han

Abstract Reliability of the mine hoist braking system is directly related to the safety of staff in the pit. For the sake of improving the accuracy of the fault diagnosis of the hoist braking system, aradial basis function (RBF) neural network diagnostic method based on improved particle swarm optimization (PSO) algorithm is proposed. Then, the hoist braking system fault diagnosis model is established, which uses some kinds of braking system fault characteristic parameters as input variables and adopts several kinds of main fault types as output ones. In view of the strong global convergence of the genetic algorithm (GA), the idea of crossover and mutation is introduced into PSO and the paper employs to optimize the parameters of hidden layer of RBF neural network. The simulation results show that the improved diagnosis strategy improves fault diagnostic speed and precision of the hoist braking system.

Keywords Mine hoist braking system · Fault diagnosis · RBF · GA · PSO

1 Introduction

As the main equipment of mine automation in the production process, mine hoist is the channel that connects ground and underground which is responsible for the lifting of the material, equipment and staffs [1]. Mine hoist is generally composed of spindle device, reducer, hoist braking system and electric traction device. Braking system is a very important part of the Mine hoist. According to incomplete statistics, the braking system failure rate accounted for more than half of the mine hoist fault approximately. Once the brake failure, brake oil cylinder sticks and the other failures occur, which will bring not only economic losses, but also more serious the personal safety accidents to staffs. Hence, it is even more important to

L. Yao · F.Z. Wang (✉) · S.M. Han
School of Electrical Engineering and Automation, Henan Polytechnic University, Jiaozuo 454000, China
e-mail: wangfzh@hpu.edu.cn

Y. Jia et al. (eds.), *Proceedings of 2017 Chinese Intelligent Systems Conference*, Lecture Notes in Electrical Engineering 459,
https://doi.org/10.1007/978-981-10-6496-8_4

improve the reliability of the hoisting braking system. In recent years, experts at home and abroad have adopted a variety of methods for the fault diagnosis of hoist braking system, such as artificial neural networks [2–6], immune optimization algorithm [7], support vector machine [8–10], fuzzy fault tree analysis method [11–13]. Due to the characteristics of particle swarm algorithm for parameter adjustment is simple, it hasbeen widely used in the fields of fault diagnosis, intelligent identification, etc. In literature [14], the RBF neural network was used in the fault diagnosis of hydraulic drilling rigs and the implicit layer parameters of RBF were optimized by PSO algorithm, the classification of hydraulic drilling rig has achieved good results. In the literature [15], a model of transformer fault diagnosis of support vector machine was established and PSO was used for optimizing the parameters of SVM model, which improved the accuracy of transformer fault diagnosis. In the literature [16], the author extracted the energy entropy of the high voltage circuit breaker which was used as the input parameter of the RBF neural network by wavelet packet transform. Finally, PSO algorithm was used for optimizing the parameters of the fault diagnosis model of the high voltage circuit breaker, and the accuracy of fault diagnosis was improved. However, PSO algorithm has its shortcomings. Aiming at the disadvantage that PSO algorithm is easy to fall into local minimum [17], the paper proposes a improved PSO algorithm, considering the feature of GA strong global convergence which puts the crossover and mutation of GA into the particle swarm algorithm.

This paper is organized as follows: Sect. 2 presents fault feature parameters of the hoist braking system. Section 3 introduces the fault diagnosis model based on RBF neural network. Section 4 establishes the fault diagnosis model based on improved PSO algorithm. The analysis and computational results are included in Sect. 5. Section 6 presents the concluding remarks.

2 Fault Feature Parameters of the Hoist Braking System

At present, most of mine hoist braking system has adopted hydraulic brake. Through a large number of reference data and field investigation, the common failure of hoist braking system is analyzed. Main fault characteristic parameters of the braking system are divided into seven kinds: brake positive pressure x_1, brake oil pressure x_2, residual pressure of the hydraulic pressure station x_3, brake oil pressure x_4, brake shoe opening and closing state x_5, wear overrun determine hydraulic oil pressure x_6 and working pressure of hydraulic station x_7. The main failure of mine hoist braking system are low friction coefficient y_1, brake shoe wear too much y_2, brake spring failure y_3, brake cylinder stuck y_4, excessive residual pressure y_5 and the loss of pressure of the hydraulic pressure station y_6. That is, $x_1 \sim x_7$ represent the 7 kinds of main characteristic parameters of hoist braking system and $y_1 \sim y_6$ represent 6 kinds of main fault of hoist braking system.

3 Fault Diagnosis Model based on RBF Neural Network Fault Diagnosis

As RBF neural network which has strong biological background is a feed forward neural networks, it has been widely used in the fields of pattern recognition, fault diagnosis, etc. A typical network structure of RBF neural network is $n-h-m$. In other words, it has n inputs nodes, h hidden layer nodes and m outputs nodes. The structure of the established RBF neural network is shown in Fig. 1.

The transformation from the input layer to hidden layer is nonlinear in this model, so we adopt the radial basis function as the activation function. Gaussian function is frequently used as activate function, which has simple form and strong data analysis, so we choose it as activation function of the ith hidden layer node, in order to provide activation response for input layer. The radial basis function has radial symmetry in the center of a n-dimensional space. Moreover, the father neuronal input is from the center, the lower activated ability it is. So the data center is represented by c_i, which represents the ith hidden layer nodes of a network data center. The gaussian function can be calculated by Eq. (1).

$$\Phi_i(\mathrm{t}) = e^{-\frac{t^2}{\delta_i^2}} \tag{1}$$

where δ_i is called the extension constants or width of this basis function. Obviously, if the smaller the value of δ_i is, the narrower the width of the radial basis function is. So the basis function is more selective. And the output of the RBF network layer can be calculated by Eq. (2).

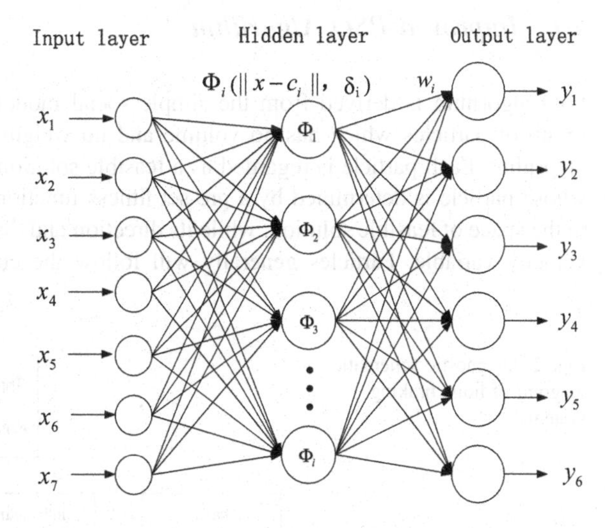

Fig. 1 Brake system of RBF network diagnosis model

$$y_k = \sum_{i=1}^{h} w_i \Phi_i(\|x - c_i\|, \delta_i) \quad (2)$$

That is, each neurongenerates a response $\Phi_i(\|x - c_i\|)$ corresponding to input x, and the output of the neural network is a weighted sum of all these responses.

According to the above introduction of hoist braking system fault, the network includes N groups of input data of which represented by $X = [x_1, x_2, \ldots, x_N]$ are set up, in which each of these elements $x_i = [x_{i1}, x_{i2}, x_{i3}, x_{i4}, x_{i5}, x_{i6}, x_{i7}]^T$ represents a group of failure data. And the paper regards these 6 kinds of typical hoist braking system fault as output of the model, and its corresponding output data is $Y = [y_{i1}, y_{i2}, y_{i3}, y_{i4}, y_{i5}, y_{i6}]^T$, which is coded with numbers 1 to 6 respectively.

4 RBF Neural Network Diagnosis Model Based on Improved Particle Swarm Optimization

Since the fault diagnosis of braking system is a complex nonlinear problem, the failure reason is influenced by multiple factors. The basic PSO algorithm which uses for diagnosing the brake system of the mine hoist is easy to fall into the local minimum value, so that the diagnosis result is not ideal. In order to overcome the flaw of basic PSO algorithm, the paper uses the improved PSO algorithm to optimize hidden layer center parameters and connection weights of RBF neural network. Figure 2 is the diagnostic schematic diagram of hoist braking system:

4.1 Improved PSO Algorithm

PSO algorithm is derived from the simple social model. Randomly initializing a group of particles which has no volume and no weight is the basic idea of PSO algorithm. Each particle is regarded as a feasible solution of optimization problem, whose particle is determined by a pre-set fitness function. Each particle will move in the space of feasible solution, while its direction and distance are determined by a velocity variable. Particles generally will follow the current optimum ones, and

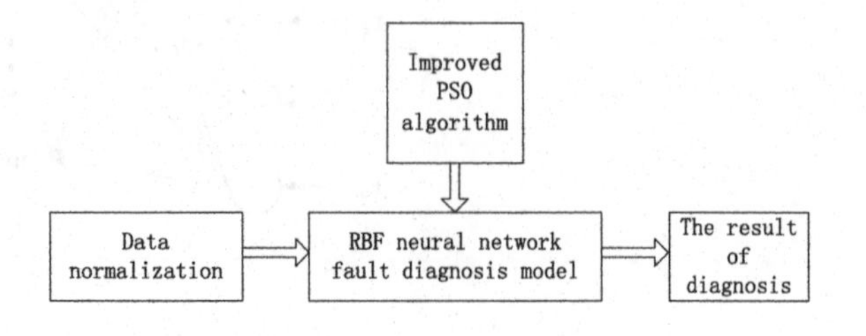

Fig. 2 Diagnostic schematic diagram of hoist braking system

eventually get the optimal solution by searching each subsequent generation. In each generation, the particle will track two extreme values: One is the optimal solution which found by the particle itself so far, and the other is the optimal solution which found by the whole group so far.

Suppose that M particles are flying at a certain velocity in the D-dimensional search space. The velocity of particle i at a certain time (t) is expressed as: $v_i^t = \left(v_{i1}^t, v_{i2}^t, \ldots, v_{id}^t\right)^T$, $v_{min,d}$ and $v_{max,d}$ are the minimum and maximum values respectively. So the velocities of the particles are limited to ensure that the velocities are not too high or too low in the iterative process. The individual best position is expressed as: $p_i^t = \left(p_{i1}^t, p_{i2}^t, \ldots, p_{iD}^t\right)^{\mathrm{T}}$, also known as p_{ibest}. The global optimal position is expressed as: $p_g^t = \left(p_{g1}^t, p_{g2}^t, \ldots, p_{gD}^t\right)^{\mathrm{T}}$, also known as g_{best}, $1 \leq d \leq D, 1 \leq i \leq M$.

The velocity and position of particles are updated by the following equations:

$$v_{id}^{t+1} = w(t)v_{id}^t + c_1 r_1 \left(p_{id}^t - x_{id}^t\right) + c_2 r_2 \left(p_{gd}^t - x_{id}^t\right) \tag{3}$$

$$x_{id}^{t+1} = x_{id}^t + v_{id}^{t+1} \tag{4}$$

where c_1 and c_2 are called learning factor, which often take random number between [0, 4] as its value. r_1 and r_2 are both random numbers which are uniformly distributed between [0, 1]. The value of t can be considered as the current evolutionary generation, the number of current iterations.

The $w(t)v_{id}^t$ in Eq. (3) is the inertial term which keeps the particle moving in the same direction as its original direction [18]. $w(t)$ is the inertia factor, which reflects the ability that the particle inherit the velocity of previous particle, whose value will affect the result of diagnosis. Hence, in order to better balance capability of the global search and the local search, the paper uses the inertial factor of linear regressive. The method is calculated as follows:

$$w(t) = w_{start} - \left(w_{start} - w_{end}\right)\left(\frac{t}{t_{\max}}\right)^2 \tag{5}$$

where w_{start} is the initial inertia weight, w_{end} is the inertia weight when the iteration is terminated, t is the current number of iteration.

Introducing the crossover can improve the global search ability of swarm. The crossover means that the two different particles are combine with each other based on a certain probability p_c. The crossover of velocity and position of particles are calculated by Eq. (6) and Eq. (7) respectively.

$$\begin{cases} v_i^{t+1} = a * v_i^t + (1-a)v_j^t \\ v_j^{t+1} = (1-a)v_i^t + a * v_j^t \end{cases} \tag{6}$$

$$\begin{cases} x_i^{t+1} = b * x_i^t + (1-b)x_j^t \\ x_j^{t+1} = (1-b)x_i^t + b * x_j^t \end{cases} \tag{7}$$

where a and b are both random number between [0, 1]. Since the crossover probability p_c will have great influence on the results, the paper adopts self-adaption adjustment strategy for p_c. The probability p_c is calculated by Eq. (8).

$$p_c = \begin{cases} \frac{(p_{c1} - p_{c2})(f_{\max} - f')}{f_{\max} - f_{avg}}, f' \geq f_{avg} \\ p_{c1}, f' < f_{avg} \end{cases} \tag{8}$$

where $f_{\max}$ is the largest fitness value within swarm. f_{avg} is the average fitness value of each generation. To compare the fitness values of the two particles crossing each other, the better one is assigned to f'. Value of p_c ranges from 0.75 to 0.9, $p_{c1} = 0.95$, $p_{c2} = 0.75$.

The mutation is to verify the fitness values of particles after the crossover. The value of traditional p_m is between 0.001 and 0.1. The value of p_m is too high that make the well-adapted swarm be destroyed easily and the value of p_m is too low to introduce new genes. Hence, the paper adopts self-adaption adjustment strategy for p_m. The p_m is calculated by Eq. (9).

$$p_m = \begin{cases} p_{m2} + \frac{(p_{m1} - p_{m2})(f_{\max} - f)}{f_{\max} - f_{avg}}, f \geq f_{avg} \\ p_{m1}, f < f_{avg} \end{cases} \tag{9}$$

where f is the individual fitness value of the mutation, $p_{m1} = 0.1$, $p_{m2} = 0.001$.

This paper selects the error of the network learning sample as the fitness function, the fitness function is calculated by Eq. (10).

$$f = \sum_{i=1}^{N} \sum_{t=1}^{M} (y_{it} - c_{it})^2 \tag{10}$$

where N is the total number of training samples, M is the number of RBF neural network output neurons, y_{it} is the ideal output of a node, c_{it} is the actual output of a node.

4.2 The Improved Particle Swarm Algorithm to Optimize RBF Neural Network

The optimization method is broadly divided into two steps. The first step is to use traditional subtraction clustering algorithm to obtain the center of the RBF neural network. The second step is to use the improved PSO algorithm to optimize the network center values, width and its connection weight between the hidden layer and the output layer. Assume that the number of nodes in the input layer, hidden layer and output layer of RBF neural network is n, h, and m respectively. The length of the central vector c_i of each neuron of the hidden layer is n, the variance σ_i is of h-dimension and the number of neurons of the output layer weights is m, hence, the dimension of the parameter vector is $n+1+m$. That is, the dimension of all neuron vectors which need to be optimized is $D=(n+1+m)*h$. Then the D-dimensional vector is used as the coding information of the particle swarm. The termination condition of the program is taken into account by the maximum number of iterations and the learning accuracy.

The steps flow chart for algorithm is shown in Fig. 3:

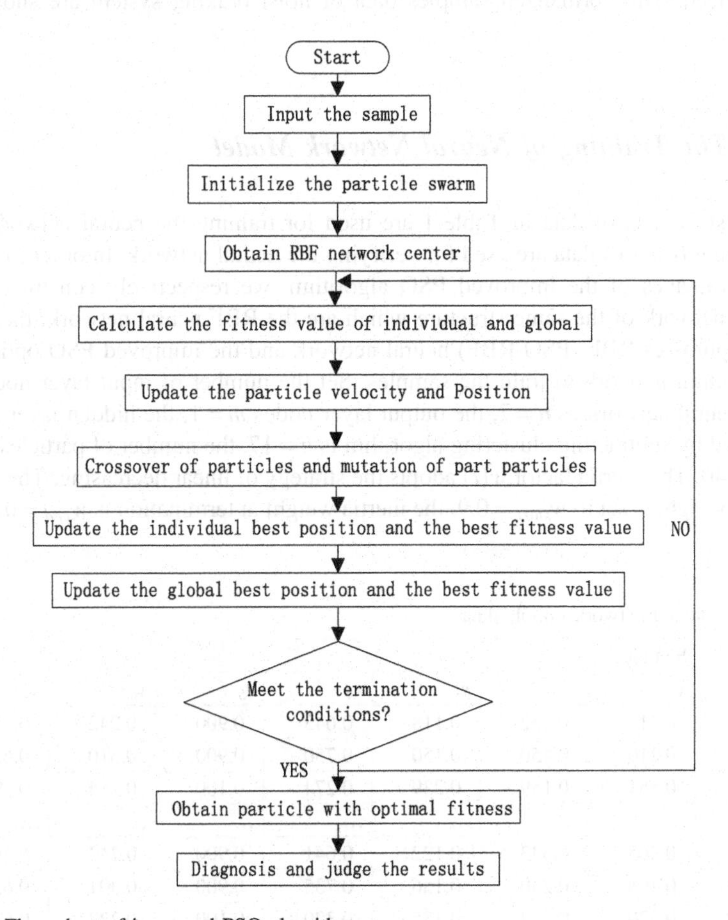

Fig. 3 Flow chart of improved PSO algorithm

5 Fault Simulation Process

5.1 Normalization of Fault Samples

The 42 sets of faulty data from the hoist braking system are selected as the simulation data in Matlab. The normalized Eq. (1) is as follows:

$$x_i = \frac{x_{acti} - x_{\min i}}{x_{\max i} - x_{\min i}} \tag{11}$$

where x_i is the input value of sample, x_{acti} is the actual value before normalized, $x_{\min i}$ is the minimum value before normalized, $x_{\max i}$ is the maximum value before normalized. The normalized samples data of hoist braking system are shown in Table 1.

5.2 The Training of Neural Network Model

The first 36 sets of data in Table 1 are used for training the neural network, the remaining 6 sets of data are used for verifying the neural network. In order to prove the advantages of the improved PSO algorithm, we respectively construct three neural network of the same structure which are the RBF neural network, the basic PSO optimizes RBF (PSO-RBF) neural network and the improved PSO optimizes RBF neural network to train the samples. Set the number of input layer nodes of RBF neural network as $n = 7$, the output layer nodes $m = 1$, the hidden layer nodes obtained by subtracting clustering algorithm is $h = 17$, the number of particles is set to $N = 40$. The inertia factor $w(t)$ adopts the strategy of linear decreasing. The initial inertia weight is set to $w_{start} = 0.9$, the inertia weight at termination is $w_{end} = 0.4$ and

Table 1 Neural network sample data

Group	Sample						
	x_1	x_2	x_3	x_4	x_5	x_6	x_7
1	0.218	0.352	0.118	0.643	0.900	0.242	0.776
2	0.210	0.450	0.150	0.750	0.900	0.310	0.620
3	0.551	0.159	0.239	0.274	0.100	0.154	0.592
…	…	…	…	…	…	…	…
37	0.265	0.343	0.122	0.641	0.900	0.242	0.765
38	0.498	0.230	0.156	0.735	0.900	0.301	0.638
39	0.520	0.224	0.185	0.320	0.100	0.227	0.642
40	0.568	0.620	0.120	0.850	0.100	0.405	0.838
41	0.300	0.524	0.825	0.450	0.900	0.440	0.550
42	0.540	0.134	0.295	0.190	0.900	0.680	0.124

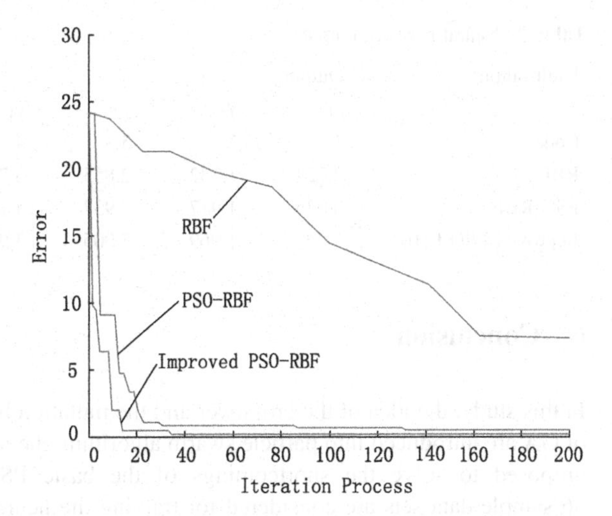

Fig. 4 Error-training frequency curve

learning factor is set to $c_1 = c_2 = 2$. Crossover probability p_c adopts self-adaption adjustment strategy, the mutation probability p_m adopts self-adaption adjustment strategy also, the maximum number of iterations is set to $t_{\max} = 200$. The error convergence accuracy is set to 0.001 and the fault output is encoded as 1–6.

Figure 4 respectively shows the systematic error variation of the three training methods.

As we can see in Fig. 4, the speed of the error convergence of PSO-RBF neural network and improved PSO-RBF neural network are both faster than the single RBF neural network in the process of the initial iteration. However, the basic PSO-RBF neural network is accompanied by a long time error value pause in the iterative convergence process, and the improved PSO-RBF neural network has already achieved the corresponding error condition at the 37th iteration. Obviously, the algorithm in the paper is superior to the other two algorithms in terms of improving the fault diagnosis effect of the hoist brake system.

5.3 Verify Simulation Results

The remaining 6 sets of fault samples data are respectively assigned to the three kinds of trained neural network model to verify the accuracy of model, and the output results are shown in Table 2.

Through the Table 2 as we can see, membership degree of diagnosis result of the improved PSO algorithm is more close to the desired output. So the improved PSO algorithm is more effective than the other two algorithms in the hoist braking system.

Table 2 Neural network output

Fault output	Output					
	y_1	y_2	y_3	y_4	y_5	y_6
Code	1	2	3	4	5	6
RBF	0.724	1.732	2.855	3.775	4.781	5.619
PSO-RBF	0.925	1.917	2.937	4.142	4.883	5.838
Improved PSO-RBF	1.009	1.969	3.005	3.947	5.001	5.981

6 Conclusion

In this study, the idea of the crossover and the mutation based on the characteristics of GA are introduced into particle swarm algorithm, the improved PSO algorithm is proposed to solve the shortcomings of the basic PSO algorithm. A total of 36 sample data sets are considered for training the neural network model, and the 6 sample data sets is used for verifying the neural network model. The results shows that the improved PSO algorithm is used for the fault diagnosis of the mine hoist braking system, not only the diagnosis accuracy is high, but also the error convergence speed is fast.

Acknowledgements Support by National Key Research and Development Program (2016YFC0600906).

References

1. Liu J, Wang F, Li Y. Fault diagnosis of hoist braking system based on neural network optimized by particle swarm. Control Eng China. 2016;23(2):294–8.
2. Liu JR, Wang SJ, Ren F, et al. Fault diagnosis of hoist braking system based on BP neural network optimized by genetic algorithm. Coal Mine Mach. 2011;05:246–8.
3. Zhang Q, Hu N, Li H. Fault diagnosis of the mine hoist brake based on GA-BP neural network. J Liaoning Tech Univ. 2016;02:155–9.
4. Jingyan LIU. Hoist fault diagnosis based on improved particle swarm neural network. J Henan Polytech Univ. 2014;03:313–7.
5. Yuwei Y, Hunju L. Hoist braking system of mine fault tree analysis based on Bayesian network. Coal Mine Mach. 2014;04:258–60.
6. Zhao M, Wenshang X, Qu Y, et al. RBF network optimization based of the hoist fault diagnosis methods. Microcomput Inf. 2012;28(09):78–9, 89.
7. Wang CJ, Xia SX, Niu Q. Artificial immune particle swarm optimization for fault diagnosis of mine hoist. Acta Electron Sinica. 2010;38(2A):94–8.
8. Dong L, Sun W, Zhao J, et al. Fault diagnosis of mine hoist braking system based on support vector machine. Mech Eng Autom. 2010;02:124–6.
9. Guo X, Ma X. Mine hoist braking system fault diagnosis based on a support vector machine. J China Univ Min Technol. 2006;(35):813–7.
10. Wang Y, Gao Y, Ma J. Fault diagnosis on braking systems of mine hoists based on ordering binary tree SVM. Min Process Equip. 2011;09:46–9.

11. Hu SF. Failure analysis on hoist braking system based on fuzzy fault tree. Min Process Equip. 2014;12:53–6.
12. Minjie J, Guo K, Zhang S, et al. Fuzzy fault tree analysis on hoist disc brake. Min Process Equip. 2013;07:53–5.
13. Xia Z, Niu Q, Zhang L. Mine hoist fault diagnosis system based on FTA. Microcomput Inf. 2008;(24):160–2.
14. Du JY, Zhang MZ. Fault diagnosis of hydraulic drilling rig based on particle swarm optimization RBF neural network. Coal Mine Mach. 2012;33(05):251–3.
15. Han S, Zhu J, Mao J, et al. Fault diagnosis of transformer based on particle swarm optimization-based support vector machine. Electr Meas Instrum. 2014;51(11):71–5, 90.
16. Xu J, Zhang B, Lin S, et al. Application of energy spectrum entropy vector method and RBF neural networks optimized by the particle swarm in high-voltage circuit breaker mechanical fault diagnosis. High Volt Eng. 2012;38(06):1299–306.
17. Lei HX, Liu N, Cui DJ, et al. Transformer fault diagnosis based on optimized FCM clustering by hybrid GA and PSO. Power Syst Prot Control. 2011;39(22):52–6.
18. Ganesan T, Vasant P, Elamvazuthy I. A hybrid PSO approach for solving non-convex optimization problems. Arch Control Sci. 2012;22(1):87–105.
19. Qian Y, Zhang H, Peng D, et al. Remote integrated fault diagnosis for generator unit using PCA and GA-PSO-RBF. J Electron Meas Instrum. 2012;26(7):600–4.
20. Cui HQ, Liu XY. Parameter optimization algorithm of RBF neural network based on PSO algorithm. Comput Technol Dev. 2009;19(12):117–9.

Attitude Control of a Quad-rotor Based on LADRC

Yong Zhang, Zengqiang Chen, Xinghui Zhang and Qinglin Sun

Abstract In this paper, the structure of the linear active disturbance rejection control (LADRC) is described in detail, including linear tracking differentiator, linear extended state observer and linear feedback control law. Typical algorithms of the each part are given as well. In order to control the attitude of the quad-rotor robot as we expected, we designed a LADRC scheme. Simulations are carried out based on Simulink. After parameters adjustment and simulating with different disturbances, the simulation results show that the LADRC can satisfy the need of control precision and speed of response. It also indicates that the LADRC has strong robustness and anti-disturbance performance, which can control the non-linear time-varying coupling system effectively.

Keywords linear active disturbance rejection control (LADRC) · Linear extended state observer · Quad-rotor robot

1 Introduction

PID is a control strategy that can eliminates the error by using the error between set point and output independent of the system mathematical model. Until now, PID holds a dominant position in the fields of aerospace control, motion control and other process control.

Y. Zhang · Z. Chen (✉) · Q. Sun
College of Computer and Control Engineering, Nankai University, Tianjin 300350, China
e-mail: chenzq@nankai.edu.cn

Y. Zhang
e-mail: zhangdayong810@163.com

Q. Sun
e-mail: sunql@nankai.edu.cn

X. Zhang
Tianjin Sino-German University of Applied Sciences, Tianjin 300350, China
e-mail: xhzhang@tute.edu.cn

Y. Jia et al. (eds.), *Proceedings of 2017 Chinese Intelligent Systems Conference*, Lecture Notes in Electrical Engineering 459,
https://doi.org/10.1007/978-981-10-6496-8_5

However, with the development of technology and science, the standards for control precision and speed are higher and higher, so that the disadvantages of PID have been revealed gradually. In order to inherit the advantages of PID control and overcome its shortcomings, Professor Han created a new control strategy, He denote this new strategy active disturbance rejection control (ADRC) [1] in 1990s.

The central idea of ADRC is to estimate and compensate both internal dynamics and external disturbances in real time, we denote the uncertainties of internal dynamics and external disturbances as total disturbance. ADRC has fast response speed, high control precision and strong anti-disturbance capability, then ADRC used widely in many fields. But the ADRC from Han is nonlinear, although the nonlinear method may be more effective, but they increase extra complexity in control algorithm implementation and parameters tuning. So Professor Gao from Cleveland State University proposed a linear active disturbance rejection control (LADRC) [2]. In this paper all of the discussions are limited to LADRC case.

Attitude control of quad-rotor is a system that multi-variable, nonlinear, strong coupling and sensitive to disturbance. It has six DOF (position and attitude) and four control input. Attitude control is the critical problem. At present, the related control methods [3–8] like PID [9], Backstepping [10], LQR [11, 12], sliding mode control [13–15], etc.

2 The Basic Principle of LADRC

Linear active disturbance rejection control is composed by linear tracking differentiator, linear extended state observer and linear error control law. The parts where inside the dotted line are LADRC, as shown in Fig. 1.

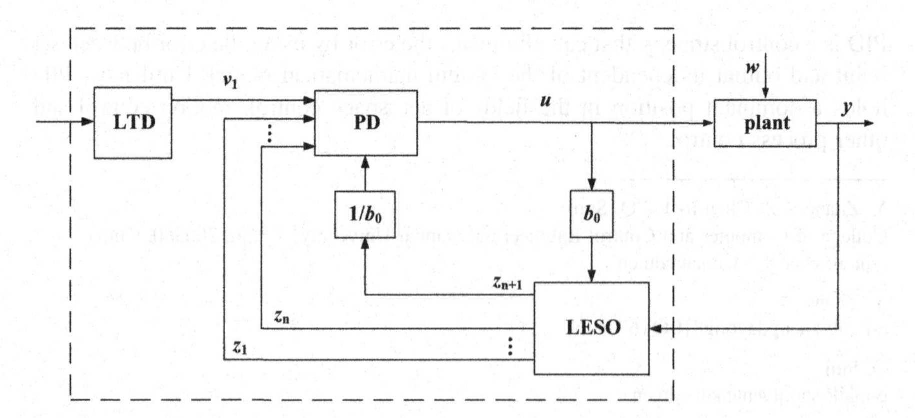

Fig. 1 Block diagram of LADRC

2.1 Linear Tracking Differentiator (LTD)

To avoid the set point jump, it is necessary to construct a transient process (i.e., LTD) to make the output of the plant can be reasonably followed.

For the sake of simplicity, consider a second order system, continuous linear tracking differentiator algorithm as follows:

$$\begin{cases} \dot{v}_1 = v_2 \\ \dot{v}_2 = -1.76 \cdot r \cdot v_2 - r^2(v_1 - v) \end{cases} \tag{1}$$

where v is input, v_1 is the tracking signal for v, v_2 is the differential signal for v. Here, the parameter, r, can determine the speed of tracking.

2.2 Linear Extended State Observer (LESO)

Both the external disturbances and the unknown internal dynamics are referred as the total disturbance as an augmented state in addition to the system states outside. The basic idea is to obtain the total disturbance and use it in the control law to simplify the plant to a cascade integrator control problem, and this is the core technology of ADRC.

Continuous linear extended state observer as follows:

$$\begin{cases} \dot{z}_1 = z_2 - \beta_1(z_1 - y(t)) \\ \dot{z}_2 = z_3 - \beta_2(z_1 - y(t)) \\ \quad \vdots \\ \dot{z}_n = z_{n+1} - \beta_n(z_1 - y(t)) + b_0 u(t) \\ \dot{z}_{n+1} = -\beta_{n+1}(z_1 - y(t)) \end{cases} \tag{2}$$

In order to represent all the parameters of the LESO with ω_o, assigning all observer eigenvalues at $-\omega_o$ as shown in the following equation:

$$s^n + \beta_1 s^{n+1} + \cdots + \beta_{n+1} s + \beta_n = (s + \omega_o)^n \tag{3}$$

here, ω_o is denoted as the bandwidth of the LESO. More importantly, the LESO tuning are reduced to adjust only one parameter ω_o.

2.3 Linear Error Control Law

The plant is simplified to a cascade integrator control system after using LESO to estimate the total disturbance, so we can reach our target with a simple linear PD controller. And the linear error control law as follows:

$$\begin{cases} u_0 = k_p(r - z_1) - k_{d_1} z_2 - \cdots - k_{d_{n-1}} z_n \\ u = u_0 - z_{n+1}/b_0 \end{cases} \quad (4)$$

where the parameters are chosen so that the closed-loop system has n poles at $-\omega_c$, as shown in the following equation:

$$s^n + k_{d_{n-1}} s^{n+1} + \cdots + k_{d_1} s + k_p = (s + \omega_c)^n \quad (5)$$

Equation (4) is used to make the closed-loop system pure n order with no zero, because this equation did not use the differentiation of the set point.

3 Attitude Control of Quad-rotor Simulation

In this paper, the simulation of the attitude control of quad-rotor based on LADRC in the environment of the MATLAB Simulink.

Step 1: Constructing the plant model in state equation form;
Step 2: According to the state equation, design the LESO with the parameter ω_o;
Step 3: Design LTD;
Step 4: Design PD controller with the parameter ω_c;
Step 5: Constructing simulation structure and tuning.

3.1 Quad-rotor Dynamics Model

The structure for attitude control system of quad-rotor with three DOF as shown in Fig. 2.

The mathematical model [16] for attitude control of quad-rotor with three DOF as follows:

$$\begin{cases} \ddot{y} = \frac{K_{tc}}{J_y}(v_f + v_b) + \frac{K_{tn}}{J_y}(v_r + v_l) \\ \ddot{p} = l\frac{K_f}{J_p}(v_f - v_b) \\ \ddot{r} = l\frac{K_f}{J_r}(v_r - v_l) \end{cases} \quad (6)$$

where y, p and r are yaw angle, pitching angle and roll angle, respectively, and v_f, v_b, v_l, v_r are the speed voltage that control front, back, left and right rotor,

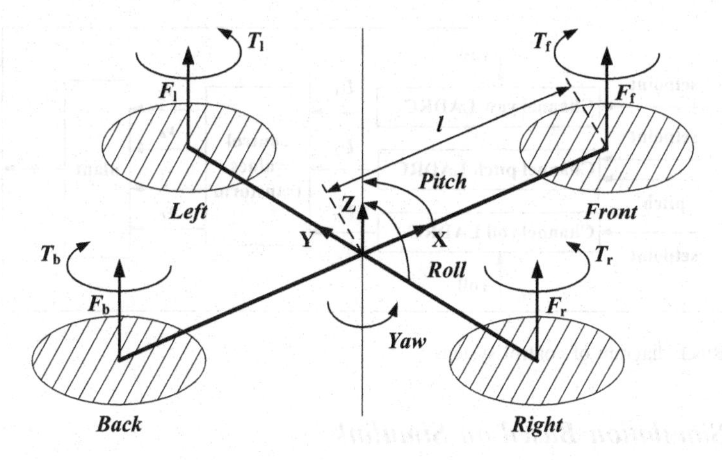

Fig. 2 Free-body diagram of quad-rotor

respectively, K_{tn} is the coefficient of clockwise screw torque, its value is 0.0036 N m/V, K_{tc} is the coefficient of counterclockwise screw torque, its value is − 0.0036 N m/V, K_f is the screw lift coefficient, its value is 0.1188 N/V. J_y, J_p, J_r are the rotational inertia of yaw shaft, pitching shaft, roll shaft and their values are 0.1104 kg m^2, 0.0552 kg m^2, 0.0552 kg m^2, respectively, l is the distance between the center of rotation and the center of propeller, its value is 0.197 m.

Then the equation can be written with all the parameters as formula (7):

$$\begin{cases} \ddot{y} = \omega_1(t) + U_1 \\ \ddot{p} = \omega_2(t) + U_2 \\ \ddot{r} = \omega_3(t) + U_3 \end{cases} \tag{7}$$

where ω_1(t), ω_2(t) and ω_3(t) are the unknown disturbances from the channels of yaw angle, pitching angle and roll angle, respectively, U_1, U_2 and U_3 are virtual control values for the three channels, i.e.:

$$U = \begin{bmatrix} U_1 \\ U_2 \\ U_3 \end{bmatrix} = \begin{bmatrix} -0.0326 & -0.0326 & 0.0326 & 0.0326 \\ 0.424 & -0.424 & 0 & 0 \\ 0 & 0 & 0.424 & -0.424 \end{bmatrix} \begin{bmatrix} v_f \\ v_b \\ v_r \\ v_l \end{bmatrix} \tag{8}$$

And

$$M = \begin{bmatrix} -0.0326 & -0.0326 & 0.0326 & 0.0326 \\ 0.424 & -0.424 & 0 & 0 \\ 0 & 0 & 0.424 & -0.424 \end{bmatrix}$$

M is the static coupling matrix.

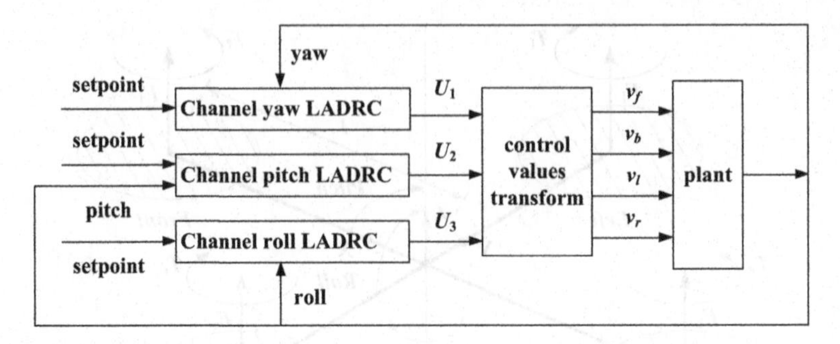

Fig. 3 Block diagram of control system

3.2 *Simulation Based on Simulink*

The attitude control system of quad-rotor with three DOF could be divided into three channels [17–19] yaw angle, pitching angle and roll angle, respectively, as shown in Fig. 3.

In Fig. 3, the transformation of the control values is a pseudo-inverse matrix of M. There are three methods to get the pseudo-inverse matrix of M, direct solving method, SVD method and QR method. In this paper, we used the direct solving method.

And

$$M^{-1} = \begin{bmatrix} -7.6687 & 1.1792 & 0 \\ -7.6687 & -1.1792 & 0 \\ 7.6687 & 0 & 1.1792 \\ 7.6687 & 0 & -1.1792 \end{bmatrix}$$

In terms of Fig. 3, we can construct a LESO for every channel. In the yaw channel, for example, the LESO is:

$$\begin{bmatrix} \dot{z}_1 \\ \dot{z}_2 \\ \dot{z}_3 \end{bmatrix} = \begin{bmatrix} -3\omega_o & 1 & 0 \\ -3\omega_o^2 & 0 & 1 \\ -\omega_o^3 & 0 & 0 \end{bmatrix} z + \begin{bmatrix} 0 & 3\omega_o \\ 1 & 3\omega_o^2 \\ 0 & \omega_o^3 \end{bmatrix} \begin{bmatrix} u \\ y \end{bmatrix} \tag{9}$$

And z_3 is the augmented state for $\omega_1(t)$.

4 Performance of LADRC

After designing the LADRC, next step is tuning the parameters. Through a series of simulation experiments, eventually determined the parameters as follows: $r = 100$, $\omega_o = 30$, $\omega_c = 20$. In the experiment, the set point of the three channels is square

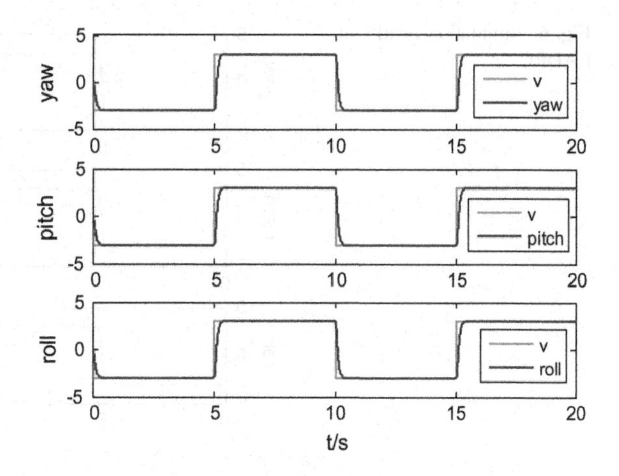

Fig. 4 Simulation results of LADRC

wave signal which the amplitude is 3 and the frequency is 0.1 Hz. The unknown disturbances ω_1(t), ω_2(t) and ω_3(t) are all sign(sin(0.9t)), and the output of the three channels with a white noise as the sensor disturbance which it's peak value is 0.1%, a 1 ms sampling period.

The initial value x_0 is [0 0 0 0 0 0], the simulation time is 20 s. Good performance is achieved through the LADRC and shown in Fig. 4.

From the Fig. 4, under the control of LADRC, tracking performance of the three channels are excellent, the response time are short and no overshoot. In order to further verify the robustness of the LADRC, we set the unknown disturbances as sign(sin(0.9t)) + cos(0.3t), the LADRC parameters are invariable. The simulation results as shown in Fig. 5.

Now we set the value of the unknown disturbances more complicated as 0.5sign (sin(0.5t)) + cos(0.3t) + 2cos(0.9t), the performance of the LADRC as shown in Fig. 6.

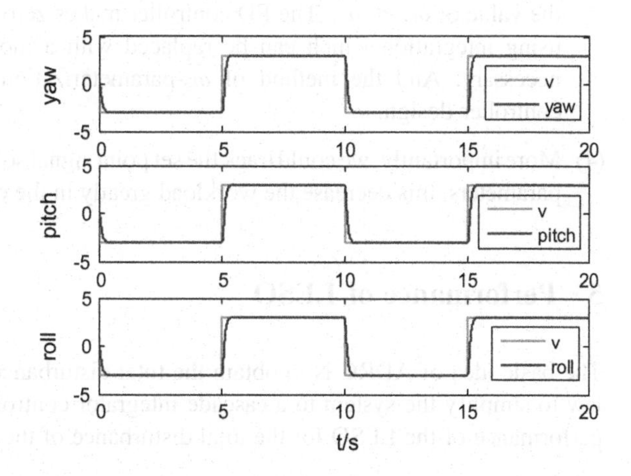

Fig. 5 Simulation results of LADRC

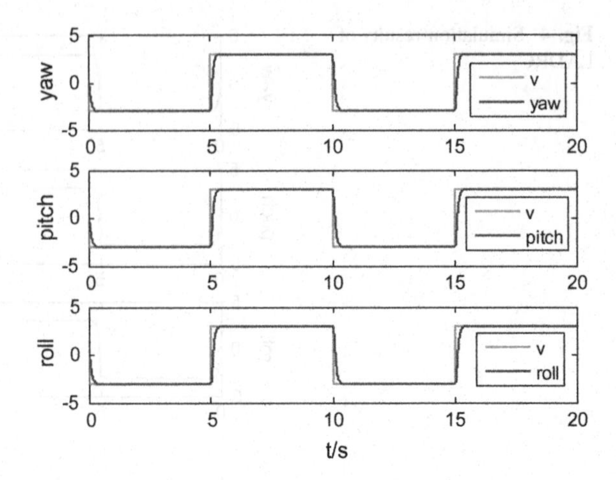

Fig. 6 Simulation results of LADRC

From the Figs. 4, 5 and 6, we can see that the LADRC has high performance to suppress the influence of the external disturbances, strong robustness and anti-disturbance ability.

Remarks:

(1) If the parameter of the LTD, *r*, bigger, then the tracking performance is better. And the LTD is not essential, the high tracking performance we would get without a LTD or TD sometimes.

(2) The further increase in ω_o doesn't improve the tracking performance greatly, but the observer can observe more noise. In general, a compromise value is made between the speed that the LESO tracks the states and its sensitivity to the noise. In fact, ESO should be work as fast as the measurement noise allows.

(3) Set an initial value of ω_c based on the bandwidth requirement from the transient response, and increase ω_c gradually until obtain good performance. In general, the value of $\omega_c < \omega_o$. The PD controller makes zero steady state error without using integration which can be replaced with a more complicated design, if necessary. And the method of ω_c-parameterization could be applied to all controller design.

(4) More importantly, we could track the set point signal so precise only by using three parameters, this decrease the workload greatly in the engineering application.

5 Performance of LESO

The basic idea of ADRC is to obtain the total disturbance and use it in the control law to simplify the system to a cascade integrator control problem. The estimating performance of the LESO for the total disturbance of the three channels are shown

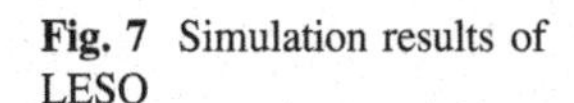

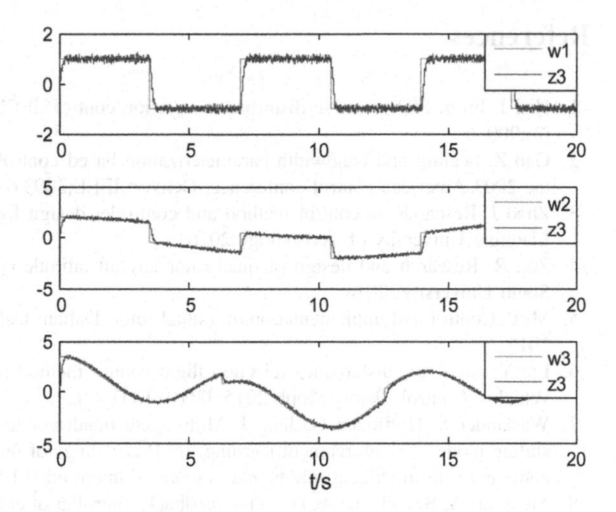

Fig. 7 Simulation results of LESO

in Fig. 7, and we set the total disturbance $\omega_1(t) = \text{sign}(\sin(0.9t))$, $\omega_2(t) = \text{sign}(\sin(0.9t)) + \cos(0.3t)$, and $\omega_3(t) = 0.5\text{sign}(\sin(0.5t)) + \cos(0.3t) + 2\cos(0.9t)$.

High estimating performance is achieved through using the LESO and examined in the following figure. From the figure we can see, z_3 estimated the disturbances accurately, and then the disturbances can be compensated in the PD controller to simplify the plant.

6 Concluding Remarks

In this paper, we designed a LADRC for the attitude control of a quad-rotor robot. This new algorithm makes controller design and parameters tuning easier and more effective, there are only three parameters we need to tuning. The method was examined successfully in MATLAB Simulink, which incorporated white sensor noise and external disturbances.

As shown in simulation results, the LADRC can satisfy the need of control precision and speed of response. It also indicates that the LADRC has strong robustness and anti-disturbance performance, which can control the nonlinear time-varying coupling system effectively.

Acknowledgements This work is supported by National Natural Science Foundation (NNSF) of China (61573197, 61573199).

References

1. Han J. From PID to active disturbance rejection control. IEEE Trans Ind Electron. 2009;56 (3):900–6.
2. Gao Z. Scaling and bandwidth-parameterization based controller tuning. In: Proceedings of the 2003 American control conference. Denver: IEEE;2003,6:4989–96.
3. Zhao J. Research on control method and controller design for quadrotor aircraft. Liaoning: Liaoning University of Technology;2016.
4. Zou R. Research and design of quad-rotor aircraft attitude control system. Hunan: Central South University; 2014.
5. Mi P. Control and implementation of a quad-rotor. Dalian: Dalian University of Technology; 2015.
6. Liu Y. An active disturbance-rejection flight control method for quad-rotor unmanned aerial vehicles. Control Theory Appl. 2015;32(10):1351–60.
7. Waslander S, Hoffmann G, Jang J. Multi-agent quadrotor test bed control design: integral sliding mode vs reinforcement learning. In: Proceedings of the 2005 IEEE/RSJ international conference on intelligent robots and systems. Edmonton: IEEE;2005. p. 3712–17.
8. Mokhtari A, Benallegue A. Dynamic feedback controller of euler angles and wind parameters estimation for a quadrotor unmanned aerial vehicle. In: Proceedings of the 2004 IEEE international conference on robotics and automation. New Orleans: IEEE;2004. vol. 3, p. 2359–66.
9. Lu W. Double loop PID control based on quad-rotor. Sci Technol Eng. 2014;14(33):127–31.
10. Bouchoucha M, Seghour S, Osmani H. Integral backstepping for attitude tracking of a quad-rotor system. Elektronika Ir Elektrotechnika. 2011;10(116):75–80.
11. Jiang Z, Han J, Wang Y, Song Q. Enhanced LQR control for unmanned helicopter in hover. Syst Control Aerosp Astronaut. 2006:1438–43.
12. Hu C, Meng Q, Liu X. Observer-based LQR control of shaping process of automobile belt. Intell Control Autom. 2004;4:3310–4.
13. Zheng E, Xiong J, Luo J. Second order sliding mode control for a quad-rotor UAV. ISA Trans. 2014;53(4):1350–6.
14. Bouabdallah S, Siegwart R, Backstepping and sliding-mode techniques applied to an indoor micro quad-rotor. In: Proceedings of the 2005 IEEE international conference on robotics and automation. Barcelona: IEEE;2005.p. 2259–64.
15. Besnard L, Shtessel Y, Landrum B. Quadrotor vehicle control via sliding mode controller driven by sliding mode disturbance observer. J Franklin Inst. 2012;349(2):658–84.
16. Yu W. Research on control of three DOF hovering system with quad-rotor. Shenyang: Northeastern University;2007.
17. Li Y, Chen Z, Liu Z. Attitude control of a quad-rotor robot based on ADRC. J Harbin Univ Technol. 2014;46(3):115–8.
18. Li J, Qi X, Han S. Attitude decoupling control for quad-rotor aircraft based on ADRC technique. Electron Opt Control. 2013;20(3):44–8.
19. Li Y, Chen Z, Sun M, Liu Z, Zhang Q. Attitude control for quad-rotor helicopter based on discrete-time active disturbance rejection control. Control Theory Appl. 2015;32(11):1470–7.

Quasi-synchronization of Chaotic Systems with Parameter Mismatches Via Aperiodically Intermittent Control

Yan Jiang and Junyong Zhai

Abstract This paper focuses on the problem of quasi-synchronization of chaotic systems, which will be dealt by aperiodically intermittent control. Different from previous results which investigated quasi-synchronization via periodically intermittent control, the method of aperiodically intermittent control removes the limit that every work time is the same and all rest time are equal. A novel sufficient condition for quasi-synchronization is established under a small error bound via using a piecewise switching time-dependent Lyapunov function, which is monotonically decreasing with respect to time. The effectiveness of the proposed approach is shown by taking Chua's circuit.

Keywords Quasi-synchronization · Aperiodically intermittent control · Piecewise Lyapunov function

1 Introduction

In recent decades, synchronization has received lots of attentions from various fields, such as signal processing, automatic control engineering, medical science and secure communication (see, for example, [1, 2] and the reference therein). In real life, the problem of parameter mismatches between drive-response systems is unavoidable. With parameter mismatches, synchronization will be reached with a small region around zero. Namely, the error system between drive-response systems will be controlled under a small region around zero, that is called quasi-synchronization. There are many results related to quasi-synchronization [3, 4]. For heterogeneous dynamic networks, the problem of quasi-synchronization was investigated in [3] via distributed impulsive control. Weak synchronization between two coupled identical chaotic systems with parameter mismatches was exploited in [4] via periodically intermittent control.

Y. Jiang · J. Zhai (✉)
Key Laboratory of Measurement and Control of CSE, Ministry of Education,
School of Automation, Southeast University, Nanjing 210096, China
e-mail: jyzhai@seu.edu.cn

Y. Jia et al. (eds.), *Proceedings of 2017 Chinese Intelligent Systems Conference*, Lecture Notes in Electrical Engineering 459,
https://doi.org/10.1007/978-981-10-6496-8_6

For regulating synchronization behavior, many control strategies have been presented, such as pinning cluster control [5], impulsive control [6], intermittent control [7] and sliding-mode control [8]. Among them, impulsive control and intermittent control have advantages of economic and efficiency, which are discontinuous control. Intermittent control is more effective than impulsive control, because intermittent control works on certain nonzero time intervals which are called work time, and impulsive control only works at some instants. Intermittent control stops on certain nonzero time intervals which are called rest time. Periodically intermittent control has been applied to lots of studies of synchronization. Such as, periodically intermittent control was utilized to study the synchronization behavior of stochastic delayed neural networks in [7].

In the frame of periodically intermittent control, work time and rest time are periodic. This condition is so restrict that may not suitable in practice, such as wind power generation which relies on the various situation of the real world. To avoid these disadvantages, it is more reasonable and necessary to study aperiodically intermittent control. In the frame of aperiodically intermittent control, work time and rest time are aperiodic, which means that aperiodically intermittent control includes periodically intermittent control. A few results considered aperiodically intermittent control. In work [9], synchronization between linearly coupled networks was investigated via a simple aperiodically intermittent controller. Aperiodically intermittent control was employed to study the intermittent H_∞ synchronization problem between reaction-diffusion neural networks in [10].

In comparison with the technique of periodically intermittent control, utilizing the technique of aperiodically intermittent control to study quasi-synchronization has the advantage of relaxing the constraint on work time and rest time. As far as we known, several results have focused on quasi-synchronization via aperiodically intermittent control. In work [11], aperiodically intermittent control combined pinning control was employed to discuss quasi-synchronization for nonlinear coupled chaotic systems. In this study, transcendental equation was presented. Is there any technique to avoid transcendental equation? Motivated by the above work, we will utilize a piecewise switching time-dependent Lyapunov function to investigate the problem of quasi-synchronization between chaotic drive-response systems via aperiodically intermittent control.

2 Systems Description and Preliminaries

Let $\mathbb{R}^n$ and $\mathbb{R}^{n\times m}$ be the n-dimensional Euclidean linear space with the norm $\|\cdot\|$ and the set of $n\times m$-dimensional real matrix, respectively. Denote $\mathbb{N}_0$ to be the set of nonnegative integer. Denote I and 0 to be identity matrix and zero matrix. For any $A\in\mathbb{R}^{n\times n}$, denote $\lambda_{\max}(A)$ ($\lambda_{\min}(A)$) to be the maximum (minimum) eigenvalue of the matrix A.

Consider the following continuous chaotic system

$$\begin{cases} \dot{x}(t) = A_1 x(t) + B_1 f(x(t)), \ t \geq 0, \\ x(0) = x_0, \end{cases} \tag{1}$$

where $x(t) \in \mathbb{R}^n$ is the state vector; $A_1 \in \mathbb{R}^{n \times n}$, and $B_1 \in \mathbb{R}^{n \times n}$ are constant matrices; $x_0 \in \mathbb{R}^n$ is the initial value, the nonlinear function $f : \mathbb{R}^n \to \mathbb{R}^n$, satisfies $f(0) = 0$ and the following assumption:

(**A1**) There exist scalars κ_i^+, κ_i^- such that for all $s_1, s_2 \in \mathbb{R}^n$,

$$\kappa_i^- \leq \frac{f_i(s_1) - f_i(s_2)}{s_1 - s_2} \leq \kappa_i^+, \ i = 1, 2, \ldots, n.$$

Let $L_f = \max\{|\kappa_i^+|, |\kappa_i^-|\}$, then

$$||f(s_1) - f(s_2)|| \leq L_f ||s_1 - s_2||.$$

We consider system (1) as the drive system, and the corresponding response system is given as follows,

$$\begin{cases} \dot{\hat{x}}(t) = A_2 \hat{x}(t) + B_2 f(\hat{x}(t)) + u(t), \ t \geq 0 \\ \hat{x}(0) = \hat{x}_0, \end{cases} \tag{2}$$

where $\hat{x}(t) \in \mathbb{R}^n$ is the state vector, $u(t) \in \mathbb{R}^n$ is the control input vector; $A_2 \in \mathbb{R}^{n \times n}$, $B_2 \in \mathbb{R}^{n \times n}$ are constant matrices; $\hat{x}_0 \in \mathbb{R}^n$ is the initial value. Since parameter dismatches exist, thus $A_1 \neq A_2$ and $B_1 \neq B_2$.

In this study, we consider the synchronization between state of drive system (1) and that of response system (2) with parameter mismatches by aperiodically intermittent output feedback control. The measured output of drive system (1) is

$$y(t) = Fx(t),$$

where $F \in \mathbb{R}^{m \times n}$ is a known matrix.

Set the time sequence $t_0 < s_0 < t_1 < s_1 < t_2 < s_2 < \cdots < t_k < s_k < t_{k+1} < \cdots$, and $\lim_{k \to \infty} t_k = \infty, k \in \mathbb{N}_0$. Then for response system (2), design the following aperiodically intermittent output feedback control law:

$$u(t) = K(t)(y(t) - F\hat{x}(t)); \tag{3}$$

with

$$K(t) = \begin{cases} K, \ t \in [t_k, s_k), \\ 0, \ t \in [s_k, t_{k+1}), \end{cases}$$

where $K \in \mathbb{R}^{n \times m}$ is the gain matrix to be determined later; and $h_{11} \leq s_k - t_k \leq h_{12}$, $h_{21} \leq t_{k+1} - s_k \leq h_{22}$, $h_{ij}, i, j = 1, 2, k \in \mathbb{N}_0$ are positive scalars.

In view of the aperiodically intermittent control law (3), then

$$\begin{cases} \dot{\hat{x}}(t) = A_2\hat{x}(t) + B_2 f(\hat{x}(t)) + K(t)F(x(t) - \hat{x}(t)), t \geq 0, \\ \hat{x}(0) = \hat{x}_0. \end{cases} \tag{4}$$

Let $\Delta A = A_2 - A_1$ and $\Delta B = B_2 - B_1$ denote the errors of parameter mismatches, denote $e(t) = \hat{x}(t) - x(t)$ as the synchronization error. We get the error system,

$$\begin{cases} \dot{e}(t) = (A_2 - K(t)F)e(t) + \Delta A x(t) \\ \qquad + B_2(f(\hat{x}(t)) - f(x(t))) + \Delta B f(x), \ t \geq 0 \\ e(0) = e_0 \triangleq \hat{x}_0 - x_0. \end{cases}$$

Let $\hat{f}(e(t)) = f(\hat{x}(t)) - f(x(t)), \ H(t) = \Delta A x(t) + \Delta B f(x(t))$. Then above error system can be rewritten as

$$\dot{e}(t) = (A_2 - KF)e(t) + B_2\hat{f}(e(t)) + H(t), \ t_k \leq t < s_k, \tag{5a}$$

$$\dot{e}(t) = A_2 e(t) + B_2\hat{f}(e(t)) + H(t), \ s_k \leq t < t_{k+1}, \ k \in \mathbb{N}_0, \tag{5b}$$

$$e(0) = e_0.$$

Notice that, for the error system (5), $e(t) = 0$ is not an equilibrium point because of the parameter mismatches term $H(t)$. Consequently, complete synchronization can not be achieved. However, in chaotic systems, the states are always contained in a bounded subset $\mathcal{R} \subseteq \mathbb{R}^n$. Therefore, it is possible to find a gain matrix K which can make the synchronization error decays to a value as small as possible. To achieve this goal, the following definition is necessary.

Definition 1 Let $\mathcal{R} = \{x \in \mathbb{R}^n; \ \|x\| \leq \varepsilon_1\}$ with $\varepsilon_1 > 0$ be a region in the phase space, which includes the chaotic attractor of drive system (1). The drive system (1) and the response system (2) are known as quasi-synchronized with an error level ε_1 over $\mathcal{R}$ if there exist $\delta_0 > 0$ and $T \geq 0$, such that if $\|\hat{x}_0 - x_0\| \leq \delta_0$, it holds that $||\hat{x}(t) - x(t)|| \leq \varepsilon_1$ for all $t \geq T$.

For some positive scalars $h_{i\ell}, \ i, \ell = 1, 2$, satisfying $h_{i1} \leq h_{i2}, \ i = 1, 2$, introducing the following admissible intermittent time sequences:

$$S(h_{11}, h_{12}, h_{21}, h_{22}) \triangleq \left\{\{(t_k, s_k, t_{k+1})\} | h_{11} \leq s_k - t_k \leq h_{12}, h_{21} \leq t_{k+1} - s_k \leq h_{22}, k \in \mathbb{N}_0\right\}.$$

For the convenience of this study, we have the following notations:

$$L_1 = \text{diag}\left\{-\kappa_1^+\kappa_1^-, -\kappa_2^+\kappa_2^-, \ldots, -\kappa_n^+\kappa_n^-\right\},$$
$$L_2 = \text{diag}\left\{\frac{\kappa_1^+ + \kappa_1^-}{2}, \frac{\kappa_2^+ + \kappa_2^-}{2}, \ldots, \frac{\kappa_n^+ + \kappa_n^-}{2}\right\},$$
$$\eta(t) = \text{col}\left\{e(t), \hat{f}(t), H(t)\right\}.$$

3 Criterion for Quasi-synchronization

In ths section, a time-varying switched Lyapunov function method will be developed to establish the quasi-synchronization criterion for the drive system (1) and response system (2). To make it, several auxiliary function sequences are introduced.

For $t \in [t_k, s_k)$, $k \in \mathbb{N}_0$, define

$$\rho_{11k}(t) = \frac{t - t_k}{s_k - t_k}, \ \rho_{12k}(t) = \frac{s_k - t}{s_k - t_k}, \ \rho_{10}(t) = \frac{1}{s_k - t_k},$$

For $t \in [s_k, t_{k+1})$, $k \in \mathbb{N}_0$, let

$$\rho_{21k}(t) = \frac{t - s_k}{t_{k+1} - s_k}, \ \rho_{22k}(t) = \frac{t_{k+1} - t}{t_{k+1} - s_k}, \ \rho_{20}(t) = \frac{1}{t_{k+1} - s_k}.$$

There exist $\varsigma_{i\ell k}(t) : \mathbb{R} \to [0, 1]$, $i, \ell = 1, 2$, $k \in \mathbb{N}_0$, such that

$$\rho_{i0}(t) = \frac{\varsigma_{i1k}(t)}{h_{i1}} + \frac{\varsigma_{i2k}(t)}{h_{i2}},$$

where $\varsigma_{i1k}(t) + \varsigma_{i2k}(t) = 1$, $i = 1, 2$, $k \in \mathbb{N}_0$. From the above definitions, one has

$$\rho_{11k}(t_k) = \rho_{12k}(s_k^-) = \rho_{21k}(s_k) = \rho_{22k}(t_{k+1}^-) = 0;$$
$$\rho_{11k}(s_k^-) = \rho_{12k}(t_k) = \rho_{21k}(t_{k+1}^-) = \rho_{22k}(s_k) = 1.$$

The quasi-synchronization criterion is described below.

Theorem 1 *For given admissible intermittent time sequences $S(h_{11}, h_{12}, h_{21}, h_{22})$, suppose that $\mathcal{R} = \{x \in \mathbb{R}^n \mid \|x\| \leq \varepsilon_1\}$ includes the attractor of the drive system* (1) *and the parameter mismatches satisfy $\|\Delta A\| + L_f\|\Delta B\| \leq \varepsilon_2$. Considering the drive system* (1) *and the response system* (2) *satisfy* (**A1**)*, for the given $m \times n$ matrix K, positive scalars ϵ, μ_1, μ_2, there exist positive matrices $P_{ij} \in \mathbb{R}^{n\times n}$, positive diagonal matrices $D_{ij} \in \mathbb{R}^{n\times n}$, positive scalars α_i, $\underline{\lambda}$, where $i, j = 1, 2$, such that the following linear matrix inequalities (LMIs) hold:*

$$\underline{\lambda} I \leq P_{ij} \tag{6}$$

$$P_{22} \leq \mu_1 P_{11}, P_{12} \leq \mu_2 P_{21} \tag{7}$$

$$\Xi_{ij\ell} \triangleq \begin{bmatrix} \Omega_{ij\ell} & P_{ij}B_2 + L_2 D_{ij} & P_{ij} \\ * & -D_{ij} & 0 \\ * & * & -\alpha_i I \end{bmatrix} < 0,\ i,j,\ell = 1,2, \tag{8}$$

where

$$\Omega_{1j\ell} = \left(\epsilon + \frac{\ln \mu_1}{h_{1\ell}}\right) P_{1j} + \frac{1}{h_{1\ell}}(P_{11} - P_{12}) + L_1 D_{1j0} + P_{1j}(A_2 - KF) + (A_2 - KF)^{\mathrm{T}} P_{1j},$$

$$\Omega_{2j\ell} = \left(\epsilon + \frac{\ln \mu_2}{h_{2\ell}}\right) P_{2j} + \frac{1}{h_{2\ell}}(P_{21} - P_{22}) + L_1 D_{2j0} + P_{2j}A_2 + A_2^{\mathrm{T}} P_{2j},$$

then the trajectory of error system (5) converges to region $\mathcal{D}$ exponentially, which contains the origin, where $\mathcal{D} = \left\{ e \in \mathbb{R}^n \mid \|e\| \leq \frac{\alpha\varepsilon_1\varepsilon_2}{\sqrt{\epsilon\mu_0\lambda}} \right\}$, in which $\mu_0 = \min\{\mu_1, \mu_2, 1\}$, $\alpha = \sqrt{\max\{\mu_1, \mu_2, 1\}\max\{\alpha_1, \alpha_2\}}$. Thus, the drive system (1) and response system (2) are said to be quasi-synchronization under error bound $\varepsilon + \frac{\alpha\varepsilon_1\varepsilon_2}{\sqrt{\epsilon\mu_0\lambda}}$ for any arbitrary small positive number ε.

Proof Consider the following piecewise switching time-dependent Lyapunov function for system (5):

$$V(t) = \begin{cases} \psi_{1k}(t)\mathrm{e}^{\epsilon t} e^{\mathrm{T}}(t) P_{1k}(t) e(t); & t_k \leq t < s_k, \\ \psi_{2k}(t)\mathrm{e}^{\epsilon t} e^{\mathrm{T}}(t) P_{2k}(t) e(t); & s_k \leq t < t_{k+1}, \end{cases} \tag{9}$$

where $\epsilon > 0$, $\mu_1 > 0$, $\mu_2 > 0$, $P_{ij} > 0$, and

$$P_{ik}(t) = \sum_{j=1}^{2} \rho_{ijk}(t) P_{ij},\ \psi_{ik}(t) = \mu_i^{\rho_{i1k}(t)},\ i,j = 1,2,\ k \in \mathbb{N}_0.$$

In view of the function sequences $\{\rho_{ijk}(t)\}$ and $\{\psi_{ik}(t)\}$, $i,j = 1,2$, $k \in \mathbb{N}_0$, it follows that

$$V(t_k) = \mathrm{e}^{\epsilon t_k} e^{\mathrm{T}}(t_k) P_{12} e(t_k),\ V(t_k^-) = \mu_2 \mathrm{e}^{\epsilon t_k^-} e^{\mathrm{T}}(t_k^-) P_{21} e(t_k^-),$$
$$V(s_k) = \mathrm{e}^{\epsilon s_k} e^{\mathrm{T}}(s_k) P_{22} e(s_k),\ V(s_k^-) = \mu_1 \mathrm{e}^{\epsilon s_k^-} e^{\mathrm{T}}(s_k^-) P_{11} e(s_k^-).$$

From condition (7), the following inequalities hold

$$V(t_k) \leq V(t_k^-),\ V(s_k) \leq V(s_k^-). \tag{10}$$

This shows that under the condition (7), $V(t)$ is non-increasing at switching instants.

For $t \in (t_k, s_k)$, the derivative of $V(t)$ is

$$\begin{aligned}\dot{V}(t) \le \psi_{1k}(t)\mathrm{e}^{\epsilon t}\left\{e^{\mathrm{T}}(t)\left[\rho_{10}(t)\ln\mu_1 P_{1k}(t) + \epsilon P_{1k}(t)\right.\right.\\ \left. + 2P_{1k}(t)(A_2 - FK) + \rho_{10}(t)(P_{11} - P_{12})\right] e(t)\\ \left. + 2e^{\mathrm{T}}(t)P_{1k}(t)B_2\hat{f}(e(t)) + 2e^{\mathrm{T}}(t)P_{1k}(t)H(t)\right\}. \quad (11)\end{aligned}$$

Let

$$D_{ik}(t) = \mathrm{diag}\left(d_{i1k}(t), d_{i2k}(t), \ldots, d_{ink}(t)\right)\sum_{j=1}^{2}\rho_{ijk}(t)D_{ij}, i,j = 1,2, k \in \mathbb{N}_0.$$

On the other hand, by the first inequality of (**A1**), we arrive at

$$\begin{aligned}0 \le& \sum_{l=1}^{n} d_{ilk}(t)\left(\kappa_l^+ e_l(t) - \hat{f}_l(e_l(t))\right)\left(\hat{f}_l(e_l(t)) - \kappa_l^- e_l(t)\right)\\ =& e^{\mathrm{T}}(t)D_{ik}(t)L_1 e(t) - \hat{f}^{\mathrm{T}}(e(t))D_{ik}(t)\hat{f}(e(t))\\ &+ 2e^{\mathrm{T}}(t)D_{ik}(t)L_2\hat{f}(e(t)),\ i = 1,2, k \in \mathbb{N}_0. \quad (12)\end{aligned}$$

Recalling that $\|x(t)\| \le \varepsilon_1$ and $\|H(t)\| \le \varepsilon_2$, it follows that for any $\alpha_i > 0$,

$$0 \le \alpha_i\psi_{ik}(t)\mathrm{e}^{\epsilon t}\left((\varepsilon_1\varepsilon_2)^2 - H^{\mathrm{T}}(t)H(t)\right), i = 1,2. \quad (13)$$

Adding the terms on the right of inequalities (12), (13) with $i = 1$ to the right side of (11) yields

$$\dot{V}(t) \le \psi_{1k}(t)\mathrm{e}^{\epsilon t}\eta^{\mathrm{T}}(t)\sum_{j=1}^{2}\sum_{\ell=1}^{2}\rho_{1jk}(t)\Xi_{1j\ell}\,\eta(t) + \alpha_1\psi_{1k}(t)\mathrm{e}^{\epsilon t}(\varepsilon_1\varepsilon_2)^2,\ t \in (t_k, s_k). \quad (14)$$

It follows from (8) with $i = 1$ that

$$\dot{V}(t) \le \alpha_1\psi_{1k}(t)\mathrm{e}^{\epsilon t}(\varepsilon_1\varepsilon_2)^2,\ t \in (t_k, s_k). \quad (15)$$

For $t \in (s_k, t_{k+1})$, using the similar technique as in the above proof, we obtain

$$\dot{V}(t) \le \psi_{2k}(t)\mathrm{e}^{\epsilon t}\eta^{\mathrm{T}}(t)\sum_{j=1}^{2}\sum_{\ell=1}^{2}\rho_{2jk}(t)\Xi_{2j\ell}\,\eta(t) + \alpha_2\psi_{2k}(t)\mathrm{e}^{\epsilon t}(\varepsilon_1\varepsilon_2)^2.$$

In view of (8) with $i = 2$, yields

$$\dot{V}(t) \le \alpha_2\psi_{2k}(t)\mathrm{e}^{\epsilon t}(\varepsilon_1\varepsilon_2)^2,\ t \in (s_k, t_{k+1}). \quad (16)$$

Combining (15) and (16) together, we have

$$\dot{V}(t) \leq \mathrm{e}^{\epsilon t}(\alpha\varepsilon_1\varepsilon_2)^2,\ t \geq 0,\ t \neq t_k,\ s_k,\ k \in \mathbb{N}_0,$$

where $\alpha = \sqrt{\max\{\mu_1, \mu_2, 1\}\max\{\alpha_1, \alpha_2\}}$. Then, taking into account (10), one has

$$V(t) \leq V(0) + \frac{1}{\epsilon}(\alpha\varepsilon_1\varepsilon_2)^2\mathrm{e}^{\epsilon t} - \frac{(\alpha\varepsilon_1\varepsilon_2)^2}{\epsilon},\ t \geq 0. \tag{17}$$

Let $\mu_0 = \min\{\mu_1, \mu_2, 1\}$, it follows from (6) and (17) that

$$\|e(t)\| \leq \frac{e_0\sqrt{\epsilon\lambda_1} - \alpha\varepsilon_1\varepsilon_2}{\sqrt{\epsilon\mu_0\underline{\lambda}}}\mathrm{e}^{-\frac{\epsilon}{2}t} + \frac{\alpha\varepsilon_1\varepsilon_2}{\sqrt{\epsilon\mu_0\underline{\lambda}}},\ t \geq 0,$$

where $\lambda_1 = \max\{\lambda_{\max}(P_{1j}), \lambda_{\max}(Q_1); j = 1, 2\}$. Therefore, the trajectory of error system (5) converges to the region $\mathcal{D}$ exponentially, which contains the origin, where

$$\mathcal{D} = \left\{ e \in \mathbb{R}^n \mid \|e\| \leq \frac{\alpha\varepsilon_1\varepsilon_2}{\sqrt{\epsilon\mu_0\underline{\lambda}}} \right\},$$

from which we can get the conclusion that the drive system (1) and response system (2) achieve quasi-synchronized under an error bound $\varepsilon + \frac{\alpha\varepsilon_1\varepsilon_2}{\sqrt{\epsilon\mu_0\underline{\lambda}}}$ for any arbitrary small positive number ε. This completes the proof.

Remark 1 Instead of employing common Lyapunov function as [4] to make stability analysis of systems which ignores the different dynamic between closed-loop systems and open loop systems, we introduce a piecewise switching time-dependent Lyapunov function (9) to analyze the stability of the error systems. This kind of Lyapunov function (9) can take more dynamic characteristic of systems in both work time $s_k - t_k$ and rest time $t_{k+1} - s_k$, and nonincrease at switching instants, thus leading a less conservative result.

4 A Numerical Example

Example 1 Consider the Chua's circuit proposed in [4] with the following parameters:

$$A_1 = \begin{bmatrix} \Phi_3 & p_1 & 0 \\ 1 & -1 & 1 \\ 0 & -p_2 & 0 \end{bmatrix},\ f(x) = \begin{bmatrix} g(x) \\ 0 \\ 0 \end{bmatrix},$$

$$B_1 = \mathrm{diag}\{(m_0 - m_1)p_1,\ 0,\ 0\},$$

Table 1 The values of b_{min}

ω (h_{11}, h_{12})	σ (h_{21}, h_{22})	Theorem 1 of [4]	Theorem 1
4 (3.6, 3.6)	0.9 (0.4, 0.4)	14.2545	0.6268
10 (8.0, 8.0)	0.8 (2.0, 2.0)	16.0102	3.2645
16 (13.6, 13.6)	0.85 (2.4, 2.4)	15.1556	2.6979
20 (15.0, 15.0)	0.75 (5.0, 5.0)	17.1383	5.7832

where $\Phi_3 = -p_1(1+m_1)$, $g(x) = -\frac{1}{2}(|x+1| - |x-1|)$, $p_1 = 9.2156$, $p_2 = 15.9946$, $m_0 = -1.24905$ and $m_1 = -0.75735$.

For simulation, we assume the response system (2) with the following parameters:

$$A_2 = \begin{bmatrix} \widetilde{\Phi}_3 & \tilde{p}_1 & 0 \\ 1 & -1 & 1 \\ 0 & -\tilde{p}_2 & 0 \end{bmatrix}, f(x) = \begin{bmatrix} g(\hat{x}) \\ 0 \\ 0 \end{bmatrix},$$

$$B_2 = \text{diag}\{(\tilde{m}_0 - \tilde{m}_1)\tilde{p}_1,\ 0,\ 0\},\ B = \begin{bmatrix} 0 & 0 & 1 \end{bmatrix},$$

where $\widetilde{\Phi}_3 = -\tilde{p}_1(1+\tilde{m}_1)$, $\tilde{p}_1 = 9.21$, $\tilde{p}_2 = 15.995$, $\tilde{m}_0 = -1.25$ and $\tilde{m}_1 = -0.758$. To compare the effectiveness of the results in Theorem 1 of [4] and our Theorem 1, we make a comparison basing the above system coefficients, then let $KF = bI$ for some positive scalars b and other suitable parameters, to compute b_{min}, i.e., the minimum values of b which can maintain the stability. Firstly, we set $h_{11} = h_{12}$, $h_{21} = h_{22}$, such that $\omega = h_{11} + h_{21}$, $\delta = h_{11}$. Choose different (ω, σ), respectively $(4, 0.9)$, $(10, 0.8)$, $(16, 0.85)$, $(20, 0.75)$, then the values of b_{min} are given in Table 1. From this table, we can directly get that our Theorem 1 is less conservative than Theorem 1 in [4].

5 Conclusion

The quasi-synchronization for continuous-time chaotic systems has been investigated via aperiodically intermittent control. A criterion which was expressed as a series linear matrix inequalities was obtained by utilizing a piecewise switching time-dependent Lyapunov function to analysis the quasi-synchronization of drive-response systems. Chua's circuits is utilized to verify the effectiveness of our result.

Acknowledgements This work is supported in part by National Natural Science Foundation of China [grant number 61473082], the Fundamental Research Funds for the Central Universities, Qing Lan Project, and PAPD.

References

1. Prakash M, Balasubramaniam P, Lakshmanan S. Synchronization of Markovian jumping inertial neural networks and its applications in image encryption. Neural Netw. 2016;83:86–93.
2. Xu D, Xu X, Yang C, et al. Spreading dynamics and synchronization behavior of periodic diseases on complex networks. Physica A: Stat Mech Appl. 2017;466:544–51.
3. He W, Qian F, Lam J, et al. Quasi-synchronization of heterogeneous dynamic networks via distributed impulsive control: error estimation, optimization and design. Automatica. 2015;62:249–62.
4. Zhang W, Huang J, Wei P. Weak synchronization of chaotic neural networks with parameter mismatch via periodically intermittent control. Appl Math Model. 2011;35(2):612–20.
5. Li T, Wang T, Yang X, Fei S. Pinning cluster synchronization for delayed dynamical networks via Kronecker product. Circ Syst Sig Process. 2013;32(4):1907–29.
6. Chen W-H, Luo S, Zheng WX. Impulsive synchronization of reaction-diffusion neural networks with mixed delays and its application to image encryption. IEEE Trans Neural Netw Learn Syst. 2016;27(12):2696–710.
7. Jiang Y, Luo S. Periodically intermittent synchronization of stochastic delayed neural networks. Circ Syst Sig Proces. 2017;36(4):1426–44.
8. Jing T, Chen F, Zhang X. Finite-time lag synchronization of time-varying delayed complex networks via periodically intermittent control and sliding mode control. Neurocomputing. 2016;199:178–84.
9. Liu X, Chen T. Synchronization of complex networks via aperiodically intermittent pinning control. IEEE Trans Autom Control. 2015;60(12):3316–21.
10. Liu L, Chen W-H, Lu X. Aperiodically intermittent H_∞ synchronization for a class of reaction-diffusion neural networks. Neurocomputing. 2017;222:105–15.
11. Liu X, Liu Y, Zhou L. Quasi-synchronization of nonlinear coupled chaotic systems via aperiodically intermittent pinning control. Neurocomputing. 2016;173:759–67.

Interactive Touch Control Method Based on Image Denoising Technology

Xueyan Chen, Lei Yu and Jun Huang

Abstract In this paper, an interactive touch control method based on image filter technology is proposed to solve the problem of noise interference in the image transmission of interactive system. Firstly, an interactive touch control system based on Kinect sensor is constructed. The intelligent interaction area is created by projection technique and dynamic capture method, and the image of Kinect is processed by Kalman filtering. Finally, the infrared operation pen is employed to simulate the mouse to achieve human-computer interaction. The experimental results show that the interaction effect of the human-computer interaction system can be well obtained.

Keywords Interactive touch control system · Kinect · Kalman filter · Projection system

1 Introduction

Traditional multimedia technology needs to rely on the mouse and keyboard to carry out human-computer interaction, rather than through the user's direct operation to achieve information input and control of the computer, so that the freedom of interaction has been limited. The emergence of large-screen interactive devices liberates the user from the keyboard and mouse constraints and allows the user to manipulate the computer directly on the interactive device with limb movements. From the human-computer interaction, it becomes more direct and natural [1–6].

Current common large-screen interactive device can be divided into five categories according to implement different technologies: pressure-sensitive-type large-screen interactive devices, electromagnetic induction type large-screen interactive equipment, ultrasonic type large-screen interactive devices, infrared-type large-screen interactive equipment, machine vision-type large-screen interactive

X. Chen · L. Yu (✉) · J. Huang
School of Mechanical and Electric Engineering, Soochow University, Suzhou 215021, China
e-mail: slender2008@163.com

Y. Jia et al. (eds.), *Proceedings of 2017 Chinese Intelligent Systems Conference*, Lecture Notes in Electrical Engineering 459,
https://doi.org/10.1007/978-981-10-6496-8_7

equipment. The large-screen interactive device based on machine vision is widely used because of its positioning accuracy, fast response, strong adaptability and natural interaction. However, as a new technology, the current domestic research is still very little. The existing human-computer interaction technology on the hardware equipment requirements are relatively high, and the system is very large, low sensitivity [3–7]. So the independent research and development based on the machine vision of large-screen interactive device touch technology are extremely important. Existing human-computer interaction system requirements for hardware equipment is relatively high, costly. So the independent research and development based on the machine vision of large-screen interactive device touch technology is extremely important.

This paper presents a human-computer interaction method based on Kinect sensor, which can realize good human-computer interaction effect on any wall [8–10]. Firstly, an interactive control system based on Kinect sensor is constructed. The intelligent interaction region is created by using projection technique and dynamic capture method, and the Kinect image is processed by Kalman filter. Finally, through the infrared pen to simulate the mouse to achieve human-computer interaction. The method is simple and practical, and can guarantee a good human-computer interaction effect. The proposed interactive touch control method based on image denoising technology provides a new way of thinking for the development of large-screen human-computer interaction technology.

2 The Structure and Realization of Interactive Touch Control System

2.1 The System Structure

In this paper, the human-computer interaction system is mainly composed of hardware part and software part. The hardware part consists of control panel, infrared operating pen, Kinect somatosensory sensor, projection equipment and other hardware components. The main function of the control panel is utilized to receive the signal transmitted by the somatosensory sensor, which is output to the screen after the software is processed. The infrared pen will emit infrared rays after being squeezed by the walls, which are reflected to the sensor via the walls. The Kinect sensor is employed to receive the infrared signal (mainly the depth signal) from the infrared pen and transfer it to the computer after processing. The projection device is used to output the video signal, and the result of the operation is displayed accurately and reliably. Figure 1 shows the interaction of the user and the touch control interactive system.

A wall-based interactive touch control system is designed by the self-developed human-computer interactive software platform. The system can not only capture the position and action of the operating pen, but also can capture the action of the

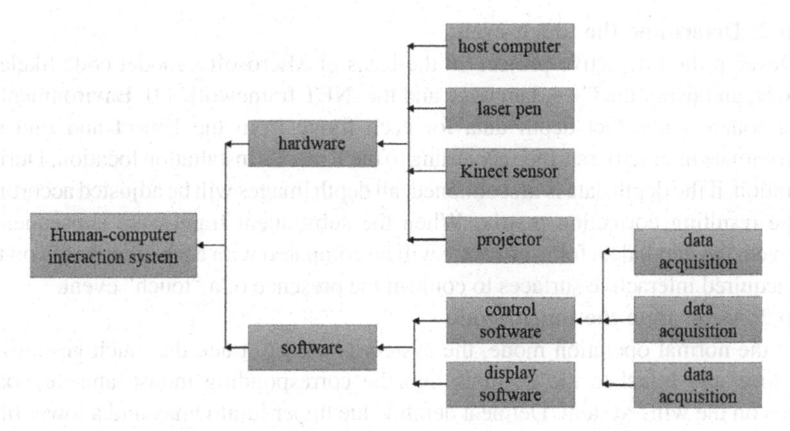

Fig. 1 System components module structure

human body. The action signal is received by the Kinect sensor and feedback to the computer, and then displayed on the wall by the projector to achieve human-computer interaction. This makes the action of the person and the action pen can replace the mouse and complete the mouse function on the wall including: click, double click, drag and so on.

The software part consists of control software and display software. The control software is composed of three modules: data acquisition, data filtering and data communication. The Kinect sensor acquires the original data and transmits it to the control host through the network cable. The data acquisition module is responsible for receiving the original data. The data filtering module first filters out the noise in the original data and then filters the data to get the induction points. Determine whether the induction point in the effective operation area, that is, whether the effective induction point. Then communicate with interactive display software for data, and get the sensor interaction; if not, then return to the data acquisition module to continue the next frame of the original data work process. The display software edits the interactive content to be displayed on the interactive surface.

The system components module structure shown in Fig. 1.

2.2 *Realization of Touch Control Interactive System*

Step 1: Create an interactive area

First create a desktop interactive platform through the 3D sensor. The Kinect needs to be installed on top of the head and make sure the Kinect's depth camera covers the entire desktop range. The minimum distance from the Kinect to the desktop is limited to 0.75 m, and the maximum distance determines the accuracy of the touch point calculation, a recommended distance value of 1 m.

Step 2: Determine the touch event

Develop the interactive project on the basis of Microsoft's model code Skeletal Viewer, and using the C++ language and the .NET framework 4.0. Environment to write source code. Get depth data for each frame from the Kinect and find the approximate interactive surface according to the Kinect's installation location. During operation, if the depth data is first obtained, all depth images will be adjusted according to the resulting correction matrix. When the subsequent frames are subsequently received, the depth data for each frame will be compared with each other based on the pre-acquired interactive surfaces to confirm the presence of a "touch" event.

Step 3: Determine the touch mode

In the normal operation mode, the system is used to track the touch gestures in real time and translate the gestures into the corresponding mouse and keyboard events on the win7 system. Define a depth value upper limit Dmax and a lower limit of the depth value Dmin. The narrow area between the two is defined as the effective area of the touch point. When the system is in idle mode, the parameters required for the gesture filter function and the classification will be initialized and ready to accept different particle information. When the number of particles is greater than 1, then it is considered a valid gesture, and all analog I/O events are reset. Use a time counter to prevent false gestures from being reset to case0 (noise jamming frames).

The system work flow is shown in Fig. 2.

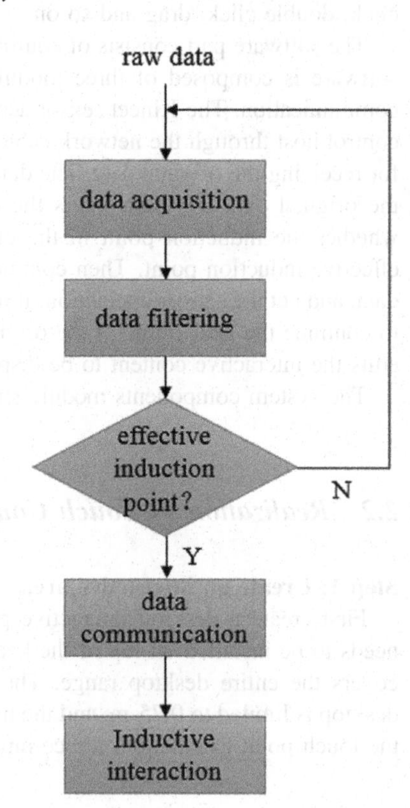

Fig. 2 System work flow chart

3 The Touch Control Interactive System Image Filtering

Kinect receives depth maps through the camera and then passes them to the computer, and the image information will be inevitably mixed with noise interference. Noise may be generated in the image signal input, transmission, output and other aspects, and may also be caused by the hardware device [2, 5, 11–13]. Noise interference can be seen directly by the jitter, breakpoint, dislocation and so on. Therefore it is necessary to filter to eliminate or reduce noise. The Kalman filter is a digital signal filter that uses the Kalman algorithm and it is a real-time recursive algorithm [3, 7, 11–15]. It obtains the current estimate by the estimated value of $k-1$ time and the current observation value. Where the input is the observed value of the system and the output is the estimated value of the current time of the system.

First, use the Kalman filter to denoise the image. There are five steps to restore the original image:

(1) Estimate any pixel in the set of pixels (0,255), assuming that:

$$X_1(i,j) = X(i,j) \tag{1}$$

(2) Estimate the horizontal direction of the pixel, whose value is the left pixel plus the first step of the pixel.

$$X_2(i,j) = a^*X(i,j) + b^*r^*X(i,j-1) \tag{2}$$

where a and b are the weighting coefficients, $a+b=1$ and r is the correlation coefficient of the horizontal direction image.

(3) Estimate $X_3(i,j)$, whose value is the sum of the right pixel and the pixel of the second step.

$$X_3(i,j) = a^*X(i,j) + b^*r^*\frac{X(i,j-1)+X(i,j+1)}{2} \tag{3}$$

(4) Estimate $X_4(i,j)$, whose value is the sum of the Kalman filter and the third step estimate.

$$X_4(i,j) = a(1-k)^*X(i,j) + b(1-k)^*r^*\frac{X(i,j-1)+X(i,j+1)}{2} \tag{4}$$

In the above formula, K is the Kalman gain.

(5) Estimate $X_5(i,j)$, whose value is the average of the vertical direction and the horizontal direction in the fourth step.

$$X_5(i,j) = \frac{X_{4x}(i,j) + X_{4y}(i,j)}{2} \tag{5}$$

where $X_{4x}(i,j)$ is the fourth-step horizontal direction estimate and $X_{4y}(i,j)$ is the fourth-step vertical direction estimate. From the above formula,

$$X^* = A^*X + K(Y - Y^*) \tag{6}$$

$$Y^* = B^*X^* \tag{7}$$

In the above equations, X is the state vector, A is the state matrix, B is the output matrix, * represents the predicted valuet.

$$X = [X(i,j-1) \quad X(i,j) \quad X(i,j+1)] \tag{8}$$

$$A = \begin{bmatrix} 0 & 0 & 0 \\ b^*(1-K)^*\frac{r}{2} & a^*(1-K) & b^*(1-K)^*\frac{r}{2} \\ 0 & 0 & 0 \end{bmatrix} \tag{9}$$

$$C = [0 \quad 1 \quad 0] \tag{10}$$

After the above calculation steps, we can restore each pixel to the noise-free pixels, and thus achieve the entire tracking process image denoising.

4 Experimental Results and Analysis

In this section, the experimental platform is set up to test the interactive effect. The experimental environment is shown in Fig. 3: Kinect sensor is placed above the projection surface (any wall can be), and the projector is in front, both connected with the computer. The experimenter holds the infrared pen to click, drag, and write on the projection screen.

First, use the unfiltered interaction method to simulate the interaction. The results are shown in Figs. 4 and 5. When the operating pen moves on the wall, the projected image is disturbed by noise and generates a lot of jitter points, which can not be accurately positioned. When Kinect receives the infrared signal of the infrared pointer, it will also be disturbed by the infrared light, so even if the operation pen is still, there will be jitter on the screen.

Next, add the Kalman filter to the interactive algorithm and simulate the interactive operation again. The results are shown in Figs. 6 and 7. It can be seen that the system effectively removes the interference of the external noise signal. From the comparison of the results before and after filtering, it can be seen that the unfiltered system is not accurate by the external noise interference, and the interaction effect is not good. After filtering the system positioning accuracy, interactive effect can be well obtained. The interactive control system can be used for teaching display, for example, use the wall as a whiteboard with a infrared pen directly

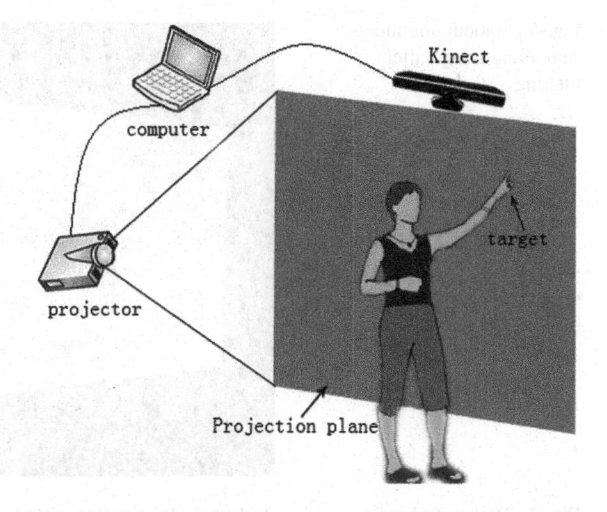

Fig. 3 Interactive touch system schematic diagram

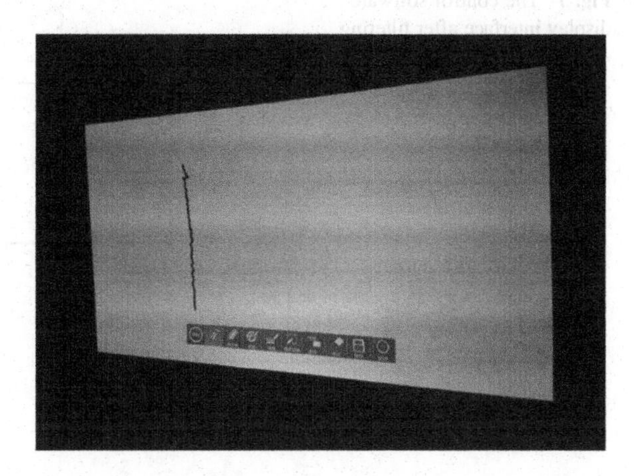

Fig. 4 Projection plane with jitter point before filtering

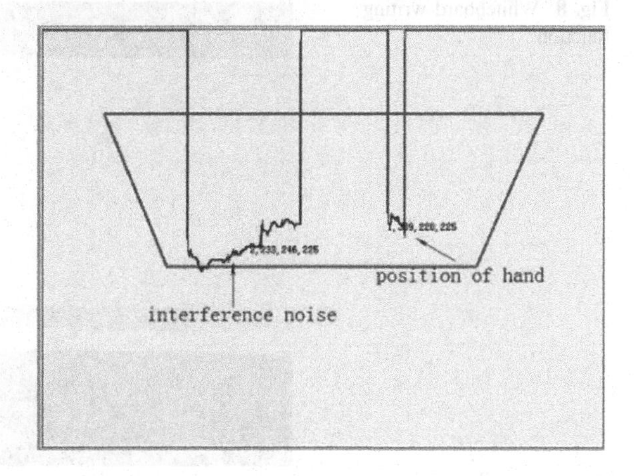

Fig. 5 The control software display interface before filtering

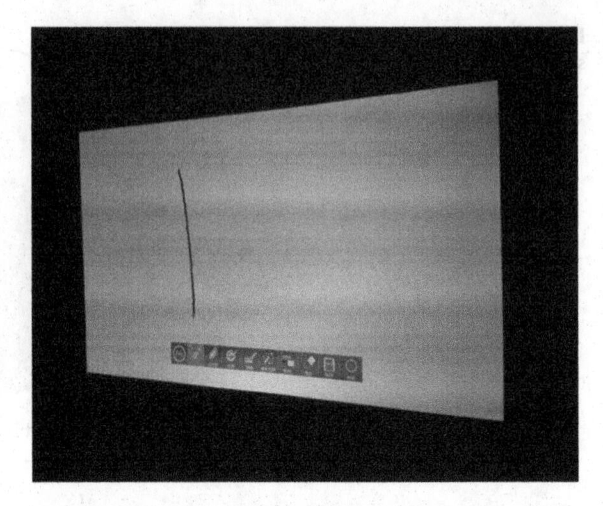

Fig. 6 Smooth continuous smoothing image after filtering

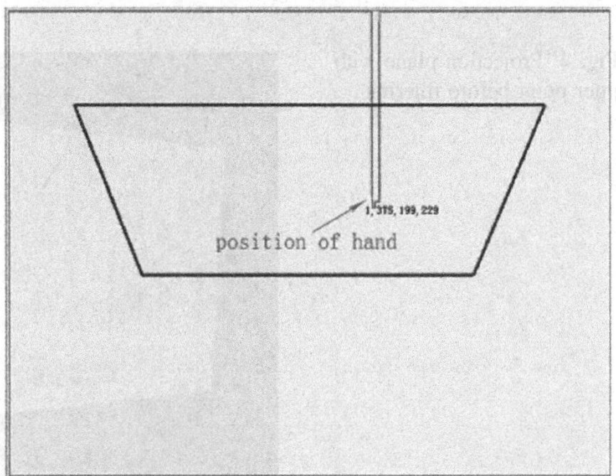

Fig. 7 The control software display interface after filtering

Fig. 8 Whiteboard writing function

writing on it, and after writing the picture can be saved. The whiteboard function is shown in Fig. 8.

5 Summary

In this paper, a set of touch control interactive system based on Kinect depth information is designed and used to create intelligent interactive region by projection technology and dynamic capture method. The influence of disturbance signal on the system is eliminated by Kalman filter. The system can be operated with an infrared pen or human gestures to perform a series of functions on the projection plane, including: click, double click, drag and so on. The system equipment is simple and easily operated. The experimental results show that the system can guarantee the good human-computer interaction effect, and provides a new way of thinking for the development of large-screen human-computer interaction technology.

Acknowledgements The work is supported by the National Natural Science Foundation of China (Nos. 61403268 and 61403267); Natural Science Fund for Colleges and Universities in Jiangsu Province (No. 16KJB120005); The Key Technology Program of Suzhou, China (No. SYG201639); The Project was Supported by the open fund for Jiangsu Key Laboratory of Advanced Manufacturing Technology (No. HGAMTL-1602).

References

1. Hutchinson TE, White KP, Martin WN, et al. Human-computer interaction using eye-gaze input. IEEE Trans Syst Man Cybern. 1989;19(6):1527–34.
2. Paravati G, Gatteschi V. Human-computer interaction in smart environments. Sensors. 2015;15(8):19487–94.
3. Manuel PJ, Saguees C, Montijano E, et al. Human-computer interaction based on hand gestures using RGB-D sensors. Sensors. 2013;13(9):11842–60.
4. Atkins J, Sharma DP. Visualization of babble-speech interactions using Andrews curves. Circuits Syst Signal Process. 2016;35(4):1313–31.
5. Alexandre LA. Gender recognition: A multiscale decision fusion approach. Pattern Recognit Lett. 2010;31(11):1422–7.
6. Hasan H, Abdul-Kareem S. Human-computer interaction using vision-based hand gesture recognition systems: a survey. Neural Comput Appl. 2014;25(2):251–61.
7. Qi WJ, Zhang P, Deng ZL. Robust sequential covariance intersection fusion kalman filtering over multi-agent sensor networks with measurement delays and uncertain noise variances. Acta Automatica Sinica. 2014;40(11):2632–42.
8. Kastaniotis D, Theodorakopoulos I, Theoharatos C, et al. A framework for gait-based recognition using Kinect. Pattern Recognit Lett. 2015;68:327–35.
9. Chen G, Li J, Wang B, et al. Reconstructing 3D human models with a Kinect. Comput Animat Virtual Worlds. 2016;27(1):72–85.
10. Erol B, Karabacak C, Gurbuz SZ. A Kinect-based human micro-doppler simulator. IEEE Aerosp Electron Syst Mag. 2015;30(5):6–17.

11. Xiao J, Stolkin R, Oussalah M, et al. Continuously adaptive data fusion and model re-learning for particle filter tracking with multiple features. IEEE Sens J. 2016;16(8):2639–49.
12. Best MC, Gordon TJ, Dixon PJ. An extended adaptive Kalman filter for real-time state estimation of vehicle handling dynamics. Veh Syst Dyn. 2000;34(1):57–75.
13. Hajiyev C, Soken HE. Robust adaptive Kalman filter for estimation of UAV dynamics in the presence of sensor/actuator faults. Aerosp Sci Technol. 2013;28(1):376–83.
14. Menegaz H, Ishihara JY, Borges GA, et al. A systematization of the unscented Kalman filter theory. IEEE Trans Autom Control. 2015;60(10):2583–98.
15. Ray LR, Ramasubramanian A, Townsend J. Adaptive friction compensation using extended Kalman-Bucy filter friction estimation. Control Eng Pract. 2001;9(2):169–79.

Average Weight Based Branching Heuristic Strategy for Satisfiability Solvers

Rong Hu, Huimin Fu and Xiaolong Li

Abstract The number of clauses and the number of variables in real life are very huge for satisfiability problem. Branching heuristic strategy plays an important role in SAT, a solver based on conflict-driven clause-learning, which is completely. In other words, efficient branching heuristics are the key to solving the satisfiability problem quickly. In this paper, we propose a new branching heuristic inspired by a bit-encoding phase selection policy used to improve Glucose 2.1 and exponential recent weighted average algorithm used to solve the bandit problem. The weight of the current clause is the weight of the new conflict clause plus the weight of the historical clauses. We let the weight of literal correspond to a binary value in each conflicting clause. Its advantage is the use of the weight of the historical clauses. The main idea of this strategy is to give priority to the literals in recent conflict clause, and also to use feedback information from historical clauses.

Keywords Satisfiability solvers · Branching heuristic · Weight of the current conflict clause

1 Introduction

Propositional satisfiability is the problem of determining whether there is a true value of its propositional variable assignment which makes the propositional formula is true, usually abbreviated as SAT. SAT is a core problem in many fields including artificial intelligence (AI), bounded model checking (BMC) of hardware

R. Hu (✉) · H. Fu · X. Li
National-Local Joint Engineering Laboratory of System Credibility Automatic Verification, Southwest Jiaotong University, Chengdu 610031, China
e-mail: 806080468@qq.com

H. Fu
e-mail: fhm6688@my.swjtu.edu

X. Li
e-mail: 1845253023@qq.com

Y. Jia et al. (eds.), *Proceedings of 2017 Chinese Intelligent Systems Conference*, Lecture Notes in Electrical Engineering 459,
https://doi.org/10.1007/978-981-10-6496-8_8

and software, intelligent planning, intelligent decision-making, theorem proving, circuit diagnostics and optimization calculations.

Stephen. A. Cook has proved that the SAT problem was NP-complete in 1971. NP problem refers to the problem that a deterministic algorithm can be checked or verified in a polynomial time, that is, these problems cannot be determined in a polynomial time using a deterministic algorithm. In recent years, SAT solvers are also diversified (e.g. GRASP [1], Chaff [2], Berkmin [3], MiniSat [4], SATO [5], real-sat [6], Glucose [7], Lingeling [8]). In general, SAT solvers are classified into conflict-driven, look-ahead and random search. We study conflict-driven clause-learning (abbreviated as CDCL) type solvers.

CDCL solvers include many sides such as pre-treatment technology, branching heuristic, Boolean Constraint Propagation (abbreviated as BCP), conflict analysis, clause learning, nonchronological backtracking, restarts and its lazy database maintenance. A key element in the success of CDCL SAT solvers is the branching heuristic. Branching strategy is a sequential assignment based on the score of variables or literals. Specifically, according to the formula of the variables and clauses information to give them an initial score. When a conflict occurs, the variable in the conflict clause adds the score and then assignment is based on the score. In nearly 16 years, the most successful branching strategy is Variable State Independent Decaying Sum (abbreviated as VSIDS). Because VSIDS plays an important role in the efficiency of the solvers, the current branching heuristics are mostly improved on the basis of it, such as Berk Min heuristic or clause-move-to-front, normalized VSIDS (NVSIDS) [9], VMTF [10], ACIDS (average conflict-index decision score) scheme [11]). STATIC, and exponential VSIDS (EVSIDS).

The importance of VSIDS is reflected in the introduction of conflict clauses, and the weight of the literals appears in the conflict clauses added to the counter. So we continue to study the counter of conflict clauses. Let the literal in the conflict clause correspond to a binary value and base on the average of the historical conflict clause weight together, and then generate a new branch heuristic. We compare it with the newly proposed branch heuristics average conflict-index decision score, and they can follow the choice of variables as the principle of decision-making.

The paper is organized as follows. Firstly, Sect. 2 introduces VSIDS branching heuristic and ACIDS branching heuristic. Afterwards, in Sect. 3 our branching heuristic will be showed and it will describe our branching heuristic and average conflict-index decision score in an example. Section 4 concludes the paper and the future work. Reference is the end section.

2 Background

In this section, we will introduce some of the terms in the propositional logic formula and branching heuristics also will be involved.

Propositional variables are denoted $x_1, \ldots, x_n$. For variable x_i, symbol x_i and $\neg x_i$ are its literals. x_i is called positive literal and $\neg x_i$ is negative one. If L is the literal, then the opposite of L is denoted by ¬L, defined as follows:

$$\neg L = \begin{cases} \neg x_i, & \textit{when } L = x_i \\ x_i & \textit{when } L = \neg x_i \end{cases}$$

Every variable can be assigned truth values false (or 0) or true (or 1). The truth values assigned to a variable x is denoted by $\nu(x)$. We can denote $\nu(x) = 0$ or $\nu(x) = 1$.

Logical symbols include and (Λ), or (V). Λ represents the minimum of both and V represents the maximum of both. Clauses are literals that are joined together by logical words V. CNF (conjunctive normal form) formula is clauses that are joined together by logical words Λ. For example, $F = (x_1 \vee x_2 \vee x_3) \wedge (\neg x_1 \vee x_3) \wedge \neg x_3$, which is a formula. For a CNF formula consists of a set of variables $X = \{x_1, \ldots, x_n\}$, the function $v{:}X \rightarrow \{0,1\}^n$ is used to represent the truth values assignment on the set of variables. Satisfiability (SAT) problem is to determine whether a given CNF formula is satisfied.

VSIDS. In 2001, VSIDS was proposed in solver Chaff to solve satisfiable problem. The concrete realization of VSIDS is as follows:

(1) Each literal has a counter with an initial value of zero.
(2) When you add a clause to the database, the counter for each literal in this clause is incremented by one. When there is a conflict, conflict derived clause adds to clause set and increases the counter value of the corresponding literal in the conflict derived clause, which is also called the bump.
(3) According to the counter of unassigned literal is sorted, we choose the unassigned literal with the highest counter at each decision level. The solver randomly select a literal from literals which has the same maximum score as a decision variable.
(4) Periodically, all counters are divided by a constant (>1).

There are two reasons to explain why VSIDS is efficient. First of all, in the decision process, ignoring clauses are satisfying. The initial score of the literals will not be subtracted from the number of the corresponding literals even if the clauses are satisfied, which is a trade-off between efficiency and time. Secondly, learning clause mechanism is added.

ACIDS. The branching heuristic was proposed by Armin Biere and Andreas Frohlich in 2016. In the ACIDS scheme, their counters keep a score for each literal. Whenever a literal is bumped, its score is updated to be

$$s' = (s + i)/2 \tag{1}$$

where i represents the conflict-index.

3 A New Branching Heuristic

Consider a simple example: there are n cards, whose numbers are from 1 to n, we arbitrarily draw a piece, then their expectations (abbreviated as E) are

$$\mathrm{E} = \sum_{r_i=1}^{n} \frac{1}{n} r_i \tag{2}$$

We call 1/n as the weight of r_i, the weight of r_i is equal. In the satisfiability problem, because the contribution of the new conflict clause is larger than the historical clauses, we hope that the weight of the newly discovered conflict clause is greater than the weight of the historical clause. Exponential recent weighted average [12] (ERWA) is a simple technique to incrementally estimate the moving average by giving more weight to more recent results, which can deal with this problem.

Giving a string of numbers, it is arranged in the form of a sequence as (b_1, b_2, b_3, …, b_n). b_n is the last number. We calculate the exponential recent weighted average for these numbers according to the following formula:

$$E = \sum_{i=1}^{n} \alpha(1-\alpha)^{n-i} b_i \tag{3}$$

Here $\alpha \in [0, 1]$ is a step parameter which controls the relative weights between the most recent and past datas.

We command $\alpha = 1/2$ and use (3) to compute exponential recent weighted average for two sequences (2,3,4,5) and (5,4,3,2) respectively. The exponential recent weighted average of (2,3,4,5) is 4.125. The exponential recent weighted average of (5,4,3,2) is 2.5625. Even if they are the same number, as long as the order of occurrence is different, the impact of the current score will relatively large. For the satisfiability problem, the literal appears in the order of the conflicting clause (we convert it to weight) also determines the choice of the next decision branch variable. They have similar functions, as long as the interpretation of the coefficients in exponential recent weighted average can be applied to satisfiability problem of branch heuristic.

Bit-encoding scheme [13] was proposed by Jingchao Chen in 2014. This strategy is the basic thought that makes the phase at each decision level with corresponding to a bit value of the binary representation of a counter. Let n represents the value of the counter. Formula is calculated as (4):

$$n = b_k 2^k + b_{k-1} 2^{k-1} + \cdots b_1 2 + b_0 \tag{4}$$

The phase of a variable at the k-th decision level is equal to b_k.

Now explaining our branching heuristic:

(1) If there is no conflict clauses in propositional logic formula, we use VSIDS heuristic to select decision variables.
(2) If there are conflict clauses in propositional logic formula, Firstly, we use VSIDS heuristic to record the number of the literals occurrence in propositional logic formula. Then we record the sequence of literals in the conflict clauses. The literals in the N-th conflict clause is denoted as b_n. For each literal, we can get a corresponding sequence (b_1, b_2, …,b_{n-1}, b_n). While for any i, the b_i is either 0 or 1. When the literal appears in the i-th conflict clause, b_i is 1, otherwise it is 0. Even if some of the literal has never appeared in the conflict clauses, the default is that its sequence is all 0.
(3) When the last bit of the corresponding sequence of words is 1 (that is, $b_n = 1$), the literal (denoted x) can be bumped $Q_n[x]$, and bonus points are as follows:

$$Q_n[x] = b_n + b_{n-1}\frac{1}{2} + b_{n-2}\left(\frac{1}{2}\right)^2 + \cdots + b_1\left(\frac{1}{2}\right)^{n-1} \quad (5)$$

For $i < n$, if $b_i = 1$ and $(1/2)^{n-i} < (1/2)^5$

$$Q_n[x] = b_n + b_{n-1}\frac{1}{2} + b_{n-2}\left(\frac{1}{2}\right)^2 + \cdots + b_{i+1}\left(\frac{1}{2}\right)^{n-i-1} \quad (6)$$

Let's assume that the number of literal x in the last decision level score is S. When b_n is equal to 0. the score in this decision level is S/2.

(4) Let's assume that the number of literal x in propositional logic formula is S_0. The literal x can be bumped as Q[x]. The literal x now has a score of S.

$$S = S_0 + Q_n[x] \quad (7)$$

We describe our branching heuristic in the CDCL algorithm.

```
1 if(Unit propagation( )==conflict)
2    then return UNSAT;
3  blevel=0;
4  conflict number n=0;
5  While (not All Variables Assigned )
6  { pick branching variable( );
7     blevel=blevel+1;
8     if(Unit propagation( )==conflict)
9     then
10       { blevel=analyze conflict( );
11         conflict number n=conflict number n+1;
12         S=S +Q_n[x];
13  The score of other variables that do not appear in the learning clause
14       S=S/2;
15       }
16          If( blevel<0)
17            then
18              return UNSAT;
19                else Backtrack(blevel);
20 }
21 return SAT;
```

Next we describe our branching heuristics and average conflict-index decision score in the same example.

The number of literal x_1 in propositional logic formula is 10. Case one: it doesn't appear in the first conflict clause, it is recorded as (0). It appears in the second conflict clause, it is recorded as (0,1). It appears in the third conflict clause, it is recorded as (0,1,1). Case two:it appears in the first conflict clause, it is recorded as (1). It does not appear in the second conflict clause, it is recorded as (1,0). It appears in the third conflict clause, it is recorded as (1,0,1). We show the score value in the Table 1.

From the above results we can see that this weight-based score satisfies the requirement for variable scoring. In the 1-th, the literal x_1 in case one does not appear, the literal x_1 in case two appears. So the weight of case two is greater than the weight of case one. In the 2-th, the literal x_1 in case one appears, the literal x_1 in case two does not appear. At the same time we want to give the new conflict clause in the text of greater weight. So the weight of case one is greater than the weight of case two. In the 3-th, the literal x_1 in case one appears, the literal x_1 in case two also appears, the literal x_1 in case one also appears in the second conflict clause. But the

Table 1 The score value of literal x_1

Case one: x_1 in conflict clause?	Score	Case two: x_1 in conflict clause?	Score
1-th no	5	1-th yes	11
2-th yes	6	2-th no	5.5
3-th yes	7.5	3-th yes	6.75

Table 2 The score value of literal x_1

Case one: x_1 in conflict clause?	Score	Case two: x_1 in conflict clause?	Score
1-th no	10	1-th yes	5.5
2-th yes	6	2-th no	2.75
3-th yes	4.5	3-th yes	2.875

literal x_1 in case two does not appear in the second conflict clause. So the weight of case one is greater than the weight of case two.

From (5) we can conclude that if the text x appears simultaneously in the n-th and (n + 1)-th conflict clauses (i. e. $b_n = 1$, $b_{n+1} = 1$), we can derive a relationship about $Q_n[x]$ and $Q_{n+1}[x]$

$$Q_{n+1}[x] = 1 + \frac{1}{2} Q_n[x] \tag{8}$$

The above formula illustrates how we give a larger weight to the new conflict clause, and also on the basis of the weight of the historical conflict clause.

For the relatively new branch heuristic ACIDS, its purpose is similar to this article. We show the score value in the Table 2.

4 Conclusions

On the basis of previous studies, we have merged some of our ideas. Theoretically explores a large weighting plan for new conflict clauses and a scoring scheme based on historical conflict clauses. At the same time, the article also cited a small example to explain. Whether from theory or example, we can see that this idea is feasible.

References

1. Sakallah MSKA. GRASP—a new search algorithm for satisfiability. 1996.
2. Malik S, Zhao Y, Madigan CF, et al. Chaff: engineering an efficient SAT solver. In: Proceedings of the design automation conference. IEEE; 2001. p. 530–5.
3. Goldberg E, Novikov Y. BerkMin: a fast and robust sat-solver. IEEE Computer Society; 2002. p. 0142.
4. Eén N, Sörensson N. An extensible SAT-solver. In: Theory and applications of satisfiability testing. Berlin, Heidelberg: Springer; 2003. p. 502–18.
5. Zhang H. SATO: an efficient propositional prover. In: International conference on automated deduction. Springer; 1997. p. 272–5.
6. RJB Jr, Schrag R. Using CSP look-back techniques to solve exceptionally hard SAT instances. In: International conference on principles and practice of constraint programming, Cambridge, Massachusetts, USA, August. DBLP; 1996. p. 46–60.

7. Audemard G, Simon L. Predicting learnt clauses quality in modern SAT solvers. In: IJCAI 2009, proceedings of the international joint conference on artificial intelligence, Pasadena, California, USA, July. DBLP; 2009. p. 399–404.
8. Biere A, Lingeling, Plingeling, PicoSAT and PrecoSAT at SAT race 2010. In: SAT race: system description. 2010.
9. Biere A. Adaptive restart strategies for conflict driven SAT solvers. 2008. p. 28–33.
10. Ryan L. Efficient algorithms for clause-learning SAT solvers. Simon Fraser University; 2004.
11. Biere A, Fröhlich A. Evaluating CDCL variable scoring schemes. In: Theory and applications of satisfiability testing—SAT 2015. Springer International Publishing; 2015. p. 405–22.
12. Liang J, Ganesh V, Poupart P, Czarnecki K. Exponential recency weighted average branching heuristic for sat solvers. In: AAAI conference on artificial intelligence. AAAI Publications; 2016.
13. Chen JC. A bit-encoding phase selection strategy for satisfiability solvers. In: Theory and applications of models of computation. Springer International Publishing; 2014. p. 158–67.

EEG Multi-fractal De-trended Fluctuation Mental Stress Analysis

Xin Li, Erjuan Cai and Jiannan Kang

Abstract In this paper, mental stress is evaluated by algorithm based on Multi-fractal De-trended Fluctuation Analysis. The key parameters of Multi-fractal De-trended Fluctuation Analysis that is singular index, Hurst index are discussed. Based on the optimal selection of the parameters, the EEG mental stress is evaluation based on EEG signal analyses. We record electroencephalogram (EEG) of 14 students and ensured the optimal order being [−5 5] via comparing the relationship between fractal indices and order, then achieved the estimate of mental stress with the β wave in EEG. The results show that Hurst index and quality index of the EEGs under mental stress are greater than those in the relaxing state, and with the increase of order, quality index is amplified and the variation of the singular index is more obvious while Hurst index decreases and tends a constant. We also compare the width of singular spectrum of the EEGs of different mental state, discovering that the features of multi-fractal spectrum of different states are different and the width of singular spectrum of the EEGs under mental stress are greater than the relax condition.

Keywords EEG · Multi-fractal de-trended fluctuation · Singular spectrum width · Singular index

X. Li · E. Cai
Institute of Biomedical Engineering, Yanshan University, Qinhuangdao 066004
Hebei Province, China

X. Li · E. Cai
Measurement Technology and Instrumentation Key Laboratory, Qinhuangdao 066004
Hebei Province, China

J. Kang (✉)
College of Electronic & Information Engineering, Hebei University, Baoding 071000
Hebei Province, China
e-mail: kangjiannan81@163.com

J. Kang
Key Laboratory of Digital Medical Engineering of Hebei Province, Baoding 071000
Hebei Province, China

Y. Jia et al. (eds.), *Proceedings of 2017 Chinese Intelligent Systems Conference*, Lecture Notes in Electrical Engineering 459,
https://doi.org/10.1007/978-981-10-6496-8_9

1 Introduction

EEG is the neurophysiologic measurement of the electrical activity of the brain. EEG signals which include abundant information of pathological, physiological and mental are known to be non-stationary, highly noisy, and irregular and vary significantly from person to person. EEG signals are almost pre-processed before any further feature extraction analysis.

Many excellent EEG feature extraction algorithms have been put forward so far. In general, those algorithms can be categorized as: (1) time-frequency analysis method as wavelet transform which based on multi-resolution achieved the analysis of EEG; (2) energy analysis method, such as complexity and approximate entropy (AE), and the former reflect the complex degree of time series by analyzing the order of time series, while the later reflect the complexity of time series efficiency, and the more complex is, the AE greater. Wavelet entropy and power spectral entropy explain the degree of confusion of multi-frequency time series, and the energy distribution of EEG signal in frequency; (3) Multi-Fractal analysis: this method which is fundamentally more complex and inhomogeneous than monofractals describe time series features via very irregular dynamics, with sudden and intense bursts of high-frequency fluctuations. This method exploits a spectrum function to introduce the growth characteristics of fractal with different levels. The multifractal analysis of EEG data would be more appropriate, e.g. MFDFA is used to illustrate the fractal feature of EEG by analyzing fractal structure of EEG signal.

Recently several papers analyzing the multifractal properties of EEG signals were published. Multi-fractal detrended fluctuation analysis of the EEG under pressure and non-pressure was done in this paper. Fractal concept was proposed by Mandelbrot in his paper published in science about how long is the coastline of England. The multifractal was designed to find the long-range-correlation and the power-law index law features of EEG signals [1]. C.K. Peng et al. were the pioneers in this field and introduced the detrended fluctuation analysis (DFA) methodology to consider the properties of DNA sequences [2]. DFA revealing the fractal feature of EEG by analyzing the scale exponents of EEG signal during rest, was used to distinguish the normal and depression state, indicating that depressed EEG signals had mutifractal properties, but the long-range-correlation was shorter than the normal, and half of them appeared opposite correlation [3]. DFA reveal the difference of fractal structure in low-frequency and high-frequency in EEG signal [4]. The scaling exponent of resting EEG signal was 1.26 by discussing the relativity of EEG signal based on DFA.

The multifractal detrended fluctuation analysis (MFDFA) was first conceived by Kantelhardt as a generalization of the standard DFA [5]. MFDFA was capable of determining multifractal scaling behavior of non-stationary time series, and applied successfully to study various non-stationary time series. Zou et al. researched the change of EEG signal of eyes open, eyes closed, normal people, no epileptic ictal and seizure based on MFDFA and discovered that the singular spectrum width and Hurst index of the EEG signal could distinguish different conditions of the brain [6].

The multi-fractal parameters could act as biological marks to discriminate the normal and epilepsia and the multifractal degree of epilepsia is much higher than normal. Just as for the normal, the degree of multifractal is also different between eyes open and closes [7].

In this study, the MFDFA was used to analyze the multifractal features of pressure and non-pressure EEG behind the optimal fractal order confirmed and achieved the estimate of mental pressure.

2 Multifractal Detrended Fluctuation Analysis

The construction of MFDFA is divided into seven steps; the first three steps are essentially identical to the conventional DFA procedure. Converting time series into a random walk time series is a preliminary step for MFDFA. We supposed that is a time series of length N, and the values of time series is nonzero. The time series that we analyzed is one-dimensional structures.

Step 1 Convert into random walk time series.

$$Y(i) = \sum_{k=1}^{i} [x_k - \bar{x}] \quad i = 1, 2, 3, \ldots, N \tag{1}$$

The $\bar{x}$ is the average of the whole time series.

Step 2 Divide $Y(i)$ into $N_s = N/s$ no overlapping segments of each length s. If N_s is an integer, then continue step 3, if not, the same procedure is repeated starting from the opposite end.

Step 3 Calculate the local trend for each segment by subtract a least-square fit of the series y_v, then determine the variance:

$$F^2(s, v) = \frac{1}{s} \sum_{i=1}^{s} \{Y[(v-1)s + i] - y_v(i)\}^2 \tag{2}$$

for each segment $v, v = 1, 2, \ldots, N_s$

$$F^2(s, v) = \frac{1}{s} \sum_{i=1}^{s} \{Y[N - (v - N_s)s + i] - y_v(i)\}^2 \tag{3}$$

for $v = N_s + 1, \ldots, 2N_s, y_v(i)$ is the fitting polynomial in segment v. Since the detrending of the time series is done by subtracting the polynomial, different orders eliminate different trends in the time series. Thus a comparison of the results for different orders of MFDFA should be done.

Step 4 Average over all segments to get the qth order fluctuation function.

$$F_q(s) = \{\frac{1}{N_s}\sum_{v=1}^{N_s}[F^2(v,s)]^{q/2}\}^{1/q} \quad (4)$$

In general, the index variable q can take any real values except zero, and when $q = 2$, the standard DFA procedure is retrieved. We are interesting in how the generalized fluctuation functions $F_q(s)$ depend on time scale s for different values of q. Therefore we must repeat steps 2–4 for several time scales s. It is apparent that $F_q(s)$ will increase with increasing s, and also depends on the order q. By construction, $F_q(s)$ is only defined for $s \geq N_s$.

Step 5 Determine the scaling behavior of the fluctuation functions by analyzing log-log plots $F_q(s)$ versus s for each order.

$$F_q(s) \sim s^{h(q)} \quad (5)$$

For large scales, like $s > N/4$, $F_q(s)$ will become statistically unreliable because of the number of segments N_s in step 4 very small. $h(q)$ is obtained by calculating the slope of log-log plots $F_q(s)$ versus s, and $h(2)$ is the classical Hurst exponent. Larger q describe the scaling behavior of larger fluctuations as smaller q describe the scaling behavior of small fluctuations. $h(q)$ is dependent on q, that as for a multifractal time series, $h(q)$ varies with q, and this dependence is considered to be a characteristic property of the multifractal processes.

Step 6 the q-order Hurst exponent H is only one of several scaling exponents used to parameterize the multifractal structure of time series. The typical procedure in the literature of MFDFA is to firstly convert H to the q-order mass exponent $\tau(q)$:

$$\tau(q) = qh(q) - 1 \quad (6)$$

$h(q)$ defined in Step 5 is directly related to the classical multifractal scaling exponents $\tau(q)$ via Formula 6. Note that $h(q)$ is different from generalized multifractal dimensions.

Step 7 The q-order mass exponent $\tau(q)$ is used to compute the q-order singularity exponent α and the q-order singularity dimension $f(\alpha)$, denoting the dimension of the subset of the series based on singularity exponent α.

$$\alpha = \tau'(q) \quad f(\alpha) = q\alpha - \tau(q) \quad (7)$$

The width of the singular spectrum is characterized by the difference between the maximum and minimum values of α,

$$\Delta\alpha = \alpha_{\max} - \alpha_{\min} \quad (8)$$

3 Parameter of MFDFA

3.1 Data Description

Seven males and seven females of colleges with right-handed participated in a sound and light isolation experiment. Neuroscan 64 channel EEG acquisition system is used to acquire EEG signal based on the standard 10–20 international system and remove the eyes artifact, with the sampling rate of 1000 Hz. The data is filtered between 0 and 30 Hz, in the course of the experimental activity, EEG was recorded with eyes open. The denoising processing was completed by ICA (independent component analysis).

EEG signal of participants were recorded after these requirement condition finished: (1) wash and dry their hair; (2) keep a better state; (3) in a shielded room with the sound and light isolated; (4) To reduce hypoglycemia effects, the time of the experiment should be controlled within 3 h. Subjects underwent through resting 2 min with opened eyes and recited for 1 min. We collected 92 groups containing 44 for resting and 48 for short-term memory task. Given theta and gamma mainly emerged in nursling while alpha will be weak or none when eyes opened, we extracted beta wave in EEG signal to analyze and evaluate the mental stress.

The MFDFA is performed over the dataset. To obtain the optimal order, We calculated the fluctuations $F_q(s)$ for scales s ranging from 4 to 256 with the length of time series was 1000 respectively. The exponents $H(q)$ could be obtained by observing the slope of log-log plots $F_q(s)$ versus s. Figure 1 displays the relationship of the parameters versus q.

It should be noted that $H(q)$ decreased with q increased and tended to be a constant, but higher than 0.5, and the pressure was greater than no pressure. Different $H(q)$ represented different features of EEG signal, such as long-range-anticorrelated ($0 < H(q) < 0.5$), uncorrelated ($H(q) = 0.5$), and positively long-range-correlated ($0.5 < H(q) < 1$) series. The exponent $H(q)$ describes the scaling behavior of the q-order fluctuation function. Figure B showed the plot of $\tau(q)$ versus q. The curvature of the pressure was much stronger, indicating that the multifractal feature was obviously. Figure C and D represent the plot of α and$f(\alpha)$ versus q, and are both curves lines. α decreased with q increased, and finally turned into a constant. The other plot is a singlet arch. Figure E showed the plot of singular spectrum, showing the multifractal feature of EEG. In the nest part we summarize how to determine the optimal fractal order.

A standard assumption in the classical control theory is that the data transmission required by the control or state estimation algorithm can be performed with infinite precision. However, due to the growth in communication technology, it is becoming more common to employ digital limited capacity communication networks for exchange of information between system components.

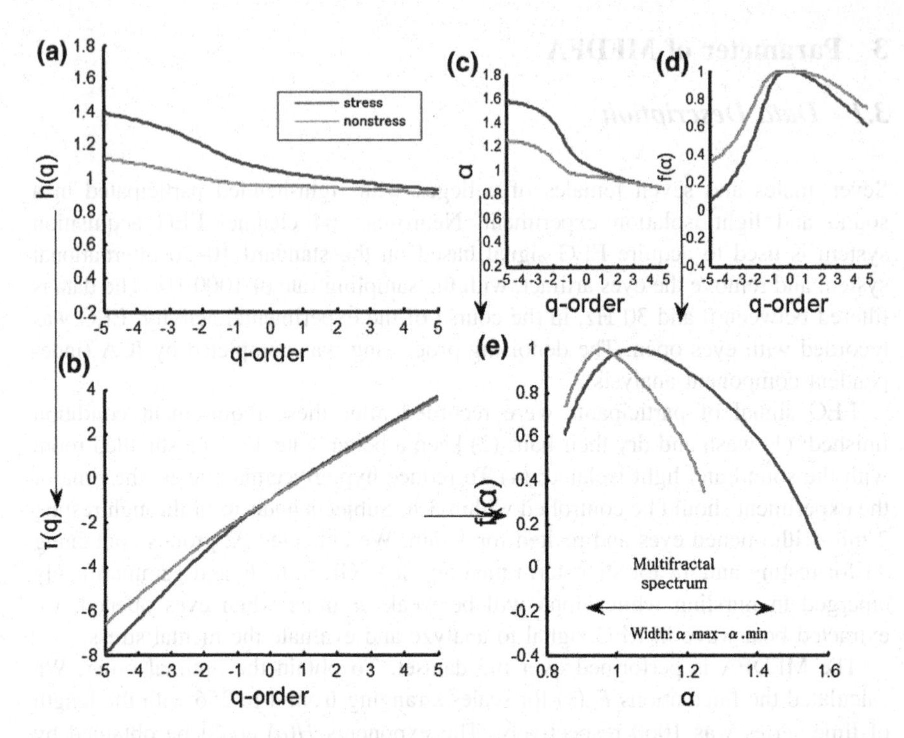

Fig. 1 Multiple representations of multifractal spectrum for stress and no stress time series. **a** q-order Hurst exponent $h(q)$, **b** Mass exponent $\tau(q)$, **c** and **d** q-order singularity exponent and q-order singularity dimension, **e** multifractal spectrum of singularity exponent and singularity dimension

3.2 Data Description

The select of order is essential for the result in MFDFA. To ascertain the optimal fractal order, the variation of $H(q)$, $\tau(q), \alpha, f(\alpha)$ versus q were analyzed. $H(q)$ is an indicator of data chaos. $H(q)$ decreased as the order increased with the rang [−5, 10], while in >10 rang, $H(q)$ was nearly equal. In the range of [0, 8], $H(q)$ is close to 1. As for $H(q)$, the order range was selected between [−5, 8] shown in Fig. 2.

Figure 3 reveals the $\tau(q)$ versus q. The variation trend was almost steadiness and tended to be equal with the q increases, but the initial value of $\tau(q)$ which affected the change of the singular dimension decreased. For $\tau(q)$, the select of order was not too big. The curvature of the curve of the pressure was little bigger than that of nonpressure. The curvature of $\tau(q)$ affected both singular index and singular dimension, so the selection of q was based on the relationship between q with α and $f(\alpha)$.

The different singular degrees of each region in complex systems were described by analyzing α. The smaller value of α is, the greater of singularity is. Figure 4 showed the plot of α versus different q-order ranges. The initial value of singularity

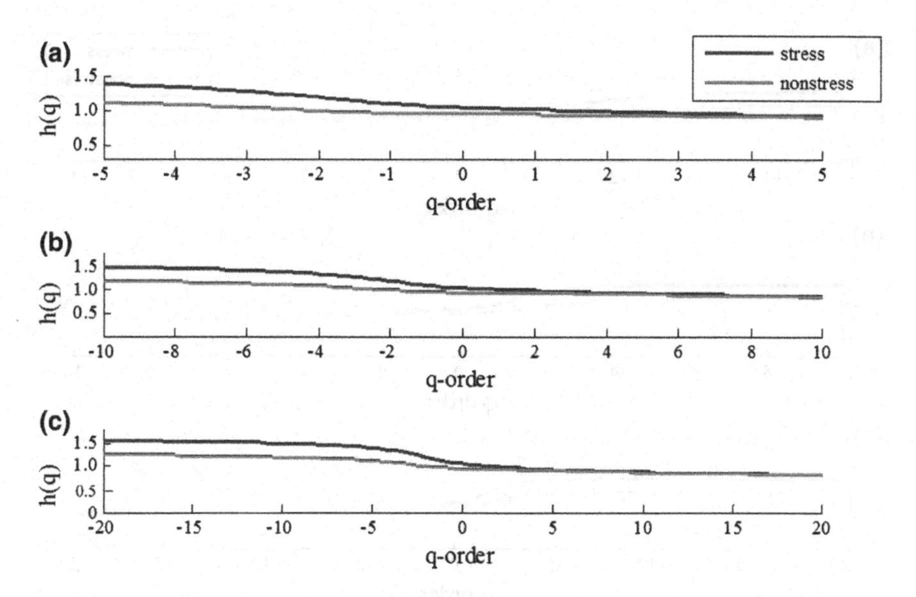

Fig. 2 Comparison of the generalized Hurst exponent h(q) with different ranges of q-order

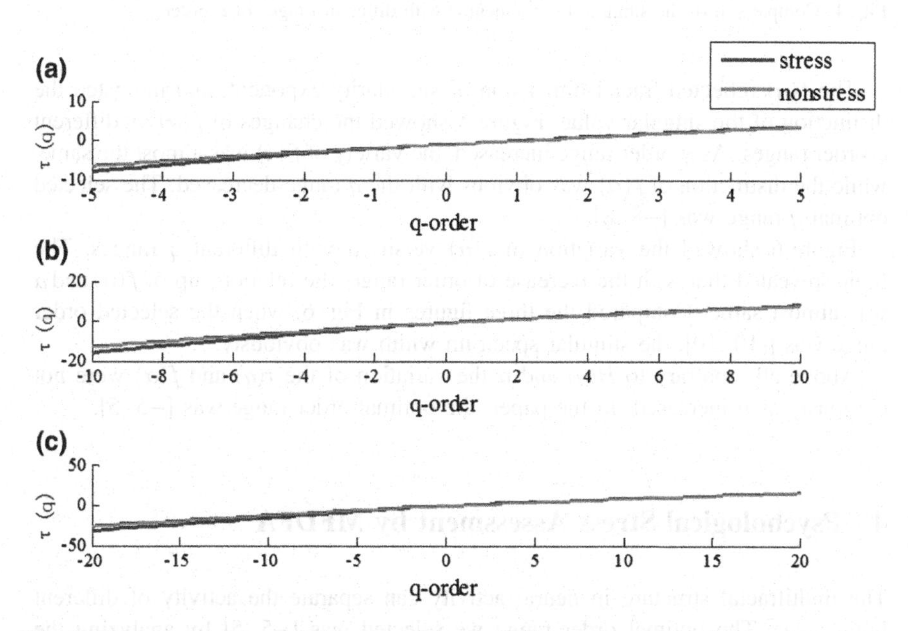

Fig. 3 Comparison of the mass exponent $\tau(q)$ with different ranges of q-order

exponent was almost the same, and in the range of >5 and <−5, the variety was not obviously with q-order ranges increase, and when q was greater than 5, the α of the EEG were not obviously to distinguish, so the best selection range was [−5, 5];

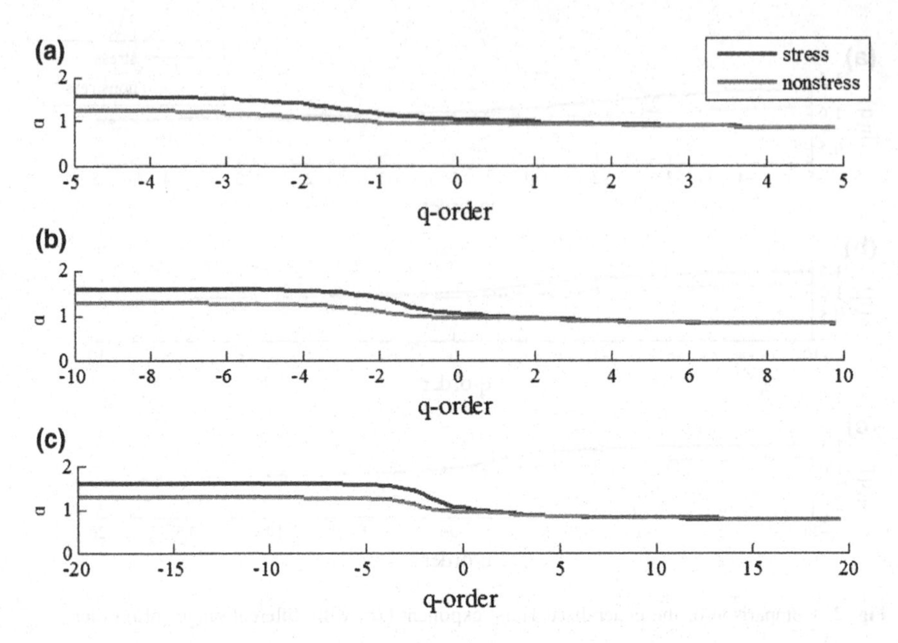

Fig. 4 Comparison of the singularity exponent α with different ranges of q-order

The $f(\alpha)$ reflected fractal dimension of singularity exponent, and indicated the distinction of the singular value. Figure 5 showed the changes of $f(\alpha)$ vs different q-order ranges. As q-order range increased, the variety of $f(\alpha)$ was almost the same, while the distinction of $f(\alpha)$ was obvious with the q range decreased. The selected optimal q range was [−8, 8].

Figure 6 showed the variation of $f(\alpha)$ versus α with different q ranges. The figure revealed that with the increase of order range, the relationship of $f(\alpha)$ and α was almost same. Compared the three figures in Fig. 6, when the selected order range was [-10, 10], the singular spectrum width was obviously.

Above all, contrary to $H(q)$ and α the variation of the $\tau(q)$ and $f(\alpha)$ were not obviously as q increased. In the paper, the optimal order range was [−5, 5].

4 Psychological Stress Assessment by MFDFA

The multifractal structure in neural activity can separate the activity of different brain areas. The optimal order range we selected was [−5, 5] by analyzing the parameter of the MFDFA method. In this order range, $H(q)$ was close to 1, indicating the long-range-correlation of EEG signal. The singular spectrum was single arch illustrating the multifractal feature of EEG signal. We selected singular spectrum width estimated the mental stress. Figure 7 showed the algorithm flow:

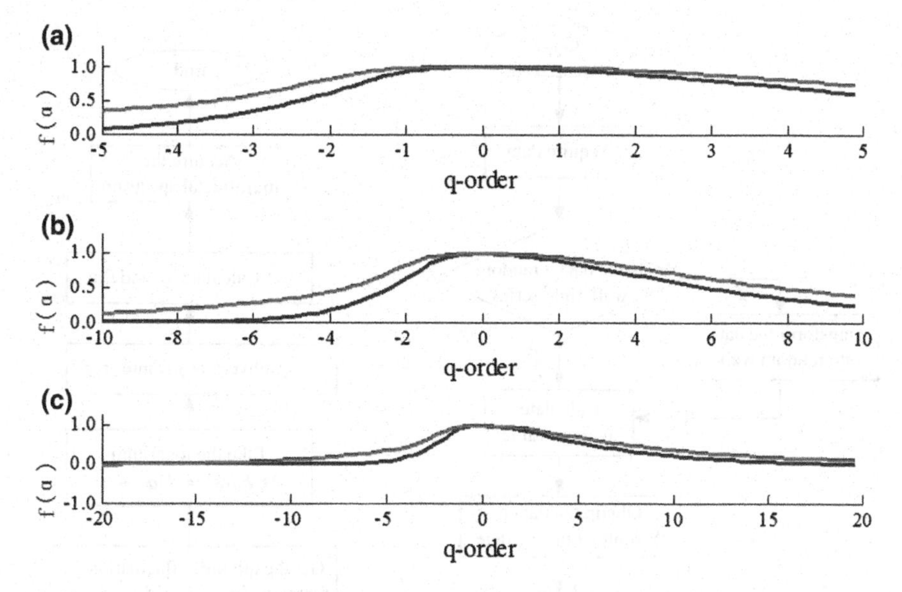

Fig. 5 Comparison of the singularity dimension $f(\alpha)$ with different ranges of q-order

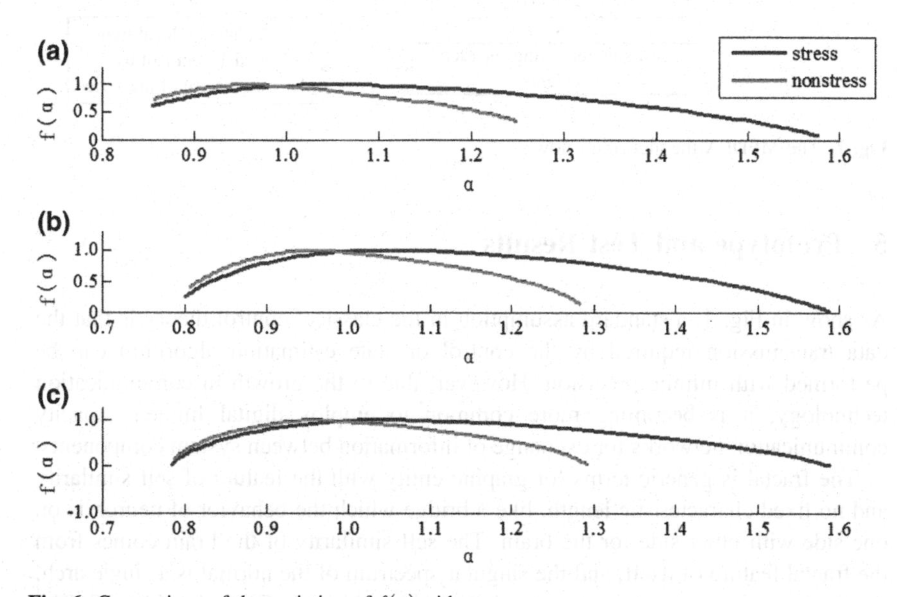

Fig. 6 Comparison of the variation of $f(\alpha)$ with α

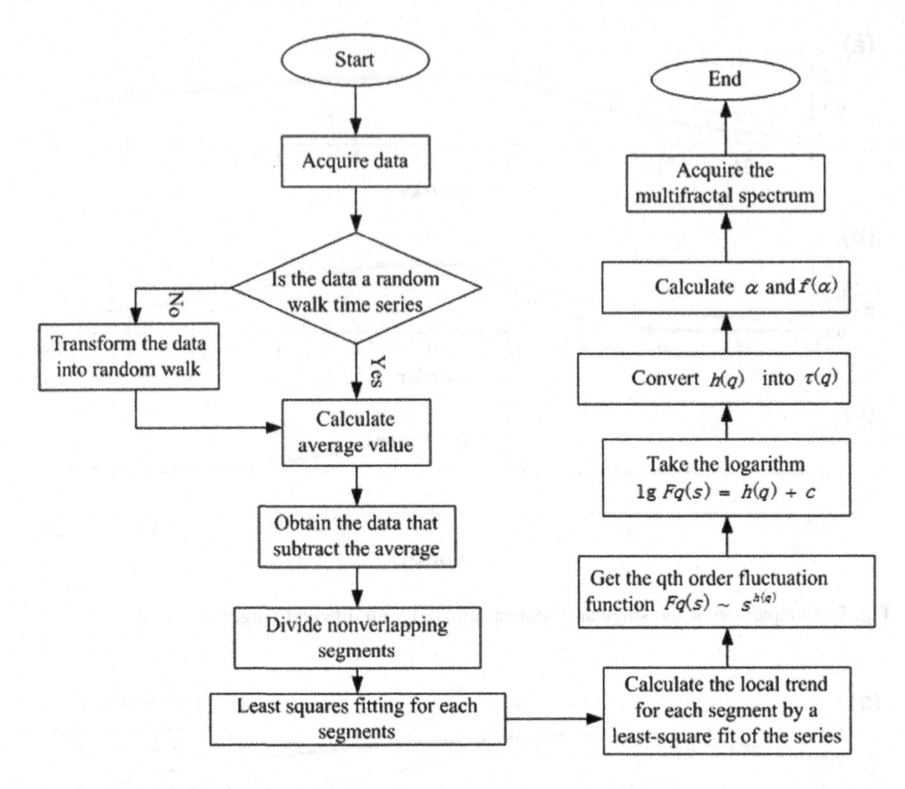

Fig. 7 The MFDFA the algorithm flow

5 Prototype and Test Results

As show in Fig. 1, a standard assumption in the classical control theory is that the data transmission required by the control or state estimation algorithm can be performed with infinite precision. However, due to the growth in communication technology, it is becoming more common to employ digital limited capacity communication networks for exchange of information between system components.

The fractal is generic terms for graphic entity with the feature of self similarity and no fixed characteristic length, like a bridge which the behavior of neuron is on one side with other side for the brain. The self similarity of the brain comes from the fractal feature of itself, and the singular spectrum of the normal is a single arch.

We found that an order corresponded to $f(\alpha)$ and α via Formulae 6 and 7. The collected EEG data was selected by the singular spectrum. Finally we chose 22 groups stress/non-stress EEG data to analyze; the result was in Table 1.

The result in Tables 2 and 3 were the statistical analysis of the features of EEG data. In Table 2, the statistical features were different. In Table 3, the value of

Table 1 The singular exponent of MFDFA with EEG

Groups	Pressure			No pressure		
	α maximum	α minimum	$\Delta\alpha$	α maximum	α minimum	$\Delta\alpha$
1	1.58	0.85	0.73	1.27	0.67	0.60
2	1.45	0.78	0.67	1.09	0.83	0.26
3	1.58	0.85	0.73	1.25	0.85	0.40
4	1.47	0.75	0.71	1.61	1.15	0.46
5	1.51	0.96	0.54	1.60	1.08	0.52
6	1.47	0.88	0.58	1.49	1.12	0.37
7	1.77	0.96	0.81	1.75	0.99	0.77
8	1.75	1.07	0.68	1.67	1.00	0.67
9	1.72	1.01	0.72	1.46	0.97	0.49
10	1.68	1.00	0.68	1.64	1.22	0.42
11	1.68	1.04	0.65	1.66	1.15	0.51
12	1.84	1.02	0.82	1.20	0.80	0.41
13	1.68	1.11	0.58	1.08	0.63	0.44
14	1.61	0.96	0.64	1.26	0.76	0.49
15	1.56	1.13	0.42	1.07	0.69	0.38
16	1.49	0.95	0.54	1.18	0.82	0.35
17	1.68	1.01	0.67	1.19	0.89	0.30
18	1.53	0.80	0.74	1.18	0.83	0.35
19	1.30	0.81	0.49	0.96	0.75	0.22
20	1.31	0.86	0.45	1.26	0.87	0.38
21	1.38	0.88	0.49	1.37	0.89	0.48
22	1.37	1.01	0.36	1.13	0.80	0.33

P indicated that the result had statistical significance. The singular spectrum was different from one person to other person owing to the response to stress was different. We selected 22 groups EEG data to analyze by subtract the larger deviation data. Calculate and compare the maximum, minimum values and mean value of α of the pressure/non-pressure, the result was showed in Table 4. The maximum and minimum value of singular spectrum width of the stress EEG data is 0.82 and 0.36, both larger than the non-pressure. Based on the analysis above, with the optimal order range, the multifractal singular spectrum width can effectively analyze the non-pressure/pressure EEG data and estimate the mental stress.

Table 2 Statistics of two groups of EEG

Classification	Groups	Mean	Standard deviation	Standard error of mean
Pressure	22	0.59	0.13	0.03
Non-pressure	22	0.46	0.16	0.03

Table 3 Independent sample t test

	Levene test of variance equation		t test of the mean equation				
	F	*p*	*t*	*df*	*p*	D-value of the mean	The value of standard error
Suppose the variance equal	0.55	0.46	3.03	42	0.004	0.13	0.04
Suppose the variance inequality			3.03	39.71	0.004	0.13	0.04

Table 4 The compare with the singular spectrum width

$\Delta\alpha$	Maximum	Minimum	Average
No pressure	0.77	0.22	0.44
Pressure	0.82	0.36	0.62

6 Conclusion

We achieved the analysis of EEG signal and the estimate of mental stress based on the multi-fractal detrended analysis method. We ensured the optimal order by analyzing the relationship between the fractal index and order. The result indicated that variation of the quality index and singular index were not obviously while Hurst index and singular index were obviously. The optimal order selected was [−5 5] which in this range, the features of EEG signal can be excavated fully.

We also compared the maximum, minimum and average value of the singular spectral width of EEG signal. The results revealed that the three values of the singular spectral width of mental stress were greater than those of relaxing state, indicating that the analysis of singular spectral width can discriminate different mental states. As based on the MFDAF method, the mental stress state can be identified. We can improve the mental stress state via some intervene and prevention methods and make a healthy life.

References

1. Halsey TC, Jensen MH, Kadanoff LP, et al. Fractl measure and their singularities. The characterization of strange sets. Phys Rev A. 1986;33:1141–51.
2. Peng CK, Buldyrev SV, Havlin S, et al. Mosaic organi-zation of DNA nucleotides. Phys Rev E. 1994;49(2):1685–9.
3. Bachmann M, Suhhova A, JASS J, et al. Detrended fluctuation analysis of EEG in depression. IFMBE Proc. 2014;41(1):694–7.

4. Marton LF, Brassai ST, BAKO L et al. Detrended fluctuation analysis of EEG signals. Proc Technol. 2014;12(1):125–32.
5. Kantelharde JW, Zschiegner SA, Koscielny BE, et al. Multifractal detrended fluctuation analysis of nonstationay time series. Phys A. 2002;316:87–114.
6. Ming Z, Yong G, XInmeng W et al. Multifractal detrended fluctuation analysis on electroencephalography. Beijing Biomed Eng. 2013;32(3):226–9.
7. Dutta S, Ghosh D, Samanta S, et al. Multifractal parameters as an indication of different physiological and pathological states of the human brain. Phys A. 2014;396(15):155–63.

The Simulation of Neural Oscillations During Propofol Anesthesia Based on the FPGA Platform

Zhenhu Liang and Cheng Huang

Abstract This paper focuses on the realization of the Pharmacokinetic (PK)-Jansen Rit neural mass model (JRNMM) for visualization neural oscillations and accelerating the calculations of complex model on hardware platform. Firstly, we set up a combined model named PK-JRNMM and produce simulated EEG-like (sEEG) signals in DSP-Builder. Then, the scheme of SOPC embedded platform is employed to reproduce the model function via Field Programmable Gate Array (FPGA). Finally, the sEEG signals can be achieved through the digital to analog conversion (DAC) and Liquid Crystal Display (LCD). The realization of this platform takes advantage of the parallel computing characteristics with FPGA processor. It will provide a new tool for the fast simulation and realization of the complex neural models in hardware platform.

Keywords Neural mass model · Simulation · DSP-Builder · Field programmable gate array · Electroencephalogram

1 Introduction

The electroencephalogram (EEG) is a macroscopic manifestation of the neurological activity in the brain and it exhibits specific neural oscillations in different brain activities such as anesthesia [1]. Not only that but EEG is the oldest and most effective method to be employed to clinical experimental research. Therefore it means a lot to understand anesthesia mechanism from EEG. In particular, using the math and physics model to produce simulated EEG-like (sEEG) signals for analyzing the mechanisms of anesthesia has great value.

Z. Liang (✉) · C. Huang
Institute of Electrical Engineering, Yanshan University, Qinhuangdao 066004, China
e-mail: zhl@ysu.edu.cn

C. Huang
e-mail: huangcheng@stumail.ysu.edu.cn

Y. Jia et al. (eds.), *Proceedings of 2017 Chinese Intelligent Systems Conference*, Lecture Notes in Electrical Engineering 459,
https://doi.org/10.1007/978-981-10-6496-8_10

In previous paper we have built the Pharmacokinetic-Jansen Rit neural mass model (PK-JRNMM) to produce the sEEG data and proved it can well distinguish different anesthesia states [2]. However it was finished in Matlab and it only stayed in the simulation phase. Therefore, in this study, a new method is proposed to reproduce the function of Pharmacokinetic-Jansen Rit neural mass model (PK-JRNMM) via Field Programmable Gate Array (FPGA) hardware platform, which linked the digital to analog conversion (DAC) and Liquid Crystal Display (LCD) to output and display the EEG signals. The PK-JRNMM was built based on the complex mathematic model [3] in DSP-Builder which is a development tool of DSP from Alter Company and it can combine QuartusII with the Matlab/Simulink together.

2 System Description

With the purpose of simulating neural oscillations during the propofol anesthesia, the model of generating sEEG signals and the model of producing the effect-site drug concentration will be needed. Therefore we choose the Jansen-Rit (JR) neural mass model and Pharmacokinetics (PK) model.

The neural oscillations activities are produced by means of the primary thought of the neural mass model that is the interaction of pyramidal cells, excitatory interneuron and inhibitory interneuron [4]. With a 6 dimensional state space's Jansen-Rit (JR) neural mass model is the basic neural mass model for producing the sEEG signals. Even if there are lots of other neural models you can see [5–7] this paper focuses on JRNNM to introduce a simple approach in order to generate sEEG data and provide a benchmark for future studies with more complex model. The neural mass model is composed of pyramidal cells, excitatory interneuron and inhibitory interneuron [8]. Figure 1 shows the dynamic evolution between excitatory interneuron, inhibitory interneuron and pyramidal cell. Firstly the presynaptic's information is converted into the information of postsynaptic.

The linear function in excitatory and inhibitory case can be expressed as following equations:

$$h_e(t) = Aate^{-at} \tag{1}$$

$$h_i(t) = Bbte^{-bt} \tag{2}$$

Secondly average membrane potentials is converted an average pulse density of action potentials by nonlinear function $S(v(t))$.

$$S(v(t)) = \frac{2e_0}{1 + \exp(r[v_0 - v(t)])} \tag{3}$$

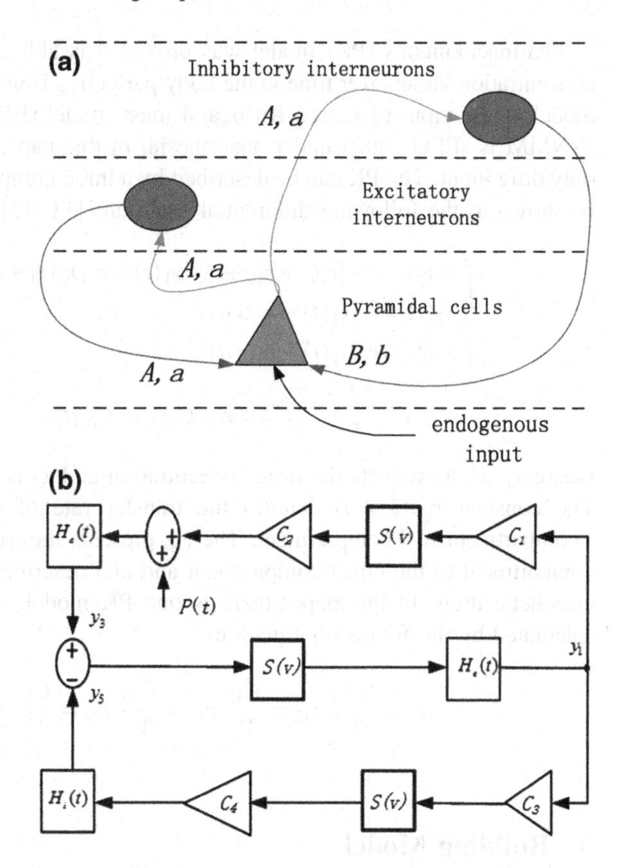

Fig. 1 **a** The dynamic evolution between pyramidal cell, excitatory interneuron and inhibitory interneuron. **b** The block diagram of the JRNMM

The neural mass model is expressed as follows [9, 10]:

$$\begin{cases} \dot{y}_1(t) = y_2(t) \\ \dot{y}_2(t) = AaS(y_3(t) - y_5(t)) - 2ay_2(t) - a^2 y_1(t) \\ \dot{y}_3(t) = y_4(t) \\ \dot{y}_4(t) = Aa[p + C_2 S(C_1 y_1(t))] - 2ay_4(t) - a^2 y_3(t) \\ \dot{y}_5(t) = y_6(t) \\ \dot{y}_6(t) = BbC_4 S(C_3 y_1(t)) - 2by_6(t) - b^2 y_5(t) \end{cases} \quad (4)$$

where C_1, C_2, C_3 and C_4 denote coupling strength between excitatory and inhibitory interneuron. The output of JRNMM is $Y = y_3 - y_5$ which is the sEEG signals. In order to simulate our brain's spontaneous activities, the Gaussian white noise was employed as the model input $P(t)$. The adjustment of model parameter is based on the physiological parameters information of standard value, and summarized as follows:

$A = 3.25$ mV, $B = 22$ mV, $a = 100\ s^{-1}$, $b = 50\ s^{-1}$,
$v_0 = 6$ mV, $e_0 = 2.5\ s^{-1}$, $r = 0.56\ mV^{-1}$, $a_d = 33\ s^{-1}$
$C_1 = 135$, $C_2 = 108$, $C_3 = 33.75$, $C_4 = 33.75$

Pharmacokinetics (PK) model was proved it is able to describe how the drug concentration varies over time in the body perfectly. Thus we take the output of PK model as the input of Jansen Rit neural mass model (JRNMM) and the output of JRNMM is sEEG signal under anaesthesia. In this paper we take propofol as the only drug input. The PK can be described by a three-compartment model which can be shown as the following differential equations [11, 12]:

$$\begin{cases} \dot{x}_1(t) = -[r_{10} + r_{12} + r_{13}]x_1(t) + r_{21}x_2(t) + r_{31}x_3(t) + u(t) \\ \dot{x}_2(t) = r_{12}x_1(t) - r_{21}x_2(t) \\ \dot{x}_3(t) = r_{13}x_1(t) - r_{31}x_3(t) \end{cases} \tag{5}$$

$$\dot{C}_{es} = r_{e0}(C_1(t) - C_{es}(t)) \tag{6}$$

where x_1, x_2, x_3 reflects the drug concentration and x_1 is the central compartment. The constant r_{ij} $(i \neq j)$ denotes the transfer rate of drug concentration from j compartment to i compartment. The r_{e0} presents the transfer ratio of the central compartment to the effect compartment and $u(t)$ describes the injection rate of the anesthetic drug. In this paper there is one PK model, and some constants were calculated by the following equations:

$$r_{10} = \frac{C_{11}}{V_1}\ r_{12} = \frac{C_{12}}{V_1}\ r_{13} = \frac{C_{13}}{V_1}\ r_{21} = \frac{C_{12}}{V_2}\ r_{31} = \frac{C_{13}}{V_2} \tag{7}$$

3 Building Model

The basic idea of hardware computing is the converting of neural mathematic model into the executable hardware description language of FPGA, using FPGA to realize the function of PK-JRNMM reproducing [13]. In this work, in order to achieve the digital simulation of the PK-JRNMM in the DSP-Builder, we should build a difference equation, which is obtained by discretizing the differential equations [14]. There are several methods to solve it such as Runge-Kutta, Euler and so on. If we choose the Runge-Kutta, we have to face the problem that complex computing process and lots of hardware sources were consumed will be needed, thus we choose Euler finally.

First of all, the linear function $h_e(t)$ and $h_i(t)$ can be presented by following differential equations:

$$\begin{cases} \dot{z}(t) = z_1(t) \\ \dot{z}_1(t) = Ggx(t) - 2gz_1(t) - g^2 z(t) \end{cases} \tag{8}$$

Secondly we discretize the ordinary differential equation and obtained the following equations:

$$\begin{cases} z(n+1) = z_1(t)^*dt + z(n) \\ z_1(n+1) = Ggx(n)^*dt + (1-2g^*dt)z_1(n) - g^2z(n)^*dt \end{cases} \tag{9}$$

Next we built PK and JRNMM module by Simulink tools based on difference Eq. (9). However there is a nonlinear functions ($S(v(t))$) [15], it is used to finish converting from average membrane potentials of the population to an average pulse density of action potentials. Therefore the thought of Look-Up-Table was employed to solve it. And then we created the module of PK (Fig. 2a) and JRNMM (Fig. 2b) so that the two simple modules can be interconnected easily. So far, most scholars hold their view that anesthetic drugs effects the brain function by improving the inhibition of the GABA neurotransmitter. Not only that but the key parameters inhibitory average synaptic gain B, average synaptic inhibitory time constants b and the input p(t) can capture the shape of EEG and B, b can increase and decrease with the drug concentration increasing respectively [16]. For this reason, we combine the output of PK with this parameters to produce sEEG signals.

4 Designing Scheme Based on FPGA

As shown in Fig. 3a, the simulation platform consists of waveform generation system and waveform control system. The waveform generation system produces sEEG data based on JRNMM which was built in DSP-Builder. Waveform control system uses SOPC embedded scheme which contains hardware and software design. In the respect of hardware, we finished system configuration by SOPC-Builder, program download and system-testing. In the respect of software, we connected the IP care, generated after compiling by Signal Compiler, with NiosII system to realize data communication with other module via Avalon Bus and realized the driver of D/A and LCD via programming in NiosIIEclipse. FPGA developed platform belongs to DE2-115 as shown in Fig. 3b.

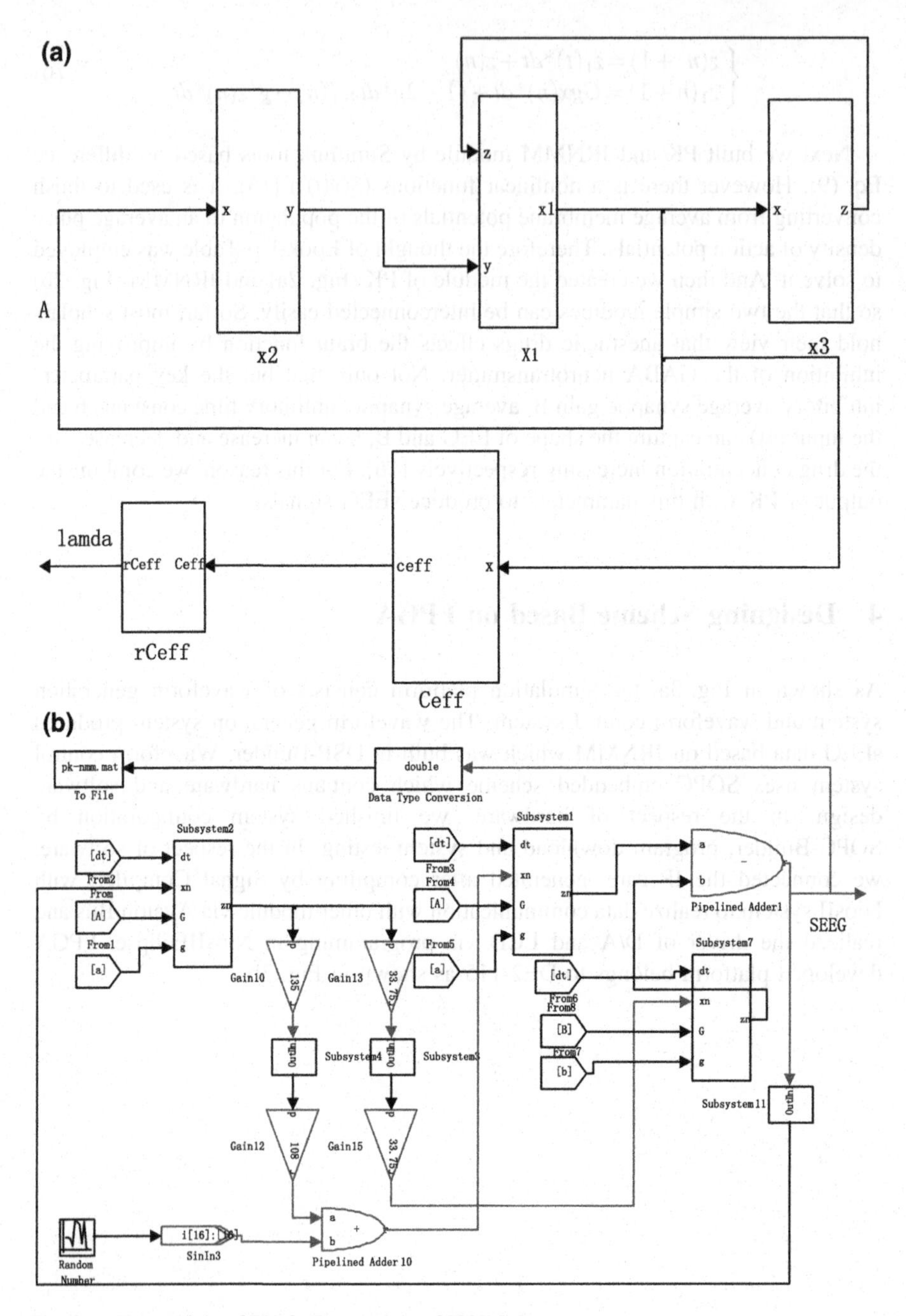

Fig. 2 **a** The module of PK. **b** The module of JRNMM

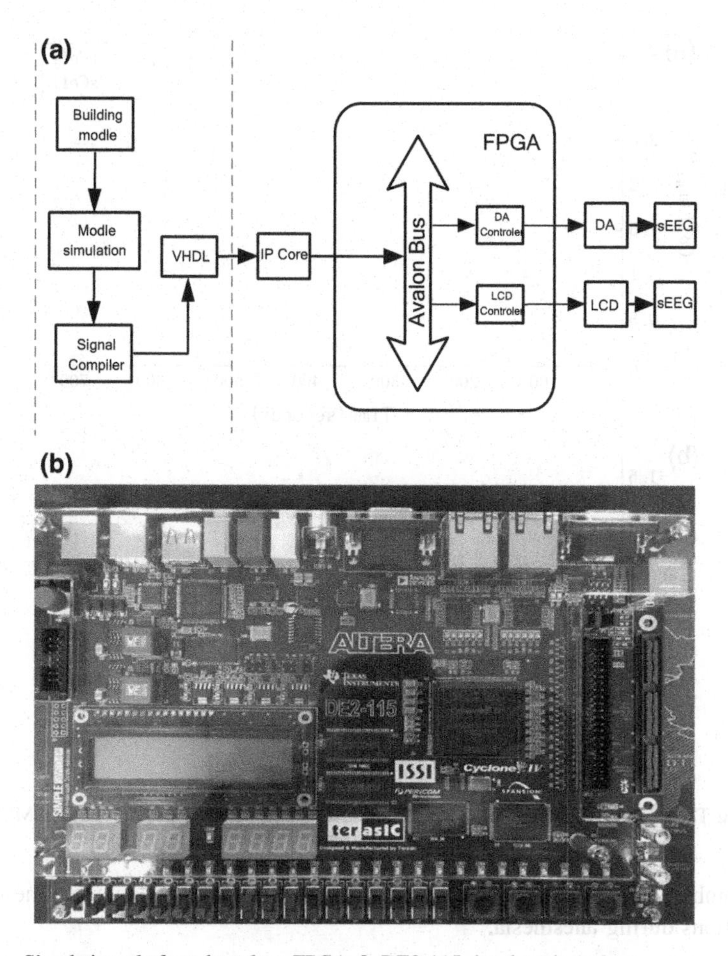

Fig. 3 **a** Simulation platform based on FPGA. **b** DE2-115 developed platform

5 Prototype and Test Results

In this paper, while, we reproduce the function of Pk and JRNMM successfully based on real anesthesia EEG data during propofol more details you can see [2]. Figure 4a shows the effect-site concentration from the Pk model in MATLAB (blue line) and DSP Builder (red line) simulation respectively. Thus we can know the simulated PK-like data in DSP Builder is parallel to MATLAB simulation. Figure 4b shows the simulated EEG-like data from the JRNMM based on physiological parameters information of standard value and we can see it is similar to normal EEG. Furthermore the unscented Kalman filter method is put into use to estimate parameters of the JRNMM to provide a theoretical basis for the fact that how the parameters varies with increasing the propofol concentration and we can

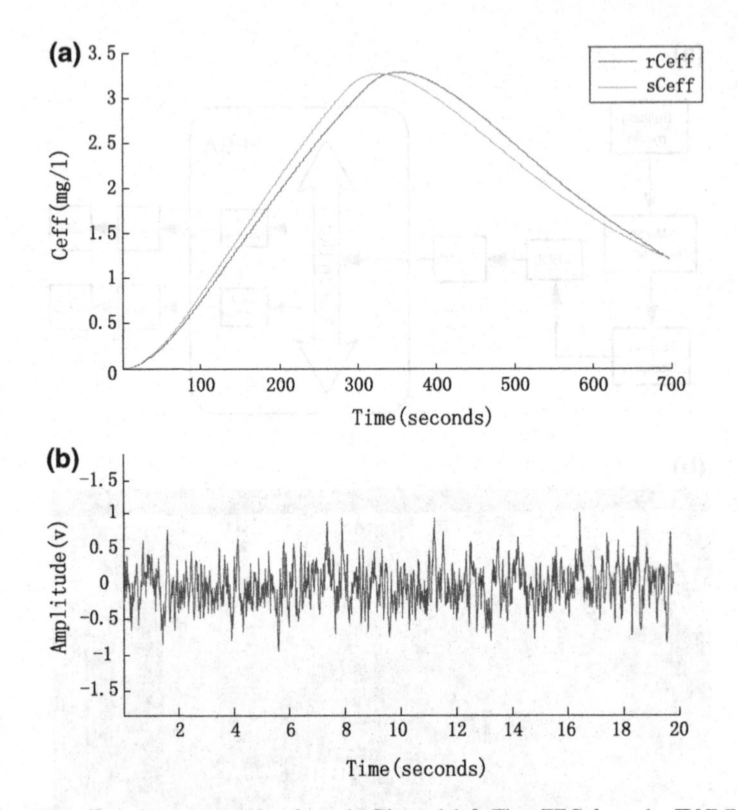

Fig. 4 **a** The effect-site concentration from the Pk model. **b** The sEEG from the JRNMM

also combine effect-site concentration with those parameters to simulate the neural oscillations during anesthesia.

6 Conclusion

In this work, JRNMM and PK model was studied and the fast computation and simulation was realized based on FPGA. The model can generate normal EEG data and effect-site concentration successfully. The combination of Simulink in MATLAB with DSP Builder in FPGA is designed to realize the effective integration of the software and hardware system, which also provides a new method and idea for computer simulation of JRNMM. Next we will focus on debugging platform so that the waveform can be output stably via D/A.

References

1. Ching SN, Brown EN. Modeling the dynamical effects of anesthesia on brain circuits. Curr Opin Neurobiol. 2014;25C(2):116–22.
2. Liang Z, Duan X, Cui S, et al. A pharmacokinetics-neural mass model (PK-NMM) for the simulation of EEG activity during propofol anesthesia. Plos One. 2015;10(12).
3. Jansen BH, Zouridakis G, Brandt ME. A neurophysiologically-based mathematical model of flash visual evoked potentials. Biol Cybern. 1993;68(3):275–83.
4. Yang DY, Rosenblau G, Keifer C, et al. An integrative neural model of social perception, action observation, and theory of mind. Neurosci Biobehav Rev. 2015;51:263–75.
5. Bojak I, Liley DT. Modeling the effects of anesthesia on the electroencephalogram. Phys Rev E. 2005;71(1):041902.
6. Mccarthy MM, Brown EN, Kopell N. Potential network mechanisms mediating electroencephalographic beta rhythm changes during propofol-induced paradoxical excitation. J Neurosci Off J Soc Neurosci. 2008;28(50):13488–504.
7. Hutt A. The anesthetic propofol shifts the frequency of maximum spectral power in EEG during general anesthesia: analytical insights from a linear model. Frontiers Comput Neurosci. 2013;7(7):2.
8. Garnier A, Vidal A, Huneau C, et al. A neural mass model with direct and indirect excitatory feedback loops: identification of bifurcations and temporal dynamics. Neural Comput. 2015;27(2):1–36.
9. Sotero RC, Trujillo-Barreto NJ, Iturria-Medina Y, et al. Realistically coupled neural mass models can generate EEG rhythms. Neural Comput. 2007;19(2):478–512.
10. Chakravarthy N, Sabesan S, Tsakalis K, et al. Controlling epileptic seizures in a neural mass model. J Comb Optim. 2009;17(1):98–116.
11. Schnider TW, Minto CF, Gambus PL, et al. The influence of method of administration and covariates on the pharmacokinetics of propofol in adult volunteers. Anesthesiology. 1998;88 (5):1170–82.
12. Ionescu CM, De Keyser R, Torrico BC, et al. Robust predictive control strategy applied for propofol dosing using BIS as a controlled variable during anesthesia. IEEE Trans Bio-med Eng. 2008;55(9):2161–70.
13. Yaghini BS, Asgharian H, Safari S, et al. FPGA implementation of a biological neural network based on the Hodgkin-Huxley neuron model. Frontiers Neurosci. 2013;8:379.
14. Liu Q, Fang JQ, Zhao G, et al. Research of Chaotic encryption system based on FPGA technology. Acta Physica Sinica. 2012;61(13):165–72.
15. Babajani-Feremi A, Soltanian-Zadeh H. Multi-area neural mass modeling of EEG and MEG signals. Neuroimage. 2010;52(3):793–811.
16. Kuhlmann L, Freestone DR, Manton JH, et al. Neural mass model-based tracking of anesthetic brain states. Neuroimage. 2016;133:438–6.

Research on CO Poisoning Risk Classification Evaluation Based SOM-AHP Method

Youlin Cai, Xiaoyi Wang and Jiping Xu

Abstract For the risk of CO poisoning risk problems in heating quarter rental housing of northern China, a risk assessment model on CO poisoning based on SOM-AHP method is proposed in this paper. On the basis of designing the index system of poisoning risk assessment, self—organizing neural network (SOM) is used to determine the division of the risk classification, and used Analytic Hierarchy Process (AHP) to obtain the relative weight of each risk index, then according to the classification rule of the hierarchical order and warning level, the risk classification of carbon monoxide poisoning has been realized. The model is applied to the risk assessment of carbon monoxide poisoning in rented houses in Chaoyang District of Beijing, and the results showed the feasibility of the model.

Keywords Carbon monoxide poisoning · The risk assessment · AHP · SOM network

1 Introduction

In recent years, with the increase of floating population, housing leasing increased. At present, China's northern suburbs of the rental housing in winter is still using coal stove heating mode, and biomass and coal is the main combustion medium. The humidity of coal is too large or the furnace, flue, wind bucket and other equipment, lack of ventilation, which easily lead to coal can not be fully burned, and lead to CO gas poisoning [1]. It is essential to scientifically assess the risk of CO poisoning and prevent the occurrence of malignant events for human's lives.

The research on the risk of CO poisoning at home and abroad mainly focuses on the field of urban meteorology. Such as Dockery DW and others have analyzed the impact of various meteorological factors on the diffusion capacity of CO in

---

Y. Cai · X. Wang (✉) · J. Xu
School of Computer and Information Engineering, Beijing Technology and Business University, Beijing 100048, China
e-mail: sdwangxy@163.com

Y. Jia et al. (eds.), *Proceedings of 2017 Chinese Intelligent Systems Conference*, Lecture Notes in Electrical Engineering 459,
https://doi.org/10.1007/978-981-10-6496-8_11

outdoor [2]; Luan Huaide and others have analyzed a serious CO poisoning incident, which occurred in Jinan in 2009, mainly for the current temperature, pressure and other meteorological conditions to explore the reason of occurrence of CO poisoning incidents [3]; Sun Yitian and others have analyzed the influence of wind pressure and temperature gradient on the diffusion of CO, and verified that meteorological conditions have a significant effect on the diffusion or aggregation of CO [4]; Xie Jingfang and others simulated the speed of the vents at different temperature difference in the indoor and outdoor with numerical simulation method, proved the influence of indoor and outdoor temperature difference on the ventilation [5]. The above researches are a study of CO poisoning from specific meteorological conditions, which has certain application value for establishing CO poisoning weather forecast. However, the occurrence of CO poisoning is a complex process, which is affected by a variety of factors, and a single meteorological conditions can not play a decisive role, so the multi-dimensional and multi-faceted analysis are needed. Because CO poisoning is a gas pollution which occur in the room, indoor furnace quality, flue plugging, wind ventilation and other facilities play a key role in the distribution and transmission of CO. According to the characteristics of indoor CO poisoning, the risk assessment index system of poisoning was constructed, and the risk assessment model of CO poisoning was constructed by the method of combining SOM network and AHP analysis. To be specific, the SOM network is used to realize the analysis and clustering of the risk assessment index information, and obtains the risk assessment level. The AHP analysis method analyzes the risk assessment index, and obtains the importance of the different level indicators and then achieve the risk assessment.

2 Design of Risk Assessment Index System

The reason of CO poisoning which is caused by rental housing winter coal stove heating is a multifaceted, a single indicator can not be a complete description of the risk of CO poisoning. According to the principle of selection of indicators, combined with the winter coal furnace heating facilities and weather factors, to determine the risk of CO risk assessment index system is divided into three levels, as shown in Fig. 1. Among them, the target layer is the level division of the risk of housing CO poisoning; the criteria layer includes two aspects of the evaluation of indoor facilities and meteorological conditions; the indicator layer is a relevant indicator of the relevance of the assessment.

Therefore, the degree of damage of the stove C1, the clogging degree C2, the bucket ventilation condition C3, the wind speed C4, the indoor and outdoor temperature difference C5, the air pressure C6 and the relative humidity C7 are determined as the risk assessment index of CO poisoning.

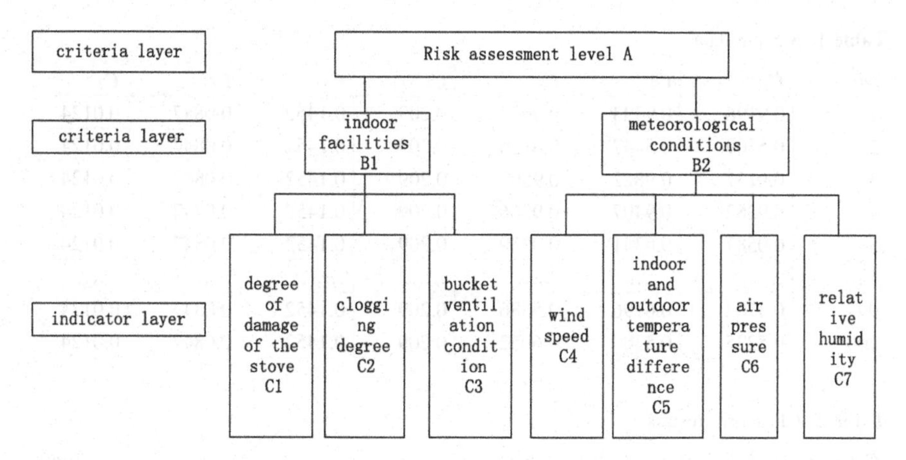

Fig. 1 CO poisoning safety risk assessment index hierarchy structure

3 Risk Assessment SOM—AHP Model Construction

Based on the SOM-AHP risk assessment model, the SOM network is used to cluster the index information, and the risk level is divided. On this basis, the AHP analysis method is used to synthesize the quantitative and qualitative indexes to obtain the risk assessment results. The basic idea of the SOM network is that the neurons of the network competition compete for the input mode, and finally only one neuron becomes the winner of the competition, and the orientation weights of those which are connected to the winning neurons are adjusted in a more favorable direction for competition, and the winning neurons represent the classification of the input patterns. In addition to the competition method, there are ways to win by the inhibition method—the network competition layer of neurons can inhibit all other neurons on the input mode of response opportunities, so that they win [6, 7]. SOM network has the characteristics of self-organization, visualization, high computational efficiency and high clustering accuracy, the biggest advantage of using it for the risk assessment of CO poisoning is that there is no need to know the prior knowledge of the categories in advance, as long as the learning can be done through the network according to the specific sampling samples, the training of the subjective factors can be overcome effectively and the classification result can be more objective and accurate. In this case, 10 households rental housing in urban and rural areas of Beijing Chaoyang District had been randomly selected one day 18:00 in November 2016 year, and obtained the index data collection, and the indicators were normalized, the data obtained as shown in Table 1.

The results of SOM network clustering are shown in Table 2.

The classification of SOM networks is divided into five categories, namely, the highest risk (category 5), greater risk (category 4), general risk (category 3), less risky (category 2) and risk minimization Category 1). Determining the weight of the

Table 1 Sample data

No.	C_1	C_2	C_3	C_4	C_5	C_6	C_7
1	0.9196	0.9393	0.9654	0.209	0.1452	0.0847	0.0124
2	0.5309	0.5447	0.6126	0.209	0.1452	0.0847	0.0124
3	0.9137	0.9822	0.9507	0.209	0.1452	0.0847	0.0124
4	0.9262	0.9407	0.9206	0.209	0.1452	0.0847	0.0124
5	0.0587	0.0341	0.0339	0.209	0.1452	0.0847	0.0124
……	……	……	……	……	……	……	……
99	0.4702	0.6196	0.5386	0.209	0.1452	0.0847	0.0124
100	0.5225	0.5309	0.6052	0.209	0.1452	0.0847	0.0124

Table 2 Clustering results

Category	House no.	Sum
1	1, 3, 4, 17, 24, 29, 51, 77, 90, 94	10
2	12, 19, 20, 22, 31, 38, 44, 53, 59, 62, 64, 68, 70, 73, 76, 81, 88, 91, 96, 97	20
3	2, 7, 8, 11, 14, 15, 16, 18, 21, 23, 25, 27, 32, 33, 34, 36, 39, 45, 47, 48, 49, 50, 52, 55, 56, 58, 60, 61, 63, 66, 67, 69, 71, 74, 75, 79, 80, 83 , 84, 85, 86, 87, 93, 100	44
4	9, 10, 13, 26, 30, 35, 40, 41, 46, 57, 65, 72, 89, 92, 95, 99	16
5	5, 6, 28, 37, 42, 43, 54, 78, 82, 98	10

Table 3 9 points a ratio judgment matrix scale table

Scaling	Scale meaning
1	Two factors are compared, Have the same importance
2	Two factors are compared, One factor is slightly more important than the other
5	Two factors are compared, One factor is more important than the other
7	Two factors are compared, One factor is more important than the other
9	Two factors are compared, One factor is more important than another factor
2, 4, 6, 8	Represents the median of the two adjacent judgments

indicator is the prerequisite for the risk assessment. After the establishment of the hierarchy, the judgment matrix of the relative importance of each index (shown in Table 4), is scored from the lowest level according to the 9-position ratio judgment matrix proposed by Saaty (shown in Table 3). The weight of the target layer A is calculated relative to the decision matrices of the criterion layers B1, B2 and B1 relative to C1, C2, C3, B2 relative to C4, C5, C6 and C7, thereby calculating the weights of the respective indexes [8, 9].

Table 4 $A-B$ Judgment matrix

	B_1	B_2	Normalized W_1
B_1	1	4	0.80
B_2	1/4	1	0.20

(1) Level single sort

According to the judgment matrix A-B to calculate $B_i(i=0,1,\ldots,n)$, the relative importance coefficient (i.e., weight) for the risk assessment level A. The calculation of the W_{bi} uses the geometric mean method, the process is to first calculate the product of each row element of the judgment matrix, that is, $W_{bi}=M\prod_{i=1}^{n} b_{ij}(i=1,2,\ldots,n)$; then calculate the geometric mean of the M_{bi}, that is, $\overline{W}_{bi}=\sqrt[n]{M_{bi}}$, $(i=0,1,\ldots,n)$; Finally, the $\overline{W}_{bi}$ is normalized, that is,

$$W_{bi}=\frac{\overline{W_{bi}}}{\sum_{i=1}^{n}\overline{W_{bi}}}(i=1,2,\ldots,n) \quad (1)$$

According to Eq. (1), the weights of the factors in the second layer B for A can be determined so that the importance of the factors in the second layer B for A can be determined. Two judgment matrices can be calculated in the same way (shown in Tables 5 and 6).

(2) The overall ranking of the hierarchy

According to the result of the second layer B and the third layer C, we can determine the relative importance weight of each factor in the third layer C for A, that is, the total order of the hierarchy. The general formula for the hierarchy is:

$$W=\sum_{i=1}^{n} W_{bi}W_{cj}^{i}(j=1,2,\ldots,n) \quad (2)$$

The results are shown in Table 7.

(3) Consistency test

In order to guarantee the reliability of the conclusion, it is necessary to carry on the consistency test to the evaluation result of the judgment matrix. In this case, the consistency test was carried out using the concept of stochastic consistency ratio

Table 5 B_1-C Judgment matrix

	C_1	C_2	C_3	Normalized W_2
C_1	1	3	4	0.62
C_2	1/3	1	1	0.21
C_3	1/4	1	1	0.17

Table 6 B_2-C Judgment matrix

	C_4	C_5	C_6	C_7	Normalized W_3
C_4	1	3	4	6	0.46
C_5	1/3	1	5	7	0.43
C_6	1/4	1/5	1	1/3	0.06
C_7	1/6	1/7	1/3	1	0.05

Table 7 Hierarchy total sorts

Level C	B_1	B_2	Level C sorting W
	0.80	0.20	
C_1	0.62	–	0.50
C_2	0.21	–	0.16
C_3	0.17	–	0.14
C_4	–	0.46	0.092
C_5	–	0.43	0.086
C_6	–	0.06	0.012
C_7	–	0.05	0.01

$(C \cdot R)$ proposed by Saaty T L. $C \cdot I$ is used as a measure of deviation from the judgment matrix.

Firstly, calculate the maximum eigenvalue of the two discriminant matrices:

$$\lambda_{\max} = \sum_{i=1}^{n} W_{bi} \left[\sum_{j=2}^{n} W_j / nW_1 \right] \tag{3}$$

Secondly, calculate the consistency indicator:

$$CI = \frac{\lambda_{\max} - n}{n - 1} \tag{4}$$

Finally, calculate the consistency ratio $CR = \frac{CI}{RI}$, according to the actual situation of each discriminant matrix, find the corresponding average random consistency index value RI as shown in Table 8.

If $CR < 0.10$, then that the matrix has a satisfactory consistency, hierarchical order is valid, otherwise it needs to be adjusted.

Taking the feature vector W_1 of matrix $A - B$ as an example:

$$\lambda_{\max} = \sum_{i=1}^{n} W_{bi} \left[\sum_{j=2}^{n} W_j / nW_1 \right] = 1.60 + 0.40 = 2 \quad CI = \frac{\lambda_{\max} - n}{n - 1} = \frac{2-2}{2-1} = 0$$

When n = 2, $RI = 0$, get $CR = 0 < 0.1$, can determine the consistency of the matrix $A - B$ can be accepted, so the weight vector W_1 can be accepted. W_2 and W_3 are judged by the above-mentioned method, and when the W_2 consistency judgment is made, $CR = 0.024 < 0.1$;

when the W_3 consistency judgment is made, $CR = 0.033 < 0.1$.

Table 8 The mean random consistency index

Order	1	2	3	4	5	6
RI value	0.00	0.00	0.58	0.90	1.12	1.24

The validity of the total sort is tested, and the result is

$$CR_{总} = \frac{\sum_{i=1}^{n} W_{b1} \cdot (CI)_1}{\sum_{i=2}^{n} W_{b1} \cdot (RI)_1} = \frac{0.00005}{0.58} < 0.10$$

State that the overall order of the hierarchy conforms to the consistency requirements.

4 Validation of Results

In order to validate the effectiveness of the risk assessment model, randomly selected 10 households rental housing in urban and rural areas of Beijing Chaoyang District as index data as shown in Table 9, using AHP analysis of its risk assessment, and the assessment results shown in Fig. 2.

Table 9 Verify the sample data

NO.	C_1	C_2	C_3	C_4	C_5	C_6	C_7
1	0.6131	0.5636	0.6452	0.209	0.1452	0.0847	0.0124
2	0.3073	0.4679	0.3776	0.209	0.1452	0.0847	0.0124
3	0.9262	0.9407	0.9206	0.209	0.1452	0.0847	0.0124
4	0.5012	0.4947	0.6091	0.209	0.1452	0.0847	0.0124
5	0.5624	0.6034	0.5427	0.209	0.1452	0.0847	0.0124
6	0.4194	0.4256	0.2134	0.209	0.1452	0.0847	0.0124
7	0.6407	0.5611	0.4017	0.209	0.1452	0.0847	0.0124
8	0.4847	0.5243	0.5991	0.209	0.1452	0.0847	0.0124
9	0.4714	0.6703	0.5358	0.209	0.1452	0.0847	0.0124
10	0.0419	0.0897	0.0671	0.209	0.1452	0.0847	0.0124

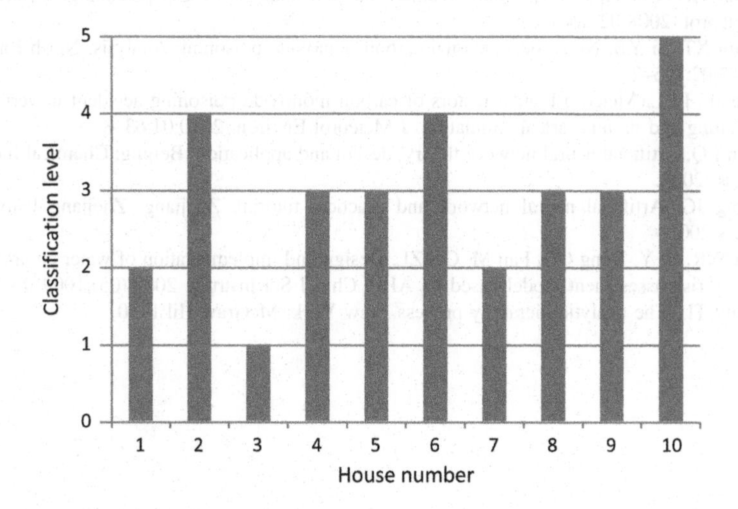

Fig. 2 CO poisoning evaluation results

According to the SOM-AHP model, the risk of CO poisoning assessment of 10 rented houses was obtained and the validity of the model was verified by field test. The relevant departments or tenants can be based on the assessment results, making the appropriate security measures for the rental housing which have been evaluated as risk assessment rating of 4, 5 class.

5 Conclusion

Based on the comprehensive analysis of the risk factors of CO poisoning, a multi-level and multi-index risk assessment index system of CO poisoning was established. Based on this, a risk assessment model of CO poisoning based on SOM-AHP was established, and the risk level of CO poisoning was quantified Analysis, the rental housing to determine the safety of housing, risk assessment for the risk of CO put forward a viable approach.

Acknowledgements This work was financially supported by National Natural Science Foundation of China (51179002), Beijing Municipal Universities Innovation Ability Improvement Program Project (PXM2014_014213_000033). Science and Technology Key Project of Beijing Municipal Education Commission (KZ201510011011). Those supports are gratefully acknowledged.

References

1. Wang J. Study on the diffusion and control strategy of CO in rural residential in sever cold region. Harbin: Harbin Institute of Technology;2013.
2. WHO International Agency for Research on Cancer Press Release. IARC classifies formaldehyde as carcinogenic to humans. J Women s Health. 2004.
3. Luan HD, Liu HP, LI H. Jinan weather situation analysis of CO poisoning. J Shandong Meteorol. 2008;02:22–4.
4. Shun XT, Si YB. Non—occupational carbon monoxide poisoning Analysis. North Environ. 2012;02:195–7.
5. Xie JF, Li L. Meteoro logical factors of carbon monoxide poisoning accident in vernacular dwelling and its num erical simulation. J Meteorol Environ. 2010;01:63–8.
6. Han LQ. Artificial neural network theory, design and application. Beijing: Chemical Industry Press; 2004.
7. Yang JG. Artificial neural network and practical tutorial. Zhejiang: Zhejiang University Press;2001.
8. Shi WR, Li Y, Deng CG, Fan M, Cai ZL. Design and implementation of water environment safety risk assessment model based on AHP. Chin J Sci Instrum. 2009;(05):1009–13.
9. Saaty TL. The analytic hierarchy process. New York: McGraw-Hil;1980.

Decision Making in Multi-agent Systems Based on the Evolutionary Game with Switching Probabilities

Zimin Xu, Jianlei Zhang, Qiaoyu Li and Zengqiang Chen

Abstract Much attention has been paid on exploring the solutions for cooperative dilemma in multi-agent systems. Thereinto, the evolutionary game theory which describes cooperative dilemma is seen as an effective approach. Notably, many of previous works are based on the ideal hypothesis that individuals can feasibly obtain their neighbours' payoffs to update strategies. Considering the difficulty of getting the exact information about payoffs, we propose the switching probabilities between strategies which do not require the payoffs. Here the evolutionary dynamics driven by the switching probabilities in a three-strategy game model is established. Results show that the steady state of the gaming system is closely related with the switching probability matrix. These findings give a novel account about the decision making process in the gaming systems, when a strategy updating rule weakening the ideal assumption about payoffs is established.

Keywords Multi-agent systems · Evolutionary game dynamics · Switching probability · Decision making · Stationary distribution

1 Introduction

In multi-agent systems which require the altruistic collaboration of the group members, free riding others' contributions can be seen as a temptation from the game theory [1]. From Darwinist perspective, it is puzzling to understand that the maintenance of cooperative behaviors among selfish individuals when defection is the advantageous strategy for self-interest agents [2–4]. Many researchers in different fields have made enormous efforts to explore solutions for this problem [5–8]. Evolutionary game theory can effectively model the mentioned cooperation dilemma by the aid of many concrete games, such as the prisoner's dilemma game (PDG),

Z. Xu · J. Zhang (✉) · Q. Li · Z. Chen
Department of Automation, College of Computer and Control Engineering,
Nankai University, Tianjin 300071, China
e-mail: jianleizhang@nankai.edu.cn

Y. Jia et al. (eds.), *Proceedings of 2017 Chinese Intelligent Systems Conference*, Lecture Notes in Electrical Engineering 459,
https://doi.org/10.1007/978-981-10-6496-8_12

the snowdrift game (SDG), the stag-hunt game (SHG) and the public goods game (PGG), etc. [9–12].

For the infinite and well-mixed populations and neglecting mutations, the deterministic replicator dynamics equation $\dot{x}_i = x_i(\pi_i - \langle\pi\rangle)$ has provided a lot of insights for exploring the strategy evolution dynamics [13–15]. Here, x_i is the fraction of strategy i in the population, π_i is the payoff or fitness of this strategy and $\langle\pi\rangle$ is the average payoff in the whole population. If the payoff of strategy i is below the average payoff, its density will decrease. If the payoff is above the mean payoff, then the corresponding density will increase. In general, x_i depends on the strategy placement of population, that is to say, on the proportions of all other strategies x_j. Then, the average payoff $\langle\pi\rangle$ is the quadratic equation of the fraction x_j. While stochastic evolutionary game based on the theory of stochastic processes verifies its effectiveness in finite populations.

A main research interest is the microscopic patterns of interactions among individuals. Until now many mechanisms show their effectiveness in promoting cooperation, such as direct and indirect reciprocity, kin selection, group interaction, spatial and networking reciprocity, teaching activity, individual rationality and individual aspiration [16, 17]. Especially, recent years have witnessed the combination between the evolutionary game theory and complex network theory. In this framework, the population structure is represented by a network, the nodes of which represent the individual agents while the links correspond to the possible interactions.

As for the strategy updating, many studies consider that players imitate or replicate their neighbors' strategy with a certain probability p, which depends on payoff comparison [18–20]. The switching probability p could be a linear function of the payoff difference [21] (e.g., $p = \frac{1}{2} + \omega_1 \frac{\pi_f - \pi_r}{\Delta\pi}$), or indicated by non-linear functions, such as Fermi function $p = \frac{1}{1+e^{\omega_2(\pi_f - \pi_r)}}$ [22, 23]. Here, π_f and π_r denote the payoffs of the focal individual and referenced one, respectively, and $\Delta\pi$ is the maximum payoff difference. The parameter ω_1 and ω_2 denote the noise or inverse temperature which control the selection intensity and take values in the range of $[0, 1]$ and $(0, \infty)$, respectively. The situation of $\omega_1 \to 0$ or $\omega_2 \to 0$ (weak selection) manifests all information is covered by noise, yet the condition of $\omega_1 \to 1$ or $\omega_2 \to \infty$ (strong selection) signifies decided imitation rules.

Despite this extraordinary level of attention, however, the mentioned updating rules are always based on the ideal assumption about payoff information. In fact, acquiring payoffs is not easy due to individuals' bounded rationality or ability in real social systems [24–26]. Thus, the rules driving the strategy evolution is a fascinating and meanwhile key topic. Especially, how can individuals update strategies and improve their profits by not requiring the known payoffs? Our previous works have realized it by introducing the willingness that one individual shifts her current strategy to the other one, an intriguing feature of which is the absence of usually required payoff information [27–29]. In this paper, we propose an extended model covering a wider range of update situations, by analyzing the evolutionary dynamics in the framework of three-strategy games to improve its applicability.

The remainder of this paper is organized as follows: the next section describes the model and Sect. 3 represents the stationary distribution of different switching probabilities and initial settings. Section 4 shows the results on complex networks and the finally section concludes this paper.

2 Model

In this section, we employ transition probability in three-strategy game to analyze the equilibrium distribution. Specifically, we consider three pure strategies: strategy *A*, *B* and *C*. Every player in the population has three choices to play with her opponents. Here, we encode the payoffs into the intentions that one individual switches her current strategy to another one. Therefore the evolution process is based on the switching probabilities: $u_{A\to A}$, $u_{A\to B}$, $u_{A\to C}$ $u_{B\to A}$, $u_{B\to B}$, $u_{B\to C}$ $u_{C\to A}$, $u_{C\to B}$ and $u_{C\to C}$ that are not related with payoff comparisons. Specifically, $u_{A\to B}$ indicates the probability with which an *A*-player will adopt strategy *B* in the next step. Similarly, $u_{A\to C}$ means the probability by which an *A*-player will alter to be a *C* player. The remaining seven switching probabilities are demonstrated in Fig. 1.

It is quite clear that every individual, irrespective of her current strategy, has three alternative strategies to choose. Thus we can get that:

$$\begin{cases} u_{A\to A} + u_{A\to B} + u_{A\to C} = 1 \\ u_{B\to A} + u_{B\to B} + u_{B\to C} = 1 \\ u_{C\to A} + u_{C\to B} + u_{C\to C} = 1 \\ 0 \le u_{i\to j} \le 1, i, j = A, B, C \end{cases} \tag{1}$$

And in the next step every individual's strategy is only related to the current strategy, and has no relation with with the previous strategy, i.e.,

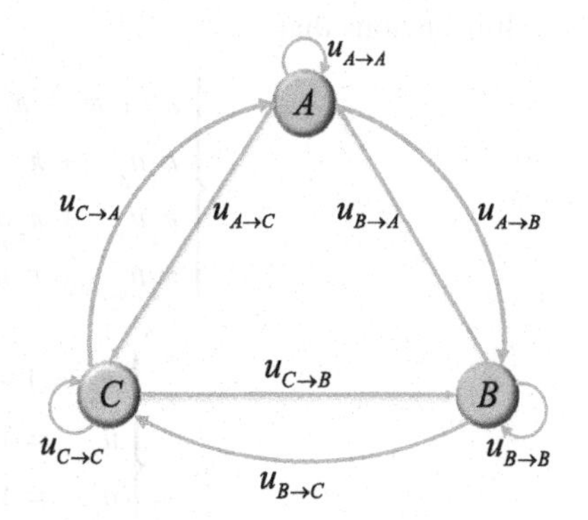

Fig. 1 The switching probability between strategies. The *A*-player will adopt strategy *B* with the probability $u_{A\to B}$ in the next step. The rest can be described in the same way

$$P\{X(k+1) = s_{k+1}|X(k) = s_k, X(k-1) = s_{k-1}, ..., X(1) = s_1, X(0) = s_0\} \\ = P\{X(k+1) = s_{k+1}|X(k) = s_k\} \quad (2)$$

Here, $X(t), t \in T$ $(T = \{0, 1, 2, 3, ...\})$ is the strategy sequence of a group member in the system, and $s_t \in I$ $(I = \{A, B, C\})$, which define a finite state Markov process with a 3×3 state transition matrix U:

$$U = \begin{bmatrix} u_{A\to A} & u_{A\to B} & u_{A\to C} \\ u_{B\to A} & u_{B\to B} & u_{B\to C} \\ u_{C\to A} & u_{C\to B} & u_{C\to C} \end{bmatrix} \quad (3)$$

3 Stationary Distribution

Definition 1 Let $\{X_n, n \geq 0\}$ be a Markov chain, the state space be I and switching probability be $u_{i\to j}$ $(i, j \in I)$. Call $\pi_j, j \in I$ be a stationary distribution if it satisfies:

$$\begin{cases} \pi_j = \sum_{i\in I} \pi_i u_{i\to j} \\ \sum_{j\in I} \pi_j = 1, \ 0 \leq \pi_j \leq 1 \end{cases} \quad (4)$$

Here, x_A, x_B and x_C respectively denote the probability of A, B and C players. The equilibrium distribution is set as $\pi = [\pi_A \quad \pi_B \quad \pi_C]$. It is easy to get that if the initial frequencies of these three kinds of individuals are equal to π_A, π_B and π_C, respectively, then the probability distribution at each subsequent step will remain. That is to say, once the probability distribution reaches the stationary one, the frequencies of A, B and C agents will remain fixed.

It is obvious that

$$\begin{cases} \pi_A + \pi_B + \pi_C = 1 \\ \pi_A u_{A\to A} + \pi_B u_{B\to A} + \pi_C u_{C\to A} = \pi_A \\ \pi_A u_{A\to B} + \pi_B u_{B\to B} + \pi_C u_{C\to B} = \pi_B \\ \pi_A u_{A\to C} + \pi_B u_{B\to C} + \pi_C u_{C\to C} = \pi_C \end{cases} \quad (5)$$

$$\begin{cases} \pi_A = 1 - \pi_B - \pi_C \\ u_{A\to A} = 1 - u_{A\to B} - u_{A\to C} \\ u_{B\to A} = 1 - u_{B\to B} - u_{B\to C} \\ u_{C\to A} = 1 - u_{C\to B} - u_{C\to C} \end{cases} \quad (6)$$

Therefore, if the first condition and any two of the next three terms in Eq. 5 are satisfied, the stationary distribution can be solved.

Then, we sort the relative relationships among the three strategies into five categories: (1) $A > B > C$; (2) $A > B, B > C, C > A$; (3) $A > B = C$; (4) $A < B = C$; (5) $A = B = C$. Next we analyze the stationary distribution in the framework of these five types.

3.1 $A > B > C$

In this case, strategy A is dominant for B and C, and B is better than C. So that $u_{A\to B} < u_{B\to A}$, $u_{A\to C} < u_{C\to A}$ and $u_{B\to C} < u_{C\to B}$. Next we take two examples to analyze.

(1) The switching probability matrix is:

$$U = \begin{bmatrix} 0.6 & 0.3 & 0.1 \\ 0.55 & 0.3 & 0.15 \\ 0.7 & 0.2 & 0.1 \end{bmatrix} \tag{7}$$

In order to calculate the stationary distribution,

$$\begin{cases} \pi_A + \pi_B + \pi_C = 1 \\ 0.6\pi_A + 0.55\pi_B + 0.7\pi_C = \pi_A \\ 0.3\pi_A + 0.3\pi_B + 0.2\pi_C = \pi_B \end{cases} \tag{8}$$

So $\pi = [\pi_A \quad \pi_B \quad \pi_C] = [\frac{40}{67} \quad \frac{58}{201} \quad \frac{23}{201}]$. Then we compute the proportions as time evolves with different initial percentages of A, B and C players by simulation, in which the number of individuals is $N = 100000$. Figure 2 shows the results of calculations and simulations with different initial proportions. The initial frequencies of these three types of individuals are: $[x_A \quad x_B \quad x_C] = [0 \quad 0 \quad 1]$ in Fig. 2a, b and $[x_A \quad x_B \quad x_C] = [0.0009 \quad 0.3319 \quad 0.6672]$ in Fig. 2c, d. A clear information is that, irrespectively of the initial probability distributions, computation and simulation results can reach the stationary distribution quickly. Considering the consistence between calculations and simulations, only the simulation results are shown in the following.

(2) The transition probability matrix:

$$U = \begin{bmatrix} 0.3 & 0.5 & 0.2 \\ 0.55 & 0.2 & 0.25 \\ 0.6 & 0.3 & 0.1 \end{bmatrix} \tag{9}$$

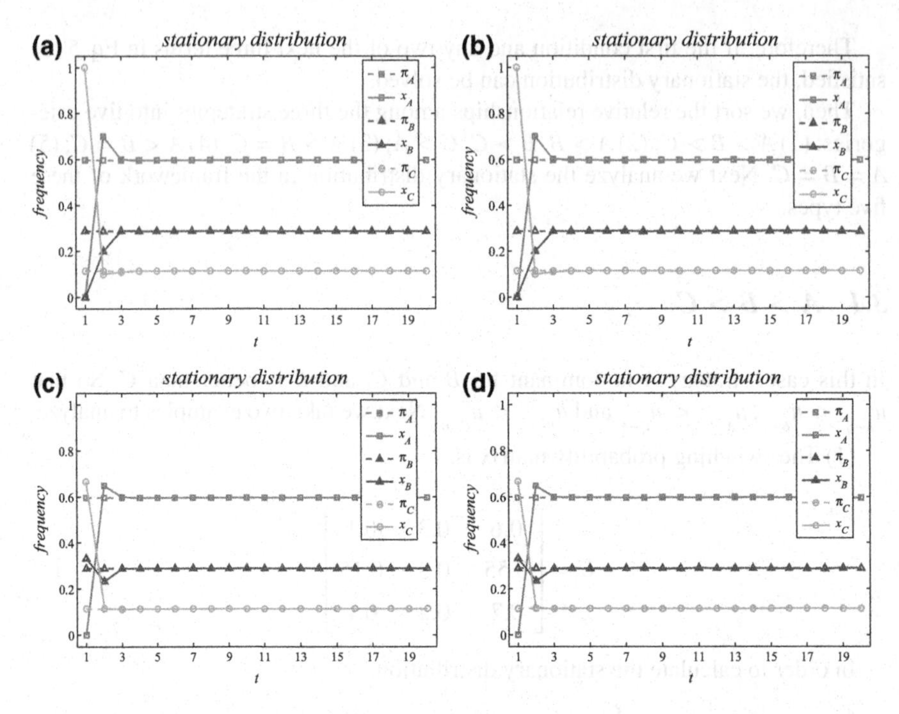

Fig. 2 The fractions of *A*, *B* and *C* players as games proceed. *Left panel* results by calculations; *right panel* simulation results. The imaginary lines mean the stationary distribution π_A, π_B and π_C, while the *solid lines* indicate the computation and simulation results. Comparison shows the consistence of the calculations and simulations with the stationary distribution

$$\begin{cases} \pi_A + \pi_B + \pi_C = 1 \\ 0.3\,\pi_A + 0.55\,\pi_B + 0.6\,\pi_C = \pi_A \\ 0.5\,\pi_A + 0.2\,\pi_B + 0.3\,\pi_C = \pi_B \end{cases} \tag{10}$$

$$\Downarrow$$

$$\pi = [\pi_A \quad \pi_B \quad \pi_C] = [\frac{43}{96} \quad \frac{17}{48} \quad \frac{19}{96}]$$

Similar with Figs. 2 and 3 also manifests that regardless what the initial settings are, even if the numbers of some strategies are zero, the system will evolve into the equilibrium point finally. On the condition of $A > B > C$ and $u_{A\to B} < u_{B\to A}$, $u_{A\to C} < u_{C\to A}$, $u_{B\to C} < u_{C\to B}$, the stationary distribution π satisfies $\pi_A > \pi_B > \pi_C$.

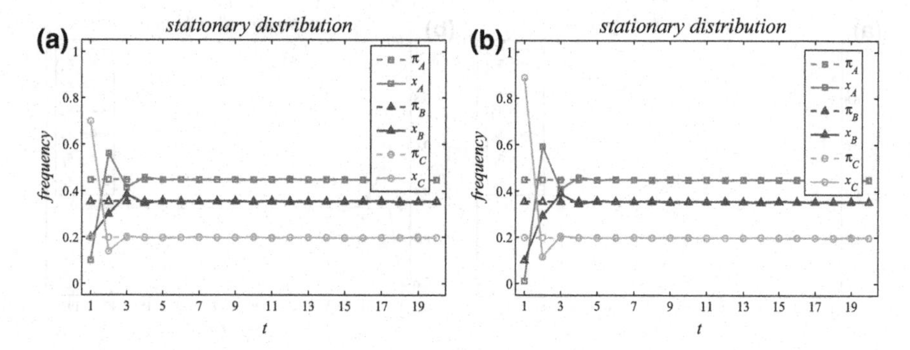

Fig. 3 The evolving proportions of strategy A, B and C with time, both of which are simulation results. The initial frequencies of A, B and C agents are $[x_A \quad x_B \quad x_C] = [0.1 \quad 0.2 \quad 0.7]$ and $[x_A \quad x_B \quad x_C] = [0.01 \quad 0.1 \quad 0.89]$, respectively

3.2 $A > B, B > C, C > A$

Under this circumstance, strategy A has advantages over B and strategy B is better than C, however, strategy C is superior to A. As a consequence, $u_{A\to B} < u_{B\to A}$, $u_{B\to C} < u_{C\to B}$, $u_{C\to A} < u_{A\to C}$. This case is similar to the rock-paper-scissors game, in which each player simultaneously forms one of three shapes with an outstretched hand. The game has only two possible outcomes other than a tie: a player who decides to play rock will beat another player who has chosen scissors ("rock crushes scissors"), but will lose to one who has played paper ("paper covers rock"); a play of paper will lose to a play of scissors ("scissors cut paper"). If both players choose the same strategy, the game is tied and usually immediately replayed to break the tie. So it is impossible to gain an advantage over a truly random opponent.

(1) The differences between any pair of strategies are the same. Thus, $u_{A\to A} = u_{B\to B} = u_{C\to C}$, $u_{A\to B} = u_{B\to C} = u_{C\to A}$ and $u_{A\to C} = u_{B\to A} = u_{C\to B}$.

$$U = \begin{bmatrix} 0.3 & 0.2 & 0.5 \\ 0.5 & 0.3 & 0.2 \\ 0.2 & 0.5 & 0.3 \end{bmatrix} \tag{11}$$

$$\begin{cases} \pi_A + \pi_B + \pi_C = 1 \\ 0.3\,\pi_A + 0.5\,\pi_B + 0.2\,\pi_C = \pi_A \\ 0.2\,\pi_A + 0.3\,\pi_B + 0.5\,\pi_C = \pi_B \end{cases} \tag{12}$$

$$\Downarrow$$

$$\pi = [\pi_A \quad \pi_B \quad \pi_C] = [\tfrac{1}{3} \quad \tfrac{1}{3} \quad \tfrac{1}{3}]$$

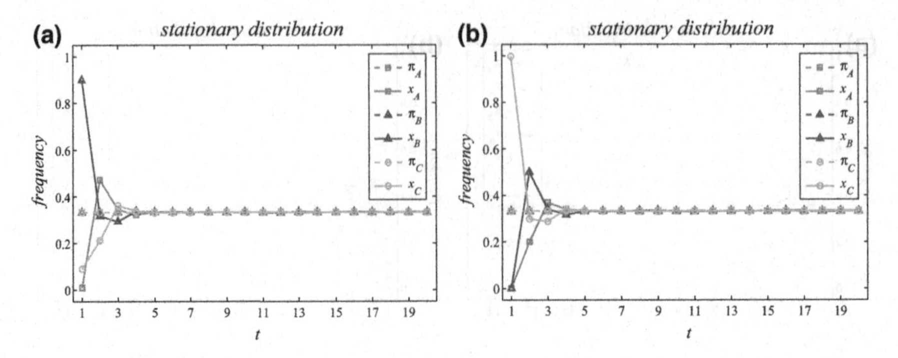

Fig. 4 The proportions of *A*, *B* and *C* players in the system as game proceeds. The simulation results nicely match the equilibrium distribution by theoretical analysis, and they arrive the steady state at a fast pace

Figure 4 reveals the varying frequencies of the three strategies. The initial frequencies of *A*, *B* and *C* of agents in Fig. 4a, b are $[x_A \quad x_B \quad x_C] = [0.01 \quad 0.9 \quad 0.09]$ and $[x_A \quad x_B \quad x_C] = [0.01 \quad 0.0008 \quad 0.9981]$, respectively. If the distinctions between *A* and *B*, *B* and *C* and *A* and *C* are the same, the gaming system will finally evolve to the coexistence state evenly divided by the three strategies.

(2) The difference of each pair of strategies are not equal in this case:

$$U = \begin{bmatrix} 0.1 & 0.2 & 0.7 \\ 0.5 & 0.4 & 0.1 \\ 0.4 & 0.3 & 0.3 \end{bmatrix} \tag{13}$$

$$\begin{cases} \pi_A + \pi_B + \pi_C = 1 \\ 0.1\pi_A + 0.5\pi_B + 0.4\pi_C = \pi_A \\ 0.2\pi_A + 0.4\pi_B + 0.3\pi_C = \pi_B \end{cases} \tag{14}$$

$$\Downarrow$$

$$\pi = [\pi_A \quad \pi_B \quad \pi_C] = [\frac{39}{118} \quad \frac{35}{118} \quad \frac{22}{59}]$$

Figure 5 shows the simulation results in dependence on two different initial placements of strategies *A*, *B* and *C* in the system. The initial probability in Fig. 5a is $[x_A \quad x_B \quad x_C] = [0.1 \quad 0.15 \quad 0.75]$, and $[x_A \quad x_B \quad x_C] = [0.8 \quad 0.1 \quad 0.1]$ in Fig. 5b.

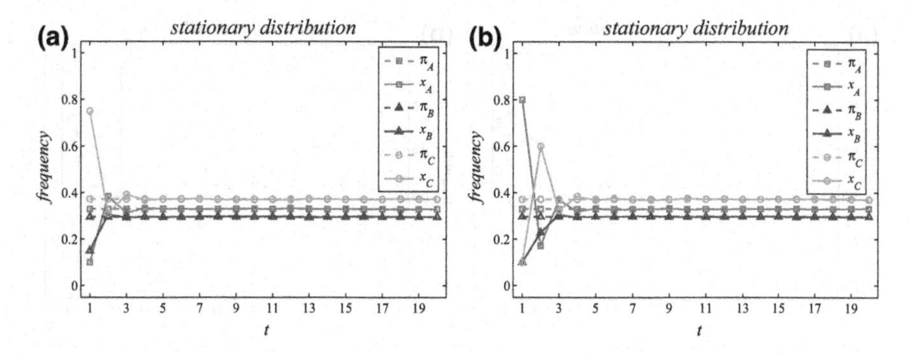

Fig. 5 The evolution process of frequencies of *A*, *B* and *C* agents with different initial settings

3.3 $A > B = C$

This situation is similar to the first case, only the difference is that strategy *B* and *C* are in the same status. So that $\pi_A > \pi_B$, $\pi_A > \pi_C$ and $\pi_B = \pi_C$.

From Figs. 6 and 7, it is easy to get that $u_{A\to B} < u_{B\to A}$, $u_{A\to C} < u_{C\to A}$ and $u_{B\to C} = u_{C\to B}$, as a consequence, strategy *A* is dominant to both strategy *B* and *C* at the same time. At the steady state, the amount of *A*-players is greater than that of *B* and *C* players.

(1) The transition matrix here is given by

$$U = \begin{bmatrix} 0.6 & 0.2 & 0.2 \\ 0.7 & 0.15 & 0.15 \\ 0.7 & 0.15 & 0.15 \end{bmatrix} \tag{15}$$

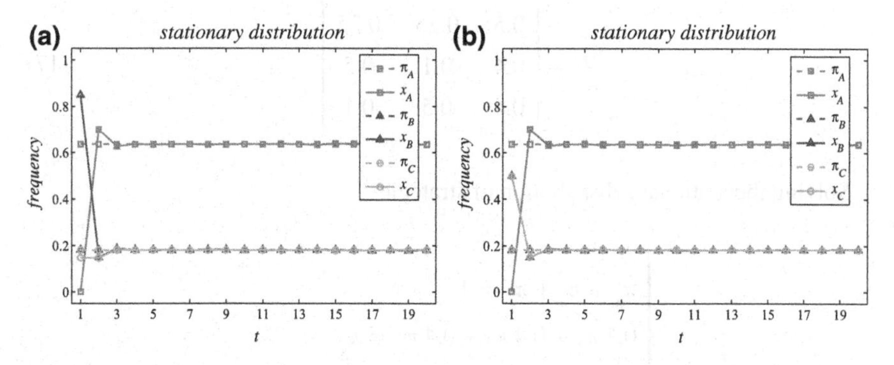

Fig. 6 The proportions of these three types of agents as games proceed. The initial probability distributions are: **a** $[x_A \quad x_B \quad x_C] = [0.001 \quad 0.85 \quad 0.149]$, **b** $[x_A \quad x_B \quad x_C] = [0.0009 \quad 0.5005 \quad 0.4985]$

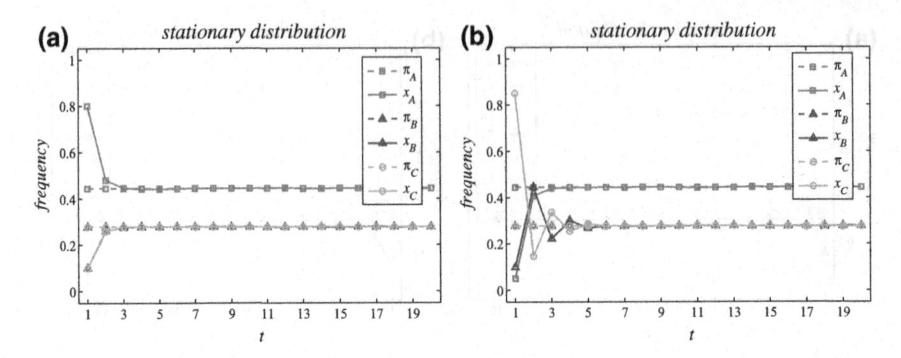

Fig. 7 The frequencies of *A*, *B* and *C* agents in the systems as games proceed. The initial fractions are: **a** $[x_A \quad x_B \quad x_C] = [0.8 \quad 0.1 \quad 0.1]$, **b** $[x_A \quad x_B \quad x_C] = [0.05 \quad 0.1 \quad 0.85]$

And then determining this steady-state distribution

$$\begin{cases} \pi_A + \pi_B + \pi_C = 1 \\ 0.6\pi_A + 0.7\pi_B + 0.7\pi_C = \pi_A \\ 0.2\pi_A + 0.15\pi_B + 0.15\pi_C = \pi_B \end{cases} \tag{16}$$

$$\Downarrow$$

$$\pi = [\pi_A \quad \pi_B \quad \pi_C] = [\frac{7}{11} \quad \frac{2}{11} \quad \frac{2}{11}]$$

(2) Given switching probability matrix U:

$$U = \begin{bmatrix} 0.5 & 0.25 & 0.25 \\ 0.4 & 0.1 & 0.5 \\ 0.4 & 0.5 & 0.1 \end{bmatrix} \tag{17}$$

Solving the stationary distribution of strategies:

$$\begin{cases} \pi_A + \pi_B + \pi_C = 1 \\ 0.5\pi_A + 0.4\pi_B + 0.4\pi_C = \pi_A \\ 0.25\pi_A + 0.1\pi_B + 0.5\pi_C = \pi_B \end{cases} \tag{18}$$

$$\Downarrow$$

$$\pi = [\pi_A \quad \pi_B \quad \pi_C] = [\frac{4}{9} \quad \frac{5}{18} \quad \frac{5}{18}]$$

3.4 $A < B = C$

The collective dilemma situation here is opposite to the case of (3). Strategy A is dominated by B and C, hence $u_{A \to B} > u_{B \to A}$, $u_{A \to C} > u_{C \to A}$ and $u_{B \to C} = u_{C \to B}$.

(1) The transitions probability matrix is

$$U = \begin{bmatrix} 0.2 & 0.4 & 0.4 \\ 0.1 & 0.45 & 0.45 \\ 0.1 & 0.45 & 0.45 \end{bmatrix} \quad (19)$$

In order to solve the steady-state distribution

$$\begin{cases} \pi_A + \pi_B + \pi_C = 1 \\ 0.2\pi_A + 0.1\pi_B + 0.1\pi_C = \pi_A \\ 0.4\pi_A + 0.45\pi_B + 0.45\pi_C = \pi_B \end{cases} \quad (20)$$

$$\Downarrow$$

$$\pi = [\pi_A \quad \pi_B \quad \pi_C] = [\tfrac{1}{9} \quad \tfrac{4}{9} \quad \tfrac{4}{9}]$$

(2) Given another matrix

$$U = \begin{bmatrix} 0.4 & 0.3 & 0.3 \\ 0.2 & 0.3 & 0.5 \\ 0.2 & 0.5 & 0.3 \end{bmatrix} \quad (21)$$

Then solving the steady-state distribution

$$\begin{cases} \pi_A + \pi_B + \pi_C = 1 \\ 0.4\pi_A + 0.2\pi_B + 0.2\pi_C = \pi_A \\ 0.3\pi_A + 0.3\pi_B + 0.5\pi_C = \pi_B \end{cases} \quad (22)$$

$$\Downarrow$$

$$\pi = [\pi_A \quad \pi_B \quad \pi_C] = [\tfrac{1}{4} \quad \tfrac{3}{8} \quad \tfrac{3}{8}]$$

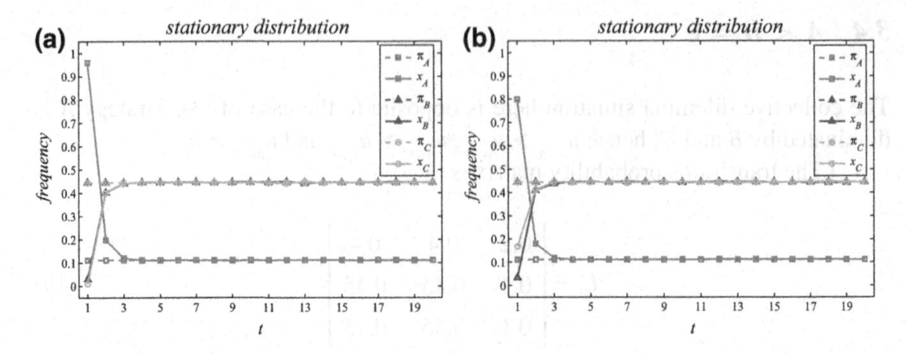

Fig. 8 The frequencies of A, B and C individuals. The population begins with the state of **a** $[x_A \quad x_B \quad x_C] = [0.96 \quad 0.03 \quad 0.01]$, **b** $[x_A \quad x_B \quad x_C] = [0.7985 \quad 0.0341 \quad 0.1674]$

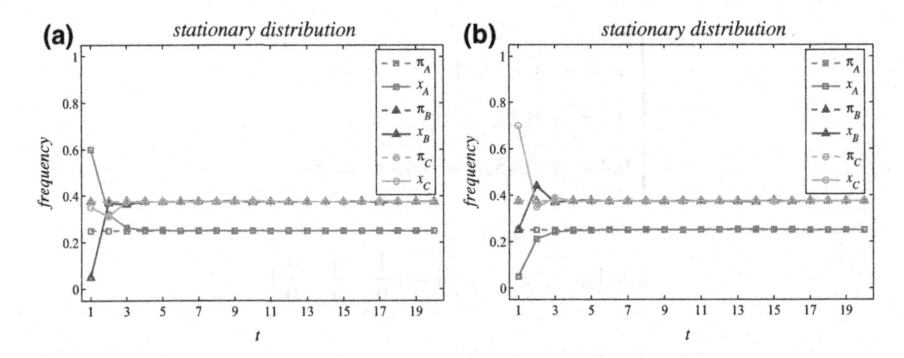

Fig. 9 The evolution process of the strategy distribution of the gaming system, and the whole population is composed of **a** $[x_A \quad x_B \quad x_C] = [0.6 \quad 0.05 \quad 0.35]$, **b** $[x_A \quad x_B \quad x_C] = [0.05 \quad 0.25 \quad 0.7]$

Figures 8 and 9 show the evolution of strategy distribution (A, B and C players) of the system, in dependence on different transition probability matrixes and initial probability distributions. If the probabilities meet the conditions that $u_{A \to B} > u_{B \to A}$, $u_{A \to C} > u_{C \to A}$ and $u_{B \to C} = u_{C \to B}$, the system will evolve to the steady state that $x_A < x_B = x_C$, irrespective of the switching probabilities and initial strategy placement.

3.5 $A = B = C$

In this special situation, $u_{A \to B} = u_{B \to A} = u_{A \to C} = u_{C \to A} = u_{B \to C} = u_{C \to B}$ and $u_{A \to A} = u_{B \to B} = u_{C \to C}$. In the following, we explore the stationary distribution under this condition.

Results summarized in Figs. 10 and 11 show that in the presence of no difference among the three strategies, the system will evolve into the coexistence state with equally distributed strategies.

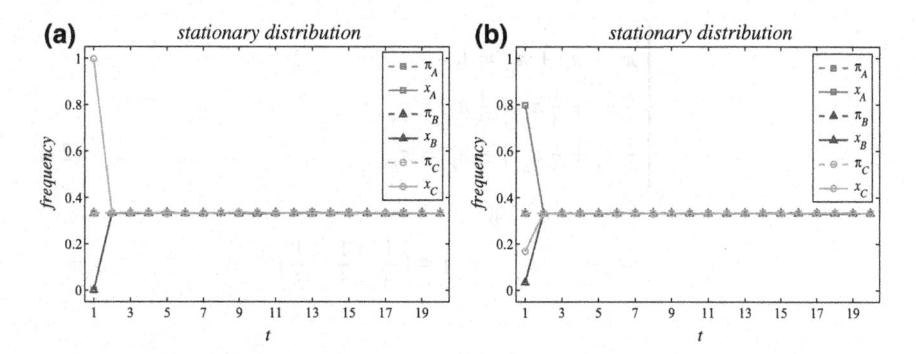

Fig. 10 The strategy distribution (*A*, *B* and *C*) in the system as game proceeds. The initial frequencies are $[x_A \quad x_B \quad x_C] = [0.001 \quad 0.001 \quad 0.998]$ and $[x_A \quad x_B \quad x_C] = [0.7981 \quad 0.0335 \quad 0.1684]$, respectively. The system quickly converges to the coexistence state with equally distributed strategies

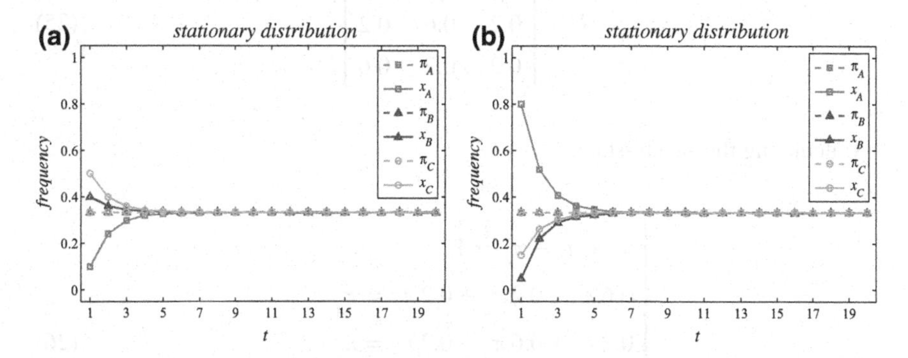

Fig. 11 As games proceed, the proportions of three strategy, x_A, x_B and x_C in the system. The initial frequencies of **a** and **b** are $[x_A \quad x_B \quad x_C] = [0.1 \quad 0.4 \quad 0.5]$ and $[x_A \quad x_B \quad x_C] = [0.8 \quad 0.05 \quad 0.15]$, respectively. The state finally reaches the steady distribution of the strategies

(1) In this special special, each probability is the same,

$$U = \begin{bmatrix} \frac{1}{3} & \frac{1}{3} & \frac{1}{3} \\ \frac{1}{3} & \frac{1}{3} & \frac{1}{3} \\ \frac{1}{3} & \frac{1}{3} & \frac{1}{3} \end{bmatrix} \tag{23}$$

The equilibrium distribution is $\pi = [\frac{1}{3} \quad \frac{1}{3} \quad \frac{1}{3}]$,

$$\begin{cases} \pi_A + \pi_B + \pi_C = 1 \\ \frac{1}{3}\pi_A + \frac{1}{3}\pi_B + \frac{1}{3}\pi_C = \pi_A \\ \frac{1}{3}\pi_A + \frac{1}{3}\pi_B + \frac{1}{3}\pi_C = \pi_B \end{cases} \tag{24}$$

$$\Downarrow$$

$$\pi = [\pi_A \quad \pi_B \quad \pi_C] = [\frac{1}{3} \quad \frac{1}{3} \quad \frac{1}{3}]$$

(2) It is meaningful to investigate the results led by normal situation, i.e., the values of the main diagonal are not equal to that of the off-diagonal.

$$U = \begin{bmatrix} 0.6 & 0.2 & 0.2 \\ 0.2 & 0.6 & 0.2 \\ 0.2 & 0.2 & 0.6 \end{bmatrix} \tag{25}$$

Calculating the steady-state:

$$\begin{cases} \pi_A + \pi_B + \pi_C = 1 \\ 0.6\pi_A + 0.2\pi_B + 0.2\pi_C = \pi_A \\ 0.2\pi_A + 0.6\pi_B + 0.2\pi_C = \pi_B \end{cases} \tag{26}$$

$$\Downarrow$$

$$\pi = [\pi_A \quad \pi_B \quad \pi_C] = [\frac{1}{3} \quad \frac{1}{3} \quad \frac{1}{3}]$$

4 Simulations

The above analysis does not take into account the information of neighbours, who may influence the player's decision making. Therefore, we perform simulations on complex networks by adopting the switching probabilities. We employ homogeneous networks (square lattice) to study the strategy evolution process. Here, the situations of Sect. 3.1 is studied for comparison with the gained analysis results. The snapshots of spatial configurations in the evolution process can help acquire more information of evolution dynamics. So the evolution process of A, B and C players in the lattice whose number of individuals is $N = 1024$, i.e., 32×32 is presented here. At the initial step, every player employs one of strategy A, B and C randomly.

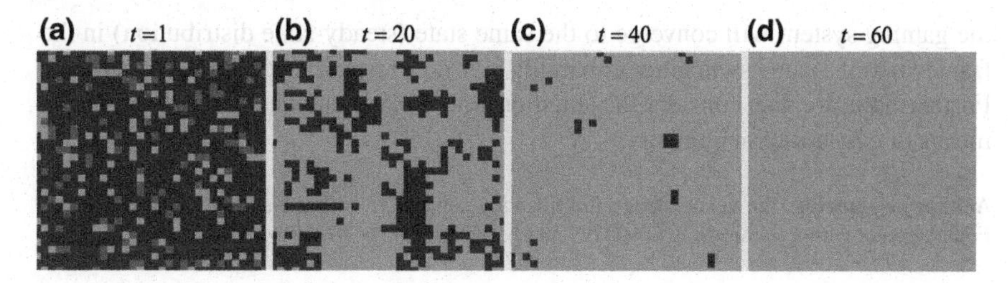

Fig. 12 The spatial distribution of A-players (*light blue*), B-players (*purple*) and C-players (*red*) that situate in the square lattice of size $N = 1024$, depending on $u_{A\to A} = 0.6$, $u_{A\to B} = 0.3$, $u_{A\to C} = 0.1$, $u_{B\to A} = 0.55$, $u_{B\to B} = 0.3$, $u_{B\to C} = 0.15$, $u_{C\to A} = 0.7$, $u_{C\to B} = 0.2$ $u_{C\to C} = 0.1$. Results show that strategy A quickly takes over the system by the aid of the condition $A > B > C$

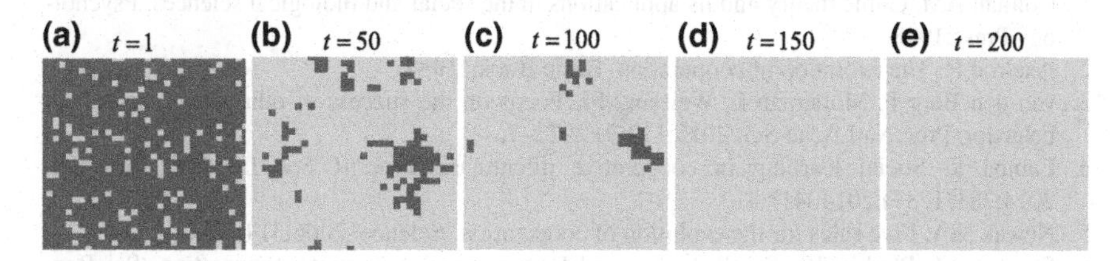

Fig. 13 The snapshots of A, B and C agents in the lattice. It is related to the switching probabilities: $u_{A\to A} = 0.3$, $u_{A\to B} = 0.5$, $u_{A\to C} = 0.2$, $u_{B\to A} = 0.55$, $u_{B\to B} = 0.2$, $u_{B\to C} = 0.25$, $u_{C\to A} = 0.6$, $u_{C\to B} = 0.3$ $u_{C\to C} = 0.1$. Similar with the Fig. 12, the best strategy A can dominate the system finally

Figures 12 and 13 provide results depending on the transition probability matrix. Here, players not only refer to the switching probabilities but also consult their neighbours for strategy updating. It is obvious that a small amount of A-players can make a large number of C-players and a few B-players extinct rapidly, and the whole population is finally occupied by A-players.

5 Conclusions

In this paper, the switching probabilities provide a new framework for studying the evolutionary game dynamics in the three-strategy game models. The motivation is inspired by the difficulty of obtaining specific payoffs of individuals in real systems. We keep away from this difficulty and encode the payoffs into the probabilities that any player shifts her current strategy to the other one. So that there is no need to get the concrete information about payoffs, but instead only consider the switching probabilities $u_{A\to A}$, $u_{A\to B}$, $u_{A\to C}$, $u_{B\to A}$, $u_{B\to B}$, $u_{B\to C}$, $u_{C\to A}$, $u_{C\to B}$ and $u_{C\to C}$.

Specifically, here we apply the switching probabilities to the three-strategy games to analyze the steady state of the evolving system. Results show that the steady fraction of better strategy in the system is larger than the relatively poor one. Moreover, the gaming system will converge to the same state (steady-state distribution) in the framework of a same switching probability matrix, irrespective of the initial settings. Furthermore, we also consider the situation where players update their strategies by means of consulting neighbours.

Acknowledgements We acknowledge the financial support from the National Natural Science Foundation of China (Grant Nos. 61603199 and 61603201 and 61573199).

References

1. Colman AM. Game theory and its applications in the social and biological sciences. Psychology Press;1998.
2. Axelrod R. The evolution of cooperation. Basic Books;1984.
3. van den Berg P, Molleman L, Weissing FJ. Focus on the success of others leads to selfish behavior. Proc Natl Acad Sci. 2015;112(9):2912–7.
4. Lamba S. Social learning in cooperative dilemmasr. Proc R Soc Lond B Biol Sci. 2014;281(1787):20140417.
5. Nowak MA. Five rules for the evolution of cooperation. Science. 2006;314:1560–3.
6. Stewart AJ, Plotkin JB. Small groups and long memories promote cooperation. Sci Rep. 2016;6(s 1–3):26889.
7. Stivala A, Kashima Y, Kirley M. Culture and cooperation in a spatial public goods game. Phys Rev E. 2016;94(3):032303.
8. Scott B. Coordination vs. voluntarism and enforcement in sustaining international environmental cooperation. Proc Nat Acad Sci. 2016;113(51):14515–22.
9. Smith JM. Evolution and the theory of games. Cambridge University Press;1982.
10. Cimini G, Sánchez A. Learning dynamics explains human behaviour in prisoners dilemma on networks. J R Soc Interface. 2014;11(94):20131186.
11. Poncela J, Gómez-Gardeñes J. Cooperation in scale-free networks with limited associative capacities. Phys Rev E. 2011;83(5).
12. Sasaki T, Okada I. Cheating is evolutionarily assimilated with cooperation in the continuous snowdrift game. BioSystems. 2015;131:51–9.
13. Abbass H, Greenwood G, Petraki E. The n-player trust game and its replicator dynamics. IEEE Trans Evol Comput. 2016;20(3):470–4.
14. Bomze IM. Lotka-volterra equation and replicator dynamics: new issues in classification. Biol Cybern. 1995;72:447–53.
15. Taylor P, Jonker L. Evolutionarily stable strategies and game dynamics. Math Biosci. 1978;40:145–56.
16. Nowak MA, Sigmund K. Evolution of cooperation by multilevel selection. Proc Natl Acad Sci. 2006;103(29):10952–5.
17. Vainstein MH, Arenzon JJ. Spatial social dilemmas: dilution, mobility and grouping effects with imitation dynamics. Physica A. 2014;394:145–57.
18. Fosco C, Mengel F. Cooperation through imitation and exclusion in networks. J Econ Dyn Control. 2011;35:641–58.
19. Galla T. Imitation, internal absorption and the reversal of local drift in stochastic evolutionary games. J Theoret Biol. 2011;269:46–56.
20. Sánchez A, Vilone D, Ramasco J, San Miguel M. Social imitation versus strategic choice, or consensus versus cooperation, in the networked prisoners dilemma. Phys Rev E. 2014;90(2):022810.
21. Traulsen A, Claussen JC, Hauert C, San M. Coevolutionary dynamics: from finite to infinite populations. Phys Rev Lett. 2005;95:238701.

22. Blume LE. The statistical mechanics of best-response strategy revision. Games Econ Behav. 1995;5(387):111–45.
23. Hauert C, Szabo G. Game theory and physics. Am J Phys. 2005;73(5):405–14.
24. Helbing D. A stochastic behavioral model and a microscopic foundation of evolutionary game theory. Theory Decis. 1996;40:149–79.
25. Blume LE. How noise matters. Games Econ Behav. 2003;44(2):251–71.
26. Zhang J, Zhang C, Chu T. The evolution of cooperation in spatial groups. Chaos Solitons Fractals. 2011;44:131–6.
27. Zhang J, Zhang C, Cao M, Weissing FJ. Crucial role of strategy updating for coexistence of strategies in interaction networks. Phys Rev E. 2015;91(4):042101.
28. Zhang J, Chen Z. Contact-based model for strategy updating and evolution of cooperation. Physica D. 2016;323–3242:27–34.
29. Xu Z, Zhang J, Zhang C, Chen Z. Fixation of strategies driven by switching probabilities in evolutionary games. EPL. 2016;116(5):58002.

Fault Diagnosis of Rolling Bearing Based on Wavelet Packet and Extreme Learning Machine

Xiaoquan Tang, Yi Chai, Yongfang Mao and Junjie Ji

Abstract The paper presents a new method using wavelet packet analysis and Extreme Learning Machine (ELM) with the following steps. First, the signal is decomposed by wavelet packet, and the root-mean-square (RMS) and energy of the decomposed subband component signals are extracted. Secondly, the fault classification model of rolling bearing is established based on the Extreme Learning Machine (ELM). Finally, the eigenvector composed of the characteristic parameters of the decomposed sub-signals is used as the model input to diagnose the fault of the rolling bearing. The results indicate that this method will be effectively applied to fault diagnosis of rolling bearings.

Keywords Rolling bearing · Wavelet packet · Fault diagnosis · Extreme learning machine (ELM)

1 Introduction

Due to their high energy conversion efficiency, easy installation, rolling bearings, as a part of rotating machinery, have gained extensive applications [1]. Rolling bearings are important for mechanical equipment, whose working condition affects the overall performance as well as the technical indicators of machinery and equipment [2]. Data shows that 30% of the mechanical equipment failures are related to bearing damage [3]. Fault diagnosis technology for rolling bearing has been a hotspot in this field of fault diagnosis. Therefore, it is of great significance to carry out the research on the new method of fault diagnosis of rolling bearing.

In essence, the rolling bearing fault diagnosis process is a problem of pattern classification, the most critical part of which is the feature extraction for signal

---

X. Tang · Y. Chai · Y. Mao (✉) · J. Ji
College of Automation, Chongqing University, Chongqing 400044, China
e-mail: yongfangmaocqu@163.com

X. Tang · Y. Chai · Y. Mao · J. Ji
School of Automation, Chongqing University, Chongqing 400044, China

Y. Jia et al. (eds.), *Proceedings of 2017 Chinese Intelligent Systems Conference*, Lecture Notes in Electrical Engineering 459,
https://doi.org/10.1007/978-981-10-6496-8_13

obtained and state recognition. Among former studies on this problem, Ref. [4] uses EMD and SVM to identify the fault state of rolling bearing [4]. In Ref. [5], the energy characteristics of different frequency bands after double tree complex wavelet transform are obtained and SVM is used to classify the faults of rolling bearing [5]. In Ref. [6], the multi-scale entropy in the seven states of the bearing is extracted for the BP neural network to diagnose the bearing failure [6]. However, in the above mentioned studies, the selection of the penalty coefficient and kernel structure parameter in SVM is based on experience, and the use of BP requires to set a large number of network training parameters, which will lead to degraded accuracy of algorithm classification.

In my paper, a method about fault diagnosis of rolling bearing is proposed. Firstly, wavelet packet is used for better feature extraction of different fault information of rolling bearing. The original signal is divided into sub-bands which is in accordance with the local characteristics of the signal obtained by wavelet packet decomposition. Then the root mean square value and energy of each sub-band component signal are extracted to be served as the eigenvector of bearing fault diagnosis. Secondly, the fault classification model of the rolling bearing of ELM is established. Finally, the eigenvector consisted with the root mean square value and the energy of component signal after the decomposition by wavelet packet is used for the ELM model to finish fault classification of rolling bearings.

2 Methodology

2.1 Wavelet Packet

The overall resolution accuracy of wavelet packet is also higher than wavelet analysis. The wavelet packet analysis continues to divide the high frequency component that can not be subdivided in wavelet analysis, and thus achieves a higher frequency resolution for the high frequency band. Therefore, wavelet packet analysis is particularly important in dealing with engineering practical problems [7].

The scale relation between the known scale function and the wavelet function is [8]:

$$\phi(\mathrm{t}) = \sqrt{2}\sum_{\mathrm{q}} \mathrm{h}_0 q\phi(2t - q) \tag{1}$$

$$\psi(t) = \sqrt{2}\sum_{\mathrm{q}} \mathrm{h}_1 \mathrm{q}\phi(2t - q) \tag{2}$$

Where, h_0 and h_1 are coefficients of the filter.

Define the following recursive relations in order to generalize the two-scale equation:

$$w_{2n}(t) = \sqrt{2}\sum_{q \in Z} h_0 q w_n(2t - q) \quad (3)$$

$$w_{2n+1}(t) = \sqrt{2}\sum_{q \in Z} h_1 q w_n(2t - q) \quad (4)$$

Wavelet packet decomposition has arbitrary multi-scale characteristics, which avoids the defect of fixed time-frequency in wavelet decomposition. It provides better solution for time-frequency analysis, thus reflecting the nature and characteristics of the signal more accurately [9].

For the purpose of clear distinction, this article refers to Z_j^n instead of u_j^n:

$$z_j^n = u_j^n = u_{j+1}^{2n} + u_{j+1}^{2n+1} \quad (5)$$

The expression of decomposition by wavelet packet can be obtained by recursion:

$$\begin{cases} Z_j = u_{j+2}^2 \oplus u_{j+1}^3 \\ Z_j = u_{j+2}^4 \oplus u_{j+2}^5 \oplus u_{j+2}^6 \oplus u_{j+2}^7 \\ \cdots \\ Z_j = u_{j+k}^{2^k} \oplus u_{j+k}^{2^k+1} \oplus \ldots \oplus u_{j+k}^{2^k+m} \end{cases} \quad \begin{cases} m = 0, 1, 2, \ldots, 2^k - 1 \\ k = 1, 2, 3\ldots \\ j = 1, 2, 3\ldots \end{cases} \quad (6)$$

The wavelet packet system recursion formula is:

$$\begin{cases} d_k^{j+1,2n} = \sum_1 h_0(2l - k) d_1^{j,n} \\ d_k^{j+1,2n+1} = \sum_1 h_1(2l - k) d_1^{j,n} \end{cases} \quad (7)$$

The corresponding reconstruction formula is:

$$\begin{aligned} d_1^{j,n} &= \sum_k \left[h_0(p - 2k) d_k^{j+1,2n} + h_1(p - 2k) d_k^{j+1,2n+1} \right] \\ &= \sum_k^k g_0(p - 2k) d_k^{j+1,2n} + g_1(p - 2k) d_k^{j+1,2n+1} \end{aligned} \quad (8)$$

From the Eqs. (7) and (8), it can be seen that the decomposition process is the further decomposition of all the bands obtained by the upper-layer decomposition, however, the wavelet decomposition only decomposes the low frequency. It is evident that the wavelet packet transform perfectly complements the defect of wavelet transform.

2.2 Feature Parameter Extraction

After decomposing the discrete sequence, the total length of the decomposition coefficient is equal to the length of the original and discrete sequence. The difference is that its components are rearranged according to frequencies. The new subsequence has the ability to concentrate the coefficients to facilitate the extraction of the essential features. In this paper, the statistical indexes selected in the frequency domain analysis are the root mean square value and wavelet packet energy of sub-band signal components after wavelet packet decomposition.

The instantaneous amplitude of the vibration signal is constantly changing over time, and the root mean square value of the signal reflects the intensity of the vibration of the bearing during the corresponding period. Suppose there is a time discrete sequence {X1, X2, …, Xn}, where N represents number of sampling points, the corresponding root mean square could be expressed as:

$$x_{rms} = \sqrt{\frac{1}{\mathrm{N}}\sum_{\mathrm{t}=1}^{N} x^2(t)} \tag{9}$$

From the energy point of view, wavelet packet breaks the signal energy into different time-frequency planes. Wavelet packet is featured with orthogonality. After decomposing the signal f (x) by wavelet packet, energy of the signal in i-th band of j-scale is expressed as:

$$E_j^i = \int_{-\infty}^{\infty} f^2(t)d_t \tag{10}$$

2.3 Extreme Learning Machine (ELM)

ELM is an easy-to-use and effective algorithm which is proposed by Professor Huang in 2004 [10]. ELM reduces the numbers of parameters to one relying on training to set. In the process of algorithm implementation, it is not necessary to adjust the weight of network input. Without iteration, it can produce a unique optimal solution that overcomes the shortcomings of local optimal solution. Therefore, ELM is superior in its simplicity of structure, rapidity of solution and generalization performance [11].

You can think of ELM as a network shown in Fig. 1, and the internal connection is also indicated. The number of neurons in the input layer, the hidden layer and the output layer is *n, l, m* respectively.

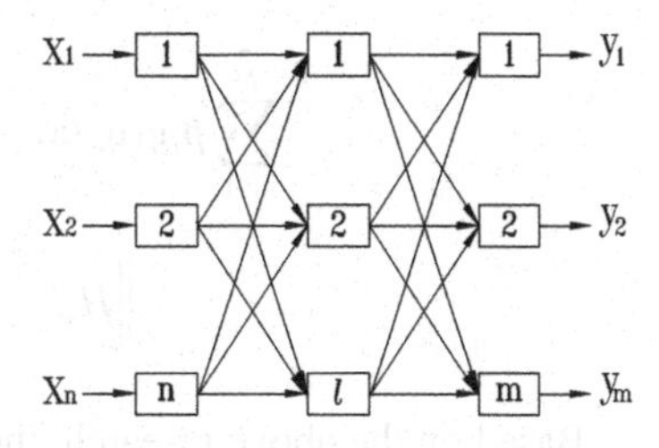

Fig. 1 ELM network diagram

The theory of ELM algorithm is as below:

Let the training sample be (x_j, y_j), where $x_j = [x_{j1}, x_{j2}, \ldots, x_{jn}]^T \in R^n$ is the input vector for training and $y_j = [x_{j1}, x_{j2}, \ldots, x_{jm}]^T \in R^m$ is the desired output vector for training.

With G (x) as the excitation function, the ELM model is as follows:

$$\sum_{j=1}^{n} \beta_j G(a_j x_t + b_j) = o_t, t = 1, 2, \ldots, n \tag{11}$$

where $a_j = [a_{j1}, a_{j2}, \ldots, a_{jn}]^T$ is the weight between hidden layer node and input layer node; b_j is the offset value of hidden layer node; $\beta_j = [\beta_{j1}, \beta_{j2}, \ldots, \beta_{jn}]^T$ is the weight between hidden layer node and output layer node.

The matrix of the N equations of (11) can be written as Hβ = Y, where H is nominated as the output matrix of ELM.

$$H(u_1, \ldots, u_{\tilde{N}}, v_1, \ldots, v_{\tilde{N}}, x_1, \ldots, x_N) = x \begin{bmatrix} g(u_1 x_1 + v_1) & \ldots & g(u_{\tilde{N}} x_1 + v_{\tilde{N}}) \\ \ldots & \ldots & \ldots \\ g(u_1 x_N + v_1) & \ldots & g(u_{\tilde{N}} x_N + v_{\tilde{N}}) \end{bmatrix}_{N \times \tilde{N}} \tag{12}$$

$$\beta = \begin{bmatrix} \beta_1^T \\ \ldots \\ \beta_N^T \end{bmatrix}_{N \times m} \tag{13}$$

$$Y = \begin{bmatrix} y_1^T \\ \ldots \\ y_N^T \end{bmatrix}_{N \times m} \tag{14}$$

E (w) is used to represent the sum of the squares of errors of the ELM network, the essence of solution is to find the optimal network weight W = F (u, v, β), so that the value of cost function E (w) is minimal, the mathematical model (11) can be rewritten as Hβ = E + Y and the following equation can be deducted [12]:

$$\sum_{i=1}^{\widehat{N}} \beta_i g(\mathrm{u}_i \cdot x_j + \mathrm{v}_i) - y_j = \varepsilon_j, j = 1, 2, \ldots N \tag{15}$$

$$\left\| H\widehat{\beta} - Y \right\| = \min_{\beta} \| H\beta - Y \| \tag{16}$$

Based on the above research, the training process of ELM network structure can be concluded to an optimization problem, which is the key to solve the minimum value of the above equation. Set the nerve parameters of ELM randomly, and the

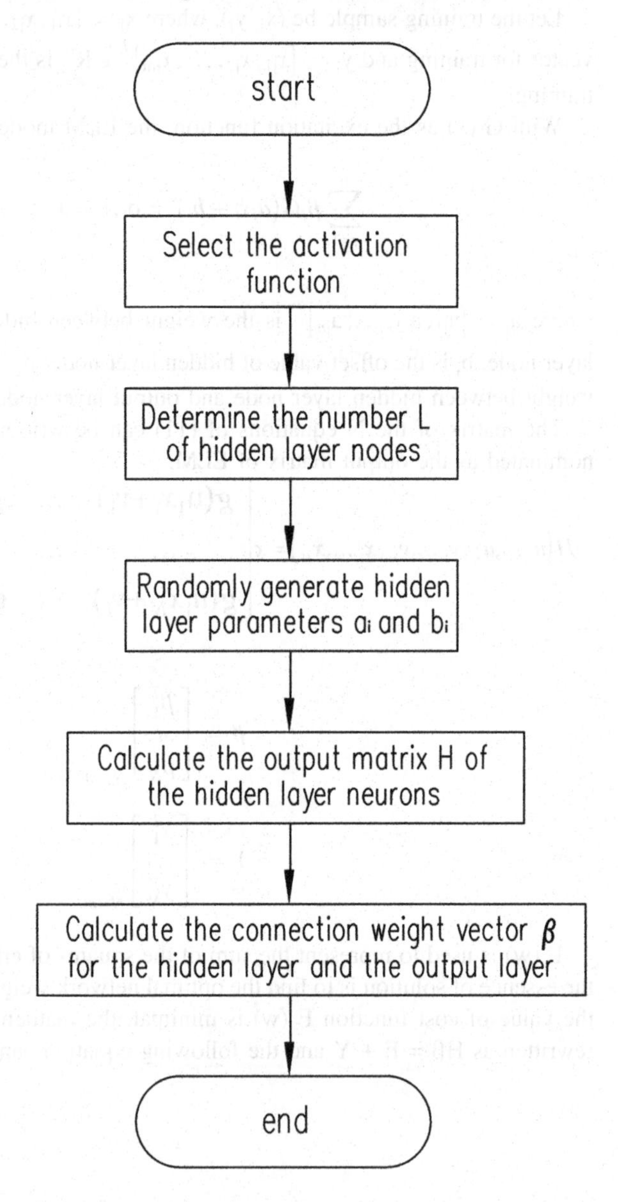

Fig. 2 ELM algorithm flowchart

output weight value can be obtained through training. Hence, the matrix H is transformed into a constant matrix, and the equations are transformed into linear equations, so that the solution $\widehat{\beta} = \arg\min\|H\beta - Y\| = H^{+}\beta$ of the equations can be obtained. Where $H^{+} = (H^{T}H)^{-1}H^{T}$. The process of training is completed after obtaining β.

ELM algorithm flowchart shown in Fig. 2.

3 Rolling Bearing Fault Diagnosis Experiment

3.1 Introduction to the Experiment

The experimental data is from the simulation test bed for the rolling bearing fault diagnosis laboratory in Case Western Reserve University, United States. The model used is rolling bearings with a rotating motor load power of 735.5 w, rolling bearing speed of 1797 r/min. The pitting corrosion of the inner, outer ring and rolling body of the bearing has been produced by electrical discharge machining. Using the acceleration sensor in the 12 kHz sampling frequency, the four kinds of signal waveforms are collected.

3.2 Eigenvector Construction

The db3 wavelet, commonly used in fault diagnosis, is chosen as the basis function of wavelet packet, whose decomposition layer is chosen as 3. The vibration signal is decomposed by wavelet packet to obtain eight component signals, from p0 to p7, followed by low frequency to high frequency arrangement.

Table 1 shows the root-mean-square values of the component signals of the sub-bands in the frequency domain analysis after the decomposition of vibration signals of the four operating states by wavelet packets.

Table 2 shows the energy values in the frequency domain analysis after the vibration signals of the four operating states are decomposed by wavelet packet.

Table 1 Root-mean-square values of each band component signal after wavelet packet decomposition

Fault type	Root mean square value							
	p0	p1	p2	p3	p4	p5	p6	p7
Normal	0.054	0.042	0.001	0.026	0.001	0.002	0.003	0.006
Rolling body	0.048	0.030	0.020	0.018	0.013	0.026	0.021	0.047
Inner	0.081	0.133	0.078	0.130	0.014	0.040	0.063	0.056
Outer	0.056	0.079	0.095	0.065	0.017	0.061	0.086	0.045

Table 2 Energy values of each band component signal after wavelet decomposition

Fault type	Energy value							
	p0	p1	p2	p3	p4	p5	p6	p7
Normal	6.514	3.191	0.231	1.385	0.003	0.018	0.028	0.098
Rolling body	4.374	1.696	0.832	0.697	0.396	1.133	1.084	4.755
Inner	13.293	35.390	12.257	34.004	0.423	3.204	8.014	6.295
Outer	6.337	12.503	18.326	8.665	0.626	7.562	14.969	4.061

The RMS value of the signal reflects the intensity of the vibration of the bearing during this period. The energy can reflect the strength of the signal. The bearing of different working states has different root mean square and energy in different frequency bands. The root mean square and the energy of the sub-band eight component signals are as the characteristic parameters to construct the corresponding eigenvector.

3.3 *Fault Identification*

In this experiment, the "Tribas" is selected as an excitation function. Normal state and 3 kinds of fault state, including inner ring, outer ring and rolling body failure, are respectively labeled with {1, 2, 3, 4}, which are used in the following experiment. In each test, 50 samples are included in every bearing state. Training samples and test samples are selected in accordance with the ratio of 2: 3. The eigenvector is designated to the ELM for the training of acquisition the desired prediction model. Secondly, the corresponding feature input vector is calculated for the current test

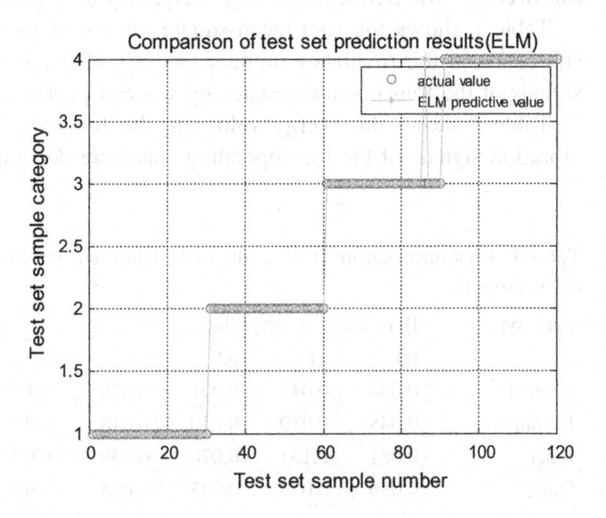

Fig. 3 ELM troubleshooting classification chart

Table 3 Test accuracy of different classifiers

Fault Type	Classification accuracy/%		
	BP	SVM	ELM
Normal	89.16	90.8	100
Rolling body	87.5	86.67	100
Inner ring	83.33	84.16	99.17
Outer ring	84.16	87.57	100

sample, and is inputted to the trained prediction model. The output of the ELM classifier can elicit diagnosis result. At the same time, classifier of BP and SVM is used to train and test the samples. ELM test classification results are shown in Fig. 3; accuracy of the three classifier is shown in Table 3.

In Fig. 3, the abscissa 0–30 indicates the 30 tag data in the normal state. The result of the ELM model has 30 points falling on "1", that is, 0 samples are judged incorrectly. The abscissa 30–60 indicates the 30 tag data of the inner ring fault. The calculated result has 30 points falling on "2", which means 0 samples are judged wrong. The abscissa 60–90 indicates the 30 tag data of the outer ring fault. The result of the calculation has 29 points falling on "3", that is, one sample is wrong. The abscissa 90–120 indicates the 30 tag data of the rolling element failure.

The result of the calculation has 30 points falling on "4", that is, 0 samples are judged wrong.

From Fig. 3 and Table 3, ELM classifier can effectively identify different working states of the rolling bearing, and it exhibits higher fault classification accuracy compared with BP and SVM.

In order to avoid the problem of contingency, 100000 data points are selected in each state and 2000 data points are selected in each test. 50 experiments were carried out, and 50 groups of experimental results were obtained. In addition, the average classification accuracy was obtained by the classification accuracy of multiple tests. The accuracy rate of using the ELM classifier to obtain the average classification is still the highest, reaching 98.82%. Meanwhile, it is featured with faster learning speed and better classification efficiency.

4 Conclusion

Monitoring and diagnosing the working status of rolling bearing is of great significance because of its wide application in various mechanical equipment. In my paper, a new method is proposed. According to the simulation results above, these conclusions could be drawn: (1) The wavelet packet decomposition method decomposes the vibration signal, and the root mean square and wavelet packet energy of the component signal of each sub-band in the frequency domain analysis can be used as effective characteristic parameters to construct the corresponding eigenvector that can accurately estimate the running state of the rolling bearing.

(2) The ELM classifier can effectively classify and identify different fault states of rolling bearings. Compared with BP and SVM, the accuracy of classification of ELM is higher, and the classification efficiency of ELM is better and the learning speed is faster. Therefore, the method of fault diagnosis in this paper is effective.

Acknowledgements The research is supported by "the Fundamental Research Funds for the Central Universities (No.106112016CDJXY170003)".

References

1. Wang GY, Xu BJ. Application of RBF neural network in fault diagnosis of rotating machinery. Mech Des Manuf. 2008;09:57–8.
2. Qu LS, Zhang XN, Shen YZ. Mechanical fault diagnosis theory and method. Beijing: Mechanical Industry Press; 2009.
3. Zhang H. Rolling bearing fault diagnosis technology research. Hunan Agric Mach Acad Ed. 2010;37(5):93–4.
4. Xiang D, Cen J. Fault diagnosis method of rolling bearing based on EMD entropy feature fusion. J Aeronaut Astronaut. 2015;05:1149–55.
5. Xu YG, Meng ZP, Lu M. Fault diagnosis of rolling bearing based on double tree complex wavelet transform and SVM. J Aeronaut Astronaut. 2014;01:67–73.
6. Zhang L, Zhang L, Xiong G et al. i. Frequency fault diagnosis of rolling bearings based on multi-scale entropy and neural network. J Mech Des Res. 2014;(05):96–98 + 105.
7. Zhang SH, Ju G. A real-coded adaptive genetic algorithm and its application research in thermal process identification. Proc CSEE China. 2004;24(2):210–4.
8. Hemmati F, Orfali W, Gadala MS. Roller bearing acoustic signature extraction by wavelet packet transform, applications in fault detection and size estimation. Appl Acoust. 2016;104:101–18.
9. Fang S, Wei Z, Fang S, et al. Rolling bearing fault diagnosis based on wavelet packet and RBF neural network. China Control Conf. 2007;451–5.
10. Cao J, Lin Z, Huang GB, et al. Voting based extreme learning machine. Inf Sci. 2012;185 (1):66–77.
11. Huang GB, Wang DH, Lan Y. Extreme learning machines: a survey. Int J Mach Learn Cybernet. 2011;2(2):107–22.
12. Wong PK, Yang Z, Vong CM, et al. Real-time fault diagnosis for gas turbine generator systems using extreme learning machine. Neurocomputing. 2014;128:249–57.

Event-Triggered Control for Multi-agent System with a Smart Leader

Bin Zhao, Zhongxin Liu and Zengqiang Chen

Abstract This paper studies the consensus problem of the leader-following system. First of all, we propose a new kind of leader, smart leader, by adding the velocity states of the followers to the leader's control algorithm. We also use the event-triggered control to make the leader only use the feedback at necessary time. Then a distributed control law for both the leader and followers is designed. Compared with the multi-agent system with traditional leader, the system with a smart leader can reach the consensus more quickly, smoothly and have a smaller control energy consumption. Moreover, we propose a suffcient condition which can ensure the system can reach the consensus. Finally, some simulation examples are presented for illustration.

Keywords Multi-agent system · Leader-following structure · Smart leader · Event-triggered control

1 Introduction

Nowadays, the consensus problem of multi-agent system is becoming more and more popular for its widespread applications. The applications of multi-agent system can be found in many fields such as mobile autonomous robots, unmanned

B. Zhao · Z. Liu (✉) · Z. Chen
College of Computer and Control Engineering, Nankai University, Tianjin 300350, China
e-mail: lzhx@nankai.edu.cn

B. Zhao
e-mail: 2573202427@qq.com

Z. Chen
e-mail: chenzq@nankai.edu.cn

B. Zhao · Z. Liu · Z. Chen
Tianjin Key Laboratory of Intelligent Robotics, Nankai University, Tianjin 300350, China

Y. Jia et al. (eds.), *Proceedings of 2017 Chinese Intelligent Systems Conference*, Lecture Notes in Electrical Engineering 459,
https://doi.org/10.1007/978-981-10-6496-8_14

aerial vehicle formation, surveillance, hazardous material handling and distributed sensor networks in [1, 2].

Among all the studies of consensus problems, leader-following consensus problem is one of the important research topics. For example, the authors in [3] analyzed the robustness of a second-order system with leader-following structure in fixed topology. When the leader-following system has a switching topology, authors in [4] proposed and solved the consensus problem of this system. A control algorithm which includes a desired input was given in [5] and the authors also analyzed its consensus performance. Besides, the authors studied the leader-following system with a dynamic leader in [6].

However, there is one disadvantage in the above leader-following systems that they rely on the leader too much. The consensus of the system may not be guaranteed if the communication between the leader and followers are failed for some reasons.

To solve this problem, the authors in [7] proposed a new kind of leader, smart leader. The leader can access its neighboring followers' position information, and use it to regulate the control algorithm at some special times. It guarantees that even if there are some faults happening on some followers, the system with the smart leader can keep a bounded tracking-error under fixed or switching topology and the system can reach the consensus eventually.

Based on the paper [7], we study the consensus problem of second-order system with a smart leader. In this paper, we need the smart leader not only use the feedback from the follower but also ask it only uses the feedback at necessary time. Then the event-triggered control is used in this paper to determine whether the leader will use the feedback and which to use.

The event-triggered control can date back to papers [8, 9]. The researchers often use it to deal with the congestion of the communication channels in the multi-agent system and improve the control efficiency. A new kind of event-triggered control which is based on the information from local neighbors is designed in [10, 11]. More information for event-triggered control can be found in paper [12], the authors gave an overview of event-triggered control in recent years.

Motivated by above studies, we design a multi-agent system with smart leader and event-triggered control. First of all, we design the distributed control laws for both leader and followers. Besides, we propose a threshold for the system. If the state errors between the leader and followers are beyond the threshold, the leader will use the feedbacks from the neighbors. Otherwise the leader will not use the feedback. It satisfies the demand that the leader only uses the feedback from the follower at necessary time. Compared with the traditional leader-following system, the system with a smart leader also reaches the consensus more quickly and smoothly. Meantime, since the leader and followers can track the desire velocity together, the system can reduce the energy consumption.

The main contributions of this paper are listed as follows.

1. Design a multi-agent system with a smart leader which can use the feedback from its neighboring follower to adjust its control law. It improves the deficiencies that the traditional leader-following system is effected by the leader to a great extent. Besides, it can reduce the velocity errors between the leader and followers and make the system reach consensus more quickly.
2. Use the event-triggered control in the multi-agent system. By using the event-triggered control, the smart leader can switch its control law to reduce the tracking errors during the convergency. It satisfies the demand of the smart leader that only uses the followers' feedbacks at necessary time.
3. For the followers and the smart leader can track the desire velocity together. The smart leader has a lower energy consumption than the traditional leader.

The remainder of this paper is organized as follows. Basic definitions of the algebraic graph theory and some lemmas are given in Sect. 2. Main results for the system with a smart leader under fixed topology are discussed in Sect. 3. Illustrative examples are presented in Sect. 4. Finally, the conclusion of the whole paper is addressed in Sect. 5.

2 Backgrounds and Preliminaries

Consider the multi-agent system composed of n followers and *1* leader. The topology of the system is represented by graph G and the communication network among the followers is represented by the undirect graph$\bar{G}_f = \{V, E, A\}$, where $V = \{v_1, v_2, \ldots, v_n\}$ is the set of the follower *1* to follower n. $E \subseteq v \times v$ is the set of edges, if $(v_i, v_j) \in E$, then we call the agent j is a neighbor of agent i. Define the neighbor set of agent i as $N_i = \{j\colon (v_i, v_j) \in E, \forall j \neq i\}$. It means that the information is flowing from agent j to agent i. The weighted adjacency matrix $A = (a_{ij}) \in R^{n \times n}$, where $a_{ii} = 0$ and $a_{ij} = a_{ji} \geq 0$. The matrix L is Laplacian matrix and the elements of it are denoted as $l_{ii} = \sum_{j=1}^{n} a_{ij}$ when $i = j$ and $l_{ij} = -a_{ij}$ when $i \neq j$.

To denote the communication between the leader and followers, we define a vector $B = [b_1, b_2, \ldots, b_n]^T$, where b_i is a binary number. $b_i = 1$ means that the leader can communicate with follower i, otherwise $b_i = 0$. In this paper, we assume there is at least one $b_i > 0$. It means that there is at least one follower connecting with the leader. Then the topology among the leader and the followers can be denoted by the diagonal matrix $D = diag\{b_1, \ldots, b_n\}$ and the system matrix here can be represented as $H = L+D$.

Next, we will propose some notations and lemmas which will be used in the following paper.

Notation 2.1 The matrix I is the identity matrix with appropriate dimension. $\lambda_{\min}(A)(\lambda_{\max}(A))$ means the minimum (or maximum) eigenvalue of the corresponding matrix A. The vector $\mathbf{1} = [1, 1 \ldots, 1]^T$ and matrix $\mathbf{0}$ means the all-zeros matrix with appropriate dimension. And $||A||_2$ means the 2-norm of the matrix A with appropriate dimension.

Lemma 2.2 ([13]) *If the entire graph $\bar{G}_f$ is connected, then the symmetric matrix $H = L + D$ associated with $\bar{G}_f$ is positive definite, where the matrix L is the Laplacian matrix of $\bar{G}_f$ and matrix $D = diag\{b_1, b_2, \ldots, b_n\}$.*

Lemma 2.3 ([7]) *Let matrix S be a symmetric matrix*

$$S = \begin{bmatrix} S_{11} & S_{12} \\ S_{12}^T & S_{22} \end{bmatrix} \in R^{n\times n}$$

where both S_{11} and S_{22} are square. Assume S_{22} is positive definite, then the following properties are equivalent:

(1) S is positive definite.
(2) $S_{11} - S_{12}S_{22}^{-1}S_{12}^T$ is positive definite.

Lemma 2.4 ([14]) *Suppose that $\Phi \in R^{n\times n}$ is a positive definite matrix, $a > 0$ is a scalar and $X, Y \in R^n$. Then we have $2X^TY \le X^T\Phi^{-1}X + Y^T\Phi Y$, and as a special case $2X^TY \le aX^TX + a^{-1}Y^TY$ where a is any positive constant.*

Lemma 2.5 ([15]) *According to the Lyapunov stability theorem, it follows that if a common Lyapunov function can be constructed from each subsystems, the switching system can switch arbitrarily and the stability of the switching system can be guaranteed.*

3 Main Results

For we have proposed a new kind of leader, smart leader, in this paper. Here, we will introduce it in detail. The purpose of the smart leader is that it will use the feedbacks from the followers at necessary time. For example, when the state errors between the leader and followers are beyond some threshold values, the smart leader will use the feedbacks to reduce the error. Then the multi-agent system is designed as follow.

Consider the multi-agent system consisting of n followers and one leader. Our goal is to design a proper control algorithm to make the followers track the leader asymptotically. The dynamic of the follower i is represented by:

$$\begin{cases} \dot{x}_i(t) = v_i(t) \\ \dot{v}_i(t) = u_i(t) \end{cases} \quad i = 1, 2, \ldots, n \tag{1}$$

where $x_i(t), v_i(t), u_i(t) \in R^m$ are the position, velocity states and the input of agent i respectively.

The dynamic of the leader can be represented as follows:

$$\begin{cases} \dot{x}_0(t) = v_0(t) \\ \dot{v}_0(t) = u_0(t) \end{cases} \tag{2}$$

where $x_0 \in R^m$ and $v_0 \in R^m$ are the position state and velocity state of the leader and $u_0 \in R^m$ is leader's control input.

In this system, the follower is designed as it only use the information from itself and its neighboring agents. Then the control algorithm for follower i is proposed as follows:

$$u_i = -k[\sum_{j=1}^{n} a_{ij}(x_i(t) - x_j(t)) + b_i(x_i(t) - x_0(t))] - kr[\sum_{j=1}^{n} a_{ij}(v_i(t) - v_j(t)) + b_i(v_i(t) - v_0(t))] \tag{3}$$

where $r, k > 0$ are control parameters which are determined by the system matrix H.

To achieve the demand of the multi-agent system with a smart leader. Here, the smart leader is designed as that it will use the feedback from the neighboring follower when the event-triggered conditions are satisfied. Otherwise, it will only use the states itself and the desire input to control the whole system. Then the control law of the smart leader is designed as follow.

$$u_0(t) = v_d - v_0 + \eta \sum_{i=1}^{n} c_i(v_i(t) - v_0(t)) \tag{4}$$

where v_d is the desire velocity. The switching parameter $\eta = 0$ or 1. $\eta = 0$ means that the leader only uses the desire input, while $\eta = 1$ means that the leader uses the feedback from follower to adjust the control algorithm. For the leader only uses one neighboring follower's feedback every time in this system, we define a column vector $C = [0, \ldots, c_i, \ldots, 0]^T$, where $c_i = 1$ means the leader adopts the follower i's feedback, otherwise $c_i = 0$. The choice of the follower i will be discussed later.

Definition 3.1. The smart leader here can use the feedbacks from the neighboring followers to adjust its control law. Compared with the traditional leader, the system with a smart leader has a faster convergency rate and less energy consumption. The simulations in Sect. 4 can also verify this conclusion. This is the reason why we call it the smart leader.

Denote $\bar{x}_i(t) = x_i(t) - x_0(t)$, $\bar{v}_i(t) = v_i(t) - v_0(t)$, $\hat{v}(t) = v_d - v_0(t)$, using protocols (3) and (4), the systems (1) and (2) can be rewritten as

$$\begin{cases} \dot{\bar{x}}(t) = \bar{v}(t) \\ \dot{\bar{v}}(t) = -kH \cdot \bar{x}(t) - krH \cdot \bar{v}(t) - \hat{v}(t) \cdot \mathbf{1} - (\eta \otimes \mathbf{1}) \cdot C^T \cdot \bar{v}(t) \\ \dot{\hat{v}}(t) = -\hat{v}(t) \cdot \mathbf{1} - (\eta \otimes \mathbf{1}) \cdot C^T \cdot \bar{v}(t) \end{cases} \tag{5}$$

or

$$\dot{\omega}(t) = F \cdot \omega(t) \tag{6}$$

where $\omega(t) = \begin{bmatrix} \bar{v}^T(t) & \bar{x}^T(t) & \hat{v}^T(t) \end{bmatrix}^T$ and

$$F = \begin{bmatrix} -rkH - (\eta \otimes \mathbf{1}) \cdot C^T & -kH & -I \\ I & 0 & 0 \\ -(\eta \otimes \mathbf{1}) \cdot C^T & 0 & -I \end{bmatrix} \tag{7}$$

Definition 3.2 If the states of the agent satisfy the following two conditions:

$$\begin{cases} \lim_{t \to \infty} (x_i(t) - x_0(t)) = 0 \\ \lim_{t \to \infty} (v_i(t) - v_0(t)) = 0 \end{cases} \tag{8}$$

then the multi-agent systems (1) and (2) are said to reach a consensus asymptotically.

To satisfy the demand that the leader only uses the follower's feedback at necessary time. In this paper, we use the event-triggered control to determine whether to use and which follower's feedback to use. Next, we will define the event-triggered functions for the system.

Definition 3.3 Consider that the leader can receive n_c followers' feedbacks at the same time. Define the threshold $\varepsilon > 0$. Then the minimum and maximum tracking errors are $f_{\min} = \min_i \|v_i - v_0\|_2$, $f_{\max} = \max_i \|v_i - v_0\|_2$ and the difference between them is $e = f_{\max} - f_{\min}$

Next, we will define the Switching Rule of the control algorithm.

Switching Rule 3.4 We assume $\eta = 1$ in (4) when the tracking error $e = f_{\max} - f_{\min} \geq \varepsilon$. It means that the leader will use the feedback from follower $i_c = \arg\max_i \|v_i - v_0\|_2$ to adjust its control algorithm. Similarly we let $\eta = 0$ when the tracking error $e = f_{\max} - f_{\min} < \varepsilon$. In this situation, the leader doesn't use the follower's feedback, then the algorithm (4) degenerates to $u_0(t) = v_d - v_0$.

Remark 3.5 The threshold ε here is different from the threshold ζ in traditional event-triggered control. In previous work, the researchers use the event-triggered control to reduce the updating of the control law and then reduce the consumption, so they should prove that there is a lower limit $\zeta > 0$ to avoid the Zeno Behavior. However, in this paper, the threshold ε and the event-triggered control is used to switch the control algorithm rather than reduce the updating of the control law. Then according to the functions (9) and Switching Rule 3.4, the threshold ε here can be chosen arbitrarily. The following theorem will also prove this conclusion

Construct the matrix $P = \begin{bmatrix} I & r^{-1}I & 0 \\ r^{-1}I & I & -r^{-1}I \\ 0 & -r^{-1}I & k_2 I \end{bmatrix}$. From Lemma 2.3, we can get P is positive definite if $k_2 > 0$ and $r > sqrt(1 + 1/k_2)$. Then we construct the function

$$V(t) = \omega^T(t)P\omega(t) \tag{9}$$

Next, we will propose some theorems to prove that the system can reach the consensus with a smart leader.

Assumption 3.6 ([7]) Assume that the graph considered in this paper is connected. The connection among the followers are undirected and the connection between the smart leader and followers are directed. The leader is a neighbor of at least one follower, which means that all the followers can be affected by the smart leader directly or indirectly.

Theorem 3.7 *For any $k_2 > 1$, suppose the Assumption 3.6 holds. When the switching parameter $\eta = 0$, we have $\dot{V}_1(t) < 0$ with the multi-agent system (1)–(4), if $r > (5 + sqrt(41))/8$ and $k > p_1/\lambda_{\min}$, where $\lambda_{\min}$ is the minimum eigenvalue of the matrix H, and $p_1 = r^2k_2/(4r^2k_2 - 4k_2 - r(1/r + 1)^2)$.*

Proof The matrix (7) can be simplified as $F = F_1 = \begin{bmatrix} -rkH & -kH & -I \\ I & 0 & 0 \\ 0 & 0 & -I \end{bmatrix}$ when $\eta = 0$.

Differentiating the function (9) with respect to t, we have

$$\dot{V}_1(t) = \omega^T(t)(PF_1 + F_1^TP)\omega(t) = \omega^T(t)Q\omega(t) \tag{10}$$

where

$$Q = PF_1 + F_1^TP = \begin{bmatrix} -2krH + 2/rI & -2kH + I & (-1/r)I - I \\ -2kH + I & (-2k/r)H & 0 \\ (-1/r)I - I & 0 & -2k_2I \end{bmatrix}$$

For any $k_2 > 1$, according to Lemma 2.3, we can get the matrix P is positive definite when $r > (5 + sqrt(41))/8 > sqrt(1 + 1/k_2)$. Next, we will prove that the matrix Q is negative definite.

For we have $k,\ r > 0,\ k_2 > 1$, then according to the Lemma 2.3, the matrix Q is negative definite when

$$\frac{1}{k}H^{-1} < \frac{4r^2 - 4 - \frac{r}{k_2}(\frac{1}{r} + 1)^2}{r^2}I \tag{11}$$

Construct the function $f(r) = 4r^2 - 4 - r(1/r+1)^2$, then we have $f(r) > 0$ when $r > r_0 = (5 + sqrt(41))/8$. Because we have $k_2 > 1$, then it must have $4r^2 - 4 - (r/k_2)(1/r+1)^2 > 0$ when $r > r_0$.

Then we can get the matrix Q is negative when $k > r^2/[\lambda_{\min}(H) \cdot (4r^2 - 4 - (r/k_2)(1/r+1)^2)]$ according to the Lemma 2.3. Thus, we have $\dot{V}_1(t) < 0$ with the multi-agent system (1)–(4) when $\eta = 0$.

Remark 3.8 In traditional leader-following system, the control algorithm of leader is always $u_0(t) = 0$. It means the leader will move with a constant velocity and many researchers have studied this problem. However, in the system with a smart leader, the smart leader will use the feedback from the follower to adjust its control law. It means the leader will adjust its velocity to reach the consensus. In other words, the traditional leader is a special case of the smart leader.

From the above theorem, we can get $\dot{V}_1(t) < 0$ when the leader only uses the desire input. Next theorem will show that we can have the similar conclusion when the leader use the desire input as well as the feedback from follower.

Theorem 3.9 *For any $k_2 > 0.5$, suppose the Assumption 3.6 holds. When the parameter $\eta = 1$, we have $\dot{V}_2(t) < 0$ with the multi-agent system (1)–(4), if $r > T/6 + s_4 + (s_1^2 + 2/3)/s_4 + 1/6$ and $k > p_2/\lambda_{\min}$, where $p_2 = r^2/(4r^2 - 4 - 2(2k_2 - 1)r - 4nr(2k_2 - 1)^{-1} - 2r(\frac{1}{r} + 1)^2)$, s_1 and s_4 will be defined in the following provement.*

Proof The matrix (7) can be simplified as

$$F = F_2 = \begin{bmatrix} -rkH - \mathbf{1} \cdot C^T & -kH & -I \\ I & 0 & 0 \\ -\mathbf{1} \cdot C^T & 0 & -I \end{bmatrix} \tag{12}$$

when $\eta = 1$.

Then the Eq. (6) can be rewritten as

$$\dot{\omega}(t) = F_1 \omega(t) + T \tag{13}$$

where

$$T = \left[\left(\sum_{i=1}^{n} c_i(-\bar{v}_i)\right) \cdot \mathbf{1}_\mathbf{n}^T \quad 0_{1 \times n} \quad \left(\sum_{i=1}^{n} c_i(-\bar{v}_i)\right) \cdot \mathbf{1}_\mathbf{n}^T \right]^T.$$

From function (9) we can get

$$\dot{V}_2(t) = \omega^T(t) Q \omega(t) + 2\omega^T(t) \cdot P \cdot T \tag{14}$$

According to Lemma 2.4, we have

$$2\omega^T(t) \cdot P \cdot T \leq a\bar{v}^T\bar{v} + a\hat{v}^T\hat{v} + 2na^{-1}\bar{v}_i^T\bar{v}_i \tag{15}$$

where a is any positive constant and we will define it later.

Using (15), the Eq. (14) can be rewritten as

$$\dot{V}_2(t) \leq \omega^T(t)Q^{'}\omega(t) \tag{16}$$

where $Q^{'} = Q + diag\{aI + 2a^{-1}nD_c, \mathbf{0}, aI\}$ and $D_c = diag\{C\}$.

We have the matrix $Q^{'}$ is negative definite when

$$2r - \frac{2}{r} - (a + 2a^{-1}n) - (\frac{1}{r} + 1)^2(\frac{1}{2k_2 - a}) > \frac{r}{2k}H^{-1} \tag{17}$$

Let $2k_2 - a = 1$, $a = 2k_2 - 1$. The in Eq. (17) can be simplified as

$$\frac{1}{k}H^{-1} < \frac{4r - \frac{4}{r} - 2a - 4a^{-1}n - 2(\frac{1}{r} + 1)^2}{r} \tag{18}$$

Similar to the proof in Theorem 3.7. Let $T = a + 2a^{-1}n$, we can the matrix $Q^{'}$ is negative definite when $r > r_1 = T/6 + s_4 + (s_1^2 + 2/3)/s_4 + 1/6$ and $k > r^2/(\lambda_{\min}(H)$ $(4r^2 - 4 - 2(2k_2 - 1)r - 4nr(2k_2 - 1)^{-1} - 2r(1/r + 1)^2))$ where n is the dimension of the system matrix H and s_1, s_2, s_3, s_4 are only influenced by the parameter T. Then we have the Lyapunov function $\dot{V}_2(t) < 0$ with the multi-agent system (1)–(4) when $\eta = 1$.

Theorem 3.10 *For any $k_2 > 1$, with the Assumption 3.6, the multi-agent system (1)–(4) can reach consensus asymptotically under the Switching Rule 3.4, when $r > \max\{(5 + sqrt(41))/8, T/6 + s_4 + (s_1^2 + 2/3)/s_4 + 1/6\}$ and $k > \max\{p_1/\lambda_{\min}, p_2/\lambda_{\min}\}$.*

Proof From the above theorem, we have the matrix P is positive definite, and $V(t) = \omega^T(t)P\omega(t)$ is the Lyapunov function of the system (1)–(4).

From the function (9), we can get $\dot{V}(t) = \omega^T(t)[PF + F^TP]\omega(t)$.

According to the Switching Rule 3.4, we can prove when the switching parameter $\eta = 0$, the Lyapunov function $\dot{V}_1 < 0$. Because the conditions proposed in Theorem 3.10 can satisfy the demands of the Theorem 3.7.

Similarly, when $\eta = 1$, we can get $\dot{V}_2 < 0$, because the conditions proposed here can also satisfy the demands of the Theorem 3.9.

Then according to the Lemma 2.5, the system can reach consensus asymptotically when the control parameters k_2, k and r satisfy the conditions.

Remark 3.11 Through the theorems proposed above, we can get the systems can reach the consensus asymptotically with a smart leader. Meanwhile, consider that the smart leader can adopt the desire input as well as the follower's feedback.

We consider that the smart leader system has a lower energy consumption than traditional leader system. The simulation in Sect. 4 validate the assumption.

4 Simulation

In this section, we will give some simple examples to verify the performance of the system with smart leader.

For having a quantitative research on the performance of the smart leader, we denote that the system reach consensus asymptotically if the following conditions are satisfied.

$$\begin{cases} \|x_i(t) - x_0(t)\| \leq 0.05 \\ \|v_i(t) - v_0(t)\| \leq 0.01 \end{cases} \tag{19}$$

Next, we will research the consensus performance and the energy cost by the following simulation.

Suppose a system which is consisting of one leader and six followers labelled with $\{0, 1, 2, \ldots, 6\}$. The topology is represented in Fig. 1.

In this system, the leader has two neighboring followers and we design its event-triggered function as:

$$f(x) = \|v_1 - v_0\|_2 - \|v_4 - v_0\|_2 \tag{20}$$

Based on the Switching Rule 3.4, we define the leader uses the Follower 1's feedback to adjust its algorithm when $f(x) \geq \varepsilon$, while it uses the Follower 4's feedback to adjust its algorithm when $f(x) \leq -\varepsilon$. Otherwise, the control law of the leader will be degenerated to $u_0(t) = v_d - v_0$.

Here, we can get the eigenvalues of the matrix H, $\lambda_{\min} = 0.2679$, $\lambda_{\max} = 3.7321$, then we can choose the control parameters $r = 2$, $k_2 = 2$ and $k = 2$ through the conditions proposed in Theorem 3.10. We propose the desire velocity $v_d = 6$. To guarantee the smart leader can switch the control law more efficiently, we choose

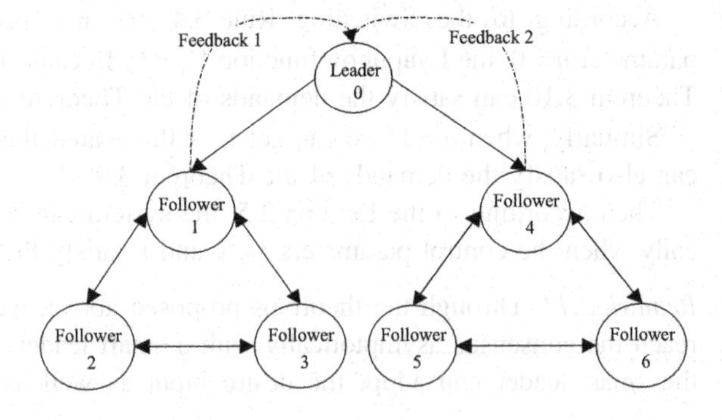

Fig. 1 Graph G with fixed topology

Table 1 The convergence time of two systems

	Smart leader	Traditional leader
Shortest time/s	14.5	25.0
Longest time/s	23.0	34.0
Average time/s	20.2	31.4

the threshold $\varepsilon = 0.05$ here. The initial positions and velocities are randomly chosen within $x \in [-20, 20]$ and $v \in [-10, 10]$. Then through the simulation, we can get the following tables and figures.

From Table 1, we can get the shortest, longest and the average time when the system reach the consensus. It shows that the system with a smart leader needs less time to make the system reach the consensus. The left figure in Fig. 2 is the velocity states of the multi-agent system with smart leader and the right one is the traditional leader, we can get that the system with a smart leader is smoother and stabler during the convergence. In Fig. 3, we let different values represent the different control laws. It shows that the system here reaches our demands that the leader will use the feedback from the follower at a necessary time. Otherwise it will only use the state itself which we let the value equals to zero in Fig. 3.

Besides, the energy consumption of two kinds of systems is also taken into consideration. The defination of the energy consumption of the system is as follow.

Definition 4.1 Define the energy consumption of the leader on time t_0 is $E_i(t_0) = \int_{t_0}^{t_0+\Delta t} u_i^2(t)dt$ and the energy consumption of the follower i is $E_i(t_0) = \int_{t_0}^{t_0+\Delta t} u_i^2(t)dt$. So the energy consumption of the system during $[t_0, t_1]$ is $E = \sum_{i=0}^{i=n} (E_i(t_0) + E_i(t_0 + \Delta t) + \cdots + E_i(t_1))$.

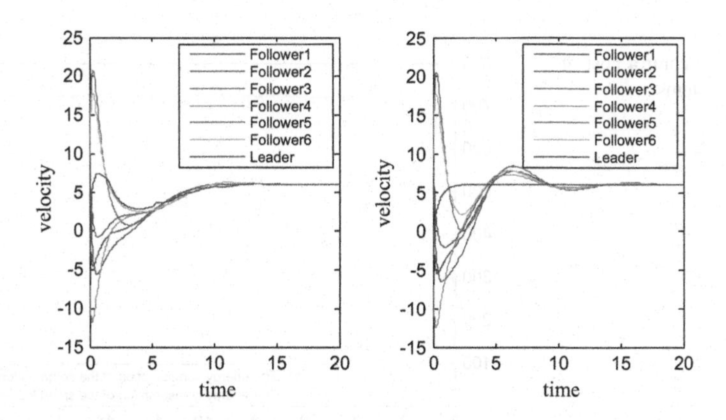

Fig. 2 Curves of velocity states for multi-agent system with smart leader and traditional leader

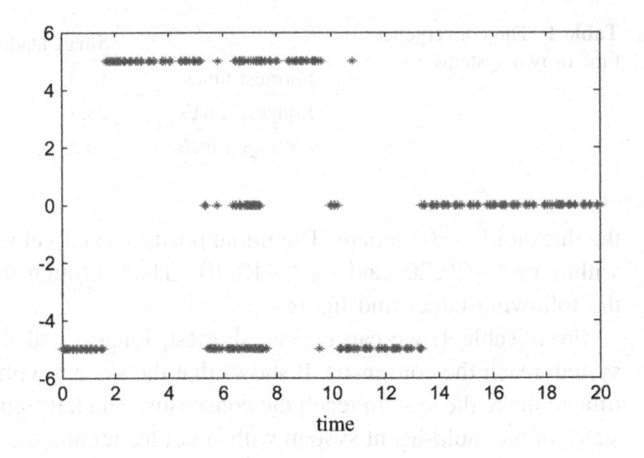

Fig. 3 The situations of different control laws which the smart leader uses

Table 2 The comparison of energy consumption between the smart leader and traditional leader

	Smart leader	Traditional leader
Lowest consumption	2.45×10^2	2.30×10^2
Highest consumption	1.93×10^3	1.95×10^3
Average consumption	8.75×10^2	8.90×10^2

Then, from Table 2 and Fig. 4, we can get the energy consumption comparison of two systems.

From Table 2, we can get the lowest, highest and the average consumption of two kinds of systems. It shows that the system with a smart leader has a lower consumption than traditional leader. From Fig. 4, we can get the same conslusion. The smart leader has a lower consumption than the traditional leader.

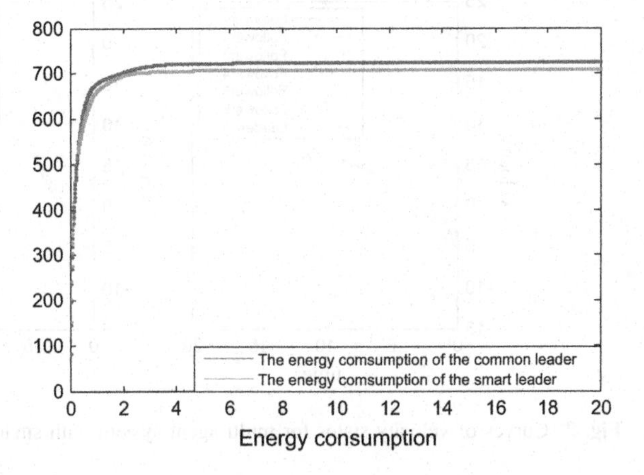

Fig. 4 The control cost of the two systems

5 Conclusion

In this paper, we have studied the consensus problem of leader-following system. First, we design a new kind of leader, smart leader, by adding the feedback of the follower's into leader's control algorithm. We design the smart leader which can use the neighboring follower's feedback and accordingly adjust the control algorithm when the event-triggered functions are satisfied. For the system, a sufficient condition that can guarantee the stability of the system is proposed. For the smart leader can reach the consensus together with followers, it has a lower energy consumption than the system with a traditional leader. Finally, the simulation results have also verified that the smart leader has a better performance than the traditional leader.

References

1. Olfati-Saber R, Murray RM. Consensus problems in networks of agents with switching topology and time-delays. IEEE Trans Autom Control. 2004;49(9):1520–33.
2. Yu W, Chen G, Cao M. Some necessary and sufficient conditions for second-order consensus in multi-agent dynamical systems. Automatica. 2010;46:1089–95.
3. Yang X, Wang JZ, Tan Y. Robustness analysis of leader–follower consensus for multi-agent systems characterized by double integrators. Syst Control Lett. 2012;61:1103–15.
4. Guo WL, Lü JH, Chen SH, Yu XH. Second-order tracking control for leader–follower multi-agent flocking in directed graphs with switching topology. Syst Control Lett. 2011;60:1051–58.
5. Ren W. Second-order consensus algorithm with extensions to switching topologies and reference models. In: Proceedings of the 2007 American control conference Marriott Marquis Hotel at Times Square New York City, USA; 11–13 July 2007.
6. Hong Y, Hu J, Gao L. Tracking control for multi-agent consensus with an active leader and variable topology. Automatica. 2006;42(7):1177–82.
7. Ma ZG, Liu ZX, Chen ZQ. Leader-following consensus of multi-agent system with a smart leader. Neurocomputing. 2016;214:401–8.
8. Gupta S. Increasing the sampling efficiency for a control system. IEEE Trans Autom Control. 1963;8(3):263–4.
9. Liff A, Wolf JK. On the optimum sampling rate for discrete-time modeling of continuous-time systems. IEEE Trans Autom Control. 1966;11(2):288–90.
10. Åström KJ, Bernhardsson B. Comparison of Riemann and Lebesque sampling for first order stochastic systems. In: Proceedings of the Conference on Decision and control, LasVegas, NV, USA;2002. p. 2011–16.
11. Dimarogonas D, Frazzoli E, Johansson KH. Distributed event-triggered control for multi-agent systems. IEEE Trans Autom Control. 2012;57(5):1291–97.
12. Heemels W, Johansson KH, Tabuada P. An introduction to event-triggered and self-triggered control. In: Proceedings of the Conference on Decision and Control, Maui, HI, USA;2012. p. 3270–85.
13. Årzén KE. A simple event-based pid controller. In: Proceedings of the IFAC, Beijing, China;1999. p. 423–8.
14. Boyd S, El Ghaoui L, Balakrishnan V. Linear matrix inequalities in system and control theory. Philadelphia: SIAM; 1994.
15. D. Liberzon, Switched systems. In: Handbook of networked and embedded control systems, Springer;2005. p. 559–74. doi:10.1007/0-8176-4404-0_24.

Visual Based Abnormal Target Annotation with Recurrent Neural Networks

Buyi Yin and Yingmin Jia

Abstract Most traffic accidents are caused by abnormal driving behaviors. Timely annotate the abnormal targets could effectively reduce the accident risk. This paper has proposed a system used on autopilot vehicles, to evaluate the targets' caution level. The target with high caution level will be treated as being abnormally driven so attract more attention from the autopilot algorithm. In this paper, a learning based relative position prediction algorithm is proposed by applying CW-RNN method on digital video data. And by modeling the vehicle dynamics and the camera parameters, a position-caution level mapping is built. The system is demonstrated in experiment with the data from Caltech Pedestrian Dataset. An analysis is done to explore the relation between the caution level and the parameters of the vehicle.

Keywords Autopilot · Caution value · Clockworks RNN · Position prediction

1 Introduction

Achieving autopilot needs to sense the vehicle's environment and react to it by making control decisions. Video camera is one of the most common used sensors on autopilot vehicles. It records the front view in real time and deliver the digital image sequence to the processor. Machine learning algorithm is unavoidable for autopilot nowadays. Most of them treat this observation-reaction system as a whole problem, by which we get directly the control actions. But the result is not good enough to

B. Yin
Ecole Centrale Pekin, Beihang University (BUAA), Beijing 100191, China
e-mail: mathieu_yin@126.com

Y. Jia (✉)
The Seventh Research Division and the Center for Information and Control,
School of Automation Science and Electrical Engineering,
Beihang University (BUAA), Beijing 100191, China
e-mail: ymjia@buaa.edu.cn

Y. Jia et al. (eds.), *Proceedings of 2017 Chinese Intelligent Systems Conference*, Lecture Notes in Electrical Engineering 459,
https://doi.org/10.1007/978-981-10-6496-8_15

replace the driver due to the problem's complexity. A specialized system concentrates on the emergency situations that gives the driver or the control algorithm an alert in advance would be a more realistic and economical choice for now.

This paper has proposed a system to evaluate the caution value of every target quantitatively. The system includes 4 parts, target recognition, prediction model training, movement generating and caution value evaluation. The target recognition is achieved by sparse optical-flow [1] and clustering method [2]. Clockworks RNN [3] is applied to process the time related data and train a model to generating the possible movement sequence in future moments. To achieve the caution level evaluation, a distribution is designed to describe the caution value for every point in the video. By combining the movement generating and the distribution, we can make a judgment of target with abnormal behaviors.

In this paper, we will make an analysis for the problem in Sect. 2, including the precise objectives of the sub-systems and the difficulties we are facing to. The design details of the system will be explained in Sect. 3, including the precise applied methods, the relations between the different parts, and the structures of the models we used. In Sect. 4, an experiment to this system has been done with some specially modified data from Caltech Pedestrian Dataset. An analysis will be made by simulation based on the experiment result. Finally in Sect. 5, we will give out some comments to this system and share some ideas which may improve its performance. We will also make a forecast for the research in the last section.

2 Problem Description

Just like the human does, the judgment of the emergency situation for a autopilot car is usually made in a self-centered way. In this way, we make the judgment according to the relative movements between the environment objects and us. We get objects' movement information by camera settled at the front of our car. These information will be the data source of our system.

The first problem to solve is extracting the position sequence of the target in a dynamic background. This kind of algorithms can be influenced by lots of interference in the environment such as illumination, similarity between the targets and sheltering, etc.

The second problem is predicting the movement of the target according to its position sequence and figure out whether the target will be dangerous or not. To achieve this function, a common movement pattern model will be needed to support the prediction. The prediction accuracy will be influenced by the design structure of the model, the data quality, the training method, etc. A camera-vehicle relation model will be needed to evaluate the targets around in order to support the judgment of the emergency situation. The evaluation result should be related with the camera parameters, the vehicle's speed and the dynamic parameters of the vehicle.

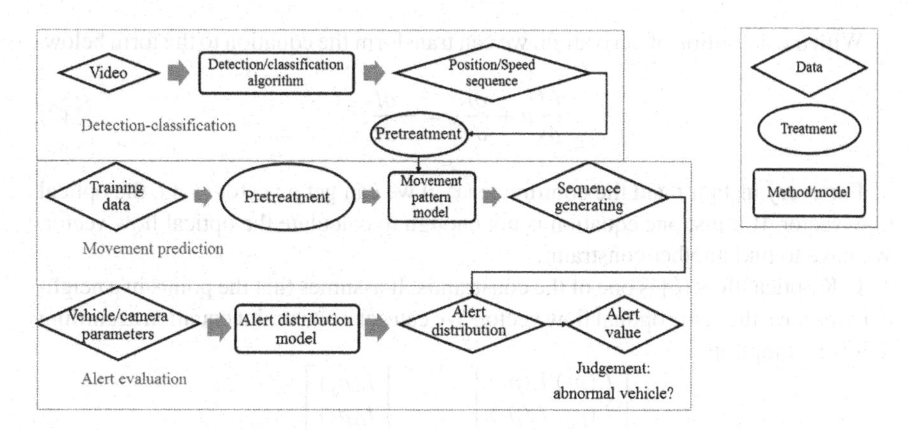

Fig. 1 System structure

3 System Design

The system is designed as Fig. 1.

The whole system can be divided into 3 sub-systems. The Detection-Classification part transforms the image sequences into targets coordinates sequences. The movement prediction part trains a model with the training data, in order to process the inputs from the Detection-Classification part. The alert evaluation part evaluate the dangerous level for every point in image and combines it with the output of the movement prediction. The combination will finally give out an alert value and we should set a dynamic threshold for it to make the abnormal target judgment.

3.1 Detection-Classification

The dynamic background is the main difficulty we should deal with in the detection step. Optical flow [4] is a method designed for extracting the movement information of image sequence and it could deal with the dynamic background. It extracts the image's movement features, which means the speed vectors of every point in the image. The base of optical flow is the gray value constancy assumption. It has been assumed that the gray value of a pixel is not changed by a small displacement in space and time dimension [5].

$$I(x, y, t) = I(x + u, y + v, t + \Delta t) \tag{1}$$

Here $I : \mathbb{R}^3 \rightarrow \mathbb{R}$ represents the gray value in the video, (x, y, t) is the coordinates in 2-dimensions space and time. $(u(x, y, t), v(x, y, t), \Delta t)$ is a vector small enough to avoid the change of the gray value.

With the definition of derivation, we can transform the equation to the form below.

$$\frac{\partial I}{\partial x}u + \frac{\partial I}{\partial y}v = -\frac{\partial I}{\partial t} \tag{2}$$

For every instant t and the coordinate (x, y) we can get a vector (u, v), the optical flow vector. But just one equation is not enough to calculate the optical flow vector. We have to find another constraint.

L-K optical flow [6] is one of the constraints. It assumes that the points in a neighbor area have the same optical flow vector. We can get a constraint equation according to this assumption.

$$\begin{bmatrix} I_x(p_1) & I_y(p_1) \\ I_x(p_2) & I_y(p_2) \\ \cdot & \cdot \\ \cdot & \cdot \\ \cdot & \cdot \\ I_x(p_n) & I_y(p_n) \end{bmatrix} \begin{bmatrix} u \\ v \end{bmatrix} = \begin{bmatrix} I_t(p_1) \\ I_t(p_2) \\ \cdot \\ \cdot \\ \cdot \\ I_t(p_n) \end{bmatrix} \tag{3}$$

In Eq. 3, we assume there are n points in the neighbor area of the point (x, y). I_x, I_y and I_t are the partial derivative of I to x, y, and t.

If we calculate the optical flow for all the points in video, the calculation will be too heavy. We usually apply the sparse method [1] with the L-K optical flow. In this way, we only calculate the features points' optical flow and assume that in the neighbor area of a feature point we have the same optical flow vector.

The sparse L-K optical flow method allow us get a series of feature points with its optical flow vector, which represent its speed vector. The speed vector from the same target are usually similar, if the target do only translation movement. So we can classify the feature points according its speed vector by clustering algorithm [2] and finally get the targets' position and speed sequences.

3.2 Movement Prediction

We predict the movement by sequence generating, which means feeding the output as input at the next prediction step [7]. For every moment t of a target with position and size vector $X_t = (x_1, x_2, s_1, s_2)$, we attempt to predict the vector with same dimension X_{t+1}. Since the vector $X_t \in \mathbb{R}^4$ is represented by real value, it would be more complex to become output and input of a neural networks. We need to transform the calculation of X_t into a classification problem. We call it as pretreatment of the data.

The target in the video is described by its center point coordinate (x_1, x_2) and its horizontal and vertical size (s_1, s_2). These four parameters constitute a representative rectangular centered (x_1, x_2) and with length and width (s_1, s_2). In order to represent these target information in a vector prepared for neural networks, we divide the plan

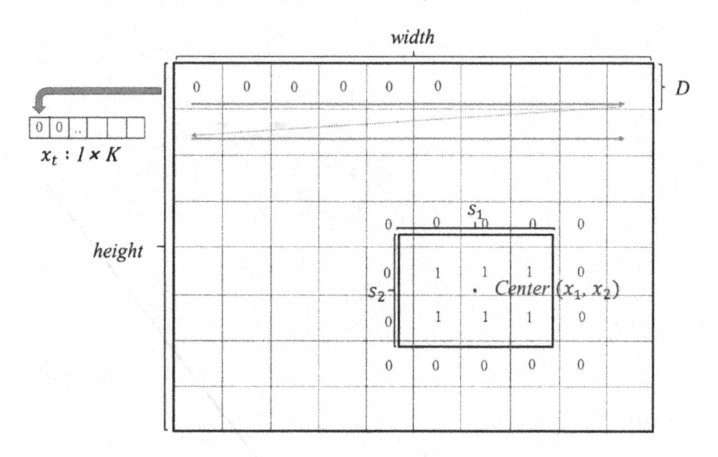

Fig. 2 Pretreatment

$\mathbb{R}^2$ in the (*width* × *height*) video into squares with fixed side length *D*, so we have totally $K = \frac{width \times height}{D^2}$ areas. For the square areas in the target rectangular, we note them as value 1, and the others as 0. After the notation, we fill the squares' values into a $1 \times K$ vector from left to right, from top to bottom. By repeating the treatment in every time step, we get a sequence of $1 \times K$ vectors (Fig. 2).

We note the $1 \times K$ vector at time t as x_t. It is obvious that x_{t+1} should be the correct answer for the prediction of x at moment t. Therefore, we set the training feature vector $X_t = x_{t-1} (t \in [2, T])$ and $X_1 = 0$, the training label vector $Y_t = x_t$ for $t \in [1, T]$, where T is the step number of the training sequence.

These sequence informations recorded by the camera like their positions and speeds is temporally related. So we usually turn to the algorithms designed to process the temporal information. Recurrent Neural Networks [8] (RNN) have the ability to store the information as a short-term memory by their recurrent connections. RNN has been already applied in target tracking [9], prediction [11] and autopilot with an acceptable performances. However, in certain situations the targets' movement pattern changes slowly and have a long time span, which requires a long-term memory to fix the information in recurrent connections and make reliable predictions. Clockworks RNN [3] (CW-RNN) is a method designed to store memories in multiple time scale. CW-RNN divides the hidden layer into several modules with different clock periods to store memories of different terms.

The structure of the CW-RNN is showed in the Fig. 3.

The networks structure is similar to classic RNN. We set $I = K$ nodes in input layer, H nodes in hidden layer, K nodes in output layer. The input layer and the hidden layer are fully connected, so do the output layer and the hidden layer. We bypass some connections in the Fig. 3 in order to show clearly the recurrent connection.

The recurrent connections of the hidden layer in CW-RNN are specially designed. The H nodes are divide into G modules, each module is related to a clock period T_g and we assume that if $i \neq j, T_i \neq T_j$ and for $i < j, T_i < Tj$. The recurrent connections

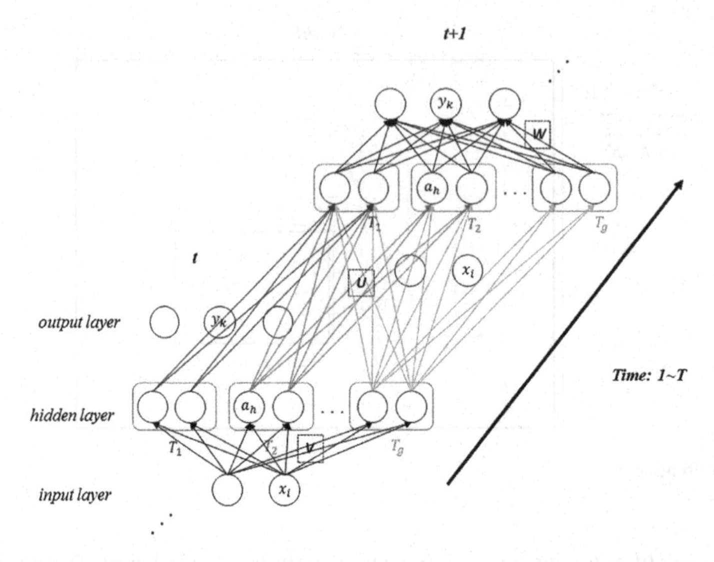

Fig. 3 Clockworks RNN

are settled only from nodes in modules with larger clock period to nodes in modules with smaller clock period. For CW-RNN, at each time step *t*, only the output of the modules *g* satisfy $mod(t, T_g) = 0$ are executed. The designer of CW-RNN has suggested [3] to take $T_g = 2^{g-1}$.

Setting the input vector $X^t = \left[x_1^t \ldots x_i^t \ldots x_I^t\right]$, the output vector $Y^t = \left[y_1^t \ldots y_k^t \ldots y_K^t\right]$, the output value of the hidden layer nodes a_h^t. We have the relations below:

$$a_h^{t+1} = f\left(\sum_{ih} x_i^{t+1} V_{ih} + \sum_{hh'} a_h^t U_{hh'}\right) \quad (4)$$

$$y_k^t = g\left(\sum_{hk} a_h^t W_{hk}\right) \quad (5)$$

Here *f* and *g* are the activation functions. We choose *sigmoid* for *f* and *g* here:

$$g(x) = f(x) = \frac{1}{1 + e^{-x}} \quad (6)$$

CW-RNN do the training and prediction in the similar way with the RNN [8], the difference is that in CW-RNN only the value a_h^t and the error of the activated module would be updated. The rest would copy from the steps before.

The CW-RNN store the short-term memory in the module with fast clock period and the long-term memory in the module with slow clock period. The target's patterns with different time scale would influence the output alternatively in both training and prediction.

The objective of the training work with data sequence from moment 1 to $T0$ is to get U, V and W. For the moment $t > T0$, we have no feature vector input X^t for the networks. The generating is made just based on the values of the hidden layer nodes a_h^t, the output of the latest moment y_k^{t-1} and the networks matrix U, V, W. In the sequence generating, before we feed the output back as input, we should find a function Pr:

$$x_g^t = Pr(y^{t-1}) \tag{7}$$

to define the relation between y^{t-1} and the input for generating at moment t, x_g^t. Usually Pr can be step function with suitable step value or normalization function.

3.3 Alert Evaluation

For every position in the image, we will set a value to evaluate its alert level to the autopilot vehicle. We set the autopilot vehicle at the center of the bottom bound, note $O = (O_x, O_y)$. The alert distribution should satisfy two properties: the point closer to O would have a higher alert level, and it should decrease when the point gets further. The decrease rate should be related to several parameters of the autopilot vehicle: the speed v, the brake acceleration a, the height of the camera to the ground h, the camera view sight angle θ, we call it the decrease property. The alert value should decrease to 0 at a position with certain value of the vertical coordinate y_0 and two positions with horizontal coordinates x_{left} and x_{right}, we call it the boundary limitation property.

The two-dimensional Gaussian distribution satisfies both the decrease property and the boundary limitation property. Since in the Gaussian distribution, the area with difference to the mean value larger than 2σ takes only 2.2%, we can regard the 2σ as the limitation boundary.

For the x_{left} and x_{right} we just chose it according to the width of the autopilot vehicle itself. In the data of this system, we set $\sigma_x = \frac{3}{16}(width\ of\ image)$. As for $y_0 = 2\sigma_y$, we set a model to make the calculation and the final function is:

$$y_0 = 2\sigma_y(v, a, h, \theta) = 4\frac{\arctan(\frac{dist}{h}) - \frac{\pi-\theta}{2}}{\theta h} \tag{8}$$

where

$$dist = \sqrt{(h\tan(\frac{\pi-\theta}{2}))^2 + \frac{v^4}{4a^2}} \tag{9}$$

Applying the σ_x and σ_y into two-dimensional Gaussian distribution, the alert distribution:

$$A(x, y, \sigma_x, \sigma_y) = exp\left(-\frac{(x - O_x)^2}{\sigma_x^2} - \frac{(y - O_y)^2}{\sigma_y^2}\right) \tag{10}$$

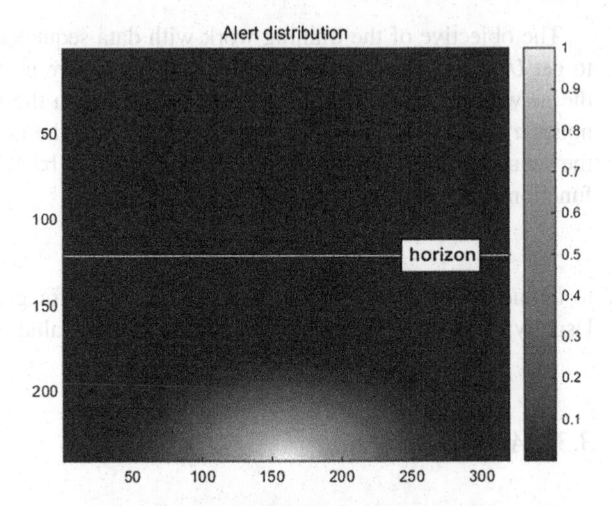

Fig. 4 Alert distribution v = 10 m/s

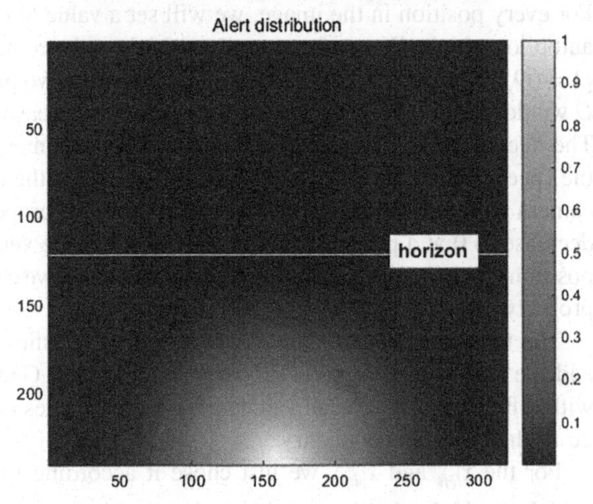

Fig. 5 Alert distribution v = 20 m/s

Figures 4 and 5 show the alert distribution in different speed. In these examples, we set $h = 1.5\,\mathrm{m}, a = 5\,\mathrm{m/s^2}, \theta = \pi/6$.

By Eq. (5), Y^t is the probabilities of the target shows in the K area at the next moment. If we make an inverse calculation of the pretreatment to Y^t, we can get the target position distribution $prob_t(x, y)$ on $\mathbb{R}^{\models}$, along the time t.

Combining A and $prob_t$, we get finally the alert value of a target:

$$alert\_value = \sum_{x,y} A(x, y) prob_t(x, y) \tag{11}$$

4 Simulation and Experiment

4.1 CW-RNN Training

The Caltech Pedestrian Dataset consists of approximately 10 h of 640×480 $30fps$ video taken from a vehicle driving through regular traffic in an urban environment. The Dataset was normally used to train the pedestrian recognition algorithm, but with the Piotr's Matlab Toolbox and Matlab labeling code, we can also label the vehicle targets in the video. The annotation is realized by a bounding box from which we could get the position and the size of the target.

After labeling the Dataset (Figs. 6 and 7) and doing the pretreatment, we could train the CW-RNN.

The data is divided into training data and test data. The test result and the parameters we have set are showed in Fig. 8.

The η_{max} is a designed maximum constraint to the update step length in Gradient Descent. The $Step - slow - coefficient$ is the coefficient to reduce the learning rate η when the Lost return in training recurrence increases. *Error* is the proportion of the square area in which the value of label and the output result are different.

4.2 Sequence Generating

In the simulation, we generate 300 frames for the test sequence, which takes 10 s in the video. The function *Pr* is a composite of normalization function with maximum 1, minimum 0 and step function with step value 0.6. The step value is chosen

Fig. 6 Labelled train data 1

Fig. 7 Labelled train data 2

Learning Rate η	η_{max}	Module Number G	Module Size	Step-slow coefficient	Error
0.002	0.1	10	16	0.9	6.12%

Fig. 8 Parameters and test result

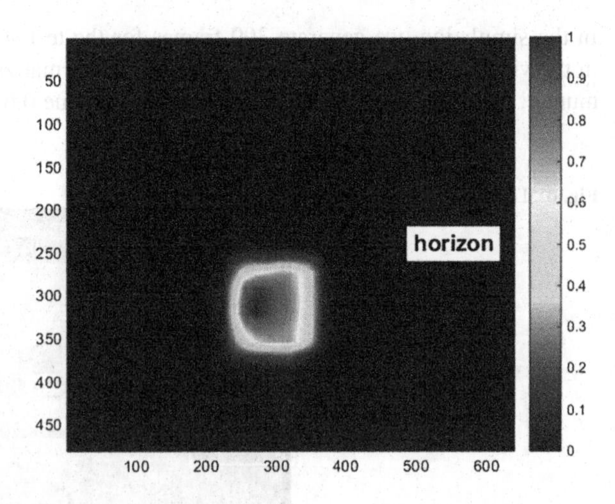

Fig. 9 Generating result at step 150

according to the analysis of test sequence. Figures 9 and 10 show the generating result at certain time steps.

The images show the probability to be occupied by the target for every point in the video view. The higher a point's value is, more possible the point will be occupied.

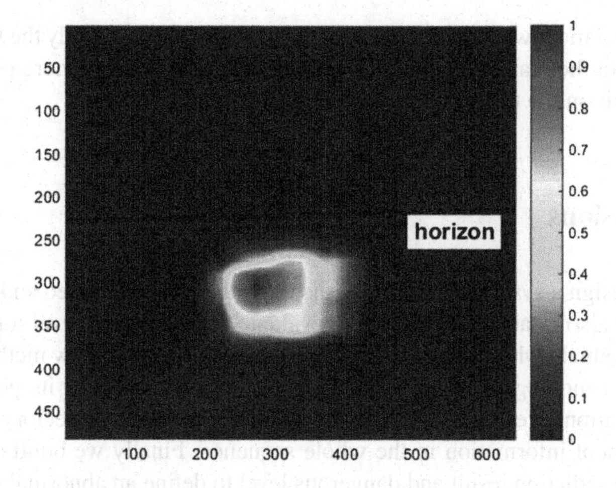

Fig. 10 Generating result at step 300

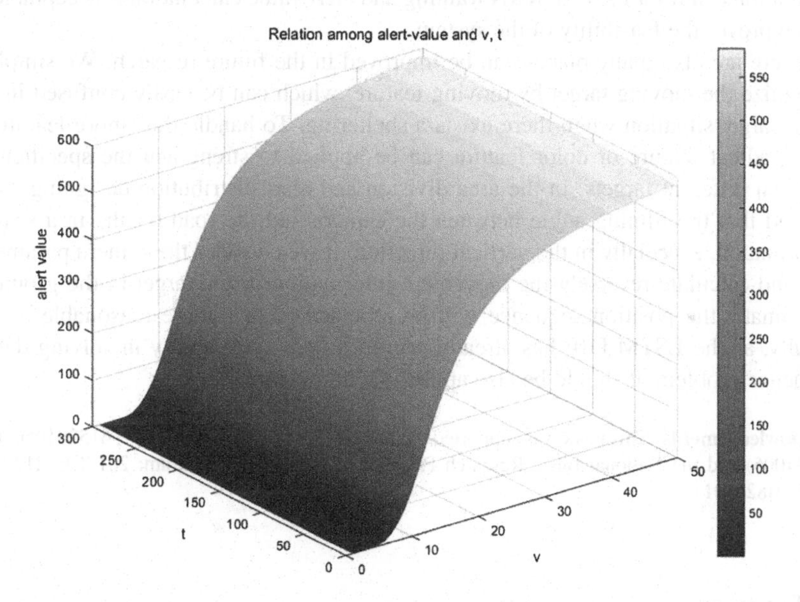

Fig. 11 Alert value

4.3 Alert Value

By combining the result of Sects. 3.3 and 4.2, the alert value can be calculated as a function of v, a, h, θ and time step t.

Assuming the probability of point (x, y) occupied by target at moment t is $prob_t(x, y)$, and $A(x, y, \sigma_x, \sigma_y)$, as in Eq. 10, the alert distribution value of point (x, y). We calculate the alert value $AV(t, \sigma_x, \sigma_y)$ for the target at moment t by Eq. 11.

The calculation result is showed in Fig. 11.

In this simulation, we find the camera's speed v influences strongly the final value. It is obvious in the real situation that the higher the speed is, the more possible the vehicle will hit on the target when there is an emergency situation.

5 Conclusions

This paper design a system to recognize the moving target in a video and predict its movement. It also establishes a model to evaluate the dangerous level for a moving target to the vehicle where the camera is on. The sparse optical flow method is used to recognize a moving target in a dynamic background and offer its position and speed information. We use CW-RNN to generating the future trajectory according to the movement information in the whole sequence. Finally we build a model to combine the prediction result and dangerous level to define an abnormal target. The simulation result for the CW-RNN training and alert value calculation is acceptable, which proves the feasibility of the system.

There are also many places can be improved in the future research. We simply recognize the moving target by moving feature, which can be easily confused in a multi-target situation when there exists a sheltering. To handle this, more features like gradient feature or color feature can be applied to strengthen the specificity between different targets. In the area division and alert distribution designing, we noticed that the filming angle between the camera and the road results in a scale deformation, especially in the vertical direction. If we consider the camera parameters and calculate reversely the movement information of the target in the ground coordinate, the position sequence will be represented in a more reasonable way. Finally, as the LSTM [10] has already proved its powerful ability in solving data sequence problem, it should be also applicable in this problem.

Acknowledgements This work was supported by the NSFC (61327807, 61521091, 61520106010, 61134005) and the National Basic Research Program of China (973 Program: 2012CB821200, 2012CB821201).

References

1. Guo K, Ishwar P, Konrad J. Action recognition using sparse representation on covariance manifolds of optical flow. In: IEEE international conference on advanced video and signal based surveillance;2010. p. 188–95.
2. Jain AK, Dubes RC. Algorithms for clustering data. Technometrics. 1988;32(2):227–9.
3. Koutnk J, Greff K, Gomez F, et al. A clockwork RNN. Comput Sci. 2014;1863–71.
4. Horn BKP, Schunck BG. Determining optical flow. Artif Intell. 1980;17(1–3):185–203.
5. Brox T, Bruhn A, Papenberg N, et al. High Accuracy Opt Flow Estim Based Theory Warping. 2004;3024(10):25–36.

6. Lucas BD, Kanade T. An iterative image registration technique with an application to stereo vision. International Joint Conference on Artificial Intelligence, Morgan Kaufmann Publishers Inc.;1981. p. 674–9.
7. Graves A. Generating Sequences With Recurrent Neural Networks;2014. arXiv:1308.0850v5 [cs.NE].
8. Graves A. Supervised sequence labelling with recurrent neural networks. Berlin: Springer;2012.
9. Milan A, Rezatofighi SH, Dick A et al. Online multi-target tracking using recurrent neural networks;2016. arXiv:1604.03635v1 [cs.CV].
10. Hochreiter S, Schmidhuber J. Long short-term memory. Neural Comput. 1997;9(8):1735.
11. Connor JT, Martin RD, Atlas LE. Recurrent neural networks and robust time series prediction. IEEE Trans Neural Netw. 1994;5(2):240–54.

Analysis of the Fluid Approximation of Stochastic Process Algebra Models

Jie Ding, Xiao Chen and Xinshan Zhu

Abstract With the advent of the Internet of Things (IoT) as a major force of change in industry, Cyber Physical Systems (CPS) is right for building the concept smart Environment. In CPS, the internal computational and physical elements generally interact, reflect and influence each other in order to obtain and analyze human behaviors and their social activities, finally to help them facilitate experiences. Nevertheless, the system complexity and scale become challenges of discrete state modelling formalisms especially in the capability issue. For the stochastic process algebra, performance evaluation process algebra (PEPA), a fluid approximation approach dealing with this problem has been developed, which approximates the continuous time Markov chain underlying a model using ordinary differential equations (ODEs). This paper establishes some basic properties for the ODE based approximation, e.g., uniqueness, existence and boundedness of ODE solutions. Our research in particular presents a convergence of the solutions for nonsynchronised models.

Keywords Convergence · Fluid approximations · PEPA

J. Ding
School of Information Engineering, Yangzhou University, Yangzhou 225127, China

J. Ding
State Key Laboratory for Novel Software Technology, Nanjing University,
Nanjing 210023, China

X. Chen
School of Computer Science and Communication Engineering, Jiangsu University,
Zhenjiang 212013, China

X. Zhu (✉)
School of Electrical and Information Engineering, Tianjin University,
Tianjin 300072, China
e-mail: xszhu@tju.edu.cn

Y. Jia et al. (eds.), *Proceedings of 2017 Chinese Intelligent Systems Conference*, Lecture Notes in Electrical Engineering 459,
https://doi.org/10.1007/978-981-10-6496-8_16

1 Introduction

Performance modelling can help system designers and managers to analyse some aspects like scalability of a system before it is deployed or tested. There are many high-level modelling languages, among them stochastic process algebras (e.g. [1, 8, 9]) are powerful and suitable which model a concurrent system in a compositional approach. Although they have enjoyed considerable success over the last twenty years, they encounter a state space explosion problem that is coursed by the large system scale.

The activity durations, defined in stochastic process algebra (SPA), are usually considered as a exponentially distributed value, which aims to model the memory-less property of a system. As a result, the underlying stochastic models are usually continuous time Markov chains (CTMC) used by quantitative analysis. By solving a balance equation, we can obtain the steady state probability distribution based on a Markov chain in order to derive performance measures, such as throughout and utilization. However, the approach cannot be applied to large scale systems as the state space explosion problem can be encountered in solving the balance equation. As a result, fluid flow approximation has been proposed, as a novel approach, to solve the problem. This approach uses a set of ODEs to approximate the underlying CTMCs, and then deriving approximate performance metrics.

The stochastic process algebra PEPA and its fluid approximation method have been successfully applied to analyze a number of large scale systems, e.g. [3–5]. However, as yet the characteristics of the ODEs generated by PEPA models have not be considered, particularly for the convergence of ODE results when the time is changing to a infinity value. This is addressed in this paper where we also consider the link between ODEs and the Markov chains for a given PEPA model.

Section 2 briefly introduce PEPA and fluid approximations of PEPA models. Some basic properties from the derived ODEs, such as uniqueness, existence and boundedness, will be detailed in Sect. 3. Section 4 demonstrates that the results of ODEs obtained from non-synchronised PEPA models converge to finite limits and these limits are related to the steady state distribution of a Markov chain. Section 5 finally concludes the whole work.

2 Fluid Approximations of PEPA Models

2.1 PEPA Syntax

PEPA (Performance Evaluation Process Algebra) [9] is a high-level modelling language which describe a concurrent system in a compositional manner. The system performance as well as functional properties can be derived or deduced by analysing the corresponding system model using PEPA. The PEPA language extends traditional process algebras, e.g. CCS or CSP through the way of an exponentially distributed value for all actions to represent their durations. So an activity of PEPA model is represented by a notation pair (α, r) in which α and r represent the action type and activity rate respectively. If a component P behaves as Q after completing activity (α, r), then this transition is denoted by $P \xrightarrow{(\alpha,r)} Q$ or $P \stackrel{def}{=} (\alpha, r) \cdot Q$.

Prefix: The prefix combinator gives a designated first action. In the usual interpretation the duration of an action is distributed exponentially based on a parameter r.

Choice: $P + Q$ represents a system that has its activity conduct and behave as P or Q. The activities of both P and Q are enabled. Due to the associated rates that they both have, a *race condition* happens between P and Q, and then the first to complete is kept while the slower one is discarded.

Hiding: For P/L, activities with their action types in L can be considered as the "private" with a notation τ. This combinator gives the type abstraction rather than affecting activity duration.

Cooperation: $P \underset{L}{\bowtie} Q$ means that component P and Q cooperates over action types in the cooperation set L. This combinator forces to synchronise P and Q with action types in set L when the activities are conducted concurrently and independently with some other enabled activities. The rate of synchronised activity depends on the slower cooperated action (see [9] for details). If $L = \emptyset$, then $P \underset{L}{\bowtie} Q$ is denoted by $P \parallel Q$.

Constant: The constant $A \stackrel{def}{=} P$ defines component A's behaviour to be the same to component P.

The structured PEPA semantics is referred to [9]. Governing by these operational rules, a PEPA model could be defined as a labelled multi-transition system with the following expression:

$$\left(\mathcal{C}, \mathcal{A}ct, \left\{ \xrightarrow{(\alpha,r)} |(\alpha, r) \in \mathcal{A}ct \right\}\right).$$

Here $\mathcal{C}$ denotes a set of components, $\mathcal{A}ct$ represents a set of activities, and the multi-relation $\xrightarrow{(\alpha,r)}$ is determined in terms of some rules.

As we have mentioned, the durations of all activities satisfy exponential distributions, leading to the underlying stochastic process of a transition system having Markov properties. So there is a CTMC underlying each PEPA model. Average performance measures of the system rely on the steady state distribution of CTMC being derived through solving the global balance equation using linear algebra. However, as we have pointed out, this is infeasible for large scale models.

For the remainder of this paper, we follow the assumptions in [10] about the considered PEPA models: the same local rates are applied for shared activities of all cooperating components.

2.2 *Fluid Approximation of PEPA Models*

To solve the state space explosion problem that encountered in calculating the steady state distribution, Hillston [10] proposed a fluid approximation approach from two perspectives:

- choosing a more abstract state representation in terms of state variables, quantifying the types of behaviour evident in the model; and
- assuming that these state variables are subject to continuous rather than discrete change.

This approach leading to a set of ODEs approximating the CTMC, which may be used for the performance evaluation of steady state measure and transient measure.

In order to derive the ODEs, We first present the *numerical vector form* that defined in [10], which is a basis of fluid approximations.

Definition 1 (*Numerical Vector Form* [10]) For an arbitrary PEPA model $\mathcal{M}$ with n component types $C_i, i = 1, 2, \ldots, n$, each with N_i distinct derivatives, the numerical vector form of $\mathcal{M}$, $\mathcal{N}(\mathcal{M})$, is a vector with $N = \sum_{i=1}^{n} N_i$ entries. The entry $N(C_{i_j})$ records how many instances of the jth local derivative of component type C_i are exhibited in the current state.

The underlying CTMC has a discrete state space, in which the entries of numerical vector form are non-negative integers and the increment or decrement in each step is one. According to [10], these steps will be relatively small in the condition of large number of components. So the movement between states can be continuously approximated.

Assume a sequential component has a local derivative D of. If D enables an activity (α, r), then (α, r) is called an *exit activity*, i.e. a transition $D \xrightarrow{(\alpha,r)}$ exists in a labelled transition system based on D. The exit activity set of D is denoted as $Ex(D)$. Moreover, the exit activity, denoted by $Ex(\alpha, r)$, is the set of local derivatives for which (α, r). For *entry activity* of D, we similarly define $En(D)$ as well as $En(\alpha, r)$.

Now we show the ODEs derived from PEPA models. Let $N(C_{i_j}, t)$ denote the number of instances used for the jth local derivative of C_i. Then after an approximation for the movement of states, i.e. replacing "δt" with "dt", it can be obtained [10]

$$\begin{aligned} \frac{dN\left(C_{i_j}, t\right)}{dt} = &- \sum_{(\alpha,r)\in Ex\left(C_{i_j}\right)} r \times \min_{C_{k_l}\in Ex(\alpha,r)} \left\{N\left(C_{k_l}, t\right)\right\} \\ &+ \sum_{(\alpha,r)\in En\left(C_{i_j}\right)} r \times \min_{C_{k_l}\in En(\alpha,r)} \left\{N\left(C_{k_l}, t\right)\right\}. \end{aligned} \tag{1}$$

The first part on the right side of Eq. (1) denotes the rates of exit activities while the second part represents the entry activities.

PEPA is the first stochastic process algebra to have a true concurrency semantics via the mapping to ODEs. The true concurrency semantics avoids the state space explosion problem and opens the door to vast applications especially for the large scale systems such as global applications.

3 Existence, Uniqueness and Boundedness of Solutions

In this section, we will show the solutions of (1) not only exist but are unique and bounded.

3.1 Features of ODEs

Let $N\left(C_{i_j}, 0\right)$ be the initial values, and denote $\sum_j N\left(C_{i_j}, 0\right) = N(C_i)$. Because the PEPA model is closed without external exchange; and for all components in any time, all the exit activity rates are equal to all entry activity rates, which leading to the following.

Proposition 1 *For all i,*

$$\sum_j \frac{dN\left(C_{i_j}, t\right)}{dt} = 0, \quad \forall t, \tag{2}$$

or

$$\sum_j N\left(C_{i_j}, t\right) = \sum_j N\left(C_{i_j}, 0\right) = N(C_i), \quad \forall t. \tag{3}$$

Remark 1 Proposition 1 demonstrates the *Conservation Law* in PEPA models and thus in the derived ODEs, i.e. eachtype of component has its amount to be constant at any time.

In Eq. (1), the "exit rates" of C_{i_j},

$$-\sum_{(\alpha,r)\in Ex\left(C_{i_j}\right)} r \times \min_{C_{k_l}\in Ex(\alpha,r)} \left\{N\left(C_{k_l}, t\right)\right\},$$

are related to the number of C_{i_j}. In fact, by the semantics of PEPA, the exit activities of C_{i_j} depend on either C_{i_j} itself or the synchronisation in which C_{i_j} takes part. Consequently $C_{i_j} \in Ex(\alpha, r)$ as long as $(\alpha, r) \in Ex(C_{i_j})$. This implies

$$\min_{C_{k_l} \in Ex(\alpha,r)} \left\{ N\left(C_{k_l}, t\right) \right\} \leq N(C_{i_j}, t). \tag{4}$$

So we have the following

Proposition 2 *For all* C_{i_j},

$$\sum_{(\alpha,r) \in Ex\left(C_{i_j}\right)} r \times \min_{C_{k_l} \in Ex(\alpha,r)} \left\{ N\left(C_{k_l}, t\right) \right\} \leq \sum_{(\alpha,r) \in Ex(\alpha,r)} r \times N(C_{i_j}, t) \tag{5}$$

These two important characteristics derived through the PEPA based fluid approximation, as shown in Propositions 1 and 2, result in good analytic results in the next subsection.

3.2 Existence, Uniqueness and Boundedness of the Solutions

Notice that $\min(\cdot)$ is a Lipschitz function, so according to classical differential equation theory, the solutions of the ODEs from a PEPA model both exist and are unique.

Theorem 1 *The solutions of (1) exist and are unique given the initial values.*

The following theorem further shows that they are bounded.

Theorem 2 *Let* $N\left(C_{i_j}, t\right)$ *satisfy ODEs (1). If the initial values are nonnegative, then the solution is nonnegative, i.e.,*

$$0 \leq N\left(C_{i_j}, t\right) \leq N(C_i), \quad \forall t. \tag{6}$$

If the initial values are positive, then the solutions are positive, i.e.,

$$0 < N\left(C_{i_j}, t\right) \leq N(C_i), \quad \forall t. \tag{7}$$

Proof Notice that $\sum_j N\left(C_{i_j}, t\right) = N(C_i)(\forall t)$, by Proposition 1. The left work is to prove that $N(C_{i_j}, t)$ is nonnegative or positive, given nonnegative or positive initial conditions. The proof includes two different cases.

Case 1: All initial values are assumed to be positive, i.e. $\min_{i_j} \left\{ N(C_{i_j}, 0) \right\} > 0$. It is said that, for all $t \geq 0$,

$$\min_{i_j} \left\{ N(C_{i_j}, t) \right\} > 0.$$

Otherwise, if a $t > 0$ exists such that $\min_{i_j}\left\{N(C_{i_j},t)\right\} \leq 0$, then a point $t > 0$ definitely exists such that

$$\min_{i_j}\left\{N(C_{i_j},t)\right\} = 0.$$

Denote

$$t^* = \inf_{t>0}\left\{\min_{i_j}\left\{N(C_{i_j},t)\right\} = 0\right\},$$

then $0 < t^* < \infty$. Without loss of generality, it is assumed that $N(C_{1_1},t)$ reaches 0 at t^*, i.e.,

$$N(C_{1_1},t^*) = 0, N(C_{i_j},t^*) \geq 0 \ \ (i_j \neq 1_1)$$

and

$$N(C_{i_j},t) > 0, \quad t \in [0,t^*), \quad \forall i,j.$$

Thus, for $t \in [0,t^*]$, by Proposition 2,

$$\begin{aligned}
\frac{dN(C_{1_1},t)}{dt} &= -\sum_{(\alpha,r)\in Ex\left(C_{1_1}\right)} r \times \min_{C_{k_l}\in Ex(\alpha,r)} \left\{N\left(C_{k_l},t\right)\right\} \\
&\quad + \sum_{(\alpha,r)\in En\left(C_{1_1}\right)} r \times \min_{C_{k_l}\in En(\alpha,r)} \left\{N\left(C_{k_l},t\right)\right\} \\
&\geq -\sum_{(\alpha,r)\in Ex\left(C_{1_1}\right)} r \times N(C_{1_1},t) \\
&= -N(C_{1_1},t) \sum_{(\alpha,r)\in Ex\left(C_{1_1}\right)} r.
\end{aligned}$$

Let $R = \sum_{(\alpha,r)\in Ex\left(C_{1_1}\right)} r$, then

$$\frac{dN(C_{1_1},t)}{dt} \geq -RN(C_{1_1},t),$$

or

$$\frac{dN(C_{1_1},t)}{N(C_{1_1},t)} \geq -Rdt,$$

which implies

$$N(C_{1_1},t^*) \geq N(C_{1_1},0)e^{-Rt^*} > 0.$$

This contradicts to $N(C_{1_1}, t^*) = 0$. Thus,

$$0 < N\left(C_{i_j}, t\right) \leq N(C_i), \quad \forall t.$$

Case 2: Suppose $\min_{i_j} \left\{ N(C_{i_j}, 0) \right\} = 0$. Let $u_\delta(i_j, 0) = N(C_{i_j}, 0) + \delta$ and $u_\delta(i_j, t)$ satisfy (1). By the proof of Case 1, $u_\delta(i_j, t) > 0 (\forall t \geq 0)$. Noting $\min(\cdot)$ is a Lipschitz function, by the Fundamental Inequality in [11] (page 14), we have

$$|u_\delta(i_j, t) - N(C_{i_j}, t)| \leq \delta e^{Kt}, \tag{8}$$

where K is a Lipschitz constant. So for any given $t \geq 0$,

$$N(C_{i_j}, t) \geq u_\delta(i_j, t) - \delta e^{Kt} > -\delta e^{Kt}. \tag{9}$$

Let $\delta \downarrow 0$ in (9), then we have $N(C_{i_j}, t) \geq 0$, which completes the proof.

Remark 2 Theorem 2 shows that each type of components just changes its states while keeping the total number unchanged. The component amount of a particular state for any time is between 0 and the total number. The method of the proof can also be applied to the ODEs in [2, 6], which are derived by different semantic mappings but for which Propositions 1 and 2 still hold.

4 Convergence of Solutions: Without Synchronisation

Regrading some particular PEPA models, i.e., they have no synchronisations, we will show that the ODE solutions converge to finite limits when the time grows to infinity.

Without loss of generality, for this class of models, we may suppose that each system has unique component type C. Factually, if a system has several types of components, then ODEs regarding these types of components are treated independently and thus be separated, because of no interactions between the different kinds of components. Therefore, just the unique component C type is assumed in the system, and C is supposed to have k states: $C_1, C_2, \ldots, C_k$. Then Eq. (1) becomes

$$\frac{dN\left(C_j, t\right)}{dt} = -\sum_{(\alpha,r)\in Ex(C_j)} r \times N\left(C_j, t\right) + \sum_{\substack{(\alpha,r)\in En(C_j)\\ C_l\in En(\alpha,r)}} r \times N\left(C_l, t\right), \tag{10}$$

where $j = 1, 2, \ldots, k$. Since (10) generates linear ODEs, so Eq. (10) can be rewritten with a new matrix form:

$$\frac{d\left(N(C_1,t),\ldots,N(C_k,t)\right)}{dt} = \left(N(C_1,t),\ldots,N(C_k,t)\right)Q, \tag{11}$$

where $Q = \left(q_{ij}\right)$ is a $k \times k$ matrix. The coefficient matrix Q has the following properties.

Proposition 3 *$Q = \left(q_{ij}\right)_{k\times k}$ in (11) is an infinitesimal generator matrix, satisfying*

1. $0 \le -q_{ii} < \infty$ *for all* i;
2. $q_{ij} \ge 0$ *for all* $i \ne j$;
3. $\sum_{j=1}^{k} q_{ij} = 0$ *for all* i.

Proof It is clear to see that item 1 and item 2 both hold with Proposition 2. We only need to prove item 3. According to Proposition 1,

$$\frac{dN\left(C_1\right)}{dt} + \frac{dN\left(C_2\right)}{dt} + \cdots + \frac{dN\left(C_k\right)}{dt} = 0, \quad \forall t,$$

which means

$$N(C_1,t)\sum_{j=1}^{k} q_{1j} + N(C_2,t)\sum_{j=1}^{k} q_{2j} + \cdots + N(C_k,t)\sum_{j=1}^{k} q_{kj} = 0, \quad \forall t, \tag{12}$$

which implies $\sum_{j=1}^{k} q_{ij} = 0$ for all i. The proof is completed.

The generator matrix $Q_{k\times k}$ demonstrated in Proposition 3 may not be the generator matrix of the "original" CTMCs. The distinction is very clear. Because the system has N components, the state space size of the original CTMC is k^N, and thus the dimension of associated generator matrix is $k^N \times k^N$, in contrast to the dimension $k \times k$ of this generator matrix $Q_{k\times k}$, which corresponds to the CTMC of the model having only "one" component. To make a difference, this CTMC refers to as a "standard" CTMC.

Remark 3 This proposition illustrates the relationship between the ODEs and the infinitesimal generator matrix of a model component. If the system has only "one" component, then (11) becomes the probability distribution evolution equations of the CTMC which has the rate transition matrix Q. This allows us to use the Markov theory to prove analytic results.

We know, by Theorem 2, that the solutions of (10) are bounded between 0 and the total number of instances of component C. Furthermore, we may ask whether these solutions converge as time tends to infinity. A positive answer can be obtained from the theorem given below.

Theorem 3 *Suppose $N\left(C_j, t\right)$ ($j = 1, 2, \dots, k$) satisfy (10), then for any given initial values $N\left(C_j, 0\right) \geq 0$($j = 1, 2, \dots, k$), we have*

$$\lim_{t\to\infty} N(C_j, t) = N(C_j, \infty), \quad j = 1, 2, \dots, k. \tag{13}$$

Proof Construct a CTMC[1] which has state space $S = \{C_1, C_2, \dots, C_k\}$, infinitesimal generator matrix Q in (11) and initial probability distribution $\pi(C_j, 0) = \frac{N(C_j,0)}{N}$($j = 1, 2, \dots, k$). Based on Markov theory, the transient probability distribution $\pi(C_j, t)$($j = 1, 2, \dots, k$) obtained from the new CTMC fits the following differential equation:

$$\frac{d\left(\pi(C_1, t), \dots, \pi(C_k, t)\right)}{dt} = \left(\pi(C_1, t), \dots, \pi(C_k, t)\right) Q. \tag{14}$$

The new CTMC is irreducible, so it has a steady state distribution, i.e.,

$$\lim_{t\to\infty} \pi(C_j, t) = \pi(C_j, \infty), \quad j = 1, 2, \dots, k. \tag{15}$$

Note that $\frac{N(C_j,t)}{N}$ also satisfies (14) and the initial value $\frac{N(C_j,0)}{N}$ equals $\pi(C_j, 0)$, on the basis of the unique solutions of (14), it can be obtained

$$\frac{N(C_j, t)}{N} = \pi(C_j, t), \quad j = 1, 2, \dots, k, \tag{16}$$

and hence,

$$\lim_{t\to\infty} N(C_j, t) = N\pi(C_j, \infty), \quad j = 1, 2, \dots, k. \tag{17}$$

Remark 4 If the system has m types of components: $C_1, C_2, \dots, C_m$, each with $k_1, k_2, \dots, k_m$ component states respectively, since there is no interaction between different types of components the derived ODEs can be written in the following forms:

$$\frac{d\left(N(C_{i_1}, t), \dots, N(C_{i_{k_i}}, t)\right)}{dt} = \left(N(C_{i_1}, t), \dots, N(C_{i_{k_i}}, t)\right) Q_{k_i \times k_i}, \tag{18}$$

where $i = 1, 2, \dots, m$. Moreover, each $Q_{k_i \times k_i}$ ($i = 1, 2, \dots, m$) is an infinitesimal generator matrix. By Theorem 3, for type C_i we have

$$\lim_{t\to\infty} N(C_{i_j}, t) = N\pi(C_{i_j}, \infty), \quad j = 1, 2, \dots, k_i, \tag{19}$$

where $\{\pi(C_{i_j}\}_{j=1,2,\dots,k_i}$ ($i = 1, 2, \dots, m$), are the corresponding steady state distributions.

[1]This CTMC is essentially the same to the "standard" CTMC.

Remark 5 IT was pointed out in [7], for some special PEPA models, the consistence between derived ODEs and its corresponding CTMCs, in the sense that the equilibrium solutions coincide with the steady state probability distributions. Here we reveal in this theorem that this kind coincidence is universal for PEPA models without synchronisation.

5 Conclusions

This paper provides some basic outcomes regarding a fluid approximation on PEPA. The solutions of derived ODEs are demonstrated to be existent, unique, nonnegative and bounded. Moreover, we also present the convergence of the solutions for non-synchronised cases as well as relationships with the Markov chain associated steady state distribution.

Acknowledgements The authors acknowledge the financial support by the NSF of China under Grant No 61472343, and the NSF of Jiangsu Province of China under Grant BK20151314 and BK20160543.

References

1. Bernardo M, Gorrieri R. A tutorial on EMPA: a theory of concurrent processes with nondeterminism, priorities, probabilities and time. Theor Comput Sci. 1998;202:1–54.
2. Bradley JT, Hillston J. Quantitative analysis of PEPA models through continuous state-space approximation. LFCS, School of Informatics, University of Edinburgh;2007.
3. Chen X, Wang LM. A cloud-based trust management framework for vehicular social networks. IEEE Access. 2017;5:2967–80. doi:10.1109/ACCESS.2017.2670024.
4. Chen X, Wang LM. Exploring fog computing based adaptive vehicular data scheduling policies through a compositional formal method-PEPA. IEEE Commun Lett. 2017;21(4):745–8.
5. Ding J. A comparison of fluid approximation and stochastic simulation for evaluating content adaptation systems. Wirel Pers Commun. 2015;84(1):231–50.
6. Geisweiller N, Hillston J, Stenico M. Relating continuous and discrete PEPA models of signalling pathways. Theor Comput Sci. 2008;404(1–2):97–111.
7. Gilmore S. Continuous-time and continuous-space process algebra. In: Process algebra and stochastically timed activities (PASTA'05);2005.
8. Götz N, Herzog U, Rettelbach M. TIPP—a language for timed processes and performance evaluation. Technical Report 4/92, IMMD7, University of Erlangen-Nörnberg, Germany, Nov 1992.
9. Hillston J. A compositional approach to performance modelling (PhD thesis). Cambridge University Press;1996.
10. Hillston J. Fluid flow approximation of PEPA models. In: International conference on the quantitative evaluation of systems (QEST'05). IEEE Computer Society;2005.
11. Hubbard JH, West BH. Differential equations: a dynamical systems approach (higher-dimensional systems). In: Texts in applied mathematics, No. 18. Springer;1990.

A Wall Interactive System Based on Infrared Electronic Pen

Haonan Xu, Lei Yu, Le Zhang and Jinhong Liu

Abstract In this paper, a new method of interactive system based on infrared electronic pen is proposed to achieve accurate tracking of the target trajectory with anti-jamming ability. Projection technology, infrared sensor, dynamic capture, image processing and other technologies are synthesized in this system. The interactive area is created intelligently and the touch event is determined, and the Improved Mean Shift algorithm is utilized for image information tracking. The simulation of the mouse function can make any wall into a touch screen. With the integration of whiteboard technology, a good human-computer interaction effect can be achieved.

Keywords Interactive system · Target trajectory · Projection technology · Infrared sensor · Improved Mean Shift algorithm

1 Introduction

Human-computer interaction technology is the technology that achieves the natural interaction between human and computer. With the rapid development of projection technology and multimedia technology, human-computer interaction system is more and more applied in our lives [1–9]. Human-computer interaction technology in museum and exhibition allows people to better accept the popular science knowledge and commodity information. However, with the improvement of people's living standards, the requirements on performance of the human-computer interaction system get higher. Now the major technology giants are focused on human-computer interaction, a hot area of computer science, it will dominate the development trend of the computer in the future. Infrared technology, touch and somatosensory technology can make human-computer interaction simplify and direct [8, 9]. The Xbox360 peripherals Kinect which is developed by Microsoft's

H. Xu · L. Yu (✉) · L. Zhang · J. Liu
School of Mechanical and Electric Engineering, Soochow University, Suzhou, China
e-mail: slender2008@163.com

Y. Jia et al. (eds.), *Proceedings of 2017 Chinese Intelligent Systems Conference*, Lecture Notes in Electrical Engineering 459,
https://doi.org/10.1007/978-981-10-6496-8_17

game department inspires us to use the infrared technology, touch and somatosensory technology to achieve small screen and immobile times [10, 11]. The touch projection device based on human-computer interaction control comes into being which is Portable, removable and large screen. However the current human-computer interaction system tracking the target track is not accurate, and anti-interference ability is poor, so it can't meet real-time accurate human-computer interaction goals.

In this paper, the hardware selects 3D sensors (Kinect) and infrared electronic pen, projection technology, infrared sensor, dynamic capture, image processing and other technologies are mixed in this system. The interactive area is created intelligently and determine the touch event. The Improved Mean Shift algorithm is used for image information tracking, and the simulation of the mouse function can make any wall into a touch screen. Through the interactive software platform simulation, it is found that the wall interactive system based on infrared electronic pen can accurately track the positioning of infrared electronic pen, with high precision and anti-jamming. The interactive experience with the integration of electronic Whiteboard technology has comfortable interactive effect.

2 Working Principle

2.1 Introduction to System Principles

The system function diagram is shown in Fig. 1.

The hardware of system is based on the Kinect 3D somatosensory camera (developed by "Project Natal"), and it imports real-time dynamic capture, image recognition, microphone input, voice recognition, community interaction and other functions to accurately identify human dynamics and capture in real time [10, 11]. The infrared pen operates on the interactive plane and forms an infrared spot on the interaction plane. Then the Kinect depth of the camera will scan the interactive plane, and scan the signal to the computer. The computer will filter the received

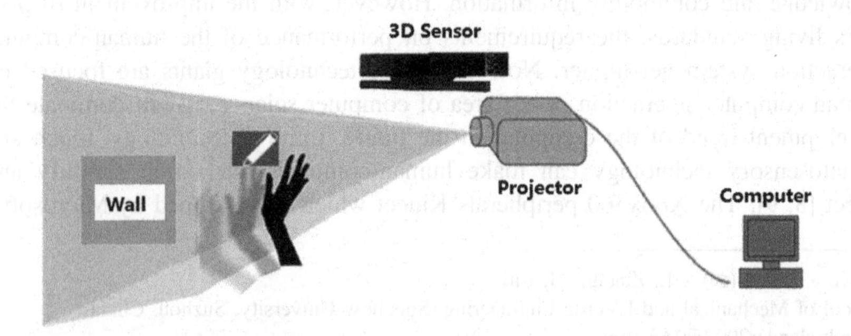

Fig. 1 Schematic diagram of the system

data first, and then transfer the processed data to the control software. System software will control interactivity on interactive plane in order to achieve human-computer interaction.

2.2 The Working Principle of Infrared Electronic Pen

The Infrared electronic-pen is a set of mouse-operated and hand-written authentic inputs in a new input device (Fig. 2). It can be written on any plane media and operation which uses infrared technology to pinpoint, not needing a dedicated writing board.

The Infrared electronic pen system consists of three main components: electronic pen, computer and Kinect. Electronic pen is for transmitting a specific wavelength of the infrared signal; Kinect is employed to receive and deal with infrared signals; Computer electronic pen processing software is used to deal with the information received by Kinect. The processed information will convert into electronic-pen writing position information, and will display on the computer.

Its working principle is described as follows:

(1) The communication between the computer and Kinect is through the USB interface; The communication between electronic pen and Kinect is through infrared technology.
(2) Electronic pen uses infrared technology to locate the pen trajectory accurately for accurate capture.

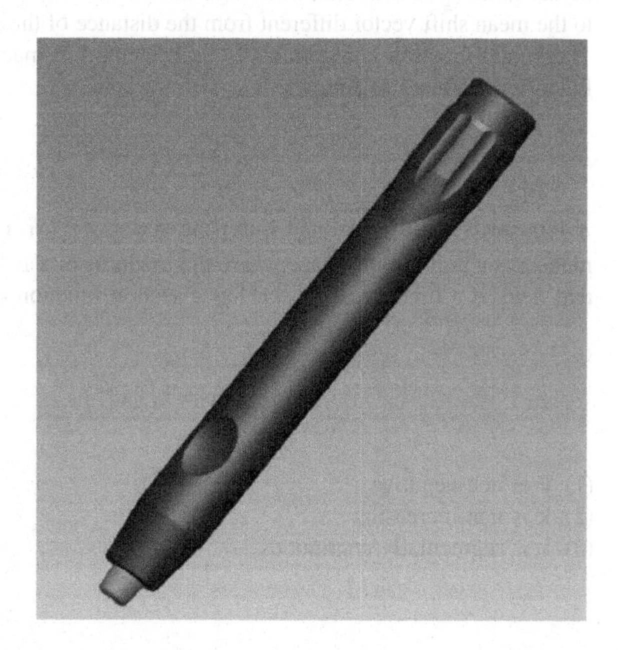

Fig. 2 Infrared electronic-pen

(3) Press the electronic pen button, and this will wake up the electronic pen work. Then the electronic pen will launch single pulse wave infrared signal in accordance with the agreed infrared format.
(4) The infrared signal which received by Kinect is converted into a single saw-tooth pulse after the amplification, filtering, shaping later, and then transfers to the microprocessor time capture pin, the microprocessor would capture the infrared time value.
(5) The electronic pen software on the computer translates the information of the pen's coordinates into points.
(6) Electronic pen will continuously shoot single pulse signal in accordance with the agreed infrared interval after the start of work, the system repeated the above (2)–(5) process. If you write with an electronic pen, then the computer screen will display almost the same words as the original handwriting.

3 Improved Mean Shift Algorithm

The current human-computer interaction system is not accurate to track the target trajectory, and its anti-interference ability is poor, so it can not achieve the goal of real-time accurate human-computer interaction. In this paper, the modified Mean Shift algorithm is used to track the image information. The Mean Shift algorithm, also known as the mean shift algorithm, is widely used in clustering, image smoothing, segmentation and video tracking [12–16]. The purpose of introducing a kernel function in the Mean Shift algorithm is to make the contribution of the offset to the mean shift vector different from the distance of the sample to the offset point. The kernel function is a commonly used method in machine learning. The kernel function is defined as follows:

$$K(x) = k\left(\|x\|^2\right)$$

X represents a d-dimensional European space. $x = \{x_1, x_2, x_3, \ldots, x_d\}$, in this formula, x is a point in the space where the modulus of x is $||x||^2 = xx^T$. R represents a real field. If a function $K: X \to R$ has a section function k: $[0, \infty] \to R$. that is

$$K(x) = k\left(\|x\|^2\right) \tag{1}$$

(1) k is non-negative
(2) k is non-increasing
(3) k is segmentally continuous

Then, the function $K(x)$ is called a kernel function.

The Gaussian kernel function is a commonly used kernel function. The Gaussian kernel function is as follows:

$$N(x) = \frac{1}{\sqrt{2\pi}h} e^{-\frac{x^2}{2h^2}} \tag{2}$$

For the Mean Shift algorithm, it is an iterative step by first calculating the offset mean of the current point, moving the point to the offset mean, and then using this as the new starting point to continue moving until the final condition is satisfied. In the Mean Shift algorithm, the most critical is to calculate the offset mean for each point, and then update the position of the point based on the newly calculated offset mean.

For n sample points $x_i, i = 1, \ldots, n$, in a given d-dimensional space R^d, the basic form of the Mean Shift vector for point x is:

$$M_h(x) = \frac{1}{k} \sum_{x_i \in S_h} (x_i - x) \tag{3}$$

where, S_h is a high-dimensional sphere with a radius of h. S_h is defined as:

$$S_h(x) = \left(y \middle| (y - x)(y - x)^T \leq h^2\right) \tag{4}$$

There is a problem with this basic Mean Shift form: in the area of S_h, the contribution of each point to x is the same. In fact, this contribution is related to the distance between x and each point. At the same time, for each sample, its importance is not the same. Based on the above considerations, the kernel function and the sample weight are added to the basic Mean Shift vector form, and the following modified Mean Shift vector form is obtained:

$$M_h(x) = \frac{\sum_{i=1}^{n} G_H(x_i - x) w(x_i)(x_i - x)}{\sum_{i=1}^{n} G_H(x_i - x) w(x_i)} \tag{5}$$

Among them:

$$G_H(x_i - x) = |H|^{-\frac{1}{2}} G\left(H^{-\frac{1}{2}}(x_i - x)\right) \tag{6}$$

$G(x)$ is a unit of kernel function. H is a positive definite symmetric $d \times d$ matrix, called the bandwidth matrix, which is a diagonal matrix .$w(x_i) \geq 0$ is the weight of each sample. The form of the diagonal matrix H is:

$$H = \begin{pmatrix} h_1^2 & 0 & \cdots & 0 \\ 0 & h_2^2 & \cdots & 0 \\ \vdots & \vdots & & \vdots \\ 0 & 0 & \cdots & h_d^2 \end{pmatrix}_{d \times d} \tag{7}$$

The above Mean Shift vector can be rewritten as:

$$M_h(x) = \frac{\sum_{i=1}^{n} G(\frac{x_i - x}{h_i}) w(x_i)(x_i - x)}{\sum_{i=1}^{n} G(\frac{x_i - x}{h_i}) w(x_i)} \tag{8}$$

Mean Shift vector $M_h(x)$ is a normalized probability density gradient. In the Mean Shift algorithm, the probability density is actually used to obtain the local optimal solution of the probability density. For a probability density function $f(x)$, the known d-dimensional space of n sampling points $x_i, i = 1, \cdots, n$, the kernel function of $f(x)$ is estimated:

$$\hat{f}(x) = \frac{\sum_{i=1}^{n} K(\frac{x_i - x}{h}) w(x_i)}{h^d \sum_{i=1}^{n} w(x_i)} \tag{9}$$

where, $w(x_i \geq 0)$ is a weight assigned to the sample point x_i, and $K(x)$ is a kernel function. The estimate of the gradient $\nabla f(x)$ of the probability density function $f(x)$ is:

$$\nabla \hat{f}(x) = \frac{2 \sum_{i=1}^{n} (x - x_i) k'\left(\left\| \frac{x_i - x}{h} \right\|^2 \right) w(x_i)}{h^{d+2} \sum_{j=1}^{n} w(x_i)} \tag{10}$$

Let $g(x) = -k'(x)$, $G(x) = g\left(\|x\|^2 \right)$, there are

$$\begin{aligned} \nabla \hat{f}(x) &= \frac{2 \sum_{i=1}^{n} (xi - x) G\left(\left\| \frac{x_i - x}{h} \right\|^2 \right) w(x_i)}{h^{d+2} \sum_{i=1}^{n} w(x_i)} \\ &= \frac{2}{h^2} \left[\frac{\sum_{i=1}^{n} G(\frac{x_i - x}{h}) w(x_i)}{h^d \sum_{i=1}^{n} w(x_i)} \right] \cdot \left[\frac{\sum_{i=1}^{n} (x_i - x) G\left(\left\| \frac{x_i - x}{h} \right\|^2 \right) w(x_i)}{\sum_{i=1}^{n} G(\frac{x_i - x}{h}) w(x_i)} \right] \end{aligned} \tag{11}$$

Among them, the second square brackets is the Mean Shift vector, which is proportional to the probability density gradient. Mean Shift vector correction results are as follows:

$$M_h(x) = \frac{\sum_{i=1}^{n} G\left(\left\| \frac{x_i - x}{h} \right\|^2 \right) w(x_i) x_i}{\sum_{i=1}^{n} G(\frac{x_i - x}{h}) w(x_i)} - x \tag{12}$$

Mind: $m_h(x) = \frac{\sum_{i=1}^{n} G\left(\left\|\frac{x_i - x}{h}\right\|^2\right) w(x_i) x_i}{\sum_{i=1}^{n} G(\frac{x_i - x}{h}) w(x_i)}$, then the above formula becomes:

$$M_h(x) = m_h(x) + x \tag{13}$$

This is consistent with the gradient rise process.
The algorithm of the Mean Shift algorithm is as follows:

- Calculate $m_h(x)$
- Let $x = m_h(x)$
- If $\|m_h(x) - x\| < \varepsilon$, end the loop, otherwise repeat the above steps.

4 The Wall Interactive System Design and Simulation

4.1 *Create Interactive Area*

Firstly, we use the 3D sensor (Kinect) to create a desktop interactive platform [10, 11]. After testing, the smallest distance of Kinect to the wall is limited to 0.75 m, and the maximum distance determines the calculation accuracy of the touch point. A recommended distance is 1 m. Also need to ensure that Kinect will not be moved after installation and will not be disturbed. Even if the small vibration will cause interference to the depth value, reduce the accuracy of interaction. If we take a wall as an interactive surface, the above rules are still applicable, but we must be very careful to confirm Kinect installation location. Kinect face the view of the interactive plane, and it can not be completely blocked by the user's shelter.

4.2 *Determine the Touch Event*

To Develop the interactive project based on the Microsoft template code SkeletalViewer, we use C++ language and .NET framework4.0. Environment to prepare the source code. Through the particle analysis algorithm, we can confirm which points are real "touch", which is interference. At this point, the data should no longer contain noise. Having found a "touch" point, we need to confirm the corresponding location of these point the in the actual space. By touching the touch point to the affine transformation, we can transform the touch point of the depth space into a desktop touch point. Get the original point from Kinect and then make some conversions to determine the scope of the interaction based on the range of interactions you know in advance. Finally, a filter is set up to determine frame by frame whether the gestures represented by these points represent the keyboard or which mouse events, respectively. This set of systems can replace the mouse function, select, single and double, drag, the basic ability to simulate all the mouse operation. As shown in Fig. 3, the pen can achieve the mouse drag and drop function. From the red circle marked in the figure, you can see that the pen tip and the mouse cursor coincide. The error is minimal.

With the action pen replacing the role of the mouse, we can greatly improve the friendly human-computer interaction, highlight the interaction between man-machine interaction. So that users can be more intuitive to take control of the system, take the passive information into a user's active knowledge and acceptance, as shown in Fig. 4.

The more important use of the human-computer interaction system equipment is that it is a real "pen". In the projection of the PPT, the user can directly use the pen in the above mark, draw the key, modify, write, as shown in Fig. 5. It is more intuitive than the mouse, and the demo is better, breaking the mouse's limit. It can be seen in the following toolbar that this "pen" has many features. It can not only be transferred into the highlighter, adjust the color of the pen, and can also adjust the thickness of the pen, which is more conducive to the use of the user. The content written on the screen can also be easily erased by the use of the rubber function of the toolbar. It has the area erase and the screen as a whole clear function, which can quickly eliminate the written writing. In addition, when temporarily need to write something to save, open the following whiteboard. After writing, choose whether to save these things wrote. When need to use the mouse function, we can point the following mode to switch this way to achieve the perfect conversion of the pen and mouse functions.

The system is the perfect alternative to electronic whiteboard, which affected by environmental factors, and the size of the whiteboard can not control constraints.

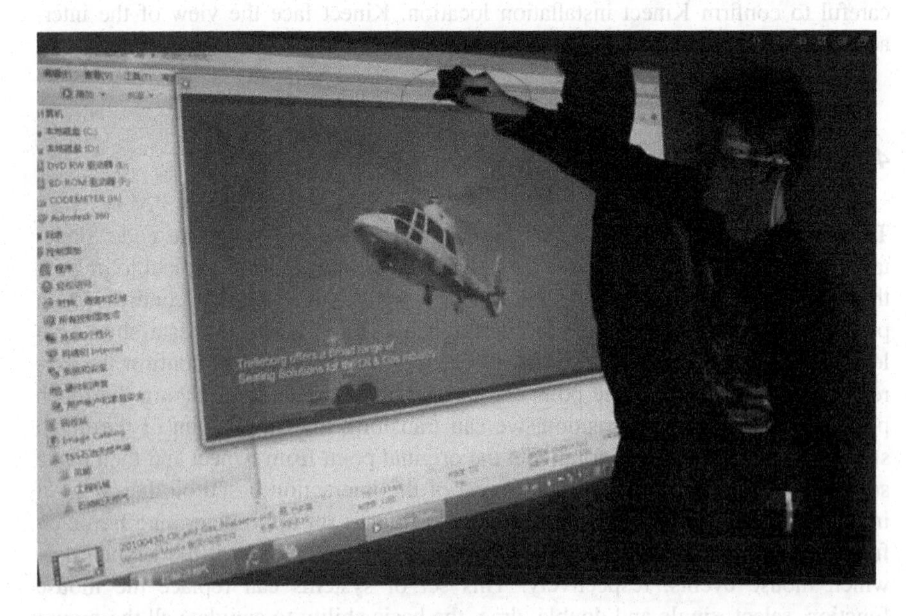

Fig. 3 Pen can achieve the mouse drag and drop function

Fig. 4 Electronic pen to replace the mouse operation diagram

Fig. 5 Text marked function map

The man-machine interaction system is not affected by environmental factors, it can turn any wall into a display area, in a certain range for users to switch size, and it is more accurate than the traditional large-screen positioning. Therefore, it can be widely used in classroom interaction, business meeting interaction, interactive exhibition and other interactive occasions.

5 Summary

In this paper, aiming at the problem that the real-time accuracy of the target trajectory tracking is not high for the human-computer interaction system, the modified Mean Shift algorithm is used to track the image information. The projection technology, infrared sensor, dynamic capture and image processing are used to simulate the mouse function to turn any wall into a touch screen. The set of systems can replace the mouse function, select, single and double, drag, the basic ability to simulate all the mouse operation. With the action pen replacing the role of the mouse, we can greatly improve the friendly human-computer interaction, highlight the interaction between man-machine interaction. So that users can be more intuitive to take control of the system, and take the passive information into a user's active knowledge and acceptance to achieve a good human-computer interaction effect.

Acknowledgements The work is supported by the National Natural Science Foundation of China (Nos. 61403268); The Key Technology Program of Suzhou, China (No. SYG201639); Natural Science Fund for Colleges and Universities in Jiangsu Province (No. 16KJB120005).

References

1. Grudin J. Human-computer interaction. Ann Rev Inf Sci Technol. 2011;45(1):367–430.
2. Maqueda AI, del-Blanco CR, Jaureguizar F, García. N. Human–computer interaction based on visual hand-gesture recognition using volumetric spatiograms of local binary patterns. Comput Vis Image Underst. 2015;141:126–37.
3. Ramamoorthy A, Vaswani N, Chaudhury S, et al. Recognition of dynamic hand gestures. Pattern Recogn. 2003;36(9):2069–81.
4. Oka K, Sato Y, Koike H. Real-time fingertip tracking and gesture recognition. IEEE Comput Graphics Appl. 2002;22(6):64–71.
5. Cheng H, Yang L, Liu Z. Survey on 3D hand gesture recognition. IEEE Trans Circuits Syst Video Technol. 2016;26(9):1659–73.
6. Kim J, Jung H, Kang M, et al. 3D human-gesture interface for fighting games using motion recognition sensor. Wirel Pers Commun. 2016;89(3):927–40.
7. Chen KB, Savage AB, Chourasia AO, et al. Touch screen performance by individuals with and without motor control disabilities. Appl Ergon. 2013;44(2):297–302.
8. Hsu MH, Shih TK, Chiang JS. Real-time finger tracking for virtual instruments. In: 7th international conference on ubi-media computing and workshops (UMEDIA). IEEE;2014. p. 133–38.
9. Lu W, Tong Z, Chu J. Dynamic hand gesture recognition with leap motion controller. IEEE Signal Process Lett. 2016;23(9):1188–92.
10. Chen G, Li J, Wang B, et al. Reconstructing 3D human models with a Kinect. Comput Anim Virtual Worlds. 2016;27(1):72–85.

11. Hsu SC, Huang JY, Kao WC, et al. Human body motion parameters capturing using kinect. Mach Vis Appl. 2015;26:919–32.
12. Yu K, Watson NR, Arrillaga J. An adaptive Kalman filter for dynamic harmonic state estimation and harmonic injection tracking. IEEE Trans Power Deliv. 2005;20(2):1577–84.
13. Kalal Z, Mikolajczyk K, Matas J. Tracking learning detection. IEEE Trans Pattern Anal Mach Intell. 2012;34(7):1409–22.
14. Peng D, Guo Y, Xue A, et al. An interacting multiple model algorithm for a 3D high maneuvering target tracking. Control Theory Appl. 2008;25(5):831–6.
15. Medeiros H, Park J, Kak AC. Distributed object tracking using a cluster-based kalman filter in wireless camera networks. IEEE J Sel Top Signal Process. 2008;2(4):448–63.
16. Hsia CH, Liou YJ, Chiang JS. Directional prediction CamShift algorithm based on adaptive search pattern for moving target tracking. J Real-Time Image Proc. 2016;12(1):183–95.

The Design of New Energy-Saving DC Brake Device

Cai Xia Gao, Zi Yi Fu and Fu Zhong Wang

Abstract In order to meet the needs of the market, such as small size, low power consumption, and low cost, the finite element analyzing software is used to design DC brake. The optimization design of DC electromagnet using magnet is introduced, including the determination of the electromagnet material, air-gap length, etc. According to the characteristics of DC brake, which are needing larger initial suction and maintaining smaller suction, the control circuit is designed, which is strong excitation starting, weak excitation maintaining. This circuit is simple, reliable and has achieved high efficiency and energy saving purposes. The results of experiments indicate that the DC brake has characteristics of small starting current, simple structure, rapid action, high reliability and durability.

Keywords DC brake · Optimum design · Control circuit · Finite element analysis

1 Introduction

DC electromagnet has advantages of small starting current, rapid action, simple structure, reliability and durability, and widely used in industrial equipment and automation device. But due to its special shape and irregular, magnetic material nonlinear and distributed, the design methods of electro-magnet are still not fully developed. Reference [1], the simplified model and by mastering some empirical data and actual data, with a certain test to complete the design, this method requires

C.X. Gao (✉) · Z.Y. Fu · F.Z. Wang
School of Electrical Engineering and Automation, Henan Polytechnic University, Jiaozuo 454000, China
e-mail: gcx81@126.com

Z.Y. Fu
e-mail: gcx@hpu.edu.cn

F.Z. Wang
e-mail: 254636920@qq.com

Y. Jia et al. (eds.), *Proceedings of 2017 Chinese Intelligent Systems Conference*, Lecture Notes in Electrical Engineering 459,
https://doi.org/10.1007/978-981-10-6496-8_18

the designer to have considerable experience in engineering, but also to prolong the time of design. Reference [2], it uses empirical formula to preliminarily design, and then determine the final design of the parameters through many experiments. The cost of design is very high. Reference [3], according to that the mathematical model of DC electromagnet structure parameter optimization design was carried out, and the model is on the assumption of linear magnetic circuit which is established under the condition of not in accord with the actual working condition of the electromagnet. The design of electromagnet dynamic and static property is poor; finally through the experiment of parameter optimization, parameters change is not convenient. In order to improve the precision of the DC electromagnet design, the author put forward the design method based on the finite element analysis software Magnet DC electromagnet, which also based on the magnetic circuit analysis method to get the structure parameters of the DC electromagnet, and energy-saving electromagnet coil design improvement; and then in the Magnet model to iron optimization design, the design of DC electromagnet is made into the sample and the sample test.

2 The Preliminary Design of DC Electromagnet

2.1 Design Requirements

The electromagnet main technical parameters: Rated suction is 2800 N, rated stroke is 3 mm, rated voltage of 220 V (DC), an ambient temperature of −20 to 20 °C, using location altitude does not exceed 2000 m, 20 MΩ insulation resistance.

DC electromagnet work requirements: Electromagnetic brake, an electromagnet coil is de-energized, under the action of the spring back in situ, drives the brake to finish the brake action; a braking electromagnet, coil 5 is energized, core 8 to 4 through a spring near the armature, electromagnetic force to the brake device, brake release, keeping time is 0.025 s.

2.2 Structure Design of DC Electromagnet

According to the general method of Electromagnet Design [5], and some experiences and reference data, a preliminary design to determine the DC electromagnet structure is as shown in Fig. 1. In order to reduce the influence of eminence, the armature is pasted on the surfaces of nonmagnetic thin copper [5]. DC electromagnet shell is designed into a closed, being capable of properly increasing the operating frequency, not easy to burn out the coil, brake and cohesion by straight push type structure, so not only guarantee the overall rigidity and machining precision, but also can improve the reliability and the service life of the whole brake.

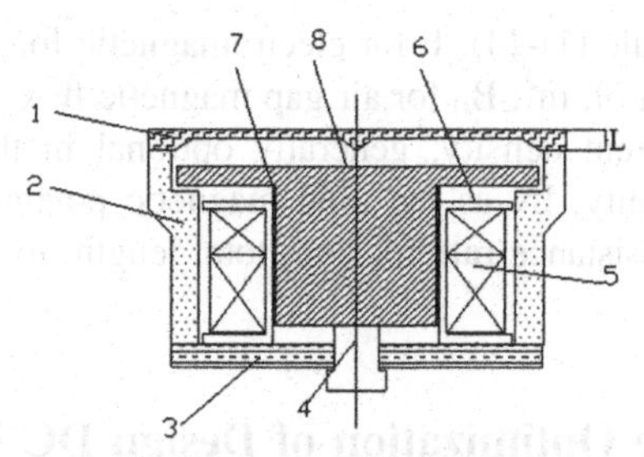

Fig. 1 Structure of DC electromagnet

1—End cover; 2—Iron yoke; 3—Base; 4—Armature; 5—Coil; 6—Coil skeleton; 7—Thin copper strip; 8—Iron core

2.3 The Design of Coil

In order to achieve high efficiency and energy saving, the design of the electromagnet is used in coil form (starting coil, maintaining coil), i.e., weak incentive to maintain strong excitation starting. In order to meet the requirements of initial suction, and electromagnet starting large magnetic force, the armature is pulled in. Starting after the end of armature and core is required only in very small magnetic potential and maintained under suction. Because the maintenance of current is very small, the coil loss is very small so as to achieve the purpose of energy saving.

Based on magnetic circuit structure, determine the coil height was H = 58 mm, coil outer radius of maximum only for 67 mm, otherwise easily causing the coil between the casing and the dielectric withstand voltage breakdown.

According to the initial suction and travel requirements, ignoring the working air gap and the iron core magnetic pressure the magnetic potential, by the formula (1), formula (2) can get the total magnetic potential [3]. According to the above magnetic potential and the size of the coil, by formula (3) can be obtained by using the coil diameter DC, according to calculation of the value choice in the vicinity of the standard diameter, can determine the starting coil diameter is 0.695 mm, maintain the coil diameter is 0.28 mm. Finally by formula (4) to determine the number of turns of coil, the starting coil turns 2236, placing the coil skeleton inside; to keep the number of turns of coil for 4336 turns, placing the coil skeleton of the lateral.

$$F = \frac{S_0 B_0^2}{2\mu_0} = \frac{10^7}{8\pi} S_0 B_0^2 \tag{1}$$

$$IN = \Phi \times R_m \tag{2}$$

$$d_c = \sqrt{\frac{4\rho l_p IN}{0.85 U_N}} \tag{3}$$

$$N = 1.28 \frac{(IN)}{0.85 j d_c^2} \tag{4}$$

Formula (1)–(4): F for electromagnetic force; S_0 for air gap magnetic pole in a total area of, m^2; B_0 for air gap magnetic flux density; Rm as the total reluctance; j wire current density, generally optional in the 2–4 A/mm^2 range selection; air permeability; IN as the total magnetic potential; magnetic flux Φ core; for conductor resistance rate for coil; total length; to add to the coil rated field voltage.

3 The Optimization of Design DC Electromagnet Based on Magnet

There are many factors influencing electromagnet static characteristic. It uses Magnet software here [6, 7], from the initial design of the DC electromagnet suction type on the model to optimize, electromagnet: including the coil skeleton material determination, the end cap size optimization, optimization of the length of air gap. Figure 2 double-winding electromagnet physics model diagram.

3.1 Determination of Coil Skeleton Material

According to the analysis of electromagnet suction process, considering the materials properties, the end cap instead of a magnetic conductive, but must have a certain degree of hardness, the selection of 3003 aluminum alloy. In order to improve the suction, yoke, base must be with high magnetic conductivity, solid selection of 10 steel. Armature selection of 3003 aluminum alloy. The following magnet software through the electromagnet is shown in Fig. 2 two dimensional physical model simulations, determine the coil skeleton material.

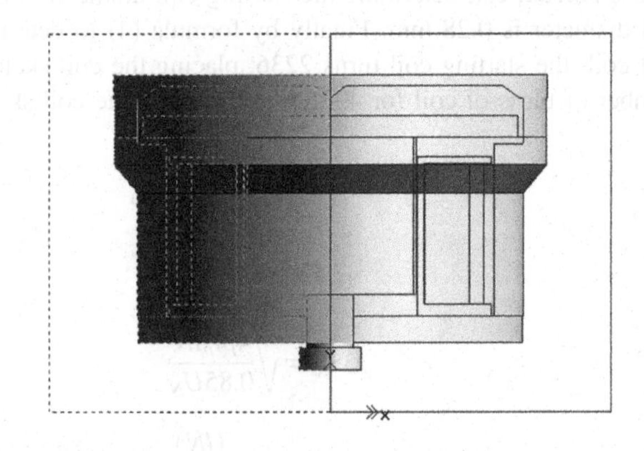

Fig. 2 Physics model of electromagnet

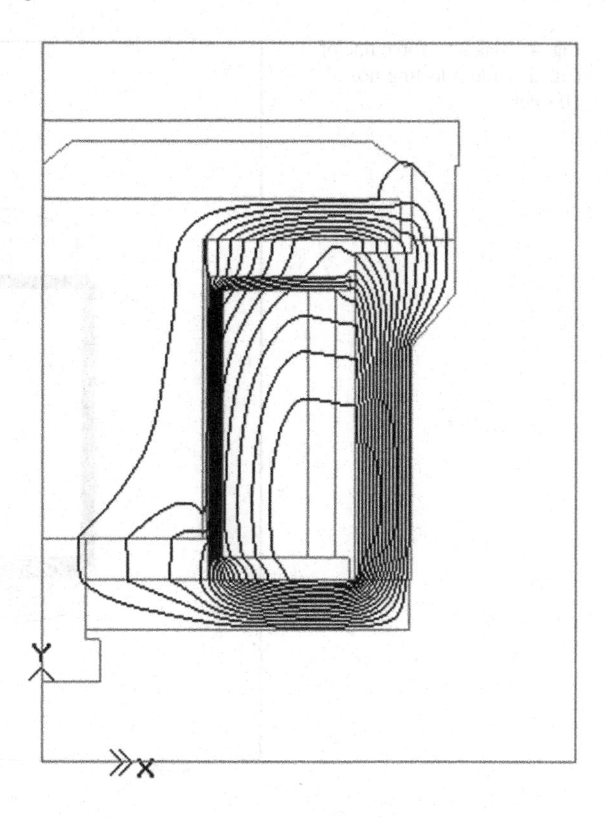

Fig. 3 Magnetic force line when coil, skeleton is 3003 aluminum alloy

Figure 3 the magnetic field distribution when coil skeleton selects 3003 aluminum alloy, the suction is 2845.77 N. Figure 4 is the Magnetic Force Line when Coil Skeleton Selecting No. 10 Steel force, from figure, it can be seen that through the magnetic lines of force most of iron yoke 2 form a loop, so the winner of the working air gap magnetic flux generated by the end face of the main suction is reduced to 1800 N. Therefore, coil skeleton should be selected in 3003 aluminum alloy.

3.2 The Optimization of Air Gap Length

The core and the end cover of air gap between the L1 affect the main air gap flux path of electromagnetic force. Different L1 simulation data is as shown in Table 1. With the increase of L1, suction increases gradually, when L1 increases to 4 mm will continue to increase suction decreased, therefore, the best air gap size L1 for 4 mm.

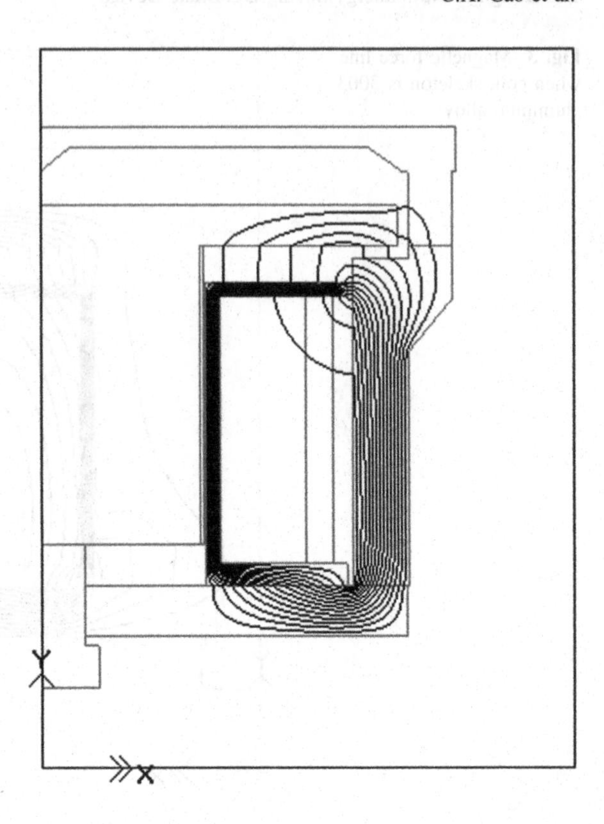

Fig. 4 Magnetic force line of coil skeleton selecting no. 10 steel

Table 1 The calculated simulation data of different air gap length L1

σ (mm)	1.5	2	2.5	3	3.5	4	4.5	5
Electromagnetic force (N)	2819	2853	2880	2901	2920	2932	2890	2820

4 The Control Circuit Design of DC Brake

DC brake control circuit is as shown in Fig. 5. The working principle is: electricity through a resistor R1 capacitors C2 charging, when the capacitor voltage is lower than the C2 VD1 diode and thyristor voltage and VT1 positive, VT1, VT2 conduction, the holding coil and the starting coil at the same time, the electromagnet has generated considerable initial suction; when the capacitor voltage reaches VD1 C2 diode and thyristor VT1 positive voltage and time, VT1 conduction, VT2 cut off, only to keep the coil at work, at this time because the electromagnet working gap is very small, keep the coil to generate electromagnetic force is large enough to allow the electromagnet to maintain in working state. This design achieves reducing the size of the product, increasing the initial force, reducing the electromagnet coil temperature rise, and reducing the energy consumption.

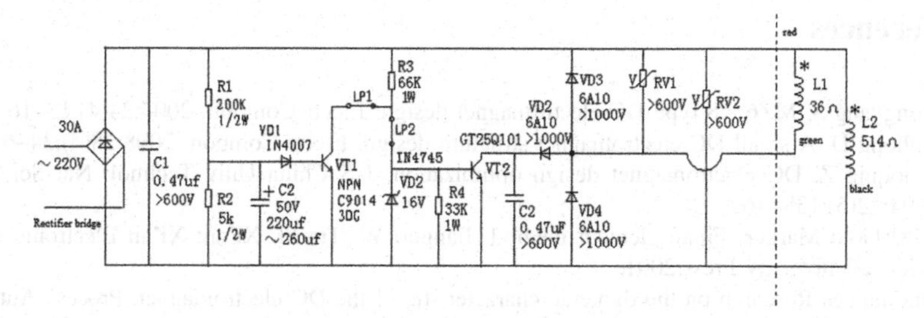

Fig. 5 Control circuit principle diagram of DC brake

5 Experimental Analysis

Finally, aiming to the above design of DC electromagnet, we have produced samples, and the samples were measured test. Starting current is 4.5 A, suction is 2967 N, action is fast and reliable; maintaining current is 0.514 A, suction is 606 N, it can be reliably maintained; starting winding resistance is 35.5 Ω, keeping the winding resistance is 529 Ω. The test results show that the design of the electromagnet is not only the indicators meet the design requirements, and high efficiency and energy saving, obvious optimization effect. Practice has proved that, compared with traditional magnetic circuit analysis method, finite element method which as calculation accuracy, high stability, is suitable for using in engineering design and optimization design.

6 Conclusions

This article used the Magnet software to optimize the electromagnet component parameters and the results were verified through experiments. Research shows that, the use of 3003 aluminum alloy coil framework 6, end cap size L = 29.1 mm, gap size L1 4 mm generates electromagnetic force. In addition, in order to achieve the purposes of high efficiency and energy saving, the electromagnet excitation mode should be used in coil form (starting coil, coil), i.e., weak incentive in order to maintain strong excitation starting. Practice has proved that, compared with traditional magnetic circuit analysis method, finite element method which as calculation accuracy, high stability, is suitable for using in engineering design and optimization design.

References

1. Dongyang S. MZ6 - 8 type DC electromagnet design. Electr Compon. 2004;24(4):13–16.
2. Yuhong D. A small DC electromagnet structure design. Electr Compon. 2008;28(4):34–6.
3. Xiaoqing Z. DC electromagnet design optimization. J S China Univ Technol: Nat Sci Ed. 1994;22(5):138–46.
4. [US] Gold Marker. Finite element method. Jianguo W, Transl. Xi'an: Xi'an Electronic and Science University Press;2001.
5. Qiuming S. Research on the dynamic characteristic of the DC electromagnet. Process Autom Instrum. 2007;5:20–3.
6. Nitta K, Watanabe K, Kagenaga K. Three-dimensional magnetic field analysis of electrodes for VCBs. IEEE Trans Power Deliv. 1997;12(4):1520–5.
7. Lianhui H. Magnetic circuit and ferromagnetic device. Beijing: Higher Education Press;1982.

Data Missing Process by Extended Kalman Filter with Equality Constraints

Hao Li and Yingmin Jia

Abstract In this paper, the extended Kalman filter (EKF) is used to estimate the position of the feature points when data missing occurs, taking the feature extraction of plane moving robot ceiling-based positioning as background and the coordinates of the feature points in the image plane as objects. Firstly, the acceleration model of the feature points in the image plane is established as the motion equation, and the motion information of the feature points is extracted by filter. Then, the equality constraints of the feature points are added to the filter scheme to increase the measurement information. In the case where there is a loss of data, that is, the feature points are lost partly, and the predicted values of the lost points are estimated as the true value. By comparing the filtering results, it shows that the addition of equality constraints can not only enhance the filtering effect, but also can estimate the loss points more effectively. Finally, the validity of the filtering scheme is verified by a numerical example.

Keywords Extended Kalman filter · Data missing · Equality constraints

1 Introduction

Image matching technology is one of the key steps to realize image fusion, image correction, image mosaic and target recognition and tracking, mainly divided into pixel-based and feature-based methods [1], has been widely used in image recognition. Feature points are important image feature, and feature points extraction and matching technology have become the mainstream of the application direction, for

H. Li · Y. Jia (✉)
The Seventh Research Division and the Center for Information and Control, School of Automation Science and Electrical Engineering, Beihang University (BUAA), Beijing 100191, China
e-mail: ymjia@buaa.edu.cn

H. Li
e-mail: lihao111813@126.com

Y. Jia et al. (eds.), *Proceedings of 2017 Chinese Intelligent Systems Conference*, Lecture Notes in Electrical Engineering 459,
https://doi.org/10.1007/978-981-10-6496-8_19

the calculation quantity is simplified, the matching precision is improved, and the operation speed is accelerated.

Ceiling-based visual positioning is one of the most important means of indoor positioning for its low computational burden and high accuracy, by treating the corner points on the ceiling as feature points [2]. The feature points extraction plays an important role in visual positioning and tracking object [3], and the accuracy of feature point extraction often determines the accuracy of the target position measurement. The vast majority of the literatures are based on the assumption that the feature points are not disturbed, such as literature [4]. Nevertheless, the actual position measurement data of the feature points will encounter the following problems inevitably: the location of the feature points will fluctuate due to changes in light intensity, and loss of feature points due to movement range limits or encounter of obstructions on the ceiling. These problems will lead to increased fluctuations of observed data, or even data missing, seriously affecting the accuracy of feature extraction.

EKF is one of the most popular filtering methods, and has been widely applied in many fields such as location [5], tracking [6], navigation [7] and control [8]. EKF is effective and practical in actual use and it can not only to estimate the past and current state of the signal, but also can estimate the future state. Furthermore, it can be carried out even in some unknown circumstances using dynamic equations. Specifically, when the position information or the measurement data is missing, it is very convincing to predict the state of the next moment using the state equation based on the current state on the condition that the model is established accurately.

There are some effective means that have been proposed to improve the performance of traditional EKF scheme, for example the EKF with equality constraints [9]. Compared to plain EKF, the addition of equality constraints can improve the estimation performance and filter accuracy for making use of the constraint information contained in the scene reasonably. There are some sub-methods of equality constraints filter, among them the pseudo measurement method is a main approach due to its simplicity. It is, however, a bit imperfect due to its computational complexity and numerical ill-posed problem, for which, literature [10] presented two forms of algorithms which can solve the abovementioned problems to a certain extent.

The core of this paper is estimating the coordinates of the feature points in image plane using EKF with equality constraints when data missing occurs. Specifically, an improved filtering scheme of EKF with nonlinear equality constraints is proposed to improve the estimation accuracy, and then a simple and intuitive filtering scheme with data missing is proposed, by using the dynamic equation of the established model to predict the missing state.

This paper is organized as follows. Section 2 formulates the state estimation problem with the equality constraint information. Section 3 introduces filtering schemes. Section 4 presents the filtering scheme with data missing. Section 5 provides a supporting numerical example. Section 6 gives concluding remarks.

2 Problem Formulation

In an indoor environment, assume that the floor is flat and parallel to the ceiling. A CCD camera is mounted on the top of the mobile robot working on the floor and the optical axis of CCD is vertically facing the ceiling. Assume that the shape of the lamp on the ceiling is arbitrary but is known in advance, and the corners of the lamp have been successfully extracted as feature points. The positive pentagon lamp shade has been chosen in this article, as shown in Fig. 1.

Assuming that the image coordinates of the feature points in the image plane have been obtained through the camera aperture model [11] are denoted by $p_i = (u_i, v_i)^T$. And the image coordinates of each feature point p_i are taken as one object. In order to make the algorithm applicable to any motion, the acceleration model is used to simulate the target motion, and the speed Vp_i and the acceleration Ap_i are denoted by $(Vu_i, Vv_i)^T$ and $(Au_i, Av_i)^T$. So the motion equation is given as follows:

$$x_k = Fx_{k-1} + G\omega_{k-1} \quad k = 1, 2, \ldots \tag{1}$$

where x is the state of the acceleration model, which is:

$$x = \left[p_1^T, Vp_1^T, Ap_1^T, p_2^T, Vp_2^T, Ap_2^T, \ldots, p_N^T, Vp_N^T, Ap_N^T\right]^T \in \mathbb{R}^{6N}$$

N denotes the number of feature points, $E(x_0) = \bar{x}_0$, $\text{cov}(x_k) = P_k > 0$ and process noise ω_{k-1} is zero-mean white noise with $\text{cov}(\omega_k) = Q_k$.

The state transition matrix F and the noise driven matrix G are determined at the same time:

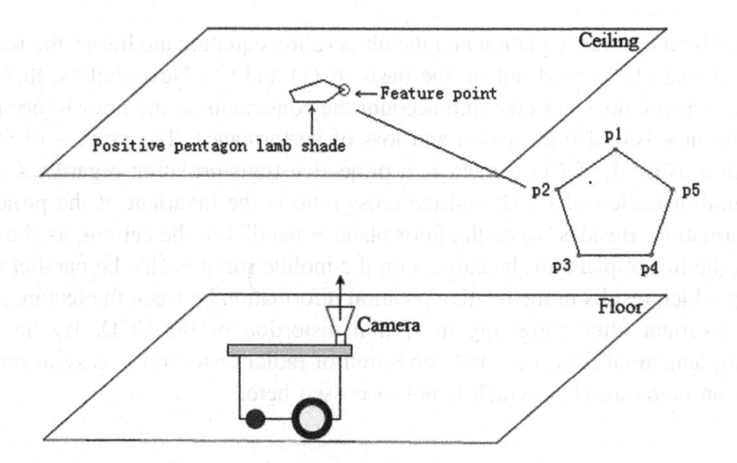

Fig. 1 Model of ceiling-based indoor mobile robot

$$F = diag(F_{c1}, F_{c2}, \ldots, F_{cN})_{6N\times 6N}$$

$$G = diag(G_{c1}, G_{c2}, \ldots, G_{cN})_{6N\times 2N}$$

where

$$F_{ci} = \begin{pmatrix} 1 & 0 & T & 0 & T^2/2 & 0 \\ 0 & 1 & 0 & T & 0 & T^2/2 \\ 0 & 0 & 1 & 0 & T & 0 \\ 0 & 0 & 0 & 1 & 0 & T \\ 0 & 0 & 0 & 0 & 1 & 0 \\ 0 & 0 & 0 & 0 & 0 & 1 \end{pmatrix} G_{ci} = \begin{pmatrix} T^2/2 & 0 \\ 0 & T^2/2 \\ T & 0 \\ 0 & T \\ 1 & 0 \\ 0 & 1 \end{pmatrix} \quad i = 1, 2, \ldots N$$

The observation model can be written in a compact fashion as following:

$$z_k^{(1)} = Hx_k + v_k \quad k = 1, 2, \ldots \tag{2}$$

where v_k is zero-mean white noise, $\text{cov}(v_k) = R_k^{(1)} > 0$, and $z_k^{(1)}$ denotes the coordinates of the feature points actually obtained through the camera in the image plane, so:

$$z_k^{(1)} = [Ou_1(k), Ov_1(k), Ou_2(k), Ov_2(k), \ldots, Ou_N(k), Ov_N(k)]_{N\times 1}^T$$

where

$$H = diag(H_{c1}, H_{c2}, \ldots, H_{cN})_{2N\times 6N}$$

$$H_{ci} = \begin{pmatrix} 1 & 0 & 0 & 0 & 0 & 0 \\ 0 & 1 & 0 & 0 & 0 & 0 \end{pmatrix} \quad i = 1, 2, \ldots N$$

Since both the state equation and the observation equation are linear, the Kalman filter (KF) can be carried out on the basis of (1) and (2). Nevertheless, this plain filtering scheme does not take into account the constraints of the prior information, resulting in waste of information and loss of performance. The process of feature extraction using the CCD camera is a projective transformation regardless of the tangential distortion of CCD, and the cross-ratio is the invariant of the projective transformation. Besides, since the floor plane is parallel to the ceiling, as shown in Fig. 1, the image plane of the camera on the mobile robot is also be parallel to the ceiling, which results in the relative position information between the feature points being invariant after correcting the radial distortion of the CCD. By the way, ignoring tangential distortion and correction of radial distortion is easy to proceed basing on literature [12], which is not discussed here.

In order to improve the accuracy of the estimation, the nonlinear equality constraints including cross-ratio constraints and relative position constraints are extracted as follows:

$$h(x_k) = 0 \quad k = 1, 2, \ldots$$

The specific details of $h(x)$ will be discussed in Sect. 5.

Take $z_k^{(2)} = 0$ as pseudo measurement, so the equality constraints can be expressed into a noise-free measurement as:

$$z_k^{(2)} = h(x_k) \quad k = 1, 2, \cdots \tag{3}$$

So (1), (2) and (3) constitute the dynamic system of the state space description. And the corresponding filtering strategy will be discussed in the next section.

3 Filtering Scheme

Two forms are proposed in literature [13] to improve numerical ill-posed problems and computational complexity, which are common in pseudo-measurement methods. Nevertheless, the equality constraints in [13] are linear and are not applicable here. In this section, an improved filtering scheme with nonlinear equality constraints is propose.

If the state equation is the same as (1), the observation equation is the same as (2), and the equality constraints are all linear, as:

$$z_k^{(2)} = H_k^{(2)} x_k \tag{4}$$

The algorithm in literature [13] is as follows:

Prediction: use the state of the previous cycle to predict the current state, which is the same as KF prediction process.

$$\hat{x}_{k|k-1} = F_{k-1} \hat{x}_{k-1|k-1} \tag{5}$$

$$P_{k|k-1} = F_{k-1} P_{k-1|k-1} F_{k-1}^T + G_{k-1} Q_{k-1} G_{k-1}^T \tag{6}$$

If the equality constraints are taken as noise-free measurement, the pseudo measurement method can be used to improve the estimation accuracy by regarding the noise-free measurement as the ideal measurement without noise, which means the noise covariance of $z_k^{(2)}$ equal to zero. This original pseudo measurement method is relatively simple in designing, but the computational complexity will increase due to the expansion of the observed dimension and the numerical ill-posed problem will occur due to the covariance matrix of the measured noise is singular.

For the above mentioned questions, literature [13] proposes that the state is updated by the noisy measurement and noise-free measurement separately.

Update by the noisy measurement:

$$\hat{x}_{k|k}^{(1)} = \hat{x}_{k|k-1} + P_{k|k-1}\left(H_k^{(1)}\right)^T \left(S_k^{(1)}\right)^{-1} \left(z_k^{(1)} - H_k^{(1)}\hat{x}_{k|k-1}\right) \tag{7}$$

$$P_{k|k}^{(1)} = P_{k|k-1} - P_{k|k-1}\left(H_k^{(1)}\right)^T \left(S_k^{(1)}\right)^{-1} H_k^{(1)} P_{k|k-1} \tag{8}$$

$$S_k^{(1)} = H_k^{(1)} P_{k|k-1}\left(H_k^{(1)}\right)^T + R_k^{(1)} \tag{9}$$

Update by the noise-free measurement:

$$\hat{x}_{k|k} = \hat{x}_{k|k}^{(1)} + P_{k|k}^{(1)}\left(H_k^{(2)}\right)^T \left(S_k^{(2)}\right)^{+} \left(z_k^{(2)} - H_k^{(2)}\hat{x}_{k|k}^{(1)}\right) \tag{10}$$

$$P_{k|k} = P_{k|k}^{(1)} - P_{k|k}^{(1)}\left(H_k^{(2)}\right)^T \left(S_k^{(2)}\right)^{+} H_k^{(2)} P_{k|k}^{(1)} \tag{11}$$

$$S_k^{(2)} = H_k^{(2)} P_{k|k}^{(1)}\left(H_k^{(2)}\right)^T \tag{12}$$

Nevertheless, the equality constraints this paper used (3) are nonlinear, and the above algorithm in literature [13] is no longer applicable. The above algorithm can be improved by the Taylor expansion linearization method, which can be extended to non-linear constraints. Then, the pseudo-measurement method with non-linear equality constraint is given as follows:

Prediction: the same as (5)–(6)

Update by the noisy measurement: the same as (7)–(9)

Update by the noise-free measurement:

$$\hat{x}_{k|k} = \hat{x}_{k|k}^{(1)} + P_{k|k}^{(1)}\left(\tilde{H}_k^{(2)}\right)^T \left(S_k^{(2)}\right)^{+} \left(z_k^{(2)} - h\left(\hat{x}_{k|k}^{(1)}\right)\right) \tag{13}$$

$$P_{k|k} = P_{k|k}^{(1)} - P_{k|k}^{(1)}\left(\tilde{H}_k^{(2)}\right)^T \left(S_k^{(2)}\right)^{+} \tilde{H}_k^{(2)} P_{k|k}^{(1)} \tag{14}$$

$$S_k^{(2)} = \tilde{H}_k^{(2)} P_{k|k}^{(1)}\left(\tilde{H}_k^{(2)}\right)^T \tag{15}$$

where:

$$\tilde{H}_k^{(2)} = \frac{\partial h}{\partial x}\Big|_{x=\hat{x}_{k|k}^{(1)}}$$

On the basis of the above discussions, the filtering algorithm with data missing is given in the next section.

4 Estimation Scheme with Data Missing

The following assumptions are made: the mobile robot move forward with uniform acceleration; the tangential distortion of the CCD has been ignored; the radial distortion of the CCD has been corrected; feature matching has been completed. The estimation scheme with data missing is carried out under such assumptions.

In the case of accurate model, the predicted state obtained by dynamic equation filtering is very close to the real state for the effectiveness and accuracy of the filtering algorithm. So it's feasible to predict the position of feature points using dynamic equation. Furthermore, when the position information or the measurement data is unavailable, the dynamic equation and the current state of the system are used to predict the state of the next moment is very credible. Specifically, when the data missing of the measurement occurs, the predicted value of this moment is taken as the true value of this moment. The corresponding algorithm is described as follows:

If the observation data of feature point i is missing in period m to n, that means the elements $Ou_i(k), Ov_i(k)$ in $z_k^{(1)}$ $(k = m, m+1, \ldots n)$ is unavailable in period m to n, then regard the predicted state $\hat{x}_{k|k-1}$ as the real state of the system, so the $H_k^{(1)}\hat{x}_{k|k-1}$ is regarded as the observations.

5 Numerical Examples

In the following, the above estimation scheme is verified through a numerical example. Consider the ceiling-based mobile robot working in the plane floor, as shown in Fig. 1. And assume that the acceleration of the mobile robot is constant, which can be described using the form of (1) and (2).

The lamp shade on the ceiling is selected as the positive pentagon, which means the number of feature points is $N = 5$ and the following two kinds of equality constraints can be extracted: positional invariant and projective invariant.

Positional invariant constraints: As the motion of mobile robot in the model is assumed as a uniform acceleration linear motion, the relative position of the marker is invariant. For example, $\overrightarrow{p_2p_5}//\overrightarrow{p_3p_4}$, the corresponding equality constraint can be described as:

$$p_5 - p_2 - k(p_4 - p_3) = 0$$

where k is determined by the positive pentagon, and can be easily got: $k = 2\sin 54°$. Similarly, the other four parallel constraints can be obtained, so a total of ten coordinate relationships can be obtained.

Projective invariant constraints: The camera imaging process is the central projective process and the cross ratio of the five points is the invariant of the projective transformation. That means the cross ratio is constant, which can be described as:

$$\frac{\sin\angle p_2p_1p_4}{\sin\angle p_3p_1p_4} \div \frac{\sin\angle p_2p_1p_5}{\sin\angle p_3p_1p_5} = C$$

where C is constant and can be calculate from the positive pentagon. Similarly, the other four cross ratio constraints can be obtained.

Set positional invariant constraints together with projective invariant constraints, the equality constraints of system are obtained. Furthermore, and the noise-free measurements (3) are got.

Set observation period as $T = 0.01s$, set the length of filtering as $L = 618$, set the variance of the observation noise as $\mathrm{cov}(v_k) = 25I_{10\times10}$, and set the variance of the process noise as $\mathrm{cov}(\omega_k) = 0.01I_{10\times10}$.

In order to simulate the data missing, the following assumptions are made: p_1 disappears in period 300 to 320, p_2 disappears in period 500 to 520, p_3 disappears in period 350 to 380, and p_4 disappears in period 290 to 310.

Numerical example is carried out 100 times, and the RMSE position error is chosen as the evaluation index to evaluate the different filter algorithms. The results are shown in Figs. 2 and 3.

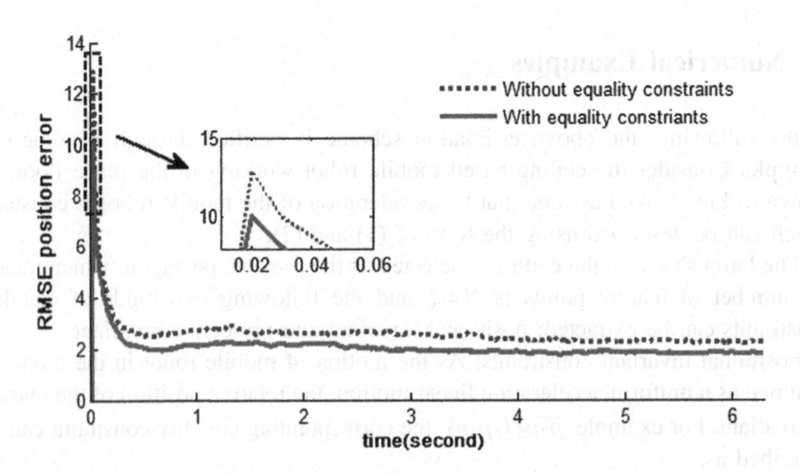

Fig. 2 REMS position error comparison of filtering scheme without data missing

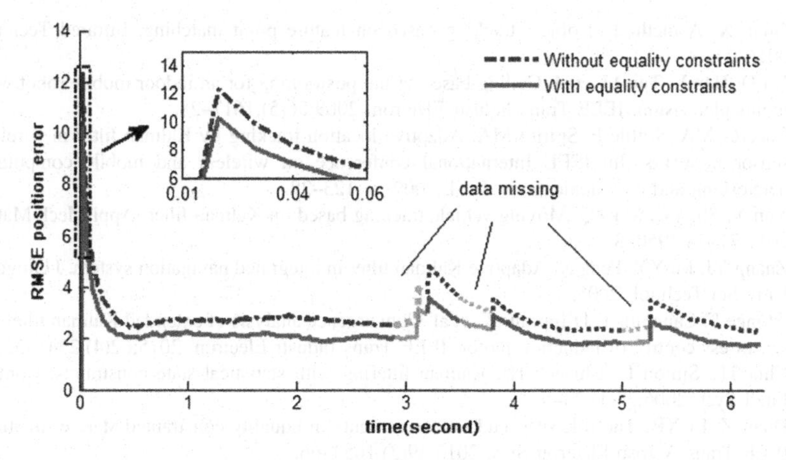

Fig. 3 REMS position error comparison of filtering scheme with data missing

From the simulation, it can be seen that the filtering scheme with equality constraints can effectively predict the state of the system in a certain period. Furthermore, the estimation error of the filtering scheme with equality constraints is reduced in both cases, so the performance of filtering scheme with equality constraints is better, regardless of whether data missing occurs.

6 Conclusion

In this paper, a filtering scheme of EKF with equality constraints is proposed to predict the coordinates of the feature points in image plane when data missing occurs. An improved filtering scheme of EKF with nonlinear equality constraints is proposed to improve the estimation accuracy, and a simple and intuitive filtering scheme with data missing is proposed to predict the state of the system. And the simulation results confirm the effectiveness of the filtering scheme.

Acknowledgements This work was supported by the NSFC (61327807, 61521091, 61520106010, 61134005) and the National Basic Research Program of China (973 Program: 2012CB821200, 2012CB821201).

References

1. Lowe DG. Distinctive image features from scale-invariant keypoints. Int J Comput Vis. 2004;60(2):91–110.
2. Chen X, Jia Y. Indoor localization for mobile robots using lampshade corners as landmarks: visual system calibration, feature extraction and experiments. Int J Control Autom Syst. 2014;12(6):1313–22.

3. Guan X. A method of object tracking based on feature point matching. Infrared Technol. 2016.
4. Xu D, Han L, Tan M, et al. Ceiling-based visual positioning for an indoor mobile robot with monocular vision. IEEE Trans Industr Electron. 2009;56(5):1617–28.
5. Caceres MA, Sottile F, Spirito MA. Adaptive location tracking by Kalman filter in wireless sensor networks. In: IEEE International conference on wireless and mobile computing, networking and communications. IEEE;2009. p. 123–28.
6. Yan Y, Shi YC, Ma ZQ. Moving vehicle tracking based on Kalman filter. Appl Mech Mater. 2011;71–78:3950–3.
7. Zhang YJ, Jin YY, Peng X. Adaptive Kalman filter in integrated navigation system. J Qingdao Univ Sci Technol. 2005.
8. Alonge F, Cangemi T, D'Ippolito F, et al. Convergence analysis of extended Kalman filter for sensorless control of induction motor. IEEE Trans Industr Electron. 2015;62(4):2341–52.
9. Chia TL, Simon D, Chizeck HJ. Kalman filtering with statistical state constraints. Control Intell Syst. 2006;34(1):73–9.
10. Duan Z, Li XR. The role of pseudo measurements in equality-constrained state estimation. IEEE Trans Aerosp Electron Syst. 2013;49(3):1654–66.
11. Potmesil M, Chakravarty I. A lens and aperture camera model for synthetic image generation. In: Conference on computer graphics and interactive techniques. ACM;1981. p. 297–305.
12. Zhang Z. A flexible new technique for camera calibration. IEEE Trans Pattern Anal Mach Intell. 2000;22(11):1330–4.
13. Duan Z, Li XR. Best linear unbiased state estimation with noisy and noise-free measurements. In: International conference on information fusion. IEEE Xplore;2009. p. 2193–200.

Design and Simulation of Fuzzy PID Controller Based on Variable Output Domain of Discourse

Zhiwei Chen, Huanyu Zhao and Jincheng Liu

Abstract The traditional variable domain of discourse fuzzy PID controller has complex structure and difficulty of expansion factor designing. This paper puts forward a new variable output domain of discourse fuzzy PID controller, and designs a new type of function expansion factor. The new variable output domain of discourse fuzzy PID controller only needs to adjust the output domain of the fuzzy controller according to the error, and not changing the input domain of the fuzzy controller. The structure of the new variable domain of discourse fuzzy PID controller is simpler and the control precision is more accurate compared with the new one. The new function expansion factor is smaller, which further improves the real-time performance of the controller. The simulation is constructed based on the platform of MATLAB. The results show that the variable domain of discourse fuzzy PID controller has better real-time, robustness and anti-jamming ability.

Keywords Domain of discourse · Fuzzy-PID · Function expansion factor · Output domain

1 Introduction

The fuzzy controller of the nonlinear time-varying system does not need precise mathematical model, thus has a wider range of applications [1, 2]. The field of input and output variables in the fuzzy controller is fixed, so when the error is changed in a small range but the corresponding fuzzy rules are not changed, the accuracy will still be low. To obtain a more accurate result, more the control rules should be added, resulting in numerous calculating [3]. However, Li Hongxing designed variable domain of discourse fuzzy control to solve this problem, by making out a stretch factor controller to adjust the system input and output domain [4]. Under the premise that the rules and the membership function are unchanged, the input and

Z. Chen · H. Zhao (✉) · J. Liu
Faculty of Automation, Huaiyin Institute of Technology, Huai'an 223003, Jiangsu, China
e-mail: hyzhao@163.com

Y. Jia et al. (eds.), *Proceedings of 2017 Chinese Intelligent Systems Conference*, Lecture Notes in Electrical Engineering 459,
https://doi.org/10.1007/978-981-10-6496-8_20

output domain will shrink (increase and expand) as the error decrease, which is equivalent to increasing the number of rules, and the control precision is improved.

The scaling factor is the key to the design of variable domain of discourse fuzzy controller [5]. Li Hongxing created the axiomatic definition of the scaling factor, and designed the proportional type scaling factor and exponential scaling factor [6]. Subsequently, a variety of intelligent algorithms, such as fuzzy algorithm [7, 8], neural network algorithm [9], ant colony algorithm [10] and so on are used to design the scaling factor. These intelligent algorithms avoid the workload to determine the scaling factor parameters and thus achieve satisfactory performances. However, these intelligent algorithms need to be adjusted in real time, which may induce a large amount of computation and thus deteriorate the real-time control performance. This is very unfavorable and the use of these algorithms would lower the stability of the system. However, the function type expansion factor can simultaneously guarantee the real-time capability and stability of systems [11]. Therefore, this paper will design the scaling factor based on the function model.

2 Design of Variable Domain Fuzzy PID Controller

Li Hongxing is the first one who proposed variable domain of discourse fuzzy control [12], the key is the changes of fuzzy control system domain in response to the error changes e in the system. The scaling controller adjusts the domain of the fuzzy control system according to the different time's error e and the rate of error change e_c, so that the fuzzy controller can create more control rules near the expected control point.

Figure 1 is a schematic extension of the domain. In Fig. 1, (b) is the initial domain of a fuzzy controller and its division, the domain of discourse is divided into negative NB, negative NM, negative NS, zero ZO, positive small PS, positive medium PM and positive big PB, total all seven subjects of even distribution of triangles. The scaling factor allows the domain to expand (shrink) as the error increases (decreases), as shown in (a) and (c) in the Fig. 1.

The traditional variable domain of discourse fuzzy controller uses a three-stage controller structure [7], as shown in Fig. 2. Fuzzy controller based on the e and e_c to adjust PID controller, the scaling factor controller gain the input domain of the expansion factor α and the output domain of the expansion factor β based on the e and e_c, respectively, used to adjust the fuzzy controller quantization factor K_e, K_{ec}, and the scale factor K_p, K_i, K_d, to achieve the telescopic changes of the input and output domain. The adjustment rules are: amplifying quantization factor is equivalent to contracting input domain, and reducing quantization factor is equivalent to expanding input domain. Shrinking scaling factor is equivalent to the shrinking output domain and amplifying scaling factor is equivalent to expanding output domain [13].

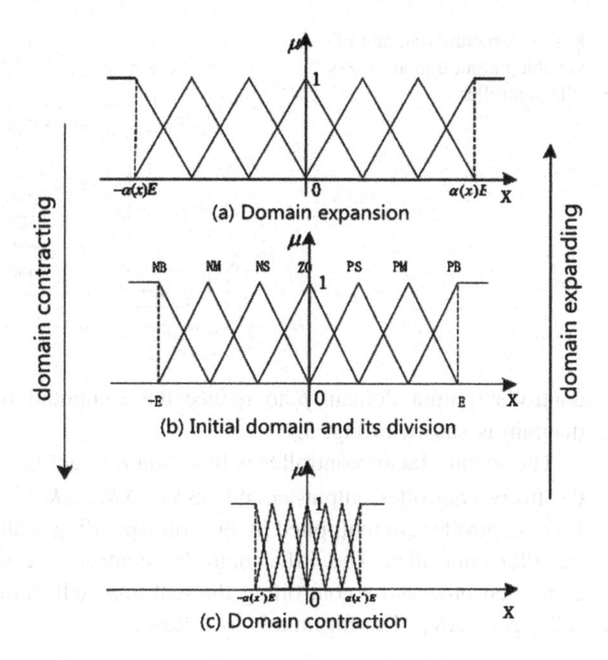

Fig. 1 Schematic extension of the domain of discourse

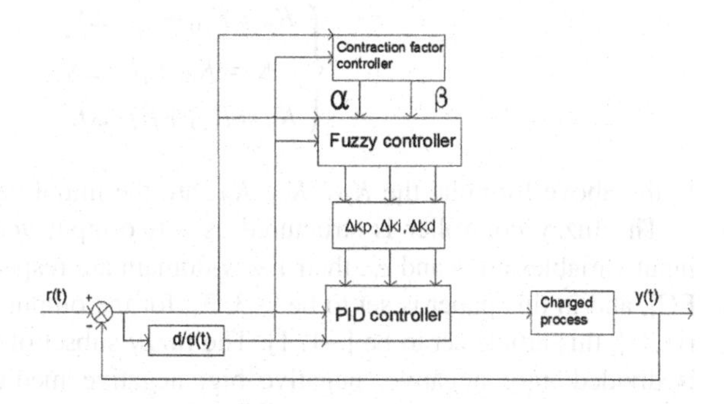

Fig. 2 Structure diagram of variable domain of discourse fuzzy PID controller

The traditional variable domain of discourse fuzzy controller needs to design the expansion factor of the input domain of discourse α and the output domain of discourse β. But the design of the expansion factor is complicated, which brings difficulties to the designers. The excessive expansion factor makes the controller structure complicated and lots of calculation, thus lowering the real-time control of the system. In this paper, we design a fuzzy PID controller with variable output domain, which is shown in Fig. 3. After abundant theoretical research and multiple simulation experiments, we find that the input theory domain of fuzzy controller has little effect on the result of variable domain of discourse fuzzy PID controller after shrinkage or expansion, and so it is in the progress of fuzzy, fuzzy reasoning and solution fuzzy after that. Therefore, we only add the scaling factor of the fuzzy

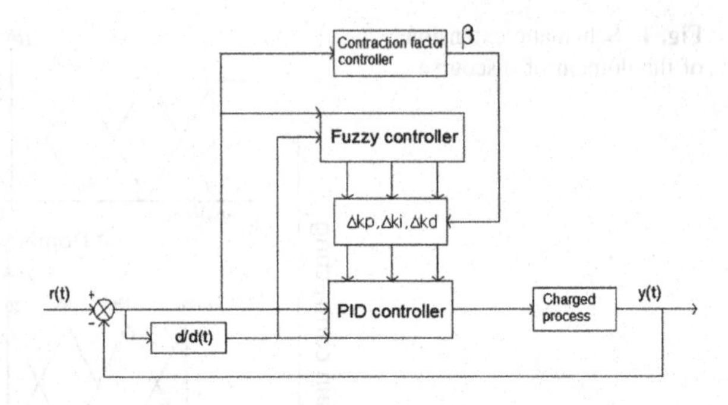

Fig. 3 Structure diagram of variable output domain fuzzy PID controller

controller output domain β to reduce the computational complexity. The block diagram is shown in Fig. 3.

The scaling factor controller is based on e to get the scaling factor β_p, β_i, β_d of the fuzzy controller output-variable ΔK_P, ΔK_i, ΔK_d. The output variables of the fuzzy controller are multiplied by the corresponding scaling factor and then input to the PID controller. The PID controller controls the closed-loop control of the controlled process by controlling the real-time self-tuning of the three parameters ΔK_P, ΔK_i, ΔK_d. The formula is as follows:

$$\begin{cases} K_p = K_{p0} + \beta_p \cdot \Delta K_p \\ K_i = K_{i0} + \beta_i \cdot \Delta K_i \\ K_d = K_{d0} + \beta_d \cdot \Delta K_d \end{cases} \tag{1}$$

In the above formula, the K_{P0}, K_{i0}, K_{d0} are the initial parameters.

The fuzzy controller is structured as a two-input and a three-output. The two input variables are e and e_c, their fuzzy domain are respectively [−E, E] and [−EC, EC], and in this paper is set to be [−3, 3]; for the output variable, its fuzzy domain [U, U], this article set to be [−1, 1]. The fuzzy subset of input and output variables is divided into: negative, negative big, negative medium, negative small, zero, positive small, positive medium, positive big, expressed as {NB, NM, NS, ZO, PS, PM, PB}.

According to the experience of the Li Hongxing, combined with the principle of PID control parameter-setting, we get the fuzzy control rule table, as shown in Table 1. And it obeys the triangular membership function curve distribution. In this paper, fuzzy reasoning method is adopted. Using the weighted average method to solve the fuzzy factor, that is, take the weighted average of the output variable y as the exact value, and the formula is as follows:

$$y_0 = \sum_{i=1}^{m} y_i k_i / \sum_{i=1}^{m} k_i \tag{2}$$

The scale factor k_i in the formula depends on the actual situation.

Table 1 Table of the fuzzy rules

$\Delta K_P/\Delta K_i/\Delta K_d$		EC						
		NB	NM	NS	ZO	PS	PM	PB
E	NB	PB/NB/PS	PB/NB/NS	PM/NM/NB	PM/NM/NB	PS/NS/NB	ZO/ZO/NM	ZO/ZO/PS
	NM	PB/NB/PS	PB/NB/NS	PM/NM/NB	PS/NS/NM	PS/NS/NM	ZO/ZO/NS	NS/ZO/ZO
	NS	PM/NB/ZO	PM/NM/NS	PM/NS/NM	PS/NS/NM	ZO/ZO/NS	NS/PS/NS	NS/PS/ZO
	N	PM/NM/ZO	PM/NM/NS	PS/NS/NS	ZO/ZO/NS	NS/PS/NS	NM/PM/NS	NM/PM/ZO
	PS	PS/NM/ZO	PS/NS/ZO	ZO/ZO/ZO	NS/PS/ZO	NS/PS/ZO	NM/PM/ZO	NM/PB/ZO
	PM	PS/ZO/PB	ZO/ZO/NS	NS/PS/PS	NM/PS/PS	NM/PM/PS	NM/PB/PS	NB/PB/PB
	PB	ZO/ZO/PB	ZO/ZO/PM	NM/PS/PM	NM/PM/PM	NM/PM/PS	NB/PB/PS	NB/PB/PB

The scaling factor is the key in the design of variable domain of discourse fuzzy controller. In this paper, we design a new scalability factor based on the function model. The formula is as follows:

$$\beta(x) = \left(1 + |x| - \frac{x^2 + 1}{|x| + 1}\right)^{\tau}, 0 < \tau \leq 1\cdot \quad (3)$$

Given the fixed fuzzy controller, the input and output domain is X = [−E, E], Y = [−U, U], where E and U are positive real numbers. Relatively to the variable domain, X and Y is the initial domain, in this paper is set to be E = 3, U = 1. According to the definition of the scaling factor, the scaling factor must satisfy the duality, zero-preserving, monotony, coordination and regularity [14]. Prove as follows:

(1) Duality: Since Eq. (3) is an even function, it is a constant and bigger than 0, obviously satisfying the duality.
(2) Zero-preserving: when $x = 0$ is substituted into the formula (3), then we can get $\beta(x) = 0$, so it meets the zero preserving.
(3) Monotony: when we derivative the formula (3), we find $\beta^{'}(x) > 0$ the $\beta(x)$ monotonically increasing, so it meets the monotony.
(4) Coordination: according to $(\forall x \in U)(|x| \leq \beta(x)U)$, and Fig. 4 shows that $\beta(x)$ are on the top of the curve $y = x$, so it meets the coordination.
(5) Regularity: from the Fig. 4 we can see that it definitely meets the regularity.

For the scaling factor of the output domain, the selection principle is that the scaling factor should have a monotonic consistency with the deviation, and should have a monotonic opposite with the deviation [13], in this paper $\beta_p = \beta_d = 1 + |e| - \frac{e^2+1}{|e|+1}, \beta_{\mathrm{i}} = \frac{1}{|e|+0.7}$.

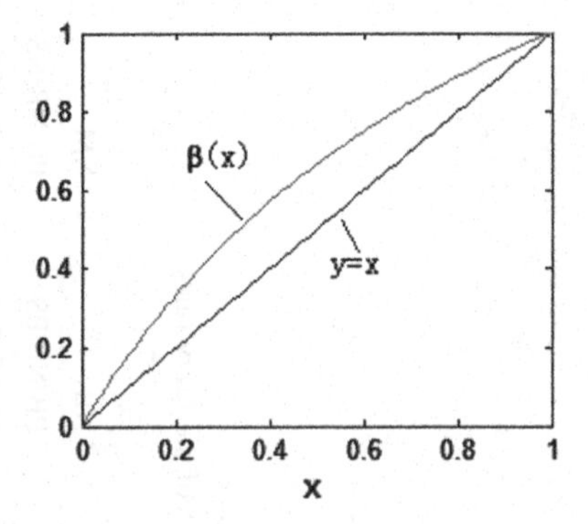

Fig. 4 Comparison of curves

3 Simulation

The software of MATLAB is used to build the system simulation model, and the simulation circuit diagram is drawn by a professional tool. The fuzzy PID control of variable output domain is carried out by using the scaling factor in this paper and the performance have been compared with that of Refs. [14] and [15].

In [14], the fuzzy PID control is compared with the traditional PID control. The object transfer function is $G(s) = \frac{6.7}{3.6s^2 + 23s + 1}$, $K_{P0} = 0.25$, $K_{i0} = 0.01$, $K_{d0} = 0.15$, and the step signal amplitude is taken by 1. Under the same conditions, the simulation results of the variable output domain fuzzy PID controller in this paper are compared with that in [14]. It can be seen from Fig. 5 that the control performance of the variable-output domain fuzzy PID controller has the following advantages over the fuzzy PID controller and so it is over the conventional PID controller: 1 faster regulation; 2 better dynamic perform; 3 smaller overshoot.

In [15], the traditional variable domain of discourse fuzzy PID controller to control the system of air-compressor, and fuzzy rule table is used to design the expansion and contraction factor. The control object transfer function is $G(s) = \frac{12}{0.04s^2 + 1.04s + 1}$, $K_{P0} = 0.4$, $K_{i0} = 0.32$, $K_{d0} = 0.02$, and the step signal amplitude is set to 0.65. Under the same conditions, the results of variable output domain fuzzy PID controller in this paper are compared with that in [15]. And the simulation results of the comparison in Fig. 6 shows that the two have almost the same performance, but this paper adopts the new control structure and function type expansion factor, so it cuts down the workload of calculation and get a better real-time performance.

In order to illustrate the robustness of the fuzzy PID controller in this paper, Fig. 7 shows the simulation results when the parameters of the control object transfer function in the paper [15] are changed to $G(s) = \frac{10}{0.08s^2 + 1.08s + 1}$, and the system only produces an overshoot of 3.07%. Figure 8 shows that when we add a

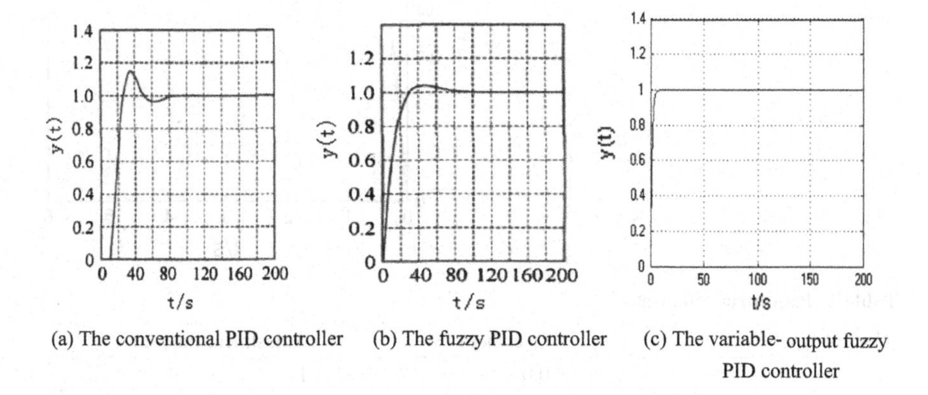

(a) The conventional PID controller (b) The fuzzy PID controller (c) The variable- output fuzzy PID controller

Fig. 5 Simulation results of three different PID controllers

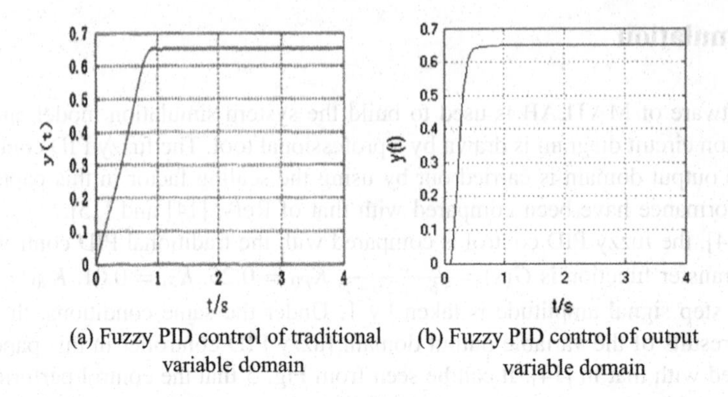

(a) Fuzzy PID control of traditional variable domain

(b) Fuzzy PID control of output variable domain

Fig. 6 The variable output domains of two controllers

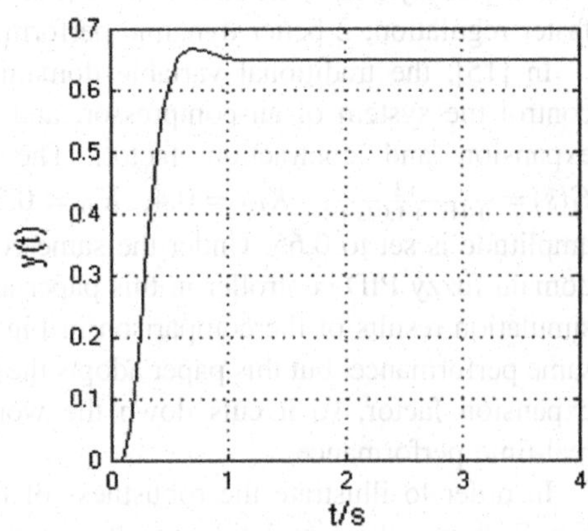

Fig. 7 Simulation results after parameter changes

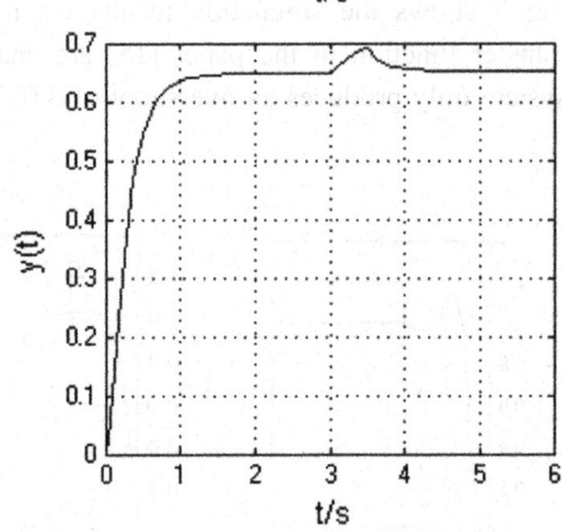

Fig. 8 Simulation results after adding interference

Table 2 Performance indicators

Control type	Overshoot (%)	Adjustment time (s)
A	0.0	1
B1	0.0	1
B2 (Parameters changing)	3.07	1.25

step amplitude of 0.05 where t = 3 to 3.5 s, the controller can quickly return to the set value, the Performance indicators show in the Table 2, where A means the traditional variable domain of discourse fuzzy controller, B1 and B2 represent variable output domain fuzzy controller (B2 represent the effect after changing the control-object parameters).

From Figs. 6, 7, 8 and Table 2, we can find that this variable output domain fuzzy PID controller has a stronger robustness and anti-jamming capability.

4 Conclusions

Based on the theory of variable domain of discourse and the fuzzy PID control strategy, this paper has simplified the structure of fuzzy PID controller with traditional variable domain of discourse, proposed a design method of fuzzy PID controller with variable output domain, and designed a new function type expansion factor. The results have shown that it can adjust the fuzzy PID controller output domain according to the size of error in real-time, and not change the fuzzy controller input domain, thus reduce the amount of workload. The simulation results have shown that the fuzzy PID controller of the variable output domain is better than the conventional fuzzy PID control, and the dynamic response performance is better than that of the conventional fuzzy PID control. Compared with the traditional variable domain of discourse fuzzy PID controller, this controller has the advantage of small calculation, good real-time response, good robustness and anti-jamming capability.

Acknowledgements This work was supported by the Natural Science Foundation of Jiangsu Province of China (Grant No. BK20151290), Six Talent Peaks Project in Jiangsu Province (Grant No. 2015-DZXX-029).

References

1. Dongli C. Study on network control system based on fuzzy PID controller. Central South University;2012.
2. Xian W. Study on self-tuning PID controller of feed servo system based on fuzzy control. Tianjin University;2014.
3. Haigang G. Several new methods of adaptive fuzzy control in variable domain of discourse. Dalian University of Technology;2013.
4. Li HX. Adaptive fuzzy controllers based on variable domain of discourse. Sci China Ser E-Technol Sci. 1999;42:10–20.
5. Cheng S, Xiwen D, Xiaofang W. Study on the selection method of the scaling factor of fuzzy controller in variable domain of discourse. Inf Control. 2010;39:536–41.
6. Li HX, Miao ZH, Lee ES. Variable domain of discourse stable adaptive fuzzy control of a nonlinear system. Comput Math Appl. 2002;44:799–815.

7. Niuyun Y, Yinghe Z. Design and simulation of variable fuzzy adaptive PID controller. China Water Transp. 2010;1:63–5.
8. Haicheng L, Qilong J, Dong L. Additional adaptive fuzzy PID control of active magnetic levitation system. Comput Meas Control. 2015;23:1165–7.
9. Li LF, Liu XY, Chen WF. A variable domain of discourse fuzzy control algorithm based on fuzzy neural network. In: Proceedings of the 7th world congress on intelligent control and automation;2008. p. 4352–356.
10. Yuntao Z, Jing W, Xinliang X, et al. Variable domain of discourse fuzzy control based on multi-layer ant colony algorithm. Pattern Recogn Artif Intell. 2009;22:794–8.
11. Yongping P, Qinruo W. Variable domain of discourse adaptive fuzzy PID controller design. Electr Autom. 2007;293:9–11.
12. Hongxing L. Mathematical essence of fuzzy control and design of a class of high precision fuzzy controller. Control Theory Appl. 1997;14:868–76.
13. Yan Z, Junping G. Study and simulation of adaptive fuzzy PID method for variable domain of discourse. J Air-Force Eng Univ (Nat Sci Ed). 2005;05:11–3.
14. Yixing Y, Dalian C, Aijun Z. Fuzzy adaptive PID controller and simulation. Ship Electron Eng. 2010;4:127–30.
15. Wenhan Z. Variable domain of discourse fuzzy PID in the application of air compressor control research. Kunming University of Science and Technology;2013.

Robust H_∞ Control of Non-cooperative Rendezvous Based on θ-D Method

Mingdong Hou and Yingmin Jia

Abstract This paper mainly investigates the space non-cooperative rendezvous control problem by robust H_∞ method which is an effective approach to attenuate the influence of parametric uncertainties and external interference in system. First of all, the 6-DOF coupling translation-rotation model established in line-of-sight (LOS) coordinate system is developed. And it is proved that when a Hamilton-Jacobi-Inequality (HJI) is satisfied, non-cooperative rendezvous system is asymptotic stable at origin with a desired H_∞ performance. Subsequently, the H_∞ controller is proposed using the solutions to the HJI. To get an approximate solution of the HJI effectively, the modified θ-D method is introduced. Finally, the numerical simulations prove the availability of the proposed control scheme.

Keywords Non-cooperative rendezvous · 6-DOF coupled system · Robust H_∞ control · Modified θ-D method

1 Introduction

Space rendezvous means two or more spacecrafts run to a same point on the orbit at the same time. It is the precondition for plenty of space tasks such as space docking technology, space junk recycling and so on. Many design theories for spacecraft rendezvous control are proposed. The C-W and T-H equations are used to establish the orbit models [1, 2]. The attitude equations are often built based on euler angle or quaternion [3, 4]. And kinds of control methods such as sliding-mode control, optimal control, adaptive control and robust H_∞ are used for spacecraft rendezvous control problem [5–7].

M. Hou · Y. Jia (✉)
The Seventh Research Division and the Center for Information and Control,
School of Automation Science and Electrical Engineering,
Beihang University (BUAA), Beijing 100191, China
e-mail: ymjia@buaa.edu.cn

M. Hou
e-mail: 603483879@qq.com

Y. Jia et al. (eds.), *Proceedings of 2017 Chinese Intelligent Systems Conference*, Lecture Notes in Electrical Engineering 459,
https://doi.org/10.1007/978-981-10-6496-8_21

In the early years, the rendezvous targets in researches are cooperative. However, when the targets of rendezvous are non-cooperative, such as enemy satellites, space junk and small planet, the result above can not be applied any more. Non-cooperative targets mean the space objects that can't exchange information with tracing vehicle actively. To obtain the motion information of non-cooperative target, the sensor on chasing spacecraft should be aimed at the non-cooperative target when running on the orbit. To meet the requirements of sensing, the translation and rotation have to be adjusted at the same time. Therefore, the coupling of attitude and orbit is not ignorable in non-cooperative rendezvous mission. In [8], the 6-DOF coupling motion model on the line-of-sight (LOS) coordinate system is proposed for non-cooperative rendezvous system. But the parametric uncertainties and external disturbance are still not considered, which may lead to the instability of system and the failure to rendezvous. To reduce the influence of parametric uncertainties and external interference, robust control is the most effective method to be chosen.

In space rendezvous field, kinds of robust control methods for instance optimal control [9], adaptive control [7] and H_∞ control [1] make contributions to the researches. And robust H_∞ control occupies the mainstream status in robust control field all the time. It can restrain the influence of parametric uncertainties and external interference by assuming the H_∞ norm of system parameters is no more than a certain range. And H_∞ control has been used to build linearized 3-DOF orbit equation for cooperative rendezvous [10]. But non-cooperative rendezvous model is a 6-DOF coupled nonlinear system, and its H_∞ controller is difficult to get. As far as we know, there are few researches about robust H_∞ control of non-cooperative rendezvous problem in the existing literatures.

In this paper, translation motion model under LOS and rotation motion model based on quaternion are established considering parametric uncertainties and external interference. And it is demonstrated that the non-cooperative rendezvous system has H_∞ performance. To get the H_∞ controller for the nonlinear system, the Hamilton-Jacobi-Bellman (HJB) equation has to be solved. Because the parametric uncertainties and external interference are considered, the HJB equation turns to be the Hamilton-Jacobi inequality (HJI). Since there is still not an effective to get the analytical solution of HJI, the modified θ-D method [11] is used to obtain an approximate solution to HJI. Finally, the numerical simulations shows that the designed controller can restrain the influence of parametric uncertainties and external interference.

2 Mathematical Model of Non-cooperative Rendezvous

2.1 Relative Orbit System

To express the relative orbit system accurately, the earth centered internal (ECI) frame, ascending node orbital (ANO) coordinate system, body coordinate system

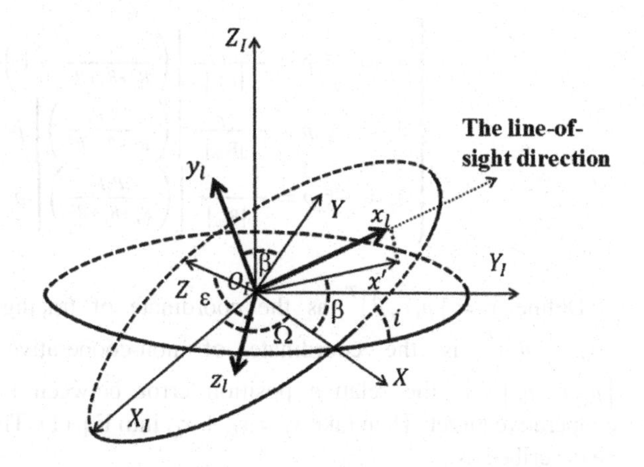

Fig. 1 Reference coordinate frame

and LOS are introduced. The ECI is defined as $O_IX_IY_IZ_I$ with X_I-axis aimed at the vernal point and Z_I-axis aimed at the north pole. The ANO is defined as O_IXYZ with X-axis towards the ascending node of spacecraft orbit plane. The body coordinate system is defined as $OX_bY_bZ_b$ with X_b-axis towards forward along vertical axis of spacecraft structure. The LOS of tracking spacecraft is defined as $O_lx_ly_lz_l$ with x_l-axis towards the line-of-sight direction. Thus the transformational relation among those coordinate systems is shown in Fig. 1, where Ω is the longitude of the ascending node. i is the orbital angle of inclination. ε is the angle between x_l-axis and the projection x' form x_l to plane XOY. β is the angle between projection x' and X-axis. It is assumed that ε, β satisfy: $\varepsilon \in \left(-\frac{\pi}{2}, \frac{\pi}{2}\right), \beta \in (-\pi, \pi)$. Then the transfer matrix from ECI to ANO is

$$T_{ir} = T_x(i)\,T_z(\Omega)$$

The transfer matrix from ANO to LOS is

$$T_{rl} = T_y(-\varepsilon)\,T_z(\beta)$$

The relative orbit system model is shown as follows:

$$\begin{cases} \ddot{\rho} - \rho\left(\dot{\varepsilon}^2 + \dot{\beta}^2\cos^2\varepsilon\right) - g_x = \frac{F_x}{m_0+\triangle m} \\ \rho\ddot{\varepsilon} + 2\dot{\rho}\dot{\varepsilon} + \rho\dot{\beta}^2\sin\varepsilon\cos\varepsilon - g_y = \frac{F_y}{m_0+\triangle m} \\ -\rho\ddot{\beta}\cos\varepsilon - 2\dot{\rho}\dot{\beta}\cos\varepsilon + 2\rho\dot{\varepsilon}\dot{\beta}\sin\varepsilon - g_z = \frac{F_z}{m_0+\triangle m} \end{cases} \quad (1)$$

where

$$\begin{cases} g_x = -\xi_x \rho = -\frac{\mu}{|\mathrm{R_{ch}}|^3}\left[\left(\frac{3R_x^2}{R_x^2+R_y^2+R_z^2} - 1\right)\right]\rho \\ g_y = -\xi_y \rho = -\frac{\mu}{|\mathrm{R_{ch}}|^3}\left[\left(\frac{3R_xR_y}{R_x^2+R_y^2+R_z^2}\right)\right]\rho \\ g_z = -\xi_z \rho = -\frac{\mu}{|\mathrm{R_{ch}}|^3}\left[\left(\frac{3R_xR_z}{R_x^2+R_y^2+R_z^2}\right)\right]\rho \end{cases}.$$

.

Define $x_l = [\rho, \varepsilon, \beta]^T$ as the coordinate of tracing vehicle in LOS. $x_{ld} = [\rho_d, \varepsilon_d, \beta_d]^T$ is the coordinate of non-cooperative target. $x_{le} = x_l - x_{ld} = [\rho_e, \varepsilon_e, \beta_e]^T$ is the relative position error between tracing vehicle and non-cooperative target. Then take $x_l = x_{le} + x_{ld}$ into Eq. (1). The orbit system model can be described as

$$\begin{aligned} &M_l(x_{le}, x_{ld})\ddot{x}_{le} + D_l(x_{le}, x_{ld})\dot{x}_{le} + K_l(x_{le}, x_{ld})x_{le} \\ &+L_l(x_{le}, x_{ld}) = B_{l1}w_l + B_{l2}u_l \end{aligned} \tag{2}$$

where

$$M_l(x_{le}, x_{ld}) = \begin{bmatrix} 1 & 0 & 0 \\ 0 & \rho_e + \rho_d & 0 \\ 0 & 0 & -(\rho_e + \rho_d)\cos(\varepsilon_e + \varepsilon_d) \end{bmatrix},$$

$$D_l(x_{le}, x_{ld}) = \begin{bmatrix} 0 & 0\,0 \\ 2(\dot{\varepsilon}_e + \dot{\varepsilon}_d) & 0\,0 \\ -2(\dot{\beta}_e + \dot{\beta}_d)\cos(\varepsilon_e + \varepsilon_d) & 0\,0 \end{bmatrix},$$

$$K_l(x_{le}, x_{ld}) = \begin{bmatrix} -(\dot{\varepsilon}_e + \dot{\varepsilon}_d)^2 - (\dot{\beta}_e + \dot{\beta}_d)^2\cos^2(\varepsilon_e + \varepsilon_d) + \xi_x & 0\,0 \\ (\dot{\beta}_e + \dot{\beta}_d)^2\sin(\varepsilon_e + \varepsilon_d)\cos(\varepsilon_e + \varepsilon_d) + \xi_y & 0\,0 \\ 2(\dot{\varepsilon}_e + \dot{\varepsilon}_d)(\dot{\beta}_e + \dot{\beta}_d)\sin(\varepsilon_e + \varepsilon_d) + \xi_z & 0\,0 \end{bmatrix},$$

$$L_l(x_{le}, x_{ld}) = M_l(x_{le}, x_{ld})\ddot{x}_{ld} + D_l(x_{le}, x_{ld})\dot{x}_{ld} + K_l(x_{le}, x_{ld})x_{ld},$$

$$B_{l1} = B_{l2} = \frac{1}{m_0 + \Delta m} \cdot I_{3\times 3}.$$

Let $e_1 = x_{le}, e_2 = \dot{x}_{le} = \dot{e}_1$. Due to $\varepsilon \in \left(-\frac{\pi}{2}, \frac{\pi}{2}\right)$, $\beta \in (-\pi, \pi)$, the matrix M_l is invertible. Then relative orbit system can be rewritten as

$$\begin{aligned} \dot{e}_1 &= e_2 \\ \dot{e}_2 &= -M_l^{-1} D_l e_2 - M_l^{-1} K_l e_1 - M_l^{-1} L_l + M_l^{-1} B_{l1} w_l + M_l^{-1} B_{l2} u_l \end{aligned} \quad (3)$$

2.2 Relative Rotation System Model

Let $q = \left(\bar{q}^T, q_4\right)^T$ be unit quaternion and ω be the angular velocity of the tracing spacecraft. q_d and ω_d denote those of non-cooperative target. The attitude and angular velocity errors between tracing vehicle and non-cooperative target are given by $q_e = q_d^{-1} \circ q$, $\omega_e = \omega - C\omega_d$, where

$$\begin{aligned} q_e &= \left(\bar{q}_e^T, q_{4e}\right)^T \\ \bar{q}_e &= q_{4d}\bar{q} - q_4\bar{q}_d + \bar{q}^\times \bar{q}_d, \\ q_{4e} &= q_{4d}q_4 + \bar{q}_d^T \bar{q} \end{aligned}$$

$$C\left(q_e\right) = \left(q_{4e}^2 - \bar{q}_e^T \bar{q}_e\right) I_{3\times3} + 2\bar{q}_e\bar{q}_e^T - 2q_{4e}\bar{q}_e^\times,$$

$$p^\times = \begin{bmatrix} 0 & -p_3 & p_2 \\ p_3 & 0 & -p_1 \\ -p_2 & p_1 & 0 \end{bmatrix}.$$

$C\left(q_e\right)$ is the rotation matrix of q_e and $\|C\| = 1$, $\dot{C} = -\omega_e^\times C$. $p^\times$ denotes the cross product operator of a vector $p = \left[p_1, p_2, p_3\right]^T$. Then the relative rotation model can be described as

$$\begin{aligned} \dot{\bar{q}}_e &= \tfrac{1}{2}\left(q_{4e} I_{3\times3} + \bar{q}_e^\times\right)\omega_e \\ \dot{\omega}_e &= -J_0^{-1}\left(\omega_e + C\omega_d\right)^\times J_0\omega_e - J_0^{-1}\left(\omega_e + C\omega_d\right)^\times J_0 C\omega_d \\ &- \left(C\dot{\omega}_d - \omega_e^\times C\omega_d\right) + \triangle\Phi + \left(J_0^{-1} + \delta J\right) d + \left(J_0^{-1} + \delta J\right)\tau \end{aligned} \quad (4)$$

where

$$\delta J = -J_0^{-1} \triangle J \left(I_{3\times3} + J_0^{-1} \triangle J\right)^{-1} J_0^{-1},$$

$$\begin{aligned} \triangle\Phi = &-\delta J\omega_e^\times J\omega_e - J_0^{-1}\omega_e^\times \triangle J\omega_e - \delta J\omega_e^\times JC\omega_d \\ &-J_0^{-1}\omega_e^\times \triangle JC\omega_d - \delta J\left(C\omega_d\right)^\times J\omega_e - J_0^{-1}\left(C\omega_d\right)^\times \delta J\omega_e. \\ &-\delta J\left(C\omega_d\right)^\times J\left(C\omega_d\right) - J_0^{-1}\left(C\omega_d\right)^\times \triangle JC\omega_d \end{aligned}$$

2.3 6-DOF Coupled Tracking Error Model

From the results above, the relative orbit model is presented in LOS F_l and relative rotation model is presented in body frame F_b. Since the control input is in F_b, the transfer matrix from F_b to F_l should be considered. The transfer matrix is $T_b^l = T_i^l \left(T_i^b\right)^T$, where $T_i^l = T_{ir} \cdot T_{rl} = T_x(i) T_z(\Omega) T_y(-\varepsilon) T_z(\beta)$, $T_i^b = \bar{q}_e \bar{q}_e^T + \left(q_{4e} I_{3\times3} - \bar{q}_e^{\times}\right)^2$.

Define the state vector $x = \left[e_1, e_2, \bar{q}_e, \omega_e\right]^T$, the coupled error system of non-cooperative target rendezvous is given as

$$\begin{aligned} \dot{x} &= f(x) + \triangle f(x) + \left(B_2(x) + \triangle B_2(x)\right) u + \left(B_1(x) + \triangle B_1(x)\right) w \\ z &= Cx + Du \end{aligned} \tag{5}$$

where $u = \left[u_l^T, \tau^T\right]^T$ is the control input. $w = \left[w_l^T, d^T\right]^T$ is the external disturbance. z is the output vector. $C^T C = Q_0, D^T(C, D) = \left(0, R_0\right), Q_0$ and R_0 are positive definite matrices.

$$f(x) = \begin{bmatrix} e_2 \\ -M_l^{-1} D_l e_2 - M_l^{-1} K_l e_1 - M_l^{-1} L_l \\ \frac{1}{2}\left(q_{4e} I_{3\times3} + \bar{q}_e^{\times}\right)\omega_e \\ -J_0^{-1}\left(\omega_e + C\omega_d\right)^{\times} J_0 \omega_e \\ -J_0^{-1}\left(\omega_e + C\omega_d\right)^{\times} J_0 C\omega_d \\ -\left(C\dot{\omega}_d - \omega_e^{\times} C\omega_d\right) \end{bmatrix}, \triangle f(x) = \begin{bmatrix} 0 \\ 0 \\ 0 \\ \triangle\Phi \end{bmatrix},$$

$$B_2(x) = \begin{bmatrix} 0 & 0 \\ \frac{T_b^l M_l^{-1} I_{3\times3}}{m_0} & 0 \\ 0 & 0 \\ 0 & J_0^{-1} \end{bmatrix}, \triangle B_2(x) = \begin{bmatrix} 0 & 0 \\ -\frac{\triangle m}{m_0(m_0+\triangle m)} T_b^l M_l^{-1} I_{3\times3} & 0 \\ 0 & 0 \\ 0 & \delta J \end{bmatrix},$$

$$B_1(x) = \begin{bmatrix} 0 & 0 \\ \frac{M_l^{-1} I_{3\times3}}{m_0} & 0 \\ 0 & 0 \\ 0 & J_0^{-1} \end{bmatrix}, \triangle B_1(x) = \begin{bmatrix} 0 & 0 \\ -\frac{\triangle m}{m_0(m_0+\triangle m)} M_l^{-1} I_{3\times3} & 0 \\ 0 & 0 \\ 0 & \delta J \end{bmatrix}.$$

Assumption 1 The norms of uncertain parameters $\| \triangle m\|$ and $\| \triangle J\|$ are bounded.

Definition 1 If the non-cooperative rendezvous system satisfies:

(1) The rendezvous system (5) is asymptotic stable.

(2) The inequality is satisfied when it is given zeros initial condition and constant $\gamma > 0$:

$$J = \int_{t_0}^{\infty} x^T Q_0 x + u^T R_0 u - \gamma^2 \omega^T \omega dt \leq 0 \tag{6}$$

Then, it is concluded that the non-cooperative rendezvous system (5) has H_∞ performance.

3 Main Results

3.1 H_∞ Performance and HJI

Assumption 2 The system uncertain terms $\triangle f(x)$ and $\triangle B_2(x)$ are norm bounded and can be expressed as the following form

$$\triangle f(x) = F(x)\Lambda_1(x),\ \triangle B_2(x) = G(x)\Lambda_2(x),\ \triangle B_1(x) = H(x)\Lambda_3(x)$$

where $F(x), G(x), H(x) \in R^{12\times12}$ are known terms, and $\Lambda_1(x) \in R^{12\times1}$, $\Lambda_2(x)$, $\Lambda_3(x) \in R^{12\times6}$ are unknown and uncertain terms.

Assumption 3 The uncertain terms $\Lambda_1(x)$, $\Lambda_2(x)$ and $\Lambda_3(x)$ satisfy the norm constraints $\|\Lambda_1(x)\| \leq \|E_1 x\|$ [12] and $\|\Lambda_2(x)\| \leq \|E_2\|$, $\|\Lambda_3(x)\| \leq \|E_3\|$.

Theorem 1 *Define $V(x) = \min_u \int_{t_0}^{\infty} x^T Q_0 x + u^T R_0 u - \gamma^2 w^T w dt$ and $V_x = \frac{\partial V(x)}{\partial x}$. The non-cooperative rendezvous system* (5) *has robust H_∞ performance when there are smooth scalars $\lambda_1 > 0$, $\lambda_2 > 0$ and $\lambda_3 > 0$ in the following HJI*

$$\Phi(x) = V_x f(x) - \frac{1}{4} V_x \widetilde{G}(x) \widetilde{R} \widetilde{G}^T(x) V_x^T + x^T \widetilde{Q} x \leq 0 \tag{7}$$

where

$$\widetilde{G}(x) = \left[B_1(x)P^{-1}, F(x), G(x), H(x), B_2(x)\right],$$

$$\widetilde{R} = diag\left\{-\frac{1}{\gamma^2}I_3, -\lambda_1 I_3, -\lambda_2 I_3, -\lambda_3 I_3, R_2^{-1}\right\}, R_2 = \frac{1}{\lambda_2}E_2^T E_2 + R_0,$$

$$\widetilde{Q} = \frac{1}{\lambda_1}E_1^T E_1 + Q_0.$$

Proof If there is a positive definite solution $V(x)$ to the HJI, then

$$\begin{aligned}\Psi &= \tfrac{dV(x)}{dt} + \|z\|^2 - \gamma^2 \|w\|^2 \\ &= V_x\dot{x} + x^T C^T Cx + u^T D^T Du - \gamma^2 w^T w \\ &= V_x f(x) + V_x B_2(x) u + V_x B_1(x) w \\ &+ V_x F(x) \Lambda_1(x) x + V_x G(x) \Lambda_2(x) u + V_x H(x) \Lambda_3(x) w \\ &+ x^T C^T Cx + u^T D^T Du - \gamma^2 w^T w\end{aligned} \tag{8}$$

Let $M = I - \frac{1}{\gamma^2 \lambda_3} E_3^T E_3 = P^T P$. From $2ab \le a^2 + b^2$ and Assumption 3, it has

$$\begin{aligned}&\Psi \le V_x f(x) + \tfrac{1}{4} V_x \{\tfrac{1}{\gamma^2} B_1(x) P^{-1} \left(P^{-1}\right)^T B_1^T(x) \\ &+ \lambda_1 F(x) F^T(x) + \lambda_2 G(x) G^T(x) + \lambda_3 H(x) H^T(x)\} V_x^T \\ &+ x^T \left(\tfrac{1}{\lambda_1} E_1^T E_1 + C^T C\right) x + u^T \left(\tfrac{1}{\lambda_2} E_2^T E_2 + D^T D\right) u + \tfrac{1}{\lambda_3} w^T E_3^T E_3 w \\ &+ V_x B_2(x) u - \gamma^2 \left\| Pw - \tfrac{1}{2\gamma^2} \left(P^{-1}\right)^T B_1^T(x) V_x^T \right\|^2 \\ &= \Phi(x) - \gamma^2 \left\| Pw - \tfrac{1}{2\gamma^2} \left(P^{-1}\right)^T B_1^T(x) V_x^T \right\|^2\end{aligned} \tag{9}$$

Then we have

$$\frac{dV(x)}{dt} + z^T z - \gamma^2 \omega^T \omega \le -\gamma^2 \left\| Pw - \frac{1}{2\gamma^2} \left(P^{-1}\right)^T B_1^T(x) V_x^T \right\|^2 \tag{10}$$

If the interference equal to 0, it is satisfied that $\frac{dV(x)}{dt} \le -z^T z$. Because Q_0 is positive definite, C is column full rank. So we can get that the rendezvous system (5) is zeros-state observable. Then, we can get that the system is asymptotic stable at the equilibrium. Furthermore, by integrating both sides of (10) from 0 to ∞, we have the H_∞ performance index (6):

$$\int_0^\infty x^T Q_0 x + u^T R_0 u dt \le \gamma^2 \int_0^\infty \omega^T \omega dt \tag{11}$$

It shows that 6-DOF system has H_∞ performance and the ability to resist the influence from parametric uncertainties and external interference. To get the robust H_∞ controller, HJI (7) has to be solved. But there is still not an effective method to obtain its analytical solution. To obtain the controller, the modified θ-D method is introduced in the following section, which is effective and feasible to get the approximate solution of the HJI.

3.2 The Controller of Rendezvous System

The θ-D method is effective to sovle the HJB equation approximatively. However, non-cooperative rendezvous system is a highly nonlinear system with parametric uncertainties and external interference. The controller is composed of the solution to HJI. So the modified θ-D method is proposed to solve the HJI approximatively.

First some perturbations should be added in Eq. (6) and get

$$J = \int_0^\infty x^T \left\{ Q_0 + \sum_{i=1}^{\infty} D_i \theta^i \right\} x + u^T R_0 u dt < \gamma^2 \int_0^\infty w^T w dt \tag{12}$$

Meanwhile, the HJI (7) turn into the form as follows

$$\Phi(x) = V_x f(x) - \frac{1}{4} V_x \widetilde{G}(x) \widetilde{R} \widetilde{G}^T(x) V_x + x^T \left(\widetilde{Q} + \sum_{i=1}^{\infty} D_i \theta^i \right) x \le 0 \tag{13}$$

Rewrite $f(x)$ and $\widetilde{G}$ as a linearized structure $f(x) = \left\{ A_0 + \theta \frac{A(x)}{\theta} \right\} x$ and $\widetilde{G}(x) = \widetilde{g}_0 + \theta \frac{\widetilde{g}(x)}{\theta}$, where $A_0 = A_0(0)$, $F_0 = F(0)$, $G_0 = G(0)$, $H_0 = H(0)$, $B_{20} = B_2(0)$, $\widetilde{g}_0 = \left[B_1 P^{-1}, F_0, G_0, H_0, B_{20} \right]$. And let (A_0, B_{20}) is a controllable pair at the same time. Then the solution $V(x)$ can be described as

$$V_x^T = \sum_{i=1}^{\infty} T_i(x, \theta) \theta^i x \tag{14}$$

where $T_i(x, \theta) = T_i^T(x, \theta)$. Take V_x^T into rewritten HJI (13) and get

$$\begin{aligned} &\textstyle\sum_{i=0}^{\infty} x^T \theta^i T_i \left\{ A_0 + \theta \frac{A(x)}{\theta} \right\} x + x^T \left\{ \left(\frac{1}{\lambda_1} E_1^T E_1 + Q_0 \right) + \sum_{i=1}^{\infty} D_i \theta^i \right\} x \\ &- \tfrac{1}{4} \textstyle\sum_{i=0}^{\infty} x^T \theta^i T_i \left\{ \widetilde{g}_0 + \theta \frac{\widetilde{g}(x)}{\theta} \right\} \widetilde{R} \left\{ \widetilde{g}_0^T + \theta \frac{\widetilde{g}^T(x)}{\theta} \right\} \sum_{i=0}^{\infty} T_i \theta^i x \le 0 \end{aligned} \tag{15}$$

Let the coefficient of $\theta^0 x^T x \le 0$. Then

$$x^T T_0 A_0 x - \frac{1}{4} x^T T_0 \widetilde{g}_0 \widetilde{R} \widetilde{g}_0^T T_0 x + x^T \left(\frac{1}{\lambda_1} E_1^T E_1 + Q_0 \right) x \le 0 \tag{16}$$

Transpose both sides of Eq. (16) and plus to (16)

$$T_0 A_0 + A_0^T T_0 - \frac{1}{2} T_0 \widetilde{g}_0 \widetilde{R} \widetilde{g}_0^T T_0 + 2 \left(\frac{1}{\lambda_1} E_1^T E_1 + Q_0 \right) \le 0 \tag{17}$$

The coefficient of $\theta^1 x^T x = 0$. Then

$$\begin{array}{l} T_1\left(A_0-\frac{1}{2}\widetilde{g}_0\widetilde{R}\widetilde{g}_0^T T_0\right)+\left(A_0^T-\frac{1}{2}T_0\widetilde{g}_0\widetilde{R}\widetilde{g}_0^T\right)T_1= \\ -T_0\frac{A(x)}{\theta}-\frac{A^T(x)}{\theta}T_0+\frac{1}{2}T_0\left(\widetilde{g}_0\widetilde{R}\frac{\widetilde{g}^T(x)}{\theta}+\frac{\widetilde{g}(x)}{\theta}\widetilde{R}\widetilde{g}_0^T\right)T_0-2D_1 \end{array} \tag{18}$$

The coefficient of $\theta^n x^T x = 0$. Then

$$\begin{array}{l} T_n\left(A_0-\frac{1}{2}\widetilde{g}_0\widetilde{R}\widetilde{g}_0^T T_0\right)+\left(A_0^T-\frac{1}{2}T_0\widetilde{g}_0\widetilde{R}\widetilde{g}_0^T\right)T_n= \\ -T_{n-1}\frac{A(x)}{\theta}-\frac{A^T(x)}{\theta}T_{n-1}+\frac{1}{2}[\sum_{j=0}^{n-1}T_j\widetilde{g}_0\widetilde{R}\widetilde{g}_0^T T_{n-j}+ \\ \sum_{j=0}^{n-1}T_j\left(\widetilde{g}_0\widetilde{R}\frac{\widetilde{g}^T(x)}{\theta}+\frac{\widetilde{g}(x)}{\theta}\widetilde{R}\widetilde{g}_0^T\right)T_{n-j-1}+\sum_{j=0}^{n-2}T_j\frac{\widetilde{g}(x)}{\theta}\widetilde{R}\frac{\widetilde{g}^T(x)}{\theta}T_{n-j-2}] \end{array} \tag{19}$$

Note that inequality (17) is an algebraic Riccati inequality, which can be transformed

$$T_0A_0+A_0^T T_0-\frac{1}{2}T_0B_{20}R_2^{-1}B_{20}T_0+\bar{Q}\leq 0 \tag{20}$$

where $\bar{Q}=\frac{1}{2\gamma^2}T_0B_1B_1^T T_0+\frac{\lambda_1}{2}T_0F_0F_0^T T_0+\frac{\lambda_2}{2}T_0G_0G_0^T T_0+\frac{\lambda_3}{2}T_0H_0H_0^T T_0+2\left(\frac{1}{\lambda_1}E_1^T E_1+Q_0\right)>0$. Because $\bar{Q}$ is a positive matrix, $\left(A_0,\bar{Q}^{\frac{1}{2}}\right)$ is a observable pair. And since $\left(A_0,B_{20}\right)$ is a controllable pair, we can get a positive definite solution of the inequality (17) by LMI tool in Matlab. And lyapunov equations (18) and (19) have the same coefficient matrix $A_0-\frac{1}{2}\widetilde{g}_0\widetilde{R}\widetilde{g}_0^T T_0$, which is constant matrix. Then, Eqs. (18) and (19) can be easily solved. Thus, control input can be expressed as

$$u=-R_2^{-1}B_2^T(x)V_x^T=-R_2^{-1}B_2^T(x)\sum_{i=0}^{\infty}T_i(x,\theta)\theta^i x \tag{21}$$

And D_i can be presented as the following pattern

$$\begin{array}{l} D_1=\frac{k_1}{2}e^{-l_1 t}\left[-T_0\frac{A(x)}{\theta}-\frac{A^T(x)}{\theta}T_0+\frac{1}{2}T_0\left(\widetilde{g}_0\widetilde{R}\frac{\widetilde{g}^T(x)}{\theta}+\frac{\widetilde{g}(x)}{\theta}\widetilde{R}\widetilde{g}_0^T\right)T_0\right] \\ \vdots \\ D_n=\frac{k_n}{2}e^{-l_n t}[-T_{n-1}\frac{A(x)}{\theta}-\frac{A^T(x)}{\theta}T_{n-1}+\frac{1}{2}\sum_{j=0}^{n-2}T_j\frac{\widetilde{g}(x)}{\theta}\widetilde{R}\frac{\widetilde{g}^T(x)}{\theta}T_{n-j-2} \\ +\frac{1}{2}\sum_{j=1}^{n-1}T_j\widetilde{g}_0\widetilde{R}\widetilde{g}_0^T T_{n-j}+\frac{1}{2}\sum_{j=0}^{n-1}T_j\left(\widetilde{g}_0\widetilde{R}\frac{\widetilde{g}^T(x)}{\theta}+\frac{\widetilde{g}(x)}{\theta}\widetilde{R}\widetilde{g}_0^T\right)T_{n-j-1} \end{array}$$

where k_i and $l_i>0, i=1,\ldots,n$ are adjustable parameters, which satisfies that the right side of Eqs. (18) and (19) are semi-positive.

4 Simulation Results

In this section, the numerical simulation of non-cooperative rendezvous system shows the effectiveness of the H_∞ controller. The state variables need to satisfy $\varepsilon\in\left(-\frac{\pi}{2},\frac{\pi}{2}\right)$, $\beta\in(-\pi,\pi)$. $m_0+\triangle m$ is the mass of spacecraft. $J_0+\triangle J$ is the

Table 1 System parameters of simulation

Initial state	$e_1(0) = [30, \pi/3, \pi/6]^T$
	$e_2(0) = [0.1, -0.1, 0.1]^T$
	$q_e(0) = [0.3, -0.2, 0.4, 0.84]^T$
	$\omega_e(0) = [-1.2, 2.0, -1.7]^T$
Desired state	$e_{d1} = [10, 0, 0]^T$
	$e_{d2} = [0, 0, 0]^T$
	$q_d = [0, 0, 0, 1]^T$
	$\omega_d = [0, 0, 0]^T$
Spacecraft	$m_0 = 100$ kg
	$J_0 = [40, 1.2, 0.9; 1.2, 17, 1.4; 0.9, 1.4, 15]$
Earth	$i = 0.6973$ rad
	$\Omega = 0.15242$
	$M_e = 5.97 \times 10^{24}$ kg
	$G = 6.67 \times 10^{-11} \mathrm{m}^3/\mathrm{kgs}^2$
Uncertainties	$\Delta J = diag[0.25sin(t), 0.25sin(2t), 0.25sin(3t)]$
	$\Delta m = 5sin(t)$
	$w_l = d = [0.25sin(t), 0.25sin(2t), 0.25sin(3t)]^T$
Weighting matrix	$Q_0 = 8I_{12}, R_0 = 0.1I_6$

mass inertia matrix. i and Ω donate the orbital angle of inclination and the longitude of the ascending node respectively. w_l and d are the external disturbance in orbit and attitude model. And when we let the attenuation level $\gamma = 0.1$, the performance index J satisfies $J \leq 0$. The parameters of non-cooperative rendezvous system are shown in Table 1.

Because ρ is in meters and ε, β are in degrees. So the fluctuation range of ρ is much bigger than that of ε and β. To show the variation trend of state clearly. ρ is presented in Fig. 2. ε and β is in Fig. 3. According to the relative orbit tracking error shown in Figs. 2 and 3, the desired track state is arrived in 25 s. And the target attitude is reached in 15 s with the relative attitude tracking error shown in Fig. 4. In Fig. 5, the input forces and torque are large relatively at the beginning. Then the required input decreases to zero gradually when the non-cooperative rendezvous system tends to be stable.

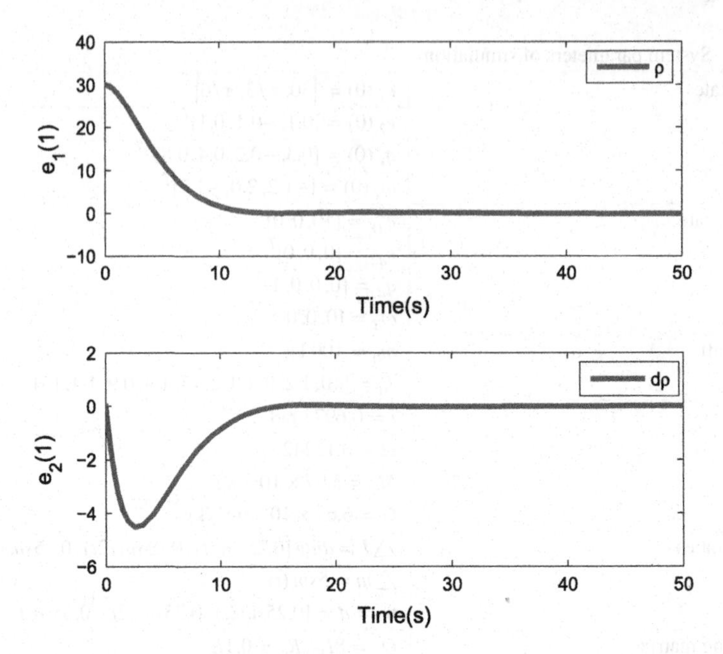

Fig. 2 Relative distance and velocity error

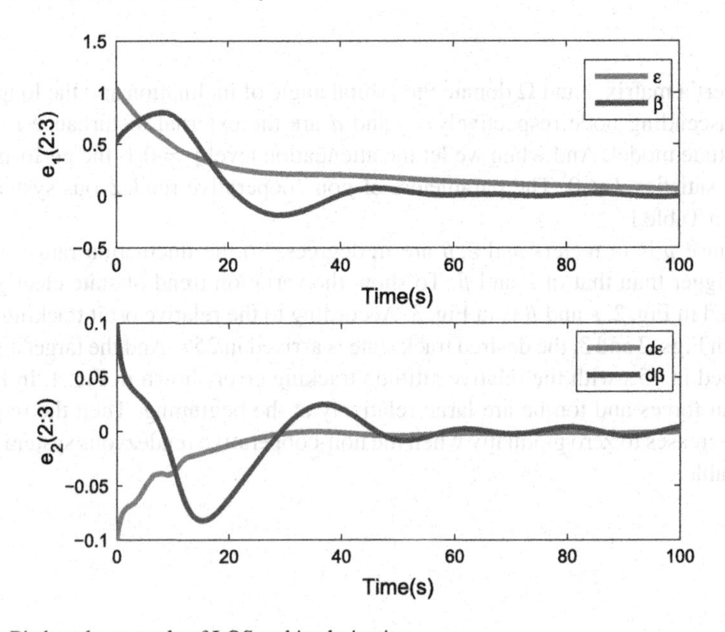

Fig. 3 Pitch and yaw angle of LOS and its derivative

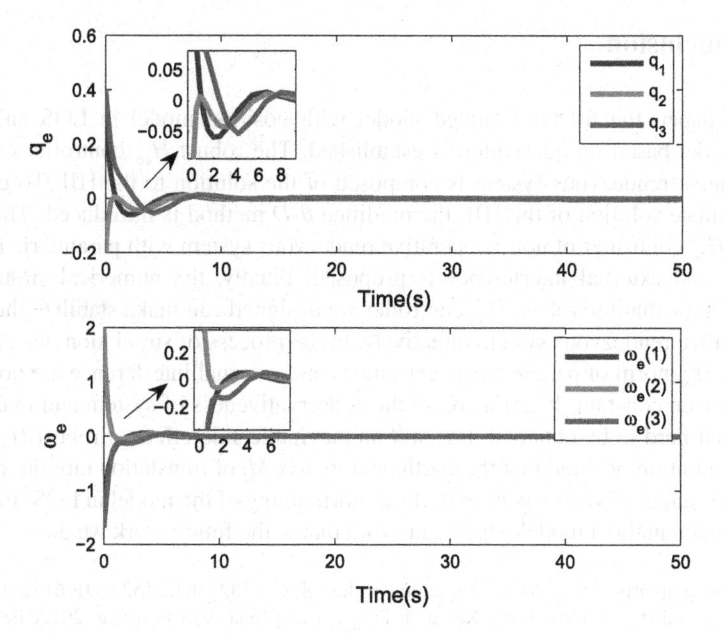

Fig. 4 Attitude and angular velocity tracking error

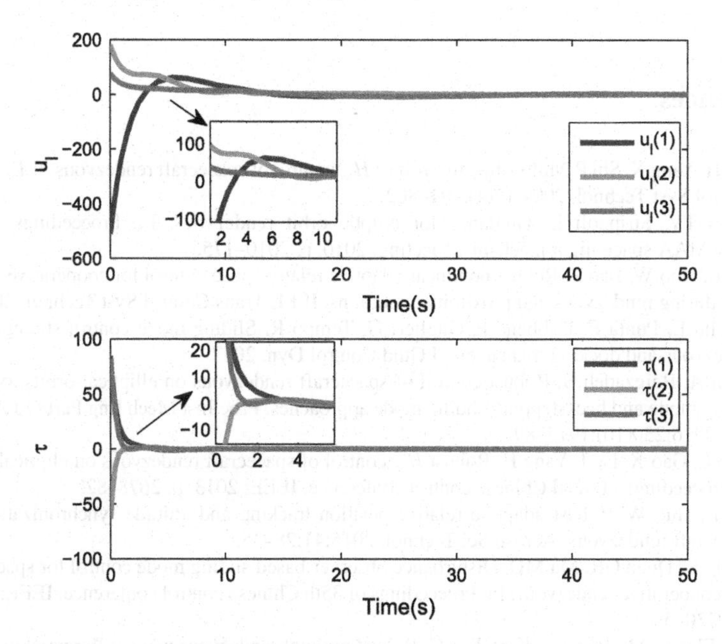

Fig. 5 Control input of orbit and attitude system

5 Conclusion

In this paper, the 6-DOF coupled model with position model in LOS and attitude model based on quaternion is established. The robust H_∞ controller of non-cooperative rendezvous system is composed of the solution to the HJI. To get the approximate solution of the HJI, the modified θ-D method is introduced. Then the robust H_∞ controller of non-cooperative rendezvous system with parametric uncertainties and external interference is proposed. Finally, the numerical simulation curves show that the robust H_∞ controller we designed can make stabilize the non-cooperative rendezvous system effectively. In the process of simulation, we assume that the H_∞ norm of parametric uncertainties and external interference are no more than a given constant. It can increase the conservativeness of system and make the controller hard to be obtained. It is still an inevitable side-effect of robust H_∞ control. In addition, we find that the coefficient matrix M_l of translation motion in LOS becomes singular when ρ goes to 0. It's a shortcoming of the model in LOS. Finding a new mathematical model which can avoid that is the future work we do.

Acknowledgements This work was supported by the NSFC (61327807, 6152 1091, 61520106010, 61134005) and the National Basic Research Program of China (973 Program: 2012CB821200, 2012CB821201)

References

1. Gao H, Yang X, Shi P. Multi-objective robust H_∞ control of spacecraft rendezvous. IEEE Trans Control Syst Technol. 2009;17(4):794–802.
2. Peters TV, Strippoli L. Guidance for elliptic orbit rendezvous. In: Proceedings of the AAS/AIAA space flight mechanics meeting; 2010. p. 2010–175.
3. Sun L, Huo W, Jiao Z. Robust nonlinear adaptive relative pose control for cooperative spacecraft during rendezvous and proximity operations. IEEE Trans Control Syst Technol. 2016.
4. Capello E, Punta E, Dabbene F, Guglieri G, Tempo R. Sliding-mode control strategies for rendezvous and docking maneuvers. J Guid Control Dyn. 2017.
5. Imani A, Beigzadeh B. Robust control of spacecraft rendezvous on elliptical orbits: optimal sliding mode and backstepping sliding mode approaches. Proc Inst Mech Eng Part G J Aerosp Eng. 2016;230(10):1975–89.
6. Shao L, Gao X, Lv J, Yang H. Robust H_∞ control of spacecraft rendezvous on eliptical orbit. In: Proceedings of 32nd Chinese control conference. IEEE; 2013. p. 2678–82.
7. Sun L, Huo W. Robust adaptive relative position tracking and attitude synchronization for spacecraft rendezvous. Aerosp Sci Technol. 2015;41:28–35.
8. Zhang K, Duan GR, Ma MD. Disturbance observer-based sliding mode control for spacecraft non-cooperative rendezvous. In: Proceedings of 35th Chinese control conference. IEEE; 2016. p. 10730–5.
9. Dell'Elce L, Martinusi V, Kerschen G. Robust optimal rendezvous using differential drag. In: AIAA/AAS astrodynamics specialist conference; 2014. p. 4161.
10. Gao X, Teo KL, Duan GR. Robust H_∞ control of spacecraft rendezvous on elliptical orbit. J Franklin Inst. 2012;349(8):2515–29.
11. Huang Y, Jia Y, Matsuno F. Robust H_∞ control for spacecraft formation flying with coupled translational and rotation dynamics. In: American control conference. IEEE; 2016. p. 4059–64.

12. Shen T, Zang H, Tamura K. Riccati equation approach to robust L_2-gain synthesis for a class of uncertain nonlinear systems. Int J Control. 1996;64(6):1177–88.

Adaptive Controller for Flexible-Joint Robot

Lei Zhang, Yingmin Jia and Xuhui Lu

Abstract An adaptive controller is proposed for flexible-joint robot subject to parameter uncertainties. The backstepping control framework has been used to obtain the virtual controller and the actual control input. Besides, the derived controller is based on non-certainty-equivalent adaptive control methodology. A set of two-order filters are also embedded into the corresponding attractive manifold design. It is proved that the position of flexible-joint robot can stabilize towards the desired position, and moreover the estimate of uncertain parameters can converge to the real values to some degrees.

Keywords Flexible-joint robot · Parameter uncertainies · Non-certainty-equivalent adaptive controller · Second-order linear filters

1 Introduction

Many robot manipulators and the corresponding robot controllers have been successfully applied into several fields, according to the assumption that the joints of the robot manipulators are rigid [1]. However, in some cases, joint flexibility of robots is inevitable due to shaft wind-up, application of harmonic drives, and gear elasticity, etc. It has been verified by a series of experiments that the control performance of the robots would be reduced, if the joint flexibility is overlook in the controller design [2].

To cope with this problem, Spong [3] proposes a model to depict the rigid-link-flexible-joint robot. Besides, when the flexible stiffness goes to infinity, the coupling model will be simplified to be a rigid-link robot model. Thereafter, several control methods have been developed in this field. Albu-Schaffer et al. A master-slave control scheme for multiple flexible-joint robots is proposed in [4] where the

L. Zhang · Y. Jia (✉) · X. Lu
The Seventh Research Division and the Center for Information and Control,
School of Automation Science and Electrical Engineering,
Beihang University (BUAA), Beijing 100191, China
e-mail: ymjia@buaa.edu.cn

Y. Jia et al. (eds.), *Proceedings of 2017 Chinese Intelligent Systems Conference*, Lecture Notes in Electrical Engineering 459,
https://doi.org/10.1007/978-981-10-6496-8_22

velocity and acceleration information are estimated by model-based observer. Nuno et al. [5] also design a decentralized energy-shaping controller to drive multiple flexible-joint robots to the same position without the need of velocity information. The Lyapunov-Krasovskii function is also incorporated into the backstepping-based controller design in [6] to deal with communication time delay between flexible-joint robots.

However, the uncertainties can not be neglected during the operation of flexible-joint robots. Once the uncertainties are not taken in consideration in the controller design, the robustness of the closed-loop system will be deteriorated. Several methods have been designed to handle uncertainties in flexible-joint robots, including direct model reference adaptive control [7], sliding-mode control [8], backstepping-based control [9], adaptive inverse dynamics control [10]. Recently, a novel adaptive control methodology named attractive manifold method has been proposed which belongs to non-certainty-equivalent adaptive control framework [11]. In this control framework, the estimation of uncertain parameters in the dynamic equations is attributed into the attractiveness and invariance of a specific manifold in the state space, and thus some degrees of convergence of parameter estimation error is ensured. The idea of attractive manifold control has been applied into various fields such as spacecrafts [12], Missiles [13], stratospheric airship [14], etc. However, notice that flexible-joint robot is a typical under-actuated control system without cascade structure, which complicates the adaptive controller design.

In this paper, a novel non-certainty-equivalent adaptive controller has been derived to stabilize the flexible-joint robot. Due to the under-actuated feature of flexible-joint robot, here the backstepping-based control framework is utilized. Besides in order to estimate uncertain parameters, a set of two-order linear filters are incorporated into the attractive manifold design, such that some degrees of convergence of parameter estimation error is guaranteed.

This rest of this paper is organized as follows. Preliminaries are provided in Sect. 2, including modeling of flexible-joint robot, problem formulation and assumption. Non-certainty-equivalent adaptive controller is developed in Sect. 3 with rigorous proofs. Simulation and the results are shown in Sect. 4. Conclusions are drawn in Sect. 5.

2 Preliminaries

For an n-link flexible-joint robot, the equation of motion is obtained as [3]:

$$M(q)\ddot{q} + C(q,\dot{q})\dot{q} + g(q) = K(\theta - q), \tag{1a}$$

$$J\ddot{\theta} + K(\theta - q) = u, \tag{1b}$$

where $q \in \mathbb{R}^n$ is robot link angles, $\theta \in \mathbb{R}^n$ is motor angles, $M(q) \in \mathbb{R}^{n \times n}$ is link inertia matrix, diagonal matrix $J \in \mathbb{R}^{n \times n}$ is motor inertia matrix, matrix $C(q,\dot{q})\dot{q} \in \mathbb{R}^n$ is related to Coriolis and centripetal force, $g(q) \in \mathbb{R}^n$ is gravity force, diagonal matrix $K \in \mathbb{R}^{n \times n}$ represents joint stiffness, and $u \in \mathbb{R}^n$ is actuator control torque. Note that matrices $M(q)$ and J are both positive definite. This means that there exist positive constants λ^M_{min}, λ^M_{max}, λ^J_{min} and λ^J_{max}, such that $\lambda^M_{max} I_3 > M(q) > \lambda^M_{min} I_3$ for any $q \in \mathbb{R}^n$ and $\lambda^J_{max} I_3 > J > \lambda^J_{min} I_3$.

In this paper, the robot link angles q is controlled to move toward desired position q_d. Denote $e_1 = q - q_d$, $e_2 = \theta - q$, and the motion equations of flexible-joint robot (1a) and (1b) can be rewritten as

$$M(q)\ddot{e}_1 + C(q, \dot{q})\dot{e}_1 + g(q) = Ke_2, \tag{2a}$$

$$J\ddot{e}_2 + J\ddot{q} + Ke_2 = u. \tag{2b}$$

Correspondingly, a controller should be propose to ensure $\lim_{t\to\infty} e_1(t) = \mathbf{0}_n$ and $\lim_{t\to\infty} \dot{e}_1(t) = \mathbf{0}_n$, $\lim_{t\to\infty} e_2(t) = \theta^*$ and $\lim_{t\to\infty} \dot{e}_2(t) = \mathbf{0}_n$, where $\theta^* = K^{-1} g(q_d)$ is the desired angle error between link angle and motor angle. It should be pointed out that the corresponding physical parameters in $M(q)$, $C(q, \dot{q})$, $g(q)$ and J are uncertain. In order to approach this problem, the following assumption is introduced [4].

Assumption 1 The signals q, $\dot{q}$, $\ddot{q}$, $q^{(3)}$, $q^{(4)}$ are uniformly bounded all the time.

According to the link dynamics (1a), together with Assumption 1, it is obtained that θ, $\dot{\theta}$ and $\ddot{\theta}$ are all bounded. Correspondingly, e_1, $\dot{e}_1$, $\ddot{e}_1$, $e_1^{(3)}$, $e_1^{(4)}$, e_2, $\dot{e}_2$ and $\ddot{e}_2$ are also bounded.

3 Controller Design and Stability Analysis

Since the flexible-joint robot consists two subsystems (2a) and (2b), we employ backstepping methodology in controller design. First a virtual controller is designed in Sect. 3.1, and control input u is then derived in Sect. 3.2. The controller is formulated in Sect. 3.3 and its control performance is also verified in Sect. 3.4 with rigorous proof.

3.1 Virtual Controller Design

Consider link angle error dynamics (2a). After adding same terms in both sides of Eq. (2a), we obtain

$$k_2[M(q)e_1] + k_1[M(q)e_1]' + [M(q)e_1]'' = W\alpha + Ke_2, \tag{3}$$

where $k_1 > 0$, $k_2 > 0$, $\alpha \in \mathbb{R}^{m1}$ is a vector of uncertain parameters in link error dynamics (2a), and $W \in \mathbb{R}^{n \times m1}$ is the according regressor matrix defined as

$$W\alpha = \ddot{M}(q)e_1 + 2\dot{M}(q)\dot{e}_1 + k_1 M(q)\dot{e}_1 + k_1 \dot{M}(q)e_1 + k_2 M(q)e_1 - C(q,\dot{q})\dot{e}_1 - g(q). \tag{4}$$

Then we introduce two second-order filters as

$$\ddot{W}_f + k_1 \dot{W}_f + k_2 W_f = W, \tag{5a}$$

$$\ddot{e}_f + k_1 \dot{e}_f + k_2 e_f = e_2. \tag{5b}$$

Substituting two filters (5a) and (5b) into (3) gives

$$\ddot{\epsilon}_1(t) + k_1 \dot{\epsilon}_1(t) + k_2 \epsilon_1(t) = 0, \tag{6}$$

where $\epsilon_1(t) = M(q)e_1 - W_f \alpha - K e_f$. Note that $\epsilon_1(t)$ will exponentially converge to zero according to (6). Therefore like [12] the signal $\epsilon_1(t)$ is neglected in the subsequent controller design, and we have

$$M(q)e_1 - W_f \alpha - K e_f = 0. \tag{7}$$

Based on the filters (5b) and Eq. (7), three extra filters are also derived as

$$\ddot{e}_{fc} + k_1 \dot{e}_{fc} + k_2 e_{fc} = e_c, \tag{8a}$$

$$\ddot{e}_{fr} + k_1 \dot{e}_{fr} + k_2 e_{fr} = e_r, \tag{8b}$$

$$\dot{e}_{f1} + \bar{k} e_{f1} = e_1 \tag{8c}$$

where $\bar{k} > 0$, $e_{fc} + e_{fr} = e_f$ and correspondingly $e_c + e_r = e_2$. Substituting (8a) and (8c) into (7) yields

$$M(q)\dot{e}_{f1} = -\bar{k} M(q) e_{f1} + W_f \alpha + K(e_{fc} + e_{fr}). \tag{9}$$

In (8a), e_{fc} is regarded as virtual controller of (7), and is proposed as $e_{fc} = -K^{-1} W_f(\hat{\alpha} + \beta)$. $\hat{\alpha} + \beta$ is the estimate of the uncertain parameter vector α, where $\beta = \gamma_1 W_f^T e_{f1}$ with $\gamma_1 > \frac{1}{\bar{k}}$, and $\hat{\alpha}$ obeys the following dynamics

$$\dot{\hat{\alpha}} = -\gamma_1 \dot{W}_f^T e_{f1} + \gamma_1 W_f^T \bar{k} e_{f1}. \tag{10}$$

Substituting e_{fc} into (9) yields

$$\dot{e}_{f1} = -\bar{k} e_{f1} - M^{-1}(q) W_f z + M^{-1}(q) K e_{fr}, \tag{11}$$

where $z = \hat{\alpha} + \beta - \alpha$ is estimation error of α. Differentiating z, and substituting (10) and (11) into the resulting equation gives

$$\dot{z} = \dot{\alpha} + \dot{\beta} = -\gamma_1 W_f^T M^{-1}(q) W_f z + r_1 W_f^T M^{-1}(q) K e_{fr}. \tag{12}$$

Here we provide a Lyapunov candidate function V_1 as $V_1 = \frac{1}{2} z^T z + \frac{1}{2}\lambda_{min} e_{f1}^T e_{f1}$, where λ_{min} is set in Property 1. Differentiating V_1 along with (11) and (12), and utilizing Young's inequality yields

$$\dot{V}_1 \leq -\frac{3}{4}(\gamma_1 - \frac{1}{\bar{k}}) z^T W_f^T M^{-1}(q) W_f z - \frac{\bar{k}\lambda_{Mmin}}{3} e_{f1}^T e_{f1} + \lambda e_{fr}^T e_{fr}, \tag{13}$$

where positive constant λ satisfies $(\gamma_1 + \frac{3}{4\bar{k}}) K^T M^{-1}(q) K \leqslant \lambda I$.

3.2 Input Controller Design

Consider motor angle error dynamics (2b). Set $e_b = e_2 - k_2 e_{fr}$ and $u = \bar{u} + K e_2$. After adding same terms in its both sides of Eq. (2b), we have

$$J\ddot{e}_{fr} + k_1 J\dot{e}_{fr} + k_2 J e_{fr} = \frac{\bar{u}}{k_2} + N\phi, \tag{14}$$

where $\phi \in \mathbb{R}^{m2}$ is vector of uncertain parameters in (2b), and $N \in \mathbb{R}^{n\times m2}$ is the corresponding regressor matrix defined as

$$N\phi = -\frac{J\ddot{q}}{k_2} - \frac{J\ddot{e}_b}{k_2} + k_1 J\dot{e}_{fr} + k_2 J e_{fr}. \tag{15}$$

Then denote k_3 and k_4 are two eigenvalues of equation $x^2 + k_1 x + k_2 = 0$. Since $k_1 > 0$ and $k_2 > 0$ defined before, it is obtained that $k_3 > 0$ and $k_4 > 0$. Therefore two filters on N and u are constructed as

$$\dot{N}_f + k_4 N_f = N, \tag{16a}$$

$$\dot{u}_f + k_4 u_f = \bar{u}, \tag{16b}$$

where the filter (16b) is introduced only for the controller design. Substituting (16a) and (16b) into (14) gives

$$\dot{\epsilon}_2(t) + k_4 \epsilon_2(t) = 0, \tag{17}$$

where $\epsilon_2(t) = J\dot{e}_{fr} + k_3 J e_{fr} - N_f\phi - \frac{u_f}{k_2}$. Due to the exponential convergence of $\epsilon_2(t)$, the variable ϵ_2 can be also omitted here, and we have

$$J\dot{e}_{fr} + k_3 J e_{fr} - N_f\phi - \frac{u_f}{k_2} = 0 \tag{18}$$

Then based on Eq. (18), the signal u_f is derived as

$$u_f = -k_2 N_f(\hat{\phi} + \eta). \tag{19}$$

In (19), $\hat{\phi} + \eta$ is the estimate of uncertain parameter vector ϕ, where $\eta = \gamma_2 N_f^T e_{fr}$ with $\gamma_2 > 0$, and the variable $\hat{\phi}$ satisfies the following dynamic equation

$$\dot{\hat{\phi}} = -\gamma_2 \dot{N}_f^T e_{fr} + \gamma_2 N_f^T k_3 e_{fr}. \tag{20}$$

Substituting u_f defined by Eq. (19) into Eq. (18) yields

$$\dot{e}_{fr} = -k_3 e_{fr} - J^{-1} N_f \tilde{\phi}, \tag{21}$$

where $\tilde{\phi} = \hat{\phi} + \eta - \phi$ denotes estimation error of ϕ. Then differentiating $\tilde{\phi}$ and substituting (20) and (8b) into the resulting equation gives

$$\dot{\tilde{\phi}} = \dot{\hat{\phi}} + \dot{\eta} = -\gamma_2 N_f^T J^{-1} N_f \tilde{\phi}. \tag{22}$$

Here we propose a Lyapunov candidate function V_2 as $V_2 = \frac{1}{3}\lambda_{Jmin}\frac{\gamma_2}{k_3} e_{fr}^T e_{fr} + \frac{1}{2}\tilde{\phi}^T\tilde{\phi}$, where $0 \leqslant \lambda_{Jmin} \leqslant \|J\| \leqslant \lambda_{Jmax}$. Differentiating V_2 and substituting Eqs. (18) and (22) into it yields:

$$\dot{V}_2 \leq -\frac{4}{9}\lambda_{min}^J \gamma_2 e_{fr}^T e_{fr} - \frac{\gamma_2}{2}\tilde{\phi}^T N_f^T J^{-1} N_f \tilde{\phi} \tag{23}$$

Here we introduce the Lyapunov candidate function of the closed-loop system:

$$V_3 = V_1 + sV_2 \tag{24}$$

where $s = \frac{9\lambda}{2\lambda_{min}^J \gamma_2}$. Differentiating V_3 and substituting (13) and (23) into the resulting equation yields

$$\dot{V}_3 \leq -\pi, \tag{25}$$

where

$$\begin{aligned} \pi = {} & \frac{3}{4}(\gamma_1 - \frac{1}{\bar{k}}) z^T W_f^T M^{-1}(q) W_f z + \frac{\bar{k}\lambda_{min}^M}{3} e_{f1}^T e_{f1} \\ & + \frac{2s}{9}\lambda_{min}^J \gamma_2 e_{fr}^T e_{fr} + \frac{s\gamma_2}{2}\tilde{\phi}^T N_f^T J^{-1} N_f \tilde{\phi}. \end{aligned} \tag{26}$$

It can be seen that $\pi \geqslant 0$ once $\gamma_1 > \frac{1}{\bar{k}}$.

3.3 Controller Formulation

Before the main result of this paper is provided, the adaptive controller should be formulated. Substituting the signal u_f (19) into (16b) gives

$$u = -k_2 N(\hat{\phi} + \eta) - k_2 \gamma_2 N_f N_f^T (\dot{e}_{fr} + k_3 e_{fr}) + K e_2, \tag{27}$$

where the signals related to motor angle error dynamics (2b) are designed as

$$\dot{N}_f = -k_4 N_f + N, \tag{28a}$$

$$N\phi = -\frac{J\ddot{q}}{k_2} - \frac{J\ddot{e}_b}{k_2} + k_1 J\dot{e}_{fr} + k_2 J e_{fr}, \tag{28b}$$

$$\dot{\hat{\phi}} = -\gamma_2 \dot{N}_f^T e_{fr} + \gamma_2 N_f^T k_3 e_{fr}, \tag{28c}$$

$$e_{fr} = e_f - e_{fc}, \tag{28d}$$

$$\dot{e}_{fr} = \dot{e}_f - \dot{e}_{fc}, \tag{28e}$$

$$e_b = e_2 - k_2 e_{fr}, \tag{28f}$$

$$\ddot{e}_b = \ddot{e}_2 - k_2 \ddot{e}_{fr} = \ddot{e}_2 - k_2(\ddot{e}_f - \ddot{e}_{fc}), \tag{28g}$$

$$\eta = \gamma_2 N_f^T e_{fr}, \tag{28h}$$

$$\ddot{e}_f = -k_1 \dot{e}_f - k_2 e_f + e_2. \tag{28i}$$

Besides, the according signals for link angle error dynamics (2a) are derived as

$$e_{fc} = -K^{-1} W_f(\hat{\alpha} + \beta), \tag{29a}$$

$$\ddot{W}_f = -k_1 \dot{W}_f - k_2 W_f + W, \tag{29b}$$

$$\begin{aligned} W\alpha = {} & \ddot{M}(q)e_1 + 2\dot{M}(q)\dot{e}_1 + k_1 M(q)\dot{e}_1 + k_1 \dot{M}(q)e_1 + k_2 M(q)e_1 \\ & - C(q,\dot{q})\dot{e}_1 - g(q), \end{aligned} \tag{29c}$$

$$\beta = \gamma_1 W_f^T e_{f1}, \tag{29d}$$

$$\dot{\hat{\alpha}} = -\gamma_1 \dot{W}_f^T e_{f1} + \gamma_1 W_f^T \bar{k} e_{f1}, \tag{29e}$$

$$\dot{e}_{f1} = -\bar{k} e_{f1} + e_1, \tag{29f}$$

where the first and second derivatives of e_{fc} are

$$\dot{e}_{fc} = -K^{-1} \dot{W}_f(\hat{\alpha} + \beta) - K^{-1} W_f \gamma_1 W_f^T e_1, \tag{30a}$$

$$\begin{aligned} \ddot{e}_{fc} = {} & -K^{-1} \ddot{W}_f(\hat{\alpha} + \beta) - 2K^{-1} \dot{W}_f \gamma_1 W_f^T e_1 \\ & - K^{-1} W_f \gamma_1 \dot{W}_f^T e_1 - K^{-1} W_f \gamma_1 W_f^T \dot{e}_1. \end{aligned} \tag{30b}$$

3.4 The Convergence Analysis

Based on the proposed controller, the main results of this paper are given below.

Theorem 1 *For the flexible-joint robot* (1a) *and* (1b) *with parameter uncertainties, Consider the proposed adaptive controller* (27), (28a)–(28i), (29a)–(29f), (30a) *and* (30b). *Then for the corresponding closed-loop system, the following properties hold:*

(P1): the signals z, e_{f1}, e_{fr}, $\tilde{\phi}$, W_f, N_f, e_f, e_{fc}, $\ddot{e}_b$ *are bounded.*

(P2): we have $\lim_{t\to\infty} e_1(t) = \lim_{t\to\infty} \dot{e}_1(t) = \mathbf{0}_n$, $\lim_{t\to\infty} e_2(t) = K^{-1}g(q_d)$, $\lim_{t\to\infty} \dot{e}_2(t) = \mathbf{0}_n$, *and* $\lim_{t\to\infty} W_f(t)z(t) = \lim_{t\to\infty} N_f(t)\tilde{\phi}(t) = \mathbf{0}_n$.

Proof First we prove Property 1. From (25), it is obtained that $V_3(t)$ is bounded and besides

$$\int_0^t \pi ds \leqslant V_3(0) - V_3(t) \leqslant V_3(0). \tag{31}$$

Therefore we have $e_{f1}, z, e_{fr}, \tilde{\phi}$ are bounded and $e_{f1} \in \mathcal{L}_2, W_f z \in \mathcal{L}_2, e_{fr} \in \mathcal{L}_2, N_f\tilde{\phi} \in \mathcal{L}_2$. Next we will prove that $\dot{\pi}$ is also bounded.

According to Assumption 1, we notice that $e_1, \dot{e}_1, \ddot{e}_1, e_1^{(3)}$ and $e_1^{(4)}$ are all bounded. Therefore e_{f1} and $\dot{e}_{f1}$ are bounded based on (29f), and besides the regressor matrix W is bounded in view of the definition of this matrix in (29c). This meaning that W_f, $\dot{W}_f$, $\ddot{W}_f$ are all bounded according to (29b). Besides $e_2 = \theta - q$ is also bounded in view of Assumption 1, thus e_f, $\dot{e}_f$ and $\ddot{e}_f$ are also bounded according to (28i). Based on the boundedness of W_f, $\tilde{\phi}$ and $\hat{\phi} + \eta$, the virtual controller e_{fc} (29a), its first-order derivative $\dot{e}_{fc}$ (30a) and the corresponding second-order derivative $\ddot{e}_{fc}$ (30b) are all bounded. Correspondingly, e_{fr}, $\dot{e}_{fr}$, $\ddot{e}_{fr}$ and $\ddot{e}_b = \ddot{e}_2 - k_2\ddot{e}_{fr}$ are also bounded. Therefore recalling the definition of regressor matrix N (28b), we obtain that N is bounded, and correspondingly N_f and $\dot{N}_f$ are also bounded from (28a).

Based on Property 1, we turn to proving Property 2. Based on the definition of η and $\hat{\phi}$ (28c), we have $\dot{\eta} + \dot{\hat{\phi}} = \dot{\tilde{\phi}} = \gamma_2 N_f^T(\dot{e}_{fr} + k_3 e_{fr})$, which means that $\dot{\tilde{\phi}}$ is bounded. Following this similar procedure, we can also obtain that $\dot{\hat{\alpha}} + \dot{\beta} = \dot{z} = \gamma_1 W_f^T e_1$ is also bounded. Remind that e_{f1}, $\dot{e}_{f1}$, e_{fr}, $\dot{e}_{fr}$, W_f, $\dot{W}_f$, N_f, $\dot{N}_f$, z, $\tilde{\phi}$. Therefore the boundedness of $\dot{\pi}$ is proved, based on the structure of π (26). Utilizing the barbalat's lemma, we have $\lim_{t\to\infty} \pi(t) = 0$, and correspondingly $\lim_{t\to\infty} e_{f1}(t) = \lim_{t\to\infty} e_{fr}(t) = \lim_{t\to\infty} W_f(t)z(t) = \lim_{t\to\infty} N_f(t)\tilde{\phi}(t) = \mathbf{0}_n$. Besides, since $\dot{e}_1$ is bounded, implying that $\ddot{e}_{f1}$ is bounded. Based on Barbalat's lemma, it is also obtained that $\lim_{t\to\infty} \dot{e}_{f1}(t) = \mathbf{0}_n$ and correspondingly $\lim_{t\to\infty} e_1(t) = \mathbf{0}_n$. Recall that $\ddot{e}_1$ is bounded according to Assumption 1, we can also get $\lim_{t\to\infty} \dot{e}_1(t) = \mathbf{0}_n$.

Besides, we will verify that $\lim_{t\to\infty} e_2(t) = K^{-1}g(q_d)$ and $\lim_{t\to\infty} \dot{e}_2(t) = \mathbf{0}_n$. Utilizing $\lim_{t\to\infty} e_1(t) = \mathbf{0}_n$ and $\lim_{t\to\infty} \dot{e}_1(t) = \mathbf{0}_n$, we can obtain that $\lim_{t\to\infty} W(t)\alpha = -g(q_d)$ according to (29c). Correspondingly, post-multiplying both sides of Eq. (29b) by α yields $\lim_{t\to\infty} W_f(t)\alpha = -\frac{1}{k_2}g(q_d)$. As for e_{fc} and e_f, it means that

$$\lim_{t\to\infty} e_f(t) = \lim_{t\to\infty} e_{fc}(t) = \lim_{t\to\infty} [-K^{-1}W_f(\alpha + z)] = \frac{K^{-1}g(q_d)}{k_2}, \quad (32)$$

where $\lim_{t\to\infty} e_{fr}(t) = \mathbf{0}_n$ is utilized. Since $\dot{e}_2$ is bounded according to Assumption 1, it can be seen from (28i) that $\dot{e}_f$, $\ddot{e}_f$ and $e_f^{(3)}$ are all bounded. Therefore utilizing Barbalat's lemma, it is obtained that $\lim_{t\to\infty} \dot{e}_f(t) = \lim_{t\to\infty} \ddot{e}_f(t) = \mathbf{0}_n$. Then in view of (28i), we have

$$\lim_{t\to\infty} e_2(t) = \lim_{t\to\infty} (\ddot{e}_f + k_1\dot{e}_f + k_2 e_f) = K^{-1}g(q_d). \quad (33)$$

Besides, since $\ddot{e}_2$ is bounded according to Assumption 1, we have $\lim_{t\to\infty} \dot{e}_2(t) = \mathbf{0}_n$ based on Barbalat's lemma.

4 Simulation

Numerical simulation results are given in this section to demonstrate the effectiveness of the adaptive control method. The simulation is based upon a two-link flexible-joint space robot, and the system is defined by the following matrices:

$$M(q) = \begin{bmatrix} p_1 + 2p_2 cosq_2 & p_3 + p_2 cosq_2 \\ p_3 + p_2 cosq_2 & p_3 \end{bmatrix}, \quad (34a)$$

$$C(q,\dot{q}) = -p_2 sinq_2 \begin{bmatrix} \dot{q}_2 & \dot{q}_1 + \dot{q}_2 \\ -\dot{q}_1 & 0 \end{bmatrix}, \quad (34b)$$

$$g(q) = \begin{bmatrix} 0 \\ 0 \end{bmatrix}, \quad (34c)$$

$$K = \begin{bmatrix} 500 & 0 \\ 0 & 500 \end{bmatrix}, \quad (34d)$$

$$J = \begin{bmatrix} p_4 & 0 \\ 0 & p_5 \end{bmatrix}, \quad (34e)$$

where $\alpha = [p_1, p_2, p_3]^T = [50.8, 15.3, 10.1]^T$, $\phi = [p_4, p_5]^T = [1, 1]^T$. Besides, $q(0) = \theta(0) = [0.1, 0.1]^T$, $\dot{q}(0) = \dot{\theta}(0) = [0, 0]^T$.

The initial values of the control variables are $W_f(0) = 0$, $\dot{W}_f(0) = 0$, $\hat{\alpha}(0) = [50, 17, 9.5]^T$, $e_{f1}(0) = 0$, $N_f(0) = 0$, $\hat{\phi}(0) = [0.9, 1.1]^T$, $e_f(0) = 0$, $\dot{e}_f(0) = 0$. Besides, the control parameters are $k_1 = 0.6$, $k_2 = 0.09$, $k_3 = 0.3$, $k_4 = 0.3$, $\gamma_1 = 0.5$, $\gamma_2 = 6$, $\bar{k} = 2$.

Set the desired position $q_d = [0, 0]^T$. The simulation results are given in Figs. 1, 2, 3 and 4.

In Fig. 1,we recognize that the robot link angles and motor angles converge to the desired position. In Fig. 2, angular velocity of robot link and motor derive to zero. As shown by the Fig. 3, parameter estimation error converge into the neighborhood of the desired equilibrium. Figure 4 shows the virtual controller and control input are bounded.

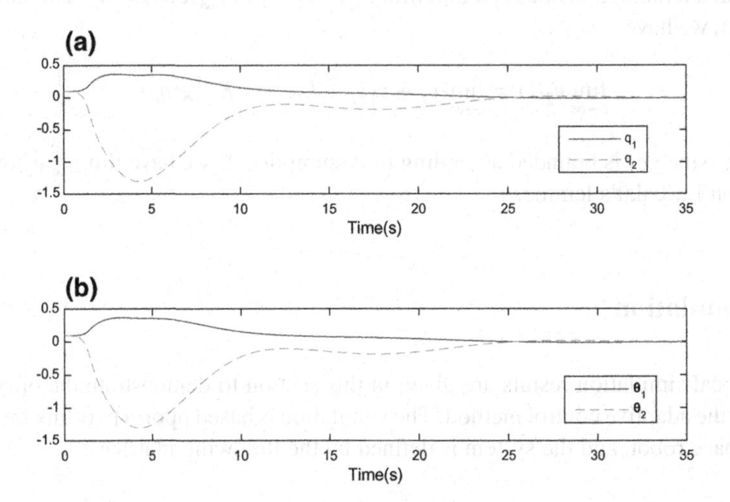

Fig. 1 Angles **a** robot link angles **b** motor angles

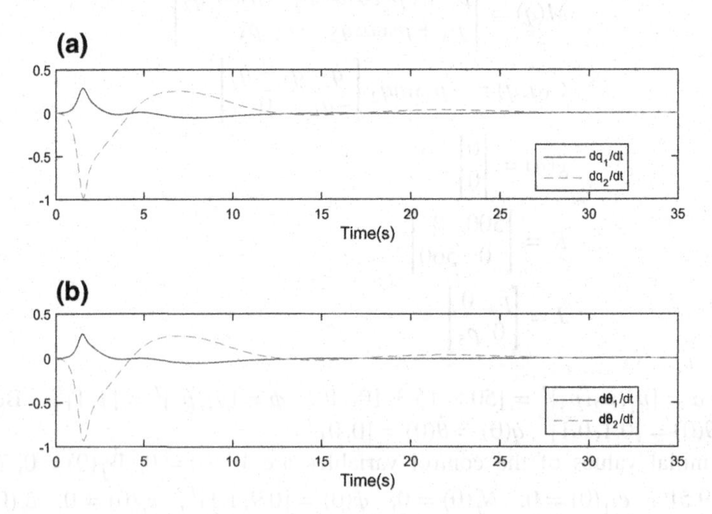

Fig. 2 Angular velocity **a** robot link angular velocity **b** motor angular velocity

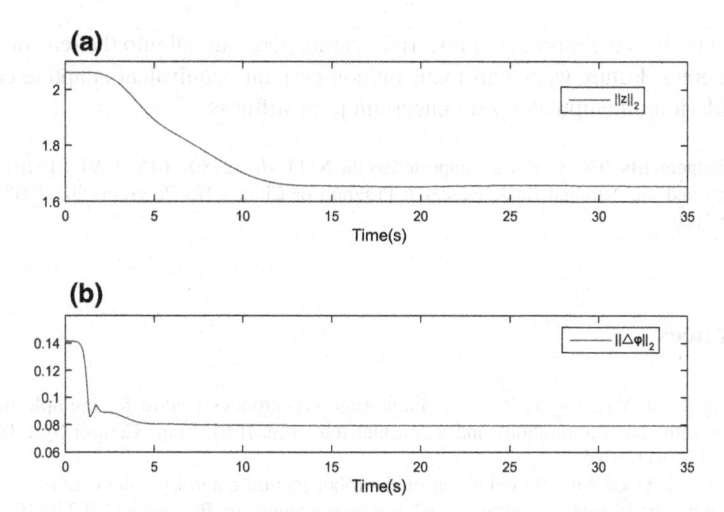

Fig. 3 Parameter estimation error **a** rigid parameter estimation error **b** flexible parameter estimation error

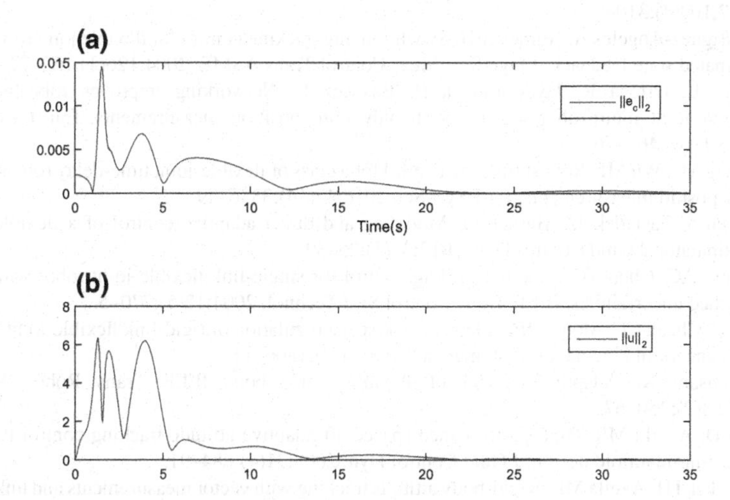

Fig. 4 Controller **a** virtual controller **b** control input

5 Conclusion

A non-certainty-equivalent adaptive controller is proposed to achieve stabilization of flexible-joint robot. The backstepping control framework is utilized in the controller design. To handle parameter uncertainties, a set of two-order linear filters have been introduced in the attractive manifold design. Based on the derived controller, it is proved that the position of flexible-joint robot can be driven to the desired position, and besides the estimate of uncertain parameters can fall into the real values to some degrees. Future work will focus on non-certainty-equivalent adaptive control of flexible-joint manipulator with uncertain joint stiffness.

Acknowledgements This work was supported by the NSFC (61327807, 6152 1091, 61520106010, 61134005) and the National Basic Research Program of China (973 Program: 2012CB821200, 2012CB821201).

References

1. Zhang B, Jia YM, Du JP, Zhang J. Finite-time synchronous control for multiple manipulators with sensor saturations and a constant reference. IEEE Trans Control Syst Technol. 2014;22(3):1159–65.
2. Sweet LM, Good MC. Re-definition of the robot motion control problem: Effects of plant dynamics, drive system constraints, and user requirements. In: Proceedings of 23th IEEE conference on decision and control, Las Vegas, NV; 1984. p. 724–32.
3. Spong M. Modeling and control of elastic joint robots. J Dyn Syst Meas Control-Trans ASME. 1987;109(4):310–9.
4. Rodriguez-Angeles A, Nijmeijer H. Synchronizing tracking control for flexible joint robots via estimated state feedback. J Dyn Syst Meas Control-Trans ASME. 2004;126(1):162–72.
5. Nuño E, Ortega R, Jayawardhana B, Basañez L. Networking improves robustness in flexible-joint multi-robot systems with only joint position measurements. Eur J Control. 2013;19(6):469–76.
6. Chang YC, Wu MF. Robust tracking control for a class of flexible-joint time-delay robots using only positionmeasurements. Int J Syst Sci. 2016;47(14):3336–49.
7. Ulrich S, Sasiadek JZ, Barkana I. Modeling and direct adaptive control of a flexible-joint manipulator. J Guid Control Dyn. 2012;35(1):25–39.
8. Huang AC, Chen YC. Adaptive sliding control for single-link flexible-joint robot with mismatched uncertainties. IEEE Trans Control Syst Technol. 2004;12(5):770–5.
9. Liu C, Cheah CC, Slotine JE. Adaptive task-space regulation of rigid-link flexible-joint robots with uncertain kinematics. Automatica. 2008;44(7):1806–14.
10. Khorasani K. Adaptive control of flexible-joint robots. IEEE Trans Robot Autom. 1992;8(2):250–67.
11. Seo D, Akella MR. High-performance spacecraft adaptive attitude-tracking control through attracting-manifold design. J Guid Control Dyn. 2008;31(4):884–91.
12. Mercker TH, Akella MR. Rigid-body attitude tracking with vector measurements and unknown gyro bias. J Guid Control Dyn. 2011;34(5):1474–84.
13. Lee KW, Singh SN. Noncertainty-equivalent adaptive missile control via immersion and invariance. J Guid Control Dyn. 2010;33(3):655–65.
14. Sun L, Zheng ZW. Nonlinear adaptive trajectory tracking control for a stratospheric airship with parametric uncertainty. Nonlinear Dyn. 2015;82(3):1419–30.

Dynamic Modeling and Analysis of the Micro-spacecraft Ejection Separation System

Xiaodong Zhao, Yingmin Jia, Jianheng Ling and Yi Li

Abstract Separation velocity and angular velocity are two key factors affecting the success of space ejection separation. In this paper, a coupled dynamic model of micro-spacecraft ejection separation mechanism is built based on the Newton-Euler method. With this model as the research object, the influences of different system parameters on the separation velocity and the angular velocity of the ejection separation mechanism are simulated and analyzed. The results indicate that the physical parameters of separating springs and the installation position of space ejection separation mechanism are the main factors affecting the separation velocity and angular velocity respectively.

Keywords Micro-spacecraft ejection separation mechanism · Dynamic model · Separation velocity · Angular velocity

1 Introduction

With the development of the space technology, the application of spacecraft is becoming more and more extensive. However, the increasing weight and volume raise additional risk to spacecrafts orbit operations. Due to this, micro-spacecraft is paid more and more attention now [1]. Comparing to the normal spacecraft, the micro-spacecraft has characteristics of less mass, small volume lower cost and good, i.e., which has more safety and flexibility [2]. Separation system determines the initial attitude of this type spacecraft and has an important influence on the overall performance. In order to ensure safe and efficient operation of the micro-spacecraft,

X. Zhao · Y. Jia (✉) · J. Ling · Y. Li
The Seventh Research Division and the Center for Information and Control,
School of Automation Science and Electrical Engineering,
Beihang University (BUAA), Beijing 100191, China
e-mail: ymjia@buaa.edu.cn

X. Zhao
e-mail: 749834174@qq.com

Y. Jia et al. (eds.), *Proceedings of 2017 Chinese Intelligent Systems Conference*, Lecture Notes in Electrical Engineering 459,
https://doi.org/10.1007/978-981-10-6496-8_23

it is necessary to analyse micro-spacecraft ejection separation system. During separation, excessive angular velocity and small relative velocity can cause damage to the micro-spacecraft ejection separation [3]. The large angular velocity is may unable to adjust the attitude of payload and small relative may produce collision. The purpose of the dynamics analysis of separation system is to study factors of angular velocity and linear velocity.

There are three kinds of separation systems: steam separation, electromagnetic separation and spring separation. Steam separation is difficult to control and electromagnetic separation can be affected by the magnetic field. Compared to steam separation and electromagnetic separation, spring separation is easy to control, not affected by space magnetic field and does not have exhaust pollution [4]. In the literature [5], micro-spacecraft on-orbit ejection separation researches are on special case that the payload launch along the center axis of platform. In the literature [6], the separation springs and limit switches are mathematically modeled. Jiang [7] analyzed centroid biased on orbit satellite separation in inertial reference frame. However, the installation methods of payload and platform are varied and direction of rail is random in practical situation.

This paper analyzes micro-spacecraft ejection separation system with helical compression spring mechanism and arbitrary direction of the rail [8]. Dynamic model is established in the translational inertial reference frame. The numerical simulations are taken by MATLAB software, and the main factor affecting the linear velocity and angular velocity are obtained from the simulation results.

We organize this paper as follows: In the second part, the ejection device is described and the reference coordinate system is established. In the third part, the dynamic model of the separation system is established by mathematical derivation. In the fourth part, the factors influencing the linear velocity and the angular velocity are obtained by analyzing the simulation results of the dynamic model.

2 Problem Configuration

The ejection separation system is the important component of research on micro-spacecraft. The ejection separation system with springs and rail can be seen, as shown in Fig. 1. Pulleys are installed between the payload and the rail to reduce the friction and deflection of the payload. Springs are mounted on the rod to ensure proper direction of force. When separation device releases in the platform, payload ejects from platform through rail to achieve separation. The process of separation should be divided into two parts: one is ejection with spring force, the other is movement under inertia.

As shown in Fig. 2, the translational inertia reference system *iXYZ* is established with payload and platform as centroid, spacecraft coordinate frame *pxyz* is established with platform as centroid. Considering the general model, the direction of rail

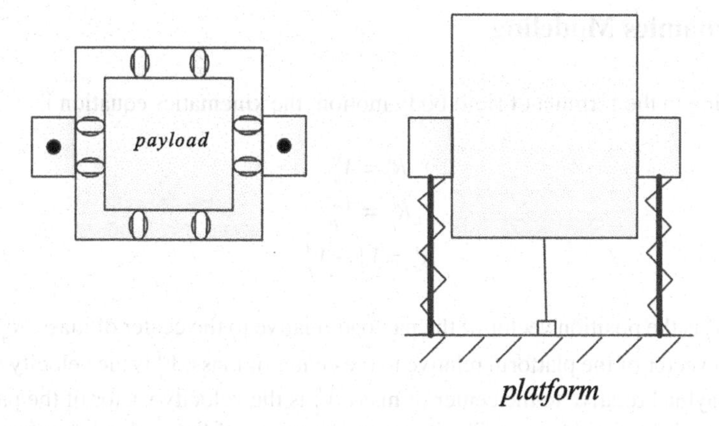

Fig. 1 Ejection separation system

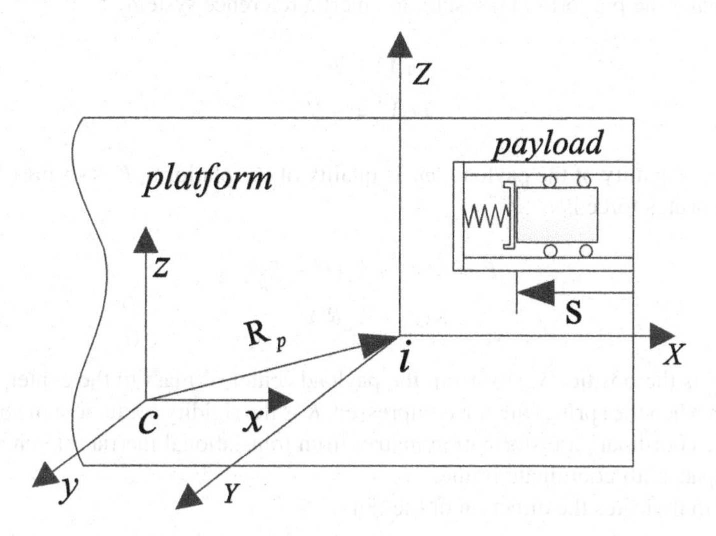

Fig. 2 Separation coordinate system diagram

and relative position between payload and platform should be random [9]. Compared to orbit radius, relative displacement between payload and platform can be neglected. And the influence of the atmospherical drag and the earth gravity will not be taken into account during separation [10]. Platform and rail are considered as a rigid body, and separation is assumed as relative motion between two rigid bodies. The contact surface is assumed to be smooth so that friction is ignored.

3 Dynamics Modeling

According to the formula of rigid body motion, the kinematics equation is

$$\dot{R}_l^i = V_l^i$$
$$\dot{R}_p^i = V_p^i \quad (1)$$
$$\dot{R}_r^i = V_l^i - V_p^i \quad (2)$$

where R_l^i is the position vector of the payload relative to the center of mass, R_p^i is the position vector of the platform relative to the center of mass, V_l^i is the velocity vector of the payload relative to the center of mass, V_l^i is the velocity vector of the payload relative to the center of mass, R_r^i is the position vector of the payload center relative to the center the platform in translational inertia reference system.

$$m_l \dot{V}_l^i = F$$
$$m_p \dot{V}_p^i = -F \quad (3)$$

where m_l is quality of the payload, m_p is quality of the platform, F is springs force.

The springs force is

$$F = K[r_0^c - A_{ic}(R_l^i - R_p^i)]$$
$$= K(r_0^c - A_{ic}R_r^i). \quad (4)$$

where r_0^c is the position vector from the payload center of mass to the center of the platform when the springs are not compressed, K is the rigidity coefficient of springs, A_{ic} is the coordinate transformation matrix from translational inertia reference system to spacecraft coordinate frame.

Let l to indicates the direction of the rail

$$r_0^c - A_{ic}(R_l^i - R_p^i) = \Delta xl \quad (5)$$

where Δx is spring deformation.

The derivative of speed can be expressed as

$$\dot{V}_r^i = \dot{V}_l^i - \dot{V}_p^i$$
$$= \frac{m_l + m_p}{m_l m_p} K\Delta xl \quad (6)$$

Take the derivative of Eq. (5) twice [11]

$$-\Delta\dot{x}l = -\,\omega_r^c \times A_{ic}R_r^i + A_{ic}V_r^i \tag{7}$$

$$\begin{aligned}-\Delta\ddot{x}l = &-\,\dot{\omega}_r^c \times A_{ic}R_r^i + \omega_r^c \times \omega_r^c \times A_{ic}R_r^i \\ &-2\omega_r^c \times A_{ic}V_r^i + A_{ic}\dot{V}_r^i\end{aligned} \tag{8}$$

Kinematic and dynamic equations of centroid are

$$\begin{aligned}\dot{q}_l^i &= \frac{1}{2}q_l^i \times \omega_l^i, \\ \dot{q}_p^i &= \frac{1}{2}q_p^i \times \omega_p^i\end{aligned} \tag{9}$$

$$\begin{aligned}J_l\dot{\omega}_l^c + \omega_l^c \times (J_l \cdot \omega_l^c) &= T \\ J_p\dot{\omega}_p^c + \omega_p^c \times (J_p \cdot \omega_p^c) &= r \times F - T\end{aligned} \tag{10}$$

where q_l^i, q_p^i are attitude quaternion in translational inertia reference system, ω_l^i, ω_p^i are angular velocities of payload and platform respectively, J_l, J_p are moments of inertia in payload and platform respectively, T is the torque of the rail on the payload [12].

In the process of ejection, relative rotation does not exist between the platform and the payload, we have

$$q_l^i = q_p^i = q_r^i \tag{11}$$

$$\omega_l^c = \omega_p^c = \omega_r^c \tag{12}$$

And because

$$\begin{aligned}r &= A_{ic}(R_l^i - R_p^i) \\ &= A_{ic}R_r^i\end{aligned} \tag{13}$$

kinematic equation of rotation can be expressed as

$$\dot{q}_r^i = \frac{1}{2}q_r^i \times \omega_r^i \tag{14}$$

dynamics equation of rotation can be expressed as

$$(J_p + J_l)\dot{\omega}_r^c + \omega_r^c \times ((J_p + J_l) \cdot \omega_r^c) = A_{ic}R_r^i \times K\Delta xl \tag{15}$$

In summary, dynamic model is organized into

$$M(X,t)\dot{X} = H(X,t) \tag{16}$$

where

$$X = [R_r^{iT}, V_r^{iT}, \Delta x, \Delta \dot{x}, q_r^{iT}, \omega_r^{cT}]$$

$$M(X,t) = \begin{bmatrix} I_{3\times3} & 0 & 0 & 0 & 0 & 0 \\ 0 & I_{3\times3} & 0 & 0 & 0 & 0 \\ 0 & 0 & 1 & 0 & 0 & 0 \\ 0 & 0 & 0 & 1 & 0 & 0 \\ 0 & 0 & 0 & 0 & I_{4\times4} & 0 \\ 0 & 0 & 0 & 0 & 0 & J_l + J_p \end{bmatrix}$$

$$H(X,t) = \begin{bmatrix} V_r^i \\ \frac{m_l+m_p}{m_l m_p} K \Delta x l \\ l^T(\omega_r^c \times A_{ic} R_r^i - A_{ic} V_r^i) \\ l^T(\dot{\omega}_r^c \times A_{ic} R_r^i - \omega_r^c \times \omega_r^c \times A_{ic} R_r^i + 2\omega_r^c \times A_{ic} V_r^i - A_{ic} \dot{V}_r^i) \\ \frac{1}{2} q_r^i \times \omega_r^i \\ -\omega_r^c \times ((J_p + J_l) \cdot \omega_r^c) + A_{ic} R_r^i \times K \Delta x l \end{bmatrix}$$

Equation (16) can be written in the form

$$\dot{X} = M^{-1}(X,t) H(X,t) \tag{17}$$

Given inertial parameters (m_l, m_p, J_l, J_p), the rigidity coefficient of the springs K, springs deformation Δx, the direction of the rail l and the length of rail, dynamic model can be solved [13].

4 Simulation and Analysis

The quality characteristics of the platform and payload are set

$$m_l = 5\ \text{kg}, J_l = diag[0.6, 0.6, 0.6]\ \text{kg} \cdot \text{m}^2,$$
$$m_p = 500\ \text{kg}, J_p = diag[80, 80, 80]\ \text{kg} \cdot \text{m}^2.$$

Setting the length of rail $s = 0.6$, direction vector of rail $l = [1, 0, 0]^T$ and springs deformation $\Delta x = 0.2$. When stretch is on the mass center line $(R_r(y) = R_r(z) = 0)$, angular velocity can be solved $(\omega = [0, 0, 0]^T)$ which conforms to the actual situations [13].

When the rigidity coefficient of the springs are respectively set $K_1 = 2400$ N/m $K_2 = 3600$ N/m and the relative position are respectively set $R_{r1} = [0, 0, 0.4]^T, R_{r2} = [0, 0, 2]^T$, the results are shown in the Figs. 3 and 4.

As the Figs. 3 and 4 show, payload and platform movement are roughly same with different conditions. When springs force work, angular velocity and relative velocity increase with time. When springing back, angular velocity and relative velocity

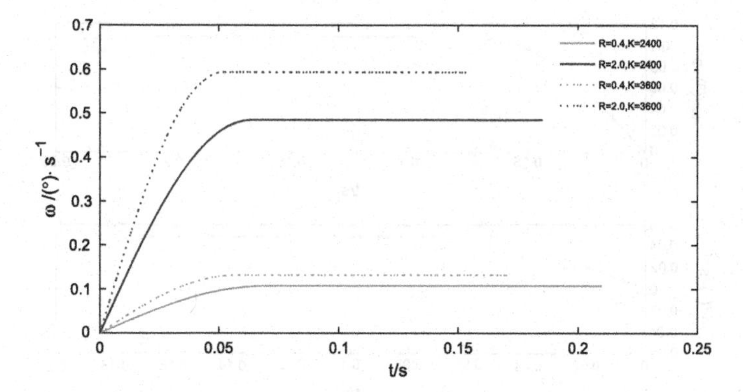

Fig. 3 Change of angular velocity as time varies

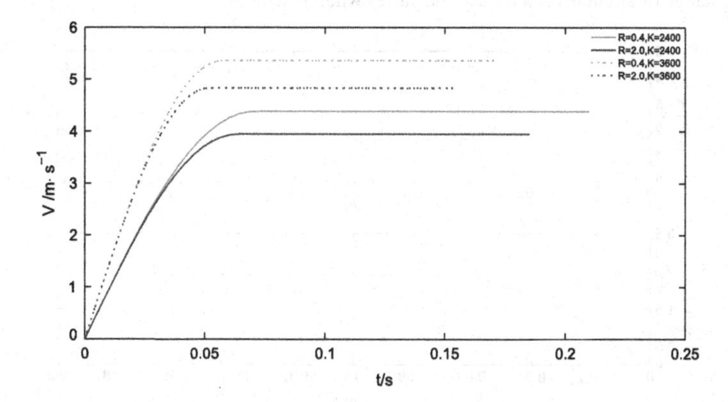

Fig. 4 Change of relative speed as time varies

remain essentially unchanged. The angular velocity and relative speed both increase as K increases. The angular velocity has changed significantly as relative position R increases. When the relative position R increases by five times, the angular velocity almost increases by five times and relative velocity increases by about one fifth.

Because the direction of rail is random, direction vector of rail is set $l = [0.4472, 0.5477, 0.7071]$. The results are shown in the Figs. 5 and 6. When direction vector of rail changes, the angular velocity and relative speed both be affected. The velocity and angular velocity components are produced in all three directions.

In short, the relative position R is the key factor affecting the angular velocity. The relative position affects the magnitude of the torque so that angular velocity increases as R increases. The elastic coefficient of the springs K has an influence of relative velocity. When K increases, relative velocity increases.

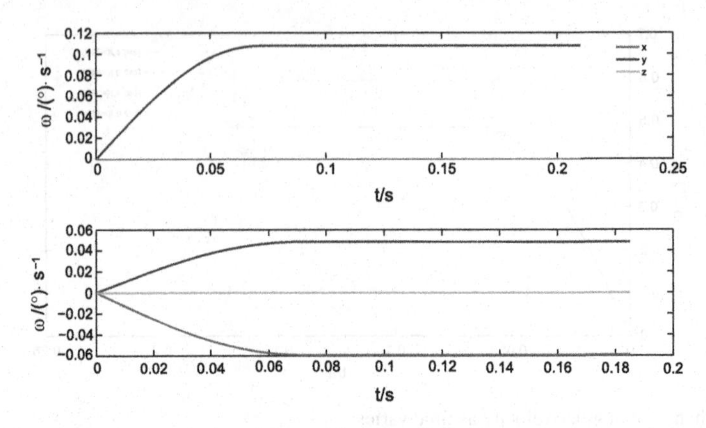

Fig. 5 Change of angular velocity as time varies when l changes

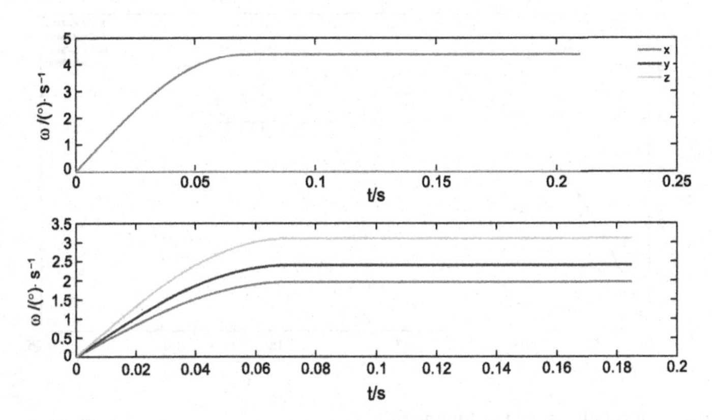

Fig. 6 Change of relative speed as time varies when l changes

5 Conclusions

Model of separation system with springs and rail is established by the basic principles of kinematics and dynamics. The elastic coefficient of the springs K determines the energy of the separation process so that relative velocity increases when K increases. The angular velocity is mainly affected by the relative installation position because the moment of force could be affected by the relative position. The angular velocity is positively related to the installation position. And the direction of rail mainly affects component of angular velocity and linear velocity in each direction and does not determine the value.

Acknowledgements This work was supported by the NSFC (61327807, 61521091, 61520106010, 61134005), the National Basic Research Program of China (973 Program: 2012CB821200, 2012CB821201) and the Fundamental Research Funds for the Central Universities (YMF-16-GJSYS-31, YMF-16-GJSYS-32).

References

1. Woellert K, Ehrenfreund P, Ricco AJ, et al. Cubesats: cost-effective science and sechnology platforms for emerging and developing nations. Adv Space Res. 2011;47(3):663–84.
2. Gongbo W, Xiaoning X. Safe on-orbit releasing velocity design for small satellites. Chin Space Sci Technol. 2007;27(3):33–8.
3. Fuxiang X. Satellite engineering. Beijing: China Astronautic Publishing House; 2004.
4. Xingzhi H. Research on design, analysis, and optimization of separation systems for small satellites. National University of Defense Technology; 2012.
5. Zhigang W, Yanan Y, Yifan D. Dynamics analysis of the spatial catation problem. Flight Dyn. 2014;32(3):239–42.
6. Xingzhi H, et al. Dynamics and transient perturbation analysis of satellite separation systems. Proc Inst Mech Eng Part G J Aerosp Eng. 2012;227(12):1968–76.
7. Jiang C, Zhaokui W, Li F, Yulin Z. Dynamics analysis of the constrained and centroid biased on-orbit satellite separation. Flight Dyn. 2010;28(1):76–9.
8. Miyamoto K, Ui K, Miyashita N, et al. Tokyo tech separation demonstration TSD as M-V rocket sub-payload for nanosatellite separation mechanism. Int Conf Eng Geol New Millennium. 2005;179:2330–8.
9. Xiaonan H, Jun W, Jinsheng P, et al. Design and experimental research of a large load and low shock release device on shape memory alloy. Sci Technol Eng. 2016;16(30):319–23.
10. Wei H. Robot dynamics and control. Beijing: Higher Education Press; 2004.
11. Xingzhi H, Xiaoqian C, Yong Z. Coordinate transform methods used in dynamic simulation of satellite separation. Adv Mater Res. 2012;569:712–7.
12. Gongbo W, Xiaoning X. Safe onorbit releasing velocity design for small satellites. Chin Space Sci Technol. 2007;27(3):33–8.
13. Qingyang L. Numercial analysis. Beijing: Tsinghua Press; 2008.

Neural Network Adaptive Control for Hysteresis Hammerstein System

Xuehui Gao, Ruiguo Liu, Bo Sun and Dawu Shen

Abstract An adaptive neural network (NN) controller is investigated for Hammerstein system with Prandtl-Ishlinskii (PI) hysteresis dynamics. The high order neural network (HONN) is applied with a new filter to performance control the system and the unknown hysteresis parameters is estimated by adaptive law. The stability of the proposed closed-loop adaptive NN control system is proved by using Lyapunov function. Finally, simulation results verify the effectiveness of the proposed adaptive controller.

Keywords HONN · Hysteresis · Hammerstein system · Adaptive control

1 Introduction

Many real systems exhibit nonlinear behavior that cannot be neglected and approximated by an appropriate linearization technique in the control synthesis. The Hammerstein model is one of the feasible option to describe these systems accurately because its special structure not only is simple, but also represents nonlinearity precisely. That structure consists of two parts: the nonlinearity section cascaded with the linearity section (Fig. 1). The dead-zone, friction, backlash and hysteresis etc. are extensively investigated nonlinearities for Hammerstein models in practice.

Hysteresis exists in smart materials, electromagnetic actuators, electromechanical devices, etc. The presence of hysteresis can degrade the control performance due to the strict nonlinearity with memory. Thus, the modeling of hysteresis dynamics has been investigated for many years, and various hysteresis models have been

X. Gao (✉) · R. Liu · B. Sun
Department of Mechanical and Electrical Engineering, Shandong University of Science and Technology, Tai'an 271019, China
e-mail: xhgao@163.com

D. Shen
Division of Information and Communications, Tai'an Public Security Bureau, Tai'an 271000, China

Y. Jia et al. (eds.), *Proceedings of 2017 Chinese Intelligent Systems Conference*, Lecture Notes in Electrical Engineering 459,
https://doi.org/10.1007/978-981-10-6496-8_24

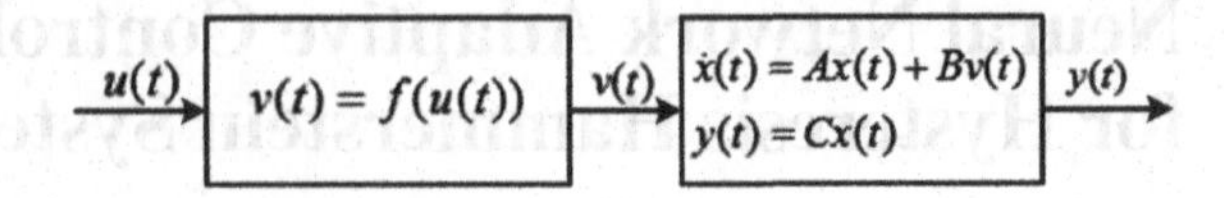

Fig. 1 The structure of Hammerstein system

suggested. Preisach model [2, 3, 10, 18, 19], Prandtl-Ishlinskii (PI) model [1, 4, 15], Jile-Atherthon model [8, 13], and Bouc-Wen model [16, 17] are the major models used in practice. In particular, PI model is one of the most important models to deal with the hysteresis nonlinearity. Jiang et al. [5] provided a new approach to model the asymmetric hysteresis nonlinearity of piezoelectric actuators with a modified PI model, which can be calculated using the recursive least-squares method. By using the inverse PI model compensator, Liu and Su [11] developed an observer-based robust adaptive output feedback controller to control the systems with hysteresis nonlinearity. An inverse rate-dependent PI model was utilized to compensate for the hysteresis nonlinearities in a piezomicropositioning stage in [1], which can suppress the rate-dependent hysteresis nonlinearities at a range of different excitation frequencies between 0.05 and 100 Hz. However, the Hammerstein system with hysteresis PI model have not been attracted sufficient attention yet.

This study adopted adaptive NN controller for a class of Hammerstein systems with hysteresis nonlinearity presented via a PI model. The feedback adaptive NN controller is designed to suppress the hysteresis nonlinearity. Different from other controllers, the proposed adaptive NN controller does not need the inverse PI model to compensate for the hysteresis and thus can simplify the control design. Simulation results are given to validate the presented methods.

This paper is organized as follows. Section 2 describes the Hammerstein system with hysteresis PI model nonlinearity. The feedback adaptive NN control is proposed in Sect. 3 and the stability of the closed loop controller is proved by Lyapunov function. The effectiveness of the proposed controller is illustrated through examples in Sect. 4. Section 5 contains the conclusions.

2 Problem Formulation

Consider a class of Hammerstein system with hysteresis nonlinearity as follows:

$$
\begin{aligned}
\dot{x}_1(t) &= x_2(t) \\
\dot{x}_i(t) &= x_{i+1}(t) \quad i = 2, \ldots, n-1 \\
\dot{x}_n(t) &= -a_1 x_1(t) - a_2 x_2(t) - \cdots - a_n x_n(t) + bv(t) \\
v(t) &= f(u(t)) \\
y(t) &= x_1(t)
\end{aligned} \tag{1}
$$

where $x(t) = [x_1, x_2, \ldots, x_n]^T \in \mathbb{R}^{n \times 1}$, $y(t) \in \mathbb{R}$ and $u(t) \in \mathbb{R}$ are system states, output and input signals, respectively. $v(t) = f(u(t))$ indicates the nonlinearity of the Hammerstein model. The structure of the Hammerstein system illustrates in Fig. 1.

In this paper, the nonlinearity $v(t) = f(u(t))$ is described by PI model established through play operator. The play operator can be defined as follows [11]:

$$\begin{aligned} \omega_0(0) &= G_r[u](0) = g_r(u(0), \omega_{m-1}(0)) \\ \omega_m(t) &= G_r[u](t) = g_r(u(t), \omega_m(t_i)) \end{aligned} \tag{2}$$

where $g_r(u, \omega_m) = max\{(u - r), min\{(u + r, \omega_m)\}\}$, $r \geqslant 0$, $t_i \leqslant t \leqslant t_{i+1}$ and ω_{m-1} is given as initial condition. Then, the PI model can be described as [11]:

$$v(t) = p_0 u(t) + \int_0^R p(r) G_r[u](t) dr \tag{3}$$

where p_0 is a positive constant, $p(r)$ means integrable density function and satisfies $\int_0^R rp(r)dr < \infty$ for $p(r) \geqslant 0$.

For the Hammerstein system (1) with hysteresis PI model (3), without loss of generality, the nonlinear parameter $p(r)$ of PI model and the linear state parameters $a_1, a_2, \ldots, a_n$ are unknown. Then, we design an adaptive NN to handle the unknown parameters.

3 Adaptive NN Control Design

3.1 Neural Network Design

A HONN will be constructed to estimate the unknown parameters. A parameterized NN approximation [9, 12] is employed over a compact set Ψ as

$$G(x) = W^{*T} \Phi(x) + \varepsilon \qquad \forall x \in \Psi \subset \mathbb{R}^n \tag{4}$$

where $W^* = [w_1^*, w_2^*, \ldots, w_L^*]^T \in \mathbb{R}^L$ is the bounded NN weights, $\varepsilon \in \mathbb{R}$ is a bounded approximation error, and $\Phi(x) = [\Phi_1(x), \Phi_2(x), \ldots, \Phi_L(x)]^T \in \mathbb{R}^L$ is the NN basis vector.

In order to design the HONN, a reference input vector is introduced as $\bar{y}_d(t) = [y_d(t), \dot{y}_d(t), \ldots, y_d^{(n-1)}(t)]^T \in \mathbb{R}^{n\times 1}$ and the tracking error as

$$e = x - \bar{y}_d \qquad s = [\Lambda^T 1]e \tag{5}$$

where $\Lambda = [\Lambda_1, \Lambda_2, \ldots, \Lambda_{n-1}]^T$ is a filter vector such that the polynomial $s^{n-1} + \Lambda_n s^{n-2} + \cdots + \Lambda_1$ is Hurwitz. Then, the tracking error e is bounded as long as the filter s is bounded on the basis of [6].

We define $\hat{x} = [x x_2, \ldots, x_n]^T \in \mathbb{R}^{(n)\times 1}$, the error $\hat{e}$ and $\hat{s}$ can be defined between the state and the reference $\bar{y}_d$ as

$$\hat{e} = \hat{x} - \bar{y}_d \qquad \hat{s} = [\Lambda^T 1]\hat{e}. \tag{6}$$

Then, the the error $\tilde{x}$ and $\tilde{s}$ can be deduced as

$$\tilde{x} = x - \hat{x} = e - \hat{e} \qquad \tilde{s} = s - \hat{s} = [\Lambda^T 1]\tilde{e} \tag{7}$$

substituting (1) into (5), we have

$$\begin{aligned} \dot{s} &= [\Lambda^T 1]\dot{e} \\ &= [0\ \Lambda^T]e - \sum_{i=1}^{n} a_i x_{i+1} + bp_0 u(t) \\ &\quad + b\int_0^R p(r)G_r[u](t)dr - y_d^{(n)} \end{aligned} \tag{8}$$

The NN approximation (4) is only valid in a compact set so that a high order NN [7, 12] is adopted for the Hammerstein system which can guarantee the approximation capacity with less neurons and computation costs. High order function $\Phi_{ik}(x) = \prod_{j\in J_k}[\sigma(x_j)]^{d_k(j)}, k = 1, 2, \ldots, L$ of HONN is used to estimate the unknown parameters, where J_k are collections of L-not ordered subsets of $\{0, 1, \ldots, n\}$, $d_k(j)$ are nonnegative integers, and $\sigma(\cdot)$ is a sigmoid function.

3.2 Controller Design

For system (1), a robust HONN controller is given:

$$\begin{aligned} u =& \frac{1}{bp_0}(-k\hat{s} - \hat{W}\Phi(X) + y_d^{(n)} \\ &- b\int_0^R \hat{p}(t, r)G_r[u](t)dr) \end{aligned} \tag{9}$$

where $k > 0$ is a positive constant, $\hat{W}$ is the estimated value of the ideal HONN weights vector W^*, $\Phi(X)$ is the basis vector of HONN with the input $X = [x, \hat{e}]$. Then, the HONN weights $\hat{W}$ can be updated as

$$\dot{\hat{W}} = \Gamma\left(\hat{s}\Phi^T(X) - \sigma_1|\hat{s}|\hat{W}\right) \tag{10}$$

where $\Gamma > 0$ and $\sigma_1 > 0$ are designed parameters.

The following Lemma is true.

Lemma 1 [12] *The HONN weights $\hat{W}$ in (10) are bounded with $\|\hat{W}\| \leqslant c_\Phi/\sigma_1$, where c_Φ is bound of the HONN basis function vector, i.e., $\|\Phi\| \leqslant c_\Phi$. If the weights* *of estimation error* $\tilde{W}$ *defines as* $\tilde{W} = W^* - \hat{W}$, *then the error* $\tilde{W}$ *is also bounded as* $\|\tilde{W}\| \leqslant c_W$, *where* $c_W = W_N + c_\Phi/\sigma_1$ *and* W_N *is positive constants which suits* $\|W^*\| \leqslant W_N$.

Considering the controller (9), $b\int_0^R \hat{p}(t,r)G_r[u](t)dr$ is used to cancel the effect of hysteresis nonlinearity of $\int_0^R p(r)G_r[u](t)dr$. Due to the integral form of $\int_0^R p(r) G_r[u](t)dr$, the boundedness cannot be assumed and thus the traditional robust control cannot be designed. However, considering the density function $p(r)$ is not a function of time, thus it can be treated as a "parameter" and an adaption law about $\hat{p}(t,r)$ can be obtained as

$$\frac{\partial}{\partial t}\hat{p}(t,r) = |\hat{s}||G_r[u](t)| + |[\Lambda^T\ 1]\tilde{x}|G_r[u](t) - \rho\hat{p}(t,r) \tag{11}$$

where ρ is positive constant.

3.3 Stability Analysis

The unknown parameters of Hammerstein system can be estimated by HONN and the error of the NN can be guaranteed by Lemma 1. The adaptive law of $\hat{p}(t,r)$ is defined in (11) and the adaptive law of HONN weight $\hat{W}$ is defined in (10). Then, the following theorem is hold.

Theorem 1 *Considering system* (1), *the controller* (9) *with adaption law* (10) *and* (11), *then, all the signals in the closed loop are uniformly ultimately bounded (UUB), and the tracking error s converges to a small compact set around zero as* $|s| \leqslant \sqrt{2(V(0)e^{-\gamma t} + \frac{\vartheta}{\gamma}(1 - e^{-\gamma t}))}$ *with* γ *and* ϑ *being positive constants defined by* (24).

Proof Considering (7), substituting (9) into (8), one can obtain

$$\begin{aligned}\dot{s} &= [0\ \Lambda^T](\hat{e} + \tilde{x}) - \sum_{i=1}^{n} a_i x_i + bp_0 u \\ &\quad + b\int_0^R p(r)G_r[u](t)dr - y_d^{(n)} \\ &= F(x,\hat{e}) - y_d^{(n)} + b\int_0^R p(r)G_r[u](t)dr \\ &\quad + [0\ \Lambda^T]\tilde{x} + bp_0 u\end{aligned} \tag{12}$$

where $F(x,\hat{e}) = [0\ \Lambda^T]\hat{e} - \sum_{i=1}^{n} a_i x_i$ is an unknown function approximated by HONN as $F(x,\hat{e}) = W^{*T}\Phi(x) + \varepsilon$.

Select the Lyapunov function as

$$V = \frac{1}{2}\left(s^2 + b\int_0^R \tilde{p}^2(t,r)dr\right) \tag{13}$$

The derivative of V can be deduced along with (7) and (12) as

$$\begin{aligned}\dot{V} &= s(W^{*T}\Phi(x) + \varepsilon - y_d^{(n)} + bp_0u + [0\ \Lambda^T]\tilde{x} \\ &\quad b\int_0^R p(r)G_r[u](t)dr + b\int_0^R \tilde{p}(t,r)\frac{\partial}{\partial t}\tilde{p}(t,r)dr\end{aligned} \tag{14}$$

Substituting (9) into (14), the derivative of V in (14) becomes

$$\begin{aligned}\dot{V} &= s(W^{*T}\Phi(x) + \varepsilon - y_d^{(n)} + [0\ \Lambda^T]\tilde{x} \\ &\quad + b\int_0^R p(r)G_r[u](t)dr - k\hat{s} - \hat{W}\Phi(x) + y_d^{(n)} \\ &\quad - b\int_0^R \hat{p}(t,r)G_r[u](t)dr) + b\int_0^R \tilde{p}(t,r)\frac{\partial}{\partial t}\tilde{p}(t,r)dr\end{aligned} \tag{15}$$

Considering Lemma 1, substituting (7) into (15), we have

$$\begin{aligned}\dot{V} &= s(\tilde{W}\Phi(x) + \varepsilon + [0\ \Lambda^T]\tilde{x} - ks - k\tilde{s} \\ &\quad - b\int_0^R \tilde{p}(t,r)G_r[u](t)dr) + b\int_0^R \tilde{p}(t,r)\frac{\partial}{\partial t}\tilde{p}(t,r)dr \\ &\leqslant -ks^2 + |s|c_W c_\Phi + \varepsilon_m + k|[\Lambda^T\ 1]\tilde{x}| + |[0\ \Lambda^T]\tilde{x}| \\ &\quad - b|s|\int_0^R \tilde{p}(t,r)G_r[u](t)dr + b\int_0^R \tilde{p}(t,r)\frac{\partial}{\partial t}\tilde{p}(t,r)dr\end{aligned} \tag{16}$$

Substituting (11) into (16), the following is hold:

$$\begin{aligned}\dot{V} &= -ks^2 + |s|(c_W c_\Phi + \varepsilon_m + k|[\Lambda^T\ 1]\tilde{x}| + |[0\ \Lambda^T]\tilde{x}|) \\ &\quad - b|s|\int_0^R \tilde{p}(t,r)G_r[u](t)dr + b|\hat{s}|\int_0^R \tilde{p}(t,r)G_r[u](t)dr \\ &\quad + b\int_0^R \tilde{p}(t,r)|[\Lambda^T\ 1]\tilde{x}|G_r[u](t)dr - b\rho\int_0^R \tilde{p}(t,r)\hat{p}(t,r)dr \\ &\leqslant -ks^2 + |s|(c_W c_\Phi + \varepsilon_m + k|[\Lambda^T\ 1]\tilde{x}| + |[0\ \Lambda^T]\tilde{x}|) \\ &\quad - b|s|\int_0^R \tilde{p}(t,r)G_r[u](t)dr + b|s|\int_0^R \tilde{p}(t,r)G_r[u](t)dr \\ &\quad - b|\tilde{s}|\int_0^R \tilde{p}(t,r)G_r[u](t)dr + b\int_0^R \tilde{p}(t,r)|[\Lambda^T\ 1]\tilde{x}|G_r[u](t)dr \\ &\quad - b\rho\int_0^R \tilde{p}(t,r)\hat{p}(t,r)dr\end{aligned} \tag{17}$$

Thus, (17) can be written as

$$\begin{aligned}\dot{V} \leqslant & -ks^2 + |s|(c_W c_\Phi + \varepsilon_m + k|[\Lambda^T \ 1]\tilde{x}| + |[0 \ \Lambda^T]\tilde{x}|) \\ & - b\rho \int_0^R \tilde{p}(t,r)\hat{p}(t,r)dr\end{aligned} \tag{18}$$

We define $c = |[\Lambda^T \ 1]|\tilde{x} \leqslant max\{\Lambda_i, 1\}\delta$ and $d = |[0 \ \Lambda^T]\tilde{x}| \leqslant max\{\Lambda_i\}\delta$ as small positive constants, and $c_1 = c_W c_\Phi + \varepsilon_m + kc + d$ is positive constant, thus, (18) can be rewritten as

$$\dot{V} = -ks^2 + c_1|s| - b\rho \int_0^R \tilde{p}(t,r)\hat{p}(t,r)dr \tag{19}$$

By applying the Young's inequality [14], we have

$$-b\rho\tilde{p}(t,r)\hat{p}(t,r) \leqslant -\frac{b\rho}{2}\tilde{p}^2(t,r) + \frac{b\rho}{2}p_{max}^2 \tag{20}$$

Integrating both sides of (20) over $[0 \ R]$ results in

$$-b\rho \int_0^R \tilde{p}(t,r)\hat{p}(t,r)dr \leqslant -\frac{b\rho}{2}\int_0^R \tilde{p}^2(t,r)dr + \frac{b\rho R}{2}p_{max}^2 \tag{21}$$

Also by Young's inequality on term $c_1|s|$, we can get

$$c_1|s| \leqslant \frac{c_1^2 + s^2}{2} \tag{22}$$

Therefore, based on (21) and (22), (19) can be rewritten further as

$$\begin{aligned}\dot{V} & \leqslant (-k + \frac{1}{2})s^2 + \frac{c_1^2}{2} - \frac{b\rho}{2}\int_0^R \tilde{p}^2(t,r)dr + \frac{b\rho R}{2}p_{max}^2 \\ & \leqslant \gamma V + \vartheta\end{aligned} \tag{23}$$

where γ and ϑ are positive constants defined as

$$\begin{aligned}\gamma &= min\{2k - 1, \rho\} \\ \vartheta &= \frac{c_1^2}{2} + \frac{b\rho R}{2}p_{max}^2\end{aligned} \tag{24}$$

According to Lyapunov theorem, V is UUB and the tracking error s is bounded. Then, that further guarantees the boundedness of $\hat{s}$. Consequently, the control signal u and the error $\hat{e}$ are all bounded.

Moreover, integrating both sides of (23) over $[0, t]$, we have

$$V \leqslant V(0)e^{-\gamma t} + \frac{\vartheta}{\gamma}(1 - e^{-\gamma t}). \tag{25}$$

Thus, considering (13), the following is hold

$$|s| \leqslant \sqrt{2(V(0)e^{-\gamma t} + \frac{\vartheta}{\gamma}(1 - e^{-\gamma t}))}. \tag{26}$$

Equation (26) illustrates the tracking error s converges to a small compact set around zero. Therefore, the control parameters should be adjusted appropriately in practical implementation.

This completes the proof.

4 Simulation Results

In this section, an example is used to verify the theoretical developments of the proposed adaptive NN controller for Hammerstein system with hysteresis nonlinearity. Consider the following nonlinear plant. The hysteresis nonlinearity see Fig. 2.

$$\begin{aligned}
\dot{x}_1 &= x_2 \\
\dot{x}_2 &= x_3 \\
\dot{x}_3 &= -2x_1 - 6x_2 - 7x_3 + 2v \\
v &= 0.25u + \int_0^{10} 0.01re^{-0.002r^2} G_r[u](t)dr \\
y &= x_1
\end{aligned} \tag{27}$$

In order to verify the performance of the proposed controller, a proportional-integral (P-I) controller and the proposed adaptive NN controller are compared with the input as $u(t) = sin(0,01\pi t) + 5sin(0.5\pi t) + sin(\pi t)$ and $K_p = 0.9, K_i = 0.3$. The control results are shown in Fig. 3a–c about the P-I controller is compared with the proposed adaptive NN controller. Figure 4 illustrates the tracking errors about the proposed adaptive NN controller compared with P-I controller. It is shown that the proposed adaptive NN controller is effective for the Hammerstein system with hysteresis and has better performance than the P-I controller. From Fig. 4, it is well known that the errors of the proposed adaptive NN control is smaller than the P-I controller. The detail data of the tracking errors are shown in the Table 1. As shown, the MAE is 0.0103 for adaptive NN control but it is 0.0169 for P-I control.

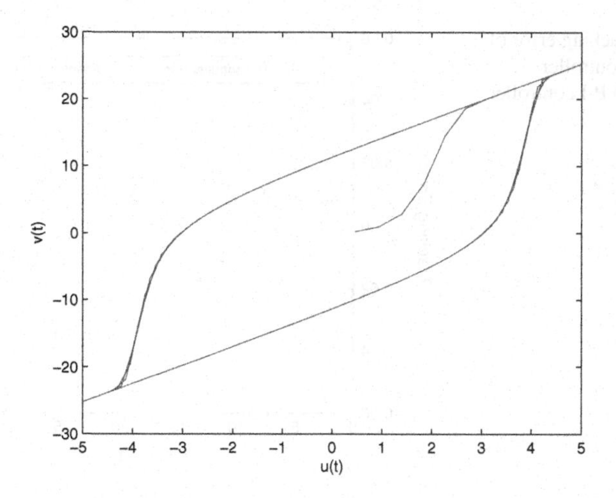

Fig. 2 The tracking error with adaptive NN controller

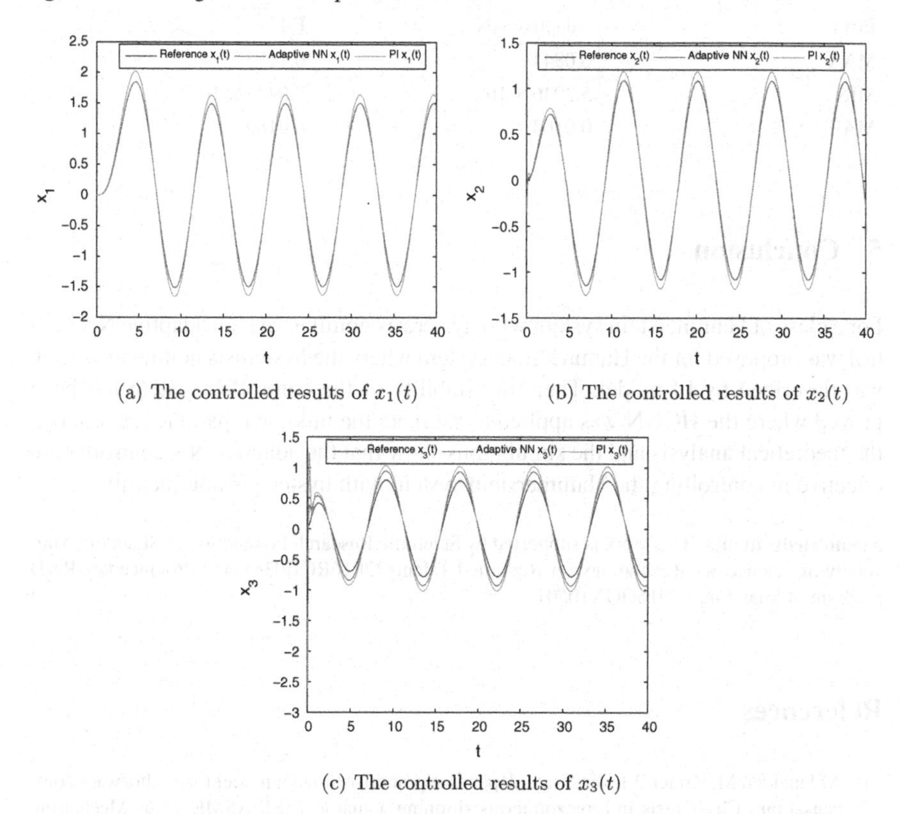

(a) The controlled results of $x_1(t)$

(b) The controlled results of $x_2(t)$

(c) The controlled results of $x_3(t)$

Fig. 3 The controlled results of adaptive NN controller compared with P-I controller

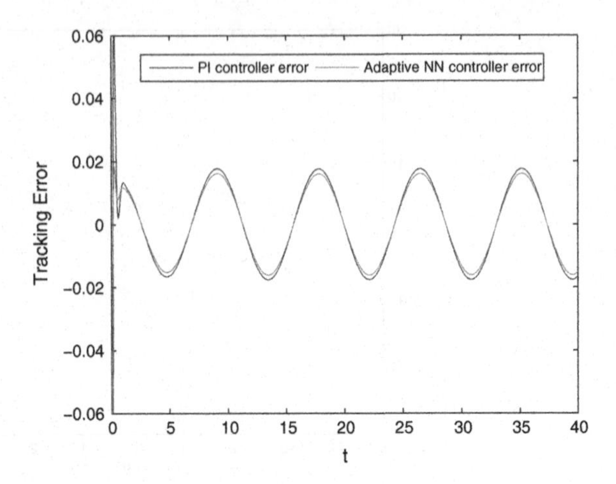

Fig. 4 The tracking error of adaptive NN controller compared with P-I controller

Table 1 The comparison of adaptive NN control errors and P-I control errors

Error	Adaptive NN	P-I
MAX	0.0844	0.1173
MIN	5.2946×10^{-6}	9.0425×10^{-6}
MAE	0.0103	0.0169

5 Conclusion

For a class of Hammerstein system with hysteresis nonlinearity, an adaptive NN control was proposed for the Hammerstein system where the hysteresis nonlinear section was described by PI model. Then, the stability of the controlled system has been proved where the HONN was applied to estimate the unknown parameters. Finally, the theoretical analysis and the simulations show that the adaptive NN controller are effective in controlling the Hammerstein system with hysteresis nonlinearity.

Acknowledgements This work is supported by Scientific Research Foundation of Shandong University of Science and Technology for Recruited Talents (2016RCJJ035), and Province key R&D program of Shandong (2016GGX105013).

References

1. Al Janaideh M, Krejci P. Inverse rate-dependent prandtl-ishlinskii model for feedforward compensation of hysteresis in a piezomicropositioning actuator. IEEE/ASME Trans Mechatron. 2013;18(5):1498–507.
2. de la Barrire O, Ragusa C, Appino C, Fiorillo F. Prediction of energy losses in soft magnetic materials under arbitrary induction waveforms and dc bias. IEEE Trans Ind Electron. 2017;64(3):2522–9 March.

3. Fallah E, Badeli V. A new approach for modeling of hysteresis in 2-d time-transient analysis of eddy current using fem. IEEE Trans Magn. 2017;53(7):1–14 July.
4. Goel V, Anderson P, Hall J, Robinson F, Bohm S. Electroless plating: a versatile technique to deposit coatings on electrical steel. IEEE Trans Magn. 2016;52(5):1–4 May.
5. Jiang H, Ji H, Qiu J, Chen Y. A modified prandtl-ishlinskii model for modeling asymmetric hysteresis of piezoelectric actuators. IEEE Trans Ultrason Ferroelectr Freq Control. 2010;57(5):1200–10 May.
6. Khalil HK. Nonlinear systems, 3rd ed. Upper Saddle River: Prentice Hall; 2002.
7. Kosmatopoulos E, Polycarpou M, Christodoulou M, Ioannou P. High-order neural network structures for identification of dynamical systems. IEEE Trans Neural Netw. 1995;6(2):422–31 Mar.
8. Lacerda Ribas J, Lourenco E, Vianei Leite J, Batistela N. Modeling ferroresonance phenomena with a flux-current jiles-atherton hysteresis approach. IEEE Trans Magn. 2013;49(5):1797–800 May.
9. Lewis FL, Yesildirak A, Jagannathan S. Neural network control of robot manipulators and nonlinear systems. Bristol: Taylor & Francis Inc; 1998.
10. Li Z, Su CY, Chai T. Compensation of hysteresis nonlinearity in magnetostrictive actuators with inverse multiplicative structure for preisach model. IEEE Trans Autom Sci Eng. 2014;11(2):613–9 April.
11. Liu S, Su CY. Inverse error analysis and adaptive output feedback control of uncertain systems preceded with hysteresis actuators. IET Control Theory Appl. 2014;8(17):1824–32.
12. Na J, Ren X, Zheng D. Adaptive control for nonlinear pure-feedback systems with high-order sliding mode observer. IEEE Trans Neural Netw Learn Syst. 2013;24(3):370–82 March.
13. Raghunathan A, Klimczyk P, Melikhov Y. Application of jiles-atherton model to stress induced magnetic two-phase hysteresis. IEEE Trans Magn. 2013;49(7):3187–90 July.
14. Ren B, Ge S, Su CY, Lee TH. Adaptive neural control for a class of uncertain nonlinear systems in pure-feedback form with hysteresis input. IEEE Trans Syst Man Cybern Part B Cybern. 2009;39(2):431–43 April.
15. Taretto K, Soldera M, Koffman-Frischknecht A. Material parameters and perspectives for efficiency improvements in perovskite solar cells obtained by analytical modeling. IEEE J Photovoltaics. 2017;7(1):206–13 Jan.
16. Wang Z, Zhang Z, Mao J. Precision tracking control of piezoelectric actuator based on bouc-wen hysteresis compensator. Electron Lett. 2012;48(23):1459–60 November.
17. Xu R, Zhou M. Sliding mode control with sigmoid function for the motion tracking control of the piezo-actuated stages. Electron Lett. 2017;53(2):75–7.
18. Xuehui G, Xuemei R, Xing'an G, Jie H. The identification of preisach hysteresis model based on piecewise identification method. In: Control conference (CCC), 2013, 32nd Chinese; 2013. p. 1680–85.
19. Zhu XGXRC, Zhang C. Identification and control for Hammerstein systems with hysteresis non-linearity. IET Control Theory Appl. 2015.

Microblog Query Expansion Based on Ontology Expansion and Borda Count Rank

Hao Gong, Junping Du and Wei Wang

Abstract Most of the traditional microblog query methods based on keywords matching lead to low query efficiency. It is difficult for users to get the real information they want without using the semantic expansion. With the study of microblog query expansion algorithm based on ontology and local query feedback, in this paper, we propose a microblog query expansion algorithm based on ontology expansion and Borda count rank. We use the semantic connection provided by ontology knowledge base to expand the initial query words, combine the local query feedback to filter the final query expansion words, and use Borda count method to rearrange the query results. The experimental results show that Microblog query expansion algorithm based on ontology expansion and Borda count rank has better recall and precision rate than keyword-based query expansion algorithm, ontology-based query expansion algorithm and query expansion algorithm based on ontology and local query feedback.

Keywords Microblog · Query expansion · Ontology · Borda count rank

1 Introduction

Microblog and other online social network platforms have a large number of active users. Thus, online social networks have accumulated a large number of users' data. How to query from these huge data to meet the needs of users became a problem that microblog retrieval needs to be resolved [1]. The query technology on which the information retrieval relies is still based on keyword-based mechanical symbol

H. Gong · J. Du (✉)
Beijing Key Laboratory of Intelligent Telecommunication Software and Multimedia, School of Computer Science, Beijing University of Posts and Telecommunications, Beijing 100876, China
e-mail: junpingdu@126.com

W. Wang
SINA Corporation, Beijing, China

Y. Jia et al. (eds.), *Proceedings of 2017 Chinese Intelligent Systems Conference*, Lecture Notes in Electrical Engineering 459,
https://doi.org/10.1007/978-981-10-6496-8_25

matching method. Query expansion is a good solution to this problem. At present domestic and foreign experts and scholars put forward a lot of query expansion technology. Wu et al. [2] proposed a query expansion based on local analysis. Liu et al. [3] proposed a query expansion based on association rules. Tannebaum et al. [4] proposed a query expansion based on user's logs. Ontology as a good concept modeling tool describes the relationship among concepts well, so ontology-based query expansion [5] has become a hot topic in recent years. Wan et al. [6] proposed a query expansion algorithm based on ontology and local context analysis for the museum knowledge base system query. Li [7] proposed a query expansion algorithm based on ontology and user query intent. Duranti et al. [8] proposed a query expansion algorithm using business domain ontology for Internet search and semantic query expansion. These domain ontologies greatly improve the efficiency and accuracy of information retrieval in their field.

Most of the above ontology-based query expansion methods are based on long text corpus, and there are few query expansions for social network information such as microblog texts. In order to improve the query efficiency of microblog query, we propose microblog query expansion algorithm based on ontology expansion and Borda count rank. This algorithm takes microblog as the research object, uses the semantic connection provided by ontology knowledge base to expand the initial query word, combines the local query feedback and Borda count method to improve microblog retrieval performance.

The structure of this paper is as follows. In Sect. 2, we establish a security domain ontology. In Sect. 3, we propose microblog query expansion algorithm based on ontology expansion and Borda count rank. In Sect. 4, we compare the algorithm we proposed with three existing algorithms, and give the experimental results and analysis. Finally, we conclude full paper in Sect. 5.

2 Establishment and Expansion of the Security Domain Ontology

We establish a security domain ontology for microblog query of national security events using Protégé 5.1. We get more than 30,000 security-related microblog texts as a microblog texts set (*D*) from Sina Weibo. Combined with ontology expansion method based on the semantic relevance of words proposed by He et al. [9], we expand the security domain ontology we have established before. We select some terminology related to natural disasters, accident disasters, public health events and social security events as the seed ontologies of our experiment, and define three relationships between concepts, which are synonymous relationship, hyponym relationship and related relationship.

We expand the security domain ontology according to the word semantic relevance rules, and select the candidate concept whose semantic correlation value is

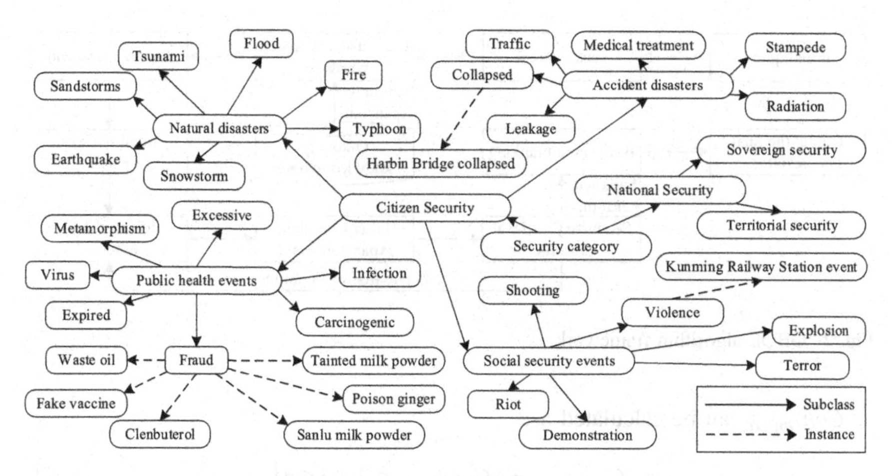

Fig. 1 Part of the security domain ontology

bigger than λ as the to-be-expanded concept of seed ontology. Then we use include analysis rules to find semantic relations between seed ontology concept and to-be-expanded concept. Part of the security domain ontology is shown in Fig. 1.

3 Microblog Query Expansion Based on Ontology Expansion and Borda Count Rank (OBQE)

3.1 OBQE Algorithm Framework

In this paper, we expand the initial query of the user with the established security domain ontology, reasoning knowledge according to the ontological semantic relationship. The framework of microblog query expansion algorithm based on ontology expansion and Borda count rank (OBQE) is shown in Fig. 2.

3.2 Determination of Candidate Expansion Words

Candidate expansion words set can be generated based on the established ontology knowledge base. Mapping the initial query words set to the ontology knowledge base, we expand q_i, we get the set $OntExtSet = \{qe_{i,j} | 1 \leq i, j \leq n\}$.

Calculating the connection value $Con(qe_i, q)$ between qe_i and q, we select the expansion words which connection value is bigger than λ_1 to join the candidate expansion words set Con_{OntExt}.

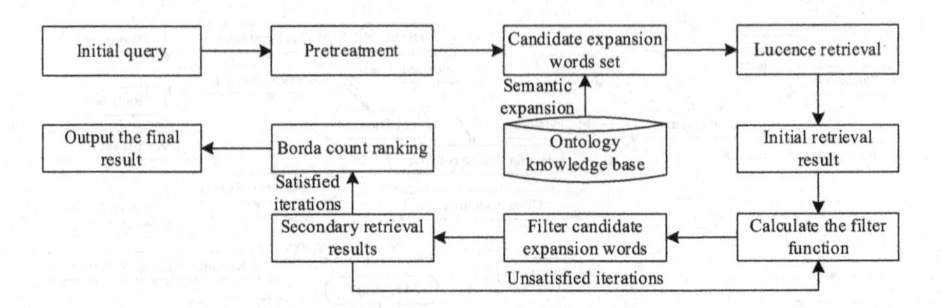

Fig. 2 OBQE algorithm framework

Con_{OntExt} can be calculated as:

$$Con_{OntExt} = \{Con(qe_i, q) | 1 \leq i \leq n\} \tag{1}$$

$$Rel(qe_i, q) = tf(q_i, doc) \times \ln(\frac{m_i + M}{M}) \tag{2}$$

$$Con(qe_i, q) = \frac{Rel(qe_i, q)}{max(Rel(qe_i, q))} \tag{3}$$

where $tf(q_i, doc)$ is the frequency of q_i in microblog texts doc, m_i is the number of query words mapping to the ontology knowledge base, and M is the total number of words in the query words set.

3.3 Filter of Candidate Expansion Words

For the results of the initial retrieval, we calculate the microblog heat H_i of each microblog text, select top N microblog texts to establish a local microblog text set (S).

H_i can be calculated as:

$$H_i = \alpha \times \log(lCount_i + fCount_i + cCount_i) + \beta \tag{4}$$

where α is the heat base, β is the heat weight, and $lCount_i$, $fCount_i$, $cCount_i$ are the number of like, forward and comment of the microblog text.

$CoFreq(qe_i, q|doc)$ is defined as the co-occurrence frequency of candidate expansion words qe_i and query word q in microblog text doc. It can be calculated as:

$$CoFreq(qe_i, q|doc) = \frac{n_c(qe_i|doc) + \alpha}{n_c(q|doc) + 2\alpha} \tag{5}$$

where $n_c(*|doc)$ is the occurrence number of concept * in microblog text doc. In order to avoid data overflow, we set the smoothing factor α.

The average co-occurrence frequency $CoFreq(qe_i, q|S)$ can be calculate as:

$$CoFreq(qe_i, q|S) = \frac{\sum_{doc \in S} CoFreq(qe_i, q|doc)}{n} \quad (6)$$

where n is the number of the microblog text which candidate expansion word and query concept appears in at the same time.

Consider the connection value at ontology expansion and the co-occurrence frequency at local query feedback, the filter function $F(qe_i, q)$ is defined as:

$$F(qe_i, q) = Con(qe, q) \times CoFreq(qe_i, q|S) \quad (7)$$

Let the candidate expansion word satisfying the condition $F(qe_i, q) > S\lambda_2$ be the final expansion word.

3.4 Borda Count Rank

According to Borda count rank, each voter has their own preference list of candidates. For each voter, the top first candidate obtains n points, the top second candidate obtains $n - 1$ points, and the top third candidate obtains $n - 2$ points, and so on. The sum value of obtained points of each voter gives the final points to each candidate. There are few candidates that are unranked by a voter (candidate terms selection method); then remaining points are divided among the unranked candidates. The candidate that has high points wins [10].

3.5 OBQE Algorithm Description

Through the iterative operation, the retrieval result meet user's requirements better. The number of iterations is set to 5. Microblog query expansion algorithm based on ontology expansion and Borda count rank is shown as follows.

Input: Initial query $Q = \{q_i | 1 \le i \le n\}$

Output: Ranked microblog texts sequence

(1) Map the initial query words set to the domain ontology knowledge base, and get *OntExtSet*;
(2) Calculate $Con(qe_i, q_i)$, and add the expansion word whose $Con(qe_i, q_i) > \lambda_1$ to Qry_{ExtSet};
(3) Repeat step (2) until all the candidate expansion words of query words in Q are found;

(4) Use Lucene to get the initial results;
(5) Select top N microblog texts to establish S, and calculate $CoFreq(qe_i, q|S)$;
(6) Calculate $F(qe_i, q)$ to filter candidate expansion words;
(7) Use Lucene for secondary retrieval;
(8) Repeat step (5) to step (7) until the iterations is completed;
(9) Use Borda count rank method to rearrange the query results;
(10) Output the final retrieval result.

4 Experimental Results and Analysis of OBQE Algorithm

4.1 Experimental Methods

The main purpose of this paper is to improve the performance and quality of microblog retrieval significantly. Thus, the experimental part is mainly used to test whether the algorithm proposed in this paper can improve the information retrieval performance compared with the traditional expansion method. This paper establishes a simple query expansion system to experiment.

The ontology used in the experiment is the security domain ontology established by Protégé 5.1. In order to combine the experiment with the security domain ontology, we use "train station" as the key words, crawling the related microblog of 34 provincial-level administrative units of Sina Weibo from February 27, 2014 to March 5, 2014, and get 32479 microblog texts. We use the ICTCLAS as the word segmentation tool, and use Lucene as the information retrieval platform.

4.2 Evaluation Indexes

Recall rate (R), precision rate (P) and F-measure (F) are used as the evaluation indexes. The F-measure that combines recall rate and precision rate can better reflect the retrieval performance. When recall rate and precision rate increase, F-measure increases. The larger the F-measure, the better the retrieval performance. The F-measure is defined as:

$$F = \frac{2 \times R \times P}{R + P} \tag{8}$$

4.3 Comparison and Analysis of Query Efficiency of OBQE Algorithm in Microblog Query

We compare our algorithm (OBQE) with keyword-based query expansion algorithm (KQE), ontology-based query expansion algorithm (OQE) and query expansion algorithm based on ontology and local query feedback (OFQE). In order to increase the accuracy and reliability of the experimental results, Seven queries are made on the experimental dataset using the three words of "fear", "violence" and "terror" and their combinations, and we take the average of seven queries as experimental results. The results are shown in Table 1 and Fig. 3.

As Fig. 3 shows that OBQE algorithm proposed in this paper is optimal in recall rate, precision rate and F-measure. Because the expansion query words include "riot", "explosion" and other keywords, query efficiency of OQE algorithm has been significantly improved compared with KQE algorithm. OFQE algorithm optimizes the final query words set. Through secondary retrieval and multiple iterations, query efficiency of OFQE increases again. OBQE algorithm uses Borda count rank to optimize the result sequence, so query efficiency of OBQE is highest of four algorithms. The average precision rate of Top N results of four algorithms are shown in Table 2 and Fig. 4.

It is shown in Fig. 4 that the precision rate of OBQE algorithm proposed in this paper is better than KQE algorithm, OQE algorithm and OFQE algorithm under different indexes. As Fig. 4 shows that the precision rate of the four algorithms shows a decreasing trend with the increase of N. This is because the high matching query results often appear in the top of the return sequence. With the increasing number of query results, the decreased precision is unavoidable. Meanwhile, We

Table 1 Recall rate, precision rate and F-measure of four algorithms

Algorithm	Recall rate	Precision rate	F-measure
KQE	0.218	0.624	0.323
OQE	0.434	0.704	0.537
OFQE	0.509	0.735	0.601
OBQE	0.537	0.782	0.637

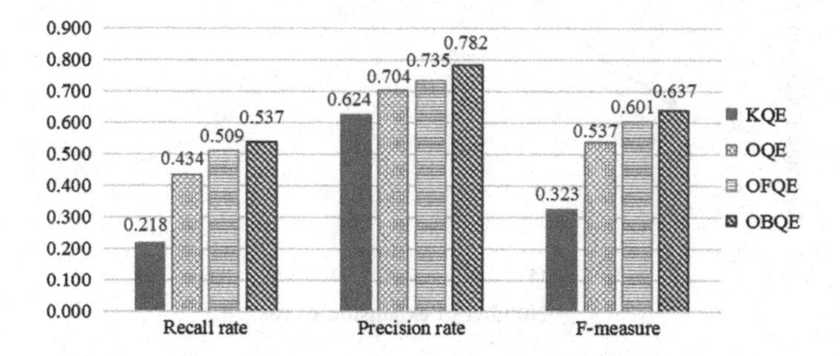

Fig. 3 Recall rate, precision rate and F-measure of four algorithms

Table 2 Average precision rate of top N results of four algorithms

Algorithm	P@20	P@50	P@100	P@200	P@500
KQE	0.714	0.663	0.634	0.603	0.591
OQE	0.807	0.759	0.658	0.632	0.620
OFQE	0.857	0.821	0.698	0.681	0.665
OBQE	0.893	0.843	0.776	0.704	0.688

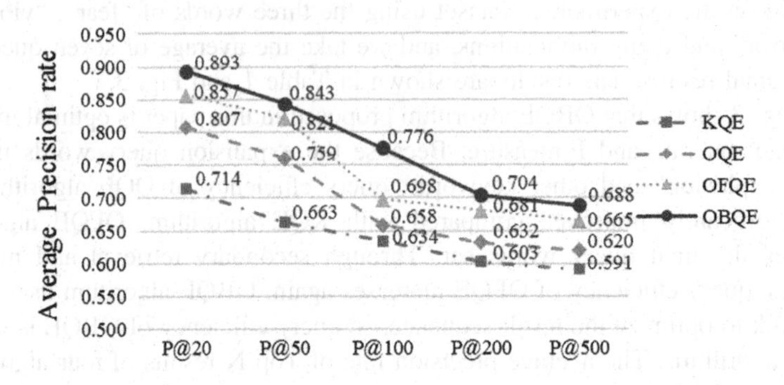

Fig. 4 Average precision rate of top *N* results of four algorithms

remove some microblogs start with ## topic but not related to the user's query artificially, which partly reduce the precision rate of microblog retrieval.

In query expansion, the number of expansion words is also an important factor affecting query performance. Thus, we carry out a comparative experiment of the number of expansion words. The relationship between the scale of expansion words and precision rate is shown in Fig. 5. OBQE algorithm can obtain a higher precision rate when the number of expansion words close to 20. When the number exceeds 40, the precision rate is significantly reduced, and may even be lower than the rate of non-expansion.

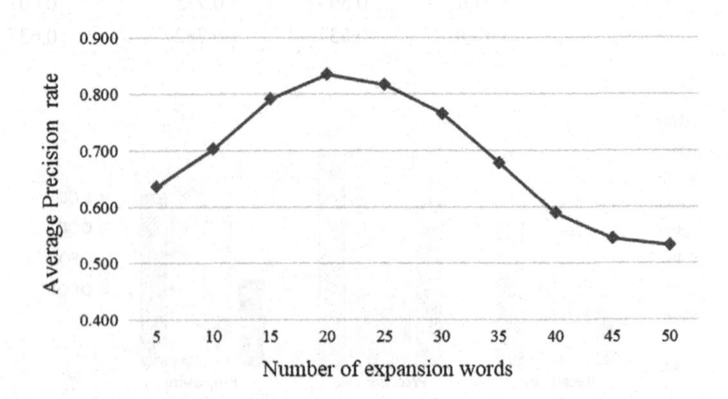

Fig. 5 The relationship between the scale of expansion words and precision rate

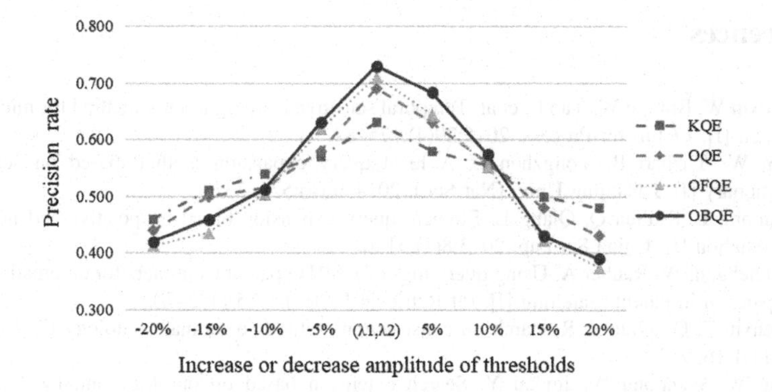

Fig. 6 Precision rate of different thresholds

The two parameters λ_1 and λ_2 are the main factors that determine the performance of our algorithm. We use percentage to express the increase or decrease amplitude of the two parameters. Precision rate of different thresholds is shown in Fig. 6. The values of λ_1 and λ_2 are optimal determined by the optimal query performance, and the positive and negative percentages represent the amount of increase or decrease compared to the optimal value. As Fig. 6 shows that OBQE is superior to OFQE, which proves the validity of Borda count rank for the rearrangement of experimental results. When the thresholds is too small or too larger, the precision rate of OBQE is lower than that of KQE and OQE, which means that OBQE is sensitive to the values of λ_1 and λ_2.

5 Conclusions

In this paper, we proposed a microblog query expansion algorithm based on ontology expansion and Borda count rank. Firstly, we establish a security domain ontology, realize semantic expansion of the initial query with the domain ontology and filter the candidate expansion words by filter function. After secondary retrieval and iterative operation, we use Borda count rank to rearrange and optimize the final results. The experimental results show that the algorithm proposed in this paper improves the recall rate and precision rate of microblog query. As the algorithm relies on the co-occurrence weight of the microblog short text, we will optimize microblog short text processing to improve the recall rate and precision rate in the future.

Acknowledgements This work is supported by the National Natural Science Foundation of China (No. 61320106006, No. 61532006, No. 61502042).

References

1. Shuxin W, Bingjie W, Yao L, et al. Temporal sensitive learning to rank method for microblog search [J]. J Chin Inf Process. 2015;29(4):175–82.
2. Qin W, Yuzhao B, Yongzhen L. A local query expansion method based on semantic dictionary [J]. J Nanjing Univ. (Nat Sci.). 2014;50(4):526–33.
3. Caihong L, Ruihua Q, Qiang L. Efficient query expansion based on positive and negative association [J]. China Sci Pap. 2013;8(1):51–6.
4. Tannebaum W, Rauber A. Using query logs of USPTO patent examiners for automatic query expansion in patent searching [J]. Inf Retrieval J. 2014;17(5):452–70.
5. Wenxiu Z, Qinghua Z. Research on construction methods of domain ontology [J]. Libr Inf. 2011;1:16–9.
6. Jing W, Wencong W, Junkai Y. Search extension based on ontology and local context analysis [J]. Control Eng China. 2013;20(3):558–61.
7. Aiming L. Research on query expansion method based on ontology and user query intention [J]. Inf Sci. 2015;5:68–71.
8. Duranti CM, Almeida FCD. Selection of online news for competitive intelligence: use of business domain ontology for internet search semantic query expansion [J]. Glob J Comput Sci Technol. 2015;15(6):10–25.
9. Wei H, Xiaoping Y. Extending ontology based on semantic relatedness between words from a text [J]. Comput Appl Softw. 2011;28(11):73–6.
10. Singh J, Sharan A. Relevance feedback based query expansion model using Borda count and semantic similarity approach [J]. Comput Intell Neurosci. 2015;2015(1):1–13.

Microblogging Event Search Based on LSTM Model

Wenzhen Zheng, Junping Du, Jincai Lai, Meiyu Liang and Ang Luo

Abstract With the rapid development of the online social media, microblogging events surveillance has become a major research topic. Traditional search method does not consider the characteristics of the events, the search algorithm has its limitations. To solve this problem, we proposed a microblogging event search method based on Long Short Term Memory (LSTM) networks called MESL. Using training corpus to extract the common characteristics of microblogging events. The establishment of event search model effectively improves the microblogging event search quality. Experimental results on the real microblogging datasets show that MESL model is better than the traditional methods for microblogging event search.

Keywords LSTM · Microblogging · Event search

1 Introduction

With the development of Web 2.0 social network, microblogging has become the main way to spread the network information because of its convenient advantages. In June 2016, the active users of Sina Weibo, which is the largest microblogging company in China, has grown up to 280 million [1]. With the huge active user base, we get massive amounts of microblogging information every day. In the microblogging network, everyone is the information producers and communicators, information dissemination is very convenient, which makes microblogging become the fastest media of information dissemination. Lots of the social events, often first appeared on the microblogging platform, with its quickly spread, causing a wide

W. Zheng · J. Du (✉) · J. Lai · M. Liang
Beijing Key Laboratory of Intelligent Telecommunication Software and Multimedia,
School of Computer Science, Beijing University of Posts and Telecommunications,
Beijing 100876, China
e-mail: junpingdu@126.com

A. Luo
SINA Corporation, Beijing, China

Y. Jia et al. (eds.), *Proceedings of 2017 Chinese Intelligent Systems Conference*, Lecture Notes in Electrical Engineering 459,
https://doi.org/10.1007/978-981-10-6496-8_26

range of social resonance, and then spread to traditional media such as news, forums, blogs. How to retrieve emergencies from these rich contexts is an urgent problem to be solved by microblogging unexpected events. Microblogging event search is different to traditional web search in many aspects. Firstly, on the retrieval content, microblogging text is short, with keywords and a variety of emoticon and URL address and so on. These features make the microblogging search become more abundant. Secondly, as the microblogging user query is usually related to the current occurrence of the event, so compared to the traditional search query microblogging search is more time-sensitive.

This paper attempts to use the deep learning method to automatically explore the characteristics of microblogging event data, and according to the characteristics of the event data to carry out its deep analysis. In this paper we proposed a microblogging event search algorithm called MESL. We improve the existing Long Short Term Memory(LSTM) neural network model, and use the model automatically extract the microblogging event feature to improve microblogging event search results. This paper is organized as follows: In Sect. 2 we describe the related works. In Sect. 3, we describe our MESL microblogging event search model. Section 4 describes the experiment and results. Finally, we conclude full paper in Sect. 5.

2 Related Work

At present, the influence of social network has been increasing, and microblogging search has become a hot research topic, and many research methods have been adopted.

Mei et al. [2] proposed a method based on semantic linking. Efron et al. [3] studied the microblogging search method based on the hash (Hashtag). Zhao et al. [4] explored a weighted based on the traditional Web method, which is limited for the sorting application based on list learning. Tang [5] proposed the use of the theme-sensitive PageRank algorithm to reorder the results of the user model based on the subject preference, with the aim of better showing the user's real interest and shortening the query feedback time and streamlining the calculation.

Li [6] using the deep learning tool Word2vec [7] for the event detection, the use of Word2vec training all the words in the training set of K-dimensional vector characterization, based on the word vector for the similarity between the vocabulary calculation, and thus achieve lexical clustering The keywords of the document. D. He et al. [8] and Wei et al. [9] have proposed an acceleration-based physical idea to find the burst mode more frequently for burst detection.

Long et al. [10] introduced the use of hashtag on the event detection. Hashtag, the information entropy factor to extract the keywords that represent the emergencies, and construct the covariance graphs to apply the clustering algorithm to get the microblogging events. Zhao et al. [11] used the relative word frequency and word frequency growth rate to extract the subject of event, based on the covariance between the words clustering. Yao [12] detected events in microblogs by monitoring the Hashtag word change marked by the user-generated information.

The above search algorithm is not for the characteristics of emergencies, so the effect of the incident search in general. Therefore, this paper presents a microblogging event search algorithm based on LSTM model to improve the search results of microblogging emergencies.

3 Microblogging Event Search Based on LSTM Model (MESL)

3.1 MESL Algorithm Framework

The framework of MESL model is shown as Fig. 1. In the word embedding layer, we get the word vectors of user queries and microblogging documents, and send them into the LSTM layer. In the LSTM layer, we get the word feature automatically by the LSTM networks. Then we get the mean pooling output of the LSTM networks and send them into the ranking layer. Beside the ranking layer, we

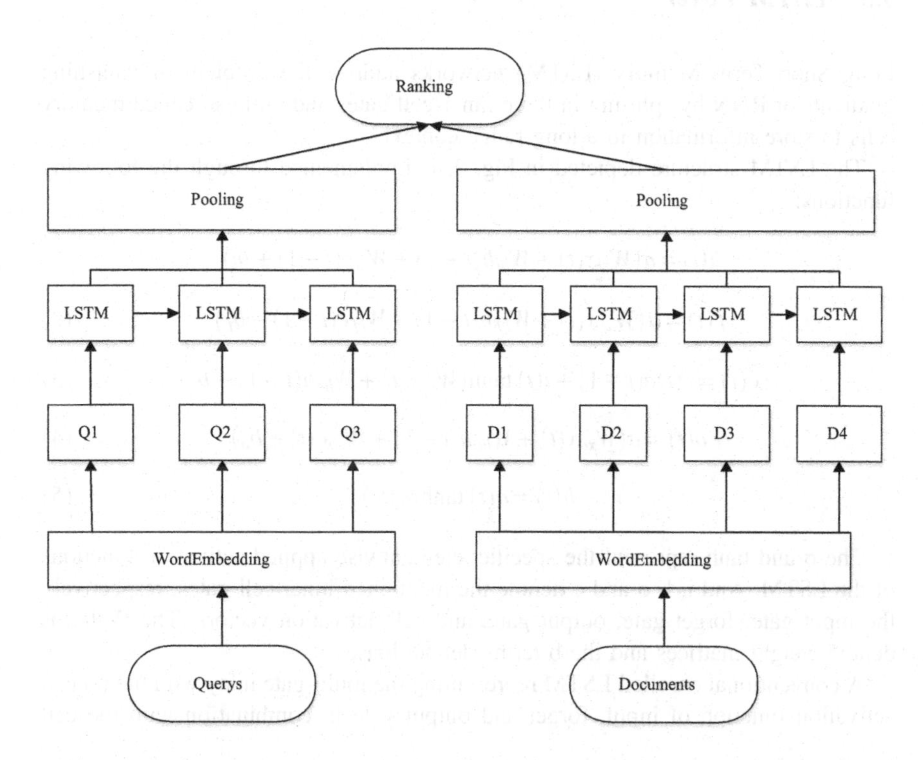

Fig. 1 MESL algorithm framework

calculate the score of each document in the microblogging documents, return the documents with high score. The final output of ranking layer is our search result by the user queries.

3.2 Word Embedding Layer

In the deep learning model, the initialization of the parameters has a great influence on the model results. In the process of model training, the word vector is also used as the model parameter. The initial effect on the model is also very important. There are two main word vector initialization methods often used: one is a language model based on corpus pre-training word vector; and the other is using random initialization word vector. From the previous research work, it is found that the model results based on the pre-trained word vector are often superior to the result of random initialization. Therefore, in this paper, we first initialize the word vector using the word2vec model and we select the dimension of the vector to be 200.

3.3 LSTM Layer

Long Short-Term Memory (LSTM) networks address the problem of vanishing gradients of RNN by splitting in three inner-cell gates and build so called memory cells to store information in a long range context.

The LSTM structure depicted in Fig. 2. is implemented through the following functions:

$$i(t) = \sigma(W_{xi}x(t) + W_{hi}h(t-1) + W_{ci}c(t-1) + b_i) \tag{1}$$

$$f(t) = \sigma\left(W_{xf}x(t) + W_{hf}h(t-1) + W_{cf}c(t-1) + b_f\right) \tag{2}$$

$$c(t) = f(t)c(t-1) + i(t)\tanh(W_{xc}x(t) + W_{hc}h(t-1) + b_c) \tag{3}$$

$$o(t) = \sigma(W_{xo}x(t) + W_{ho}h(t-1) + W_{co}c(t) + b_o) \tag{4}$$

$$h(t) = o(t)\tanh(c(t)) \tag{5}$$

The σ and tanh represent the specific, elementwise applied activation functions of the LSTM. And i, f, o and c denote the mentioned inner-cell gates, respectively the input gate, forget gate, output gate, and cell activation vectors. The W terms denote weight matrices and the b terms denote biases.

A conventional enrolled LSTM representing the multi-gate inlay with the related activation function of input, forget and output gate in combination with the cell

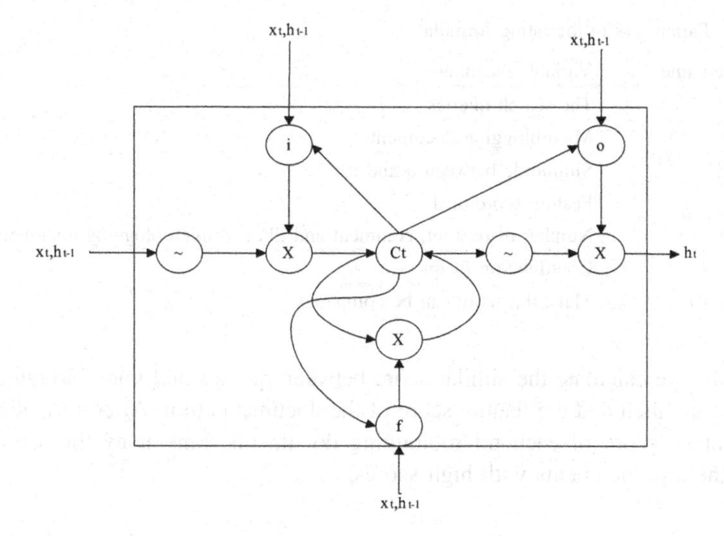

Fig. 2 Structure of LSTM node

block. Hidden outputs of one LSTM block are transported through the whole chain to ensure information storage in a long rage context.

In our MESL model, we use the basic LSTM network. The input of our model is the word embeddings of user queries and microblogging documents constructed by the word embedding layer. The output of our LSTM layer is a word vector calculated by the mean pooling of all the word in the outputs.

3.4 Ranking Layer

The main rating formula of the ranking layer is shown in Eq. 6.

$$score(q,d) = \frac{sim(q,d) + feat(d) \ * coord}{norm(q,d)} \tag{6}$$

where sim (q, d) is calculated as shown in Eq. 7. And feat is calculated as shown in Eq. 8.

$$sim(q,d) = \frac{v(d) \cdot v(q)^T}{||v(d)|| \cdot ||v(q)} \tag{7}$$

$$feat(d) = n_t * \beta + n_c \ * \gamma + n_l \ * \delta \tag{8}$$

The meaning of the parameters in the above formula is shown as Table 1.

Table 1 Parameters in the rating formula

Variable name	Variable meaning
q	The search queries
d	Microblogging documents
sim(q, d)	Similitude between q and d
feat(d)	Feature score of d
n_t, n_c, n_l	Number of retweet, comment and likes of microblogging documents
coord	Coordination factor
Norm(q, d)	Make the score can be compared

Firstly, we calculate the similar score between queries and microblogging documents, and then add the feature score of the documents to it. After normalization, we compare score of each microblogging documents, rank it by the score, and return the top documents with high scores.

4 Experimental Results and Analysis of MESL Algorithm

4.1 Dataset

We crawled microblogging data through Sina Weibo application program interface (Weibo API). We collect Random microblogging users microblogging documents between January 1, 2014 to May 1, 2014. Finally, we obtained about 25 thousands original microblogging documents.

4.2 Parameters Setting

For the word embedding layer, we set the dimension of the word vector to be 128. This word embedding set is trained on a large Chinese news corpus and microblogging data we collected with word2vec model. In this paper, we adopt the Python implement of word2vec using gensim toolset. Since the microblogging dataset is quite limited, we do not update these word embeddings in the training processing, which greatly reduces the model parameters to be learned. In the LSTM network layer, we set the hidden units to be 128, it is the same as the word embeddings. The initial learning rate is 0.05 and batch size is 50.

4.3 Evaluation Indexes

There are lots of evaluation in comparing search results on sorting algorithms, such as ERR (expected reciprocal rank), NDCG (normalized discount cumulative gain), P@N (precision at position) and MAP (mean average precision), etc. In this paper, we use the P@N, NDCG and MAP to evaluate the effect of the search result. We compare our method to BM25 and Feature Score algorithm used by Srijith et al. [13]

4.4 Comparison and Analysis of MESL Algorithm

We use different user queries for different results. And judged the results by different people to get a mean evaluation for every result. The final result is shown as below.

As it can be seen from Tables 2 and 3, because the search algorithm of this paper considers the event keywords and other features, and therefore is a significant improvement on the event search results. At K = 50, the search results of our model reached 0.82 of the P@K value, 0.21 higher than BM25 algorithm, 0.10 higher than FeatureScore algorithm, and in the value of the comparison MAP, when K is 50, the MAP@50 value of our model reached 0.88, 0.08 higher than BM25 algorithm, 0.01 higher than FeatureScore algorithm. The results showed that in the search for multiple events the MAP value of MESL model is higher than the other two algorithm.

We use different query words to see the performance of different algorithm in long query words and short query words. We use short query words such as "Kunming Railway station", "Kunming violence", "Kunming attack" and so on. On the other hand, we use long query words like "Kunming Railway Station violent terrorist incident" for instance. The comparison of three algorithms on different length of event query words is shown as Figs. 3 and 4.

Table 2 P@K value of the three algorithms

Model	P@K			
	K = 5	K = 10	K = 30	K = 50
BM25	0.65	0.67	0.68	0.65
FeatureScore	0.75	0.78	0.79	0.76
MESL	**0.82**	**0.81**	**0.83**	**0.82**

Table 3 MAP@K value of the three algorithms

Model	MAP@K			
	K = 5	K = 10	K = 30	K = 50
BM25	0.76	0.75	0.75	0.80
FeatureScore	0.86	0.85	0.87	0.87
MESL	**0.90**	**0.89**	**0.89**	**0.88**

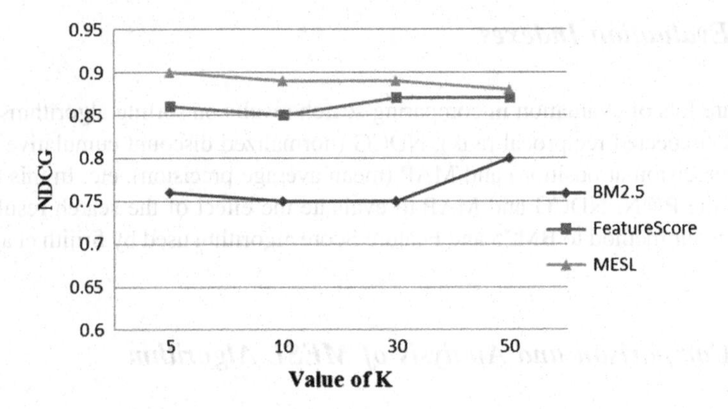

Fig. 3 NDCG value comparison for long query

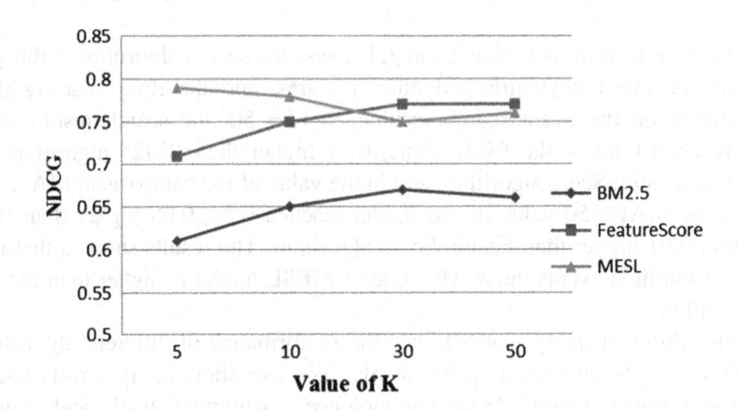

Fig. 4 NDCG value comparison for short query

In Fig. 3, we can see that MESL get better results in long query words than other two methods. Because of LSTM networks are explicitly designed to avoid the long-term dependency problem. Remembering information for long periods of time is practically their default behavior, not something they struggle to learn. So the use of LSTM networks can get more information in long query words, so the performance of long query words is better.

As Fig. 4 shows that the performance of MESL in short query words is not better than above in the long query words, but it still has good performance. With the value of K grows, the NDCG@K value of MESL is not better than FeatureScore algorithm, but the performance is better than BM2.5.

Therefore, it can be seen from the experiment that the algorithm in this paper works well in microblogging event search.

5 Conclusions

Microblogging event search is facing the problem that too much microblogging data and too many search results to match. In this paper, an event search algorithm based on the LSTM model is proposed. The experimental results show that this method can improve the accuracy of microblogging event search, and can get better results in a large number of microblogs. In the future experiment we will improve our model to get better performance in short user query words.

Acknowledgements This work is supported by the National Natural Science Foundation of China (No. 61502042, No. 61320106006, No. 61532006).

References

1. SinaWeibo.2016.8. http://tech.sina.com.cn/i/2016-08-09/doc-ifxutfpc4848750.shtml.
2. Meij E, Weerkamp W, Rijke MD. Adding semantics to microblog posts. In: Proceedings of the 5th ACM international conference on web search and data mining. New York: ACM Press;2012. p. 563–72.
3. Efron M. Hashtag retrieval in a microblogging environment. In: Proceedings of the 5th ACM international conference on web search and data mining. New York: ACM Press;2012. p. 563–72.
4. Zhao LL, Zeng Y, Zhong N. A weighted multi-factor algorithm for microblog search. In: Zhong N, Callaghan V, Ghorbani A, Hu, editors. Proceedings of the 7th international conference on AMT 2011. Berlin: Springer;2011. p. 153–61.
5. Tang J, Wang K, Shao L. Supervised matrix factorization hashing for cross-modal retrieval. IEEE Trans Image Process. 2016; 25(7):3157–66.
6. Yuepeng L, Cui J, Junchuan J. A keyword extraction algorithm based on Word2vec. E-sci Technol Appl. 2015;4:54–9.
7. Mikolov T, Chen K, Corrado G, et al. Efficient estimation of word representations in vector space. arXiv:1301.3781;2013.
8. He D, Parker D. Topic dynamics: an alternative model of bursts in streams of topics. In: Proceedings of the 16th ACM SICKDD international conference on knowledge discovery and data mining;2010 p. 443–52.
9. Xie W, Zhu F, Jiang J, et al. Topicsketch: real-time bursty topic detection from twitter. In: 2013 IEEE 13th international conference on data mining (ICIM). IEEE;2013. p. 837–46.
10. Long R, Wang H, Chen Y, et al. Towards effective event detection, tracking and summarization on microblog data. In: Web-age information management. Berlin, Heidelberg: Springer;2011. p. 652–63.
11. Zhao W, Hou X. News topic recognition of chinese microblog based on word co-occurrence graph. CAAI Trans Intell Syst. 2012;7(5):444–9.
12. Yao J, Cui B, Huang Y, et al. Bursty event detection from collaborative tags. World Wide Web. 2012;15(2):171–95.
13. Ravikumar S, Talamadupula K, et al. RAProp: ranking tweets by exploiting the tweet/user/web ecosystem and inter-tweet agreement. In: CIKM'13, Oct 2013, San Francisco, CA, USA.

Microblog Search Method Based on Neural Network Language Model

Jincai Lai, Junping Du, Wenzhen Zheng and Wei Wang

Abstract Deep neural network language model has gained significant development among natural language processing (NLP) in recent years. In this paper, we focused on using neural language model (NNLM) to enhance microblog search. This paper proposed a microblog search method based on neural network language model (NBSM). Firstly, we train neural network language model based on microblog data, so as to get the distributed representation of words which may contain internal express model of microblog. Then, we use the distributed representation of words to get the expanding words of users' searching words. Finally, we re-rank microblog search results combining deep sematic text similarity and social signal features. The method we proposed can effectively obtain microblog express model, and its search result can reflect the social hot-topics of the topic related to users searching words. Experiment results show that the proposed method yields significant improvements over state-of-arts methods and significantly improves the user's search experience.

Keywords Microblog search · Neural network · Word vector · Social network

1 Introduction

According to Teevan's research [1], the user expects the microblogging to obtain the contents of the following aspects: timeliness information, such as celebrity news and major events; topic or event development, microblogging users tend to repeat

J. Lai · J. Du (✉) · W. Zheng
Beijing Key Laboratory of Intelligent Telecommunication Software and Multimedia, School of Computer Science, Beijing University of Posts and Telecommunications, Beijing 100876, China
e-mail: junpingdu@126.com

W. Wang
SINA Corporation, Beijing, China

Y. Jia et al. (eds.), *Proceedings of 2017 Chinese Intelligent Systems Conference*, Lecture Notes in Electrical Engineering 459,
https://doi.org/10.1007/978-981-10-6496-8_27

the same keywords to track the development of the topic or event; advice and other social information. According to Kelly [2] and others survey, Twitter in 40% of the garbage microblog. Microblog short unique text features for microblog search brings a different challenge from the web search, but also contains new opportunities. How to make full use of rich social characteristics and time and space characteristics of microblog, enhance the user's search experience, more and more become the academic and industrial research hot spots.

The related research on microblog search mainly focuses on the use of microblog attention or fans relationship, user's behavior model for user-oriented personalized search recommendation [3], combined with microblog time and space attributes to re-rank microblog search results [4], which could improve the relevance and novelty between searching results and user's query. To resolve microblog rapid update features and big data characteristics, [5] proposed the relevant cache and index optimization research methods. Severyn and Moschitti [6] used the convolutional neural network language model to model sentence and applied them to enhance microblogging search, and the use of more advanced language model to improve the search algorithm is becoming a popular research topic nowadays.

Early studies of language problems using neural network models were performed by Bengio et al. [7], who completed the construction of the framework of this field of study and compared it with the traditional language model. The most significant study was done by Mikolov et al. [8]. He carefully designed the framework of neural network, and carried out a series of simplification and training aspects of optimization. After that, neural network language model became the most significant research project, and has been widely used in all aspects of natural language processing. The recent research on information retrieval based on deep neural network has the following aspects: [9, 10] focuses on the study of word embedding using deep neural network model, while [6] focuses on the similarity of short text. These studies had made great progress and encouraged the researchers' confidence in the research of information retrieval based on the deep neural network to a certain extent.

In this paper, we propose a method to model the microblog by using the neural network language model, and then get the word vector and use the word vector to expand the search keywords. Finally, we resort the search results combine with social signals. Compared with the traditional microblog search algorithms, the method we proposed in this paper has several advantages. Firstly, this paper introduces the neural network language model based on the traditional keyword search, and can make full use of the microblog language expression mode, make the search result more accurate. It can more accurately understand the user's search intent, better meet the user's information needs. Second, this method we proposed consider the impact of social signals in the microblog ranking function, which make more popular microblog at the top of the searching results, and that might help users more accurately grasp hot spots of the relevant topic.

The rest of the paper is organized as follows: in Sect. 2, we state details of the microblog search method we proposed, Sect. 3 is the experiment and analysis of our methods, and Sect. 4 is conclusions of this paper.

2 The Framework of Microblog Search Method

In this paper, we proposes a microblog search method based on neural network language model (NBSM). The searching process is shown in Fig. 1.

In our pre-process work, we use corpus (microblog data) to train the neural network language model and then obtain the word vector representation of each word. Also, we construct the microblog index database. As shown in Fig. 1, in the searching process of our microblog search system, when we get the query keywords of the user, we first use the word vector data to get the first k expanding words of the first query word, and then these expanded words are concatenated together and input into the search system as the query words. According to the semantic similarity, get top n relevant microblog, then we calculate the text similarity and social popularity of the microblog to re-rank them, and finally we get search results of corresponding user query.

2.1 Neural Network Language Model

This section introduces the concept of neural network language model. Mikolov et al. [8] proposed a well know Neural Network Language Model framework of learning the distributed representation of words, that is, the Word2Vec framework. In this paper, the skip-gram model in Word2vec framework is used as our neural network language model. Given a set of training words $w_1, \ldots, w_T$, the training goal is to maximize the average log probability:

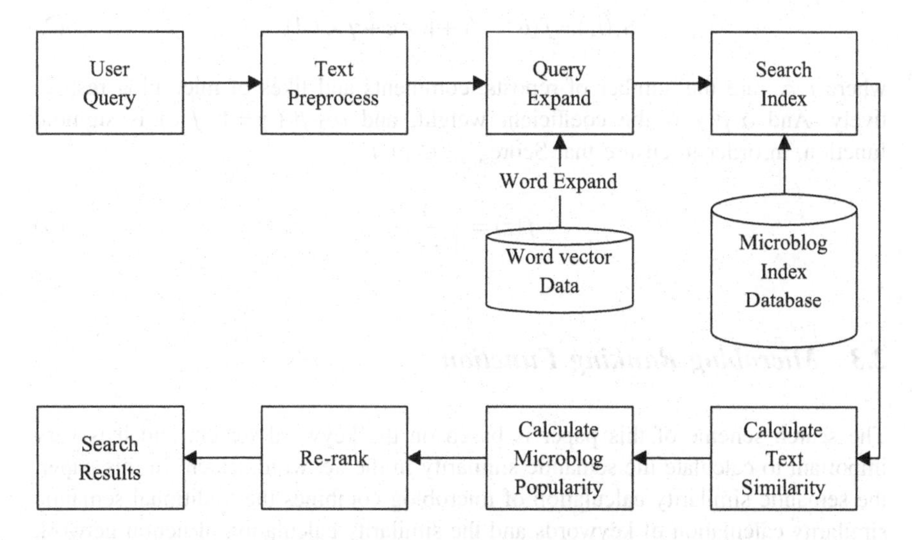

Fig. 1 The searching process of microblog search method (NBSM)

$$\max_{\theta} \frac{1}{T} \sum_{t=1}^{T} \sum_{-c \leq j \leq c} \log(p(w_{t+j}|w_t;\theta)) \tag{1}$$

where c is the size of the training corpus, T is the length of the training sequence, w_t is the training word in time t, and p is the log probability function. Through training, θ is the maximum parameter estimate and the word eigenvector are output as a result.

The similarity of the two words in the neural network language model is calculated by the distance of the represent vector of the words (using cosine distance). Calculated the distance between word vectors, it is possible to find a few words that are most similar to a given word. These words can be understood as words that have the less semantic distance to a given word in corpus, and these semantic similar words can be used to expand the search keywords. On the other hand, the similarity between the two words in the neural network can be added to the search sort function as part of the semantic similarity of the microblog and the query keywords, which makes the microblog search more accurate.

2.2 Calculation of Microblog Popularity

Each microblog contains certain social signals, including the number of likes, the number of comments, and the number of reposts. In the microblog information flow, the rapid accumulation of the number of reposts and comments are often associated with the popularity of the topic. This paper uses different ways to calculate the weight of these social features, as follows:

$$S_s(t_k) = f(\alpha \times \sqrt{r} + \beta \times c + \gamma \times \sqrt{l}) \tag{2}$$

where r, c, l are the number of reposts, comments and likes of microblog respectively. And α, β, γ is the coefficient weight, and $\alpha+\beta+\gamma=1$. $f(z)$ is sigmoid function, in order to ensure that $Score_{social} \in [0, 1]$

$$f(z) = \frac{1}{1+e^{-z}} \tag{3}$$

2.3 Microblog Ranking Function

The search scheme of this paper is based on the keyword search, and it is very important to calculate the semantic similarity in the keyword search. In this paper, the semantic similarity calculation of microblog combines the traditional semantic similarity calculation of keywords and the similarity calculation of neural network language model. Assuming that the input query word is q, the microblog document to be sorted is t_k, text similarity between microblog and query word can be expressed as follows:

$$S_r(t_k, q) = \prod_{i=i}^{m} ((1-\lambda)(\beta \sum\nolimits_{w \in t_k} P_{ml}(w|t_k) \cdot y_w + (1-\beta)P_{ml}(q_i|t_k)) + \lambda P_{ml}(q_i|Coll)) \quad (4)$$

where $y_w \in [0, 1]$ represents the word vector spaces distance of two words, $\lambda \in [0, 1]$ is a smoothing factor, $\beta \in [0, 1]$ is the coefficient of semantic similarity according to NNLM. $P_{ml}(q_i|t_k)$ is the maximum likelihood probability between query words q_i and microblog documents t_k, $P_{ml}(q_i|Coll)$ is maximum likelihood probability in the corpus of the query word q_i.

$$P_{ml}(q_i|t_k) = \frac{tf_{q_i, t_k}}{\sum_{j=1, q_j \in t_k}^{m} tf_{q_j, t_k}} \quad (5)$$

$$P_{ml}(q_i|Coll) = \frac{tf_{q_i, Coll}}{\sum_{j=1, q_j \in Coll}^{m} tf_{q_j, Coll}} \quad (6)$$

Finally, we calculate the microblog sort score, combined with microblog text similarity and microblog popularity, the ranking function can be expressed as:

$$r(t_k, q) = \alpha_1 S_r(t_k, q) + (1-\alpha_1) S_s(t_k) \quad (7)$$

where $S_r(t_k, q)$, $S_s(t_k)$ are the text similarity and microblog popularity respectively, according to formula (2) and (4).

3 Experiments and Results

In this paper, we use the search algorithm described in Sect. 2, take the neural network language model as the core, supplemented by the social network characteristics of microblog search results resort, carried out a series of microblogging search experiment.

The benchmark of this search experiment is based on the BM25 algorithm of text similarity ranking. (NSM) based on the neural network language model, and the search algorithm (NBSM) of this paper, which combine neural network sematic similarity with social network popularity to resort microblog. These experiments are mainly used to verify the effectiveness of neural network language model to expand search keywords, and resort microblogs combined neural network semantic similarity with social network signals. As a comparative experiment, the RAProp algorithm in [3] has been implemented on our microblog data set and compared with the algorithm of this paper. RAProp algorithm mainly considers the influence of user and microblog authority on the score of microblogging search.

3.1 Data Set and Evaluation

The corpus of the experiment includes two aspects. One is the general corpus of Sogou news as a general corpus, which can be used as a broad corpus to train the neural network language model; another is sina weibo coupus, which we crawl from the interface API of Sina Weibo, about 1 million microblog, as particular corpus of our training neural network language model. Microblog corpus is divided into two parts, 75% microblog as a neural network language model's training corpus, another 25% microblog as a test set for the search system search validation test.

The evaluation criteria used in this paper include the values of the MAP (Mean Average Precision) and the NDCG (Normalized Discounted Cumulative Gain) values to evaluate the search results. We evaluate multiple search results, calculate the score, and finally take the average as the final score. The larger the value of MAP and NDCG, the better the search results.

3.2 Experiment Results and Analysis

We perform a number of microblog search experiment. In our experiment, we enter the keyword, and then the system returns the microblog results to score. We take n = 5, 10, 20, …, 50, that is, the top n results of the search results, calculate and their average NDCG value and MAP value. Average NDCG values for multiple experiments are shown in Table 1, and the corresponding figure is shown in Fig. 2. Average MAP values for multiple experiments is shown in Table 2, and the corresponding figure is shown in Fig. 3.

As shown in Table 1, Fig. 2 and Table 2, Fig. 3, the proposed method NBSM has a significant improvement over existing microblog algorithms in the search accuracy. Furthermore, when the value of n is small, that is, only consider a few top results, the proposed method NBSM performance to a very high level, which means that the microblog search method we proposed could meet the user's information needs very well. When the value of n increase, that is, consider more searching

Table 1 The comparison results of different methods over NDCG value

NDCG	@5	@10	@20	@30	@40	@50
BM25	0.2	0.24	0.32	0.31	0.31	0.32
NSM	0.41	0.41	0.44	0.44	0.45	0.46
NBSM	0.63	0.52	0.55	0.53	0.51	0.50
RAProp	0.36	0.37	0.39	0.42	0.41	0.42

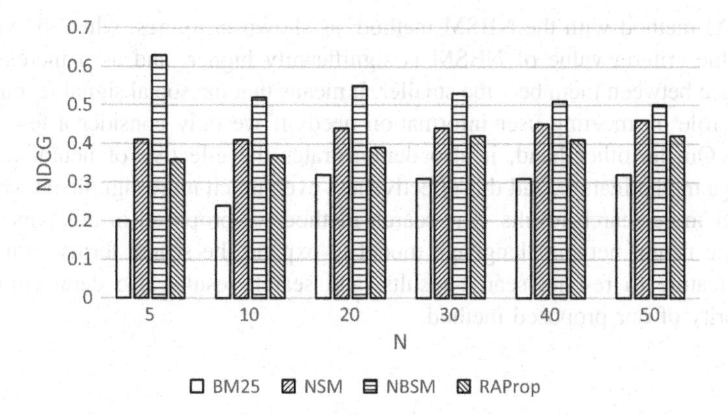

Fig. 2 The comparison results of different methods over NDCG value

Table 2 The comparison results of different methods over MAP value

MAP	@5	@10	@20	@30	@40	@50
BM25	0.21	0.25	0.28	0.29	0.27	0.26
NSM	0.41	0.42	0.44	0.44	0.45	0.44
NBSM	0.59	0.52	0.54	0.53	0.52	0.51
RAProp	0.45	0.43	0.37	0.35	0.34	0.34

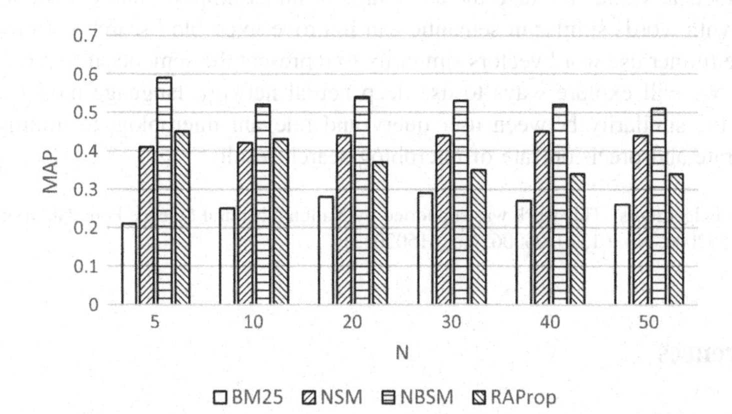

Fig. 3 The comparison results of different methods over MAP value

results, the value of evaluation criteria of the proposed method NBSM is also higher than that of other algorithms.

We comparing the search results (NSM), which are extended by the neural network language model, with the search results of BM25 and RAProp methods. As shown in figures, in each of the values of n, the search accuracy of the NSM method is higher than the search accuracy of BM25, which indicates that the neural network language model does improve the search accuracy. Compared the search result of the NSM method with the NBSM method, as shown in figures, when the value is small, the criteria value of NBSM is significantly higher, and as n increase, the difference between them become smaller. It means that the social signal feature play a great role on meeting user information needs if we only consider a few search results. On the other hand, it also demonstrates the effective of neural network language model feature, and this effectiveness even much more significant when we consider more search results. The search method we proposed in this paper combines the neural network language model to expand the search terms with social signal feature to re-rank search results, and search results also demonstrate the superiority of our proposed method.

4 Conclusions

In this paper, we proposed a microblog search method based on neural network language model (NBSM), which combines neural network language model with social signal features to improve search accuracy. Experiment results show that the method we proposed has a significantly better performance. Furthermore, we demonstrate the effectiveness of neural network language model feature and social signal features on improving microblog search accuracy, which can bring significant practical value. We take the advantage of an assumption that expanding user query with words similar in semantic can improve microblog search performance, and we further use word vectors similarity to represent the sematic similarity. In the future, we will explore ways to use deep neural network language model to calculate the similarity between user query and relevant microblog, to improve the recall rate and precision rate of microblog search results.

Acknowledgements This work was supported by National Natural Science Foundation of China (No. 61532006, No. 61320106006, No. 61502042).

References

1. Teevan J, Ramage D, Morris MR. # TwitterSearch: a comparison of microblog search and web search. In: Proceedings of the fourth ACM international conference on Web search and data mining. ACM;2011. p. 35–44.
2. Kelly R. Twitter study reveals interesting results about usage. PearAnalytics. 2009; (8).
3. Ravikumar S, Talamadupula K, Balakrishnan R, Kambhampati S. RAProp: ranking tweets by exploiting the tweet/user/web ecosystem and inter-tweet agreement. In: Proceedings of the 22nd ACM international conference on information and knowledge management. ACM;2013. p. 2345–50.
4. Mouratidis K, Li J, Tang Y, Mamoulis N. Joint search by social and spatial proximity. IEEE Trans Knowl Data Eng. 2015;27(3):781–93.

5. Busch M Gade, K, Larson B, Lok P, Luckenbill S, Lin J. Earlybird: real-time search at twitter. In: 2012 IEEE 28th international conference on data engineering (ICDE). IEEE;2012. p. 1360–69.
6. Severyn A, Moschitti A. Learning to rank short text pairs with convolutional deep neural networks. In: Proceedings of the 38th international ACM SIGIR conference on research and development in information retrieval. ACM;2015. p. 373–82.
7. Bengio Y, Ducharme R, Vincent P. A neural probabilistic language model. J Mach Learn Res. 2003, 3(Feb):1137–55.
8. Mikolov T, Karafiat M, Burget L, Cernocky J, Khudanpur S. Recurrent neural network based language model. Interspeech. 2010;2:3.
9. Mitra B, Nalisnick E, Craswell N, Caruana R. A dual embedding space model for document ranking. arXiv:1602.01137;2016.
10. Zheng G, Callan J. Learning to reweight terms with distributed representations. In: Proceedings of the 38th international ACM SIGIR conference on research and development in information retrieval. ACM;2015. p. 575–84.

Scenic Negative Comment Clustering Based on Balance Weighted Comment Topic Model

Zijian Lin, Junping Du, Yang Li, Lingfei Ye and Ang Luo

Abstract The scenic comment information from visitors often hidden the different aspects of the recommendations and expectations of the attractions, the extraction of these key information will help the spot managers find their own shortcomings and improve themselves. In this paper, we improved the author topic model and proposed a model of clustering the negative comments of the scenic spots. There are two improvements from our proposed model. Firstly, we added the importance of the comment category to the text clustering. Secondly, in order to prevent the stop words accumulating in the sampling process, we introduced the balance weight to the proposed model. Experiments showed that the model could not only effectively cluster these data, but also could extract the rich information related to different comment categories from the clustering results, which could help the managers of the scenic spots to better manage the attractions and attract tourists.

Keywords Scenic negative comment • Topic model • Clustering analysis • Data mining

1 Introduction

Each negative tourist comment from visitors may contain one or more categories. Such as the negative tourist comment for the "Many people go to the Forbidden City to visit on the National Day, the toilets are not enough, the tourist mobile interpreters are most broken", this comment contains some categories which are scenic crowded and supporting facilities. That the toilets are not enough is related to

Z. Lin · J. Du (✉) · Y. Li · L. Ye
Beijing Key Laboratory of Intelligent Telecommunication Software and Multimedia, School of Computer Science, Beijing University of Posts and Telecommunications, Beijing 100876, China
e-mail: junpingdu@126.com

A. Luo
SINA Corporation, Beijing, China

Y. Jia et al. (eds.), *Proceedings of 2017 Chinese Intelligent Systems Conference*, Lecture Notes in Electrical Engineering 459,
https://doi.org/10.1007/978-981-10-6496-8_28

the comment category of scenic crowded, and that the tourist mobile interpreters are most broken is related to the comment category of supporting facilities.

This paper designs the balance weighted comment topic model which uses the topic model to analyse the negative comments and their related comment categories and dig out richer information.

2 Related Work

The clustering algorithm has always been an important method for the analysis of unlabeled data. The general method extracts the text feature, constructs the vector representing the text, and then uses the similarity calculation between vectors [1] for text clustering. The work of extracting the text feature is the basis of text clustering, and good features are especially important for the model. Bharti and Singh [2] has analyzed the feature selection of text clustering in detail.

The text topic model is a kind of clustering method on the potential topic level of the text, which has been widely used in various fields in recent years. The improved variant algorithms based on the traditional LDA topic model are endless. Lee et al. [3] proposed the weight topic model WTM and the balance weight topic model BWTM have better performance in the small-scale training set and are suitable as a feature of text classification. Hashimoto et al. [4] used LDA model to analyze the needs of the affected people after the earthquake in Japan, Liu et al. [5] used the improved LDA model to extract the hot topic, Ding et al. [6] applied the topic model to the sorting algorithm, Yang et al. [7] studied parametric and nonparametric perceived user emotion topic models. The improved topic model based on LDA is also used to discover similar TV viewers and recommended TV shows [8]. The user topic model can mine the interest of the microblog users [9]. The semi-supervised thematic model also plays a role in multi-label text classification [10] and activity pattern discovery [11].

The author topic model [12] is a probability generation model that simplifies the writing process of the text into a series of probabilistic steps that not only discover the topic distribution of the text, but also the related authors of each topic.

The balance weighted comment topic model proposed in this paper draws on the author topic model, but takes into account the importance of the relevant comment categories and introduces the parameter factor λ to prevent the large aggregation of the stop words in the topic, so that the final clustering results are superior to other models. In order to prevent the sparseness of data from affecting the clustering process, we can use the method of modeling Twitter short text [13] to merge comments in the same day into a document.

3 Establishment of the Balance Weighted Comment Topic Model (BWCTM)

3.1 Definition of the Balance Weighted Comment Topic Model

A comment of a scenic spot can contain multiple categories, each comment has a relevant category distribution, each category has its own topic distribution, each topic also has its own word distribution. The generation process of words in comment is like this, according to the category distribution of the comment, we select an category randomly, then use the topic distribution of this category to get a topic, and then under the word distribution of the topic we generate the word.

Figure 1 shows the graph model of the balance weighted comment topic model, the shadowed variables are observable, and the shadowless variables represent hidden variables. The arrow indicates the conditional dependency between the variables, and the box indicates the repeated sampling and the variable at the bottom of the box is the number of samples in the model. The words w in the comment and the comment category information set L of comments are observable variables.

There are K different comment categories in the comment category set, and each comment has a polynomial distribution l of the relevant comment category. Parameter θ indicates that each comment category has a polynomial distribution (The number of topic is T). ϕ is that each topic has a polynomial distribution of words. And α and β are the hyperparameters of the Dirichlet priori of polynomial distributions θ and ϕ, respectively. For each word w in the comment, a comment category l is randomly selected in the comment category set L according to the polynomial distribution of the relevant comment category, and then a topic z is sampled from the topic polynomial distribution of the comment category l, and the word w is finally sampled from the word polynomial distribution of the topic z. This process can be repeated in order to generate the whole comment.

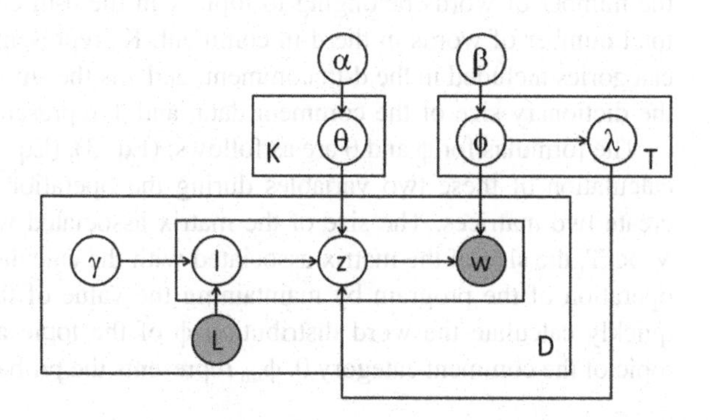

Fig. 1 Balance weighted comment topic model

3.2 Balance Weighted Comment Topic Model Parameter Estimation

Parameter λ is the reverse total number of words which assigned to the same topic used to balance the topic weight. If the larger the total number of words assigned to the k-topic, the smaller the balance factor λ_k is, the probability that the current word is assigned to the topic k becomes smaller during the gibbs sampling process, So that the word is prevented from being assigned to a single topic, and (formula 1) represents the formula for λk. Parameter N_d represents the total number of words of the comment, parameter $N_{k,v}$ represents the number of word v under the k comment category.

$$\lambda_k = \frac{\sum_{d=1}^{D} N_d}{\sum_{v=1}^{V} N_{k,v}} \tag{1}$$

The topic distribution of the comment category and the word distribution of each topic are unknown parameters in the model. The posterior distribution of the parameters can be sampled using the Gibbs sampling and the Markov chain Monte Carlo algorithm.

For each word in the comment, we can use (formula 2) to sample the topic and comment category associated with it.

$$P(z_i = j, l_i = k | w_{d,i} = m, z_{\neg i}, l_{\neg i}) \propto \frac{C_{mj}^{WT} + \beta}{\sum_{m'} C_{m'j}^{WT} + V\beta} \frac{C_{kj}^{LT} + \alpha}{\sum_{j'} C_{kj'}^{LT} + T\alpha} \frac{N_{dk} + \gamma}{N_d + K_d \gamma} \lambda_k \tag{2}$$

$z_i = j$ and $x_i = k$ denote the i-th word in the comment belongs to the topic j and the category k, $w_{d,i} = m$ denotes the i-th word in the d-th comment is the m-th word in dictionary, $z_{\neg i}$ and $l_{\neg i}$ indicate that the calculations for multiple distributions of all topics and categories do not contain the i-th word. The total number of word m allocated to the topic j is denoted by CWT mj except that the current term. The total number of category k allocated to the topic j is denoted by CLT k_j. N_{dk} represents the number of words belonging to topic k in the d-th comment, N_d represents the total number of words in the d-th comment, K_d represents the number of comment categories included in the d-th comment, and γ is the smoothing factor. V represents the dictionary size of the comment data, and T represents the number of topics.

The formulas for ϕ and θ are as follows: (Eq. 3), (Eq. 4), in order to facilitate the calculation of these two variables during the operation of the algorithm, we can create two matrices. The size of the matrix associated with the calculation of ϕ is V × T, the size of the matrix associated with the calculation of θ is K × T. In the operation of the program by maintaining the value of these two matrices, we can quickly calculate the word distribution ϕ of the topic and the distribution of the topic of the comment category θ. ϕ_{mj} represents the probability of the word m under the topic j, θ_{kj} is the probability of the topic j in the category k. (Eq. 5) lists the formula for the distribution of the comment categories for each comment.

$$\phi_{mj} = \frac{C_{mj}^{WT} + \beta}{\sum_{m'} C_{m'j}^{WT} + V\beta} \tag{3}$$

$$\theta_{kj} = \frac{C_{kj}^{LT} + \alpha}{\sum_{j'} C_{kj'}^{LT} + T\alpha} \tag{4}$$

$$l_{dk} = \frac{N_{dk} + \gamma}{N_d + K_d\gamma} \tag{5}$$

3.3 *Evaluation of the Different Topic Model Based on Perplexity*

There are two main methods of evaluating the quality of the language model, which are practical method and theoretical method respectively. Practical method is evaluated by the performance of the model in practical application (such as spelling checker, machine translation). The advantages are intuitive and practical, the shortcomings are lack of pertinence, not objective. And the basic idea of theoretical method is to give the test set to a higher probability of the language model is better. The indicator of content perplexity is widely used in the evaluation of the topic model. Lower the perplexity is, the better performance the topic model will have. The calculation method is shown as a formula 6 where $r_m = w_1, w_2, \ldots, w_n$. By the formula we can see that the smaller the degree of perplexity, the greater the probability of the sentence, the better the topic model.

$$perplexity(R) = \exp\left\{ - \frac{\sum_{m=1}^{M} \log\left(\prod_{i=1}^{n} \sum_{j=1}^{K} P(w_i|z_i = j) P(z_i = j|r_m)\right)}{\sum_{m=1}^{M} N_m} \right\} \tag{6}$$

4 Experimental Results and Analysis

4.1 *Experimental Data*

Through the crawling the tourism website under the topic of the tourist attractions comment data, a series of preprocessing work is performed on the raw data obtained by crawling. Use the dictionary to filter the stop words of comment, delete the punctuation mark in the text, in order to make the final clustering results contain fewer stop words [2]. The use of word segmentation and part-of-speech tagging tools only retains the nouns, verbs, adjectives and adverbs for each comment, and finally deletes the comments of the length of the word less than 5. The total number of comment data is 5430.

4.2 Comparison of Different Topic Model Perplexity

In order to test the performance of different topic models in the experimental data set of this paper, we calculate the perplexity of different models respectively. The model proposed in this paper is the balance weighted comment topic model (BWCTM), the perplexity score from different topic models under different topic numbers shown in Table 1. The comparative models in the experiment include LDA [14] (Latent Dirichlet Allocation), WTM (Weighted Topic Model), BWTM (the balance weighted topic model) and ATM (Author Topic Model). The WTM assigns different weights to words according to the degree of importance of the word in the process of clustering because that the LDA topic model gives the same weight to each word is unreasonable. The balanced weights topic model the adds balance parameter to the model in order to prevent clustering result of WTM from being affected by the stop words. In the experiment, the number of topic K is set to 10, 30 and 50 respectively. The hyperparameters α and β of Dirichlet distribution are set to 0.16 and 0.01 respectively, and the smoothing factor γ is set to 1.

From the results of Table 1, we can see that the LDA topic model has the highest perplexity score in the case of different number of topic in the same data set. The perplexity score of BWCTM proposed in this paper is the lowest in the different number of topic, and the perplexity score of BWCTM when K is 30 is less than the number of the topic of 10 and 50. BWCTM has the best text modeling results for the negative comments of the attractions, better than the other four topic models used for comparison. It can also be seen from Table 1 that most of the models can achieve the lowest perplexity score when the number of topic is 30, so the optimal number of topics is 30 for the data set.

Figure 2 is the number of topic set to 30, the perplexity score of different topic models with the number of iterations. From the Fig. 2 the proposed BWCTM compared to the traditional Latent Dirichlet Allocation topic model LDA and the balance weighted topic model converges faster in the process of model training, and the final model has a lower degree of perplexity.

Table 1 Perplexity score of the different topic models under different topic numbers

Perplexity score	K = 10	K = 30	K = 50
LDA	398.62	395.34	414.40
WTM	354.77	350.28	373.25
BWTM	301.64	299.06	316.88
ATM	312.21	308.52	319.56
BWCTM	**286.87**	**255.26**	**260.63**

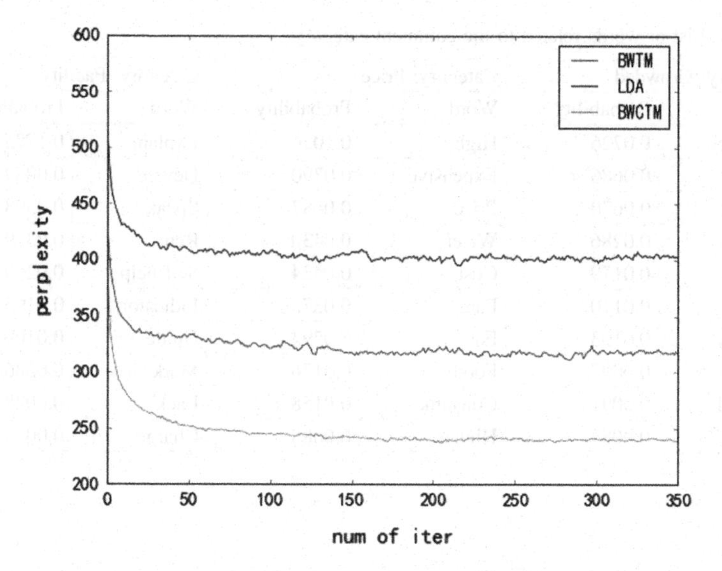

Fig. 2 Perplexity score from different models with the number of iterations

Table 2 Negative comment category information

Id	Category	Id	Category
1	Crowded	7	Tourist guide
2	Traffic	8	Management
3	Price	9	Entertainment
4	Landscape	10	Facility
5	Environment	11	Ticket
6	Security		

4.3 Scenic Negative Comment Clustering Based on Balance Weighted Comment Topic Model

The negative comments in the balance weighted comment topic model are divided into 11 categories, as shown in Table 2.

For each comment category, we can find the relevant maximum probability topic first, and then use the word probability distribution under the topic to describe the topic, lists some words which are most likely belong to the topic.

Table 3 lists the relevant words of different comment categories, through the related words set of category, we can find the relevant information of comment category. For example, by analyzing the collection of words under the crowded category, we can know that the peak flow of people in the scenic area generally appears on holidays. The crowded places in the scenic area are generally located in

Table 3 List of words related to the comment category

Category: Crowded		Category: Price		Category: Facility	
Word	Probability	Word	Probability	Word	Probability
Crowded	0.0726	High	0.1026	Explain	0.1775
Crowd	0.0686	Expensive	0.0790	Device	0.0089
Security	0.0670	Price	0.0487	Speak	0.0398
Very	0.0286	Water	0.0430	Rent	0.0339
People	0.0179	Cost	0.0424	Self-help	0.0200
Hard	0.0110	Fare	0.0373	Indicator	0.0195
Holiday	0.0103	Eat	0.0294	Guide	0.0109
Toilet	0.0097	Food	0.0176	Mark	0.0086
Festival	0.0091	Consume	0.0158	Lack	0.0078
Person	0.0087	Higher	0.0088	Charge	0.0074

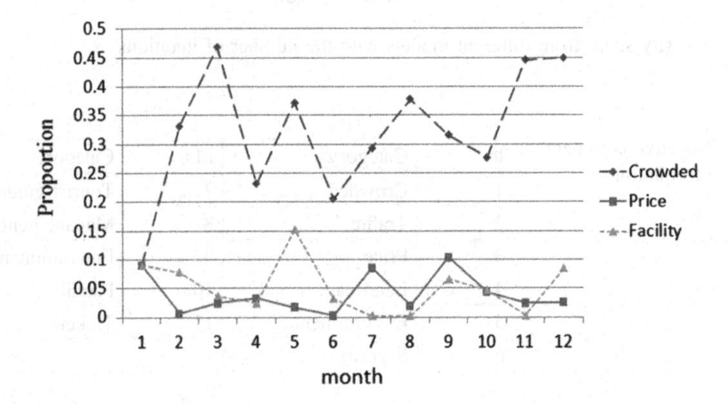

Fig. 3 Proportion of Beijing tourism comment category data in different months in 2015

security and toilet places. The relevant words under the price category are "high", "price", "water", "fares", "food", "expensive", which tells scenic managers the price of water and food sold in the area is too high. This is obvious that the self-help explain devices, indicators and logo facilities in scenic spots are not perfect from the collection of words under the facility category.

Figure 3 shows the proportion of words from different comment categories in different months, we can see from the figure, the proportion of the category of crowded in scenic areas generally higher. It is clear that the crowded phenomenon of the prevalence of scenic spots, but increased significantly when the holidays. Prices in the off-season has low level of concern, and the peak of the category of the Facility occurred in the May 1 Golden Week.

5 Conclusions

Experiments show that the balance weighted comment topic model mentioned in this paper has a better performance than the other text clustering models on the evaluation index of the clustering results. We can dig out richer information from the results of clustering, these information is convenient for tourists and scenic area managers to use these attractions negative comments more easily. Most of the evaluation information on the Internet is unlabeled data, and the design of efficient clustering analysis model is a very challenging work.

Acknowledgements This work is supported by the National Natural Science Foundation of China (No. 61320106006, No. 61532006, No. 61502042).

References

1. Kavitha Karun A, Mintu P, Lubna K. Comparative analysis of similarity measures in document clustering. In: International conference on green computing, communication and conservation of energy;2013. p. 857–60.
2. Bharti KK, Singh PK. Hybrid dimension reduction by integrating feature selection with feature extraction method for text clustering. Expert Syst Appl Int J. 2015;42(6):3105–14.
3. Lee S, Kim J, Myaeng S H. An extension of topic models for text classification: a term weighting approach. In: International conference on big data and smart computing. IEEE;2015. p. 217–24.
4. Hashimoto T, Chakraborty B, Aramvith S, et al. Affected people's needs detection after the East Japan Great Earthquake—time series analysis using LDA. In: Asia-Pacific signal and information processing association, 2014 summit and conference. IEEE;2014. p. 1–6.
5. Liu G, Xu X, Zhu Y, et al. An improved latent Dirichlet allocation model for hot topic extraction. In: IEEE fourth international conference on big data and cloud computing. IEEE;2014. p. 470–6.
6. Ding Y, Yan S, Xiao Y, et al. Ranking algorithm based on relational topic model. In: International joint conference on neural networks. IEEE;2015. p. 1–8.
7. Yang Z, Kotov A, Mohan A, et al. Parametric and non-parametric user-aware sentiment topic models. In: International ACM SIGIR conference on research and development in information retrieval. ACM;2015. p. 413–22.
8. Pyo S, Kim E, Kim M. LDA-based unified topic modeling for similar TV user grouping and TV program recommendation. IEEE Trans Cybern. 2014;45(8):1476–90.
9. Li HE, Jia Y, Han W, et al. Mining user interest in microblogs with a user-topic model. Commun China. 2014;11(8):131–44.
10. Lu Y, Okada S, Nitta K. Semi-supervised latent Dirichlet allocation for multi-label text classification. In: Recent trends in applied artificial intelligence;2013. p. 351–60.
11. Huynh T, Fritz M, Schiele B. Discovery of activity patterns using topic models. In: UBICOMP 2008: ubiquitous computing, international conference, UBICOMP 2008, Seoul, Korea, 21–24 Sept 2008, Proceedings. DBLP;2008. p. 10–9.

12. Ichise R, Fujita S, Muraki T, et al. Research mining using the relationships among authors, topics and papers. In: International conference information visualization. IEEE Computer Society;2007. p. 425–30.
13. Cai H, Yang Y, Li X, et al. What are popular: exploring twitter features for event detection, tracking and visualization. ACM international conference on multimedia. ACM;2015. p. 89–98.
14. Blei DM, Ng AY, Jordan MI. Latent Dirichlet allocation. J Mach Learn Res. 2003;3:993–1022.

Scale Adaptive Kernelized Correlation Filter with Scale-Invariant Feature Transform

Xueting Qiao and Yingmin Jia

Abstract In order to solve the problem of scale variation in Kernelized Correlation Filter (KCF) tracker, a scale adaptive tracking method based on Scale-Invariant Feature Transform (SIFT) is proposed. Firstly, it uses SIFT to extract and match keypoints between two successive frames to estimate the new scale of the target. Secondly, it utilizes keypoints information to resist strong disturbance of complex scenes, so that the method this paper proposes can be more robust. The method is tested in standard tracking library and compared with original tracking method in center location error and the overlap rate. These results illustrate that our tracking method with SIFT scale compensation improves the performance effectively.

Keywords Object tracking · Kernelized correlation filter · Scale-invariant feature transform · Scale estimation

1 Introduction

Recently, several tracking-by-detection methods have succeeded in object tracking field with high efficiency and accuracy. The basic idea is to find the most possible position of the target in the instant frame by a discriminative classifier trained by labeled samples. Struck (Structured output tracking with kernel) [1], based on structured Support Vector Machines (SVM), predicts directly the deviation of target position between frames. Tracking-Learning-Detection (TLD) [2] learns from not only the tracker, but also the tracker's error to improve the performance.

X. Qiao
Ecole Centrale Pekin, School of Automation Science and Electrical Engineering,
Beihang University (BUAA), Beijing 100189, China
e-mail: xueting.qiao@outlook.com

Y. Jia (✉)
The Seventh Research Division and the Center for Information and Control,
School of Automation Science and Electrical Engineering,
Beihang University (BUAA), Beijing 100189, China
e-mail: ymjia@buaa.edu.cn

Y. Jia et al. (eds.), *Proceedings of 2017 Chinese Intelligent Systems Conference*, Lecture Notes in Electrical Engineering 459,
https://doi.org/10.1007/978-981-10-6496-8_29

Multiple Instance Learning (MIL) [3] use bags of instance instead of instances to train classifier for superior results. Minimum Output Sum of Squared Error (MOSSE) [4] is initialized by a single frame which results in a higher tracking speed. Henriques proposed Circulant Structure Kernel (CSK) [5] which uses circulant matrices to collect more samples in order to improve classifier performance. Besides, circulant matrices divert the problem to Fourier analysis which turns out to be faster in learning and detecting. Based on CSK tracker, Kernelized Correlation Filter (KCF) [6] improves both accuracy and robustness by appling Histogram of oriented Grandient (HoG) feature instead of raw pixel features.

Different from original KCF tracker, our tracking method solves scale variation problem. Firstly, use SIFT to extract and match feature points between two successive pictures. Afterwards, calculate the position of geometric center of all feature points, which is then used to evaluate the deviation of the frame. Finally, estimate the new target scale by comparing two deviations. While the amount of keypoints is too small, the scale estimation is sometimes inaccurate. So a threshold is needed, below which the update of target scale is unacceptable.

2 KCF Tracker

The inspiration of KCF Tracker results from successful adaption of correlation filters in object tracking field. As the convolution of two image patches is equal to an element-wise product in the Fourier domain, KCF Tracker is very competitive at the processing speed level. On the training side, it shifts the object patch to generate a pool of samples to train discriminative classifier for a better performance in tracking.

2.1 Sample Representation

In order to avoid the redundancy problem of traditional sparse sampling methods, KCF tracker shifts the base sample (positive example) to generates a large number of virtual samples (negative examples) utilizing the cyclic matrix theory. All samples are made to train the classifier.

$x = [x_1, x_2, \ldots, x_n]^T$, an $n * 1$ vector represents the base sample. And shiftings of x are generated by the permutation matrix P, which is:

$$P = \begin{bmatrix} 0 & 0 & 0 & \cdots & 1 \\ 1 & 0 & 0 & \cdots & 0 \\ 0 & 1 & 0 & \cdots & 0 \\ \vdots & \vdots & \ddots & \ddots & \vdots \\ 0 & 0 & \cdots & 1 & 0 \end{bmatrix} \tag{1}$$

The sample shifted by 1 element is noted Px, and $Px = [x_n, x_1, \ldots, x_{n-1}]^T$. So the assemble of training sample is $\{P^l x | l = 0, 1, \ldots, n-1\}$. And each sample composes a row of data matrix X:

$$X = C(x) = \begin{bmatrix} x_1 & x_2 & x_3 & \cdots & x_n \\ x_n & x_1 & x_2 & \cdots & x_{n-1} \\ x_{n-1} & x_n & x_1 & \cdots & x_{n-2} \\ \vdots & \vdots & \ddots & \ddots & \vdots \\ x_2 & x_3 & x_4 & \cdots & x_1 \end{bmatrix} \quad (2)$$

So far, we've got a circulant matrix, which according to some papers, such as [7] [8] has several useful properties. For example, all circulant matrices can be diagonalized by the Discrete Fourier Transform (DFT):

$$X = F diag(\hat{x}) F^H \quad (3)$$

where constant matrix F is the DFT matrix, F^H is the Hermitian transpose of F, $\hat{x}$ is the DFT of the vector x, and $\hat{x} = \mathcal{F}(x)$.

Every sample $P^i x$ corresponds to a regression target y_i. y_i obeys Gaussian distribution, whose range is $[0, 1]$. All these labels make up the regression target vector $y = [y_0, y_1, \ldots, y_{n-1}]^T$.

2.2 *Classifier Training*

In many practical problems, Regularize Least Square classifier (RLS) [9] performs as well as SVM, with the advantage of easy to implement. In this case, the only thing is to solve a linear equation $f(z) = w^T z$ which minimizes cost function:

$$\min_w \sum_i (f(x_i) - y_i)^2 + \lambda \parallel w \parallel^2 \quad (4)$$

where λ is a regularization parameter to prevent overfitting. The equation above is also called Redge Regression problem, and the minimizer has a closed-form:

$$w = (X^T X + \lambda I)^{-1} X^T y \quad (5)$$

where I is an identity matrix. And transfer the problem into the Fourier domain:

$$w = (X^H X + \lambda I)^{-1} X^H y \quad (6)$$

As mentioned, in the Fourier domain, we can largely exploit the properties of circulant matrix. So take the term $X^H X$, and replace X by Eq. 3:

$$X^H X = Fdiag(\hat{x}^*)F^H Fdiag(\hat{x})F^H \tag{7}$$

Since for a diagonal matrix $diag(\hat{x})$, its Hermitian transpose is equivalent to its complex-conjugate $diag(\hat{x}^*)$, and element-wise product can be defined as $\odot$. $F^H F = I$ can be eliminated:

$$X^H X = Fdiag(\hat{x}^* \odot \hat{x})F^H \tag{8}$$

Taking into consideration all the diagonalized expressions, finally:

$$\hat{w} = \frac{\hat{x}^* \odot \hat{y}}{\hat{x}^* \odot \hat{x} + \lambda} \tag{9}$$

2.3 Kernel Trick

Non-linear regression function $f(z)$ allows the classifier more powerful. The key is to map the inputs of a linear problem to a non-linear high-dimensional feature-space with a mapping function $\varphi(x)$, whose advantage is that the regression problem can be kept linear.

First, express w as a linear combination of the samples with coefficients α_i:

$$w = \sum_i \alpha_i \varphi(x_i) \tag{10}$$

Then, because of kernel trick: the dot-product of two sample features in high-dimensional feature-space can be get from a kernel function κ in original space, we get: $\varphi^T(x)\varphi(x') = \kappa(x, x')$. General choice for kernel function is Gaussian kernel or Polynomial kernel.

Kernel matrix K, an $n * n$ matrix storing dot-product between different samples, whose elements are:

$$K_{ij} = \kappa(P^i x, P^j x) \tag{11}$$

When given a single testing sample z, response of the classifier is:

$$f(z) = w^T z = \sum_{i=0}^{n-1} \alpha_i \kappa(z, P^i x) \tag{12}$$

In the kernel domain, the solution of regression problem turns to optimize coefficient α, instead of w. As is proved in [9]:

$$\alpha = (K + \lambda I)^{-1} y \tag{13}$$

Since K is circulant, finally:

$$\hat{\alpha} = \frac{\hat{y}}{\hat{k}^{xx} + \lambda} \tag{14}$$

where $\hat{k}^{xx}$ is the first row of the kernel matrix $K = C(k^{xx})$.

2.4 Fast Detection

When detecting target position in potential area, the idea is to generate several candidate samples and to compare which one is the most possible to be the target by calculating the kernel correlation response. And determine the translation of target position.

Candidate samples are generated by cyclic shift to maintain the whole problem in circulant matrix domain. Next, we calculate the translation by evaluating the regression function $f(z)$, as mentioned in Eq. 12. Since there're several candidate samples $z_i = P^i z, f(z)$ becomes a $n * 1$ vector, with f_i the response for z_i. Equation 12 is transformed into:

$$f_i = w^T z_i = \sum_{j=0}^{n-1} \alpha_i \kappa(P^i z, P^j x); i = 0, 1, \cdots, n-1 \tag{15}$$

Similarly with the training section, K^z is defined as the kernel matrix between training samples and candidate samples, with elements $\kappa(P^i z, P^j x)$. K^z can be obtained in an easier way:

$$K^z = C(\hat{k}^{xz}) \tag{16}$$

where $\hat{k}^{xz}$ is the kernel correlation of base sample x and base testing sample z.

From now on, Eq. 12 is reformed with:

$$f(z) = (K^z)^T \alpha \tag{17}$$

After diagonalization, we obtain

$$\hat{f}(z) = \hat{k}^{xz} \odot \hat{\alpha} \tag{18}$$

Convert $\hat{f}(z)$ back to the time domain. The area corresponding to the maximum value of the response is considered to be the target's detected position.

3 Scale Invariant Feature Transform (SIFT)

Known as an excellent image matching method, SIFT [10] transforms image data into highly distinctive features and represents image with keypoints which are invariant to image scaling and rotation, and partially invariant to change of camera viewpoint in actual complex scenes. And these excellent invariances ensure keypoints to be correctly matched against database.

3.1 Keypoints Detection

The scale space of an image, $L(x, y, \sigma)$, is defined as the convolution of Gaussian function, $G(x, y, \sigma)$, with that image, $I(x, y)$. Gaussian function is the only possible scale-space kernel proved by Koenderink and Lindeberg:

$$L(x, y, \sigma) = G(x, y, \sigma) * I(x, y) \tag{19}$$

where $*$ is the convolution operation, and

$$G(x, y, \sigma) = \frac{1}{2\pi\sigma^2} exp(-\frac{x^2 + y^2}{2\sigma^2}) \tag{20}$$

Difference-of-Gaussian (DoG) scale space is defined as the difference of two scales, with k a constant:

$$D(x, y, \sigma) = (G(x, y, k\sigma) - G(x, y, k\sigma)) * I(x, y) = L(x, y, k\sigma) - L(x, y, \sigma) \tag{21}$$

Local extremum in the difference-of-Gaussian scale space, the largest or smallest among the points in the cube which is three-point-sized and centered at local extremum, is chosen as a potential keypoint. But a series of processes is also needed to get the exact positions of keypoints and to eliminate poor-performance potential keypoints: low-contrast points, edge-located points for a better stability. By now, the position and the scale of each keypoint are obtained.

Then, assigning an orientation to each keypoint using the gradient distribution of surrounding samples, that ensures SIFT to be invariant to rotation. For each keypoint, in the closest scale $L(x, y)$, calculate the gradient magnitude, $m(x, y)$, and the orientation, $\theta(x, y)$ by:

$$m(x, y) = \sqrt{((L(x + 1, y) - L(x - 1, y))^2 + (L(x, y + 1) - L(x, y - 1))^2} \tag{22}$$

$$\theta(x, y) = tan^{-1}((L(x, y + 1) - L(x, y - 1))/(L(x + 1, y) - L(x - 1, y))) \tag{23}$$

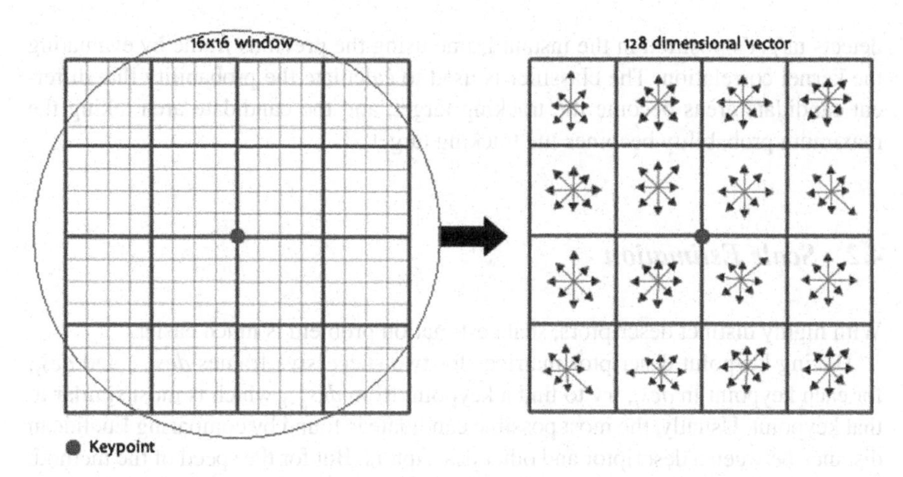

Fig. 1 Descriptor

3.2 *Descriptor*

Create a 8-bin orientation histogram to cover 360° range of orientations. Every sample around the keypoint within certain limit is added to the histogram by its weighted gradient magnitude.

Centered with each keypoint, a 16 by 16 sample window is taken into consideration. Calculate the gradient magnitude for each sample, then filter with Gaussian-weighted circular window to reduce the weight of the sample far from the center. For each 4 * 4 samples, calculate local orientation histogram. So we generate for each keypoint a 4 * 4 * 8 dimensioned descriptor, illustrated as above.

After normalization, the descriptors are invariant to illumination. For a patch, put all its m descriptors into a $128 * m$ matrix, denoted *des*.

4 KCF Combining SIFT Tracking Method

Based on KCF Tracker, this method proposes a scale estimation strategy using SIFT method so that it realizes scale adaptive tracking.

4.1 *Detection*

Normally, the target does not translate much between two successive frames. So it makes sense to select the candidate samples near the same position as the previous frame. Notice that KCF Tracker is a tracking-by-detection method, each time it detects target's position in the instant frame using the previous frame by evaluating the kernel correlation. The classifier is used to calculate the probability that different candidate areas become the tracking target, and the candidate area taking the maximum probability becomes the tracking target.

4.2 Scale Estimation

With highly distinct descriptors, scale estimation problem is much easier.

Having keypoint descriptor matrices for two successive frames des_{t-1} and des_t, for each keypoint in des_t, try to find a keypoint from des_{t-1} which is most similar to that keypoint. Usually, the most possible candidate is found by comparing Euclidean distance between a descriptor and other descriptors. But for the speed of the method, calculating dot products between descriptors is a better idea. For each keypoint in des_t, calculate:

$$dotprods = des_t(i, :) * des_{t-1} \tag{24}$$

Take inverse cosine because *acos* of dot product of unit vectors is a close approximation to the ratio of Euclidean distances for small angles. And sort results to get two nearest distances.

If two nearest distances are very different from each other, it's argued that this keypoint has matched to a keypoint in des_{t-1}. If not, it's considered that this keypoint can't be matched with any keypoint from des_{t-1}. There are several similar answers, so none of them is the best. Actually, this is realized by comparing the ratio of two nearest distance with a threshold to be set.

Firstly, for all descriptors which can be matched in des_{t-1}, calculate the geometric center position:

$$pos_{t-1} = \sum_{i=1}^{m} loc_{t-1}^{i} / num \tag{25}$$

where loc_{t-1}^{i} is the coordinate of matched descriptor, *num* is the number of matched descriptors. Then, calculate the sum of Euclidean distance between each matched descriptor and the geometric center position to estimate the deviation:

$$dev_{t-1} = \sum_{i=1}^{m} \sqrt{(loc_{t-1}^{i} - pos_{t-1})^2 / num} \tag{26}$$

Similarly, we can get pos_t and dev_t. Finally, target scale change ratio is estimated by:

$$ratio = dev_t / dev_{t-1} \tag{27}$$

4.3 Post Treatment

In many reality situations, when the tracking target has a sharp tremble, it's possible for SIFT to collect a small number of keypoints, that can lead to a matching error.

Since the amount of matched keypoints *num* can be get from previous subsection, we can set a threshold of *num* to avoid the situation above. If *num* >= *threshold_num*, it's believed that the scale estimation is accurate and update target scale. If *num* < *threshold_num*, it's argued that the scale estimation is inaccurate and the present target scale is kept.

4.4 Update

Since tracking is a dynamic process, it's also necessary to update detection coefficients after every tracking-by-detection.

Considering x' is detected target area in the frame V_t, we shift this area and train the classifier for new coefficients $\alpha_{x'}$. The update process is:

$$\begin{cases} x_t = (1-\beta)x_{t-1} + x' \\ \alpha_t = (1-\beta)\alpha_{t-1} + \alpha_{x'} \end{cases} \tag{28}$$

where β is updating coefficient, α_{t-1} and α_t are classifier coefficients for the previous frame and instant frame respectively, x_{t-1} and x_t are target models for the previous frame and instant frame respectively.

4.5 Framework

The framework of proposed method is:

Initialization:

Determine parameters. And according to the first frame, choose tracking target: initial position P_1 and initial scale s_1.

Extract base sample x_1 in the frame V_1 using P_1 and s_1.

Calculate kernel matrix K_1 using x_1 by Eq. 11.

Calculate position detection model α_1 by Eq. 14.

Get SIFT descriptor des_1.

For every frame V_i, $(i > 1)$:

Position detection:

Extract base testing sample z_i in the frame V_i using P_{i-1} and s_{i-1}.

Calculate kernel matrix K_i^z using x_{i-1} and z_i.

Calculate the response f_i using K_i^z and α_{i-1} by Eq. 18.

Find the coordinate of maximum of the response, $[i, j]$.

Update position P_i.
Scale estimation:
Get SIFT descriptor des_i.
Match two SIFT descriptor des_{i-1} and des_i.
Calculate geometric center position pos_{i-1} and pos_i by Eq. 25.
Calculate deviation dev_{i-1} and dev_i by Eq. 26.
Return the amount of matched keypoints *num*. If *num* >= *threshold_num*, calculate scale change *ratio* by Eq. 27, else set *ratio* = 1.
New scale $s_i = s_{i-1} * ratio$.
Update:
Extract base sample x' in the frame V_i using P_i and s_i.
Calculate kernel matrix K_i using x' by Eq. 11.
Calculate position classifier coefficients $\alpha_{x'}$ by Eq. 14.
Update target model x_i and classifier coefficients α_i by Eq. 28.

5 Experimentations

5.1 Experimental Setup and Evaluation Index

Experiments are implemented in MATLAB R2010a. Regularization parameter λ is set to be 10^{-4}. Updating coefficient β is set to be 0.02. And *threshold_num* is set to be 15. If the threshold is smaller than 15, post treatment could be inaccurate. In the other case, the cost of calculation increases.

There're two evaluation index and two figures: center location error (CLE) and overlap rate (OR), precision plot and success plot. CLE is the Euclidean distance between the detected position and the ground truth. When CLE is less than a certain threshold, 20 usually, tracking is considered to be successful. Precision plot shows while the threshold changes from 0 to 50, the number of frames successfully tracked as a percentage of the total number of frames. For tracking bounding box B_t and ground truth bounding box B_g, OR is defined as $OR = \frac{B_t \bigcap B_g}{B_t \bigcup B_g}$, $\bigcap$ and $\bigcup$ represent the overlap and union of the two areas. When OR is bigger than a certain threshold, 0.5 usually, tracking is successful. Success plot shows while the threshold changes from 0 to 1, the number of frames successfully tracked as a percentage of the total number of frames.

5.2 Qualitative Evaluation

To see the performance of our KCF combining SIFT tracking method, test it with several challenging sequences with obvious scale change in the benchmark database.

Fig. 2 Target scale adaptive tracking: Sequence Dog1

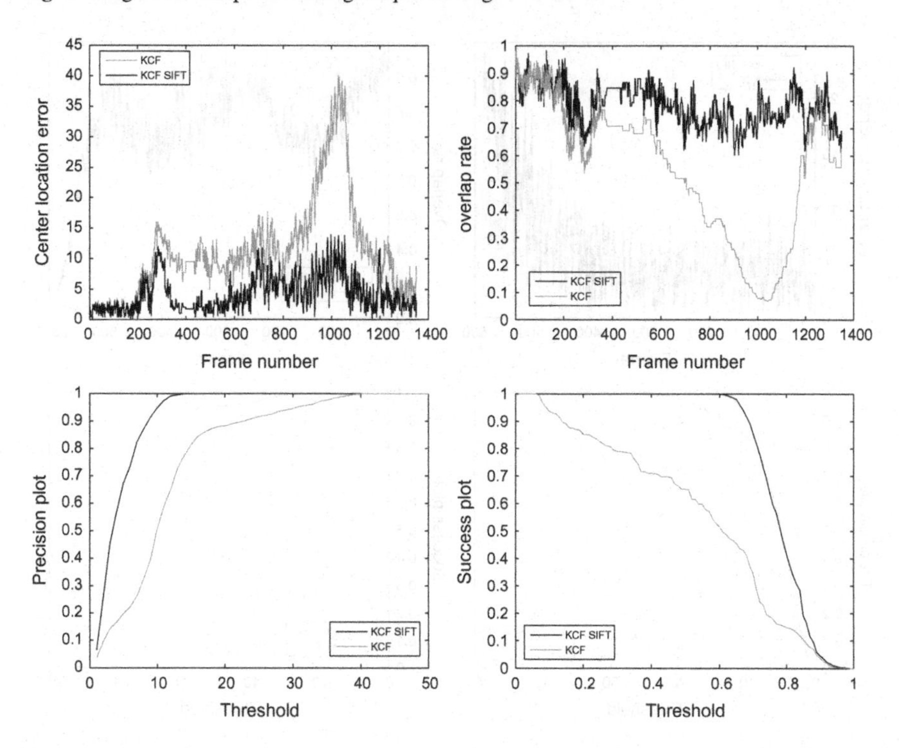

Fig. 3 CLE and OR, precision plot and success plot

Every a certain number of frames, take one frame to determine scale adaptive effect. After processing all frames, compare its evaluation index with those of original method. Firstly, take sequence Dog1 as a example.

Figure 2 shows the response of scale adaptive tracking. As can be seen from figures above, the tracking bounding box changes as target scale changes. And the center of the tracking bounding box is quite near the center of tracking area. But for original KCF tracker, with the change of target scale, two centers of are no longer near each other. That leads to failure of target tracking.

Figure 3 shows CLE and OR of every frame for sequence Dog1, and precision plot and overlap plot for sequence Dog1. Mean center location error has been reduced from 11.2864 to 4.2337. And mean overlap rate has been improved from 54.97 to 78.33%.

Fig. 4 Target scale adaptive tracking: Sequence BlurCar2

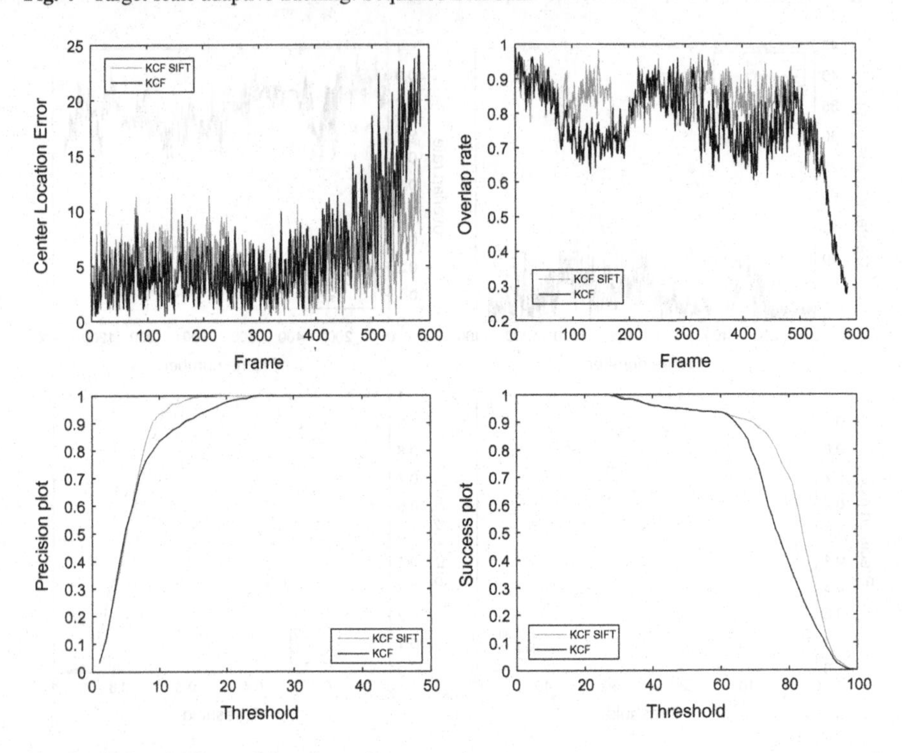

Fig. 5 CLE and OR, precision plot and success plot

Take sequence BlurCars2 as another example. In this sequence, the frames are trembling and become blurred from time to time because the camera is moving.

Figure 4 shows the response of scale adaptive tracking. And Fig. 5 shows precision plot and overlap plot for sequence BlurCar2. Mean center location error has been reduced from 6.2570 to 5.3250. And mean overlap rate has been improved from 75.74 to 80.84%.

Compared with the original algorithm, the proposed method has the ability of scale adaptability. The tracking bounding box changes as target scale changes, which assures better distinction between the background and the tracking target. And that reduces the background interference in the training classifier, so that the positive and negative sample information are more accurate to train the classifier. As a result, the proposed method is more efficient to detect the position and the scale.

6 Conclusion

In view of KCF can not solve the scale variance problem of target, this paper presents a new method donated KCF combining SIFT tracking method. Based on KCF tracker's good performance on target tracking, this method utilizes SIFT to estimate target scale change. That results in better samples for training classifier. So this method obtains a better performance both in center location error and overlap rate. And the proposed method has certain robustness to illumination, occlusion, rotation and so on, and has certain research and application value.

Acknowledgements This work was supported by the NSFC (61327807, 61521091, 61520106010, 61134005) and the National Basic Research Program of China (973 Program: 2012CB821200, 2012CB821201)

References

1. Hare S, Saffari A, Torr PHS. Struck: structured output tracking with kernels. In: International conference on computer vision, vol 23, no 5;2011. p. 263–70.
2. Kalal Z, Mikolajczyk K, Matas J. Tracking-learning-detection. IEEE Trans Pattern Anal Mach Intell. 2011;34(7):1409–22.
3. Yang M-H, Belongie S, Babenko B. Robust object tracking with online multiple instance learning. IEEE Trans Pattern Anal Mach Intell. 2011;33(8):1169–632.
4. Bolme DS, Ross Beveridge J, Draper BA, et al. Visual object tracking using adaptive correlation filters. In: Proceedings of the 2010 IEEE conference vision and pattern recognition, CVPR 2010;2010. p. 2544–50.
5. Henriqueso J, Caseiro P, Martins J, Batista FR. Exploiting the circulant structure of tracking-by-detection with kernels. In: European conference on computer vision;2012. p. 702–15.
6. Henriqueso J, Caseiro P, Martins J, Batista FR. High-speed tracking with kernelized correlation filters. IEEE Trans Pattern Anal Mach Intell. 2014;37(3):582–96.
7. Davis PJ. Circulant matrices. American Mathematical Society;1994.
8. VedaldiA Zisserman A. Efficient additive kernels via explicit feature maps. IEEE Trans Pattern Anal Mach Intell. 2012;34(3):480–92.
9. Rifkin G, Yeo T, Poggio R. Regularized least-squares classification. Nato Sci Ser Sub Ser III: Comput Syst Sci. 2003;190:131–54.
10. Lowe DG. Distinctive image features from scale-invariant keypoints. Int J Comput Vis. 2004;60:91–110.

Synchronization of Discrete-Time Delayed Neural Networks with Stochastic Missing Data: A Switching Method

Nan Xiao, Chao Liu and Yuan Ma

Abstract This paper deals with the problem of synchronization of discrete-time delayed neural networks subject to stochastic missing data. The aim of the addressed problem is to design a feedback controller for the error system, such that under unreliable communication links, the error system is guaranteed to be globally asymptotic stable. By adopting the switching techniques and constructing the corresponding Lyapunov-Krasovskii functionals, sufficient condition is established for the existence of the desired controller. The obtained criterion is in terms of LMIs which can be solved by using Matlab Toolbox. Finally, numerical example is given to show the effectiveness of the proposed method.

Keywords Discrete-time delayed neural networks · Synchronization · Missing data · Switching method · Linear matrix inequality

1 Introduction

During the past decades, neural networks have found successful applications in many engineering and scientific areas such as signal processing, model identification and optimization problem and much efforts have been done to the synchronization problem for neural networks with time delays [1–7].

On the other hand, some practical systems are influenced by additive randomly occurred nonlinear disturbances(RONs) which are caused by environmental circumstances [8–10]. In neural networks, the RONs may occur in a probabilistic way and it may be subject to random abrupt changes, such as random failures and repairs of the components, environmental disturbance and so on. In recent years, the phenomenon

N. Xiao (✉) · C. Liu
School of Mathematics and Statistics, Northeastern University at Qinhuangdao, Qinhuangdao 066004, China
e-mail: kgxnmy@aliyun.com

Y. Ma
School of Computer and Communication Engineering, Northeastern University at Qinhuangdao, Qinhuangdao 066004, China

Y. Jia et al. (eds.), *Proceedings of 2017 Chinese Intelligent Systems Conference*, Lecture Notes in Electrical Engineering 459,
https://doi.org/10.1007/978-981-10-6496-8_30

of data packet dropout has attracted much attention, and several important results have been reported [11, 12]. In these research, the RONs are modeled as mutually uncorrelated Bernoulli-distributed white noise sequences, and the stability criteria are obtained to guarantee global exponential stability of closed-loop system in the mean square sense. Meanwhile, for systems subject to some stochastic parameters, the switching method is adopted in [13–15] and stability criteria are obtained to guarantee global exponential stability of closed-loop system. To the best of authors' knowledge, the synchronization problem of discrete neural networks subject to stochastic missing data has not been investigated by utilizing the switching method before, which motivates our research.

In this paper, we investigate the synchronization problem for delayed discrete-time delayed neural networks. The unreliable communication links are modeled as stochastic dropouts which is assumed to be subject to Bernoulli distribution. By constructing suitable Lyapunov-Krasovskii functionals and considering the probability of the missing data, sufficient condition is established based on switching method, which can ensure that the error state system is globally asymptotic stable. The obtained criterion is in terms of LMIs that can be solved by using Matlab Toolbox. Finally, we give one numerical example to show the effectiveness of the obtained criterion.

Notation: In this paper, the notation $P > 0$ (≥ 0) means that P is real symmetric and positive definite (semi-definite). The superscripts T and -1 denote matrix transposition and matrix inverse. $\|\cdot\|$ stands for the Euclidean vector norm. A symmetric term in a symmetric matrix is denoted by $*$, diag$\{\cdots\}$ stands for a block-diagonal matrix. I and 0 denote the identity matrix and zero matrix with proper dimensions, respectively.

2 Problem Statement and Preliminaries

Consider the following discrete-time delayed neural networks:

$$\begin{aligned} x(k+1) &= Ax(k) + Bf(x(k)) + Cf(x(k-\tau)), \\ x(k) &= \phi_1(k), k = -\tau, -\tau+1, \ldots, 0, \end{aligned} \tag{1}$$

where $x(k) \in R^n$ is the neural state vector, $f(x(k)) = [f_1(x_1(k)), f_2(x_2(k)), \ldots, f_n(x_n(k))]^T$ represents the nonlinear activation function with the initial condition $f(0) = 0$, $A = \text{diag}\{a_1, a_2, \ldots, a_n\}$ is a positive diagonal matrix, $B = [b_{ij}]_{n\times n}$ represents the connection weight matrix, $C = [c_{ij}]_{n\times n}$ represents the delayed connection weight matrix, $\phi_1(k)$ is the initial condition and $\tau > 0$ denotes the constant transmission delay.

It is assumed that the function $f_i(\cdot)$ in (1) satisfies the following condition:

$$l_i^- \leq \frac{f_i(\zeta_1) - f_i(\zeta_2)}{\zeta_1 - \zeta_2} \leq l_i^+, \ \forall \zeta_1, \zeta_2 \in R, \ \zeta_1 \neq \zeta_2, i = 1, 2, \dots, n, \tag{2}$$

where l_i^+ and l_i^- are known constant scalars.

In this paper, we consider neural network (1) as the master system, the slave system is described as:

$$\begin{aligned} y(k+1) &= Ay(k) + Bf(y(k)) + Cf(y(k-\tau)) + u(k), \\ y(k) &= \phi_2(k), k = -\tau, -\tau+1, \dots, 0, \end{aligned} \tag{3}$$

where $y(k) \in R^n$ is the neural state vector, $\phi_2(k)$ is the initial condition and $u(k)$ is the control input.

By defining the synchronization error as $e(k) = y(k) - x(k)$, the synchronization error system can be described as follows:

$$\begin{aligned} e(k+1) &= Ae(k) + Bg(e(k)) + Cg(e(k-\tau)) + u(k), \\ e(k) &= \phi_2(k) - \phi_1(k), k = -\tau, -\tau+1, \dots, 0, \end{aligned} \tag{4}$$

where $g(e(k)) = f(y(k)) - f(x(k))$, $g(e(k-\tau)) = f(y(k-\tau)) - f(x(k-\tau))$. According to (2) we have that, for any $i = 1, 2, \dots, n$, the nonlinear function $g_i(e_i(k))$ satisfies

$$l_i^- \leq \frac{g_i(e_i(k))}{e_i(k)} \leq l_i^+. \tag{5}$$

Since unreliable communication links exist in practical systems, the stochastic approach is used to model data loss phenomena and control input is given by

$$u(k) = \gamma(k)Ke(k), \tag{6}$$

where $K \in R_{n \times n}$ is the controller gain matrix to be determined. The stochastic variable $\gamma(k)$, which is used to describe the existence of unreliable communication links, as $\gamma(k) = 0$ when the transmission fails and $\gamma(k) = 1$ when the transmission is successful. The variable $\gamma(k)$ is assumed to be subject to Bernoulli distribution taking values on zero and one. It satisfies

$$\Pr\{\gamma(k) = 1\} = \gamma, \ \Pr\{\gamma(k) = 0\} = 1 - \gamma, \ \gamma \in [0, 1]. \tag{7}$$

Substitute (6) into (4), the closed-loop error dynamic system is obtained as:

$$\begin{aligned} e(k+1) &= (A + \gamma(k)K)e(k) + Bg(e(k)) + Cg(e(k-\tau)), \\ e(k) &= \phi_2(k) - \phi_1(k), k = -\tau, -\tau+1, \dots, 0. \end{aligned} \tag{8}$$

The purpose of this paper is to design the feedback controller (6) for the error system (8), such that in the presence of missing data, the closed-loop error system (8) is asymptotically stable, i.e. $\lim_{k\to+\infty}\|e(k)\|=0$.

Before proceeding further, the following assumption and lemmas are given.

Assumption 1 For integer constants k_i, a time sequence $k_0 < k_1 < \ldots < k_i < \ldots (i \in N)$ denotes the switching instants of the transmission fails and success alternatively. Without loss of generality, we assume that the transmission is successful during the time interval $[k_{2i-1}, k_{2i}], i = 1, 2, \ldots$, and define $\bar{T}$ as the minimum time interval between k_{2i-1} and k_{2i+1}, i.e. $k_{2i+1} - k_{2i-1} \geq \bar{T}, \forall i \in N^+$.

Lemma 1 *([2]) Given constant symmetric matrices S_{11}, S_{12}, S_{22} satisfying $S_{11} = S_{11}^T$ and $S_{22} = S_{22}^T > 0$, then $S_{11} + S_{12}^T S_{22}^{-1} S_{12} < 0$ if and only if*

$$\begin{bmatrix} S_{11} & S_{12}^T \\ S_{12} & -S_{22} \end{bmatrix} < 0, \quad or \quad \begin{bmatrix} -S_{22} & S_{12} \\ S_{12}^T & S_{11} \end{bmatrix} < 0.$$

Lemma 2 *([3]) Suppose that $D = diag\{d_1, d_2, \ldots, d_n\}$ is a positive-semidefinite diagonal matrix. Let $v = [v_1, v_2, \ldots, v_n]^T \in R^n$ and $H(v) = [h_1(v_1), h_2(v_2), \ldots, h_n(v_n)]^T$ be a continuous nonlinear function satisfying*

$$m_i^- \leq \frac{h_i(v)}{v} \leq m_i^+, v \neq 0, v \in R, i = 1, 2, \ldots, n.$$

with m_i^+ and m_i^- being constant scalars. Then

$$\begin{bmatrix} v \\ H(v) \end{bmatrix}^T \begin{bmatrix} D\Gamma_1 & -D\Gamma_2 \\ * & D \end{bmatrix} \begin{bmatrix} v \\ H(v) \end{bmatrix} \leq 0,$$

where $\Gamma_1 = diag\{m_1^+ m_1^-, m_2^+ m_2^-, \ldots, m_n^+ m_n^-\}$, $\Gamma_2 = diag\{\frac{m_1^+ + m_1^-}{2}, \ldots, \frac{m_n^+ + m_n^-}{2}\}$.

3 Main Results

In this section, we establish our asymptotic stability criterion for the error system (8) by resorting to the switching method. For simplicity, in the following we denote

$$\eta(k) = [e^T(k) \ e^T(k-\tau) \ g^T(e(k)) \ g^T(e(k-\tau))]^T,$$
$$G_1 = [A \ 0 \ B \ C], \ G_2 = [A + K \ 0 \ A \ B],$$
$$L_1 = \text{diag}\{l_1^+ l_1^-, l_2^+ l_2^-, \ldots, l_n^+ l_n^-\}, L_2 = \text{diag}\{\frac{l_1^+ + l_1^-}{2}, \frac{l_2^+ + l_2^-}{2}, \ldots, \frac{l_n^+ + l_n^-}{2}\}.$$

We now present the synchronization condition as follows.

Theorem 1 *For given scalars $\gamma > 0, \tau > 0$ and $\mu_1 > 1$, choosing $0 < \alpha < 1, \beta > 0$ and $\bar{T} > 0$ such that* (12) *is satisfied, the error system* (8) *is asymptotically stable if there exist symmetric positive matrices P_i, Q_i, R_i, positive diagonal matrices S_i, T_i, and matrices W, such that the following LMIs* (9)–(11) *hold:*

$$\Pi_1 = \begin{bmatrix} \Upsilon_{11} & 0 & S_1L_2 & 0 & A^TP_1 \\ * & \Upsilon_{12} & 0 & T_1L_2 & 0 \\ * & * & \Upsilon_{13} & 0 & B^TP_1 \\ * & * & * & \Upsilon_{14} & C^TP_1 \\ * & * & * & * & -P_1 \end{bmatrix} < 0, \tag{9}$$

$$\Pi_2 = \begin{bmatrix} \Upsilon_{21} & 0 & S_2L_2 & 0 & A^TP_2 + W^T \\ * & \Upsilon_{22} & 0 & T_2L_2 & 0 \\ * & * & \Upsilon_{23} & 0 & B^TP_2 \\ * & * & * & \Upsilon_{24} & C^TP_2 \\ * & * & * & * & -P_2 \end{bmatrix} < 0, \tag{10}$$

$$P_i \le \mu_1 P_j, Q_i \le \mu_1 Q_j, R_i \le \mu_1 R_j, \ \forall i,j \in \{1,2\}, \tag{11}$$

$$\mu^{\frac{1}{\bar{T}}} \bar{\alpha}^{\gamma} \bar{\beta}^{(1-\gamma)} < 1. \tag{12}$$

where

$$\begin{aligned} &\bar{\alpha} = (1-\alpha), \ \ \bar{\beta} = (1+\beta), \mu_2 = (\bar{\beta}/\bar{\alpha})^{\tau-1}, \mu = \mu_1^2\mu_2, \\ &\Upsilon_{11} = -\bar{\beta}P_1 + Q_1 - S_1L_1, \Upsilon_{12} = -\bar{\beta}^{\tau}Q_1 - T_1L_1, \Upsilon_{13} = R_1 - S_1, \Upsilon_{14} = -\bar{\beta}^{\tau}R_1 - T_1, \\ &\Upsilon_{21} = -\bar{\alpha}P_2 + Q_2 - S_2L_1, \Upsilon_{22} = -\bar{\alpha}^{\tau}Q_2 - T_2L_1, \Upsilon_{23} = R_2 - S_2, \Upsilon_{24} = -\bar{\alpha}^{\tau}R_2 - T_2. \end{aligned}$$

Moreover, the controller gain matrix in (6) can be given by $K = P_2^{-1}W$.

Proof We construct the following Lyapunov-Krasovskii functionals for system (8), in case that the variable $\gamma(k) = 0$ and $\gamma(k) = 1$, respectively:

$$\begin{aligned} V_1(e(k)) = {} & e^T(k)P_1e(k) + \sum\nolimits_{i=k-\tau}^{k-1}(1+\beta)^{k-1-i}e^T(i)Q_1e(i) \\ & + \sum\nolimits_{i=k-\tau}^{k-1}(1+\beta)^{k-1-i}g^T(e(i))R_1g(e(i)), \end{aligned} \tag{13}$$

$$\begin{aligned} V_2(e(k)) = {} & e^T(k)P_2e(k) + \sum\nolimits_{i=k-\tau}^{k-1}(1-\alpha)^{k-1-i}e^T(i)Q_2e(i) \\ & + \sum\nolimits_{i=k-\tau}^{k-1}(1-\alpha)^{k-1-i}g^T(e(i))R_2g(e(i)). \end{aligned} \tag{14}$$

Then taking the forward difference of Lyapunov functional (13–14) along the trajectory of system (8), we obtain

$$\Delta V_1(k) = e^T(k+1)P_1e(k+1) - \bar{\beta}e^T(k)P_1e(k) + e^T(k)Q_1e(k) + \beta V_1 \\ + g^T(e(k))R_1g(e(k)) - \bar{\beta}^\tau e^T(k-\tau)Q_1e(k-\tau) - \bar{\beta}^\tau g^T(e(k-\tau))R_1g(e(k-\tau)), \tag{15}$$

$$\Delta V_2(k) = e^T(k+1)P_2e(k+1) - \bar{\alpha}e^T(k)P_2e(k) + e^T(k)Q_2e(k) - \alpha V_2 \\ + g^T(e(k))R_2g(e(k)) - \bar{\alpha}^\tau e^T(k-\tau)Q_2e(k-\tau) - \bar{\alpha}^\tau g^T(e(k-\tau))R_2g(e(k-\tau)). \tag{16}$$

It follows from (5) and Lemma 2 that there exist positive diagonal matrices S_i and T_i $(i = 1, 2)$, such that

$$\begin{bmatrix} e(k) \\ g(e(k)) \end{bmatrix}^T \begin{bmatrix} S_iL_1 & -S_iL_2 \\ * & S_i \end{bmatrix} \begin{bmatrix} e(k) \\ g(e(k)) \end{bmatrix} \leq 0, \tag{17}$$

and

$$\begin{bmatrix} e(k-\tau) \\ g(e(k-\tau)) \end{bmatrix}^T \begin{bmatrix} T_iL_1 & -T_iL_2 \\ * & T_i \end{bmatrix} \begin{bmatrix} e(k-\tau) \\ g(e(k-\tau)) \end{bmatrix} \leq 0. \tag{18}$$

Then, for the case $\gamma(k) = 0$, from (15) and (17–18) for i = 1 we can get

$$\Delta V_1 - \beta V_1 \leq \eta^T(k)(\Omega_1 + G_1^TP_1G_1)\eta(k), \tag{19}$$

where

$$\Omega_1 = \begin{bmatrix} Y_{11} & 0 & S_1L_2 & 0 \\ * & Y_{12} & 0 & T_1L_2 \\ * & * & Y_{13} & 0 \\ * & * & * & Y_{14} \end{bmatrix}. \tag{20}$$

For the case $\gamma(k) = 1$, from (16)–(18) for i = 2 we get

$$\Delta V_2 + \alpha V_2 \leq \eta^T(k)(\Omega_2 + G_2^TP_2G_2)\eta(k), \tag{21}$$

where

$$\Omega_2 = \begin{bmatrix} Y_{21} & 0 & S_2L_2 & 0 \\ * & Y_{22} & 0 & T_2L_2 \\ * & * & Y_{23} & 0 \\ * & * & * & Y_{24} \end{bmatrix}. \tag{22}$$

Let $P_2K = W$, by using Lemma 1, we can see that the hold of $\Omega_i + G_i^TP_iG_i < 0$ $(i = 1, 2)$ is guaranteed by LMIs (9–10). Then we can get the following results for given $k \in [k_{2n-1}, k_{2n}]$:

$$V_1(k+1) < \bar{\beta}V_1(k),\ k \in [k_{2j}, k_{2j+1}],\ j = 1, 2, \ldots, n-1,$$
$$V_2(k+1) < \bar{\alpha}V_2(k),\ k \in [k_{2j-1}, k_{2j}],\ j = 1, 2, \ldots, n.$$

From (11) we have

$$V_2(k) \le \mu_1 V_1(k),\ \ V_1(k) \le \mu_1\mu_2 V_2(k).$$

Together with above inequalities we can get

$$\begin{aligned} V_2(k) &< (1-\alpha)V_2(k-1) < \cdots < (1-\alpha)^{k-k_{2n-1}}V_2(k_{2n-1}) \\ &< \mu_1(1-\alpha)^{k-k_{2n-1}}V_1(k_{2n-1}) \\ &< \mu_1(1-\alpha)^{k-k_{2n-1}}(1+\beta)^{k_{2n-1}-k_{2n-2}}V_1(k_{2n-2}) \\ &< \mu_1^2\mu_2(1-\alpha)^{k-k_{2n-1}}(1+\beta)^{k_{2n-1}-k_{2n-2}}V_2(k_{2n-2}) \\ &< \cdots < \mu_1^{2n}\mu_2^{n}(1-\alpha)^{T_\downarrow}(1+\beta)^{T_\uparrow}V_1(k_0), \end{aligned}$$

where $T_\downarrow$ represents the total length of time interval for $\gamma(k) = 1$, and $T_\uparrow$ represents the total length of time interval for $\gamma(k) = 0$.

From (7) we have

$$\lim_{kto+\infty} \frac{T_\uparrow}{T_\downarrow} = \frac{1-\gamma}{\gamma},$$

hence we know that $\forall\ \epsilon > 0$, there $\exists\ K > k_0$, such that

$$\frac{T_\uparrow}{T_\downarrow} < \frac{1-\gamma}{\gamma} + \epsilon, \text{for } k > K_0.$$

Since $T_\uparrow + T_\downarrow = k - k_0$, we can get

$$T_\downarrow > \frac{\gamma}{1+\epsilon\gamma}(k-k_0),\ T_\uparrow < \frac{1-\gamma+\epsilon\gamma}{1+\epsilon\gamma}(k-k_0).$$

Notice that from the definition of $\bar{T}$ we have $n \le [(k-k_0)/\bar{T}]$, from above inequalities we can conclude that

$$V_2(k) < (\mu^{\frac{1}{\bar{T}}}\bar{\alpha}^{\gamma}\bar{\beta}^{(1-\gamma)})^{(k-k_0)}V_1(k_0). \tag{23}$$

Then from (12) and (23) we have $e(k) \to 0$ as $k \to +\infty$. Similar results can be obtained for the case $k \in [k_{2n}, k_{2n+1}]$. This completes the proof.

Remark 1 The criterion we obtained above is delay-dependent. For given γ and τ, from (12) we can see that the admissible lower bound of $\bar{T}$ is affected by the parameters μ_1, α and β. Different choice of the parameters could get different results by using Theorem 1. Thus we need to choose the parameters appropriately to find admissible lower bound of $\bar{T}$ for given δ and τ.

4 Numerical Example

In this section, numerical simulation example is given to show the effectiveness of our obtained criterion.

Example 1 Consider the system (8) with the following parameters:

$$A = \begin{bmatrix} 0.9 & 0 \\ 0 & 0.6 \end{bmatrix}, B = \begin{bmatrix} 0.02 & 0.2 \\ 0 & 0.4 \end{bmatrix}, C = \begin{bmatrix} -0.1 & 0.1 \\ -0.2 & -0.1 \end{bmatrix}, f_i(s) = tanh(0.9s) - 0.2s.$$

It can be calculated that $L_1 = \text{diag}\{-0.14, -0.14\}$, $L_2 = \text{diag}\{0.25, 0.25\}$. By using Theorem 1, the feedback controller can be designed for system (8) with different parameters, some numerical simulation results are listed below in Table 1. From which we can see that, the time interval between two successful transmissions can be allowed smaller for a larger γ or smaller τ.

Remark 2 From inequality (12) we can have that

$$\bar{T} > \frac{(\tau - 1)ln(\bar{\beta}/\bar{\alpha}) + 2ln\mu_1}{\gamma ln(\bar{\beta}/\bar{\alpha}) - ln\bar{\beta}}.$$

Thus for given μ_1, α and β, by increasing δ or decreasing τ, smaller value of allowable lower bound of $\bar{T}$ can be obtained.

Let $\gamma = 0.7$, $\tau = 8$, $\bar{T} = 19$, $x(0) = [-0.3, 0.6]$, $y(0) = [0.5, -0.3]$, the synchronization errors response without and with controller are depicted in Figs. 1 and 2, respectively. The time response of controller is also depicted in Fig. 3. From which we can see that the synchronization can be achieved under above condition.

Table 1 Admissible lower bound $\bar{T}$ for different γ and τ

γ	τ	(μ_1, α, β)	$\bar{T}$	K
0.9	4	(1.1, 0.3, 0.12)	6	$\begin{bmatrix} -0.9026 & -0.0538 \\ 0.0131 & -0.7099 \end{bmatrix}$
0.8	6	(1.1, 0.28, 0.12)	10	$\begin{bmatrix} -0.9040 & -0.0523 \\ 0.0074 & -0.7056 \end{bmatrix}$
0.7	8	(1.1, 0.23, 0.12)	19	$\begin{bmatrix} -0.9851 & -0.0527 \\ 0.0066 & -0.7050 \end{bmatrix}$

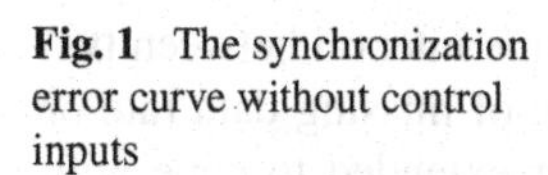

Fig. 1 The synchronization error curve without control inputs

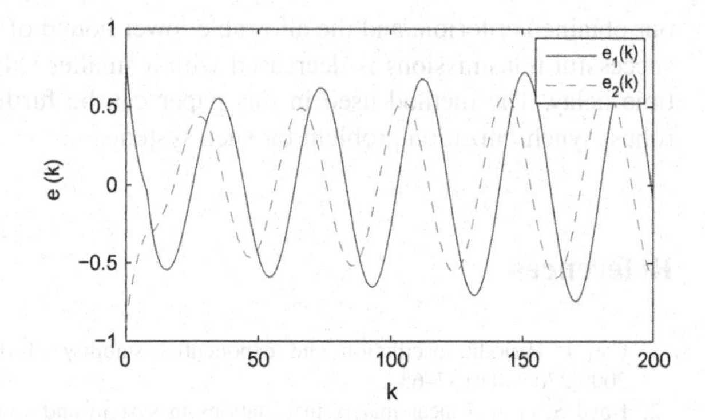

Fig. 2 The synchronization error curve with control inputs

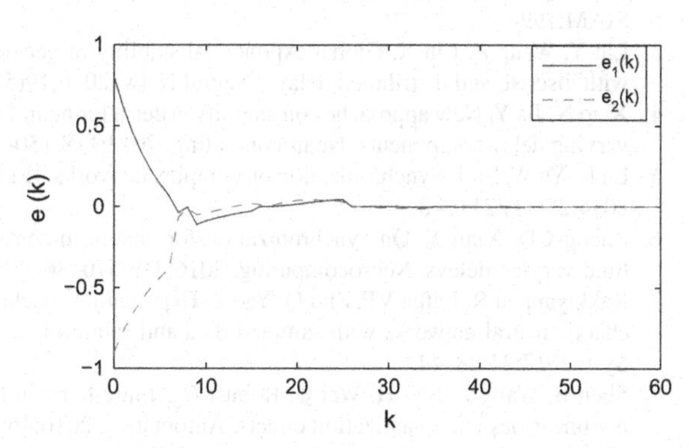

Fig. 3 The time response of controller $\gamma(k)u(k)$

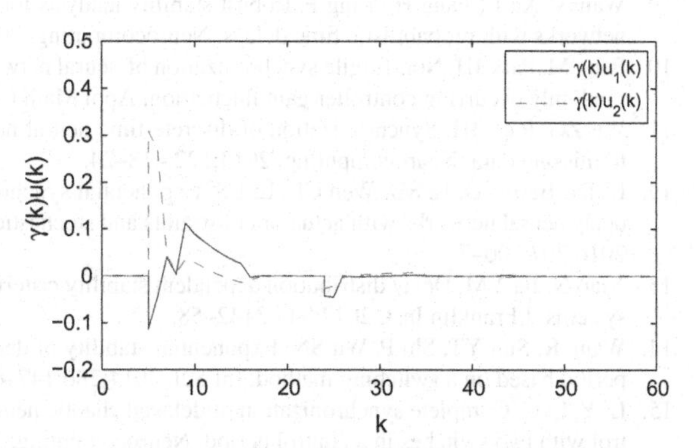

5 Conclusions

In this paper, the synchronization problem for discrete-time delayed neural networks with stochastic missing data has been investigated. Stability criterion is obtained by using switching method. The designation of feedback controller is obtained which can guarantee the error state system to be globally asymptotic stable under unreliable communication links. Numerical example is given to illustrate the effectiveness of our obtained criterion, and the allowable lower bound of time interval between two successful transmissions is decreased with a smaller value of missing data rate or time delay. The method used in this paper can be further extended to cope with robust synchronization problem for such systems.

References

1. Cao J. Periodic oscillation and exponential stability of delayed CNNs. Phys Lett A. 2000;270(3–4):157–63.
2. Boyd S, et al. Linear matrix in equations in system and control theory. Philadelphia, PA: SIAM;1994.
3. Liu Y, Wang Z, Liu X. Global exponential stability of generalized recurrent neural networks with discrete and distributed delays. Neural Netw. 2006;19(5):667–75.
4. Xiao N, Jia Y. New approaches on stability criteria for neural networks with two additive time-varying delay components. Neurocomputing. 2013;118:150–6.
5. Lu R, Yu W, Lu J. Synchronization on complex networks. IEEE Trans Neural Netw Learn Syst. 2014;25(11):2110–8.
6. Zheng CD, Xian Y. On synchronization for chaotic memristor-based neural networks with time-varying delays. Neurocomputing. 2016;216:570–86.
7. Rakkiyappan R, Latha VP, Zhu Q, Yao Z. Exponential synchronization of Markovian jumping chaotic neural networks with sampled-data and saturating actuators. Nonlinear Anal: Hybrid Syst. 2017;24:28–44.
8. Shen B, Wang Z, Shu H, Wei G. Robust H_∞ finite-horizon filtering with randomly occurred nonlinearities and quantization effects. Automatica. 2010;46:1743–51.
9. Wan X, Xu L, Fang H, Yang F. Robust stability analysis for discrete-time genetic regulatory networks with probabilistic time delays. Neurocomputing. 2014;124(26):72–80.
10. Fang M, Park JH. Non-fragile synchronization of neural networks with time-varying delay and randomly occurring controller gain fluctuation. Appl Math Comput. 2013;219:8009–17.
11. Wu ZG, Park JH. Synchronization of discrete-time neural networks with time delays subject to missing data. Neurocomputing. 2013;122:418–24.
12. Li JN, Bao WD, Li SB, Wen CL, Li LS. Exponential synchronization of discrete-time mixed delay neural networks with actuator constraints and stochastic missing data. Neurocomputing. 2016;207:700–7.
13. Xiao N, Jia YM. Delay distribution dependent stability criteria for interval time-varying delay systems. J Franklin Inst. 2012;349:3142–58.
14. Wang R, Sun YT, Shi P, Wu SN. Exponential stability of descriptor systems with large delay period based on a switching method. Inf Sci. 2014;286:147–60.
15. Li Y, Li C. Complete synchronization of delayed chaotic neural networks by intermittent control with two switches in a control period. Neurocomputing. 2016;173:1341–7.

A Real Strong Tracking Filter with Application to Frequency Estimation

Xiuling Sun, Yi Chai, Shanbi Wei, Haoyang He and Junying Hu

Abstract This paper proposes a real strong tracking filter (RSTF) and apply it for frequency estimation of distorted signals in power systems with mutation. In this method, the robustness is improved due to the smoothing ability of three sampling sine wave relational model, as well as the adaptive ability of strong tracking filter (STF). By analyzing the experiments, it is validated that the RSTF algorithm can track frequency in the presence of mutation rapidly and accurately, in addition to guaranteeing the real-time and effectiveness for online application.

Keywords Real strong tracking filter · Distorted power system signal · Parameter mutation · Frequency estimation

1 Introduction

With the complexity of power system, great efforts are made by researchers towards the operation of power system safe and stable. One of the key and challenging problems is the estimation of power signal parameters. In the case of power system, the estimation of parameters like frequency, amplitude, and phase is required not only for monitoring, control and protection. Moreover, in the process of grid-connected distributed generation system, the dynamic energy balance between power supply unit and power grid can be analyzed based on frequency. However, with the widespread use of power electronic devices in power system, the thyristor converter and the high frequency inverter are injected into the non-sinusoidal current signal, resulting in distortion of power signals in presence of noise and harmonic. Therefore, parameters estimation of distorted signals in power system with rapidity and accuracy have been a hot issue.

X. Sun (✉) · S. Wei · H. He · J. Hu
College of Automation, Chongqing University, Chongqing 400044, China
e-mail: xiuling@cqu.edu.cn

Y. Chai
State Key Laboratory of Power Transmission Equipment and System Security and New Technology, College of Automation, Chongqing University, Chongqing 400044, China

Y. Jia et al. (eds.), *Proceedings of 2017 Chinese Intelligent Systems Conference*, Lecture Notes in Electrical Engineering 459,
https://doi.org/10.1007/978-981-10-6496-8_31

Frequency is one of the most important parameters, and once it is measured accurately, the estimation of amplitude and phase is relatively easy. The well known methods to estimate the frequency include the following: fast Fourier transform (FFT) [1, 2], Newton-type algorithm [3], adaptive notch filter [4, 5], least-mean-square method (LMS) [6], Kalman filter (KF) [7, 8], wavelet transform [9], and so forth. FFT is a suitable methods for stationary signal, but it suffers from leakage and picket fence effects, along with giving erroneous results for time-varying signals [10]. KF is put forward for overcoming the influence of noise and harmonic. Whereas, the main limitation is that it loses the tracking ability of electric parameters in mutation condition. Newton-type algorithm is generally calculated via a linear predictor formulation and subsequent optimization. LMS has the drawbacks for poor convergence in addition to be failure in case of signal drifting and changing condition.

Focus on accurate frequency estimation under the conditions of noise and mutation, this paper presents a real strong tracking filter (RSTF). The proposed method is capable of improving the robustness as well as adaptive ability of tracking and estimating signal mutation, which is attributed to three sampling sine wave relational model and strong tracking filter (STF). Furthermore, it is validated to track the signal in presence of mutation rapidly and accurately by the analysis of experiment.

The rest of this paper is organized as follows. Section 2 gives the signal model via analyzing power signal. Next, Sect. 3 puts forward the calculation method of real-time parameters (including frequency, amplitude, phase). Then, simulation results are presented to illustrate the effectiveness of the proposed method in Sect. 4. Finally, followed by some concluding remarks in Sect. 5.

2 Signal Model

Consider the voltage or current signal at time k corrupted by noise [10]:

$$y_k = A\cos(k\omega T_s + \phi) + \varepsilon_k = \hat{y}_k + \varepsilon_k \tag{1}$$

where, y_k is instantaneous signal value; A is amplitude of signal; ω is angular frequency of signal; ϕ is phase of signal; k is sampling instant; T_s is fixed sampling interval; ε_k is additive noise; and $\hat{y}_k$ is the estimated value at time k.

According to analysis, the three values sampled consecutively of the single-phase sinusoidal satisfy the following equation:

$$\hat{y}_k - 2\cos\omega T_s \hat{y}_{k-1} + \hat{y}_{k-2} = 0 \tag{2}$$

Then, the state estimated vector is denoted as:

$$\hat{x}_k = \left[2\cos\omega T_s \;\hat{y}_{k-1}\; \hat{y}_{k-2}\right]^T \tag{3}$$

$$\hat{x}_{k+1} = \begin{bmatrix} 1 & 0 & 0 \\ 0 & 2\cos\omega T_s & -1 \\ 0 & 1 & 0 \end{bmatrix} \hat{x}_k \tag{4}$$

Correspondingly, the matching measurement equation can be written as:

$$\hat{y}_k = \begin{bmatrix} 0 & 2\cos\omega T_s & -1 \end{bmatrix}^T \hat{x}_k + \varepsilon_k \tag{5}$$

As Eq. (2) expressed, the constructed signal model satisfies three sampling sine wave relationship. This kind of model makes the successive three sampling values of signal mutually constrained, generating a filtering effect, which is equivalent to be an implicit filter. Consequently, this paper utilizes the function of filtering generated from the signal model to carry out a coarse adjustment, and then filters and tracks the coarse results via the RSTF algorithm, to obtain a effect similar to double-layer filter, further improve the robustness.

Essentially, the proposed method is a strong tracking filter with real state variables determined above. It is well known that STF not only just has strong robustness for uncertain models, but possess strong tracking ability for abrupt states. Moreover, it can keep track of slow and abrupt state when the system reaches a steady state. The reason is mainly attributed to adjust Kalman gain in real time by bringing in a sub-optimal fading factor, making the residual error forced to have orthogonality or approximate orthogonality for overcoming the un-modeled error of dynamic system. The details procedure of STF algorithm can be referred from [11–15]. Accordingly, RSTF algorithm also has these advantages mentioned early.

3 Frequency, Amplitude and Phase Estimation

Once the state variables of system is obtained, the frequency f, amplitude A and phase angle φ at time k can then optimally be estimated as follows.

$$f = \frac{1}{2\pi T_s} \arccos\left(\frac{\begin{bmatrix} 1 & 0 & 0 \end{bmatrix} \cdot \hat{x}_k}{2}\right) \tag{6}$$

$$A = \frac{\sqrt{\hat{y}_{k-1}^2 + \hat{y}_{k-2}^2 - 2\cos\omega T_s \cdot \hat{y}_{k-1} \cdot \hat{y}_{k-2}}}{\sin\omega T_s} \tag{7}$$

$$\varphi = \arctan\left\{ \frac{\frac{\hat{y}_{k-1}}{\hat{y}_{k-2}} \cos[(k-2)\omega T_s] - \cos[(k-1)\omega T_s]}{\frac{\hat{y}_{k-1}}{\hat{y}_{k-2}} \sin[(k-2)\omega T_s] - \sin[(k-1)\omega T_s]} \right\} \tag{8}$$

4 Analysis of Experimental Results

This section verifies the effectiveness of RSTF algorithm for frequency estimation via comparision with EKF, the variance reset extended Kalman filter (hereinafter referred to as VREKF) and unscented Kalman filter (UKF). All the algorithms are implemented in MATLAB R2012b. In order to ensure the unity of the results, the test signal is given as Eq. (9). Where, the amplitudes V_m is debugged to be 1.0 p.u., the phase φ is set as 0.2618, fundamental frequency ω is 50 Hz, sampling frequency is selected to be 2 kHz and the Gaussian white noise υ_k is 40 dB. In addition, the state estimation initialization is as follows: $\hat{x}_1 = [1.9021, 0, -1.0000]$, $\hat{P}_1 = 2I_3$, $\rho = 0.98$ and $\beta = 4$, respectively.

$$y_k = V_m \cos(k\omega T_s + \varphi) + \upsilon_k \tag{9}$$

Figure 1a presents that the frequency, amplitude and phase of test signal are all suddenly changed from 1 to 1.2 p.u., from 50 to 51.5 Hz, from $15/180 * \pi$ to $20/180 * \pi$ at $k = 51$, respectively. In terms of tracking performance, it can be reported in Fig. 1b. Obviously, the absolute error of EKF algorithm is maximum in contrast to other algorithms, followed by UKF, then VREKF, and the absolute error of RSTF is very close to zero. Therefore, the RSTF algorithm is superior to the EKF, VREKF and UKF in the performance of tracking when the signal is changed suddenly.

The results of frequency, amplitude and phase estimations of different algorithms are shown in Fig. 2. Owing to the lack of tracking ability after mutation, the parameters estimations of EKF algorithm may not be accurate, although the graphic of frequency estimation seems to be satisfactory. Hence, it is not necessary to be

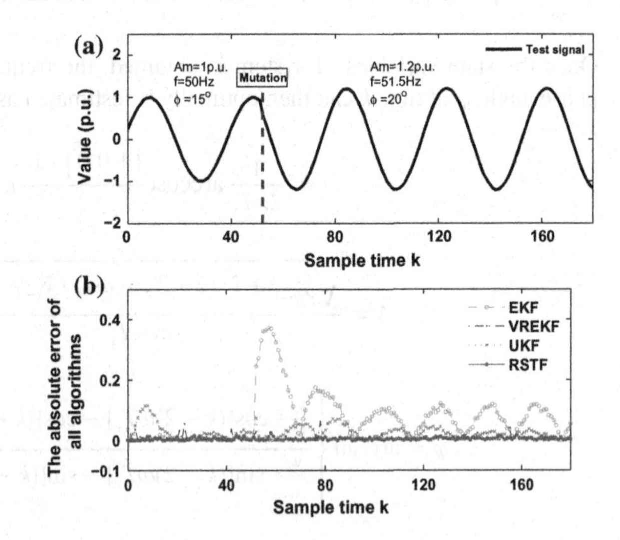

Fig. 1 Test signal (**a**) and the absolute error of all algorithms (**b**)

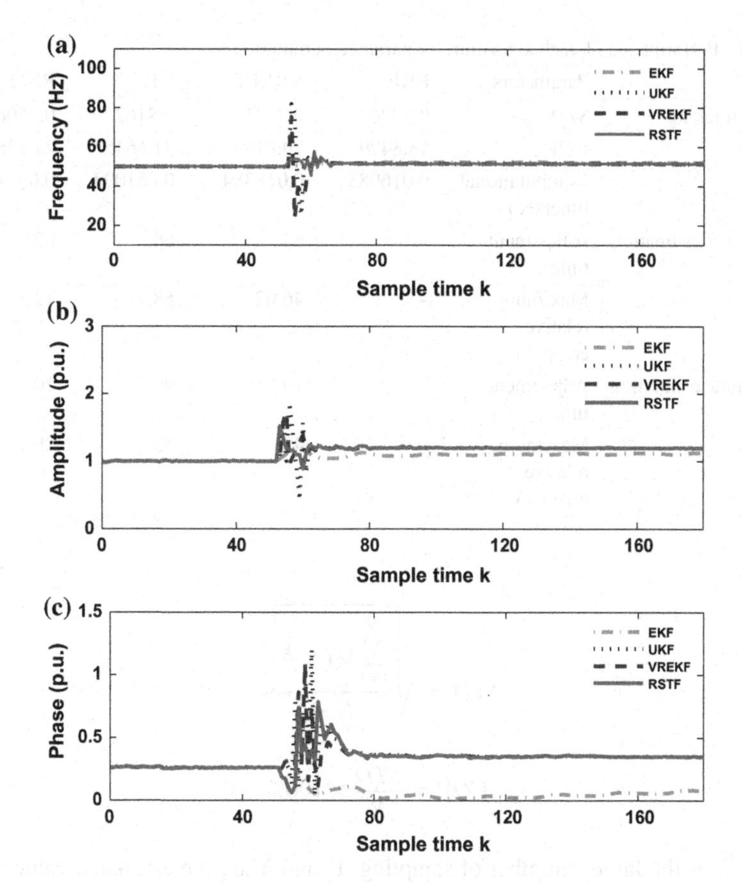

Fig. 2 Frequency, amplitude and phase estimation of all algorithms

analyzed and it may not applicable for parameter estimation in the situation of mutation. Due to the variation of parameters, it is clearly evident from the figures that RSTF algorithm is better than other three algorithms in accuracy.

Table 1 presents the comprehensive performance of each algorithm. The standard deviation (*STD*) is used to illustrate the deviation between the estimation and the mean, whereas the coefficient of variation *COV* acts as the evaluation index of the accuracy for estimated values. The smaller value of *STD* and *COV*, the better performance of the algorithm. Adjustment time is defined as the sampling time when the estimated values kept steady in the range of expected value ±2%, which reflects the rapidity of convergence. As for the computation time, despite RSTF algorithm is approximately 2.2 times and 1.9 times of that of the EKF and VREKF algorithm, respectively, it is not beyond the real-time requirements. Thus, by analyzing all the parameter except for EKF's, even though the RSTF algorithm is not optimal in all aspects, together all the parameters determine its superiority.

Table 1 Performance of each algorithm for parameters mutation

	Parameters	EKF	VREKF	UKF	RSTF
Signal tracking	*STD*	0.7026	0.7531	0.8162	0.7508
	COV	18.8479	29.8192	31.6645	27.7380
	Computational time(sec)	0.016083	0.018384	0.051093	0.035442
Frequency estimation	Adjustment time	–	64	66	62
	Maximum relative error(%)	–	46.02	58.7	12.3
Amplitude estimation	Adjustment time	–	79	90	70
	Maximum relative error(%)	–	45	55	26.7

$$STD = \sqrt{\frac{\sum_{k=1}^{N} (\hat{x}_k - \bar{X})^2}{N-1}} \tag{10}$$

$$COV = \frac{STD}{\bar{X}} \times 100\% \tag{11}$$

where, N is the largest number of sampling, $\hat{x}_k$ and $\bar{X}$ are the estimated value at the kth time point and the mean value of estimated value, respectively. The *STD* can illustrate the deviation degree between the estimation and the mean, whereas the *COV* is used as the evaluation index of the accuracy of estimating. estimating.

5 Conclusions

A real strong tracking filter (RSTF) is presented for frequency estimation of distorted signals corrupted with white Gaussian noise in this paper. According to the theoretical calculation and experimental analysis, RSTF algorithm is proved to meet the requirement of real-time. In addition, it can react to and accurately estimate the mutation in power system. Thus, the proposed method is meaningful and significant in frequency estimations of power system, especially the rapid and accurate estimation of the distorted signals. However, the weakening factor and the forgetting factor of the RSTF algorithm in this paper are empirical values, and the experiment is limited in specific situation, which causes a certain gap with the actual application. Therefore, it is necessary to do further testing in the actual application environment.

Acknowledgements This work is supported by the National Natural Science Foundation of China (61374135) and Chongqing University Postgraduates Innovation Project (CYB15051).

References

1. Lin HC, Lee CS. Enhanced FFT-based parametric algorithm for simultaneous multiple harmonics analysis. IEE Proc-Gener Transm Distrib. 2001;148(3):209–14.
2. Singh SK, Sinha N, Goswami AK, et al. Several variants of Kalman filter algorithm for power system harmonic estimation. Int J Electr Power Energy Syst. 2016;78:793–800.
3. Terzigia VV. Improving recursive Newton-typealgorithm for power system relaying and measurement. IEEE Trans Instrum Meas. 1998;145(1):877–81.
4. Mojiri M, Karimi-Ghartemani M, Bakhshai A. Estimation of power system frequency using an adaptive notch filter. IEEE Trans Instrum Meas. 2007;56(6):2470–7.
5. Shiping Z, Yongping Z, Shaoqing Z. A novel approach to measurement of power system frequency using adaptive notch filter. Proc-Chin Soc Electr Eng. 2003;23(7):81–3.
6. Tichavsky P, Nehorai A. Comparative study of four adaptive frequency trackers. IEEE Trans Signal Proces. 1997;45(6):1473–84.
7. Jang LI, Yi-wei W. A survey on the application of Kalman filtering method in power system. Power Syst Prot Control. 2014; 42(6):135–144.
8. Dash PK, Panda G, Pradhan AK, Routray, A, Duttagupta B. An extended complex Kalman filter for frequency measurement of distorted signals. In: Power engineering society winter meeting, vol. 3. IEEE;2000, pp. 1569–74.
9. Driesen J, Van Craenenbroeck T, Reekmans R, et al. Analysing time-varying power system harmonics using wavelet transform. In: Instrumentation and measurement technology conference, vol. 1;1996, p. 474–9.
10. Routray A, Pradhan AK, Rao KP. A novel Kalman filter for frequency estimation of distorted signals in power systems. IEEE Trans Instrum Meas. 2002; 51(3):469–79.
11. Shuai Ma, Rende Zhao, Xiaobo Wu. Strong tracking filter based fundamental phase real-time extraction for single-phase voltage. Proc CSEE. 2012;32(28):83–9.
12. Zhao Ren-De, Ma Shuai, Li Hai-Jian, Wu Xiao-Bo. Strong tracking filter based frequency-measuring algorithm for power system. Power Syst Prot Control. 2013;41(7):85–90.
13. Ding Zizhe, Zhang Xianda, Zhu Xiaolong. Nonlinear principal component analysis using strong tracking filter. Tsinghua Sci Technol. 2007;12(6):652–7.
14. Shuai MA, Zhao R, Wu X. A single-phase voltage fundamental phase real time extraction based on strong tracking filter. Chin J Electr Eng. 2012; 32(28):83–9.
15. Zhao Ren-De, Li Hai-Jian. A strong tracking filter algorithm for frequency measurement of power system. Power Syst Prot Control. 2013;41(7):85–90.

Sliding Mode Based He'non Mapping Control

Yunzhong Song

Abstract A new control strategy based on discrete-time variable structure systems theory is proposed to target the He'non mapping. This method enables the asymptotical stability of the control with uncertain discrete-time environments. Under auspices of the new designing strategy, the states of the system can reach the sliding manifold without chattering and converge exponentially to the zero states, in spite of the matched and mismatched perturbations. The efficacy of the suggested scheme was illustrated with the well-known He'non mapping, and results shown that it can be a promising candidate strategy to the other complex systems.

Keywords Discrete time control · Sliding mode control · Chattering · He'non mapping

1 Introduction

Discrete time variable structure control system becomes more and more powerful recent years, and this can be attributed for three reasons, the first one is the widely used computers, and the second one is the convenient tools for analysis of the discrete signals and the third one can be boiled down to the cost savings of wired lines and cables configuration fees [1]. However, discrete time variable structure control strategies are still lag behind the pace of the increment of the pressing acquirement of application of discrete time sliding mode control [2, 3]. Especially, the disturbances in the real world can be matched and mismatched ones and it is not hard finding that the matched perturbations are assumed so often, where the

Y. Song (✉)
School of Electrical Engineering and Automation, Henan Polytechnic University,
Jiaozuo 454003, China
e-mail: songhpu@126.com

Y. Jia et al. (eds.), *Proceedings of 2017 Chinese Intelligent Systems Conference*, Lecture Notes in Electrical Engineering 459,
https://doi.org/10.1007/978-981-10-6496-8_32

mismatched perturbations are seldom considered, this may be due to the complexity of reaching condition of discrete-time variable structure control [4–9].

Here, a new sliding mode control scheme for matched and mismatched perturbations was suggested. And sliding surface was made invariance as the following, that is, as the state is shifted from the prescribed margin of the surface, it will be dragged back again, and even further, the sliding mode surface can be approached again in limited number of steps.

The next several sections are listed as follows: Sect. 2 is about system description and problem formulation. Section 3 is intended to deal with sliding surface, where a class of discrete system with matched and mismatched perturbations will be explored upon. Then, at Sect. 4, chattering free sliding mode reaching condition together with sliding controller design and case study example will be provided, and finally, Sect. 5 is assigned to the conclusion.

2 System Descriptions and Problem Formulations

The sliding mode control is usually categorized into two different steps, the first one is the coined sliding function, which should be capable in guarantee both the stability of the closed-loop system, and the desired dynamic behavior of the sliding surface. The discrete-time systems are always assumed to be

$$x_1(k+1) = A_{11}x_1(k) + A_{12}x_2(k), x_2(k+1) = A_{21}x_1(k) + A_{22}x_2(k) + u(k)$$

where $x(k) \in R^n$ is assumed to be measurable, $u(k) \in R^m$, A is the known constant system matrix and $(A, \bar{B})$ is controllable, $\bar{B} = [0, I_m]^{\mathrm{T}}$. Since $(A, \bar{B})$ is controllable, from the theory of linear control system, it is not to conclude that (A_{11}, A_{12}) is controllable too, with helps of pole placement, choose the switching function as $s(k) = C_1x_1(k) + x_2(k)$, and the asymptotically stable sliding mode equation is obtained $x_1(k+1) = (A_1 - A_2C_1)x_1(k)$. The second step is about controller built to complete the reaching condition. Because of the already existed discrete system sliding mode reaching condition is like

$$s(k+1) = \alpha s(k) - \varepsilon \operatorname{sgn} s(k),\ 0 < \alpha < 1, \varepsilon > 0.$$

the $\|s(k)\|$ will be attenuated if the states of the system are made jump out of the domain: $S_\varepsilon = \{x(k); \|s(k)\| \le \varepsilon/(1-\alpha)\}$. To be evident, the already existed switching surface will be oscillated in nature.

3 Sliding Mode Surface Design

Assumed that the discrete systems can be modeled as

$$\begin{aligned} x_1(k+1) &= A_{11}x_1(k) + A_{12}x_2(k) + p_1(x,k) \\ x_2(k+1) &= A_2x(k) + f(x(k)) + u(k) + p_2(x,k) \end{aligned} \tag{1}$$

where $x = [x_1^{\mathrm{T}}, x_2^{\mathrm{T}}]^{\mathrm{T}}$, $u \in R^m$ are state vector and control input vector respectively, $x_1 \in R^{n-m}$, $x_2 \in R^m$; and $p_1(x(k),k)$ and $p_2(x(k),k)$ are the unknown matched and mismatched perturbations; $A_1 = [A_{11}, A_{12}]$ and A_2 with appropriate dimensions.

Assumption 1 $(A, \bar{B})$ is controllable, here $A = [A_1^{\mathrm{T}}, A_2^{\mathrm{T}}]^{\mathrm{T}}, \bar{B} = [0^{\mathrm{T}}, I_m]^{\mathrm{T}}$.

Assumption 2 $\lim_{x_1(k) \to 0} p_1(x(k),k) = 0$.

In order to repress the bad influence of mismatched disturbance on the sliding mode, the dynamic compensator will be introduced as

$$v(k+1) = -N(A_{11} - A_{12}C - I)x(k) + v(k) \tag{2}$$

where $C \in R^{m \times (n-m)}$ is selected to keep $A_{11} - A_{12}C$ is Hurwitz, and will be determined later. Take switching function as

$$s(k) = (C + N)x_1(k) + v(k) + x_2(k) \tag{3}$$

Let $s(k) = 0$, yields

$$x_2(k) = -(C + N)x_1(k) - v(k) \tag{4}$$

Substituting (4) into the first equation of (1), then sliding mode equation is obtained

$$x_1(k+1) = [A_{11} - A_{12}(C + N)]x_1(k) + p_1(x(k),k) - A_{12}v(k) \tag{5}$$

Now we have the following result:

Theorem 1 *If there exists* $N \in R^{m \times (n-m)}$, *such that*

$$|\lambda_{\max}(I_{n-m} - A_{12}N)| < 1 \tag{6}$$

and $p_1(k)$ *vary slowly i.e.* $p_1(k) \approx p_1(k-1)$, *then sliding mode Eq. (5) is asymptotically stable.*

Proof Definite

$$e(k) = p_1(x,k) - A_{12}v(k) \tag{7}$$

From (5) and (2), we have

$$\begin{aligned} x_1(k+1) &= [A_{11} - A_{12}(C+N)]x_1(k) + e(k) \\ e(k+1) &= A_{12}N(A_{11} - A_{12}C - I)x_1(k) + e(k) + \Delta p_1(k) \\ &\approx A_{12}N(A_{11} - A_{12}C - I)x_1(k) + e(k) \end{aligned} \tag{8}$$

Take state transformation

$$\begin{pmatrix} \bar{x}_1(k) \\ \bar{e}(k) \end{pmatrix} = \begin{pmatrix} I_{n-m} & 0 \\ A_{12}N & -I_{n-m} \end{pmatrix} \begin{pmatrix} x_1(k) \\ e(k) \end{pmatrix} \tag{9}$$

Then (8) is transformed into

$$\begin{pmatrix} x_1(k+1) \\ \bar{e}(k+1) \end{pmatrix} = \begin{pmatrix} A_{11} - A_{12}C & -I_{n-m} \\ 0 & I_{n-m} - A_{12}N \end{pmatrix} \begin{pmatrix} \bar{x}_1(k) \\ \bar{e}(k) \end{pmatrix} \tag{10}$$

Since $A_{11} - A_{12}C$ and $I_{n-m} - A_{12}N$ both are stable, therefore

$$\lim_{k\to\infty} x_1(k) = \lim_{k\to\infty} \bar{x}_1(k) = 0, \quad \lim_{k\to\infty} e(k) = \lim_{k\to\infty} \bar{e}(k) = 0$$

Using (4) and Assumption 2, on the sliding surface

$$\begin{aligned} \lim_{k\to\infty} ||x_2(k)|| &\le \lim_{k\to\infty} ||(C+N)||\,||x_1(k)|| + \lim_{k\to\infty} ||v(k)|| \\ &\le \lim_{k\to\infty} ||(C+N)||\,||x_1(k)|| + \lim_{k\to\infty} ||e(k)|| + \lim_{k\to\infty} ||p_1(x(k),k)|| \\ &= \lim_{x_1(k)\to\infty} ||p_1(x(k),k)|| = 0 \end{aligned} \tag{11}$$

So sliding mode is asymptotically stable.

Theorem 2 If $p_1(x(k),k) \in R(A_{12})$, i.e. there exists $\bar{p}_1(k)$ such that $p_1(k) = A_{12}\bar{p}_1(k)$ and $p_1(k)$ vary slowly, then sliding mode Eq. (5) is asymptotically stable. In case of brevity, the proof is omitted here.

4 Sliding Mode Controller Design

The second step in designing sliding control is to determine a control law such that the state of system (5) approaches the sliding surface and is sustained thereafter. In order to overcome the disadvantage of traditional sliding mode reaching condition, we propose a new sliding mode reaching condition as following:

$$s(k+1)=\frac{1}{2}[\alpha-\beta+(\alpha+\beta)\mathrm{sgn}(||s(k)||-\delta)]s(k) \tag{12}$$

where $0\leq\alpha<1, 0<\beta<1, \delta>0$.

Theorem 3 Under the reaching condition (12), the domain $S_\delta=\{x(k); ||s(k)||\leq\delta\}$ is reached in finite number of steps, and once into the domain, the $s(k)$ will converge exponentially to zero with chattering free.

Proof When $x(k)\notin S_\delta$, i.e. $||s(k)|| >\delta$, then $\mathrm{sgn}(||s(k)||-\delta)=1$, from (12), we have

$$s(k+1)=\alpha s(k)=\alpha^2 s(k-1)=\cdots=\alpha^{k+1}s(0) \tag{13}$$

Thus $s(k)$ is monotonously decreasing, moreover, the domain S_δ is reached in finite number of steps, say k_δ steps, and the state once into the domain, the state will always remain in it. When state reached into S_δ, then $||s(k)|| <\delta$,$\mathrm{sgn}(||s(k)||-\delta)=-1$, using (13), yields

$$s(k_\delta+h)=\cdots=(-1)^h\beta^h s(k_\delta)\to 0\ (h\to\infty) \tag{14}$$

This indicates that the $s(k)$ will across the sliding surface each step and converge exponentially to zero with chattering free. In order to ensure the existence of sliding mode, even if there is matched and mismatched perturbations exist, we need to design a dynamic compensator again as follow

$$\mu(k+1)=s(k)+\mu(k)-u_1(k) \tag{15}$$

Theorem 4 Design the controller as

$$\begin{aligned} u(k)=&-(C+N)[A_{11}x_1(k)+A_{12}x_2(k)]-A_2x(k)-f(x(k))] \\ &+N(A_{11}-A_{12}C-I)x(k)-v(k)-s(k)-\mu(k)+u_1(k) \end{aligned} \tag{16}$$

where

$$u_1=\frac{1}{2}[\alpha-\beta+(\alpha+\beta)\mathrm{sgn}(||s(k)||-\delta)]s(k)\ ,\ 0<\alpha,\beta<1 \tag{17}$$

$\mu(k)$ is state of dynamic compensator (15), then state of system (1) will reach the sliding surface in finite number of steps, and converge exponentially to zero with chattering free. In case of brevity, proof of Theorem 4 is also omitted here. As an example, we consider the well-known He'non mapping

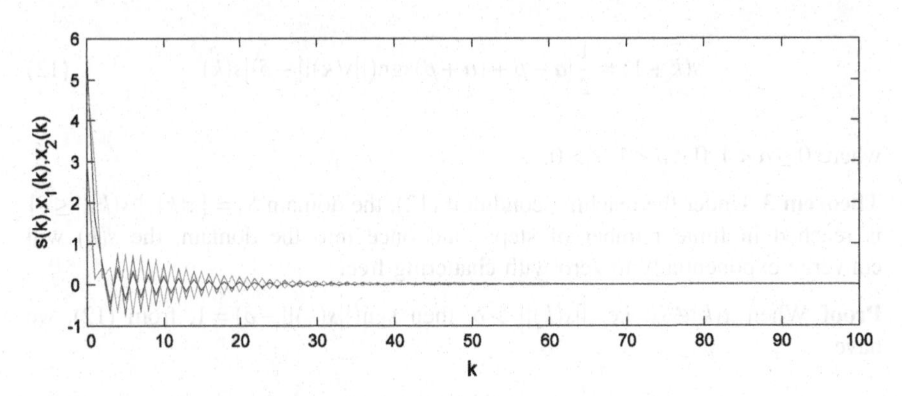

Fig. 1 Time evolution of sliding surface together sub-state variables

$$\begin{aligned} x_1(k+1) &= x_2(k) + \Delta_1(k) \\ x_2(k+1) &= (1+\Delta_2(k))x_1(k) + 1 + \Delta_3 - 1.4x_2^2 + u(k) \end{aligned} \tag{18}$$

where $\Delta_1 = 0.3\sin(0.01k)$, $\Delta_2 = 0.2\cos(0.01k)$, $\Delta_3 = 0.5 + 3\cos(0.01k)$

According to Theorem 1 choose switching function

$$s(k) = (C+N)x_1(k) + v(k) + x_2(k) = 1.5x_1(k) + v(k) + x_2(k) \tag{19}$$

Choose controller according Theorem 4

$$\begin{aligned} u(k) = &-(C+N)[A_{11}x_1(k) + A_{12}x_2(k)] - A_2x(k) - f(x(k))] \\ &+ N(A_{11} - A_{12}C - I)x(k) - v(k) - s(k) - \mu(k) + u_1(k) \\ = &-2.5x_1(k) - 1.5x_2(k) - 1 + 1.4x_2^2 - v(k) - s(k) - \mu(k) + u_1(k) \end{aligned} \tag{20}$$

The simulation results of the states and the switching surface are listed in Fig. 1. It can be seen obviously, according to the new designing strategy in this paper, the system can be asymptotically stabilized, despite the existence of the matched and mismatched perturbations.

5 Conclusions

Discrete system with matched and mismatched perturbations was touched upon here, with the new sliding mode design technique, the oscillation was mitigated. The suggested scheme was employed to deal with the He'non map control, and results verified its effectiveness.

Acknowledgements This work aims to be published in memory of Professor Wenlin Li, who used to be a great mentor and also a very good friend of mine, unfortunately, Professor Li passed away several years ago, Professor Li explored a lot in sliding mode control, especially to the discrete time sliding mode control, Professor Li had published several papers, which should to be long last meaningful in the future. The knowledge of sliding mode control, especially about the discrete time case here is attributed to Professor Li, for his kindness, generosity and his help.

This work was supported in part by National Science Foundation of China (60850004, 61340041 and 61374079), and The Project-sponsored by SRF for ROCS, SEM to Yunzhong Song.

References

1. Dra_zenovi_c B. The invariance conditions in variable structure systems. Automatica. 1969; 5:287–95.
2. Bartolini G, Ferrara A, Utkin VI. Adaptive sliding mode control in discrete-time systems. Automatica. 1995;31:769–73.
3. Furuta K. Sliding mode of a discrete system. Syst Control Lett. 1990;14:145–52.
4. Young KD, Utkin VI, Zguner UO. A control engineer's guide to sliding mode control. IEEE Trans Control Syst Technol. 1999;7:328–42.
5. Potts RB, Yu X. Discrete variable structure system with pseudo-sliding mode. J Aust Math Soc Ser B. Appl Math. 1991;32:365–76.
6. Gao W, Wang Y, Homaifa A. Discrete-time variable structure control systems. IEEE Trans Ind Electron. 1995;42:117–22.
7. Gao W, Hung JC. Variable structure control of nonlinear systems: a new approach. IEEE Trans Ind Electron. 1993;40:45–55.
8. Wang W-J, Wu G-H, Yang D-C. Variable structure control design for uncertain discrete-time systems. IEEE Trans Autom Control. 1994;39:99–102.
9. Golo Goran. Cedomir Milosavljevic Robust discrete-time chattering free sliding mode control. Syst Control Lett. 2000;41:19–28.

Approximate Solution for Three-Player Mixed-Zero-Sum Nonlinear Game via ADP Structure

Yongfeng Lv, Xuemei Ren, Jing Na, Qinqin Yang and Linwei Li

Abstract In this paper, a three-player mixed-zero-sum game situation with nonlinear dynamics is proposed, and an approximate dynamic programming (ADP) learning scheme is used to solve the proposed problem. First, the problem formulation is presented. A value function for player 1 and 2 nonzero-sum game is constructed, another value function for player 1 and 3 zero-sum game is presented for three-player nonlinear game system. Because of the difficulty to solve the nonlinear Hamilton-Jacobi (HJ) equation, the single-layer critic neural networks are used to approximate the optimal value functions. Then the approximated critic neural networks (NNs) are directly used to learn the optimal solutions for three-player mixed-zero-sum nonlinear game. A novel adaptive law with the estimation performance index is proposed to estimate the unknown coefficient vector. Finally, a simulation example is presented to illustrate the proposed methods.

Keywords Approximate dynamic programming · Neural networks · Zero-sum game · Parameter estimation

1 Introduction

Game theory [1] provides an ideal environment to study multiplayer optimal decision and control problems, and Nash differential games were originally stated by [2]. This theorem has been highly applied to a number of potential applications

The work was supported by National Natural Science Foundation of China (No. 61433003, No. 61573174, and No. 61273150).

Y. Lv · X. Ren (✉) · Q. Yang · L. Li
School of Automation, Beijing Institute of Technology, Beijing 100081, China
e-mail: xmren@bit.edu.cn

J. Na
Faculty of Mechanical & Electrical Engineering, Kunming University of Science & Technology, Kunming 650500, China

Y. Jia et al. (eds.), *Proceedings of 2017 Chinese Intelligent Systems Conference*, Lecture Notes in Electrical Engineering 459,
https://doi.org/10.1007/978-981-10-6496-8_33

in control engineering and economics [3, 4]. In a differential game, each player independently chooses an optimal policy to minimize its own performance objective, which depends on the actions of itself and all the other players [5]. ADP algorithms have been applied to solve the zero-sum game problems in the past few years [6]. Liu et al. solved the neural-network-based zero-sum game problem for discrete-time nonlinear systems, then presented an online synchronous approximate optimal learning algorithm for multiplayer nonzero-sum games.

To the best of our knowledge, three player system with both zero-sum and nonzero-sum game is not investigated in the literature. In this paper, we propose three player mixed-zero-sum differential game problem, and define the Nash equilibrium of the proposed problem. Moreover, an NN ADP learning scheme [7] is introduced to get the optimal control inputs for each player, and a novel adaptive law with performance index is proposed to estimate the unknown coefficient vector.

2 Preliminary

Consider the following multi-player game system

$$\dot{x}=f(x)+\sum_{j=1}^{2} g_j(x)u_j+k(x)d \tag{1}$$

where $x\in\mathbb{R}^n$ is game state, $u_j\in\mathbb{R}^m$ is game player policies, $d\in\mathbb{R}^q$ is another policy. $f(x)\in\mathbb{R}^n$ is system dynamic, $g_j(x)\in\mathbb{R}^{n\times m}$ and $k(x)\in\mathbb{R}^{n\times q}$ are the unknown input dynamics and disturbance injection dynamics, respectively.

To solve the game situation in this paper, the following assumption will be used throughout this paper.

Assumption 1 In the game system, player 1 u_1 is cooperative nonzero-sum with regard to player 2 u_2, simultaneously is non-cooperative zero-sum with regard to player 3 d.

Thus, the optimal performance index for player 1 and player 2 can be described as

$$V_1^*(x)=\min_{u_{1,2}}\int_t^{\infty}(x^TQ_{11}x+u_1^TR_{11}u_1+u_2^TR_{12}u_2)d\tau \tag{2}$$

where $r_1=x^TQ_{11}x+u_1^TR_{11}u_1+u_2^TR_{12}u_2$, $Q_{11}>0$ and $R_{11}>0, R_{12}>0$ are positive symmetric matrices. In the game system, we will find an optimal admissible control policy pair $\{u_1^*,u_2^*\}$ to minimize the value function (2) with players 1 and 2.

Simultaneously, the optimal performance index for player 1 and player 3 can be described as

$$V_2^*(x) = \min_{u_1} \max_{d} \int_t^{\infty} \left(x^T Q_{21} x + u_1^T R_{21} u_1 + u_2^T R_{22} u_2 - \gamma^2 \|d\|^2 \right) d\tau \tag{3}$$

where $r_2 = x^T Q_{21} x + u_1^T R_{21} u_1 + u_2^T R_{22} u_2 - \gamma^2 \|d\|^2$, $Q_{21} > 0$ and $R_{21} > 0, R_{22} > 0$ are positive symmetric matrices. We will find a solution $\{u_1^*, d^*\}$ to optimize the value function (3).

Definition 1 (*Nash Equilibrium*): In the one-to-two player game system, $\{\mu_1^*, \mu_2^*\}$ and $\{\mu_1^*, d^*\}$ constitute Nash equilibriums with regard to player 1, if the following inequalities with regard to the value function J_i are satisfied. Player 1 and player 2 satisfy [8]

$$\begin{cases} J_1^*(\mu_1^*, \mu_2^*, d^*) \le J_1(\mu_1, \mu_2^*, d^*) \\ J_2^*(\mu_1^*, \mu_2^*, d^*) \le J_2(\mu_1^*, \mu_2, d^*) \end{cases} \tag{4}$$

Simultaneously, player 1 and player 3 satisfy [9]

$$J_2(u_1^*, \mu_2^*, d) \le J(u_1^*, \mu_2^*, d^*) \le J_2(u_1, \mu_2^*, d^*) \tag{5}$$

Assume that the value functions (2) and (3) are continuous and differentiable. The differential equation for the value function can be described as

$$0 = r_i(x, u_1, u_2, d) + (\nabla V_i)^T \left(f(x) + \sum_{j=1}^{2} g_j(x) u_j + k(x) d \right), \; i = 1, 2 \tag{6}$$

where $\nabla V \triangleq \partial V / \partial x$. Then the Hamiltonian functions for each situation can be presented as

$$H_i(x, \nabla V_i, u_1, u_2, d) = r_i(x, u_1, u_2, d) + (\nabla V_i)^T \left(f(x) + \sum_{j=1}^{2} g_j(x) u_j + k(x) d \right), \; i = 1, 2 \tag{7}$$

Thus from (7), it can be obtained that

$$\frac{\partial H_i}{\partial u_i} = 0 \Rightarrow u_i^* = -\frac{1}{2} R_{ii}^{-1} g_i^T(x) \nabla V_i^*, \; i = 1, 2 \tag{8}$$

$$\frac{\partial H_2}{\partial d} = 0 \Rightarrow d^* = \frac{1}{2\gamma^2} k^T(x) \nabla V_2^* \tag{9}$$

Remark 1 In this three-player game, player 1 cooperates with player 2, but non-cooperates with player 3. Thus, we give the value function (2) and (3), respectively. It is obviously that both the zero-sum and nonzero-sum situation is included in this three player game. It should be noted there exists an equilibrium point that player 1 is with regard to both player 2 and player 3.

3 Optimal Control Policies Design

To get the optimal solution for the Nash equilibrium, one has to solve the coupled HJ Eq. (7). For the linear system, it can be solved by the coupled algebraic Riccati equations (ARE). As we known, it is difficult to obtain the optimal solution of the nonlinear coupled HJ equations for the nonlinear system [10]. In the following section, we will present a new single-layer NN ADP methods for the one-to-two player game system.

3.1 Critic NN Design

The critic neural networks of each player are given as

$$V_i^*(x) = W_{ci}^T \varphi_{ci}(x) + \varepsilon_{ci},\ i = 1, 2 \tag{10}$$

where $W_{ci} \in \mathbb{R}^K$ are the ideal unknown coefficient vectors, and $\varphi_{ci}(x) \in \mathbb{R}^K$ are the linearly independent basis function vectors with the hidden neuron number K. ε_{ci} is the residual error.

Assumption 2 [11]: Both The NN weights W_{ci} and the activation functions $\varphi_{ci}(x)$ are bounded, i.e. $\|W_{ci}\| \le b_{wi}$, $\|\varphi_{ci}(x)\| \le b_{\varphi i}$.

Let the estimation of the NN weight be $\hat{W}_{ci}$, then the estimated optimal value function can be

$$\hat{V}_i(x) = \hat{W}_{ci}^T \varphi_{ci}(x), i = 1, 2 \tag{11}$$

Accordingly, its derivative is

$$\nabla \hat{V}_i(x) = \nabla \varphi_{ci}^T(x) \hat{W}_{ci}, i = 1, 2 \tag{12}$$

The approximate optimal control policies are then given by

$$\hat{u}_i = -\frac{1}{2} R_{ii}^{-1} g_i^T(x) \nabla \varphi_{ci}^T(x) \hat{W}_{ci},\ i = 1, 2 \tag{13}$$

$$\hat{d} = \frac{1}{2\gamma^2} k^T(x) \nabla \varphi_{c2}^T(x) \hat{W}_{c2} \tag{14}$$

The approximated coupled Lyapunov HJ Eq. (7) is rewritten as

$$
\begin{aligned}
H_i(x, \nabla V_i, u_1, u_2, d) &= r_i(x, u_1, u_2, d) \\
&\quad + \left(\nabla \varphi_{ci}^T(x) \hat{W}_{ci}\right)^T (f(x) + g_1(x)u_1 + g_2(x)u_2 + k(x)d) \\
&= e_i,\ i = 1, 2
\end{aligned} \tag{15}
$$

where e_i is the residual error. It should be pointed that many ADP NN algorithms [12] get the approximation optimal solution by minimizing the corresponding squared residual error, i.e. $E_i = (1/2)e_i^T e_i$.

Denote that $\Theta_i = r_i(x, u_1, u_2, d)$ and $\Xi_i = \nabla \varphi_{ci}^T(x)(f(x) + g_1(x)u_1 + g_2(x)u_2 + k(x)d)$, then Eq. (15) can be rewritten as

$$
\Theta_i = -\hat{W}_{ci}^T \Xi_i + e_i, i = 1, 2 \tag{16}
$$

We will use a new adaptive law [13] with estimation performance to estimate the critic NN weight W_{ci}, such that $P_i \in \mathbb{R}^{K \times K}$ and $Q_i \in \mathbb{R}^K$ are defined as

$$
\begin{cases}
\dot{P}_i = -\ell P_i + \Xi_i \Xi_i^T, & P_i(0) = 0 \\
\dot{Q}_i = -\ell Q_i + \Xi_i \Theta_i, & Q_i(0) = 0
\end{cases} \tag{17}
$$

where $i = 1, 2$; $\ell > 0$. $M_i \in \mathbb{R}^K$ is deduced as

$$
M_i = P_i \hat{W}_{ci} + Q_i, i = 1, 2 \tag{18}
$$

From (17) and (18), one has $Q_i = -P_i W_{ci} - \upsilon_i$, where $\upsilon_i = -\int_0^t e^{-\ell(t-r)} e_i(r) \Xi_i^T(r) dr$ is bounded error in $\Omega \in \mathbb{R}^n$, i.e., $\|\upsilon_i\| \le \varepsilon_{\upsilon i}$ for a positive constant $\varepsilon_{\upsilon i}$. Moreover, one can obtain from (17) and (18) that

$$
M_{2i} = P_{2i} \hat{W}_{ci} + Q_{2i} = -P_{2i} \tilde{W}_{ci} + \upsilon_{2i}, i = 1, 2 \tag{19}
$$

where $\tilde{W}_{ci} = W_{ci} - \hat{W}_{ci}$ is the estimated parameter error. From (19), it can be known that M_i is obtained by $\tilde{W}_{ci}$ and υ_i. We will use the following performance index to improve the estimator as in [14]

$$
J(\hat{W}_{ci}) = \frac{1}{2} \int_0^t e^{-\beta(t-\tau)} \frac{[Q_i - \hat{W}_{ci} P_i]^T [Q_i - \hat{W}_{ci} P_i]}{m^2} d\tau + \frac{1}{2} e^{-\beta t} (\hat{W}_{ci} - W_{ci}(0))^T Q_0 (\hat{W}_{ci} - W_{ci}(0)) \tag{20}
$$

where $m^2 = I + P^T P$, $Q_0 > 0$ and $\beta > 0$. Equation (20) contains the past data discounting and the initial NN weight $W_{ci}(0)$. Thus, minimum value function can be obtained by the derivative of the parameter performance index (20)

$$\nabla J(\hat{W})=\int_0^t e^{-\beta(t-\tau)}\frac{(P_iQ_i^T+P_iP_i^T\hat{W}_{ci})}{m^2}d\tau+e^{-\beta t}Q_0(\hat{W}_{ci}-W_{ci}(0))=0 \quad (21)$$

Furthermore, it can be concluded that

$$\hat{W}_{ci}=\left(\int_0^t e^{-\beta(t-\tau)}\frac{P_iP_i^T}{m^2}d\tau+e^{-\beta t}Q_0)\right)^{-1}\left(\int_0^t e^{-\beta(t-\tau)}\frac{P_iQ_i^T}{m^2}d\tau+e^{-\beta t}Q_0W_{ci}(0)\right) \quad (22)$$

Denote $\Gamma_i(t)=(\int_0^t e^{-\beta(t-\tau)}\frac{P_iP_i^T}{m^2}d\tau+e^{-\beta t}Q_0)^{-1}$ and $A_i=\int_0^t e^{-\beta(t-\tau)}\frac{P_iQ_i^T}{m^2}d\tau+e^{-\beta t}Q_0W_{ci}(0)$. Then we have

$$\frac{d}{dt}\Gamma_i\Gamma_i^{-1}=\Gamma_i\Gamma_i^{-1}+\Gamma_i\frac{d}{dt}\Gamma_i^{-1}=0 \quad (23)$$

and

$$\Gamma_i=\beta\Gamma_i-\Gamma_i\frac{P_iP_i^T}{m^2}\Gamma_i^{-1} \quad (24)$$

Thus, the critic NN weight W_{ci} can be updated by the auxiliary vector M_i as

$$\dot{\hat{W}}_{ci}=\Gamma_iA_i+\Gamma_i\frac{dA_i}{dt}=-\frac{\Gamma_iP_iM_i}{m^2},\ i=1,2 \quad (25)$$

Theorem 1 *For critic NN* (11), *with the adaptive law* (25), *the approximation coefficient vector* $\hat{W}_{ci}$ *converges to a small neighborhood surrounding the ideal value* W_{ci}.

Proof Consider the following Lyapunov function

$$V=\sum_{i=1}^N V_i=\sum_{i=1}^N\frac{(\tilde{W}_{ci}^T\Gamma_i^{-1}\tilde{W}_{ci})}{2},\ i=1,2 \quad (26)$$

According to (19) and (25), the above Lyapunov function can be deduced as

$$\begin{aligned}\dot{V}=\sum_{i=1}^N\dot{V}_i=\sum_{i=1}^N\tilde{W}_{ci}^T\Gamma_i^{-1}\dot{\tilde{W}}_{ci}=\sum_{i=1}^N\left(-\tilde{W}_{ci}^TP_i\tilde{W}_{ci}+\tilde{W}_{ci}^T\upsilon_i\right)=\sum_{i=1}^N\left(-\sigma_i\|\tilde{W}_{ci}\|^2+\tilde{W}_{ci}^T\upsilon_i\right)\\ \le-\sum_{i=1}^N\|\tilde{W}_{ci}\|(\sigma_i\|\tilde{W}_{ci}\|-\varepsilon_{\upsilon i}),\ i=1,2\end{aligned} \quad (27)$$

Then according to Lyapunov Theorem, $\tilde{W}_{ci}$ converges to the compact set Ω_i: $\left\{\tilde{W}_{ci} \middle| \left\|\tilde{W}_{ci}\right\| \leq \varepsilon_{vi}/\sigma_i\right\}$.

3.2 Stability Analysis

Theorem 2 *For nonlinear three-player game system* (1), *under the performance index* (2) *and* (3), *with the single-layer ADP NN* (11), *the approximate optimal control policies* (13) *and* (14) *converge to a small neighborhood surrounding their ideal values, and the system states can be stabilized to zero.*

Proof The errors between the approximate controls and ideal controls can be concluded as

$$u_i - u_i^* = \frac{1}{2} R_{ii}^{-1} g_i^T(x) \nabla \varphi_{ci}^T (W_{ci} - \hat{W}_{ci}) = \frac{1}{2} R_{ii}^{-1} g_i^T(x) \nabla \varphi_{ci}^T \tilde{W}_{ci} \leq \frac{1}{2} \lambda_{\min}^{-1}(R_{ii}) b_{gi} b_{\varphi i} \left\| \tilde{W}_{ci} \right\| \tag{28}$$

$$\begin{aligned} d - d^* &= \frac{1}{2\gamma^2} [k(x)]^T \nabla \phi_{c2}^T \hat{W}_{c2} - \frac{1}{2\gamma^2} [k(x)]^T \nabla \phi_{c2}^T W_{c2} \\ &= -\frac{1}{2\gamma^2} [k(x)]^T \nabla \phi_{c2}^T \tilde{W}_{c2} \leq \frac{1}{2\gamma^2} b_k b_{\phi 2} \left\| \tilde{W}_{c2} \right\| \end{aligned} \tag{29}$$

From (28) and (29), Theorem 2 holds. This completes the proof.

4 Example

In this section, we use a simulation to illustrate the effectiveness of the proposed algorithm. The nonlinear three-player model [15] is improved as:

$$\dot{x} = f(x) + \textstyle\sum_{i=1}^{2} g_i(x) u_i + k(x) d,\ x \in \mathbb{R}^2 \tag{30}$$

where

$$f(x) = \begin{bmatrix} x_2 \\ -x_2 - 0.5x_1 - 0.25x_2(\cos(2x_1)+2)^2 + 0.25x_2(\sin(4x_1)+2)^2 \end{bmatrix}, g_1(x) = \begin{bmatrix} 0 \\ \cos(2x_1)+2 \end{bmatrix},$$
$$g_2(x) = \begin{bmatrix} 0 \\ \sin(4x_1^2)+2 \end{bmatrix},\ k(x) = \begin{bmatrix} 0 \\ \sin(4x_1)+2 \end{bmatrix},\ \text{with } \gamma = 8$$

To obtain the Nash equilibrium of the nonlinear system, the simulation parameters are set as $Q_1 = 2I$, $Q_2 = I$, $R_{11} = R_{12} = 2I$ and $R_{21} = R_{22} = I$. The initial

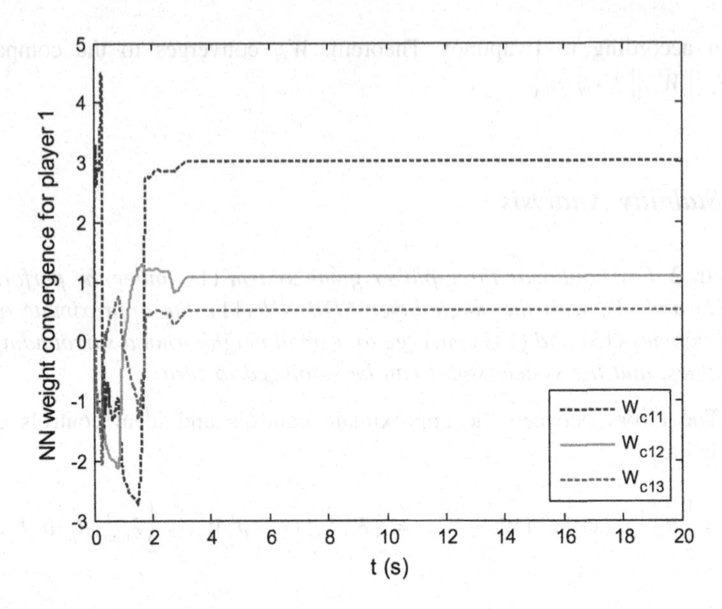

Fig. 1 Critic NN weights $\hat{W}_{c1}$

states are given as $x_1(0) = 3$, $x_2(0) = -1$, and initial critic weights are $\hat{W}_{c1}(0) = \hat{W}_{c2}(0) = 0$. Figures 1 and 2 show the convergence of the critic NN 1 and critic NN 2, which indicate that the mixed-nonzero-sum game get to Nash

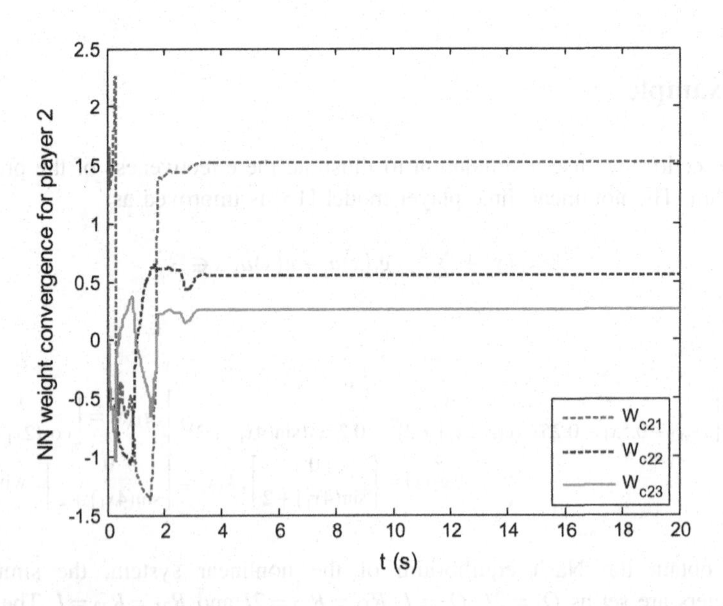

Fig. 2 Critic NN weights $\hat{W}_{c2}$

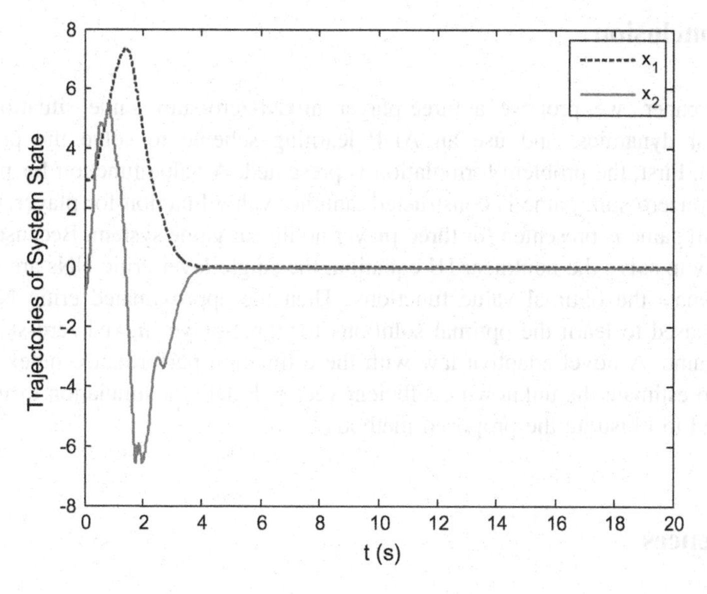

Fig. 3 Trajectories of the system states

Equilibrium for every player. Figure 3 presents the trajectories of the system states, which converge to zero with the optimal control inputs. Every player's control inputs are given as in Fig. 4. The simulation results indicate that three players in the game system get a Nash equilibrium under the given performance index.

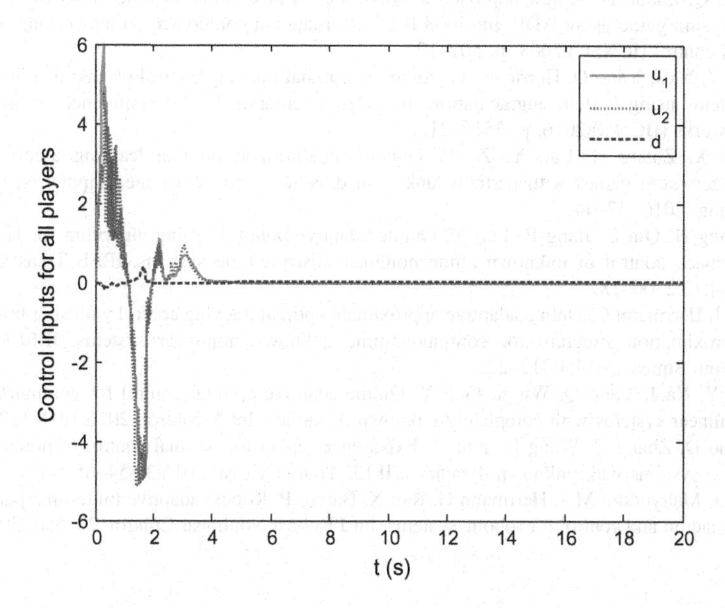

Fig. 4 Control inputs of every player

5 Conclusion

In this paper, we propose a three-player mixed-zero-sum game situation with nonlinear dynamics, and use an ADP learning scheme to solve the proposed problem. First, the problem formulation is presented. A value function for player 1 and 2 nonzero-sum game is constructed, another value function for player 1 and 3 zero-sum game is presented for three-player nonlinear game system. Because of the difficulty to solve the nonlinear HJ equation, the single-layer critic NNs are used to approximate the optimal value functions. Then the approximated critic NNs are directly used to learn the optimal solutions for three-player mixed-zero-sum nonlinear game. A novel adaptive law with the estimation performance index is proposed to estimate the unknown coefficient vector. Finally, a simulation example is proposed to illustrate the proposed methods.

References

1. Morris P. Introduction to game theory. Springer Science & Business Media; 2012.
2. Starr AW, Ho Y-C. Nonzero-sum differential games. J Optim Theory Appl. 1969;3:184–206.
3. Abou-Kandil H, Freiling G, Jank G. Necessary conditions for constant solutions of coupled Riccati equations in Nash games. Syst Control Lett. 1993;4:295–306.
4. V. M. Shah, Power control for wireless data services based on utility and pricing, 1998.
5. Liu D, Li H, Wang D. Online synchronous approximate optimal learning algorithm for multi-player non-zero-sum games with unknown dynamics. IEEE Trans Syst Man Cybern Syst. 2014;8:1015–27.
6. Wei Q, Zhang H. A new approach to solve a class of continuous-time nonlinear quadratic zero-sum game using ADP. In: 2008 IEEE international conference on networking, sensing and control (ICNSC);2008, p. 507–12.
7. Lv Y, Na J, Yang Q, Herrmann G. Adaptive optimal tracking control of unknown nonlinear systems using system augmentation. In: 2016 International joint conference on in neural networks (IJCNN);2016, p. 3516–21.
8. Cui X, Zhang H, Luo Y, Zu P. Online finite-horizon optimal learning algorithm for nonzero-sum games with partially unknown dynamics and constrained inputs. Neurocomputing. 2016; 37–44.
9. Zhang H, Qin C, Jiang B, Luo Y. Online adaptive policy learning algorithm for H∞ state feedback control of unknown affine nonlinear discrete-time systems. IEEE Trans Cybern. 2014;12:2706–18.
10. Na J, Herrmann G. Online adaptive approximate optimal tracking control with simplified dual approximation structure for continuous-time unknown nonlinear systems. IEEE/CAA J Autom Sinica. 2014;4:412–22.
11. Lv Y, Na J, Yang Q, Wu X, Guo Y. Online adaptive optimal control for continuous-time nonlinear systems with completely unknown dynamics. Int J Control. 2016;1:99–112.
12. Zhao D, Zhang Q, Wang D, Zhu Y. Experience replay for optimal control of nonzero-sum game systems with unknown dynamics. IEEE Trans Cybern. 2016;3:854–65.
13. Na J, Mahyuddin MN, Herrmann G, Ren X, Barber P. Robust adaptive finite-time parameter estimation and control for robotic systems. Int J Robust Nonlinear Control. 2015;16:3045–71.

14. Ioannou PA, Sun J. Robust adaptive control. Courier Corporation;2012.
15. Vamvoudakis KG, Lewis FL. Online solution of nonlinear two-player zero-sum games using synchronous policy iteration. Int J Robust Nonlinear Control. 2012;13:1460–83.

An Evolution Perception Shape Creation Mechanism for 3D Shapes

Lingling Zi, Xin Cong, Yanfei Peng and Pei Yang

Abstract Making use of existing shapes to create creative shapes is a challenging problem in the field of 3D modeling. To resolve this problem, we change the problem of shape creation change into the shape evolution problem and an evolution perception 3D shape creation mechanism (EPSCM) is proposed. The core idea of EPSCM is fittest survive and genetic diversity. On the one hand, we present the shape evolutionary method based on the shape components, including crossover operation, the variant operation and the phagolysis operation, which could evolve diversity shape individuals under the condition of preserving shape functions. on the other hand, we design the evolution multiplication strategies, including structural constrains and fitness selection scheme, so as to further ensure the diversity and adaptability of EPSCM. Experimental results show that EPSCM could obtain novel and creative 3D shapes under the condition limited 3D shapes.

Keywords Evolution perception · Shape creation · Shape components · Creative exploration

---

L. Zi · X. Cong (✉) · Y. Peng
School of Electronic and Information Engineering, Liaoning Technical University, Huludao 125105, Liaoning, China
e-mail: chongzi610@163.com

L. Zi
e-mail: lingling19812004@126.com

Y. Peng
e-mail: pengyf75@126.com

P. Yang
College of Science, Liaoning Technical University, Fuxin 123000, Liaoning, China
e-mail: 55298571@qq.com

Y. Jia et al. (eds.), *Proceedings of 2017 Chinese Intelligent Systems Conference*, Lecture Notes in Electrical Engineering 459,
https://doi.org/10.1007/978-981-10-6496-8_34

1 Introduction

3D shape creation is a hot issue in 3D modeling and many modeling tools have been developed [1]. However, these tools could only provide a two-dimensional display of 3D shapes and realistic and available shapes are not quickly obtained from the digital design concepts of non-professional users. So, it is nature to propose a convenient shape creation method, which could not only greatly reduce the workload of designing new shapes, but also enable non-professional users to quickly obtain professional shapes. Therefore, it is a challenging problem to study how to use existing 3D shapes to create new shapes that meet users' requirements.

Component-driven shape creation method is one of solutions and it adopts the relationships of the components of the existing shapes to create new shapes. This method could be mainly divided into two categories. One is shape creation based on the structure features of the shape components. For example, shape variations exploration method by 3D-model decomposition and part-based recombination [2], topology-varying 3D shape creation method via structural blending [3], and shape creation based on topology structure optimization [4]. These above methods could create 3D shapes with the potential topology changes of shape components. But due to the information of insufficient part semantic, the obtained shape variations are limited. The other is shape creation based on the semantic features of the shape components. For example, the semantic decomposition and reconstruction of 3D shapes [5], the shape creation by using learning part-based templates [6], shape synthesis by using the probabilistic model [7], shape creation by style transfer [8], and shape editing by deformation handles [9]. Semantic features of the shape components in the above methods are computed by using machine learning techniques. So, large training data are used to improve the accuracy of semantic in-formation, and the obtained semantic information are used to guide the creation of 3D shapes.

Through the analysis of the above methods, it is found that the existing methods mainly depend on users' imagination to construct the 3D shapes and lack the excitation of the user's creativity. So, it is necessary to propose an intelligent modeling method to assist in exploring creative shapes for non-professional users. Evolutionary theory gives us a good inspiration and its goal is to implement organic evolution by through changing the genetic. Biological genes, individuals and populations are important factors in the evolutionary theory [10, 11]. In this paper, we introduce the evolutionary theory into the shape modeling. Specifically, 3D shape corresponds to the biological individual, shape component corresponds to the biological gene, and shape class set corresponds to the biological population. The shape creation is described the process of shape evolution. Firstly, the shapes with the same semantic class are input the initial set, denoted as the initial shape population. Shape individuals in the initial shape population are processed by using the function recognition and shape segmentation to obtain shape genes. By adopting the evolution operation, shape genes are changed and diverse shape individuals are generated. Based on the obtained shape individuals, the evolution multiplication strategies are designed to ensure the diversity and adaptability of shape population. This evolution process is performed repetitively until the user' requirements are met.

2 Evolution Perception Shape Creation Mechanism

In this paper, we change the problem of shape creation to the mapping problem between 3D shapes and functional semantics. Through using the correspondence between them, a new shape could be generated, shown as following:

$$D(S) \xrightarrow{M(S)} D(F) \tag{1}$$

In which $D(S)$ denotes the three-dimensional shape domain, $D(F)$ denotes the functional semantic domain, and $M(S)$ denotes the mapping relationship between the shape domain and the functional semantic domain. $M(S)$ does not exist independently in the process of shape creation and it is a link between $D(S)$ and $D(F)$. So, the basic rules of EPSCM are shown as follows:

Rule 1: Shape creation is the process of seeking a valid combination of function for 3D shapes.
Rule 2: It is necessary to find shape structure as reasonable as possible under the condition of satisfying functional semantics.
In order to implement EPSCM, the following definitions are shown as follows:
Shape gene: A shape gene is the basic unit of EPSCM, denoted as $G(F_G, S_G)$. Where F_G represents the function semantic and S_G represents the shape component.
Shape individual: A shape individual is a combination of shape genes, denoted as $P(G, L, M)$. Where G represents the contained shape genes, L represents the relationship between the different shape genes, and M represents the extent of fitness in the process of shape evolution.
Shape population: A shape population is a set containing shape individuals, denoted as $A(P, F)$. Where P represents the contained shape individuals and F represents the common function semantics.

The framework of EPSCM are shown in Fig. 1. Firstly, we put 3D shapes with the same functional semantic into the initial shape population, and determine the shape genes in the shape population. Secondly, new shape individuals are generated by proposing the evolutionary operations, including gene crossover, gene variant and gene phagocytosis. Finally, the shape population multiplication strategies are presented, including structural constraints and fitness selection. If the shape population reaches the evolutionary expectation, the process of shape evolution ends. Otherwise, shape evolution could be performed again until the number of acceptable shape population is expected.

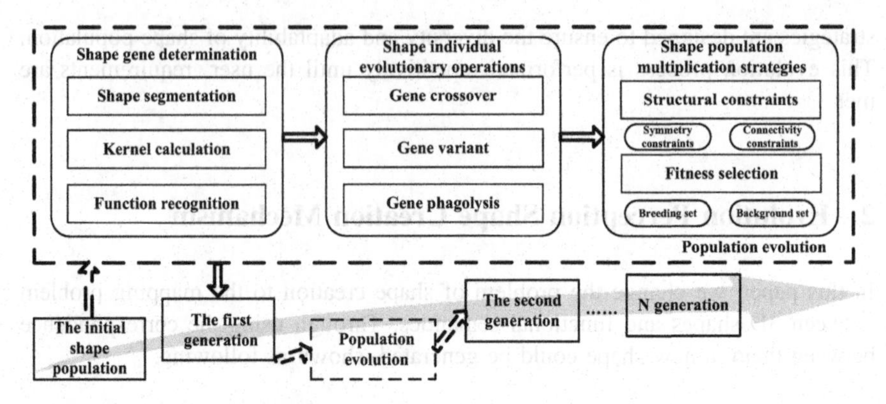

Fig. 1 The framework of EPSCM

3 The Details of EPSCM

In this section, we will elaborate shape gene determination, shape individual evolutionary operations and shape population multiplication strategies in EPSCM.

3.1 Shape Gene Determination

We segment the shape individuals in the shape population by using shape segmentation based on the approximation convexity decomposition and obtain the shape of each gene, i.e. S_G. On this basis, we qualify context relations between shape genes to calculate kernel function.

On this basis, we qualify context relations between shape genes to calculate kernel function Ψ. Given graph R (V, E) denotes the relations of shape genes in shape individuals, V denotes the nodes of shape genes, and E denotes the different type relations among the nodes. Ψ is shown as follows.

$$\psi^{l}(R_p, R_q, p_A, p_B) = \begin{cases} \psi_N(p_A, p_B) \times \sum\limits_{\substack{p_s \in AN_{Gp}(p_A) \\ p_{s'} \in AN_{Gq}(p_B)}} \psi_E(e, e')\, \psi^{l-1}(R_p, R_q, p_s, p_{s'}),\ l > 0 \\ \psi_N(p_A, p_B),\ l = 0 \end{cases} \tag{2}$$

where R_p and R_q are the relation graph for shape gene G_p and G_q, respectively. l denotes the path length of relation graph, $p_A \in G_p$, $p_B \in G_q$, $AN(x)$ denotes all adjacent nodes for x in R, e and e' are the edges which connect between p_A and p_s, p_B and $p_{s'}$. Ψ_N denotes the node kernel reflecting the similarity geometric characteristics of the two shape genes. Ψ_E denotes the edge kernel reflecting the context of different nodes.

According to the obtained kernel function, we use SVM to determine the functional semantic for each gene, i.e. F_G. Finally, we acquire shape gene $G(F_G, S_G)$.

3.2 Shape Individual Evolutionary Operations

The nature of shape individual evolutionary is to operate the genes from original shape individuals, including gene crossover, gene variant, and gene phagolysis. And methods are shown as follows.

Gene crossover. Shape genes from the different shape individuals are exchanged to generate a new shape individual P_3. GS_3 is the shape gene set of P_3 is to satisfy the following condition:

$$GS_3 = G_1^{'} \cup G_2^{'} \ s.t.\ G_1^{'} \in GS_1, G_2^{'} \in GS_2, F_1^{'} \neq F_2^{'} \tag{3}$$

Specifically, $GS_1(F_1, S_1)$ and $GS_2(F_2, S_2)$ are the gene set, in which genes are selected from different shape individuals. And S_1 and S_2 are aligned by using anisotropic scaling based on the bounding box. Through functional analysis, new individuals containing the different shape genes are generate.

Gene variant. The deformation space of shapes should be constructed to complete the task of gene variant. Specifically, we perform PCA method for shape genes, obtain the deformation space of shapes and calculate the deformation vectors. Then we map all the deformation vectors and compete the maximum and minimum values of each dimension to form a bounding box that allows deformation. Note that optimum structure should be executed to ensure the structure continuity of generated shape individuals.

Gene phagolysis. The operation of gene phagolysis should be performed under the premise of preservation of functional semantic, which contains gene splitting and gene merging, shown as follows.

$$G_1 \Rightarrow \{G_1^{'}, G_2^{'}, G_3^{'}, \ldots G_m^{'}\} \ s.t.\ F_{G1} = F_{G1}^{'} = F_{G2}^{'} = \ldots = F_{Gm}^{'} \tag{4}$$

$$\{G_1, G_2, G_3, ..G\}_n \Rightarrow G^{'} \ s.t.\ F_{G1} = F_{G2} = F_{G3} = \ldots = F_G^{'} \tag{5}$$

The process of gene splitting: firstly, we activate G_1 and find the other shape genes that have the adjacent relationship with G_1, denoted as G_2 and G_3. Then a serious of shape genes $G_i^{'}(i = 1 \ldots m)$ are interpolated and they are the varieties of G_2 or the varieties of G_3. Finally, new individuals are reconstructed by combining generated shape genes with original shape genes. Gene merging is the inverse operation of gene splitting and it synthesizes the shape genes which have both the same functional semantic and geometric symmetry.

3.3 Shape Population Multiplication Strategies

The shape individuals in the same shape population are divided into two shape sets: the breeding set R and the background set B. The former is to guarantee high quality evolution and the latter is to retain the diversity of the shape population. Specifically, starting from the initial shape population A_0, the shape individuals in A_0 are randomly put into R and B. From A_i to A_{i+1}, shape individuals in R perform the shape evolutionary operations to obtain offspring and they are checked by structural constraints. The passed shape individuals are set into the R. At the same time, the fitness selection method is used to select good-quality shape individuals. The selected shape individuals are stayed R and the eliminated shape individuals are put into B. This process is performed repeatedly until the number of acceptable shape individuals reaches the predefined value.

Structural constraints. We design structural constraints to eliminate the shape individuals which have the unstable geometry structure. Structural constraints contain symmetry constraints and connectivity constraints.

(1) Symmetry constraints. If the shape structural of paternal individuals have symmetry relationship, including translation symmetry, rotational symmetry and reflection symmetry, the shape structural of the offspring also have symmetry relationships. In the process of testing, the symmetry axis of shapes could be fixed based on the center of gravity of the paternal individuals, and then the Hausdorff distance is used to calculate the deviation values.
(2) Connectivity constraints. The created shape individuals need to maintain the physical connectivity of the parent individuals. In the process of testing, we determine association relationships between shape genes of shape individuals firstly. Then we record all the positions of shape genes in the generated individuals. Finally, we compute the contact distance between the corresponding shape genes from shape individuals. If it is larger than the distance threshold, it is considered that it does not meet the connectivity constraints and needs to be eliminated.

Fitness selection. To reflect user preferences for shapes, we record the fitness score M for each shape individual in R. Specifically, if the generated shape individual is selected by users, M is 1. If the values of M in the successive generations, it is unsuitable to breed shapes and the corresponding shape individual is input it into B.

4 Experiments Results and Discussion

We used the experimental data from Princeton University database [12] and the COSEG dataset [13], including plane, ant, chair, goblet, candlestick and bird. For each shape population, Table 1 shows some shape individuals, the input the number of shape individuals and functional semantic of each shape gene.

Table 1 Experimental data

Shape population	Examples of shape individuals	Input the number of shape individuals	Functional semantic for each shape gene
Plane		11	Body, wing, tail.
Ant		10	Tentacles, head, body, legs, tail.
Chair		18	Chair legs, back, seat.
Goblet		20	Cup body, stalk, seat.
Candlestick		24	Container, handle, flame, candle.
Bird		25	Body, tail, wings.

Figure 2 shows the percentage of valid shapes of the six shape populations. The left figure shows the number of the generated shape individuals for each generation of shape population. From this subfigure, we find that as the number of evolution generations increases, the percentage of valid shape individuals ascends. The right figure shows the comparisons of average percentage of valid shapes for the different shape population using the proposed method and the fit method. Using the fit method, controllability is defined by the selection process and the fitness function determines the generation of offsprings [14]. We find that our results are higher than the comparison results.

We invited twenty non-professional users to choose their favorite shapes. Figure 3 shows the percentage of liked shapes for the six shape populations. Form the left figure, we find that as the number of evolution generations increases, the number of shapes the users liked is increasing. The right figure shows the average percentage of liked shapes using EPSCM and Fit. It can be observed that the values obtained using the EPSCM method are higher than those using the Fit method.

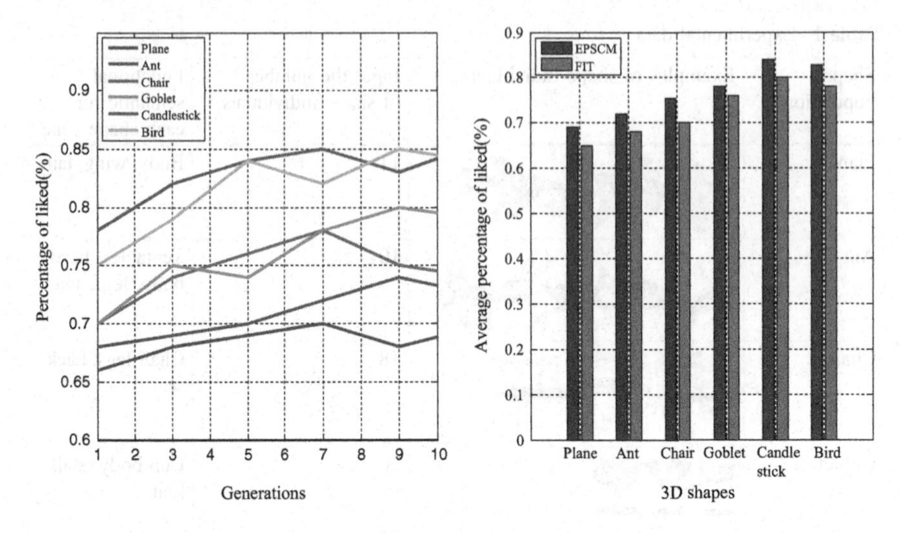

Fig. 2 The percentage of valid shapes

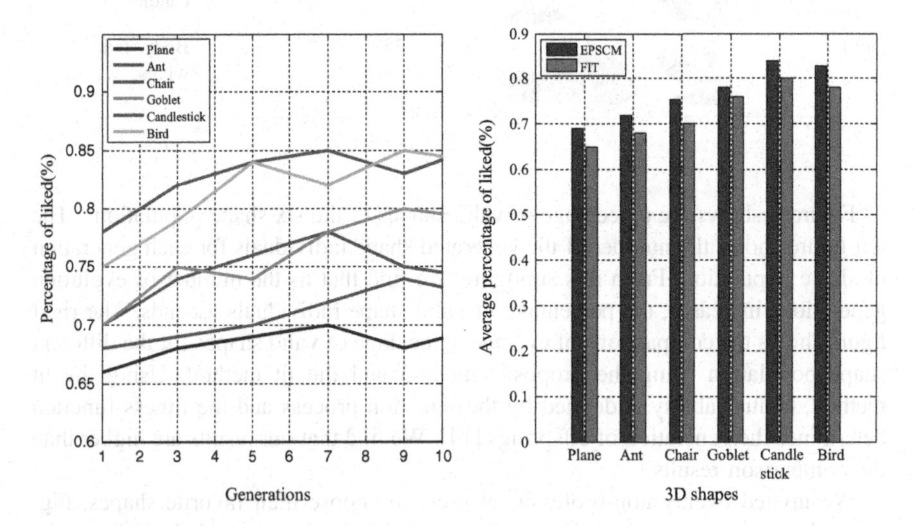

Fig. 3 The percentage of liked shapes

5 Conclusions

In this paper, we propose an evolution perception shape creation mechanism for 3D shapes to obtain novel shapes under the limited resource condition. The contributions of EPSCM are shown as follows. (1) The framework of EPSCM is proposed by using evolutionary theory and it solves the problem of shape creation from a different perspective. (2) The shape individual evolutionary operations based on shape genes are presented to evolve high quality shape offspring, which could preserve the intrinsic structure and functions of the origin shapes. (3) The shape population multiplication strategies are proposed, including structural constraints and fitness selection. On the one hand, it is suitable for users' preference, and on the other hand, it could assure the rationality of shape and the diversity of created shapes. The experimental results show that the proposed method has obtain a lot of diverse and novel 3D shapes.

Acknowledgements This work is supported by the National Natural Science Foundation of China (61702241, 61602227); The Foundation of the Education Department of Liaoning Province (L2015225, LJYL019) and the Doctoral Starting up Foundation of Science Project of Liaoning Province (201601365).

References

1. Nguyen DT, Hua BS, Tran MK, et al. A field model for repairing 3D shapes. In: IEEE conference on computer vision and pattern recognition. IEEE; 2016, p. 5676–84.
2. Jain A, Thormählen T, Ritschel T, et al. Exploring shape variations by 3D-model decomposition and part-based recombination. Comput Graph Forum. 2012;31:631–40.
3. Alhashim I, Li H, Xu K, et al. Topology-varying 3D shape creation via structural blending. ACM Trans Graph. 2014; 33(4):Article 158.
4. Christiansen A, Bærentzen J, Morten N, et al. Combined shape and topology optimization of 3D structures. Comput Graph. 2015;46:25–35.
5. Lin H, Gao J, Zhou Y, et al. Semantic decomposition and reconstruction of residential scenes from LiDAR data. ACM Trans Graph. 2013; 32(4):Article 66.
6. Kim VG, Li W, Mitra NJ, et al. Learning part-based templates from large collections of 3D shapes. ACM Trans Graph. 2013; 32(4):Article 70.
7. Kalogerakis E, Chaudhuri S, Koller D, et al. A probabilistic model for component-based shape synthesis. ACM Trans Graph. 2012; 31(4):Article 55.
8. Han Z, Liu Z, Han J, et al. 3D shape creation by style transfer. Vis Comput. 2015;31(9):1147–61.
9. Yumer ME, Chaudhuri S, Hodgins JK, et al. Semantic shape editing using deformation handles. ACM Trans Graph. 2015; 34(4):Article 86.
10. Wang R, Pujos C. Emergence of diversity in a biological evolution model. J Phys Conf Ser. IOP. 2015; 604(1):012019.
11. Jeddi I, Saiz L. Structure prediction and 3D modeling of single stranded DNA from sequence for aptamer-based biosensors. Biophys J. 2016;110(3):333a.
12. Chen X, Golovinskiy A, Funkhouser T. A benchmark for 3D mesh segmentation. ACM Trans Graph. 2009;28(3):341–52.
13. COS. The shape coseg dataset; 2012. http://web.siat.ac.cn/yunhai/ssl/ssd.htm.
14. Cohen-Or D, Zhang H. From inspired modeling to creative modelling. Vis Comput. 2016;32 (1):7–14.

Fixed-Time Consensue-Based Scheme for Economic Dispatch of Smart Grid

Peng Lei, Zhi-Wei Liu, Ming Chi, Ding-Xin He and Zhi-Hong Guan

Abstract In this paper, we introduce a new economic dispatch strategy to solve the energy management problem in a smart grid which is a distributed and cooperative method. Different from the existing works, this paper aims to achieve the optimization in a settling time. The balance between supply and demand constraint can be kept during the computing time. Then, it is demonstrated that the total cost can reach it's minimal value in the settling time with initial values satisfied the balance constraint. Finally, we verify the effectiveness of the theoretical result with a numerical simulation.

Keywords Distributed control · Consensus-based strategy · Fixed time · Economic dispatch

1 Introduction

Economic dispatch problem (EDP) is a fundamental problem in smart grids. The key to solve the EDP problem is reaching the minimum total generation cost under some practical constrains, that is meeting the power demand at the same time. Great efforts have been paid to study the EDP problem in traditional power grids by centralized optimization algorithms [1–3]. In order to operate the electric power system more economically, Attaviriyanupap proposed a hybrid method by using evolutionary programming and sequential quadratic programming [1]. The algorithm proposed in [2] solved the EDP by using a particle swarm optimization method. Different from the

This work was partially supported by the National Natural Science Foundation of China under Grants 61673303, 61572208, 61572210, and 61672244.

P. Lei · Z.-W. Liu · M. Chi (✉) · D.-X. He (✉) · Z.-H. Guan
College of Automation, Huazhong University of Science and Technology, Wuhan 430074, People's Republic of China
e-mail: chiming@hust.edu.cn

D.-X. He
e-mail: hedingxin@hust.edu.cn

Y. Jia et al. (eds.), *Proceedings of 2017 Chinese Intelligent Systems Conference*, Lecture Notes in Electrical Engineering 459,
https://doi.org/10.1007/978-981-10-6496-8_35

work in [2], the EDP with nonsmooth cost functions is proposed the particle swarm optimization method in [3], which is also solved by however, all above mentioned method is centralized algorithms, and the centralized algorithms may not suitable for smart grid today due to the following reasons. First, the centralized algorithms require global information, they will lose efficacy because of a small change and collecting the unit's detailed information may cause high cost. Then, most of energy in the system is provided by the distribute renewable energy sources, the traditional centralized algorithms are unable to meet the requirements. Due to the development of smart grid technology, many researchers have turn their attention from centralized manner to distributed manner [4–8], which enjoy the advantages of less information requirement, robustness and scalability. A distributed algorithm that converge to the global optimum without requiring a central controller is proposed in [5]. In [6], a consensus based algorithm is proposed, the algorithm enables generators to collectively learn the mismatch between demand and total amount of power generation. In [7], a distributed model predictive control is presented, which formulated and solved the combination of environmental and economic dispatch problem.

All of the above works have deeply promoted the development of smart grid, but most of the strategies are linear, the convergence speed are slow and not controllable, it is significant to propose a nonliner strategy to solve the EDP in finite time. The finite time stabilization is widely applied in many areas, the most typical is finite time consensus problem. The control object of finite-time consensus is to make the considered multi-agent system converge to an invariant manifold in finite time. In [9], it gives a detail explanation of finite time stabilization and propose time-invariant finite-time control law for the double integrator. Based on the finite time stabilization and multi-agent consensus theory, the finite-time consensus problem of multi-agent system with second-order dynamics is studied in [10]. By using the finite-time stabilization theory, the system can converge to an invariant manifold in settling time, but the time is related to the initial conditional. In [11], the fixed-time stabilization is first given which means that the settling time is independent of initial conditions. In [12], it use the fixed-time stabilization to solve the consensus tracking problem for second-order multi-agent networks. On the other hand, the distributed control method have attractive many research in multi-agent control area [13–16].

Motivated by the above works, we aim to solve the smart grid economic dispatch problem in finite time. Different from the existing works, we give the settling time when the system converge to the optimality regardless of the initial condition. Compared with the existing work, the speed of convergence in this paper is much faster.

The rest of this paper is organized as follows. In Sect. 2, preliminaries on economic dispatch problems, algebraic graph theory and some basic consensus result including some useful lemmas are provided. In Sect. 3, the main theoretical results are derived. In Sect. 4, some numerical simulations are performed to illustrate effectiveness of the proposed protocol. Finally, the conclusion is shown in Sect. 5.

2 Problem Formulation

In this section, some preliminaries on economic dispatch problems and algebraic graph theory are introduced.

The considered smart grid is supposed to be composed of n generating units. $C_i(P_{G_i})$ is the cost function of generating units i $(1 \leq i \leq n)$,

$$C_i(P_{G_i}) = \alpha_i + \beta_i P_{G_i} + \gamma_i P_{G_i}^2, \tag{1}$$

where P_{G_i} is the output of generating units i, α_i, β_i and γ_i are positive constants. The output of generating units should meet the total demand P_D in the smart grid, i.e., $\sum_{i=1}^{n} P_{G_i} = P_D$, while minimizing the total cost of the generating units $\sum_{i=1}^{n} C_i(P_{G_i})$. Therefore, the economic dispatch problem is to design $P_{G_i}(t)$ such that

$$\begin{aligned} &\min \sum_{i=1}^{n} C_i(P_{G_i}) \\ &\sum_{i=1}^{n} P_{G_i} = P_D. \end{aligned} \tag{2}$$

The generating units in the smart grid is connected by a communication network, which can be described by an undirected graph $\mathcal{G} = (v, \varepsilon, A)$, where $v = \{v_1, v_2, \ldots, v_n\}$ denote the generating unit set, $\varepsilon \subseteq v \times v$ is the set of the communication links between nodes, $A = [a_{ij}]_{n \times n}$ is the weight adjacency matrix. Denote the Laplacian graph of $\mathcal{G}(A)$ by $L(A) = [l_{ij}] \in \mathbb{R}^{n \times n}$, and

$$l_{ij} = \begin{cases} \sum_{k=1,k \neq i}^{n} a_{ik}, j = i; \\ -a_{ij}, \quad\quad j \neq i. \end{cases}$$

We have that 0 is an eigenvalue of $L(A)$ and 1_n is the associated eigenvector. If $\mathcal{G}(A)$ is undirected and connected, then $L(A)$ has the following properties:

(1) For any $x \in \mathbb{R}^n$, we have $x^T L(A) x = \frac{1}{2} \sum_{i=1}^{n} \sum_{j \in N_i} a_{ij} (x_j - x_i)^2$;

(2) The second smallest eigenvalue of $L(A)$, which is called the algebraic connectivity of $\mathcal{G}(A)$ and denoted by $\lambda_2(L(A))$, is larger than zero;

(3) If $1_n^T x = 0$ where 1_n is the unit column vector, then $x^T L(A) x \geq \lambda_2(L(A)) x^T x$.

The graph $\mathcal{G}(A)$ is said to be connected if there exists communication between any two generating units. When generating unit i can obtain information from generating unit j, the generating unit j is called the neighbor of generating unit i. Denote the neighbor set of generating unit i in $\mathcal{G}(A)$ as $\mathcal{N}_i = \{j \in v | a_{ij} > 0, j \neq i\}$.

The objective of this paper is to design a distributed approach to find the solution of economic dispatch problem (2) in a settling time. Let $x_i(t) = \beta_i + 2\gamma_i P_{G_i}(t)$, $i = 1, 2, \ldots, n$, which denotes the incremental cost of generating units i. To solve the economic dispatch problem, the following lemma will be used.

Lemma 1 *[17]* $\{p_1^*, p_2^*, \ldots, p_n^*\}$ *is the solution of economic dispatch problem (2) if the following two conditions are simultaneously satisfied:* $x_i(t) = x^*$, x^* *is a constant;* $\sum_{i=1}^{n} P_{G_i} = P_D$.

Lemma 2 *[18] For any* $\xi_1, \xi_2, \ldots, \xi_n \geq 0$,

$$\sum_{i=1}^{n} \xi_i^p \geq \left(\sum_{i=1}^{n} \xi_i \right)^p, \qquad 0 < p \leq 1;$$

$$\sum_{i=1}^{n} \xi_i^p \geq n^{1-p} \left(\sum_{i=1}^{n} \xi_i \right)^p, \qquad 1 < p \leq \infty.$$

Lemma 3 *[19] Assume there exists a scalar system*

$$\dot{y} = -\alpha y^{\frac{m}{r}} - \beta y^{\frac{p}{q}},$$

with $\alpha > 0, \beta > 0, m > r, p < q$ *and* m, r, p, q *are both positive odd integers, then for any initial state* $y(0)$*, the system is fixed-time stable with the settling time is uniformly bounded by*

$$T_{\max} < \frac{1}{\alpha} \frac{r}{m-r} + \frac{1}{\beta} \frac{q}{q-p}.$$

3 Main Result

According to Lemma 1, the key of the economic dispatch problem is to find a constant x^* which satisfied $\{x_1, x_2, \ldots, x_n = x^*\}$ while keeping $\sum_{i=1}^{n} P_{G_i}(t) = P_D$ at the same time. We design a distributed consensus-based algorithm for the economic dispatch. We will proof that the increment cost x_i will converge to the constant x^* in settling time by employing the following algorithm:

$$\begin{aligned} \dot{x}_i(t) = & 2\gamma_i \left(\alpha \sum_{j \in N_i} a_{ij} \operatorname{sign}(x_j - x_i) \left| x_j - x_i \right|^{\frac{m}{r}} \right) \\ & + 2\gamma_i \left(\beta \sum_{j \in N_i} a_{ij} \operatorname{sign}(x_j - x_i) \left| x_j - x_i \right|^{\frac{p}{q}} \right), \end{aligned} \tag{3}$$

where $\alpha > 0, \beta > 0$ and m, r, p, q are positive odd integers which are satisfied that $m > r, p < q$, $\operatorname{sign}(\cdot)$ denotes the symbol function. Let $B_1 = [a_{ij}^{\frac{2r}{m+r}}] \in \mathbb{R}^{n \times n}$ and $B_2 = [a_{ij}^{\frac{2q}{p+q}}] \in \mathbb{R}^{n \times n}$, denotes the Laplacian matrixes of undirected graph $\mathcal{G}(B_1)$ and

$\mathcal{G}(\mathrm{B}_2)$ as $L_{\omega 1}$ and $L_{\omega 2}$, denotes the second smallest eigenvalue of the Laplacian matrixes as λ_2, define $T_{\max}$ as follow

$$T_{\max} = \frac{r}{2^{\frac{r-m}{2r}} \alpha n^{\frac{r-m}{r}} \lambda_2(L_{\omega 1}) \gamma_{\min}(m-r)} + \frac{q}{2^{\frac{p-q}{2q}} \beta \gamma_{\min} \lambda_2(L_{\omega 2})^{\frac{p+q}{2q}} \lambda_2(L_{\omega 1})^{\frac{q-p}{2q}} (q-p)}.$$

Theorem 1 *If the initial condition satisfied* $\sum_{i=1}^{n} P_{G_i}(0) = P_D$, *then the optimal economic dispatch problem (2) will be solved under the designed protocol (3), the convergence time t satisfies* t $\leq T_{\max}$.

Proof Let $P(t) = \sum_{i=1}^{n} P_{G_i}(t)$, then $\dot{P}(t) = \sum_{i=1}^{n} \dot{P}_{G_i}(t)$. As $\dot{x}_i = 2\gamma_i \dot{P}_{G_i}(t), \gamma_i > 0$, one has

$$\begin{aligned} \dot{P}_{G_i}(t) = & \alpha \sum_{j \in N_i} a_{ij} \mathrm{sign}(x_j - x_i) \left| x_j - x_i \right|^{\frac{m}{r}} \\ & + \beta \sum_{j \in N_i} a_{ij} \mathrm{sign}(x_j - x_i) \left| x_j - x_i \right|^{\frac{p}{q}}. \end{aligned} \tag{4}$$

Sum the derivative of $P_{G_i}(t)$, we can get $\dot{P}(t) = 0$. According to the initial condition, the following result can be obtained: $\sum_{i=1}^{n} P_{G_i}(t) = P_D$. Let $s = \sum_{i=1}^{n} k_i x_i, k_i = \frac{1}{2\gamma_i \gamma_0}, \gamma_0 = \sum_{j=1}^{n} \frac{1}{2\gamma_j}, e_i = x_i - s$, we can get that $\dot{s} = 0$. Considering the following Lyapunov function

$$V = \frac{1}{2} \sum_{i=1}^{n} \frac{1}{\gamma_i} e_i^2.$$

Its derivative can be obtained as

$$\dot{V} = \sum_{i=1}^{n} \frac{1}{\gamma_i} e_i \dot{e}_i = \sum_{i=1}^{n} (x_i - s) \frac{\dot{x}_i}{\gamma_i}.$$

By Lemma 2, one has

$$\begin{aligned} \dot{V} &= -\alpha \sum_{i=1}^{n} \sum_{j \in N_i} \left(a_{ij}^{\frac{2r}{m+r}} (e_i - e_j)^2 \right)^{\frac{m+r}{2r}} - \beta \sum_{i=1}^{n} \sum_{j \in N_i} \left(a_{ij}^{\frac{2q}{p+q}} (e_i - e_j)^2 \right)^{\frac{p+q}{2q}} \\ &\leq -\alpha n^{\frac{r-m}{r}} \left(\sum_{i=1}^{n} \sum_{j \in N_i} a_{ij}^{\frac{2r}{m+r}} (e_i - e_j)^2 \right)^{\frac{m+r}{2r}} - \beta \left(\sum_{i=1}^{n} \sum_{j \in N_i} a_{ij}^{\frac{2q}{p+q}} (e_i - e_j)^2 \right)^{\frac{p+q}{2q}}. \end{aligned}$$

From $\sum_{i=1}^{n}\sum_{j=1}^{n} a_{ij}(x_i - x_j)^2 = 2e^T Le$, one has

$$\sum_{i=1}^{n}\sum_{j\in N_i} a_{ij}^{\frac{2r}{m+r}}(e_i - e_j)^2 = 2e^T L_{\omega 1} e \geq 2\lambda_2(L_{\omega 1})e^T e, \tag{5}$$

$$\sum_{i=1}^{n}\sum_{j\in N_i} a_{ij}^{\frac{2q}{p+q}}(e_i - e_j)^2 = 2e^T L_{\omega 2} e \geq 2\lambda_2(L_{\omega 2})e^T e. \tag{6}$$

Let $\gamma_{\min} = \min\{\gamma_i\}$ for $i = 1, 2, \ldots, n$, combining with (5) and (6),

$$\dot{V} \leq -\alpha n^{\frac{r-m}{r}}(2\lambda_2(L_{\omega 1})\gamma_{min}V)^{\frac{m+r}{2r}} - \beta(2\lambda_2(L_{\omega 2})\gamma_{min}V)^{\frac{p+q}{2q}}.$$

If $V = 0$, because of $V \geq 0$, and $\dot{V} \leq 0$ then $\dot{V} = 0$, or if $V \neq 0$, then let $y = \sqrt{\gamma_{\min}\lambda_2(L_{\omega 1})V}$, we have

$$\begin{aligned}\dot{y} &= \frac{\sqrt{\gamma_{\min}\lambda_2(L_{\omega 1})}}{2\sqrt{V}}\dot{V} \\ &\leq -2^{\frac{r-m}{2r}}\alpha n^{\frac{r-m}{r}}\gamma_{\min}\lambda_2(L_{\omega 1})y^{\frac{m}{r}} - 2^{\frac{p-q}{2q}}\beta\gamma_{\min}\lambda_2(L_{\omega 2})^{\frac{p+q}{2q}}\lambda_2(L_{\omega 1})^{\frac{q-p}{2q}}y^{\frac{p}{q}}.\end{aligned}$$

By Lemma 3, the following result can be obtain

$$\begin{aligned}T_{\max} \leq &\frac{r}{2^{\frac{r-m}{2r}}\alpha n^{\frac{r-m}{r}}\lambda_2(L_{\omega 1})\gamma_{\min}(m-r)} \\ &+ \frac{q}{2^{\frac{p-q}{2q}}\beta\gamma_{\min}\lambda_2(L_{\omega 2})^{\frac{p+q}{2q}}\lambda_2(L_{\omega 1})^{\frac{q-p}{2q}}(q-p)}.\end{aligned} \tag{7}$$

This completes the proof.

4 Simulations

In the section, numerical simulation satisfying the applicability of the distributed economic dispatch strategy is given. Suppose there are 4 generating units in the considered system, the network topology is shown in Fig. 1. The communication weights between each units are assumed to be 1.

The total power demand is 800 MW. Some parameters are: $\beta_i = 2.9, 2.62, 0.24, 1$, $\gamma_i = 0.003, 0.012, 0.016, 0.02$, $\alpha_i = 573, 241, 371, 409$ for i = 1, 2, 3, 4. The initial P_{G_i}

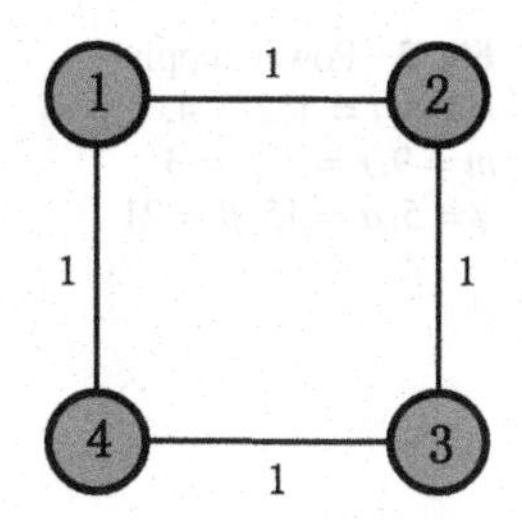

Fig. 1 Network topology of four-unit smart grid system.

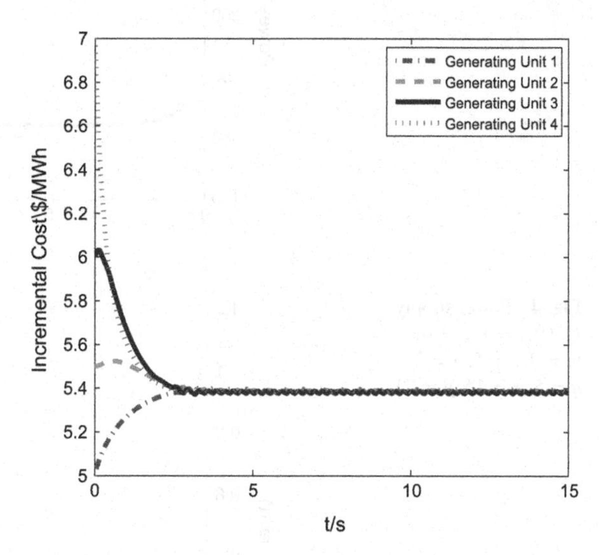

Fig. 2 Incremental cost functions $IC_i(t), i = 1, 2, 3, 4$, $m = 9, r = 5, p = 3$, $q = 5, \alpha = 15, \beta = 21$

is assumed to be $P_{G_i} = 350, 120, 180, 150\,MW$, i = 1, 2, 3, 4. From the simulation, we can get that the optimal incremental cost of the four generating units are obtained as $IC_i = 5.389$/MWh, $i = 1, 2, 3, 4$. By some simple calculation, the output power of the four units is $PG_1 = 415$ MW, $PG_2 = 115.41$ MW, $PG_3 = 160.93$ MW, $PG_4 =$ 109.5 MW. The incremental cost function $IC_i(t)$ is shown in Fig. 2, we can get that the economic dispatch problem is solved in about 3 s. From the Fig. 3, it can be obtained that the total power supply remain unchanged, and the power supply of each unit keeps unchanged after 3 s.

Defined that $err\,(t) = \left(IC_2 + IC_3 + IC_4 - 3IC_1\right)/3$, the system is said to be stability when err converge to zero. The red line in Fig. 4 denotes our result, the blue line denotes the result by using the protocol in [20], it is easily to get that the speed of convergence in our result is more faster.

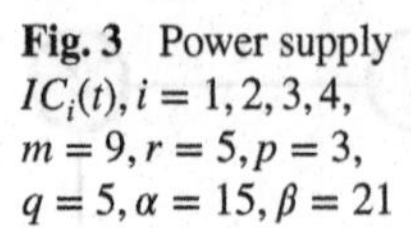

Fig. 3 Power supply $IC_i(t), i = 1, 2, 3, 4,$ $m = 9, r = 5, p = 3,$ $q = 5, \alpha = 15, \beta = 21$

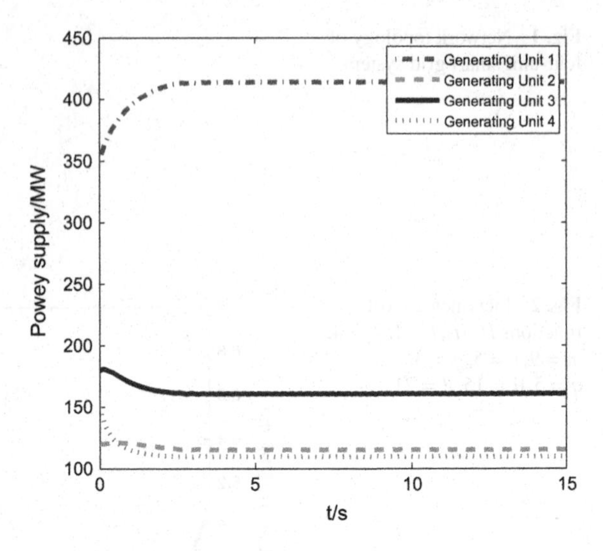

Fig. 4 Power supply $IC_i(t), i = 1, 2, 3, 4,$ $m = 9, r = 5, p = 3,$ $q = 5, \alpha = 15, \beta = 21$

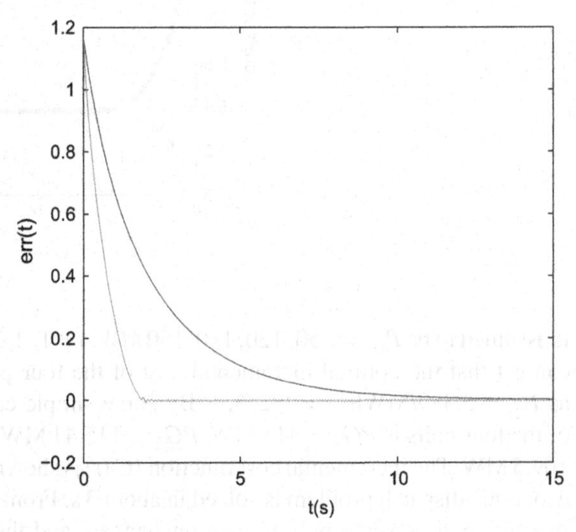

5 Conclusions

In this paper, the economic power dispatch problem in a smart grid has been discussed. If the incremental cost for all generation units can achieve consensus while the supply and demand can keep balance at the same time, the total cost for generators in a smart grid can reach its minimal value. The generating units that working cooperatively under a local neighboring area in this paper have been regarded as agents, thus the economic dispatch problem can be translated into multi-agent system consensus problem. Based on distributed consensus algorithm in multi-agent system, a novel distributed strategy has been proposed to solve the economic dispatch in settling time. It has been shown that the optimal economic dispatch problem can be solved problem in settling time under any initial condition. A simulation example have been shown to support our result.

References

1. Attaviriyanupap P, Kita H, Tanaka E, Hasegawa J. A hybrid EP and SQP for dynamic economic dispatch with nonsmooth fuel cost function. IEEE Trans Power Syst. 2002;17(2):411–6.
2. Gaing ZL. Particle swarm optimization to solving the economic dispatch considering the generator constraints. IEEE Trans Power Syst. 2003;18(3):1187–95.
3. Park JB, Lee KS, Shin JR, et al. A particle swarm optimization for economic dispatch with nonsmooth cost functions. IEEE Trans Power Syst. 2005;20(1):34–42.
4. Park JB, Lee KS, Shin JR, et al. Convergence analysis of the incremental cost consensus algorithm under different communication network topologies in a smart grid. IEEE Trans Power Syst. 2012;27(4):1761–8.
5. Rahbari-Asr N, Ojha U, Zhang Z, et al. Incremental welfare consensus algorithm for cooperative distributed generation/demand response in smart grid. IEEE Trans Power Syst. 2014;5(6):2836–45.
6. Yang SP, Tan SC, Xu JX. Consensus based approach for economic dispatch problem in a smart grid. IEEE Trans Power Syst. 2013;28(4):4416–26.
7. Alejandro J, Arce A, Bordons C. Combined environmental and economic dispatch of smart grids using distributed model predictive control. Int J Electr Power Energy Syst. 2014;54(6):65–76.
8. Tan Z, Yang P, Nehorai A. An optimal and distributed demand response strategy with electric vehicles in the smart grid. IEEE Trans Power Syst. 2014;5(2):861–9.
9. Bhat SP, Bernstein DS. Continuous finite-time stabilization of the translational and rotational double integrators. IEEE Trans Autom Control. 1998;43(5):678–82.
10. Wang X, Hong Y. Finite-time consensus for multi-agent networks with second-order agent dynamics. Proc IFAC World Congr. 2008;41(2):15185–90.
11. Polyakov A. Nonlinear feedback design for fixed-time stabilization of linear control systems. IEEE Trans Autom Control. 2012;57(8):2106–10.
12. Zuo ZY. Nonsingular fixed-time consensus tracking for second-order multi-agent networks. Automatica. 2015;54:305–9.
13. Guan ZH, Han GS, Li J, et al. Impulsive multiconsensus of second-order multiagent networks using sampled position data. IEEE Trans Neural Netw Learn Syst. 2015;26(11):2678–88.
14. Liu ZW, Yu XH, Guan ZH, et al. Pulse-modulated intermittent control in consensus of multiagent systems. IEEE Trans Syst Man Cybern Syst. 2017;47(5):783–93.
15. Fax JA, Murray RM. Information flow and cooperative control of vehicle formations. IEEE Trans Autom Control. 2004;49(9):1465–76.
16. Ren W, Atkins E. Distributed multi-vehicle coordinated control via local information exchange. Int J Robust Nonlinear Control. 2007;17(10–11):1002–33.
17. Yu WW, Li CJ, Yu XH, et al. Distributed consensus strategy for economic power dispatch in a smart grid. In: Proceedings of 10th Asian Control Conference;2015, p. 1–6.
18. Vidyasagar M. Nonlinear systems analysis. Englewood Cliffs, NJ, USA: Prentice-Hall; 1993.
19. Zuo ZY, Tie L. Distributed robust finite-time nonlinear consensus protocols for multi-agent systems. Int J Syst Sci. 2016;47(6):1366–75.
20. Wen GH, Yu WW, Yu XH, et al. Designing adaptive consensus-based scheme for economic dispatch of smart grid. In: 2016 eighth international conference on advanced computational intelligence; 2016

Stability Analysis for a Class of Caputo Fractional Time-Varying Systems with Nonlinear Dynamics

Yuxiang Guo and Baoli Ma

Abstract This paper investigates mainly stability problem of equilibrium points for a class of Caputo fractional time-varying systems with nonlinear dynamics. By employing Gronwall-Bellman's inequality, Laplace transform and estimates of Mittag-Leffler functions, when the fractional-order belongs to the interval (0, 2), several stability criterions for fractional time-varying system described by Caputo's definition are presented. Besides, some problems about the stability of fractional time-varying systems in existing literatures are pointed out. Finally, an example and corresponding numerical simulations are presented to show the validity and feasibility of the proposed stability criterions.

Keywords Fractional calculus · Fractional time-varying systems · Mittag-leffler function · Gronwall-Bellman inequality · Stability

1 Introduction

Fractional calculus, which is viewed as a generalization of the traditional integer-order differentiation and integration, goes back to more than 300 years ago in history. In the last few decades, however, some mathematical models with fractional calculus have attracted increasing attentions since it has obvious advantages in the description of memory and hereditary properties of many anomalous processes [1, 2]. As we all know, stability theory plays an important role in systems theory and engineering and, it has already been well developed in the integer-order case. Yet, the stability problems of fractional-order dynamical systems have not been solved yet and it still has a long way to go. In recent years, many scholars have been devoted to the stability analysis of fractional differential equations (systems). For example, by checking the location of the matrix eigenvalues of the system, Matignon [3] has given a well-known stability theorem for linear time invariant fractional-order systems. For the

Y. Guo · B. Ma (✉)
School of Automation Science and Electrical Engineering, Beijing University of Aeronautics and Astronautics, Beijing 100191, China
e-mail: mabaoli@buaa.edu.cn

Y. Jia et al. (eds.), *Proceedings of 2017 Chinese Intelligent Systems Conference*, Lecture Notes in Electrical Engineering 459,
https://doi.org/10.1007/978-981-10-6496-8_36

nonlinear cases, the Lyapunov's method of stability analysis for integer-order nonlinear systems has been extended to fractional-order systems [4, 5]. Besides, there are other kinds of judgment methods that have arisen in the study of stability problems. In [6], authors presented two sufficient criteria on the asymptotical stability and stabilization of a class of fractional-order nonlinear systems by using Mittag-Leffler function, Laplace transform, and the Gronwall inequality. Reference [7] is based on Krasnoselskii's fixed point theorem in a weighted Banach space, several stability criterions for nonlinear fractional differential equations are derived. Due to the tremendous efforts devoted by scholars, some valuable results have been obtained on the stability problems of fractional-order systems. More detailed information could be reported in [8], which almost covers recent contributions in this area. However, for linear time-varying fractional-order systems, the corresponding stability criteria have not been established yet. The reason for this is that there is not an effective tool to judge the stability of equilibrium points of fractional time-varying systems; that is, the state transition matrix of the system does not exist an theoretical formulation. At the same time, the roots of the eigenfunction of the system is not used to check the stability of equilibrium points of time-varying systems with either integer-order or fractional-order derivative.

Recently, based on the property of fractional derivatives of quadratic functions in the sense Caputo, Ref. [9] has given some conditions to guarantee the boundedness and convergence of fractional linear time-varying unforced equations. However, the theoretical criterions of [9] can not be applied for fractional linear time-varying system with nonlinear dynamics. In [10], authors attempted to present a new way to study stability of fractional time-varying systems by taking the lower and upper boundaries of the time matrix. For some special cases, however, there was not an efficient method to determine the stability of such a time-varying system (see, Example 1 in Sect. 4). Thus, the stability analysis of fractional time-varying systems is still an open problem. In this paper, we will solve this open problem partly.

Motivated by the previous works, this paper studies mainly stability of equilibrium points of Caputo fractional time-varying systems with nonlinear dynamics. We will present some stability criterions for fractional time-varying system described by Caputo's definition. To derive such important results on stability, Gronwall-Bellman's inequality, Laplace transform and estimates of Mittag-Leffler functions are employed. Several sufficient conditions, when the fractional order α belongs to the interval $(0, 2)$, are addressed for the asymptotical stability and stabilization of fractional time-varying systems with nonlinear dynamics. Besides, some problems about the stability of fractional time-varying systems in existing literatures are pointed out. Finally, an example and corresponding numerical simulations are presented to show the superiority and effectiveness of the proposed stability criterion.

Throughout the paper, $\mathbb{R}^+$ and $\mathbb{Z}^+$ are the set of positive real and integer numbers, respectively; while $\mathbb{R}$ and $\mathbb{C}$ denote separately the set of real and complex numbers. $\mathbb{R}^n$ represents the n-dimensional Euclidean space. For a vector $x = [x_1, x_2, \ldots, x_n]^T \in \mathbb{R}^n$, let us use $\|x\| = \sqrt{\sum_{i=1}^n |x_i|^2}$ to denote the Euclidean norm of vector x. We denote by A^T the transpose of matrix A; $\|A\| = \sqrt{\lambda_{\max}(A^T A)}$ represents the 2-norm of A, where $\lambda_{\max}(A^T A)$ is the maximum eigenvalue of the matrix $A^T A$. D^α and ${}_0I_t^\alpha$ denote the Caputo fractional derivative of fractional order $\alpha \in \mathbb{R}^+$ and Riemann-Liouville fractional integral of order $\alpha \in \mathbb{R}^+$ on $[0, t]$, respectively.

2 Preliminaries

In this section, some basic concepts of fractional operators are recalled. Without loss of generality, the lower limit of all fractional integrals and derivatives is supposed to be zero throughout the paper.

Definition 1 ([1]) Let $x(t)$ be a continuous function on an interval $[0, b]$. The Riemann-Liouville fractional integral of order $\alpha \in \mathbb{R}^+$ is defined as

$${}_0I_t^\alpha x(t) = \frac{1}{\Gamma(\alpha)} \int_0^t (t-\tau)^{\alpha-1} x(\tau) d\tau, \ (t > 0, \ \alpha > 0), \tag{1}$$

where $\Gamma(\cdot)$ is the Gamma function.

Definition 2 ([1]) The Caputo fractional derivative with order $\alpha \in \mathbb{R}^+$ of function $x(t)$ is defined by

$${}_0^C D_t^\alpha x(t) = \frac{1}{\Gamma(n-\alpha)} \int_0^t (t-\tau)^{n-\alpha-1} x^{(n)}(\tau) d\tau, \ (t > 0), \tag{2}$$

where $n - 1 < \alpha \leq n \in \mathbb{Z}^+$, $x^{(n)}(\tau)$ denotes the n-th derivative of x with respect to τ. Meanwhile, the Laplace transform (LT) of the Caputo fractional derivative are given as,

$$LT\left\{{}_0^C D_t^\alpha x(t)\right\} = s^\alpha X(s) - \sum_{k=0}^{n-1} s^{\alpha-k-1} x^{(k)}(0), \ (n - 1 < \alpha \leq n), \tag{3}$$

where $X(s)$ denotes the Laplace transform of $x(t)$, t and s are the variable in the time domain and complex-frequency domain. Furthermore, the Laplace transform of ${}_0I_t^\alpha x(t)$ takes the particularly simple form

$$LT\left\{{}_0I_t^\alpha x(t)\right\} = s^{-\alpha} X(s), \ (\alpha > 0). \tag{4}$$

In order to investigate the stability of equilibrium points of fractional-order systems, the Mittag-Leffler function is employed frequently by scholars. To this end, the Mittag-Leffler function with two parameters is defined by

$$E_{\alpha,\beta}(z) = \sum_{k=0}^{\infty} \frac{z^k}{\Gamma(\alpha k + \beta)}, \ (\alpha > 0, \beta > 0), \tag{5}$$

where $z \in \mathbb{C}$. For $\beta = 1$, one has $E_{\alpha,1}(z) = E_\alpha(z)$ and $E_{1,1}(z) = e^z$. In what follows, we introduce several results about the Mittag-Leffler function that will be used in the proof later.

Property 1 ([1]) *The Laplace transform of the Mittag-Leffler function in two parameters is*

$$LT\left\{t^{\alpha k+\beta-1}E^{(k)}_{\alpha,\beta}(\pm\lambda t^\alpha)\right\} = \frac{k!s^{\alpha-\beta}}{(s^\alpha \mp \lambda)^{k+1}}, \ (Re(s) > |\lambda|^{\frac{1}{\alpha}}, k = 0, 1, 2, \ldots), \quad (6)$$

where $E^{(k)}_{\alpha,\beta}(\pm\lambda t^\alpha) = \frac{d^k}{dt^k}E_{\alpha,\beta}(\pm\lambda t^\alpha)$, $\lambda \in \mathbb{R}$ *and* $Re(s)$ *denotes the real part of s.*

Property 2 ([1]) *Let* $0 < \alpha < 2$, $\beta \in \mathbb{C}$ *and a real number* μ *such that* $\alpha\pi/2 < \mu < \min\{\pi, \alpha\pi\}$. *If* $\mu \leq |\arg(z)| \leq \pi$, *then for an arbitrary integer* $m \geq 1$, *the following asymptotic expansion holds*

$$E_{\alpha,\beta}(z) = -\sum_{k=1}^{m} \frac{z^{-k}}{\Gamma(\beta - \alpha k)} + O(|z|^{-m-1}), \ |z| \to \infty. \quad (7)$$

As far as I know, fractional-order system is a nonlocal and infinite dimensional system or long memory system. So in this paper, it is assumed that the initialization function $x(t)$ of fractional-order systems involving Caputo's definition is a constant function of time (or null), and $x(t) = x(0^+)$ for $t \leq 0$. For a comprehensive treatment of the initialization issue of fractional operators, the reader may consult the literature [11].

3 Main Results

In this section, let us consider a class of Caputo fractional order time-varying systems with nonlinear dynamics

$$\frac{d^\alpha x(t)}{dt^\alpha} = A(t)x(t) + f(t, x(t)), \ (0 < \alpha < 2), \quad (8)$$

where $x(t) \in \mathbb{R}^n$ is the pseudo-state vector of the system, because the notion of state variables is generally not used in the context of fractional systems [11]; and $A(t) \in \mathbb{R}^{n\times n}$ is a time-varying matrix. $f(t, x) : [0, \infty) \times \mathbb{R}^n \to \mathbb{R}^n$ is a nonlinear function, which is piecewise continuous in t; further let $x = 0$ be an equilibrium point of the system, i.e. f(t, 0) = 0. The notation $\frac{d^\alpha}{dt^\alpha}$ is viewed as fractional derivative operator with Caputo's definition. In the following, we shall give our main results about stability of equilibrium points of the system (8). To derive two stability criterions of Caputo type fractional-order system ${}_0^C D_t^\alpha x(t) = A(t)x(t) + f(t, x(t))$, $(0 < \alpha < 2)$, let us denote first the initial-values $x^{(k)}(0) = x_k$, $(k = 0, 1)$ in this section.

Theorem 1 *Let $x = 0$ be an equilibrium point of system (8) involving Caputo's definition, for any $\alpha \in (0, 1]$, if*
(i) $\lim_{t\to\infty} A(t) = A$;
(ii) $f(t, 0) = 0$, and $\lim_{\|x\|\to 0} \frac{\|f(t,x)\|}{\|x\|} = 0$;
(iii) $|\arg(\mathrm{spec}(A))| > \frac{\alpha\pi}{2}$, and $\|A\| > \frac{N}{\alpha} \geq 1$.
Then, the equilibrium point $x = 0$ is locally asymptotically stable.

Proof By the condition (i) of Theorem 1, for arbitrarily small $\varepsilon_1 > 0$, there exists a positive constant T (T could be infinity), when $t \geq T$, one has $\|A(t) - A\| \leq \varepsilon_1$. Thus, the system (8) with Caputo fractional derivative operator can be expressed as

$$ {}_0^C D_t^\alpha x(t) = Ax(t) + (A(t) - A)x(t) + f(t, x(t)), \ (0 < \alpha \leq 1). \tag{9}$$

According to the Laplace transform (3) and its inverse Laplace transform (6), the solution of the system (9) with initial-value x_0 is represented by

$$x(t) = E_{\alpha,1}(At^\alpha)x_0 + \int_0^t (t-\tau)^{\alpha-1} E_{\alpha,\alpha}(A(t-\tau)^\alpha)[(A(\tau) - A)x(\tau) + f(\tau, x(\tau))]d\tau.$$

That is

$$\begin{aligned}\|x(t)\| &\leq \|E_{\alpha,1}(At^\alpha)\| \|x_0\| \\ &+ \int_0^t (t-\tau)^{\alpha-1} \|E_{\alpha,\alpha}(A(t-\tau)^\alpha)\| [\|(A(\tau) - A)x(\tau)\| + \|f(\tau, x(\tau))\|]d\tau.\end{aligned}$$

From the condition (ii) of Theorem 1, there exists a real constant $r > 0$, for any $\varepsilon_2 > 0$, such that

$$\|f(t, x)\| \leq \varepsilon_2 \|x\|, \ \forall \|x\| \leq r.$$

It then follows from Ref. [12, Corollary 1] that there exist two real positive constants D_0 and D_2 such that

$$\|x(t)\| \leq \frac{D_0\|x_0\|}{1 + \|A\|t^\alpha} + \int_0^t \frac{D_2(\varepsilon_1 + \varepsilon_2)(t-\tau)^{\alpha-1}\|x(\tau)\|}{1 + \|A\|(t-\tau)^\alpha} d\tau.$$

By Ref. [6, Lemma 2] and denoting $N = D_2(\varepsilon_1 + \varepsilon_2)$, one has

$$\begin{aligned}\|x(t)\| &\leq \frac{D_0\|x_0\|}{1 + \|A\|t^\alpha} \\ &+ \int_0^t \frac{D_0 N(t-\tau)^{\alpha-1}\|x_0\|}{(1 + \|A\|\tau^\alpha)(1 + \|A\|(t-\tau)^\alpha)} \exp\left(\int_\tau^t \frac{N(t-s)^{\alpha-1}}{1 + \|A\|(t-s)^\alpha} ds\right) d\tau\end{aligned}$$

$$\le \frac{D_0\|x_0\|}{1+\|A\|t^\alpha} + D_0N\|x_0\|\|A\|^{\frac{N}{\alpha\|A\|}-2}\int_0^t (t-\tau)^{\frac{N}{\|A\|}-1}\tau^{-\alpha}d\tau$$

$$= \frac{D_0\|x_0\|}{1+\|A\|t^\alpha} + D_0N\|x_0\|\|A\|^{\frac{N}{\alpha\|A\|}-2}\frac{\Gamma(\frac{N}{\|A\|})\Gamma(1-\alpha)}{\Gamma(\frac{N}{\|A\|}+1-\alpha)}t^{\frac{N}{\|A\|}-\alpha}.$$

Hence, for all $\|A\| > \frac{N}{\alpha}$, we can conclude that $\|x\| \to 0$ as $t \to \infty$. Namely, the equilibrium point $x = 0$ of the system (8) with Caputo's definition is locally asymptotically stable. This completes the proof.

Theorem 2 *Let $x = 0$ be an equilibrium point of system* (8) *involving Caputo's definition, for any $\alpha \in (1, 2)$, if*
(i) $\lim_{t\to\infty} A(t) = A$;
(ii) $f(t, 0) = 0$, and $\lim_{\|x\|\to 0} \frac{\|f(t,x)\|}{\|x\|} = 0$;
(iii) $|\arg(\text{spec}(A))| > \frac{\alpha\pi}{2}$, and $\|A\| > \frac{N}{\alpha-1} \ge 1$.
Then, the equilibrium point $x = 0$ is locally asymptotically stable.

Proof Similar to the proof of Theorem 1, the system (8) with Caputo fractional derivative operator can be written as

$$ {}_0^C D_t^\alpha x(t) = Ax(t) + (A(t) - A)x(t) + f(t, x(t)),\ (1 < \alpha < 2). \tag{10}$$

Consequently, we can show that the solution of the system (10) with initial-values x_0 and x_1 can be described by

$$x(t) = E_{\alpha,1}(At^\alpha)x_0 + tE_{\alpha,2}(At^\alpha)x_1$$
$$+ \int_0^t (t-\tau)^{\alpha-1}E_{\alpha,\alpha}(A(t-\tau)^\alpha)[(A(\tau) - A)x(\tau) + f(\tau, x(\tau))]d\tau.$$

And by Ref. [12, Corollary 1], there exist three real positive constants D_0, D_1, and D_2 satisfying

$$\|x(t)\| \le \frac{D_0\|x_0\|}{1+\|A\|t^\alpha} + \frac{D_1\|x_1\|t}{1+\|A\|t^\alpha} + \int_0^t \frac{D_2(\varepsilon_1+\varepsilon_2)(t-\tau)^{\alpha-1}\|x(\tau)\|}{1+\|A\|(t-\tau)^\alpha}d\tau.$$

Applying Ref. [6, Lemma 2] and denoting $N = D_2(\varepsilon_1 + \varepsilon_2)$, one gets

$$\|x(t)\| \le \frac{D_0\|x_0\|}{1+\|A\|t^\alpha} + \frac{D_1\|x_1\|t}{1+\|A\|t^\alpha}$$
$$+ \int_0^t \frac{N(D_0\|x_0\| + D_1\|x_1\|\tau)(t-\tau)^{\alpha-1}}{(1+\|A\|\tau^\alpha)(1+\|A\|(t-\tau)^\alpha)}\exp\left(\int_\tau^t \frac{N(t-s)^{\alpha-1}}{1+\|A\|(t-s)^\alpha}ds\right)d\tau$$
$$\le \frac{D_0\|x_0\| + D_1\|x_1\|t}{1+\|A\|t^\alpha} + D_0N\|x_0\|\|A\|^{\frac{N}{\alpha\|A\|}-2}\int_0^t (t-\tau)^{\frac{N}{\|A\|}-1}\tau^{-\alpha}d\tau$$

$$+D_1N\|x_1\|\|A\|^{\frac{N}{\alpha\|A\|}-2}\int_0^t (t-\tau)^{\frac{N}{\|A\|}-1}\tau^{1-\alpha}d\tau$$

$$=\frac{D_0\|x_0\|+D_1\|x_1\|t}{1+\|A\|t^{\alpha}}+D_0N\|x_0\|\|A\|^{\frac{N}{\alpha\|A\|}-2}\frac{\Gamma(\frac{N}{\|A\|})\Gamma(1-\alpha)}{\Gamma(\frac{N}{\|A\|}+1-\alpha)}t^{\frac{N}{\|A\|}-\alpha}$$

$$+D_1N\|x_1\|\|A\|^{\frac{N}{\alpha\|A\|}-2}\frac{\Gamma(\frac{N}{\|A\|})\Gamma(2-\alpha)}{\Gamma(\frac{N}{\|A\|}+2-\alpha)}t^{\frac{N}{\|A\|}+1-\alpha}.$$

Therefore, for all $\|A\| > \frac{N}{\alpha-1}$, we can conclude that $\|x\| \to 0$ as $t \to \infty$. Namely, the equilibrium point $x = 0$ of the system (8) with Caputo's definition is locally asymptotically stable. This completes the proof.

Remark 1 By Property 2, there exists an arbitrary integer $m \geq 1$ $(m < \infty)$, one has

$$\left|E_{\alpha,\beta}(z)\right| = \left|\sum_{k=1}^{m}\frac{z^{-k}}{\Gamma(\beta-\alpha k)}\right| + O(|z|^{-m-1}),\ |z| \to \infty.$$

According to the analytic continuation of Gamma function, and we can choose $D \triangleq \max\{|\Gamma(\beta-\alpha)|^{-1}, |\Gamma(\beta-2\alpha)|^{-1}, \ldots, |\Gamma(\beta-m\alpha)|^{-1}\}$, when $|z| \to \infty$, it can be concluded that

$$\left|E_{\alpha,\beta}(z)\right| \leq D\sum_{k=1}^{m}\left|z^{-k}\right| \leq D\sum_{k=1}^{\infty}\left|z^{-k}\right|.$$

So far, it follows from Ref. [1, Theorem 1.6] that $\left|E_{\alpha,\beta}(z)\right| \leq \frac{D}{1+|z|} \leq \frac{D}{|z|-1}$ as $|z| \geq 1$. Hence, there is a simple task on how to select a proper parameter N in Theorems 1 and 2, that is, $N \geq D(\varepsilon_1 + \varepsilon_2)$.

Remark 2 Assume that $A(t)$ is a constant matrix A. Then, the locally asymptotic stability of Caputo type fractional-order system ${}_0^CD_t^{\alpha}x(t) = Ax(t) + f(t, x(t))$, for any $\alpha \in (0, 2)$, with the initial conditions $x^{(k)}(0) = x_k$, $(k = 0, 1)$ has been investigated in [6, 13], we will not reiterate them here.

Remark 3 It should be pointed out in the above proofs of the obtained results, the fractional time-varying systems with nonlinear dynamics (8) is globally asymptotically stable, when f is globally Lipschitz in x on $[0, \infty) \times \mathbb{R}^n$ (i.e. $[\partial f/\partial x]$ is uniformly bounded on $[0, \infty) \times \mathbb{R}^n$).

If, let us write the system (8) as the following controlled fractional time-varying systems with nonlinear dynamics:

$$\frac{d^{\alpha}x(t)}{dt^{\alpha}} = A(t)x(t) + f(t, x(t)) + Bu(t),\ (0 < \alpha < 2), \tag{11}$$

where $u(t)$ is the control input and, the constant matrix $B \in \mathbb{R}^{n\times 1}$. For the system (11), then we are interested in the design of stabilizing pseudo-state feedback control laws of the form:

$$u(t) = Kx(t),$$

where $K \in \mathbb{R}^{1\times n}$ is a constant matrix to be determined later, and we denote here that $\hat{A}(t) = A(t) + BK$. Subsequently, our aim is to design a suitable pseudo-state feedback gain matrix K such that the controlled system (11) is asymptotically stable.

Similar to the proof of Theorems 1 and 2, we have the following results.

Corollary 1 *Let $x = 0$ be an open-loop equilibrium point of system* (11) *with Caputo fractional operator, for any $\alpha \in (0, 1]$, and the feedback gain K is selected such that $\lim_{t\to\infty} \hat{A}(t) = \hat{A}$, $|\arg(\mathrm{spec}(\hat{A}))| > \frac{\alpha\pi}{2}$, and $\|\hat{A}\| > \frac{N}{\alpha} \geq 1$.*
(i) If $f(t, x(t))$ satisfies $f(t, 0) = 0$ and $\lim_{\|x\|\to 0} \frac{\|f(t,x)\|}{\|x\|} = 0$, then the equilibrium point $x = 0$ of the closed-loop system is locally asymptotically stable.
(ii) If $f(t, x(t))$ satisfies $f(t, 0) = 0$ and is globally Lipschitz in $x(t)$ on $[0, \infty) \times \mathbb{R}^n$, then the equilibrium point $x = 0$ of the closed-loop system is globally asymptotically stable.

Corollary 2 *Let $x = 0$ be an open-loop equilibrium point of system* (11) *with Caputo fractional operator, for any $\alpha \in (1, 2)$, and the feedback gain K is selected such that $\lim_{t\to\infty} \hat{A}(t) = \hat{A}$, $|\arg(\mathrm{spec}(\hat{A}))| > \frac{\alpha\pi}{2}$, and $\|\hat{A}\| > \frac{N}{\alpha-1} \geq 1$.*
(i) If $f(t, x(t))$ satisfies $f(t, 0) = 0$ and $\lim_{\|x\|\to 0} \frac{\|f(t,x)\|}{\|x\|} = 0$, then the equilibrium point $x = 0$ of the closed-loop system is locally asymptotically stable.
(ii) If $f(t, x(t))$ satisfies $f(t, 0) = 0$ and is globally Lipschitz in $x(t)$ on $[0, \infty) \times \mathbb{R}^n$, then the equilibrium point $x = 0$ of the closed-loop system is globally asymptotically stable.

4 Illustrative Example

In a recent paper [10], a class of fractional-order nonlinear time varying systems with Caputo derivative is considered. We will give an example to illustrate that their stability criterions are not valid in general. Subsequently, it is also used to show the superiority and effectiveness of our proposed stability criterion in this paper.

Example 1 Let us consider the following Caputo type fractional-order system.

$$\begin{cases} {}_0^C D_t^\alpha x_1(t) = -x_1(t) + 8x_2(t) + 0.1x_1(t)x_2(t), \\ {}_0^C D_t^\alpha x_2(t) = (0.1 + 0.9\exp(-t)\cos(5t))x_1(t) - 0.95x_2(t), \end{cases} \tag{12}$$

where $x(t) = [x_1(t), x_2(t)]^T$ is the pseudo-state vector of the system. According to the conditions of paper [10], one has

$$A(t) = \begin{bmatrix} -1 & 8 \\ 0.1 + 0.9\exp(-t)\cos(5t) & -0.95 \end{bmatrix}, \bar{A} = \begin{bmatrix} -1 & 8 \\ 1 & -0.95 \end{bmatrix}.$$

By the equation $\det(\lambda I - \bar{A}) = 0$, one can calculate the corresponding eigenvalues as:

$$\lambda_1 = -3.8035 \text{ and } \lambda_2 = 1.8535,$$

which implies that the condition $\alpha\pi/2 < |\arg(\operatorname{spec}(\bar{A}))| < \pi$ does not hold when fractional order $\alpha \in (0, 2)$. Consequently, it follows from Theorem 1 and Theorem 3 of Ref. [10] that the zero solution of the system (12) is unstable or unidentified. However, by employing our proposed Theorem 1, one can calculate that

$$\lim_{t\to\infty} A(t) = \begin{bmatrix} -1 & 8 \\ 0.1 & -0.95 \end{bmatrix} \triangleq A, \ \|A\| = 8.1333.$$

And, the corresponding eigenvalues of the matrix A are given by

$$\lambda_1 = -1.8698, \ \lambda_2 = -0.0802,$$

which implies that $|\arg(\operatorname{spec}(A))| > \frac{\alpha\pi}{2}$ holds. Meanwhile, it yields

$$\lim_{x\to 0} \frac{\|f(x(t))\|}{\|x(t)\|} = \lim_{x\to 0} \frac{0.1\sqrt{(x_1x_2)^2}}{\sqrt{x_1^2 + x_2^2}} \le 0.05 \lim_{x\to 0} \sqrt{x_1^2 + x_2^2} = 0.$$

By Theorem 1, we can show that the zero solution of the system (12) is locally asymptotically stable, as shown in Figs. 1 and 2. Therefore, the proposed stability theorems of paper [10] are ineffective. Since the matrix $A(t)$ of the system may depend on time t, we are not going to check the location of the eigenvalues of the system matrix by selecting the upper boundary of the time matrix $A(t)$. In the usual situation, it thus is not an efficient way to judge the stability of the zero solution of the time-varying system with either integer-order or fractional-order derivative operator.

Remark 4 It is noted that the stability problem of fractional time-varying system has been solved partly in this paper. Due to the absence of the mathematical theory of fractional differential systems including the spectral estimation of matrix, the Leibniz rule and semigroup property, however, the stability of fractional time-varying system has not been completely solved yet. (See Appendix).

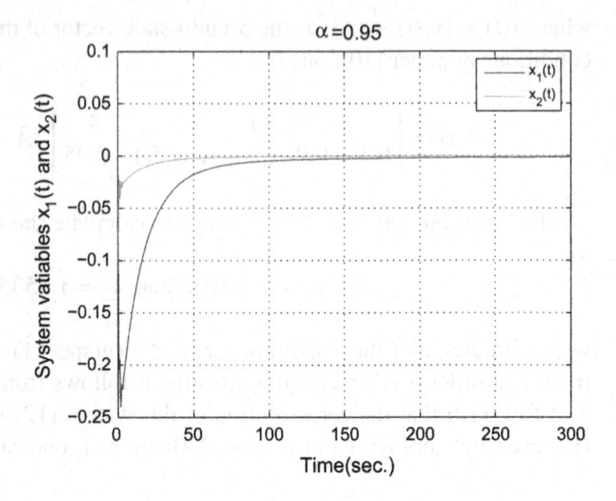

Fig. 1 Evolution of the state $x_1(t)$ and $x_2(t)$ of the system when initial-value (0.1, -0.1)

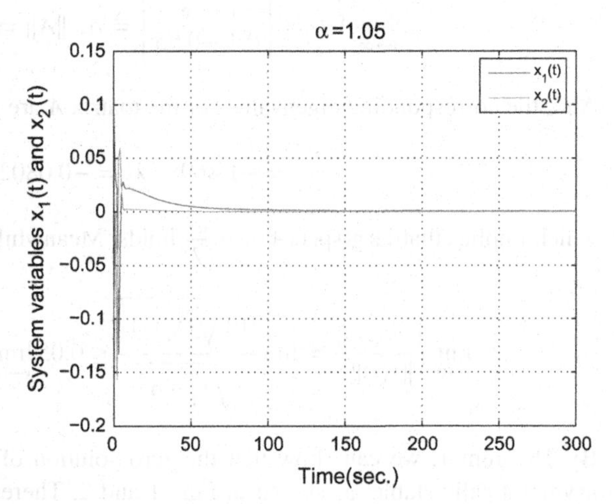

Fig. 2 Evolution of the state $x_1(t)$ and $x_2(t)$ of the system when initial-values (0.1, -0.1) and (-9, 10)

5 Conclusion

The stability problem of equilibrium points for a class of fractional time-varying systems with nonlinear dynamics is studied in this paper. By applying mathematical tools such as Gronwall-Bellman's inequality, Laplace transform and estimates of Mittag-Leffler functions, we give several stability criterions for fractional time-varying system described by Caputo's definition. Besides, some problems about the stability of fractional time-varying systems in existing literatures are pointed out. Finally, two examples and corresponding numerical simulations are presented to show the validity and feasibility of the proposed stability criterion. Because of the complexity of fractional calculus, however, the stability of fractional time-varying system has not been completely solved; it will be our future topic of research.

Acknowledgements This work was supported by the National Nature Science Foundation (No. 61327807, No. 61573034).

Appendix

Let us modify the system (12) as the following system, where $\alpha = 0.95$.

$$\begin{cases} \dfrac{d^\alpha x_1(t)}{dt^\alpha} = -x_1(t) + 10x_2(t) + 0.1x_1(t)x_2(t), \\ \dfrac{d^\alpha x_2(t)}{dt^\alpha} = (-1 + 0.9\cos(5t))x_1(t) + 0.95x_2(t), \end{cases} \tag{13}$$

where $\frac{d^\alpha}{dt^\alpha}$ stands for Caputo fractional derivative operator, $x(t) = [x_1(t), x_2(t)]^T$ is the pseudo-state vector of the system. Subsequently, one gets

$$A(t) = \begin{bmatrix} -1 & 10 \\ -1 + 0.9\cos(5t) & 0.95 \end{bmatrix}, \bar{A} = \begin{bmatrix} -1 & 10 \\ -0.1 & 0.95 \end{bmatrix}, \underline{A} = \begin{bmatrix} -1 & 10 \\ -1.9 & 0.95 \end{bmatrix}.$$

By the equations $\det(\lambda I - \bar{A}) = 0$ and $\det(\tilde{\lambda} I - \underline{A}) = 0$, one can calculate the corresponding eigenvalues as:

$$\lambda_{1,2} = \frac{-0.05 \pm 0.89303i}{2} \text{ and } \tilde{\lambda}_{1,2} = \frac{-0.05 \pm 8.49691i}{2}.$$

Then, $|\arg(\lambda_i(\bar{A}))| = 1.6267$ and $|\arg(\tilde{\lambda}_i(\underline{A}))| = 1.5767$, which implies that $\alpha\pi/2 < |\arg(\lambda_i(\bar{A}))|, |\arg(\tilde{\lambda}_i(\underline{A}))| < \pi$. Consequently, it is easy to calculate that $\|\bar{A}\| = 10.0952$, $\|\underline{A}\| = 10.1158$, $\|\bar{A} - \underline{A}\| = 1.8$, $\alpha\|\bar{A}\| = 9.5904$, $\alpha\|\underline{A}\| = 9.6100$, and $\lim_{x\to 0} \frac{\|f(x(t))\|}{\|x(t)\|} = \lim_{x\to 0} \frac{0.1\sqrt{(x_1x_2)^2}}{\sqrt{x_1^2+x_2^2}} \le 0.05\lim_{x\to 0}\sqrt{x_1^2 + x_2^2} = 0$.

Obviously, the system (13) satisfies the three conditions which are shown in the literature [10]. However, the state of system (13) cannot converge to zero as time t tends to infinity. In the following, Adams-Bashforth-Moulton predictor-corrector algorithm [14] is employed to the numerical calculation of fractional system. When the initial condition $(x_1(0), x_2(0)) = (0.1, 0.1)$, the system (13) can generate subharmonic, harmonic or almost-periodic oscillation and even yield a chaotic trajectory. Its chaotic figure is shown in Fig. 3, while Fig. 4 depicts the evolution of the trajectories $x_1(t)$ and $x_2(t)$ of system (13). Therefore, the proposed results of paper [10] are not correct. Since the matrix $A(t)$ of system may depend on time t, we cannot select the lower and upper boundaries of the time matrix $A(t)$ to determine the stability of the zero solution of the time-varying system with integer-order or fractional-order differential operator. Meanwhile, the stability criterions of our presented paper is not used to check the convergence and divergence of the system (13) as well; it will be our future topic of research.

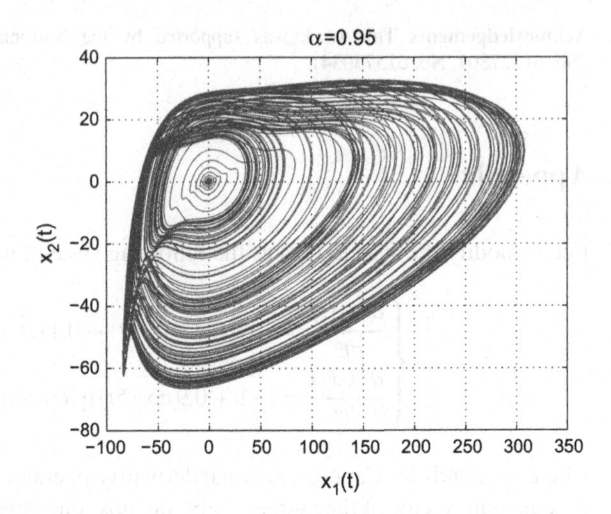

Fig. 3 Chaotic behavior of system

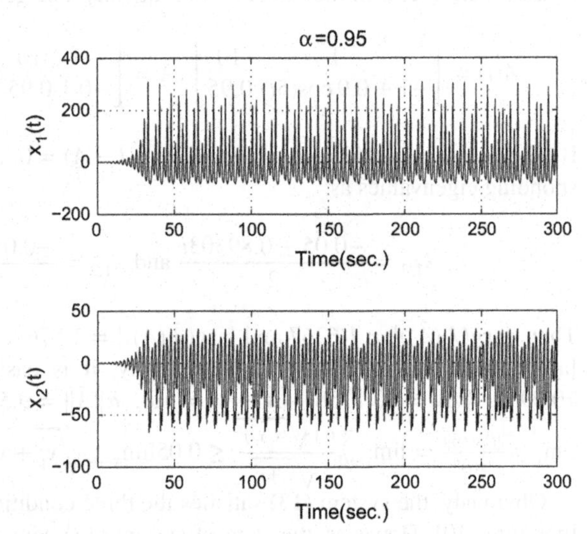

Fig. 4 Evolution of the state $x_1(t)$ and $x_2(t)$ of the system

References

1. Podlubny I. Fractional differential equations. NewYork: Academic; 1999.
2. Zhou Y (2014) Basic theory of fractional differential equations, World Scientific Publishing Co. Pte. Ltd., Hackensack, NJ
3. Matignon D. Stability results for fractional differential equations with applications to control processing. Comput Eng Syst Appl. 1996;2:963–8.
4. Trigeassou J, Maamri N, Sabatier J, Oustaloup A. A Lyapunov approach to the stability of fractional differential equations. Signal Process. 2011;91(3):437–45.
5. Li Y, Chen Y, Pudlubny I. Mittag-Leffler stability of fractional order nonlinear dynamic systems. Automatica. 2009;45(8):1965–9.

6. Chen L, Chai Y, Wu R, Yang J. Stability and stabilization of a class of nonlinear fractional-order systems with Caputo derivative. IEEE Trans Circuits Syst II, Express Briefs. 2012;59(9): 602–6.
7. Ge F, Kou C. Stability analysis by Krasnoselskii's fixed point theorem for nonlinear fractional differential equations. Appl Math Comput. 2015;257:308–16.
8. Li C, Zhang F. A survey on the stability of fractional differential equations. Eur Phys J Spec Top. 2011;193:27–47.
9. Gallegos J, Duarte-Mermoud M. Boundedness and convergence on fractional order systems. J Comput Appl Math. 2016;296:815–26.
10. Huang S, Zhang R, Chen D. Stability of nonlinear fractional-order time varying systems. ASME J Comput Nonlinear Dyn. 2016;11(3):031007.
11. Sabatier J, Farges C, Oustaloup A. Fractional systems state space description: some wrong ideas and proposed solution. J Vib Control. 2014;20(7):1076–84.
12. Wen X, Wu Z, Lu J. Stability analysis of a class of nonlinear fractional-order systems. IEEE Trans Circuits Syst II, Express Briefs. 2008;55(11):1178–82.
13. Zhang R, Tian G, Yang S, Cao H. Stability analysis of a class of fractional order nonlinear systems with order lying in (0, 2). ISA Trans. 2015;56:102–10.
14. Diethelm K, Ford NJ, Freed AD. A predictor-corrector approach for the numerical solution of fractional differential equations. Nonlinear Dyn. 2002;29(1–4):3–22.

Bipartite Containment Control of Nonlinear Multi-agent Systems with Input Saturation

Tao Yu and Lei Ma

Abstract This paper considers the bipartite containment problem for nonlinear multi-agent systems with multiple leaders over a signed directed graph. Each follower's dynamic is taken as a strict-feedback form with unknown nonlinearity and input saturation. A distributed adaptive control law is designed and analyzed. A simulation example demonstrates the proposed control algorithm.

Keywords Bipartite containment · Input saturation · Multi-agent system

1 Introduction

Over the past few decades, distributed coordination for multi-agent systems has attracted much attention due to its potential applications including natural networks, biological communities, engineering systems [1–3]. The number of leaders categorize the coordination problem into three classes, namely, the leaderless consensus, the leader-following tracking with one leader, and the containment control driving followers into a geometric space spanned by multiple leaders [4, 5]. The distributed containment control designs provide a general method for leaderless consensus and leader-following tracking, and apply to the real-world scenarios such as multi-robots obstacle-avoiding, networked Lagrange systems and biological swarming networks [6–8].

Recently, a class of consensus problems over signed graphs has an extensive concern with the coexistence of cooperative and competitive relations between agents [9, 10]. When the signed graph is structurally balanced [11], the corresponding consensus-related problems are called bipartite consensus [12, 13], bipartite tracking [14, 15], and bipartite containment control. Although there are lots of results which

T. Yu · L. Ma (✉)
School of Electrical Engineering, Southwest Jiaotong University,
Chengdu, Sichuan 610031, China
e-mail: malei@swjtu.edu.cn

T. Yu
e-mail: yutao@my.swjtu.edu.cn

Y. Jia et al. (eds.), *Proceedings of 2017 Chinese Intelligent Systems Conference*, Lecture Notes in Electrical Engineering 459,
https://doi.org/10.1007/978-981-10-6496-8_37

are presented for bipartite consensus in single-integrator dynamics [11], general linear time-invariant systems [12, 13, 15], nonlinear systems [14, 16], the research of bipartite containment control for nonlinear systems is exceedingly spare. On the other hand, input saturation is ubiquitous in realistic applications owing to the energies or physics limitations. Bipartite consensus result for multi-agent systems with input saturation has been an active research topic. Especially, the bipartite consensus for general linear systems subject to input saturation has been studied in [17, 18].

To the best of our knowledge, almost all works on bipartite consensus problems fucus on bipartite leaderless consensus or bipartite tracking with one leader. There are few efforts to investigate bipartite containment control problem for nonlinear multi-agent systems with input saturation, where the importance of containment control over signed graphs motives our present work.

2 Problem Formulation

Consider a class of multi-agent systems consisting of N followers and R leaders. Dynamics of the i-th ($i = 1, \dots, N$) following agent is described as

$$\begin{aligned} \dot{x}_{i,m} &= x_{i,m+1}, \ (m = 1, \dots, M-1) \\ \dot{x}_{i,M} &= f_i(\boldsymbol{x}_i) + S_i(u_i) + \zeta_i, \\ y_i &= x_{i,1}, \end{aligned} \tag{1}$$

where $\boldsymbol{x}_i = [x_{i,1}, \dots, x_{i,M}]^T \in \mathbb{R}^M$ denotes the state vector; unknown nonlinear function $f_i(\boldsymbol{x}_i) : \mathbb{R}^M \to \mathbb{R}$ is locally Lipschitz; $\zeta_i \in \mathbb{R}$ is an unknown bounded disturbance; $u_i \in \mathbb{R}$ and $y_i \in \mathbb{R}$ are the control input and output signal, respectively; and $S_i(u_i)$ denotes the saturation input signal defined as

$$S_i(u_i) = \begin{cases} \operatorname{sgn}(u_i)u_i^*, & \text{if } |u_i| \geq u_i^* \\ u_i, & \text{if } |u_i| < u_i^* \end{cases} \tag{2}$$

where constant $u_i^* > 0$ is the known bound of input saturation, and $\operatorname{sgn}(\cdot)$ is the sign function.

A multi-agent network consisting of N followers and R leaders can be expressed by a signed graph $\mathcal{G} = \{\mathcal{V}, \mathcal{E}\}$ with the nodes set $\mathcal{V} = \{v_1, \dots, v_{N+R}\}$ and edges set $\mathcal{E} \subseteq \mathcal{V} \times \mathcal{V}$. Let $\mathcal{N}_p = \{q \mid (v_q, v_p) \in \mathcal{E}\}$ denote the neighbors set of node p. Let $A = [a_{pq}] \in \mathbb{R}^{(N+R)\times(N+R)}$ denote the adjacency matrix, where $a_{pq} > 0$ means cooperation, and $a_{pq} < 0$ means competition; otherwise, $a_{pq} = 0$. Assume that the graph has no self-loops, i.e., $a_{pp} = 0$. Let $d_p = \sum_{q=1}^{N+R} |a_{pq}|$ and $D = \operatorname{diag}(d_1, \dots, d_{N+R})$ denote the in-degree and in-degree matrix, respectively. The Laplacian matrix is denoted as $L = D - A \in \mathbb{R}^{(N+R)\times(N+R)}$. Suppose that the followers 1 to N have at least one neighbor, and the leaders $N+1$ to $N+R$ have no neighbors. Then the Laplacian matrix L is redefined as

$$L = \begin{bmatrix} L_1 & L_2 \\ O_{R\times N} & O_{R\times R} \end{bmatrix}, \tag{3}$$

where $L_1 \in \mathbb{R}^{N\times N}$ and $L_2 \in \mathbb{R}^{N\times R}$.

A directed path from v_p to v_q is described by a sequence of successive edges in a digraph. Suppose that, for each follower, there exists at least one leader that has a directed path to the follower. Let $\bar{\mathcal{G}} = \{\bar{\mathcal{V}}, \bar{\mathcal{E}}\}$ denote the subgraph of $\mathcal{G}$, which includes the followers set $\bar{\mathcal{V}} = \{v_1, \ldots, v_N\}$ and the edges set $\bar{\mathcal{E}} \subseteq \bar{\mathcal{V}} \times \bar{\mathcal{V}}$. A signed digraph is structurally balanced if there exists a bipartition $\{\bar{\mathcal{V}}_1, \bar{\mathcal{V}}_2\}$ of the nodes with $\bar{\mathcal{V}}_1 \cup \bar{\mathcal{V}}_2 = \bar{\mathcal{V}}$ and $\bar{\mathcal{V}}_1 \cap \bar{\mathcal{V}}_2 = \emptyset$, such that $a_{ij} > 0$ when $\forall v_i, v_j$ are in the same subgroup, and $a_{ij} < 0$ when v_i and v_j are in different subgroups, $\forall i, j \in \{1, \ldots, N\}$, and denote $\theta_i = 1$ if $v_i \in \mathcal{V}_1$ and $\theta_i = -1$ if $v_i \in \mathcal{V}_2$ [11].

Assumption 1 The leader's trajectory $r_l \in C^1$, $l \in \{N+1, \ldots, N+R\}$, is known and available for the followers with $(l, i) \in \mathcal{E}$. Simultaneously, assume the weight value $b_{il} = a_{il} \geq 0$. Moreover, $r_l, \dot{r}_l \in \mathcal{L}_\infty$.

Lemma 1 ([6]) *$\Theta L_1 \Theta$ is a nonsingular M-matrix with $\Theta = diag(\theta_1, \ldots, \theta_N)$. In addition, each entry of matrix $-\Theta L_1^{-1} \Theta L_2$ is nonnegative, which all row sums equal to one. Consequently, $-\Theta L_1^{-1} \Theta L_2 \boldsymbol{r}$ is a convex hull spanned by the multiple leaders with $\boldsymbol{r} = [r_{N+1}, \cdots, r_{N+R}]^T$.*

The control objective is to design a distributed control law for the following agent (1), so that the followers' outputs of different subgroups converge to the convex hull spanned by the multiple leaders r_l, i.e.,

$$\lim_{t\to\infty} ||\Theta \boldsymbol{y} + \Theta L_1^{-1} \Theta L_2 \boldsymbol{r}|| = 0, \tag{4}$$

with $\boldsymbol{y} = [y_1, \ldots, y_N]^T$. Let $\boldsymbol{\delta} = \Theta \boldsymbol{y} + \Theta L_1^{-1} \Theta L_2 \boldsymbol{r}$ denote bipartite containment error vector, where $\boldsymbol{\delta} = [\delta_1, \ldots, \delta_N]^T$. Then we say that the multi-agent systems achieve bipartite containment consensus.

3 Adaptive Control Design

3.1 Distributed Control Law Design

In this subsection, utilizing the backstepping approach with tracking differentiator, a distributed adaptive control law is proposed. The design procedure contains M steps. Define the following error variables

$$z_{i,1} = e_i = \sum_{j=1}^{N} |a_{ij}| \big(y_i - \mathrm{sgn}(a_{ij}) y_j\big) + \sum_{l=N+1}^{N+R} b_{il}(y_i - \theta_i r_l), \tag{5}$$

$$z_{i,k} = x_{i,k} - \alpha_{i,k-1}, \ (k = 2, \dots, M-1) \tag{6}$$

$$z_{i,M} = x_{i,M} - \alpha_{i,M-1} - \varpi_i, \tag{7}$$

where $\alpha_{i,m}$ is the virtual controller that will be designed later; and ϖ_i is the auxiliary signal defined as (18) to compensate the input saturation.

3.1.1 Step 1:

Considering (1), (5) and (6), the derivative of $z_{i,1}$ is

$$\begin{aligned} \dot{z}_{i,1} &= \sum_{j=1}^{N} |a_{ij}|\left(x_{i,2} - \operatorname{sgn}(a_{ij})x_{j,2}\right) + \sum_{l=N+1}^{N+R} b_{il}(x_{i,2} - \theta_i \dot{r}_l) \\ &= d_i\left(z_{i,2} + \alpha_{i,1} - \frac{1}{d_i}\sum_{j=1}^{N} a_{ij}x_{j,2} - \frac{1}{d_i}\sum_{l=N+1}^{N+R} b_{il}\theta_i \dot{r}_l\right). \end{aligned} \tag{8}$$

The virtual controller $\alpha_{i,1}$ is designed as

$$\alpha_{i,1} = -\frac{c_{i,1}}{d_i} z_{i,1} + \frac{1}{d_i}\sum_{j=1}^{N} a_{ij}x_{j,2} + \frac{1}{d_i}\sum_{l=N+1}^{N+R} b_{il}\theta_i \dot{r}_l, \tag{9}$$

where $c_{i,1}$ is a positive design parameter.

Select the Lyapunov function candidate as

$$V_{i,1} = \frac{1}{2} z_{i,1}^2. \tag{10}$$

Substituting (8) and (9), the time derivative of $V_{i,1}$ is

$$\dot{V}_{i,1} = z_{i,1}\dot{z}_{i,1} = -c_{i,1} z_{i,1}^2 + d_i z_{i,1} z_{i,2}. \tag{11}$$

3.1.2 Step K:

At this step, to avoid the complicated differential computation, let $\alpha_{i,k-1}$ pass through a tracking differentiator defined as

$$\begin{aligned} \dot{\omega}_{i,k} &= \omega'_{i,k}, \\ \dot{\omega}'_{i,k} &= -\tau_{i,k}\operatorname{sgn}\left(\omega_{i,k} - \alpha_{i,k-1} + \frac{\omega'_{i,k}|\omega'_{i,k}|}{2\tau_{i,k}}\right), \end{aligned} \tag{12}$$

to obtain the estimate value of $\dot{\alpha}_{i,k-1}$, where $\omega_{i,k}, \omega'_{i,k} \in \mathbb{R}$ are the state variables of tracking differentiator, and $\tau_{i,k} > 0$ is a design constant. Let $\chi_{i,k} = \omega'_{i,k} - \dot{\alpha}_{i,k-1}$ denote the bounded error according to Theorem 1.1 in [19]. Hence, there exists a positive constant $\bar{\chi}_{i,k}$ such that $|\chi_{i,k}| \leq \bar{\chi}_{i,k}$.

Considering (1), (6) and (12), the derivative of $z_{i,k}$ is

$$\dot{z}_{i,k} = x_{i,k+1} - \dot{\alpha}_{i,k-1} = z_{i,k+1} + \alpha_{i,k} - \omega'_{i,k} + \chi_{i,k}. \tag{13}$$

The virtual controller $\alpha_{i,k}$ is designed as

$$\alpha_{i,k} = -c_{i,k} z_{i,k} + \omega'_{i,k}, \tag{14}$$

where $c_{i,k}$ is a positive design parameter.

Select the Lyapunov function candidate as

$$V_{i,k} = \frac{1}{2} z_{i,k}^2. \tag{15}$$

Substituting (13) and (14), the time derivative of $V_{i,k}$ is

$$\dot{V}_{i,k} = z_{i,k}\dot{z}_{i,k} = -c_{i,k} z_{i,k}^2 + z_{i,k} z_{i,k+1} + z_{i,k}\chi_{i,k}. \tag{16}$$

3.1.3 Step M:

At the last step, let $\alpha_{i,M-1}$ pass through a tracking differentiator

$$\begin{aligned} \dot{\omega}_{i,M} &= \omega'_{i,M}, \\ \dot{\omega}'_{i,M} &= -\tau_{i,M} \operatorname{sgn}\left(\omega_{i,M} - \alpha_{i,M-1} + \frac{\omega'_{i,M} |\omega'_{i,M}|}{2\tau_{i,M}} \right), \end{aligned} \tag{17}$$

to obtain the estimate of $\dot{\alpha}_{i,M-1}$, where $\omega_{i,M}, \omega'_{i,M} \in \mathbb{R}$ are the state variables, and $\tau_{i,M} > 0$ is a design constant. In addition, $\chi_{i,M} = \omega'_{i,M} - \dot{\alpha}_{i,M-1}$ is bounded by $|\chi_{i,M}| \leq \bar{\chi}_{i,M}$ with $\bar{\chi}_{i,M} > 0$.

To handle the problem of input saturation, an auxiliary compensation signal ϖ_i is defined as

$$\dot{\varpi}_i + c_{i,M}\varpi_i = S_i(u_i) - u_i, \tag{18}$$

where $c_{i,M}$ is a positive design parameter.

Considering (1), (7), (17) and (18), and using neural networks to approximate the unknown function $f_i(\boldsymbol{x}_i) = \boldsymbol{W}_i^T \boldsymbol{\varphi}_i(\boldsymbol{x}_i) + \epsilon_i$, the derivative of $z_{i,M}$ is

$$
\begin{aligned}
\dot{z}_{i,k} &= f_i(\boldsymbol{x}_i) + S_i(u_i) + \zeta_i - \dot{\alpha}_{i,M-1} - \dot{\varpi}_i \\
&= u_i + \boldsymbol{W}_i^T \boldsymbol{\varphi}_i(\boldsymbol{x}_i) - \omega'_{i,k} + c_{i,M}\varpi_i + \epsilon_i + \zeta_i + \chi_{i,M},
\end{aligned} \tag{19}
$$

where $\boldsymbol{\varphi}_i(\boldsymbol{x}_i) : \mathbb{R}^M \to \mathbb{R}^{v_i}$ abbreviated $\boldsymbol{\varphi}_i$ with neurons number v_i is a basis functions vector; $\boldsymbol{W}_i \in \mathbb{R}^{v_i}$ is the ideal output weights vector; and $\epsilon_i \in \mathbb{R}$ is the bounded approximation error [20]. Define $\rho_i = \epsilon_i + \zeta_i$, and there exists a positive constant $\bar{\rho}_i$ such that $|\rho_i| \le \bar{\rho}_i$.

Then, the actual controller u_i is designed as

$$
u_i = -c_{i,M}(z_{i,M} + \varpi_i) - \frac{1}{2} z_{i,M} \hat{w}_i \boldsymbol{\varphi}_i^T \boldsymbol{\varphi}_i + \omega'_{i,M}, \tag{20}
$$

where $\hat{w}_i$ is the estimate of $w_i = ||\boldsymbol{W}_i||^2$, and the adaptive law is designed as

$$
\dot{\hat{w}}_i = \gamma_{wi} \left(\frac{1}{2} z_{i,M}^2 \boldsymbol{\varphi}_i^T \boldsymbol{\varphi}_i - \sigma_{wi} \hat{w}_i \right), \tag{21}
$$

where γ_{wi}, σ_{wi} are positive design constants.

Select the Lyapunov function candidate as

$$
V_{i,M} = \frac{1}{2} z_{i,M}^2 + \frac{1}{2\gamma_{wi}} \tilde{w}_i^2, \tag{22}
$$

where $\tilde{w}_i = w_i - \hat{w}_i$. Substituting (19), (20) and (21), and considering the following inequality

$$
\begin{aligned}
z_{i,M} \boldsymbol{W}_i^T \boldsymbol{\varphi}_i &\le \frac{1}{2} z_{i,M}^2 ||\boldsymbol{W}_i||^2 \boldsymbol{\varphi}_i^T \boldsymbol{\varphi}_i + \frac{1}{2} \\
&= \frac{1}{2} z_{i,M}^2 w_i \boldsymbol{\varphi}_i^T \boldsymbol{\varphi}_i + \frac{1}{2},
\end{aligned} \tag{23}
$$

the time derivative of $V_{i,M}$ is

$$
\begin{aligned}
\dot{V}_{i,M} =& z_{i,M} \dot{z}_{i,M} - \frac{1}{\gamma_{wi}} \tilde{w}_i \dot{\hat{w}}_i \\
=& - c_{i,M} z_{i,M}^2 + z_{i,M} \boldsymbol{W}_i^T \boldsymbol{\varphi}_i - \frac{1}{2} z_{i,M}^2 \hat{w}_i \boldsymbol{\varphi}_i^T \boldsymbol{\varphi}_i + z_{i,M} \rho_i + z_{i,M} \chi_{i,M} \\
& - \frac{1}{2} z_{i,M}^2 \tilde{w}_i \boldsymbol{\varphi}_i^T \boldsymbol{\varphi}_i + \sigma_{wi} \tilde{w}_i \hat{w}_i \\
\le& - c_{i,M} z_{i,M}^2 + z_{i,M} \rho_i + z_{i,M} \chi_{i,M} + \sigma_{wi} \tilde{w}_i \hat{w}_i + \frac{1}{2}.
\end{aligned} \tag{24}
$$

3.2 Stability Analysis

The main result of this paper is presented by the following theorem.

Theorem 1 *Consider a multi-agent network consisting of the dynamic followers* (1) *and multiple leaders. Let Assumption 1 hold. The control law* (20)*, the virtual controller* (9) *and* (14)*, the adaptive law* (21)*, the tracking differentiator* (12) *and* (17) *are designed. For any bounded initial conditions, all signals in the closed-loop system are semi-globally uniformly ultimately bounded. In addition, the bipartite containment error converges to an arbitrarily small neighborhood of the origin by tuning the design parameters.*

Proof The Lyapunov function candidate of closed-loop system is selected as

$$V = \sum_{i=1}^{N} V_{i,1} + \cdots + V_{i,k} + \cdots + V_{i,M}. \tag{25}$$

The following inequalities hold: $z_{i,m}z_{i,m+1} \leq \frac{1}{2}z_{i,m}^2 + \frac{1}{2}z_{i,m+1}^2$, $z_{i,k'}\chi_{i,k'} \leq \frac{1}{2}z_{i,k'}^2 + \frac{1}{2}\bar{\chi}_{i,k'}^2$, $z_{i,M}\rho_i \leq \frac{1}{2}z_{i,M}^2 + \frac{1}{2}\bar{\rho}_i^2$, $\tilde{w}_i\hat{w}_i \leq -\frac{1}{2}\tilde{w}_i^2 + \frac{1}{2}w_i^2$ by the Young's Inequality. Substituting (10), (16) and (24), the time derivative of V is

$$\begin{aligned} \dot{V} \leq \sum_{i=1}^{N}\Bigg(& -\left(c_{i,1} - \frac{d_i}{2}\right)z_{i,1}^2 - \left(c_{i,2} - \frac{d_i}{2} - 1\right)z_{i,2}^2 - \left(c_{i,3} - \frac{3}{2}\right)z_{i,3}^2 - \cdots \\ & - \left(c_{i,M} - \frac{3}{2}\right)z_{i,M}^2 - \frac{1}{2}\sigma_{wi}\tilde{w}_i^2 + \sum_{k'=2}^{M}\frac{1}{2}\bar{\chi}_{i,k'}^2 + \frac{1}{2}\bar{\rho}_i^2 + \frac{1}{2}\sigma_{wi}w_i^2 + \frac{1}{2}\Bigg). \end{aligned} \tag{26}$$

The design parameters are chosen as $\bar{c}_{i,1} = c_{i,1} - \frac{d_i}{2} > 0$, $\bar{c}_{i,2} = c_{i,2} - \frac{d_i}{2} - 1 > 0$, $\bar{c}_{i,\ell} = c_{i,\ell} - \frac{3}{2} > 0$ $(\ell = 3, \ldots, M)$. Then, (26) leads to

$$\begin{aligned} \dot{V} &\leq -\sum_{i=1}^{N}\sum_{m'=1}^{M}\bar{c}_{i,m'}z_{i,m'}^2 - \sum_{i=1}^{N}\frac{1}{2}\sigma_{wi}\tilde{w}_i^2 + \mu \\ &\leq -\beta V + \mu, \end{aligned} \tag{27}$$

where β and μ are positive constants defined as

$$\begin{aligned} \beta &= \min_{1\leq i\leq N, 1\leq m'\leq M}\left\{2\bar{c}_{i,m'}, \gamma_{wi}\sigma_{wi}\right\}, \\ \mu &= \sum_{i=1}^{N}\sum_{k'=2}^{M}\frac{1}{2}\bar{\chi}_{i,k'}^2 + \sum_{i=1}^{N}\frac{1}{2}\bar{\rho}_i^2 + \sum_{i=1}^{N}\frac{1}{2}\sigma_{wi}w_i^2 + \frac{N}{2}. \end{aligned} \tag{28}$$

Multiplying (27) by $e^{\beta t}$ and integrating both sides of (27) over $[0, t]$, we have

$$\int_0^t e^{\beta\tau}\dot{V}(\tau)\mathrm{d}\tau \le -\int_0^t e^{\beta\tau}\beta V(\tau)\mathrm{d}\tau + \int_0^t \mu e^{\beta\tau}\mathrm{d}\tau. \tag{29}$$

Then, (29) results in

$$0 \le V \le \left(V(0) - \frac{\mu}{\beta}\right)e^{-\beta t} + \frac{\mu}{\beta}. \tag{30}$$

Due to $\lim_{t\to\infty} V = \mu/\beta$, V is exponentially convergent. Thus, all signals of the closed-loop system are bounded. Consequently, error vector $\boldsymbol{e} = [e_1, \ldots, e_N]^T = [z_{1,1}, \ldots, z_{N,1}]^T$ is bounded. According to $\boldsymbol{e} = L_1\Theta(\Theta\boldsymbol{y} + \Theta L_1^{-1}\Theta L_2\boldsymbol{r}) = L_1\Theta\boldsymbol{\delta}$, bipartite containment error $\boldsymbol{\delta}$ has the bounded residual error. □

4 Simulation Example

Consider a multi-agent topology with five followers and two leaders which is described as Fig. 1.

Each following agent is modeled by a single-link robotic manipulator with actuator saturation as

$$\begin{aligned}
\dot{x}_{i,1} &= x_{i,2},\ (i = 1, \ldots, 5)\\
\dot{x}_{i,2} &= -J_i^{-1}M_i g l_i \sin(x_{i,1}) - J_i^{-1}B_i x_{i,2} + J_i^{-1}S_i(u_i) + J_i^{-1}\zeta_i,\\
y_i &= x_{i,1},
\end{aligned} \tag{31}$$

where $x_{i,1}$ is the angle position; $x_{i,2}$ is the angular velocity; J_i, B_i, M_i, l_i, g are system parameters. The bound of input saturation is $u_i^* = 30$. The two leaders' signals are given as $r_6 = 0.6\sin(0.3t + \pi) + 0.65$ and $r_7 = 0.2\sin(t + 0.5) + 0.75$.

During the simulation, the initial conditions are taken as $\boldsymbol{x}_1(0) = [-0.2, 0]^T$, $\boldsymbol{x}_2(0) = [0.2, 0]^T$, $\boldsymbol{x}_3(0) = [0.4, 0]^T$, $\boldsymbol{x}_4(0) = [-0.3, 0]^T$, $\boldsymbol{x}_5(0) = [0.5, 0]^T$, $\omega_{i,2}(0) = \omega'_{i,2}(0) = 0$, $\hat{w}_i(0) = 0$. The design parameters are chosen as $c_{i,1} = 18$, $c_{i,2} = 10$, $\tau_{i,2} = 2$, $\gamma_{wi} = 5$, $\sigma_{wi} = 0.01$.

The simulation results are shown in Figs. 2 and 3. The bipartite containment trajectories of the angular positions are depicted by Fig. 2a, where the envelope curves

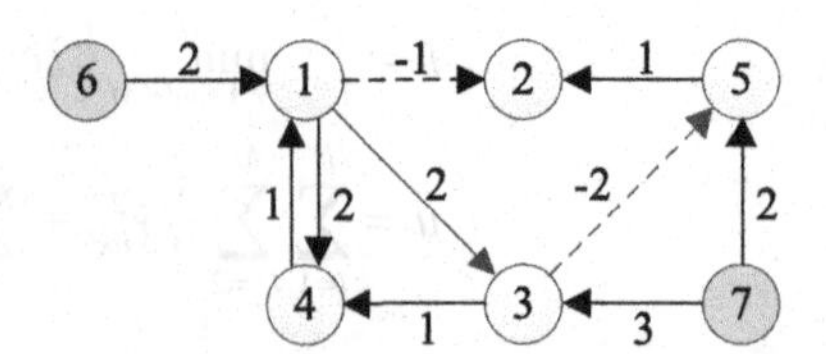

Fig. 1 Topology of the signed digraph $\mathcal{G}$

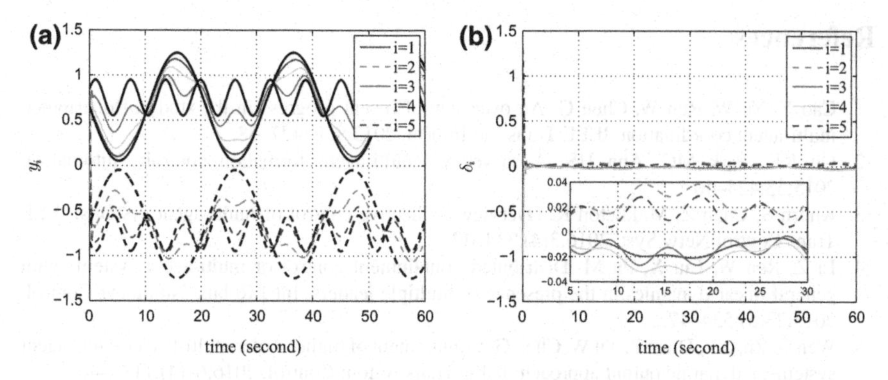

Fig. 2 **a** Profiles of the output trajectories y_i; **b** Profiles of the bipartite containment errors δ_i

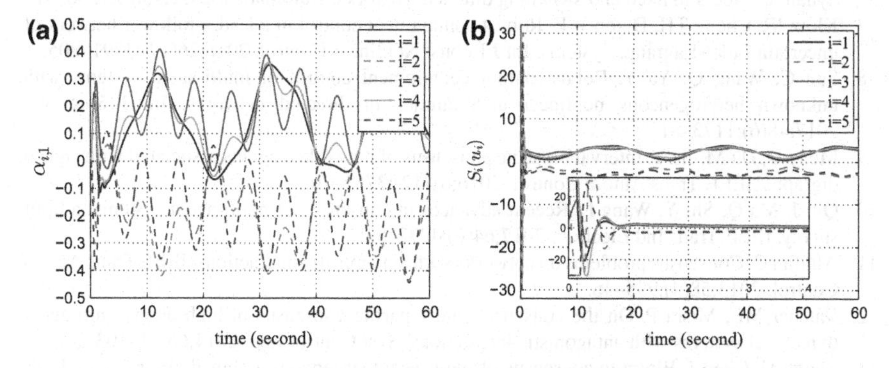

Fig. 3 **a** Evolutions of the virtual control signals $\alpha_{i,1}$; **b** Evolutions of the saturation input signals $S_i(u_i)$

are $\pm r_6$ and $\pm r_7$. Figure 2b shows that the bipartite containment errors converge to a small neighbourhood of the origin. The control signals are shown by Fig. 3, with $\alpha_{i,1}$ being the virtual control signals and $S_i(u_i)$ being the saturation input signals.

5 Conclusions

The bipartite containment control problem of nonlinear multi-agent systems with input saturation has been investigated in this paper. The multi-agent network is described by a signed digraph with multiple leaders and structurally balanced followers subgraph. In addition, a distributed control law for each following agent is proposed. Finally, simulation results justify the design algorithm.

References

1. Cao Y, Yu W, Ren W, Chen G. An overview of recent progress in the study of distributed multi-agent coordination. IEEE Trans Ind Inform. 2013;9(1):427–38.
2. Oh KK, Park MC, Ahn HS. A survey of multi-agent formation control. Automatica. 2015;53:424–40.
3. Knorn S, Chen Z, Middleton R. Overview: collective control of multi-agent systems. IEEE Trans Control Netw Syst. 2016;3(4):334–47.
4. Li Z, Ren W, Liu X, Fu M. Distributed containment control of multi-agent systems with general linear dynamics in the presence of multiple leaders. Int J Robust Nonlinear Control. 2013;23(5):534–47.
5. Wen G, Zhao Y, Duan Z, Yu W, Chen G. Containment of higher-order multi-leader multi-agent systems: a dynamic output approach. IEEE Trans Autom Control. 2016;61(4):1135–40.
6. Cao Y, Ren W, Egerstedt M. Distributed containment control with multiple stationary or dynamic leaders in fixed and switching directed networks. Automatica. 2012;48(8):1586–97.
7. Klotz JR, Cheng TH, Dixon WE. Robust containment control in a leader-follower network of uncertain Euler-Lagrange systems. Int J Robust Nonlinear Control. 2016;26(17):3791–805.
8. Sun C, Wang Q, Yu Y. Robust output containment control of multi-agent systems with unknown heterogeneous nonlinear uncertainties in directed networks. Int J Syst Sci. 2017;48(6):1173–81.
9. Meng D, Du M, Jia Y. Interval bipartite consensus of networked agents associated with signed digraphs. IEEE Trans Autom Control. 2016;61(12):3755–70.
10. Qin J, Ma Q, Shi Y, Wang L. Recent advances in consensus of multi-agent systems: a brief survey. IEEE Trans Ind Electron. 2017;64(6):4972–83.
11. Altafini C. Consensus problems on networks with antagonistic interactions. IEEE Trans Autom Control. 2013;58(4):935–46.
12. Valcher ME, Misra P. On the consensus and bipartite consensus in high-order multi-agent dynamical systems with antagonistic interactions. Syst Control Lett. 2014;66:94–103.
13. Zhang H, Chen J. Bipartite consensus of multi-agent systems over signed graphs: state feedback and output feedback control approaches. Int J Robust Nonlinear Control. 2017;27(1): 3–14.
14. Zhai S, Li Q. Practical bipartite synchronization via pinning control on a network of nonlinear agents with antagonistic interactions. Nonlinear Dyn. 2017;87(1):207–18.
15. Yaghmaie FA, Su R, Lewis FL, Olaru S. Bipartite and cooperative output synchronizations of linear heterogeneous agents: a unified framework. Automatica. 2017;80:172–6.
16. Wu Y, Hu J, Zhang Y, Zeng Y. Interventional consensus for high-order multi-agent systems with unknown disturbances on coopetition networks. Neurocomputing. 2016;194:126–34.
17. Fu W, Qin J, Zheng WX, Gao H, Shi G. Semi-global bipartite consensus for linear multi-agent systems subject to actuator saturation. In: Proc. 55th IEEE Conf. Decision Control, Las Vegas, USA;2016. pp. 1757–62
18. Qin J, Fu W, Zheng WX, Gao H. On the bipartite consensus for generic linear multiagent systems with input saturation. IEEE Trans Cybern. 2017;47(8):1948–58.
19. Guo BZ, Zhao ZL. On convergence of tracking differentiator. Int J Control. 2011;84(4):693–701.
20. Ge SS, Hang CC, Lee TH, Zhang T. Stable Adaptive Neural Network Control. New York, NY, USA: Springer; 2002.

The Development of a Brain-Controlled Lock Based on SSVEP and MI

Jingtao Guan, Zhiwen Zhang, Rensong Liu, Zengqiang Chen and Feng Duan

Abstract Ordinary locks, even fingerprint or iris locks are no longer as safe as they used to be. However, the Brain-Controlled Lock can solve this problem because of different electroencephalogram (EEG) for different people. In this paper, the lock is controlled by a Brain–Computer Interface (BCI) system. The Motor Imagery (MI) procedures for EEG process are de-noising, feature extraction, and classification. And steady state visual evoked potential (SSVEP) can judge different frequencies by canonical correlation analysis (CCA) method. The BCI system with 2 parts, each for 9 targets, chooses which part of flickers by MI and frequencies as number codes by SSVEP. 3 subjects are tested on offline and online experiments. The developed BCI system performs well in experiments, and the average accuracy is 87%.

Keywords Brain-Controlled Lock · Brain–computer interface (BCI) system · Motor imagery (MI) · Steady state visual evoked potential (SSVEP)

J. Guan · Z. Zhang · R. Liu · Z. Chen · F. Duan (✉)
College of Computer and Control Engineering, NanKai University,
Tianjin 300350, China
e-mail: duanf@nankai.edu.cn

J. Guan
e-mail: guanjingtao1@126.com

Z. Zhang
e-mail: 2120160365@mail.nankai.edu.cn

R. Liu
e-mail: liurensong@mail.nankai.edu.cn

Z. Chen
e-mail: chenzq@nankai.edu.cn

Y. Jia et al. (eds.), *Proceedings of 2017 Chinese Intelligent Systems Conference*, Lecture Notes in Electrical Engineering 459,
https://doi.org/10.1007/978-981-10-6496-8_38

1 Introduction

The security of ordinary locks become more and more serious. For some professionals or other people, it is not very difficult to unlock by the degree of wear, plasticity of tinfoil and so on. In 2015, China's public security organization investigated about two hundred thousand larcenies. The burglary of going into room takes up a large share. It is still a critical factor influencing citizens' life.

Under this situation, the locks based on fingerprint or iris have appeared. The fingerprint lock has advantage on safety compared to common locks [1]. But the fingerprint is easy to get from objects that a person touches. Besides, it might be unrecognized by fingerprint and needs to require multiple attempts. However, the iris lock can solve this problem to a certain extent [2]. But iris recognition needs to take photos towards eyes. It may unlock through the picture of one's eyes. Two types of locks are safer than ordinary lock, but are still not totally safe.

However, EEG are different for everyone [3]. This feature can be used as number codes on lock and Brain-Controlled Lock can be safer than other locks.

Also, BCI system can establish a communication system between brain and devices to transfer EEG signals into commands, which does not depend on the brain's normal output pathways of peripheral nerves and muscles [4]. The input signals for BCI contains many kinds and P300, steady-state visual evoked potentials, motor imagery are the most common types [5, 6].

In this paper, it will carry out a research for the Brain-Controlled lock based on SSVEP and MI. Choose flickering part from two parts, and then choose 3 number codes to unlock. Through different EEG of different persons, it can be safer than other locks.

2 Brain-Controlled Lock System Structure

In the experiment, two actions will be classified. One action is chosen for flickers on the upper left and the other is for the lower right part. The subjects can choose which part of flickers by imaging one action through de-noising, feature extraction and classification methods of MI.

Then, stare at one target of flickering part which the subjects choose. By the CCA method, it can judge which frequency stared at. One frequency represents one number. The Brain-Controlled Lock receives 3 number codes, and judge whether to unlock (Fig. 1).

In addition, the lock contains electronic lock, relay and Atmel SAMD21 development board. Relay is used to help unlock. The Atmel SAMD21 development board is a low power microcontroller based on 32-bit ARM® Cortex®-M0 + , with 256 KB flash memory. The procedures for lock are initializing the lock at first. Then, define an address of signal port, for waiting an unlocking command from a computer. When getting an opening command, rotate the electronic lock by the

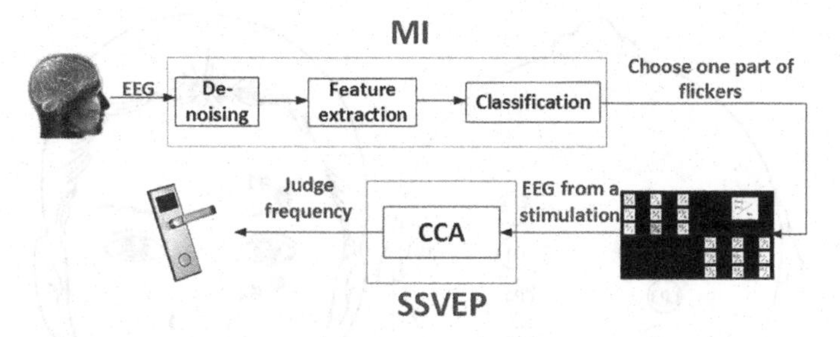

Fig. 1 Brain-Controlled Lock system structure

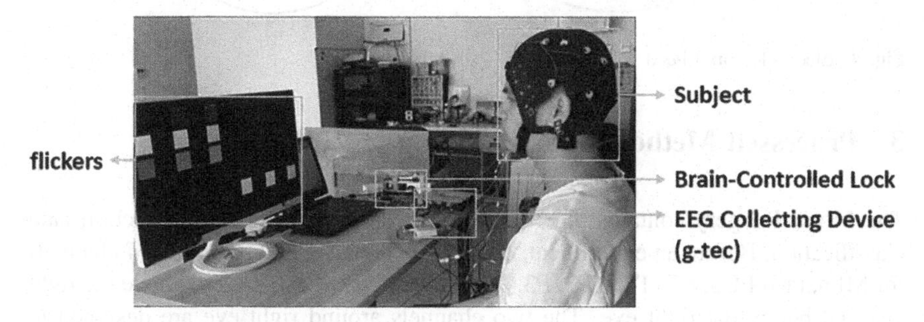

Fig. 2 The environment of experiments

corresponding pulse sent from signal port, and unlock successfully. And the photo of experiment's environment is shown as Fig. 2.

As shown in Fig. 3, the flickers are divided into two parts. One is the 9 targets on the upper left and the other is on the lower right. And every target stimulation's frequency and correspondent number are presented in Fig. 2. The upper left frequencies are 6, 6.5, 7, 7.5, 8, 8.5, 9, 9.5 and 10 Hz, which correspond to number 1, 2, 3, 4, 5, 6, 7, 8, 9. It is same as the lower right flicker besides frequencies, which are 6.2, 6.7, 7.2, 7.7, 8.2, 8.7, 9.2, 9.7 and 10.2 Hz.

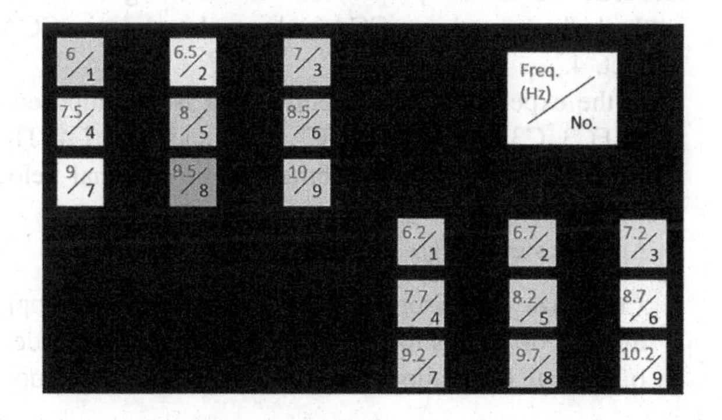

Fig. 3 Frequencies and correspondent numbers of flickers

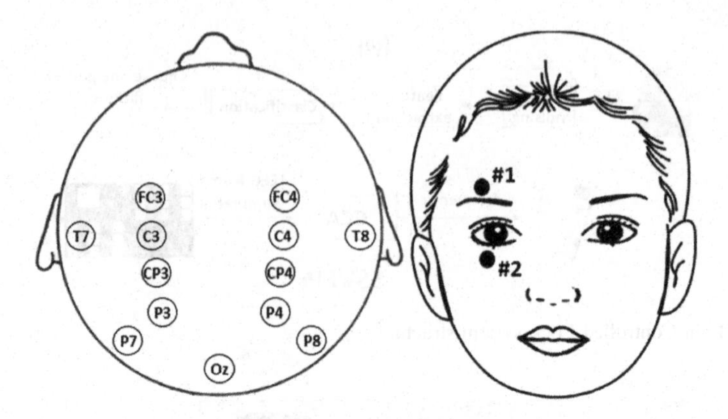

Fig. 4 Electrodes on a head

3 Processed Methods in MI

The Motor Imagery contains three procedures: de-noising, feature extraction, and classification. Before an experiment, collect 15 channels' EEG signals, 14 channels for MI named FC3, C3, T7, CP3, P3, P7, FC4, C4, T8, CP4, P4, P8, above the right eye and below the right eye. The two channels around right eye are designed to reduce ocular artifacts. And the last channel Oz is for SSVEP. The sampling frequency is 256 Hz. In offline experiments, 50 groups' data are for training and 10 groups' data are for testing. And in online experiments, 60 groups' data are for training and input one data as testing (Fig. 4).

3.1 De-Nosing

The EEG signals are easily to be interfered by irrelevant noise. In the noises, the noise of eyes called ocular artifacts interferes most, which derives from potential difference between the cornea and the fundus of the eye. For reducing the ocular artifacts, we should preprocess the EEG signals. The method taken is wCCA method which combines CCA and wavelet analysis. CCA method will be illustrated in Sect. 4.

In the experiment, one group's data is decomposed signals into two opposite parts: FC3, C3, T7, CP3, P3, P7, EOG and FC4, C4, T8, CP4, P4, P8, EOG. EOG is the difference which is subtracting EEG signal below the right eye from one above the right eye.

And the steps to go are listed:

1. Centralize the disordered EEG signals of two opposite parts, and turn the average of signals into zero. By the CCA method, decompose to get first couple of typical correlation variables. And then, decompose the typical correlation

variables with wavelet method into 5 levels. Get each level's wavelet coefficient and the last level's scale coefficient.

2. Each level's wavelet coefficient and the last level's scale coefficient are processed with hard threshold. Make zero when coefficient is above the threshold, and maintain coefficient when it is below the threshold. The threshold for each wavelet coefficient (D_1 to D_5) is:

$$Threshold = \sqrt{2\log(Di)} * |median(Di)|/0.6745 \quad (1)$$

And the threshold for last level's scale coefficient (A_5) is:

$$Threshold = \sqrt{2\log(A5)} * |median(A5)|/0.6745 \quad (2)$$

3. After wavelet coefficient and scale coefficient processed by threshold, the first couple of typical correlation variables are reconstituted by inverse wavelet transform. Then, two opposite parts are reconstituted by weight vectors from CCA method, and finally, get the EEG signals with ocular artifacts reduced.

3.2 Feature Extraction

Common Spatial Pattern (CSP) is a spatial filtering algorithm which can extract spatial features of multi-channel EEG signals. It is based on the simultaneous diagonalization of two covariance matrices [7, 8].

XA is EEG signals which have been collected about relaxation and *XB* is for right foot movement's imagination.

The covariance matrix R_A about *XA* is:

$$R_A = \frac{X_A X_A^T}{trace(X_A X_A^T)} \quad (3)$$

And the covariance matrix R_B about *XB* is similar.

Mixed covariance matrix *R*:

$$R = R_A + R_B = U_0 \Lambda U_0^T \quad (4)$$

Sort the eigenvalues of diagonal matrix Λ in descending order and U_0 corresponding to Λ is reordered as well.

Whitening matrix P:

$$P = \sqrt{\Lambda^{-1}} U_0^T \tag{5}$$

Whiten covariance matrixes R_A and R_B:

$$S_A = PR_A P^T \quad S_B = PR_B P^T \tag{6}$$

There can be a common feature matrix between S_A and S_B:

$$S_A = U\Lambda_A U^T \quad S_B = U\Lambda_B U^T \tag{7}$$

Also, the relation between Λ_A and Λ_B:

$$\Lambda_A + \Lambda_B = I \tag{8}$$

Projection matrix W:

$$W = U^T P \tag{9}$$

Extract features from an experiment data X:

$$Z = WX \tag{10}$$

3.3 Feature Classification

The k-Nearest Neighbor (KNN) algorithm is a non-parametric method used for classification. The result of classification depends on a majority vote of its neighbors among k closest training data in the feature space. The method is a type of learning based on instance. Call the function of KNN in matlab, and the parameter k is 11.

In the offline experiments, collect 60 groups' data of each action. 50 groups are for training and 10 groups are for testing. And every group's data lasts 4 s. There are 2 actions (right foot movement's imagination and relaxation) in the experiments. The data are from 3 healthy right-handed subjects. The average accuracy of discriminating two classes is 96.6%. And the confusion matrix of three subjects is shown as Table 1:

Table 1 MI offline experiments

	Predicted classification		
Actual clasification		Relaxation	Right foot movement
	Relaxation	100%	0
	Right foot movement (%)	6.3	93.3

4 CCA Method in SSVEP

To get the underlying correlation for two sets of data, CCA is a good choice. It is a multivariable statistical algorithm and extensively used. By the two multi-dimensional variables X, Y and their linear combinations $x = X^T W_x$, $y = Y^T W_y$, CCA can find the weight vectors, W_x and W_y, which can maximize the correlation between x and y. The maximum of ρ corresponds to the maximum canonical correlation [9].

$$\max_{W_x, W_y} \rho(x, y) = \frac{E[x^T y]}{\sqrt{E[x^T x] E[y^T y]}} = \frac{E[W_x^T X Y^T W_y]}{\sqrt{E[W_x^T X X^T W_x] E[W_y^T Y Y^T W_y]}} \tag{11}$$

X refers to multi-channel SSVEPs and Y relates to reference signals. Considering influence of harmonic frequencies, conference signals Y_f called sinusoidal signals are shown as the function below, where f is the stimulation frequency and N_h is the number of harmonics. In this paper, N_h is 3. CCA can calculate the canonical correlation between two sets of data, multi-channel SSVEPs and the reference signals, to get frequency by the maximal correlation [10].

$$Y_f = \begin{bmatrix} \sin(2\pi f t) \\ \cos(2\pi f t) \\ \vdots \\ \sin(2\pi N_h f t) \\ \cos(2\pi N_h f t) \end{bmatrix} \tag{12}$$

In the offline experiments, collect 27 groups' data. As a result of no high need for real time, it sets every data last 4 s to get a high accuracy, compared to 2 s data usually set. It is the same in online experiments. Subject 2 has not been trained before, so the accuracy is lower than others (Table 2).

Table 2 SSVEP offline experiments

	Right numbers	Wrong numbers	Accuracy (%)
Subject1	25	2	92.6
Subject2	23	4	85.2
Subject3	26	1	96.3
Average	24.7	2.3	91.5

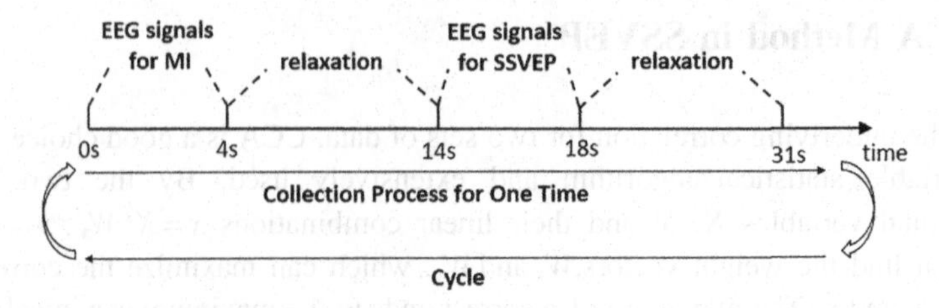

Fig. 5 Experimental paradigm

5 Online Experiments

Initialize the EEG collecting device called g-tec first, and then start collecting 15 channels' EEG signals. The passwords containing 3 number codes have been set before. Collect data for 4 s and by methods of MI, choose one part of flickers from two. After 10s' rest, on the help of CCA algorithm of SSVEP, stare at one stimulation for 4s to get one number. Have a rest of 13s, continue repeating again (Fig. 5).

When getting 3 numbers, compare passwords with inputs. If inputs are the same with passwords, unlock. If not, try another chance and clear inputs of 3 numbers.

The chances can be 3 times at most. If failing for 3 chances, the subject cannot input any number (Fig. 6).

Fig. 6 The flow chart of an experiment

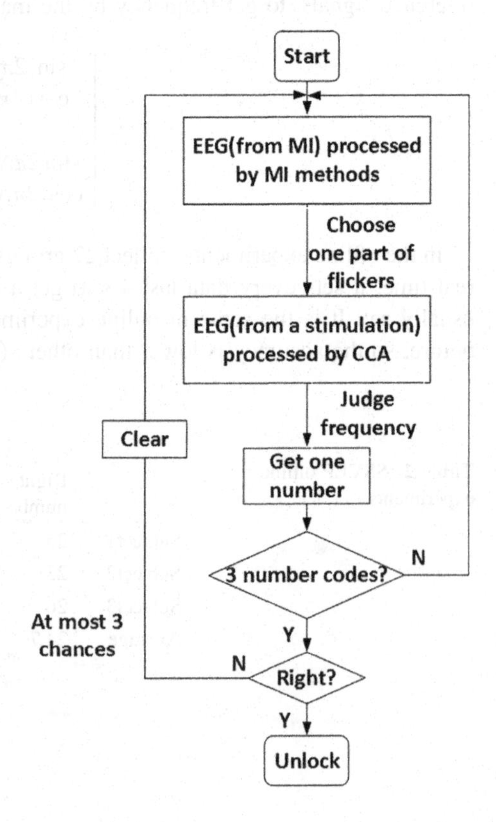

Table 3 Online experiments

	Right times	Wrong times	Accuracy (%)
Subject1	9	1	90
Subject2	7	3	70
Subject3	10	0	100
Average	8.7	1.3	87

Three subjects are tested for 10 times in the experiments and every time has 3 chances. Because there are many differences between two classifications of relaxation and right foot movement's imagination on MI, it is easy to classify the EEG signals processed by wCCA, CSP, KNN methods. The classification of SSVEP can be better through the collecting time set for 4s in every period. The reasons why the accuracy of subject 2 is lower than others is that subject 2 has not been trained for SSVEP before. Also, staring at stimulations for a long time may make him uncomfortable. Besides, Subject 1 and 3 can reach a satisfactory result. The average accuracy of experiments is about 87% (Table 3).

6 Conclusion

The multi-channel BCI system based on SSVEP and MI performs well in this paper. The procedures for EEG process of MI are de-noising, feature extraction, and classification. De-noising is an essential step to reduce ocular artifacts. Feature extraction and classification aim to classify EEG signals to choose one flickers' part.

CCA is a method to extract frequency information according to the EEG signals. The offline experiments show that when time goes longer until 4 s without high need for real time of MI and SSVEP data, the accuracies of both can get higher. The online data also last 4 s. And the online experiments achieve an average accuracy of 87%. Because of the high accuracy, Brain-Controlled Lock can be possible. In addition, Parameter optimization and channel selection are not a must for every subject. So the Brain-Controlled Lock system is more useful in practical applications.

References

1. Yang WJ. Design of hardware platform for high-performance fingerprint locks. Electron Des Eng. 2010.
2. Kim S. Security system and method by iris key system for door lock: US;2006. WO/2006/043765.
3. Hu JF. Person identification based on electroencephalogram signals. J Clin Rehabil Tissue Eng Res. 2009;13(17):3260–4.

4. Wolpaw JR, Birbaumer N, Heetderks WJ, et al. Brain-computer interface technology: a review of the first international meeting. IEEE Trans Rehabil Eng. 2000;8(2):164.
5. Farwell LA, Donchin E. Talking off the top of your head: toward a mental prosthesis utilizing event-related brain potentials. Electroencephalogr Clin Neurophysiol. 1988;70(6):510–23.
6. Birbaumer N, Kübler A, Ghanayim N, et al. The thought translation device (TTD) for completely paralyzed patients. IEEE Trans Rehabil Eng. 2000;8(2):190.
7. Kai KA, Zheng YC, Zhang H, et al. Filter bank common spatial pattern (FBCSP) in brain-computer interface. In: IEEE international joint conference on neural networks. IEEE Xplore; 2008. p. 2390–397.
8. Keng AK, Yang CZ, Wang C, et al. Filter bank common spatial pattern algorithm on BCI competition IV datasets 2a and 2b. Front. Neurosci. 2012;6:39.
9. Bin G, Gao X, Yan Z, et al. An online multi-channel SSVEP-based brain-computer interface using a canonical correlation analysis method. J Neural Eng. 2009;6(4):046002.
10. Chen X, Wang Y, Nakanishi M, et al. High-speed spelling with a noninvasive brain-computer interface. Proc Natl Acad Sci USA. 2015;112(44):E6058.

SOC Estimation of Extended Kalman Filter Based on the Model Data Optimization

Ziyi Fu, Xian Hua and Xiangwei Guo

Abstract An accurate power-battery state of charge(SOC) estimation plays an important role in battery electric vehicles(EVs). Affected by random factors such as working conditions and environment, Li-ion power battery has very strong time-varying nonlinearity in the application of EVs, the research on SOC estimation of power battery is of great theoretical significance and application value. This paper studies the SOC estimation using extended Kalman filter(EKF), which is based on the Thevenin equivalent circuit model. And then a reasonable optimization method of the parameters of the model is presented to improve the SOC estimation accuracy. The simulation results demonstrate that the optimization method can improve the SOC estimation precision remarkably with little influence on the initial error convergence by means of EKF.

Keywords Thevenin model · Data optimization · Extended Kalman filter (EKF) · State of charge(SOC)

1 Introduction

With the rapid development of electric vehicles(EVs), Lithium-iron battery has become the mainstream and preferred of the EV power battery [1]. However, the structure of battery is complex and some unstable factors can cause a serious damage to the batteries of EVs, such as the charge-discharge current, ambient

Z. Fu · X. Hua · X. Guo (✉)
School of Electrical Engineering and Automation, Henan Polytechnic University, Jiaozuo 454000, China
e-mail: gxw@hpu.edu.cn

Z. Fu
e-mail: fuzy@hpu.edu.cn

X. Hua
e-mail: 573990866@qq.com

Y. Jia et al. (eds.), *Proceedings of 2017 Chinese Intelligent Systems Conference*, Lecture Notes in Electrical Engineering 459,
https://doi.org/10.1007/978-981-10-6496-8_39

temperature and so on. To safeguard the safety performance of EV battery, an accurate state-of-charge(SOC)estimation algorithm is highly desired [2].

Currently, SOC estimation method mainly include the open circuit voltage (OCV) method, ampere-hour integral method, neutral network method and Kalman filter(KF). OCV method is sufficiently accurate but it requires a long rest time and thus cannot be used in real-time estimation; The ampere-hour integral method has high estimation accuracy under the premise of accurate initial value in a short period of time, and it accumulates error in the process of running; The neutral network method requires a large number of sample data for training, while its shortcoming is that the computational cost is high; KF algorithm used in this paper has strong error correction ability, is the most popular approach in practical research [3, 4], KF is considered an integrated method because it requires the use of a battery model and a state observe. Combined with the estimation methods above, due to the nonlinear system, the extended Kalman filter(EKF)algorithm can be used for real-time SOC estimation. The EKF has emerged as one of the practical algorithms to enhance the accuracy of SOC estimation [5].

The equivalent circuit model of the battery is used to simulate the dynamic characteristics of the battery. Currently, the Rint model, Thevenin model, PNGV model, GNL model and a multi-RC model, are used widely [6]. Comparing with the above models, the Thevenin model is accurate enough, furthermore, it is simpler in structure than the multi-order ones [7, 8].

The KF is mainly affected by the accuracy of the battery model, in order to establish a model as much accurate as possible, a optimizing method based on data parameters is used to improve the accuracy of the model. Finally, the feasibility of the optimized model and EKF algorithm are verified by simulation experiments.

2 EKF Algorithm for SOC Estimation

Kalman filter is a recursive estimator, which is widely used in estimation problems and optimal estimator for linear models. For the nonlinear time-varying characteristics of the lithium battery studied in this paper, the EKF estimation method is used as the core algorithm.

EKF, is described by the following state-space equations:

$$\begin{cases} x_{k+1} = f(x_k, u_k) + \Gamma_k w_k \\ y_k = g(x_k, u_k) + v_k \end{cases} \tag{1}$$

where x_k is the present state estimation, x_{k+1} is the next state estimation, y_k is the present output estimation, u_k is the system input, Γ_k stands for the interference matrix, w_k is the system noise, v_k is the observation noise, $f(x_k, u_k)$ is the nonlinear state transition function, $g(x_k, u_k)$ is the observation function.

$f(x_k, u_k)$ and $g(x_k, u_k)$ are linearized using a first order Taylor approach. And performing a linearization at point $\hat{x}_k$ the following functions are obtained

$$\begin{cases} f(x_k, u_k) \approx f(\hat{x}_k, u_k) + \frac{\partial f(x_k, u_k)}{\partial x_k}\Big|_{x_k = \hat{x}_k}(x_k - \hat{x}_k) \\ g(x_k, u_k) \approx g(\hat{x}_k, u_k) + \frac{\partial g(x_k, u_k)}{\partial x_k}\Big|_{x_k = \hat{x}_k}(x_k - \hat{x}_k) \end{cases} \tag{2}$$

Definition: $\hat{A}_k = \frac{\partial f(x_k, u_k)}{\partial x_k}\Big|_{x_k = \hat{x}_k}, \hat{C}_k = \frac{\partial g(x_k, u_k)}{\partial x_k}\Big|_{x_k = \hat{x}_k}$. The system state equation and observation equation can be rewritten as

$$\begin{cases} x_{k+1} \approx \hat{A}_k x_k + [f(\hat{x}_k, u_k) - \hat{A}_k \hat{x}_k] + \Gamma_k w_k \\ y_k \approx \hat{C}_k x_k + [g(\hat{x}_k, u_k) - \hat{C}_k \hat{x}_k] + v_k \end{cases} \tag{3}$$

According to the formula (3), the recursive formula of EKF algorithm for nonlinear discrete systems can be expressed as

$$\begin{cases} \hat{x}_{0/0} = E(x_0), P_{0/0} = \mathrm{var}(x_0) \\ \hat{x}_{k/k-1} = f(\hat{x}_{k-1/k-1}, u_{k-1}) \\ P_{k/k-1} = \hat{A}_{k-1} P_{k-1/k-1} \hat{A}_{k-1}^T + \Gamma_{k-1} Q_{k-1} \Gamma_{k-1}^T \\ K_k = P_{k/k-1} \hat{C}_k^T (\hat{C}_k P_{k/k-1} \hat{C}_k^T + R_k)^{-1} \\ \hat{x}_{k/k-1} = \hat{x}_{k/k-1} + K_k [y_k - g(\hat{x}_{k/k-1}, u_k)] \\ P_{k/k} = (I - K_k \hat{C}_k) P_{k/k-1} \end{cases} \tag{4}$$

where $k = 1, 2, \ldots$.

After discretization and linearization, the system state equation is shown as

$$\begin{bmatrix} SOCk+1 \\ Up,k+1 \end{bmatrix} = \begin{bmatrix} 1 & 0 \\ 0 & Ap \end{bmatrix} \begin{bmatrix} SOCk \\ Up,k \end{bmatrix} + \begin{bmatrix} -\eta T/C \\ Bp \end{bmatrix} Ik + \begin{bmatrix} w_1(k) \\ w_2(k) \end{bmatrix} \tag{5}$$

where $Ap = 1 - \Delta T / R_p C_p$, $Bp = \Delta T / C_p$, ΔT equals system sampling time, C means battery rated capacity, η stands for the coefficient of variation of capacity at different discharge rates, $w_1(k)$, $w_2(k)$ are system noise.

System observation equation can be derived as

$$[U_L(k)] = [U_{oc}] - [0 \quad 1] \begin{bmatrix} SOC(k) \\ U_p(k) \end{bmatrix} - [R_0][I(k)] + [v(k)] \tag{6}$$

From formulas (5) and (6) the desired matrix of the KF can be identified as follows:

$$x_k = \begin{bmatrix} SOC(k) \\ U_p(k) \end{bmatrix}, A_k = \begin{bmatrix} 1 & 0 \\ 0 & A_P \end{bmatrix}, B_k = \begin{bmatrix} -\eta T/C \\ Bp \end{bmatrix}, C_k = \begin{bmatrix} \frac{\partial(U_{oc})}{\partial SOC(k)} & 1 \end{bmatrix}$$

Then we can get the recursive process:

$$\begin{cases} \hat{x}_{0/0} = E(x_0)\,,\ P_{0/0} = \mathrm{var}(x_0) \\ \hat{x}_{k/k-1} = A_k x_k + B_k I(k) \\ P_{k/k-1} = A_{k-1} P_{k-1/k-1} A_{k-1}^T + Q_k \\ K_k = P_{k/k-1} C_k^T (C_k P_{k/k-1} C_k^T + R_k)^{-1} \\ \hat{x}_{k/k-1} = \hat{x}_{k/k-1} + K_k [y_k - g(\hat{x}_{k/k-1}, u_k)] \\ P_{k/k} = (I - K_k C_k) P_{k/k-1} \\ k = 1,\ 2,\ \ldots \end{cases} \tag{7}$$

3 Battery Modeling

Considering the model accuracy, complexity and use of lithium-iron battery in experiment, a Thevenin equivalent circuit model is used in this paper and the model is shown in Fig. 1 [9].

OCV U_{oc} is a battery voltage that measured when the battery is not connected to the load, I_L is the current of the battery, U_p is the voltage of the parallel R_p and C_p . R_0 is the internal resistance while R_p and C_p are the polarization resistance and capacitance respectively.

The circuit principle for the equivalent circuit model in Fig. 1 is as follows:

$$\begin{cases} I_L = \frac{U_p}{R_p} + C_p \frac{dU_p}{dt} \\ U_L = U_{oc} - U_p - I_L R_0 \end{cases} \tag{8}$$

The initial polarization voltage of the model is defined as $U_p(0)$, and the time constant is $\tau = R_p C_p$, solving the Eq. (8):

$$\begin{cases} U_p(t) = U_p(0) \cdot e^{-(t/\tau)} + I_L \cdot R_p \cdot (1 - e^{-(t/\tau)}) \\ U_L(t) = U_{oc} - U_p(0) \cdot e^{-(t/\tau)} - I_L \cdot R_p \cdot (1 - e^{-(t/\tau)}) - I_L(t) R_0 \end{cases} \tag{9}$$

4 Identification of Dynamic Parameters of the Battery Model

Taking the nominal capacity of 2.7 Ah and the rated voltage of 3.2 V lithium-iron phosphate(LiFePO4)battery as research objects, the HPPC Test(Hybrid Pulse Power Characterization Test) was held in this paper [10]. The pulse discharge

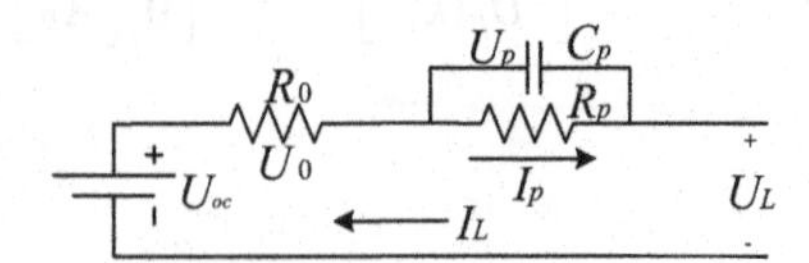

Fig. 1 Thevenin equivalent circuit

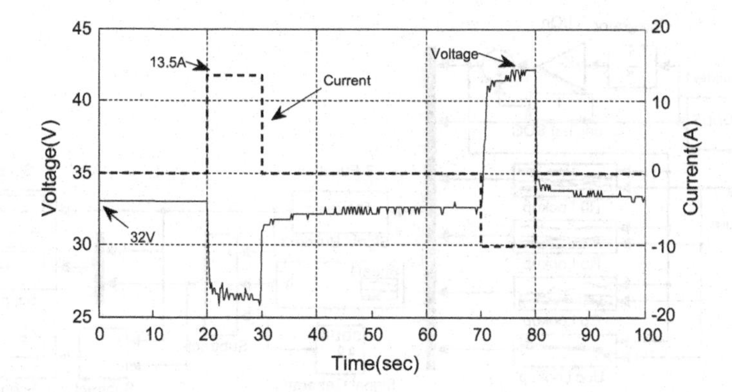

Fig. 2 HPPC pulse current and voltage responses

current is 5C, pulse charging current is 0.75 times the pulse discharge current. The input current waveform and the voltage response are shown in Fig. 2.

Considering the influence of temperature and SOC on the parameters of the model, nine SOC settings of 0.9, 0.8, 0.7, …, 0.1 are chosen for comparison of the battery model performances across various operating conditions. The parameters are identified by the completely discharge of HPPC experiment as shown in Fig. 3.

The U_{oc} of the cell has a nonlinear relationship with SOC. To obtain this function, the OCV test is conducted using the lithium-iron battery. A polynomial curve fitting is used to describe the relationship between the U_{oc} and the SOC

$$U_{oc} = 13.62\lambda^5 - 47.17\lambda^4 + 61.64\lambda^3 - 37.5\lambda^2 + 11.34\lambda + 31.39 \quad (10)$$

Other parameters of the EKF system, R_0, R_p and C_p can be obtained by two-dimensional table look-up interpolation based on the corresponding values of temperature and SOC values of the Thevenin model [11].

The experimental object is made up of 10 sections of LiFePO4 battery in series. In this paper, a battery simulation model is established in Matlab/Simulink, as shown in Fig. 4.

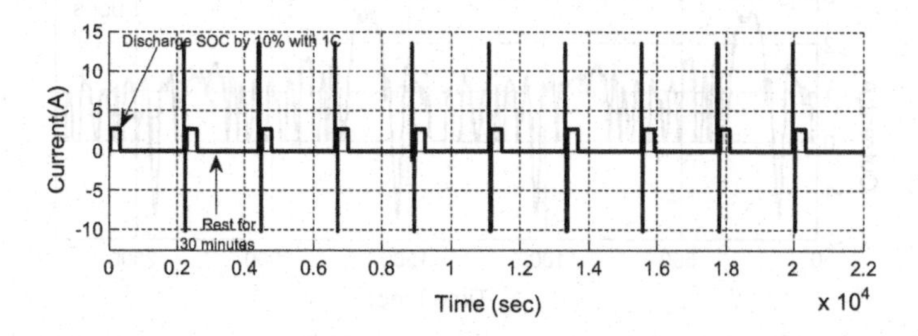

Fig. 3 Battery discharge current profile for circuit parameters identification

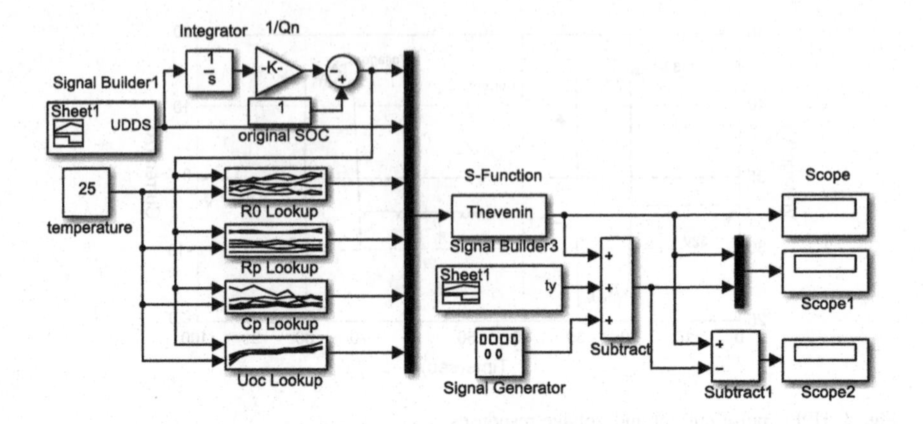

Fig. 4 Thevenin simulation model in MATLAB/Simulink

The battery model is discrete in the actual application condition, so we must handle the state equations as discrete equation:

$$\begin{cases} U_L(k) = U_{oc} - I_k R_0 - U_p(k) \\ U_p(k+1) = U_p(k) e^{-\Delta t/\tau} + I_k R_p (1 - e^{-\Delta t/\tau}) \end{cases} \quad (11)$$

The accuracy of the model has a significance influence on the precision of the SOC estimation, so it needs to be verified by comparison with the actual situation. In this paper, two kinds of input current are used to verify the accuracy of the Thevenin model by comparing the battery terminal voltage and the output voltage of the model, namely UDDS mode and ECE mode. UDDS input current is shown in Fig. 5, which lasted 1367 s a cycle.

In a UDDS cycle, collect the terminal voltage and make a comparison with the output voltage in the Simulink model, the experimental voltage and the emulate error are shown in Figs. 6 and 7 respectively.

From Fig. 7, the results show that the battery model using the identified parameters could estimate the voltage accurately. The maximum error is 0.12 V and

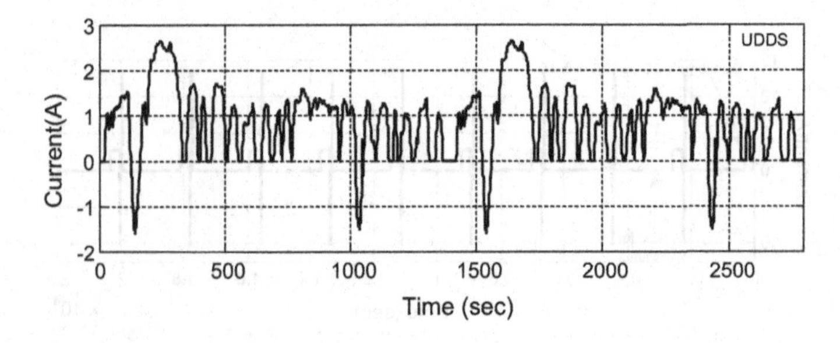

Fig. 5 The input current waveform

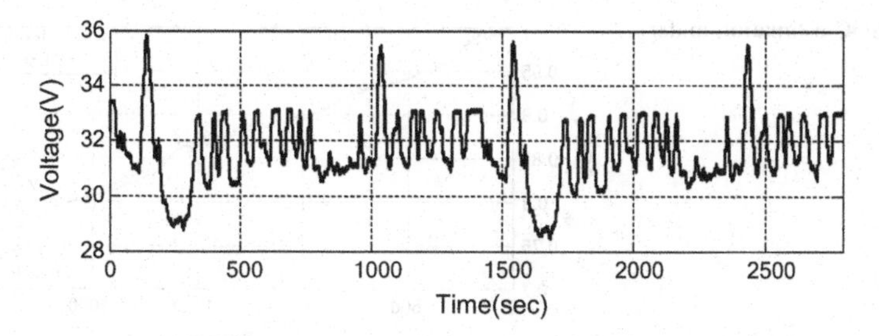

Fig. 6 Simulation of voltage waveform in UDDS mode

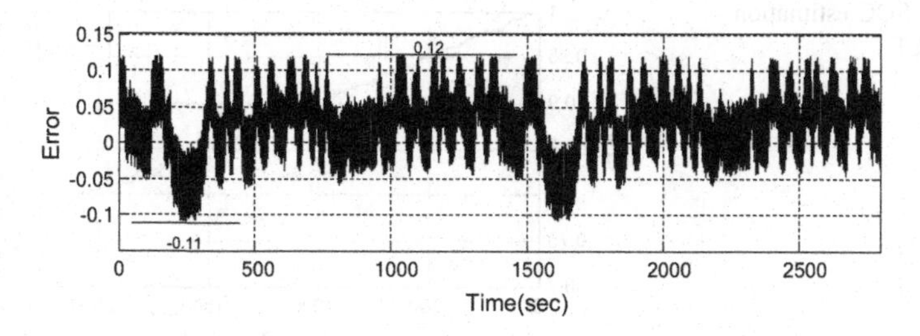

Fig. 7 Error curve between the simulated values and the measured ones

the average error is 0.04 V. Considering that the test object is connected in series with 10 cells, the error assigned to each cell is smaller, this model can be used to estimate SOC online precisely.

The verification model of SOC estimation based on the EKF algorithm is established in Matlab/Simulink environment, as shown in Fig. 8.

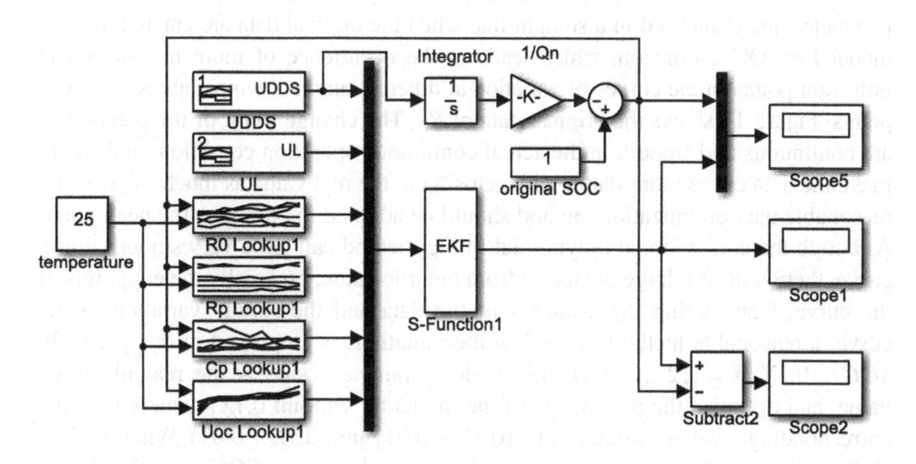

Fig. 8 Simulation model of SOC estimation based on EKF

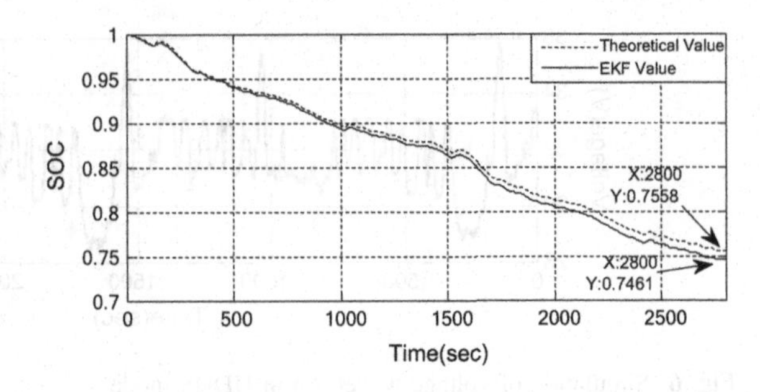

Fig. 9 SOC estimation under UDDS

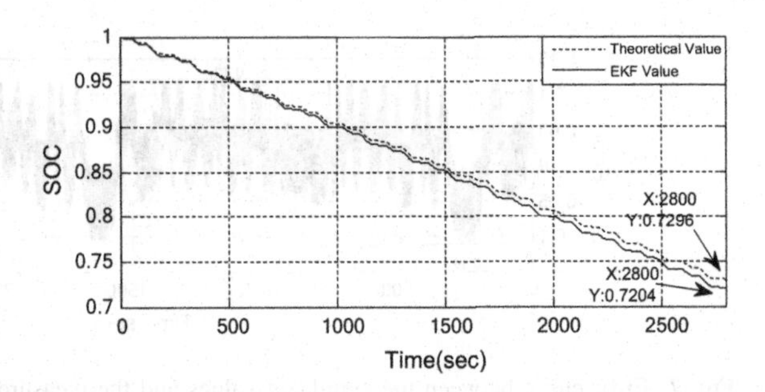

Fig. 10 SOC estimation under ECE

Simulation results show that the SOC estimation values in Figs. 9 and 10, which indicates that EKF estimation has better convergence property.

5 Model Parameters Optimization

Given the limited number of data coming from experiments, the numerical point of parameters are connected in a straight line when the original data are entered into the model for SOC estimation, which leads to the occurrence of more mutations and inflection points in the curves of variation at different ambient temperatures and SOC points. Figure 11 shows the original data of R_0. The change trend of the parameters are continuous and smooth in the actual continuous operation condition, in order to make the data curve more flatter while closing to the real value as much as possible, reasonable data optimization method should be adopted to eliminate the peak point. Although the conventional polynomial fitting method can get a very smooth fitting curve, there will be a large deviation from the true value, especially the endpoints of the curve. Considering the accuracy of the data and the slowly variation of the curve, a reasonable method to weaken the mutations is employed in this paper. In $SOC = 0.1X\,(X = 1, 2 \ldots\ldots 9)$, the model parameter value is the real identified value, and optimize the parameters of the model by setpoint $0.1X$, namely that, the corresponding value points in $(0.1X - 0.01)$ and $(0.1X + 0.01)$. When $SOC = (0.1X - 0.01)$, linear connection the values between $SOC = (0.1X - 0.01)$

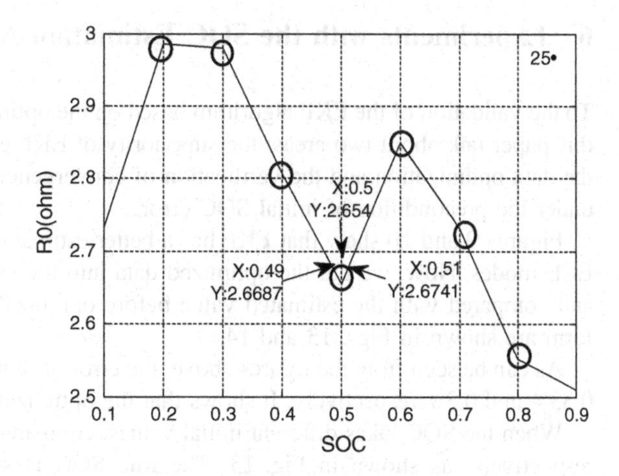

Fig. 11 The original curve

and SOC = 0.1 X. According to the linear function, we can obtain the value of model parameter when $SOC = (0.1\,X - 0.01)$, and sets its value to m; Similarly, the method has been applied to the points of $SOC = 0.1X$ and $SOC = (0.1X + 0.01)$ when $SOC = (0.1\,X + 0.01)$, calculating the parameters values, and setting it as n. The optimization of model parameters is $(m + n)/2$ at $SOC = 0.1\,X$. Although there is a little error, the curve of the model parameters are very close to the actual condition.

In 25 °C and $SOC = 0.5$, taking the measured R_0 as an example, as shown in Figs. 11 and 12, the optimized curve can be seen that the mutations and the peak points are weakened and smooth. The parameters of the model are close to the actual operating conditions of EVs.

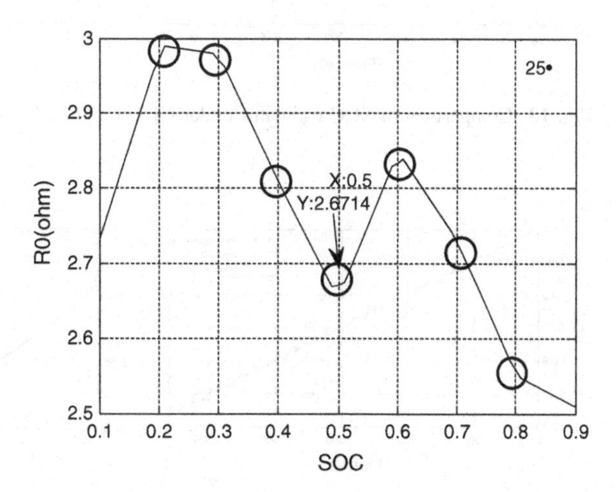

Fig. 12 The optimized curve

6 Experiments with the SOC Estimation Algorithm

To the validation of the EKF algorithm based on the optimization of the model data, this paper talk about two areas: the superiority of EKF estimation before and after the data optimization and the verification of convergence based on EKF algorithm under the precondition of initial SOC error.

Figures 9 and 10 show that EKF has a better estimation effect under UDDS and ECE modes. Now, we put the optimized data into the simulation model of Fig. 8, and compared with the estimated value before optimization, the simulation waveform are shown in Figs. 13 and 14.

As can be seen from the figures above, the error of SOC estimation is reduced by 0.33% and 0.3% respectively. It shows that the optimization method is reasonable.

When the SOC takes different initial values, compared with the theoretical ones respectively, as shown in Fig. 15. The true SOC is set as 100%, and the two measured values of initial SOC are 0.95 and 0.9 respectively.

The figure above shows that the estimated SOC converges to the real SOC value within 500 s with different SOC_0. The results manifest that the initial error does not impact the convergence of the SOC estimation using the proposed EKF. The convergence curves before and after the optimization of the model parameters are not changed, which demonstrates that the EKF model has higher accuracy.

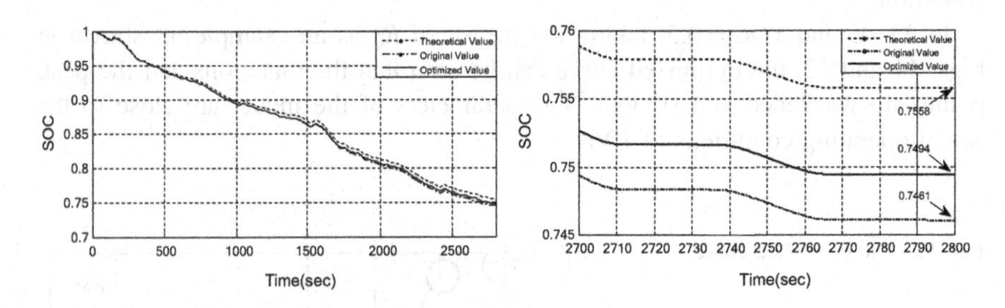

Fig. 13 Comparison of SOC estimates under UDDS and its local enlarged image

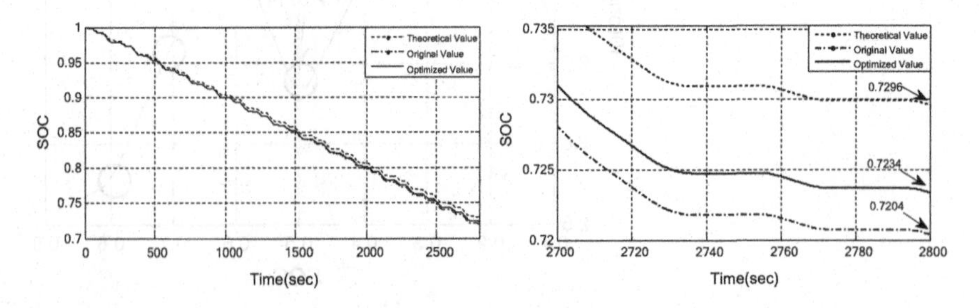

Fig. 14 Comparison of SOC estimates under ECE and its local enlarged image

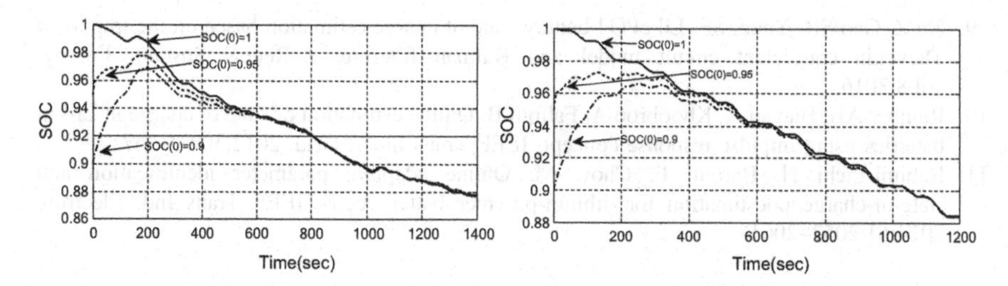

Fig. 15 SOC estimation with different initial values under UDDS and ECE mode

Considering that the convergence time is relatively small compared to the running time of EV, the convergence rate does not affect the estimation of SOC based on EKF algorithm in the practical application.

7 Conclusion

The simulation results show that, the model after parameter optimization is compared with the model without parameter optimization, without affecting the convergence of the initial error of the extended Kalman filter, has higher SOC estimation accuracy. It is verified that the original parameters of the model can be optimized reasonably, which can improve the performance of the online SOC estimation of EV battery by EKF.

References

1. Rajashekara K. Present status and future trends in electric vehicle propulsion technologies. IEEE J Emerg Sel Topics Power Electron. 2013;1:3–10.
2. Yilmaz M, Krein PT. Review of battery charger topologies, charging power levels, and infrastructure for plug-in electric and hybrid vehicles. IEEE Trans Power Electron. 2013;28:2151–2169.
3. Guo XW, Kang LY. Joint estimation of the electric vehicle power battery state of charge based on the least squares method and the kalman filter algorithm. Energies. 2016;9:100.
4. He H, Zhang X, Xiong R, Xu Y, Guo H. Online model based estimation of state-of-charge and open-circuit voltage of 10 Journal of Applied Mathematics lithium-ion batteries in electric vehicles. Energy. 2012;39(1):310–318.
5. Chen Z, Fu Y, Mi CC. State of charge estimation of lithium-ion batteries in electric drive vehicles using extended Kalman filtering. IEEE Trans Veh. Technol. 2013;62:1020–1030.
6. Corno M, Bhatt N, Savaresi SM, Verhaegen M. Electrochemical model-based state of charge estimation for Li-ion cells. IEEE Trans Control Syst. 2015;23:117–127.
7. Zhang H, Zhao L, Chen Y. A lossy counting-based state of charge estimation method and its application to electric vehicles. Energies. 2015;8:13811–13828.
8. Hariharan KS, Kumar VS. A non-linear equivalent circuit model for lithium ion cells. J Power Sources. 2013;222:210–217.

9. Xu Z, Gao SB, Yang, SF. LiFePO4 battery state of charge estimation based on the improved Thevenin equivalent circuit model and Kalman filtering. J Renew Sustain. Energy, vol.8;2016.
10. Ranjbar AH, Banaei A, Khoobroo A, Fahimi B. Online estimation of state of charge in Li-Ion batteries using impulse response concept. IEEE Trans Smart Grid. 2012;3:360–367.
11. Rahimi-Eichi H, Baronti F, Chow M. Online adaptive parameter identification and state-of-charge coestimation for lithium-polymer battery cells. IEEE Trans Ind. Electron. 2014;61:2053–2061.

The Development of a Wheelchair Control Method Based on sEMG Signals

Jinlong Shi, Xina Ren, Zhenqiang Liu, Zengqiang Chen and Feng Duan

Abstract This paper proposed a control method of the electric wheelchair based on surface electromyography (sEMG) signals. In this method, a mapping between hand motions and control commands was established. When a certain kind of hand motion was recognized from sEMG signals, corresponding control would be applied in the wheelchair. The sEMG signals was as raw material for the pattern recognition type of classifier, which promoted the accuracy rate and robustness. The fusion features of Autoregressive (AR) model coefficient and root mean square ratio (RMSR) were used as features of data of hand motions. Support vector machine (SVM) as one of state-of-the-art supervised learning models, was used as classifier. Furthermore, comprehensive real-time simulation and control experiment were implemented. The accuracy rate of hand motions recognition in real-time reached 95% and the success rate of control experiment was up to 88%, which showed the proposed method was feasible and practical.

Keywords Electric wheelchair · sEMG · Support vector machine (SVM) · Hand motion recognition

1 Introduction

There are many kinds of control methods for electric wheelchairs. Traditional methods are usually dependent on the hand to control the joystick or keypads, which are unavailable to the people who have disability in hand. Researchers have explored a variety of new control methods based on biological signals, such as tongue movement signals [1], eye movement signals [2] and EEG-SSVEP signals [3]. In addition to the human machine interface (HMI) mentioned above, sEMG also plays an important role in the field of HMI because of its non-invasive characteristics. State-of-the-art researchers have explored the characteristics and applications

J. Shi · X. Ren · Z. Liu · Z. Chen · F. Duan (✉)
College of Computer and Control Engineering, Nankai University, Tianjin 300350, China
e-mail: duanf@nankai.edu.cn

Y. Jia et al. (eds.), *Proceedings of 2017 Chinese Intelligent Systems Conference*, Lecture Notes in Electrical Engineering 459,
https://doi.org/10.1007/978-981-10-6496-8_40

of sEMG. Because sEMG can capture information of hand motion by sensors with stable robustness and accuracy rate of it, a kind of hand motions recognition systems based on sEMG has been used in controlling the prosthetic hand [4]. While sEMG is usually accompanied by noises, identifying action intent accurately is not easy [5]. It is significant to choose suitable methods of feature extraction and classification algorithms for sEMG signals.

This paper proposed a control method in which the hand motions recognized from the sEMG signals were selected to control the electric wheelchairs. In the hand motion recognition system, RMSR and AR model [6] were selected to extract time-domain features of the sEMG, and SVM [7] was chosen as a classifier to identify the hand motions.

The rest of this paper is arranged as follows. Section 2 introduces three key components of the control system which includes data segmentation, feature extraction and classification. Experiment and discussion are shown in Sect. 3. The conclusion and future work are presented at the end of this paper.

2 Methodology

The architecture of the proposed myoelectric control method was shown in Fig. 1. The system can regard as two stages. The first part recorded the features of muscle activity characteristic that were using to train SVM model. In other word, extent of muscle contraction mapped related hand motion label. Stable EMG data were difficult to obtain during switch of hand motions. Thus, we need to segment active data. After extracting features based on previous work [8, 9], we trained SVM classifier with given labels of default hand motions. Once that model was been built, the second part which is online control system for a wheelchair. According to

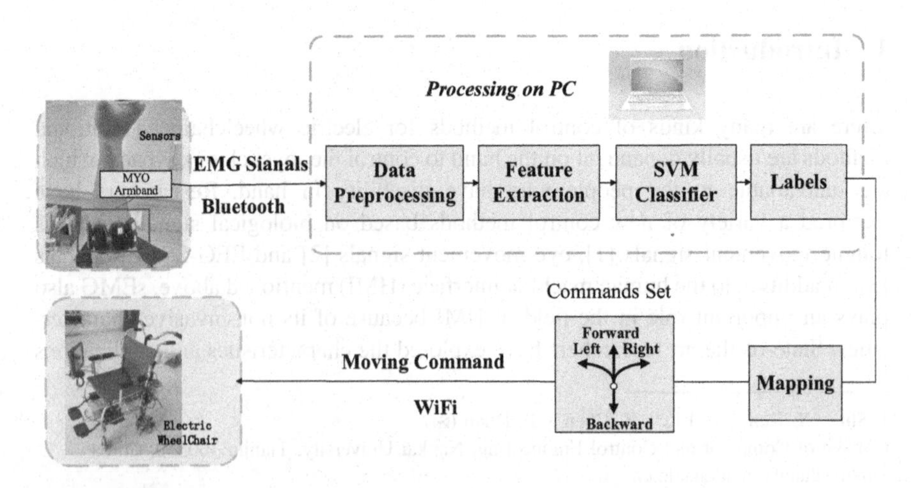

Fig. 1 The architecture of the proposed control system

the defined mapping between hand motions and move orders, control commands corresponding to the identified action will be sent to the control system of the electric wheelchair through WIFI. Finally the wheelchair moves as the user desired. The main method used in whole control strategy are described herein.

2.1 Data Segmentation

Continuous stream of inputs signals is one characterize of on-line recognition system. Information of hand motions contained in continuous signals. To distinguish between contracting state and relaxed state, it is important to choose a suitable segmentation way. Comparing with on-line system, off-line system always extract useful signals by manually as previous work, because researcher can ensure quality of signals of segmentation. Nonetheless, an intelligent recognition processing requires automatically determine the start and the end of active hand motions. Main challenge of segmentation is movement epenthesis [10, 11]. To solve this problem, this paper proposed a method to segment signals automatically using inertial measurement unit (IMU). IMU consists of accelerometer (ACC) and gyroscope, it had widely applied to dynamic detection [12]. The segmentation consists of three steps: calculated muscle energy, estimated motion amplitude, and segment using the threshold by experiments.

During movement, value of several channels of sEMG signals will amplify due to muscle contraction. The energy of sEMG reflect movement intension in time-domain. Thus, we calculated energy of sEMG at the beginning. The formula as shown in Eqs. (1) and (2). In formula (1), n is the number of channels. To smooth out short-term fluctuations, a moving average algorithm was applied. Here, w is size of window.

$$EMG_{ave}(t) = \frac{\sum_{c=1}^{n} emg_c}{n} \tag{1}$$

$$EMG_{energy}(t) = \frac{1}{w}\sum_{i=t-w+1}^{t} EMG_{ave}(i)^2 \tag{2}$$

After obtaining energy of muscle contraction, all EMG data that are linear array would been segment by this energy of EMG. There are two threshold, onset threshold and offset threshold, to avoid erroneous judgments because of data noise. The onset threshold was greater than the offset threshold. Then, we estimated motion amplitude from IMU signals. In laboratory environment experiment, most experiment defaulted subjects performed hand motion in a quiet status. Once they performed hand motion, forearm would have a shake that is easy to detect. We amplified this active shake to judge the subject start or stop to perform hand motions.

$$Motion_{gyro} = \sum (gyro_x^2 + gyro_y^2 + gyro_z^2) \qquad (3)$$

For sEMG signals and IMU signals generated at the same time index, the energy of muscle contraction could be segmented between two active shake signals, which were from IMU signals.

2.2 Feature Extraction

Feature extraction is a crucial component in pattern recognition which influents accuracy of classification to a great extent. Most literature extracted various feature based on their own acknowledge and previous research [13]. Generally, time domain (TD), frequency domain (FD), and time-frequency domain (TFD) is the traditional partition method of feature of sEMG signals. Time domain feature extraction is relatively simple, so the time domain analysis methods in the sEMG signal application have been more extensive. For example, many mature EMG products are using EMG signal characteristics of the time domain as the control signal of the prosthetic hand. In this paper, we used time domain features as an input to SVM classifier.

In time-domain category, there are quite a lot methods, such as mean absolute value (MAV), root mean square (RMS), waveform length (WL), and AR model. And RMS can indicate the effective value of sEMG signals, reflect the average power of the signal and represent the energy information. However, RMS is easy to be influenced by the power of muscle contraction. To solve this problem, we replaced RMS by RMSR. Because RMSR has the advantage over RMS and it is unaffected by different grip strength. The RMSR value reflected ratio of two channels. And the formula of it is shown as:

$$RMSR = \frac{RMS_1}{RMS_2} \qquad (4)$$

Auto regressive model, as a linear predicted model, widespread used in analysis of time series data such as electromyography. Based on n continuous data of the time series data, we can find a suitable model to express its rule and calculate required value of next data $(n+1)$. Therefore, AR model able to reflect attributes of EMG signals applied in recognizing corresponding hand motion. Four-order AR model coefficient was proved as best choice in this circumstance. A standard AR model is described in Eq. (5).

$$y(t) = \sum_{i=1}^{4} a_i{}^{*}\, y(t-1) + e(t) \qquad (5)$$

The $a_i (i \in [1, 4])$ is AR model coefficients that we used as features. And $e(t)$ is white Gaussian noise. In this paper, we used Burg algorithm to calculate AR model coefficients.

So in this paper, RMSR and coefficients of AR model are selected to extract features of sEMG signals. A nine-dimensional feature vector which includes four AR model coefficients per channel and a RMSR values is formed.

2.3 Classification

For pattern-recognition-based method myoelectric control, an appropriate classifier could improve the accuracy rate. Several types classifier were developed within this area in the past decades such as artificial neural network (ANN) [14], linear discriminant analysis (LDA), fuzzy systems, Bayesian techniques, and some hybrid algorithms. Recently, SVM as one of the most outstanding supervised classifiers attracted more and more attention and that method has showed improved results applied for myoelectric signals classification [15]. SVM is a sorting algorithm which can improve the generalization ability of machine learning by minimizing structured risk. Moreover, it can realize the minimization of empirical risk and confidence range, to achieve a good statistical law when the sample size is small.

Originally, SVM was designed for classifying binary proble. However, many researcher wanted to use SVM to deal with multiple problem depend on great performance of SVM. Thus, there are two traditional construct method to form adapted multiple problem classifier. The directed method is considering parameter computation of multiple problem into one optimization problem by modifying target function. It looks sample, but computation complexity is high. The indirected method is constructing and combining several binary classifier. There are two ordinary ways to combine binary classifier, one-against-all (OAA) and one-against-one (OAO). In OAO method, binary classifiers are trained as combined way. The OAO method has advantage of that it conducts binary classifications on all pairs of classes, and decided classification results based on the probability for each class. For example, we wanted to classified new sample into three classes. In Fig. 2, each binary classifier would output a result. A class that gains most votes is seen as the final output that here is label one. In particular condition that each class gets equal votes, we default the class which label is smallest as the final output. In this paper, we employed library LibSVM [16] as the classifier that using C-SVM and OAO method above.

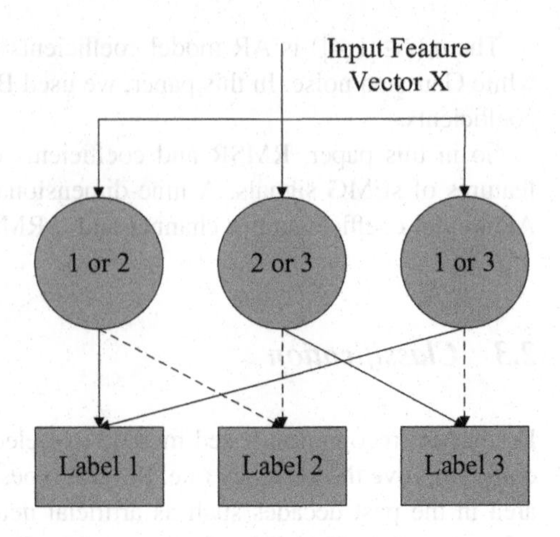

Fig. 2 The schematic diagram of SVM OAO method

3 Experiment Procedure and Result

3.1 Experiment Preparation

The hand motions recognition method of this paper uses one off-the-shelf sensors that named MYO armband (Thalmic, Inc., Firmware version 1.5.1970). The sensor integrated eight medical grade electrode and one inertial measurement units that contains three-axis gyroscope, three-axis accelerometer and three-axis magnetometer. To verified high efficiency of proposed method, we only used two channels of sEMG signals to classifying five hand status. The MYO armband was putted on the forearm of the subject, and we used the two bipolar electrode on forearm to recognize hand motions from the extensor the flexor carpi radialis muscle, and the extensor digitorum muscle approximately.

To test the proposed control method in real-world environment, this paper designed an on-line experiment and a control experiment.

In the on-line experiment, five healthy subjects who had no history of disease of neuromuscular took part in. The ratio of female is 2–3. This experiment aimed to recognize four daily hand motions (wrist extension, wrist flexion, stretch, and fist) and relax state showed in Fig. 3.

The procedural experiment contains two sections where each section consists of five hand states. Interval between two hand motions was relax state. Between each section, the subject has break time and decides when next section start. The first section was trained part. The subject was required to perform WE (wrist extension), WF (wrist flexion), S (stretch), and F (fist) in order. Each hand motion was collected for ten times as a training set. Therefore, training set of one subject includes 40 samples. During training phase, a features-to-actions map (a SVM model in this paper) was built in supervised machine learning condition and saved. Later section is test parts. The sequence of hand motions was {Relax, WE, WF F, S}. Thus, the subject needed to adjust order to comply with requested hand motions. In testing phase, the on-line experiment required each subject to perform each hand motion

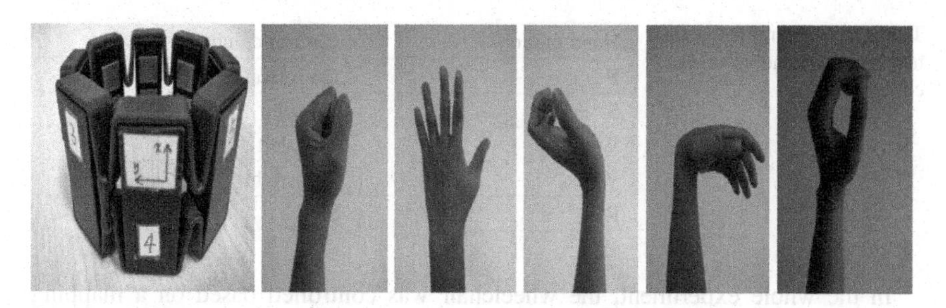

Fig. 3 The predefined hand motions and MYO armband. In *bottom line*, there were F (fist), S (stretch), WE (wrist extension), WF (wrist flexion), and R (relax) from *left* to *right*

for a certain time, during which, there contained 20 samples. In total, there were 2000 samples rely on five subjects. These samples were used to verify the performance of the saved model in training phase by analyzing accuracy rate of hand states classification.

In the control experiment, the five subjects in on-line experiment also took part in. And each subject sat on the wheelchair and controlled it to move in according with predetermined route showed in Fig. 4. Firstly, the wheelchair stopped at point A was controlled to go forwards to point B. After stopping at point B, the wheelchair turned left 90° and stopped, then, the wheelchair turned right 90° and stopped. Finally, the user controlled the wheelchair go back to point A and stop. Each subject was required to finish the ten times which was regarded as ten samples. The label of each sample was defined as success or failure depended on whether the subject completed the whole task without any control error or not.

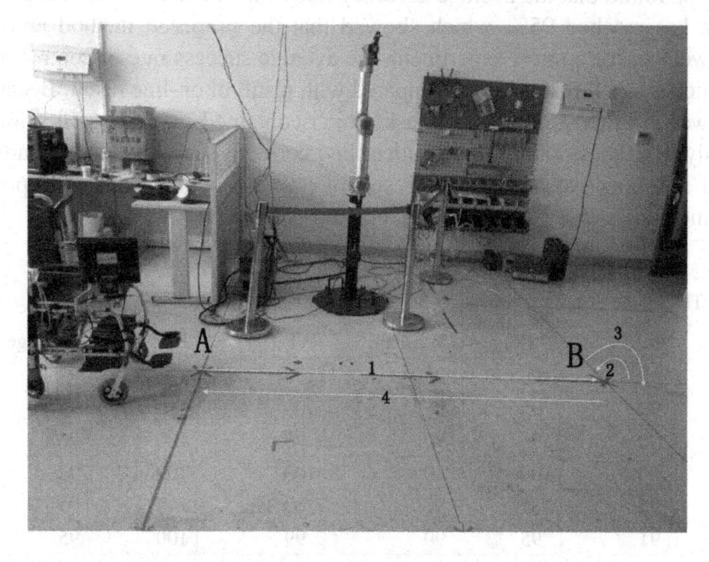

Fig. 4 Moving path of the wheelchair in experiment

Table 1 Mapping between hand motions and control commands

Hand motion	Control commands
F	Go forwards
S	Go back
WE	Turn left
WF	Turn right
R	Stop

In the whole experiment, the wheelchair was controlled based on a mapping between four kinds of hand motions as well as relaxed state and different control commands for control experiment which was shown in Table 1. The SVM models trained and saved in on-line experiment were applied in control experiment to recognize the hand motions and relax state.

3.2 Results and Discussion

Thanks to features-to-actions map had been built, this method can make system adapted to anyone quickly only based few training data. In previous off-line classify problem, we used cross validation (CV) to verification proposed method. However, real-time performance allowed us test more samples in circumstance that is close to practice environment. An improvement comparing with off-line system, we do not need to increase samples to divide sample into several pieces usually used in previous literature. Results of on-line experiment and control experiment were respectively listed in Table 2 and Table 3.

It can be found that the average accuracy rate of the hand motions recognition in real-time has reached 95%, which showed that the proposed method above performed well. In the control experiment, the average success rate of control reached 88%, which was a little bit lower compared with result of on-line result. Because the subject was required to finished a task which contained many classification samples. Only if all these samples were identified correctly, can this control sample be regarded as a positive one. So the result of 88% also showed that the proposed control method in this paper was feasible and practical.

Table 2 The result of the classification

Subject	F (%)	S (%)	WE (%)	WF (%)	R (%)	Average (%)
No. 1	90	100	95	100	100	97
No. 2	90	95	90	100	100	95
No. 3	90	95	90	95	100	96
No. 4	80	100	80	100	100	92
No. 5	95	85	95	100	100	95
AC	91	95	90	99	100	95

F fist, *S* stretch, *WE* wrist extension, *WF* wrist flexion, *R* relax, *AC* the average accuracy rate

Table 3 The success rate of the control experiment

Subject	No. 1	No. 2	No. 3	No. 4	No. 5	Average
Success rate (%)	100	80	90	80	90	88

4 Conclusion and Future Work

In this paper, we proposed a control method that controls electric wheelchairs based on the recognition of sEMG signals. Fusion time feature of AR coefficient and RMSR value are adopted for increasing classification accuracy rate. We employed SVM classifier to seek out correspondence from EMG signals to predefined hand motion. Quantitative and complete assessment have verified that proposed method could control wheelchair for a high accuracy in practice environment. In the designed on-line experiment, the accuracy rate of hand motion recognition reached 95%. The control experiment also acquired an average success rate of 88% in control.

Our ongoing work is to use multiply types of sensor for analysis hand motions as we got IMU and EMG information already. In detail, we found EMG signals would drifting after long time experience. Many reasons cause this problem including muscle fatigue, skin sweat, and sensors displacement and so on. But the IMU signals would not been influenced by muscle fatigue and skin sweat. This provided a new way that using IMU information to calibrate EMG information. In other word, we should retrained system parameter once we detected EMG signals drifting by comparing with EMG signals and IMU signals.

References

1. Huo XL, Wang J, Ghovanloo M. Wireless control of powered wheelchairs with tongue motion using tongue drive assistive technology. In: IEEE international conference on engineering in medicine and biology society, 2008. p. 4199–202.
2. Plesnick S, Repice D, Loughnane P. Eye-controlled wheelchair. In: IEEE international conference on innovations in information, embedded and communication systems, 2015. p. 1–6.
3. Turnip A, Simbolon AI, Amri MF, Utilization of EEG-SSVEP method and ANFIS classifier for controlling electronic wheelchair. In: EEE international conference on technology, informatics, management, engineering and environment, 2016. p. 143–3.
4. Pan LZ, Zhang DG, Liu JW, Sheng XJ, Zhu XY. Continuous estimation of finger joint angles under different static wrist motions from surface EMG signals. Biomed Signal Process Control. 2014;14:265–71.
5. Pau JWL, Xie SSQ, Pullan AJ. Neuromuscular interfacing: Establishing an EMG-driven model for the human elbow joint. IEEE Trans Biomed Eng. 2012;59:2586–93.
6. Luh, JJ, et al. Using time-varying autoregressive filter to improve EMG amplitude estimator. In: IEEE 17th annual conference IEEE, 1995. p. 1343–4.
7. Sun ZQ, Feng JQ, Liu W, Zhu XM. Traffic congestion identification based on parallel SVM. In: International conference on natural computation; 2012. p. 286–9.

8. Duan F, Dai L, Chang W, Chen Z, Zhu C, Li W. sEMG-Based identification of hand motion commands using wavelet neural network combined with discrete wavelet transform. IEEE Trans Industr Electron. 2016;63:1923–34.
9. Chang W, Dai L, Sheng S, Tan JTC, Zhu C, Duan F. A hierarchical hand motions recognition method based on IMU and sEMG sensors. IEEE Int Conf Robot Biomimetics (ROBIO). 2015;2015:1024–9.
10. Yang R, Sarkar S, Loeding B. Handling movement epenthesis and hand segmentation ambiguities in continuous sign language recognition using nested dynamic programming. IEEE Trans Pattern Anal Mach Intell. 2010;32:462–77.
11. Crawford H, Renaud K, Storer T. A framework for continuous, transparent mobile device authentication. Comput Secur. 2013;39:127–36.
12. Kela J, Korpipää P, Mäntyjärvi J, Kallio S, Savino G, Jozzo L, et al. Accelerometer-based gesture control for a design environment. Pers Ubiquit Comput. 2006;10:285–99.
13. Micera S, Carpaneto J, Raspopovic S. Control of hand prostheses using peripheral information. IEEE Rev Biomed Eng. 2010;3:48–68.
14. Abel E, Zacharia P, Forster A, Farrow T. Neural network analysis of the EMG interference pattern. Med Eng Phys. 1996;18:12–7.
15. Kaur G. EMG diagnosis via AR modeling and binary support vector machine classification. Int J Eng Sci Technol. 2010;2:1767–72.
16. Chang C-C, Lin C-J. LIBSVM: a library for support vector machines. ACM Trans Intell Syst Technol (TIST). 2011;2:27.

Global Adaptive Tracking for Multivariable Nonlinear Systems with Unknown Control Direction

Wanli Wang and Yan Lin

Abstract For a class of multivariable nonlinear systems with unknown control direction, the adaptive backstepping method is utilized to handle the parameter uncertainties in this paper. A new switching mechanism based on the monitoring functions(MFs) is proposed to address the control direction uncertainty. The proposed method guarantees the boundedness of the tracking error with the prescribed performance and globally makes it evolve to zero eventually. Finally, a simulation result is demonstrated to show the effectiveness of the approach.

Keywords Multivariable nonlinear systems · Switching mechanism · Unknown control direction · Global tracking · Adaptive backstepping

1 Introduction

As a powerful methodology for controlling systems with modeling uncertainties which render traditional robust control inapplicable, adaptive control has its unique superiority in vast ways. As for multivariable systems with unknown control direction, there has been a lot of research investigated with methods borrowed from SISO systems with unknown control direction [1–8]. In [1], a sliding mode control (SMC) method has been proposed to solve the problem under the assumption that a priori information about the high frequency gain matrix(HFGM) K_p is the knowledge of a matrix S such that $-K_pS$ is Hurwitz. Under the assumption that the knowledge of the signs of the leading principal minors of K_p is priori, three control parameterizations to design stable direct MRAC for the class of MIMO systems with minimum phase, diagonal interactor and arbitrary vector relative degree have been proposed in [2]. While in [3], the problem has been dealt under a difference

W. Wang · Y. Lin (✉)
School of Automation, Beihang University, Beijing 100191, China
e-mail: linyan@buaa.edu.cn

W. Wang
e-mail: wanliwang@buaa.edu.cn

Y. Jia et al. (eds.), *Proceedings of 2017 Chinese Intelligent Systems Conference*, Lecture Notes in Electrical Engineering 459,
https://doi.org/10.1007/978-981-10-6496-8_41

assumption which is that K_p satisfies $K_pS^T + SK_p^T > 0$ with an adaptive scheme. Notably, the paper [4] relaxed the restriction on the HFGM from a totally different perspective with some other assumptions.

However, the papers specifically mentioned above [1–4] are all about multivariable linear systems, and less research has been done for the nonlinear ones [5–8]. In [5], the research object was a class of multivariable linear systems with a known nonlinear function disturbed. With the Nussbaum-type functions, the papers [6] and [7] considered some other MIMO nonlinear systems, namely, distributed-like systems. Regrettably, the Nussbaum-type functions being used were in scalar form instead of vector form, which made the method a little less than perfect. Following [1, 8] has used an output-feedback unit vector SMC approach and monitoring functions to tackle the problem for multivariable nonlinear systems with the assumption about K_p relaxed to a great extent. However, the norm-observability assumption proposed in the paper would be a little restrictive.

In this paper, a new perspective is proposed to handle the dilemma caused by the unknown control direction for a class of multivariable nonlinear systems. Since we have learned that the suitable static pre-compensator S_q can be selected out of a finite index set of matrices Ω such that $-BS_q$ is Hurwitz, here we bring up a switching mechanism inspired by the prescribed performance and the MFs in [9]. With the specific mechanism, we find out the proper control direction finally in finite attempts within the options. To be emphasized, firstly, the raised approach demands much less information about the systems to be controlled. Secondly, the method provides a promising way to improve the transient performance even when the switch happens. Eventually, the method makes the output of the multivariable nonlinear systems tracks the reference signal globally and the error will evolve to zero with all signals bounded, which is verified from the simulation demonstrated.

2 Problem Statement

Consider the following class of M-input M-output nonlinear systems:

$$\dot{x}_j = x_{j+1} + \sum_{i=1}^{v} A_i \varphi_{i,j}(\bar{x}_j), j = 1, \ldots, n-1$$

$$\dot{x}_n = \sum_{i=1}^{v} A_i \varphi_{i,n}(x_1, \ldots, x_n) + Bu, \quad y = x_1, \tag{1}$$

where $x = [x_1^T, \ldots, x_n^T]^T \in \Re^{Mn}$ is the state vector with $x_i \in \Re^M$; $u \in \Re^M$ and $y \in \Re^M$ are the input and output, respectively; $\varphi_{i,j}(\bar{x}_j) \in \Re^M$ with $i = 1, \ldots, v, j = 1, \ldots, n$, $\bar{x}_j = [x_1^T, \ldots, x_j^T]^T$ are known smooth functions; $A_i \in \Re^{M\times M}, i = 1, \ldots, v$ and $B \in \Re^{M\times M}$ are unknown constant parameter matrices; and v, n, M are known integers. For convenience, we define

$$\Theta = [A_1, \ldots, A_v] \in \Re^{M\times Mv}, \qquad \varphi_i = [\varphi_{1,i}^T, \ldots, \varphi_{v,i}^T]^T \in \Re^{Mv}.$$

The control objective of this paper is to design an adaptive backstepping control based on a switching mechanism so that the output y tracks a desired trajectory y_r with a prescribed transient and steady-state performance even we have little information about the HFGM B.

For the question to be solved, we make the following assumptions:

Assumption 1 (i) For the HFGM B, there exist two known constants $0 < c_1 < c_2$ such that $c_1 \le \|B^{-1}\|_1 \le c_2$; (ii) There exists a finite index set Ω of known matrices $S_q \in R^{M\times M}$ such that $-BS_q$ is Hurwitz for some (unknown) $q \in \Omega$.

Remark 1 This assumption significantly relaxes the usual requirement of positive definiteness and symmetry of B. The existence of the finite index set Ω is guaranteed by the results presented in [8].

Assumption 2 The unknown parameter matrix Θ lies in a known bounded convex set, which means that Θ_M, an upper bound of $\|\Theta\|_\infty$, can be obtained such that $\|\Theta\|_\infty \le \Theta_M$ for any Θ. The reference signal y_r and its derivatives up to order n are known, bounded and piecewise continuous.

3 Adaptive Backstepping Control Based on the MFs

3.1 Prescribed Performance

Since the system (1) to be controlled is of strict feedback form, we shall use backstepping technique to design adaptive controller, for which we define the tracking error and other error variables as follows:

$$z_1 = y - y_r, z_i = x_i - \alpha_{i-1}, i = 2, \ldots, n, \tag{2}$$

where α_{i-1} are the virtual control signals to be designed.

The prescribed performance of the errors $z_i (i = 1, \ldots, n)$ is defined as follow:

$$z_i^T z_i < K_i^2, \forall t \ge 0, \tag{3}$$

where K_i are prescribed constants. With the help of (3), we can employ the following Barrier Lyapunov Functions (BLFs) in the backstepping design:

$$V(z_i) = \frac{1}{2} \ln \frac{K_i^2}{K_i^2 - z_i^T z_i}. \tag{4}$$

Obviously, $V(z_i)$ escapes to infinity at $z_i^T z_i = K_i^2$, which gives us a way to keep the boundness of the internal signals.

3.2 Adaptive Backstepping Control with the Proper Control Direction

Before we get the proper control direction, suppose that we already get one, say, we know an exact $S_q \in \Omega$ which makes $-BS_q$ Hurwitz. With the help of S_q, we have the following lemma.

Lemma 1 *For the multivariable nonlinear systems (1), with the known matrix S_q which makes $-BS_q$ Hurwitz and the assumptions satisfied, there exists a controller constructed by the backstepping technique*

$$\dot{\hat{P}} = -S_q^{-T} z_{K_n} \alpha_n^T, \dot{\hat{\Theta}} = \tau_n, u = \hat{P}\alpha_n, \tag{5}$$

which makes the derivative of the following Lyapunov function $V_n = \sum_{i=1}^n V(z_i) + \frac{1}{2}tr(\tilde{\Theta}\Gamma^{-1}\tilde{\Theta}^T) + \frac{1}{2}tr(\tilde{P}^T B^T S_q^T \tilde{P})$ satisfies

$$\dot{V}_n = -\sum_{i=1}^{n} z_{K_i}^T C_i z_i, \tag{6}$$

where $\tilde{\Theta}, \Gamma, \tilde{P}, C_i$ and $z_{K_i}(i = 1, \dots, n), \alpha_n, \tau_n$ are defined in detail in the Appendix.

Proof See the Appendix. □

3.3 Switching Mechanism and Monitoring Functions

Following the contents above, by integrating both sides of (6), we have

$$V_n(t) - V_n(t_0) = -\sum_{i=1}^{n} \int_0^t z_{K_i}^T(\tau) C_i z_i(\tau) d\tau \leq 0, \tag{7}$$

then we get

$$V_n(t) \leq \frac{1}{2}(\sum_{i=1}^{n} z_{K_i}^T(t_0) C_i z_i(t_0) + cons_0), \tag{8}$$

where $cons_0 = k_\Gamma M^2(\Theta_M^2 + \|\hat{\Theta}(t_0)\|_\infty^2) + k_S M(\frac{c_2^2}{c_1} + \frac{\|\hat{P}(t_0)\|_1^2}{c_1})$ and k_Γ, k_S are constants related to Γ and all the S to be chosen, respectively.

Define

$$\mu_0 = \sum_{i=1}^{n} z_{K_i}^T(t_0) C_i z_i(t_0) + \varepsilon_0 + cons_0, \tag{9}$$

where ε_0 is a small positive constant.

In the beginning, the chosen control direction is S_1, which means $S_q = S_1$, using the control law obtained from (5), then we have

$$V_n(t) \le V_n(t_0) < \frac{1}{2}\mu_0 \tag{10}$$

for some time interval. With $\frac{1}{2}\ln\frac{K_i^2}{K_i^2 - z_i^T z_i} < V_n$, we can get

$$z_i^T z_i < (1 - e^{-\mu_0})K_i^2 < K_i^2, \tag{11}$$

for all z_i.

We now construct the monitoring functions for all $z_i (i = 1, \ldots, n)$ based on the foregoing discussion. Suppose the first time of (11) being violated is t_1, then we have to change the S_q from S_1 to S_2. And we define

$$\mu_1 = \sum_{i=1}^{n} z_{K_i}^T(t_1) C_i z_i(t_1) + \varepsilon_1 + \underbrace{k_\Gamma M^2(\Theta_M^2 + \|\hat{\Theta}(t_1)\|_\infty^2) + k_S M(\frac{c_2^2}{c_1} + \frac{\|\hat{P}(t_1)\|_1^2}{c_1})}_{cons_1}, \tag{12}$$

where ε_1 is a small positive constant too, then we have

$$z_i^T z_i < (1 - e^{-\mu_1})K_i^2 < K_i^2, \tag{13}$$

for all $t \in [t_1, t_2)$. The switching instant is determined as

$$t_k := min\{t | z_i^T(t) z_i(t) = (1 - e^{-\mu_{k-1}})K_i^2, i = 1, \ldots, n\}, \tag{14}$$

where

$$\mu_{k-1} = \underbrace{\sum_{i=1}^{n} z_{K_i}^T(t_{k-1})C_i z_i(t_{k-1}) + \varepsilon_{k-1} + k_\Gamma M^2(\Theta_M^2 + \|\hat{\Theta}(t_{k-1})\|_\infty^2) + k_S M(\frac{c_2^2}{c_1} + \frac{\|\hat{P}(t_{k-1})\|_1^2}{c_1})}_{cons_{k-1}}, \quad (15)$$

for a small positive constant ε_{k-1} and all $k = 2, 3, \ldots, Q-1$. Notably, all the functions $z_i^T(t)z_i(t) = (1 - e^{-\mu_{k-1}})K_i^2, i = 1, \ldots, n$ are the MFs.

3.4 Main Results

For the all $\mu_{k-1}(k = 1, \ldots, Q-1)$, we have

$$\mu_k > 2V_n(t_k) \geq \ln \frac{K_i^2}{K_i^2 - z_i^T(t_k)z_i(t_k)}. \quad (16)$$

Since we have $z_i^T(t_k)z_i(t_k) = (1 - e^{-\mu_{k-1}})K_i^2$, then we can get

$$\mu_k > \mu_{k-1}. \quad (17)$$

Hence the main results are given as follow.

Theorem 1 *For the system (1) satisfying Assumptions 1–3, set the prescribed performance for the tracking error z_1 and the errors $z_i(i = 2, \ldots, n)$ as shown in (3) previously, respectively. With the control law (5) and the switching mechanism (14), let the initial conditions satisfy (3), the switching will be stopped within a finite number of switches, and all the closed-loop system signals are bounded with the tracking error converging to zero.*

Proof Firstly, we prove that the times of switching is finite and no more occurs thereafter. As shown in the proof of Lemma 1, all the signals will be bounded in the prescribed region once the control law with proper direction is applied to the system. Assume that the first proper control direction in the sequence is S_p, so what to do is just to prove that all the signals are bounded during the whole time interval $[t_0, t_{p-1})$.

Actually, from (14) we can deduce the following

$$z_i^T(t)z_i(t) < (1 - e^{-\mu_k})K_i^2, i = 1, \ldots, n \quad (18)$$

for $k = 0, \ldots, p-2$ on $[t_k, t_{k+1})$. Further, from (17) we have

$$(1 - e^{-\mu_0})K_i^2 < \cdots < (1 - e^{-\mu_{p-2}})K_i^2 < K_i^2, \quad (19)$$

hence, during the whole time interval $[t_0, t_{p-1})$, the following

$$z_i^T(t)z_i(t) < (1 - e^{-\mu_{p-2}})K_i^2, i = 1, \dots, n \tag{20}$$

hold firmly. So far, the first part of the theorem has been proved.

Secondly, we show the boundness of the tracking error z_1 and the errors $z_i(i = 2, \dots, n)$. From the previous part, we have achieved the boundness of all these signals on the time interval $[t_0, t_{p-1})$, so what left to do is to show the boundedness on the interval $[t_{p-1}, +\infty)$. However, from the Appendix we can obtain the boundness on the interval $[t_{p-1}, +\infty)$ with the proper control direction smoothly.

Therefore, the whole part completes the proof. □

4 Simulation Example

Consider the following uncertain nonlinear system

$$\begin{pmatrix} \dot{x}_{11} \\ \dot{x}_{12} \end{pmatrix} = \begin{pmatrix} x_{21} \\ x_{22} \end{pmatrix} + A_1 \begin{pmatrix} 3x_{11} - x_{12} \\ \sin x_{12} \end{pmatrix} + A_2 \begin{pmatrix} x_{11} \\ -\sin x_{12} \end{pmatrix}, \begin{pmatrix} y_1 \\ y_2 \end{pmatrix} = \begin{pmatrix} x_{11} \\ x_{12} \end{pmatrix}$$
$$\begin{pmatrix} \dot{x}_{21} \\ \dot{x}_{22} \end{pmatrix} = B \begin{pmatrix} u_1 \\ u_2 \end{pmatrix} + A_1 \begin{pmatrix} \sin x_{21} + x_{22} \\ \sin x_{21} + x_{12} \end{pmatrix} + A_2 \begin{pmatrix} x_{22} - x_{12} \\ 2x_{11} \end{pmatrix}, \tag{21}$$

where all signals are scalars and A_1, A_2, B are unknown constant matrices.

In the simulation, the parameter matrices are chosen to be $A_1 = \begin{bmatrix} 0.4 & 0.1 \\ -0.4 & 0.2 \end{bmatrix}$, $A_2 = \begin{bmatrix} -0.4 & 0.2 \\ 0.1 & -0.1 \end{bmatrix}$, and $B = \begin{bmatrix} 2 & 0 \\ -1 & 3 \end{bmatrix}$. For a 2-input 2-output system, the direction is chosen within

$$S_1 = \begin{bmatrix} 1 & 0 \\ 0 & 1 \end{bmatrix}, S_2 = \begin{bmatrix} -1 & 0 \\ 0 & -1 \end{bmatrix}, S_3 = \begin{bmatrix} 0 & 1 \\ -1 & 0 \end{bmatrix}, S_4 = \begin{bmatrix} 0 & -1 \\ 1 & 0 \end{bmatrix}. \tag{22}$$

Following the detailed procedure proposed before, we can get the control law. To proceed, the initial conditions of four states are $x_{11} = 0.5, x_{12} = 0.3, x_{21} = 0.2$, $x_{22} = 0.5$, respectively. The two signals to be tracked are $y_{1r} = y_{2r} = \sin(t/4)$. Other initial conditions are chosen properly. Besides, the used parameters are set to be $K_1^2 = 3, K_2^2 = 6$.

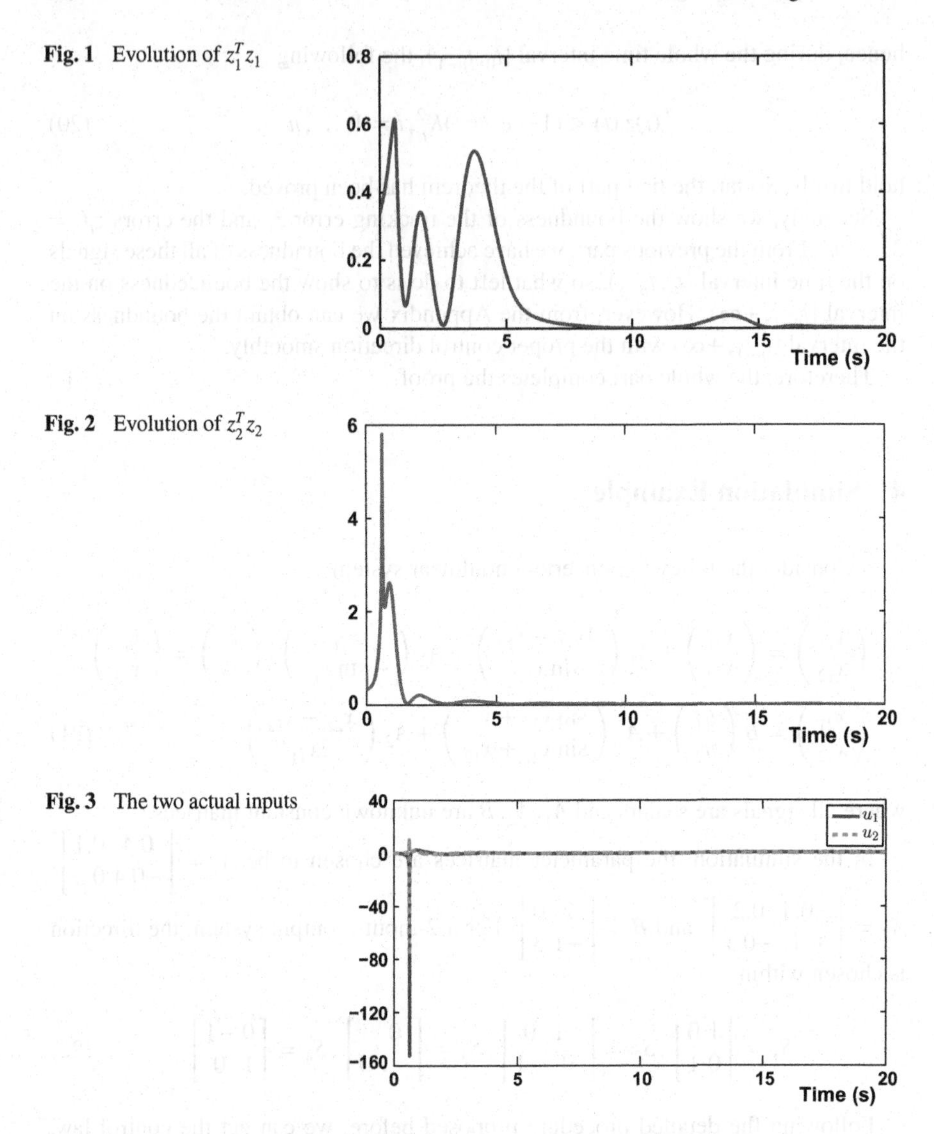

Fig. 1 Evolution of $z_1^T z_1$

Fig. 2 Evolution of $z_2^T z_2$

Fig. 3 The two actual inputs

The simulation results are shown in Figs. 1, 2, 3 and 4. Figures 1 and 2 are the evolution of the two signals $z_1^T z_1$ and $z_2^T z_2$, respectively. The two actual inputs are shown in Fig. 3. From these figures, we can observe that the switching occurs at time 0.62 s, which is much more clear from Fig. 4. Eventually, the effectiveness of the switching mechanism has been verified.

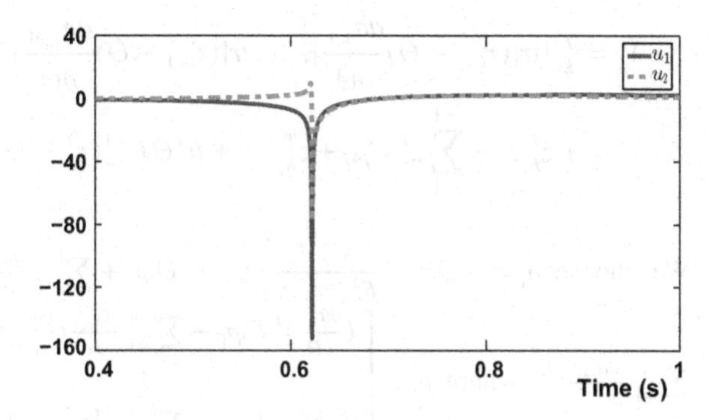

Fig. 4 Switching instant

5 Conclusion

In this paper, a new switching mechanism is proposed to tackle the tracking problem for multivariable nonlinear systems with unknown control direction with adaptive backstepping control. The novelty of this paper is that the restriction on the systems and the HFGM B is relaxed greatly with the proposed MFs-based switching mechanism. Also, the strategy can solve the problem globally.

Appendix

Proof (of Lemma 1) $\hat{\Theta}$ is defined as the estimate of Θ, and $\Gamma \in \Re^{M_V \times M_V}$ is a positive definite matrix. All the $C_i (i = 1, \ldots, n)$ below are positive definite and $\tilde{\Theta} = \Theta - \hat{\Theta}, z_{K_i} = \frac{z_i}{K_i^2 - z_i^T z_i} \in \Re^M$.

Step 1: For the given Lyapunov function $V_1 = V(z_1) + \frac{1}{2} tr(\tilde{\Theta}\Gamma^{-1}\tilde{\Theta}^T)$, we have

$$\dot{V}_1 = z_{K_1}^T (z_2 + \alpha_1 + \Theta\varphi_1 - \dot{y}_r) + tr(\tilde{\Theta}\Gamma^{-1}\dot{\tilde{\Theta}}^T). \tag{A1}$$

We choose $\alpha_1 = -C_1 z_1 - \hat{\Theta}\varphi_1 + \dot{y}_r$, then $\dot{V}_1$ becomes

$$\dot{V}_1 = -z_{K_1}^T C_1 z_1 + z_{K_1}^T z_2 + tr[\tilde{\Theta}\Gamma^{-1}(\tau_1^T - \dot{\hat{\Theta}}^T)], \tag{A2}$$

where $\tau_1^T = \Gamma \varphi_1 z_{K_1}^T$.

Step $i (i = 2, \ldots, n-1)$: With the derivative of $z_i = x_i - \alpha_{i-1}$ and $V_i = V(z_i) + V_{i-1}$, we can get

$$\dot{V}_i = z_{K_2}^T [tr[(\tau_{i-1} - \dot{\hat{\Theta}})\frac{\partial \alpha_{1,1}}{\partial \hat{\Theta}}], \ldots, tr[(\tau_{i-1} - \dot{\hat{\Theta}})\frac{\partial \alpha_{1,M}}{\partial \hat{\Theta}}]]^T + \cdots + z_{K_{i-1}}^T \delta_{i-1} \qquad \text{(A3)}$$

$$+ z_{K_i}^T \dot{z}_i - \sum_{j=1}^{i-1} z_{K_j}^T C_j z_j + z_{K_{i-1}}^T z_i + tr[\tilde{\Theta}\Gamma^{-1}(\tau_2^T - \dot{\hat{\Theta}}^T)].$$

We choose $\alpha_i = -C_i z_i - \frac{K_i^2 - z_i^T z_i}{K_{i-1}^2 - z_{i-1}^T z_{i-1}} z_{i-1} - \hat{\Theta}\varphi_i + \sum_{j=1}^{i-1} \frac{\partial \alpha_{i-1}}{\partial x_j}(\hat{\Theta}\varphi_j + x_{j+1}) + \hat{\delta}_i + \upsilon_i + \sum_{j=1}^{i} \frac{\partial \alpha_{i-1}}{\partial y_r^{(j-1)}} y_r^{(j)}$, where $\upsilon_i = \begin{bmatrix} (\frac{\partial \alpha_{1,1}}{\partial \hat{\Theta}})^T \Gamma \varphi_i - \sum_{j=1}^{i-1} \frac{\partial \alpha_{i-1}}{\partial x_j}(\frac{\partial \alpha_{1,1}}{\partial \hat{\Theta}})^T \Gamma \varphi_j \\ \vdots \\ (\frac{\partial \alpha_{1,M}}{\partial \hat{\Theta}})^T \Gamma \varphi_i - \sum_{j=1}^{i-1} \frac{\partial \alpha_{i-1}}{\partial x_j}(\frac{\partial \alpha_{1,M}}{\partial \hat{\Theta}})^T \Gamma \varphi_j \end{bmatrix} z_{K_2}$
$+ \cdots + \begin{bmatrix} (\frac{\partial \alpha_{i-2,1}}{\partial \hat{\Theta}})^T \Gamma \varphi_i - \sum_{j=1}^{i-1} \frac{\partial \alpha_{i-1}}{\partial x_j}(\frac{\partial \alpha_{i-2,1}}{\partial \hat{\Theta}})^T \Gamma \varphi_j \\ \vdots \\ (\frac{\partial \alpha_{i-2,M}}{\partial \hat{\Theta}})^T \Gamma \varphi_i - \sum_{j=1}^{i-1} \frac{\partial \alpha_{i-1}}{\partial x_j}(\frac{\partial \alpha_{i-2,M}}{\partial \hat{\Theta}})^T \Gamma \varphi_j \end{bmatrix} z_{K_{i-1}}$, so we have

$$\dot{V}_i = -\sum_{j=1}^{i} z_{K_j}^T C_j z_j + z_{K_i}^T z_{i+1} + tr[\tilde{\Theta}\Gamma^{-1}(\tau_i^T - \dot{\hat{\Theta}}^T)] \qquad \text{(A4)}$$

$$+ z_{K_2}^T [tr[(\tau_i - \dot{\hat{\Theta}})\frac{\partial \alpha_{1,1}}{\partial \hat{\Theta}}], \ldots, tr[(\tau_i - \dot{\hat{\Theta}})\frac{\partial \alpha_{1,M}}{\partial \hat{\Theta}}]]^T + \cdots + z_{K_i}^T \tilde{\delta}_i,$$

where $\tau_i^T = \tau_{i-1}^T + \Gamma \varphi_i z_{K_i}^T - \sum_{j=1}^{i-1} \Gamma \varphi_j z_{K_i}^T \frac{\partial \alpha_{i-1}}{\partial x_j}$.

Step n: The derivative of $z_n = x_n - \alpha_{n-1}$ is expressed as

$$\dot{z}_n = Bu + \Theta\varphi_n - \delta_n - \sum_{j=1}^{n-1} \frac{\partial \alpha_{n-1}}{\partial x_j}(\Theta\varphi_j + x_{j+1}) - \sum_{j=1}^{n} \frac{\partial \alpha_{n-1}}{\partial y_r^{(j-1)}} y_r^{(j)}. \qquad \text{(A5)}$$

With $V_n = V(z_n) + V_{n-1} + \frac{1}{2} tr(\tilde{P}^T B^T S_q^T \tilde{P})$, where $\tilde{P} = P - \hat{P}$, in which $P = B^{-1}$ and $\hat{P}$ is the estimate of P, then we have

$$\dot{V}_n = z_{K_n}^T \left(Bu + \Theta\varphi_n - \delta_n - \sum_{j=1}^{n-1} \frac{\partial \alpha_{n-1}}{\partial x_j}(\Theta\varphi_j + x_{j+1}) - \sum_{j=1}^{n} \frac{\partial \alpha_{n-1}}{\partial y_r^{(j-1)}} y_r^{(j)} \right) \qquad \text{(A6)}$$

$$+ \dot{V}_{n-1} + tr(\tilde{P}^T B^T S_q^T \dot{\tilde{P}}).$$

Letting

$$u = \hat{P}\alpha_n, \dot{\hat{P}} = -S_q^{-T} z_{K_n} \alpha_n^T, \dot{\hat{\Theta}} = \tau_n, \qquad \text{(A7)}$$

where α_n and τ_n are defined in the same way before, then we have $\dot{V}_n = -\sum_{i=1}^{n} z_{K_i}^T C_i z_i$. Then with the boundedness of V_n, there exists a constant $c_0 > 0$ such that

$$\int_0^t \sum_{i=1}^{n} z_{K_i}^T(\tau) C_i z_i(\tau) d\tau = -\int_0^t \dot{V}_n d\tau = V_n(0) - V_n(t) \leq c_0, \forall t \geq 0, \qquad \text{(A8)}$$

which implies that $z_i \in L^2, i = 1, \ldots, n$. The fact ensures that the error signals $z_i (i = 1, \ldots, n)$ all have finite energy. With the boundness of z_1, we could conclude that $\lim_{t \to \infty} z_1(t)$ exists and is zero. □

References

1. JoséPaulo VS, Cunha JP, Hsu L, Costa RR, Lizarralde F. Output-feedback model-reference sliding mode control of uncertain multivariable systems. IEEE Trans Autom Control. 2003;48(12):2245–50.
2. Imai José AK, Costa RR, Hsu L, Tao G, Kokotović PV. Multivariable adaptive control using high-frequency gain matrix factorization. IEEE Trans Autom Control. 2004;49(7):1152–6.
3. Ortega R, Hsu L, Astolfi A. Adaptive control of multivariable systems with reduced prior knowledge. In: IEEE Conference on Decision and Control; 2001. p. 4198–203.
4. de Mathelin M, Bodson M. Multivariable model reference adaptive control without constraints on the high-frequency gain matrix. In: IEEE Conference on Decision and Control; 1991. p. 2842–47.
5. Wu Y, Zhou Y. Output feedback control for MIMO non-linear systems with unknown sign of the high frequency gain matrix. Int J Control. 2004;77(1):9–18.
6. Li Y, Tong S, Li T. Observer-based adaptive fuzzy tracking control of MIMO stochastic nonlinear systems with unknown control directions and unknown dead zones. IEEE Trans Fuzzy Syst. 2015;23(4):1228–41.
7. Boulkroune A, Tadjine M, M'Saad M, Farza M. Fuzzy adaptive controller for MIMO nonlinear systems with known and unknown control direction. Fuzzy Sets Syst. 2010;161(6):797–820.
8. Oliveira TR, Peixoto AJ, Hsu L. Sliding mode control of uncertain multivariable nonlinear systems with unknown control direction via switching and monitoring function. IEEE Trans Autom Control. 2010;55(4):1028–34.
9. Hupo O, Yan L. Adaptive fault-tolerant control for actuator failures: a switching strategy. Automatica. 2017;81:87–95.

Algebraic Criteria for Consensus Problems of Signed Directed Networks

Mingjun Du, Baoli Ma and Wenjing Xie

Abstract This paper proposes some algebraic criteria for consensus problems and structure of signed networks whose interactions among agents are denoted by singed digraphs. Firstly, we develop a new method to obtain the left eigenvector of the Laplacian matrix associated with zero eigenvalue. With the left eigenvector, auxiliary vector is constructed and correlated with the connectivity of signed digraph. Finally, sufficient and necessary algebraic criteria for consensus problems and structure of signed graphs are provided based on auxiliary vector. Numerical instances are presented to verify the theoretical results.

Keywords Algebraic criteria · Signed digraph · Consensus problems · Left eigenvector · Auxiliary vector

1 Introduction

Recently, the research of consensus problems for networks [1–5] has attracted more and more attentions due to the wide applications of networks, such as control engineering [6, 7]. Most of the existing literatures [1–7] only consider cooperative networks whose interactions are described by nonnegative graph in which the weights of all the edges are nonnegative. Consensus on signed graph is also significant because the interaction between the nodes in network contains not only cooperative behavior but also competition phenomenon. In signed graph, positive weight of edge means cooperative relationship, and negative weight of edge suggests compete behaviors.

Due to the existence of both cooperative and antagonistic interactions in networks, many complicated behaviors can be researched deeply. Bipartite consensus problem is addressed in [8] where all the agents achieve an identical modulus but not in sign.

M. Du · B. Ma (✉)
The Seventh Research Division, Beihang University, Beijing 100191, China
e-mail: mabaoli@buaa.edu.cn

W. Xie
School of Computer and Information Science, Southwest University,
Chongqing 400700, China

Y. Jia et al. (eds.), *Proceedings of 2017 Chinese Intelligent Systems Conference*, Lecture Notes in Electrical Engineering 459,
https://doi.org/10.1007/978-981-10-6496-8_42

Interval consensus problem is studied in [9] where all the rooted agents achieve bipartite consensus and the convergent values of non-rooted agents less than the modulus of rooted agents.

From [8, 9], we realize that the known connectivity of signed graph is a base of studying consensus problems for signed graph. Due to the complexity of graph, it is difficult to obtain the connectivity of signed graph. In [9], rooted nodes play an important role in dealing with interval consensus. How to distinguish rooted nodes and non-rooted nodes from all the nodes is a difficult problem. On the other hand, the singed networks generate more complicated behaviors, such as bipartite consensus, interval consensus, stability. If we are unaware of the topology of singed graph, how to distinguish these behaviors corresponding to signed graph is also a difficult issue. As we know, although we are unaware of the structure of signed networks, Laplacian matrix can be easily obtained by the relationships among agents. So we can determine the structure of signed graph based on Laplacian matrix.

In this paper, we aim to develop algebraic criteria for solving the above issues of signed networks. Toward this end, we propose a new method to computer the left eigenvector of Laplacian matrix associated with zero eigenvalue. Based on the left eigenvector, we construct an auxiliary vector. By this auxiliary vector, some algebraic criteria for the connectivity and complex behaviors of signed networks are provided.

This paper is organized as follows. Section 2 presents some notations and preliminaries used in this paper. We give the problem statement in Sect. 2. Main results can be obtained in Sect. 4. Simulation examples are provided in Sect. 5. Finally, the conclusions are provided in Sect. 6.

2 Notations and Preliminaries

2.1 Notations

Considering a positive integer $n \in \mathbb{Z}$, we define n order identity matrix $I_n \in \mathbb{R}^{n \times n}$, and let $\mathcal{F}_n = \{1, 2, \ldots, n\}$ and column vector $1_n = [1, 1, \ldots, 1]^T \in \mathbb{R}^n$. If $G \in \mathbb{R}^{n \times n}$, $\det(G)$ represents the determinant of matrix G. $|y|$ represents the absolute value of real number y. $\mathrm{diag}\{d_1, d_2, \ldots, d_n\} \in \mathbb{R}^{n \times n}$ denotes diagonal matrix whose diagonal elements are $d_1, d_2, \ldots, d_n$ and non-diagonal elements are zero.

2.2 Signed Graph

Let $\mathcal{G} = (\mathcal{V}, \mathcal{E}, \mathcal{A})$ be a weighted signed digraph of n order with the node set $\mathcal{V} = \{v_1, v_2, \ldots, v_n\}$, the edge set $\mathcal{E} \subseteq \mathcal{V} \times \mathcal{V}$ and the adjacency weight matrix $\mathcal{A} = [a_{ij}] \in \mathbb{R}^{n \times n}$ whose elements satisfy $a_{ij} \neq 0 \Leftrightarrow (v_j, v_i) \in \mathcal{E}$ and $a_{ij} = 0$ otherwise. Assume that $\mathcal{G}$ has no self-loops, such as $a_{ii} = 0, i \in \mathcal{F}_n$, and $\mathcal{G}$ is digon sign-symmetric, such as $a_{ij}a_{ji} \geq 0, \forall i, j \in \mathcal{F}_n$. The edge (v_j, v_i) denotes that v_i can receive information from v_j and v_j is a neighbor of v_i. The neighbor set of v_i can be denoted as $N(i) = \{j : (v_j, v_i) \in \mathcal{E}\}$. The in-degree $\deg_{in}(v_i)$ and out-degree $\deg_{out}(v_i)$ of v_i in $\mathcal{G}$ are defined as follows:

$$\deg_{in}(v_i) = \sum_{j=1}^{n} |a_{ij}|, \deg_{out}(v_i) = \sum_{j=1}^{n} |a_{ji}|.$$

The signed graph $\mathcal{G}$ is said to be balanced if and only if the in-degree and out-degree of $\mathcal{G}$ are equal. We use $\triangle = \text{diag}\{\deg_{in}(v_1), \deg_{in}(v_2), \ldots, \deg_{in}(v_n)\}$ to denote the in-degree matrix of $\mathcal{G}$. According to in-degree matrix, the Laplacian matrix $L = [l_{ij}] \in \mathbb{R}^{n \times n}$ for singed graph $\mathcal{G}$ can be defined as:

$$L = \triangle - \mathcal{A}, \tag{1}$$

where l_{ij} satisfies

$$l_{ij} = \begin{cases} \sum_{j \in N(i)} |a_{ij}|, & \text{if } j = i \\ -a_{ij}, & \text{if } j \neq i. \end{cases} \tag{2}$$

The signed graph $\mathcal{G}$ is called strongly connected if there exists a directed path between and different pair of nodes. If $\mathcal{G}$ has at least one node v_i which has paths to all the other nodes, then $\mathcal{G}$ contains a spanning tree and v_i is rooted node. A path $\mathcal{P}$ in $\mathcal{G}$ from v_i to v_j is constructed by a finite sequence edges: $(v_i, v_{k_1}), (v_{k_1}, v_{k_2}), \ldots, (v_{k_{m-1}}, v_{k_j})$, where $v_i, v_{k_1}, \ldots, v_{k_{m-1}}, v_j$ are different nodes. If $v_i = v_j$, we call $\mathcal{P}$ as cycle. The product of weights of all edges in cycle $\mathcal{P}$ are positive/negative, we say that $\mathcal{P}$ is positive/negative cycle. The cycles which are formed by rooted agents are called as rooted cycles.

In addition, we will introduce the definition of structure balance which plays a dominant role in solving consensus problems for signed graph.

Definition 1 [8] The signed graph $\mathcal{G}$ is structurally balanced if its nodes set $\mathcal{V}$ can be divided into $\mathcal{V}_1$ and $\mathcal{V}_2$ which satisfy $\mathcal{V}_1 \cup \mathcal{V}_2 = \mathcal{V}, \mathcal{V}_1 \cap \mathcal{V}_2 = \emptyset$ such that $a_{ij} \geq 0$, $v_i, v_j \in \mathcal{V}_l, l \in \{1, 2\}$ and $a_{ij} \leq 0, v_i \in \mathcal{V}_l, v_j \in \mathcal{V}_q, l \neq q (l, q \in \{1, 2\})$. Otherwise, $\mathcal{G}$ is said to be structurally unbalanced.

Changing the weights a_{ij} of the signed graph as $|a_{ij}|$ results into a traditional graph $\overline{G} = (\mathcal{V}, \mathcal{E}, \overline{\mathcal{A}})$ with the Laplacian matrix $\overline{L}$, where $\overline{\mathcal{A}} = [a_{ij}] \in \mathbb{R}^{n \times n}$ and $\overline{L} = \Delta - \overline{\mathcal{A}}$. Under the condition of structure balance, the relationship between L for signed graph and $\overline{L}$ of traditional graph can be found from the following Lemma.

Lemma 1 *[8] If $\mathcal{G}$ is structurally balanced, then there exists a diagonal matrix $D = \text{diag}(\sigma_1, \sigma_2, \ldots, \sigma_n)$ with $\{\sigma_i \in \{-1, 1\}, i \in \mathcal{F}_n\}$ such that $\overline{L} = DLD$.*

3 Problem Statement

Consider signed digraph $\mathcal{G}$ which consists of n agents and the model of every agent is

$$\dot{x}_i(t) = u_i(t), i \in \mathcal{F}_n, \tag{3}$$

where $x_i(t) \in \mathbb{R}$ is the state and $u_i(t) \in \mathbb{R}$ is control protocol. According to [9], the control protocol can be designed

$$u_i(t) = \sum_{j \in N(i)} a_{ij}[x_j(t) - \text{sign}(a_{ij})x_i(t)], i, j \in \mathcal{F}_n, \tag{4}$$

where sign(·) is sign function. According to (2), we can rewrite (3) and (4) in a compact form:

$$\dot{x}(t) = -Lx(t), \tag{5}$$

where $x(t) = [x_1(t), x_2(t), \ldots, x_n(t)]^T$.

From [8] and [9], three consensus results can be found for (5) based on especial L (i.e., the structure of signed graph), see below.

The closed-loop multi-agent systems (5) reach

(I) Bipartite consensus:

$$\lim_{t \to \infty} |x_i(t)| = \theta, i \in \mathcal{F}_n, \theta > 0 \tag{6}$$

if the signed graph $\mathcal{G}$ is structurally balanced;

(II) Interval consensus:

$$\lim_{t \to \infty} |x_i(t)| \leq \theta, i \in \mathcal{F}_n, \theta > 0 \tag{7}$$

if the signed graph $\mathcal{G}$ is structurally unbalanced with all the positive rooted cycles positive, where θ is the modulus of the convergent values of root agents;

(III) Consensus with zero group value:

$$\lim_{t \to \infty} |x_i(t)| = 0, i \in \mathcal{F}_n \tag{8}$$

if the signed graph $\mathcal{G}$ has at least one negative rooted cycle.

The Laplaian matrix L plays an important role in dealing with consensus problems for signed networks and determines the final group value of systems. Although the singed graph structure is complicated in case of huge number of nodes, we can easily attain L by signed graph. By [9], det(L) reveals some properties of signed graph.

Lemma 2 *[9] For any signed digraph $\mathcal{G}$ with spanning trees,* det(L) $\neq 0$ *is equivalent to the following statements:*

(1) all the eigenvalues of L have positive real part;

(2) $\mathcal{G}$ has at least one negative rooted cycle;

(3) The signed network can converge to zero.
and $\det(L) = 0$ *is equivalent to the following statements:*
(1) L has one zero eigenvalue and all the other eigenvalues of L have positive real parts;
(2) $\mathcal{G}$ does not have negative rooted cycles.

From $\det(L)$, we can only obtain Lemma 2 for signed networks. More information about signed graph, such as structure and behaviors, can not be realized by $\det(L)$. In order to attain more information for signed graph, we introduce the left eigenvector ω of L associated with eigenvalue 0 and an auxiliary vector which is constructed by ω. By this auxiliary vector, we can establish a systematic methodology to obtain the structure and behaviors of signed graph.

4 Main Results

In this section, we develop a method to obtain the left eigenvector ω of Laplacian matrix associated with zero eigenvalue. Based on ω, we can get an auxiliary vector ε. By ω and ε, we provide algebraic criteria for connectivity and consensus problems of signed networks.

4.1 Left Eigenvector and Auxiliary Vector

Assume that connected signed graph $\mathcal{G}$ contains spanning trees and has m rooted nodes, then we rearrange the sequence of nodes. Specifically, let $v_1, v_2, \ldots, v_m$ denote rooted nodes and $v_{m+1}, v_{m+2}, \ldots, v_n$ denote non-rooted nodes. We can define anther signed digraphs $\mathcal{G}_r = (\mathcal{V}_r, \mathcal{E}_r, \mathcal{A}_r)$ and $\mathcal{G}_{nr} = (\mathcal{V}_{nr}, \mathcal{E}_{nr}, \mathcal{A}_{nr})$, where $\mathcal{V}_r = \{v_1, v_2, \ldots, v_m\}$, $\mathcal{V}_{nr} = \{v_{m+1}, v_{m+2}, \ldots, v_n\}$, $\mathcal{E}_r = \{(v_i, v_j) \in \mathcal{E} : v_i, v_j \in \mathcal{V}_r$, $\mathcal{E}_{nr} = \{(v_i, v_j) \in \mathcal{E} : v_i, v_j \in \mathcal{V}_{nr}$, $\mathcal{A}_r = [a^r_{ij}] \in \mathbb{R}^{m\times m}$ with $a^r_{ij} = a_{ij}, \forall 1 \le i, j \le m$, $\mathcal{A}_{nr} = [a^{nr}_{ij}] \in \mathbb{R}^{(n-m)\times(n-m)}$ with $a^{nr}_{ij} = a_{(i+m)(m+j)}, \forall 1 \le i, j \le n-m$. The adjacency weight matrix of $\mathcal{G}$ can be described by

$$\mathcal{A} = \begin{bmatrix} \mathcal{A}_r & 0_{m\times(n-m)} \\ \mathcal{A}_{rnr} & \mathcal{A}_{nr} \end{bmatrix},$$

where $\mathcal{A}_{rnr} = [a^{rnr}_{ij}] \in \mathbb{R}^{(n-m)\times n}$ with $a^{rnr}_{ij} = a_{(i+m)j}, \forall 1 \le i \le n-m, \forall 1 \le j \le m$. Let L_r and L_{nr} be defined in the same way as L which is defined by (1) and denote Laplcian matrices for $\mathcal{A}_r$ and $\mathcal{A}_{nr}$, respectively. The Laplacian matrix for $\mathcal{G}$ satisfies

$$L = \begin{bmatrix} L_r & 0_{m\times(n-m)} \\ -\mathcal{A}_{nr} & L_{nr} + B \end{bmatrix} = \begin{bmatrix} \begin{bmatrix} L_r^* & C \\ R & l_{mm} \end{bmatrix} & 0_{m\times(n-m)} \\ -\mathcal{A}_{nr} & L_{nr} + B \end{bmatrix}, \tag{9}$$

where $B = \mathrm{diag}\{b_1, b_2, \ldots, b_{n-m}\}$ with $b_i = \Sigma_{j=1}^{m} |a_{(i+m)j}|, \forall 1 \le i \le n-m$, $C \in \mathbb{R}^{m-1}$, $L_r^* \in \mathbb{R}^{(m-1)\times(m-1)}$, $R^T \in \mathbb{R}^{m-1}$, $l_{mm} \in \mathbb{R}$.

According to [9], for this signed graph $\mathcal{G}$, it easily follows that the eigenvalues of $L_{nr} + B$ have positive real parts. Furthermore, if $\mathcal{G}_r$ is structurally balanced (i.e., $\mathcal{G}_r$ does not contain negative rooted cycles), then $\det(L) = 0$.

Let L_{ii} (respectively, $\overline{L}_{ii}$) denote the matrix L (respectively, $\overline{L}$) with the ith row and the ith column deleted. We can establish the following lemma for the relationships of L_{ii} for signed graph and $\overline{L}_{ii}$ for traditional graph.

Lemma 3 *Assume that $\mathcal{G}$ is strongly connected and structurally balanced, then*

$$\det(L_{11}) = \det(\overline{L}_{11}), \det(L_{22}) = \det(\overline{L}_{22}), \ldots, \det(L_{nn}) = \det(\overline{L}_{nn}),$$

where $\det(L_{ii})$ and $\det(\overline{L}_{ii})$ denote the determinant of L_{ii} and $\overline{L}_{ii}$ respectively.

Proof Deleting the in-edges and out-edges of v_i constructs a subgraph $\mathcal{G}_{ii}$ of $\mathcal{G}$ and L_{ii} is the Laplacian matrix of $\mathcal{G}_{ii}$. According to Lemma 2.1 in [10], we know that $\mathcal{G}_{ii}$ is structurally balanced because $\mathcal{G}$ is structurally balanced. We can obtain that all the rooted cycles in $\mathcal{G}_{ii}$ are positive. By the definition of determinant, we can easily obtain this conclusion. The proof of Lemma 3 is complete.

Let $\overline{\omega} = [\overline{\omega}_1, \overline{\omega}_2, \ldots, \overline{\omega}_n]^T$ denote the left eigenvector of $\overline{L}$ associated with zero eigenvalue. According to [5], we have the following results about $\overline{\omega}$.

(C1): $\overline{\omega} = [\overline{\omega}_1, \overline{\omega}_2, \ldots, \overline{\omega}_n]^T = [\det(\overline{L}_{11}), \det(\overline{L}_{22}), \ldots, \det(\overline{L}_{nn})]^T$.

(C2): Suppose $\overline{\mathcal{G}}$ contains spanning trees, then $\overline{\omega}^T L = 0$. Meanwhile, the elements of $\overline{\omega}$ are nonnegative.

Consider that $\mathcal{G}$ contains spanning trees and does not have negative rooted cycle, then L has one zero eigenvalue and all the other eigenvalues of L are in the right open half. Because $\mathcal{G}_r$ is structurally balanced, then there exists a matrix D_m, $D_m = \mathrm{diag}(\sigma_1, \sigma_2, \ldots, \sigma_m)$ with $\sigma_i \in \{-1, 1\}, \forall 1 \le i \le m$, such that $D_m L_r D_m = \overline{L}_r$. Let $\omega = [\omega_1, \omega_2, \ldots, \omega_n]^T$ denote the left eigenvector of Laplacian matrix L associated with zero eigenvalue and satisfy $\omega^T L = 0$. We construct two vectors ω_r and ω_{nr} from L as follows:

$$\omega_r = [\omega_1, \omega_2, \ldots, \omega_m]^T = [\sigma_1 \det(L_{11}), \sigma_2 \det(L_{22}), \ldots, \sigma_m \det(L_{mm})]^T \quad (10)$$

$$\omega_{nr} = [\omega_{m+1}, \omega_{m+2}, \ldots, \omega_n]^T = [0, 0, \ldots, 0]^T \quad (11)$$

About the ω, we have the following Theorem.

Theorem 1 *If $\mathcal{G}$ contains spanning trees and does not have negative rooted cycle, then $\omega = [\omega_r^T, \omega_{nr}^T]^T$ is the left eigenvector of Laplacian matrix associated with zero eigenvalue.*

Proof We first prove that ω_r satisfies (10). Because $\mathcal{G}$ does not have negative rooted cycle, then $\mathcal{G}_r$ is structurally balanced and $D_m L_r D_m = \overline{L}_r$. Based on Lemma 3 and (C1), we have

$$[\det(L_{11}), \det(L_{22}), \dots, \det(L_{mm})]\overline{L}_r = 0.$$

Further,

$$[\det(L_{11}), \det(L_{22}), \dots, \det(L_{mm})]D_m L_r = 0.$$

So we have $\omega_r^T = [\det(L_{11}), \det(L_{22}), \dots, \det(L_{mm})]D_m$.

Next, we prove that ω_{nr} satisfies (11). From (9), L_r can be divided into four sub-matrices

$$L_r = \begin{bmatrix} L_r^* & C \\ R & l_{mm} \end{bmatrix} = \begin{bmatrix} l_{11} & l_{12} & \dots & l_{1m} \\ l_{21} & l_{22} & \dots & l_{2m} \\ \vdots & \vdots & \vdots & \vdots \\ l_{m1} & l_{m2} & \dots & l_{mm} \end{bmatrix} \tag{12}$$

Delete the row m of L_r

$$L_r^m = \begin{bmatrix} l_{11} & l_{12} & \dots & l_{1m} \\ l_{21} & l_{22} & \dots & l_{2m} \\ \vdots & \vdots & \vdots & \vdots \\ l_{(m-1)1} & l_{(m-1)2} & \dots & l_{(m-1)m} \end{bmatrix} = [\alpha_1, \alpha_2, \dots, \alpha_m]$$

Due to $L_r D_m 1_m = 0$, we can obtain $L_r^m D_m 1_m = 0$. Further, $\sigma_1\alpha_1 + \sigma_2\alpha_2 + \dots + \sigma_m \alpha_m = 0$. It implies that $\sigma_i\alpha_i, i \in \{1, 2, \dots, m\}$ can be represented by the other $m-1$ column vectors, so we can get $\text{rank}(L_r^m) = m-1$. Further, the rank of L_r^* is $m-1$ and L_r^* is inverse. We construct an invertible matrix W such that

$$L_r W = \begin{bmatrix} \begin{bmatrix} I_{m-1} & 0 \\ R(L_r^*)^{-1} & l_{mm} - R(L_r^*)^{-1}C \end{bmatrix} & 0_{m\times(n-m)} \\ -A_{rnr}(L_{nr}+B)^{-1} & I_{n-m} \end{bmatrix} \tag{13}$$

By (13), we know $\omega^T LW = 0$. Further, we can deduce $\omega_{m+1} = 0, \omega_{m+2} = 0, \dots, \omega_n = 0$. The proof of Theorem 1 is complete.

In this paper, the Laplacian matrix L can be obtained but D_m is unknown, so the signs of elements in ω are uncertain. However, the modulus of elements in ω can disclose some information about signed graph $\mathcal{G}$. Based on the modulus of elements, we introduce an auxiliary vector

$$\varepsilon = [\det(L_{11}), \det(L_{22}), \dots, \det(L_{nn})]^T.$$

There exists a relationship between ε and ω.

Lemma 4 *If $\mathcal{G}$ is strongly connected and structurally balanced, then ε and ω satisfy*

$$|\omega_1| = \det(L_{11}), |\omega_2| = \det(L_{22}), \ldots, |\omega_n| = \det(L_{nn}).$$

Proof By (C2) and Lemma 3, we know that elements of ε are nonnegative. So we can directly attain this conclusion by Theorem 1.

4.2 Some Algebraic Criteria

As we well known, ω plays an important role in the convergence value of signed networks. According to [9], we know

$$\begin{aligned} \lim_{t\to\infty} x(t) &= (\omega^T x(0))\eta \\ &= [x_1(0)\omega_1 + x_2(0)\omega_2 + \cdots + x_n(0)\omega_n][\eta_1, \eta_2, \ldots, \eta_n]^T \end{aligned} \tag{14}$$

where ω and η are respectively the left and the right eigenvectors of L associated with eigenvalue zero, $x(0)$ denotes the initial state of agents. From (14), $\omega_i = 0$ means that v_i does not make contribute to convergent value and v_i can be in some extent viewed as a follower. On the other hand, $\omega_j \neq 0$ implies that v_j influences the final convergent values and v_j acts as a leader. In the signed graph, leader is rooted agent and follower is non-rooted agent.

The Laplacian matrix L can be obtained by the relations of all the agent although we are unaware of the structure of signed graph. In order to obtain more information of signed digraph $\mathcal{G}$, we shall consider the following difficult problems.

(Q1) How to distinguish rooted nodes and non-rooted nodes from all the nodes?

(Q2) How to realize the connectivity of signed graph?

(Q3) How to identity the complicated behaviors of signed networks?

By ε, we provide answers to the above problems.

Theorem 2 *Consider L is Laplacian matrix of digraph connected signed graph $\mathcal{G}$ and obtain ε from L, then we have the following results:*

(I) $\det(L_{ii}) > 0$ (respectively, $\det(L_{ii}) = 0$), $i \in \{1, 2, \ldots, n\}$ if and only if v_i is rooted node (respectively, non-rooted node).

(II) all elements of ε are not equal to zero if and only if $\mathcal{G}$ is strongly connected and structurally balanced;

(III) ε has at least one zero element if and only if $\mathcal{G}$ contains spanning trees and is not strongly connected.

(IV) all elements of ε are not equal to zero if and only if signed network achieves the bipartite consensus;

(V) ε has at least one zero element if and only if signed network achieves the interval consensus or the bipartite consensus;

Remark Based on the auxiliary vector that is obtained by our newly developed method, we can realize the structure and behaviors of signed networks.

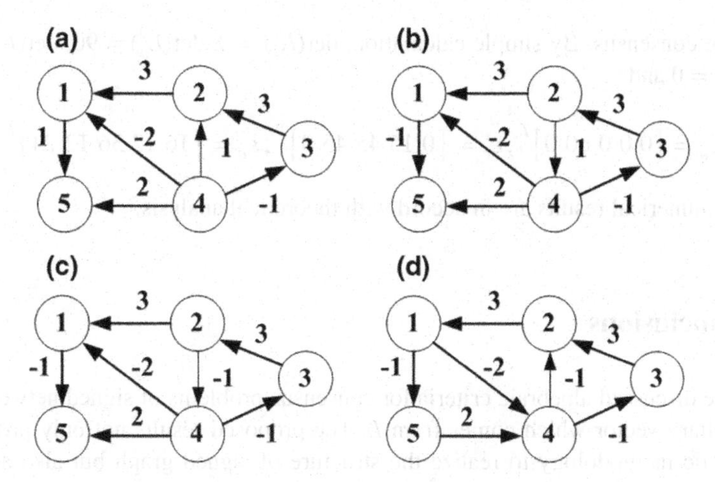

Fig. 1 Four signed graphs

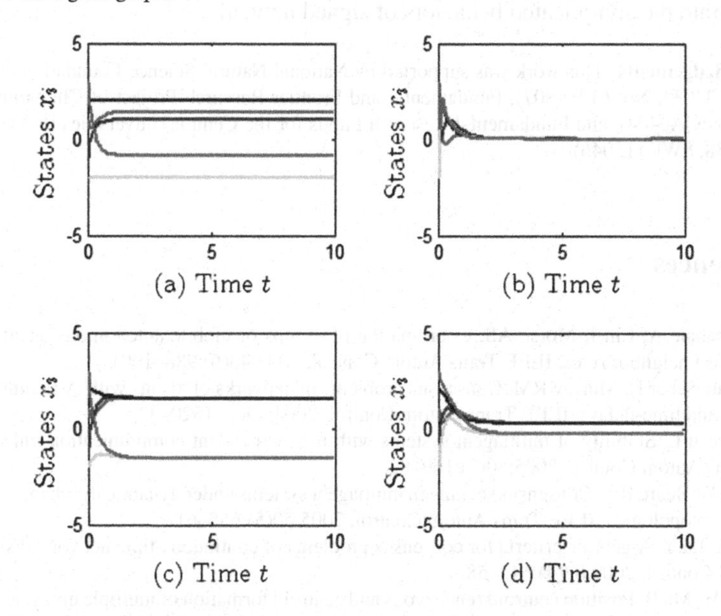

Fig. 2 Consensus of signed graphs

5 Simulations

In this section, we provide some examples to verify the validity of the results. Signed graphs are given by Fig. 1. The number of agents is five and the initial state of all agents is $x(0) = [-1, 1, -2, 2, 3]^T$. The simulation results are shown in Fig. 2. More specifically, Fig. 2a shows that all agents achieve the interval consensus; Fig. 2b shows that all agents achieve stability; Fig. 2c, d shows that all agents achieve bipartite consensus. By simple calculation, $\det(L_a) = 0$, $\det(L_b) = 90$, $\det(L_c) = 0$, $\det(L_d) = 0$ and

$$\varepsilon_a = \left[0\ 0\ 0\ 60\ 0\right]^T, \varepsilon_c = \left[0\ 15\ 45\ 45\ 0\right]^T, \varepsilon_d = \left[16\ 12\ 36\ 12\ 24\right]^T$$

The numerical results are in accord with theoretical analysis.

6 Conclusions

We have discussed algebraic criteria for consensus problems of signed networks by an auxiliary vector which comes from L. The proposed results not only provide a systematic methodology to realize the structure of signed graph but also analyze the behavior of signed network. Furthermore, these algebraic criteria provide a new insight into the complicated behaviors of signed networks.

Acknowledgements This work was supported by National Natural Science Foundation of China (No. 61573034, No. 61327807), Fundamental and Frontier Research Project of Chongqing (No. cstc2016jcyjA0404), and Fundamental Research Funds for the Central Universities (SWU XDJK 2016C038, SWU115046).

References

1. Jadbabaie A, Lin J, Morse AS. Coordination of groups of mobile autonomous agents using nearest neighbor rules. IEEE Trans Autom Control. 2003;48(6):988–1001.
2. Olfati-Saber R, Murray RM. Consensus problems in networks of agents with switching topology and time-delays. IEEE Trans Autom Control. 2004;49(9):1520–33.
3. Moreau L. Stability of multiagent systems with time-dependent communication links. IEEE Trans Autom Control. 2005;50(2):169–82.
4. Ren W, Beard RW. Consensus seeking in multiagent systems under dynamically changing interaction topologies. IEEE Trans Autom Control. 2005;50(5):655–61.
5. Li Z, Jia Y. Algebraic criteria for consensus problems of continuous-time networked systems. Int J Control. 2009;82(4):643–58.
6. Xie W, Ma B. Position centroid rendezvous and centroid formation of multiple unicycle agents. IET Control Theory Appl. 2014;17(8):2055–61.
7. Xie W, Ma B, Fernando T, Herbert HI. A new formation control of multiple underactuated surface vessels. Int J Control, (Published online). doi:10.1080/00207179.2017.1303849.
8. Altafini C. Consensus problems on networks with antagonistic interactions. IEEE Trans Autom Control. 2013;58(4):935–46.
9. Meng D, Du M, Jia Y. Interval bipartite consensus of networked agents associated with signed digraphs. IEEE Trans Autom Control. 2016;61(12):3755–70.
10. Meng D. Convergence analysis of directed digned networks via an M-matrix approach. Int J Control, (Published online). doi:10.1080/00207179.2017.1294263.

Fault Detection and Diagnosis for Servo Systems with Backlash

Fumin Guo and Xuemei Ren

Abstract This paper is concerned with the fault detection and diagnosis problem for the single motor servo systems. The continuous-time nonlinear servo system with disturbance, actuator fault and backlash is modeled. An observer based on radial basis function neural network is constructed to approximate the unknown backlash nonlinear, and a threshold is computed to detect the occurrence of fault. Then, another radial basis function neural network is provided to identify the fault information after a fault occurs. Finally, simulation results show the effectiveness and applicability of the proposed method.

Keywords Fault detection and diagnosis · Servo systems · Backlash · Neural network

1 Introduction

Fault detection (FD) has been becoming a significant research topic since the early 1970s [1–3], and up to now, there are many FD methods have been proposed, for example, event-triggered approach [4, 5] and neural network method [6, 7]. The authors in [8] provided a novel neural-network-based fault diagnosis method for the analog circuits, and the proposed method can detect and diagnose the faulty components by analyzing the time responses efficiently. For a class of nonlinear systems with the uncertainties, disturbances, actuator and sensor faults, a robust FD and isolation scheme with two modified backpropagation recurrent neural networks was proposed in [9], and the scheme did not rely on availability of full state measurements.

Backlash, as an important factor influencing a servo system dynamic performance, is a kind of nonlinear which exists in the process of gear transmission [10, 11]. The work [12] provided a fault detection scheme for mechanical transmission systems with backlash, and a non-smooth observer was designed due to the

F. Guo · X. Ren (✉)
School of Automation, Beijing Institute of Technology, Beijing 100081, China
e-mail: xmren@bit.edu.cn

Y. Jia et al. (eds.), *Proceedings of 2017 Chinese Intelligent Systems Conference*, Lecture Notes in Electrical Engineering 459,
https://doi.org/10.1007/978-981-10-6496-8_43

effect of backlash inherent in the transmission system. The authors in [13] investigated FD and isolation of the electromechanical system with backlash, and the fault indicators were given to detect and isolate some possible faults of the physical system which includes the undesirable backlash.

This paper investigates the FD and diagnosis of a nonlinear servo system with backlash, an observer based on the radial basis function (RBF) neural network is designed to approximate the unknown backlash nonlinear, and a threshold is computed to detect the occurrence of the fault. Besides, another RBF neural network is provided to identify fault information after a fault occurs. The rest of this paper is organized as follows. In Sect. 2, the mathematical modeling process of the single motor servo system is given. Section 3 shows the fault detection and diagnosis scheme. Simulation results and a conclusion are provided in Sects. 4 and 5, respectively.

2 Mathematical Model of the Single Motor Servo System

The structure of the single motor servo system with backlash is shown in Fig. 1, and the dynamic equations of the single motor servo system can be given by:

$$\begin{aligned} RI + L\dot{I} + C_e\dot{\theta}_m &= U \\ K_d I &= J_m\ddot{\theta}_m + b_m\dot{\theta}_m + kf(z) \\ i_m kf(z) &= J_d\ddot{\theta}_d + b_d\dot{\theta}_d \end{aligned} \tag{1}$$

where R is the resistance; I is the current; L is the inductance; θ_m is the angular position of motor; $\dot{\theta}_m$ is the rotate speed of motor; θ_d is the angular position of load; $\dot{\theta}_d$ is the rotate speed of load; C_e is the back electromotive-force coefficient; K_d is the electromagnetic torque coefficient; b_m is the viscosity damping coefficient of motor;

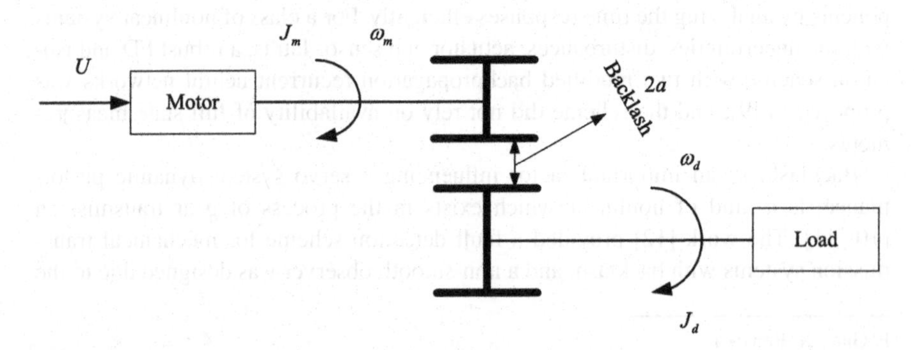

Fig. 1 Structure of the single motor servo system with backlash

i_m is the transmission ratio; k is the stiffness coefficient. The backlash $f(z)$ is a dead-zone function and can be expressed as:

$$f(z) = \begin{cases} z + \alpha, & z < -\alpha \\ 0, & -\alpha \leq z \leq a \\ z - \alpha, & z > \alpha \end{cases} \tag{2}$$

where $2\alpha > 0$ is the backlash width parameter which is assumed to be known, and $z = \theta_m - i_m\theta_d$.

According to [11], the backlash $f(z)$ an be replaced by a smooth and continuous model:

$$f(z) = z - \frac{2}{r}\alpha\left(\frac{2}{1 + e^{-rz}} - 1\right) \tag{3}$$

where r is the slope constant.

Let $x_1 = I$, $x_2 = \theta_m$, $x_3 = \dot{\theta}_m$, $x_4 = \theta_d$, $x_5 = \dot{\theta}_d$, $x = \left[x_1, x_2, x_3, x_4, x_5\right]^T$. Then, the system (1) can be described by

$$\begin{aligned} \dot{x} &= Ax + g(x, u) \\ y &= Cx \end{aligned} \tag{4}$$

where $x(t)$ is the state, and $y(t)$ is the output; A, $g(x, u)$ and C are:

$$A = \begin{bmatrix} -\frac{R}{L} & 0 & -\frac{C_e}{L} & 0 & 0 \\ 0 & 0 & 1 & 0 & 0 \\ \frac{K_d}{J_m} & 0 & -\frac{b_m}{J_m} & 0 & 0 \\ 0 & 0 & 0 & 0 & 1 \\ 0 & 0 & 0 & 0 & -\frac{b_d}{J_d} \end{bmatrix}, g(x, u) = \begin{bmatrix} \frac{u}{L} \\ 0 \\ -\frac{k}{J_m}f(z) \\ 0 \\ \frac{i_m k}{J_d}f(z) \end{bmatrix}, C = \begin{bmatrix} 0 \\ 1 \\ 0 \\ 1 \\ 0 \end{bmatrix}^T.$$

Then in the context of FD, by denoting $d(t)$ and $f_a(t)$ as the disturbance and actuator fault, the system (4) can be extended to the following model:

$$\begin{aligned} \dot{x} &= Ax + g(x, u) + d + \vartheta(t - T)f_a \\ y &= Cx \end{aligned} \tag{5}$$

The term $\vartheta(t - T) \in R$ stands for the time profile of fault and whose form is

$$\vartheta(t - T) = \begin{cases} 0, & t < T \\ 1 - e^{-\theta(t-T)}, & t \geq T \end{cases} \tag{6}$$

where $\theta > 0$ is the rate, and T is the failure occurrence time.

3 Fault Detection and Diagnosis Scheme

This section will use a RBF neural network approximator to construct a nonlinear observer in fault-free case. Then, the fault diagnosis is developed by introducing another RBF neural network.

3.1 Fault Detection Scheme

For the system (5), a FD observer can be given by

$$\begin{aligned} \dot{\hat{x}} &= A\hat{x} + \hat{g}\left(\hat{x}, u\right) + K\left(y - C\hat{x}\right) \\ \hat{y} &= C\hat{x} \end{aligned} \tag{7}$$

where $\hat{x}(t)$ is the estimated state, $\hat{y}(t)$ is the estimated output, K is the observer gain, and $\hat{g}\left(\hat{x}, u\right)$ is the estimation of $g(x, u)$.

Since the nonlinear $g(x, u)$ is unknown, we can use neural network to approximate it. According to universal approximation theorem, we have

$$g(x, u) = W^{*T}\Phi(x, u) + \varepsilon(x, u) \tag{8}$$

where $\Phi(x, u)$ is the RBF; ε is the approximation error and satisfies $\|\varepsilon(x, u)\| \leq \varepsilon_M$. Generally, the ideal weight vector W^* is unknown and need to be estimated. Let $\widehat{W}$ be estimate of W^*, then we obtain

$$\hat{g}\left(\hat{x}, u\right) = \widehat{W}^T\Phi\left(\hat{x}, u\right) \tag{9}$$

Therefore, the observer system (7) is rewritten as

$$\begin{aligned} \dot{\hat{x}} &= A\hat{x} + \widehat{W}^T\Phi\left(\hat{x}, u\right) + K\left(y - C\hat{x}\right) \\ \hat{y} &= C\hat{x} \end{aligned} \tag{10}$$

where the adaptation law of $\widehat{W}$ is

$$\dot{\widehat{W}} = \Gamma\Phi\left(\hat{x}, u\right)e \tag{11}$$

where $\Gamma > 0$ is designed by users and $e = y - \hat{y}$.

According to [14], consider the nonlinear system (5) without the fault, suppose that the matrix $\bar{A} = A - KC$ is Hurwitz matrix, thus $\widetilde{x} = x - \widehat{x}$ is uniformly bounded, and $\widehat{W}$ converges to the neighborhoods of the optimal values W^*.

After the weight vector $\widehat{W}$ approach to the optimal vector W^*, we obtain $\Gamma = 0$, and the neural network parameter is fixed and can be noted as $\widehat{W}_o$. Then, the system (10) can be given by:

$$\begin{aligned} \dot{\widehat{x}} &= A\widehat{x} + \widehat{W}_o^T \Phi(\widehat{x}, u) + K(y - C\widehat{x}) \\ \widehat{y} &= C\widehat{x} \end{aligned} \tag{12}$$

and the residual e is:

$$e = Ce^{\bar{A}t}\widetilde{x}(0) + \int_0^t Ce^{\bar{A}(t-\tau)} \left[\widetilde{W}_o^T \Phi(\widehat{x}, u) + d_{xu} + \varepsilon + d \right] d\tau \tag{13}$$

where $\widetilde{W}_o = W^* - \widehat{W}_o$, $d_{xu} = W^{*T}\left[\Phi(x, u) - \Phi(\widehat{x}, u)\right]$. If $\|\widetilde{x}(0)\| \le \varepsilon_{x0}$, we have

$$\|e\| \le \left\|Ce^{\bar{A}t}\right\| \varepsilon_{x0} + \widetilde{w}_{oM} \int_0^t \left\|Ce^{\bar{A}(t-\tau)}\right\| \left\|\Phi(\widehat{x}, u)\right\| d\tau + q \int_0^t \left\|Ce^{\bar{A}(t-\tau)}\right\| d\tau \tag{14}$$

where $\left\|\widetilde{W}_o^T\right\| \le \widetilde{w}_{oM}$ and $\|d_{xu} + \varepsilon + d\| \le d_{xuM} + \varepsilon_M + d_M = q$. Thus, we can choose the upper bound of e as:

$$e_{ub} = \left\|Ce^{\bar{A}t}\right\| \varepsilon_{x0} + \widetilde{w}_{oM} \int_0^t \left\|Ce^{\bar{A}(t-\tau)}\right\| \left\|\Phi(\widehat{x}, u)\right\| d\tau + q \int_0^t \left\|Ce^{\bar{A}(t-\tau)}\right\| d\tau \tag{15}$$

Note that since the neural network weight is fixed, the FD problem is to generate the e_{ub} satisfying the fault detection logic

$$\begin{cases} \|e\| \le e_{ub}, & \text{no fault} \\ \|e\| > e_{ub}, & \text{a fault occurs} \end{cases} \tag{16}$$

which stands for the fault symptom and provides an alarm for attracting the attention of operators.

3.2 Fault Diagnosis Scheme

After a fault occurs, an alarm in fault detection logic (16) is generated by the above FD method, then the fault formation will be identified in this section.

For the unknown fault $f_a(x,u)$, we use the following RBF neural network to approximate it:

$$\widehat{f}_a(x,u) = \Xi^{*T}\Psi(x,u) + \xi \tag{17}$$

with ideal weight Ξ^* and the bounded approximation error $|\xi| \leq \xi_M$. Consider the fault diagnosis system:

$$\begin{aligned} \dot{\widehat{x}} &= A\widehat{x} + \widehat{W}_o^T\Phi\left(\widehat{x},u\right) + K\left(y - C\widehat{x}\right) + \widehat{f}_a\left(\widehat{x},u\right) \\ \widehat{y} &= C\widehat{x} \end{aligned} \tag{18}$$

where $\widehat{f}_a$ is the estimation of f_a. Thus, the error dynamics is

$$\begin{aligned} \dot{\widetilde{x}} &= (A - KC)\widetilde{x} + \left[\widetilde{W}_o^T\Phi\left(\widehat{x},u\right) + d_{xu} + \varepsilon + d + \vartheta(t-T)f_a - \widehat{f}_a\right] \\ e &= y - C\widetilde{x} \end{aligned} \tag{19}$$

Based on (17), we have

$$\vartheta(t-T)f_a - \widehat{f}_a = \widetilde{\Xi}^T\Phi\left(\widehat{x},u\right) - \Theta(t-T)\,\Xi^{*T}\Phi(x,u) + \vartheta(t-T)\,\zeta + z_{xu} \tag{20}$$

where $\widetilde{\Xi} = \Xi^* - \widehat{\Xi}$, $\Theta(t-T) = e^{-\theta(t-T)}$, and $z_{xu} = \Xi^{*T}\left[\Phi(x,u) - \Phi\left(\widehat{x},u\right)\right]$.

It should be noted that Ξ^* is a constant vector, and the error $\Phi(x,u) - \Phi\left(\widehat{x},u\right)$ is bounded, hence $\left\|z_{xu}\right\| \leq \left\|\Xi^*\right\| \psi_m$. Based on (18) and (20), we obtain

$$\begin{aligned} \dot{\widetilde{x}} &= \overline{A}\widetilde{x} + \left[\widetilde{W}_o^T\Phi\left(\widehat{x},u\right) + d_{xu} + \varepsilon + d\right] + \left[\widetilde{\Xi}^T\Phi\left(\widehat{x},u\right) - \Theta(t-T)\,\Xi^{*T}\Phi(x,u) \right. \\ &\quad \left. + z_{xu} + \vartheta(t-T)\,\zeta\right] \\ e &= y - C\widehat{x} \end{aligned} \tag{21}$$

Then, the following adaptive law in (22) is proposed for Ξ

$$\dot{\widehat{\Xi}} = \Upsilon\left[\Phi\left(\widehat{x},u\right)e - \mu\widehat{\Xi}\right]D[e] \tag{22}$$

where Υ is a symmetric positive definite matrix; $\mu > 0$ is a small positive parameter; the dead-zone operator $D[\cdot]$ is

$$D[e] = \begin{cases} 0, & \text{if } \|e\| \leq e_{ub} \\ 1, & \text{otherwise} \end{cases} \tag{23}$$

Similarly, $e(t)$, $\widetilde{x}$ and $\widetilde{\Xi}$ are uniformly bounded according to [14].

4 Simulation Results

In this section, the following parameters are considered to illustrate the FD performance:

$$L = 50\,\mathrm{mH}, R = 2.6\,\Omega, C_e = 67.2\,\mathrm{V/KRPM}, k = 5.6\,\mathrm{Nm/rad},$$

$$K_d = 1.066\,\mathrm{N \cdot m \cdot s/A}, i_m = 1, J_m = 0.003\,\mathrm{kg \cdot m^2}, J_d = 0.0026\,\mathrm{kg \cdot m^2},$$

$$b_m = 0.015\,\mathrm{Nm \cdot s/rad}, b_d = 0.02\,\mathrm{Nm \cdot s/rad}.$$

Assume the sampling period $h = 0.005$ s, $d(k) = 0.05\sin(2\pi t)$, $\Gamma = 0.5I$, $r = 4$, $\alpha = 0.5$, the neural network architectures are respectively 10-9-5 and 6-4-1, the observer gain matrix is chosen as $K = [0; 12.5; 0; 4.5; 0]$, and the fault is

$$f_a(t) = \begin{cases} 0, & t < 2 \\ 4\sin(2\pi t)\left(1 - e^{-0.4(t-2)}\right), & t \geq 2 \end{cases}$$

The simulation results are given in Figs. 2, 3 and 4.

As shown in Fig. 2, the fault can be detected by using the proposed approach after it occurs, and the neural network can respectively approximate the backlash and fault accurately in Figs. 3 and 4, which demonstrate the effectiveness of the proposed approach.

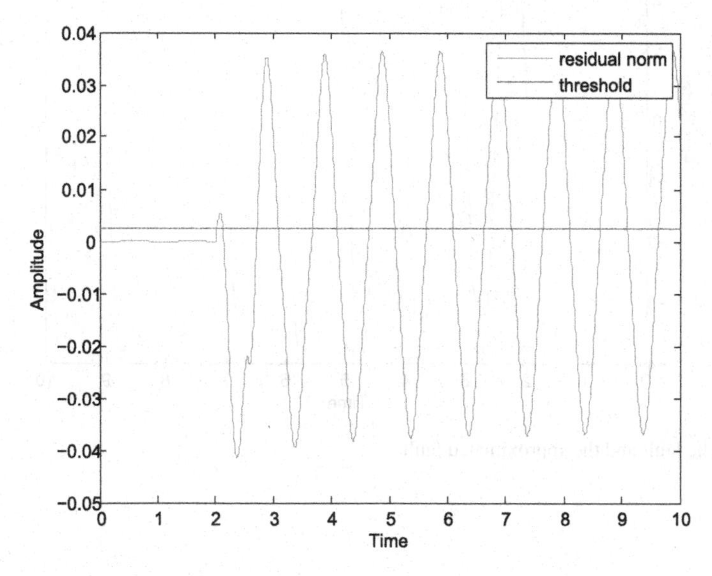

Fig. 2 The residual norm and the threshold

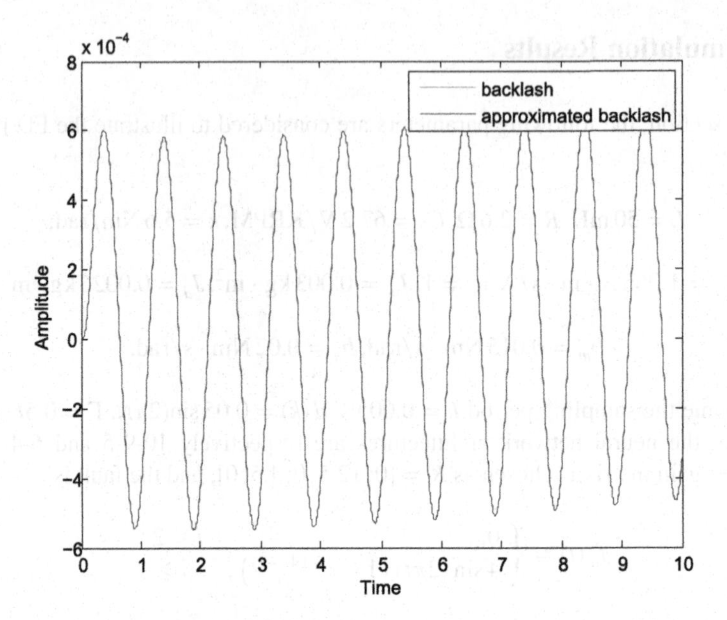

Fig. 3 The backlash and the approximated backlash

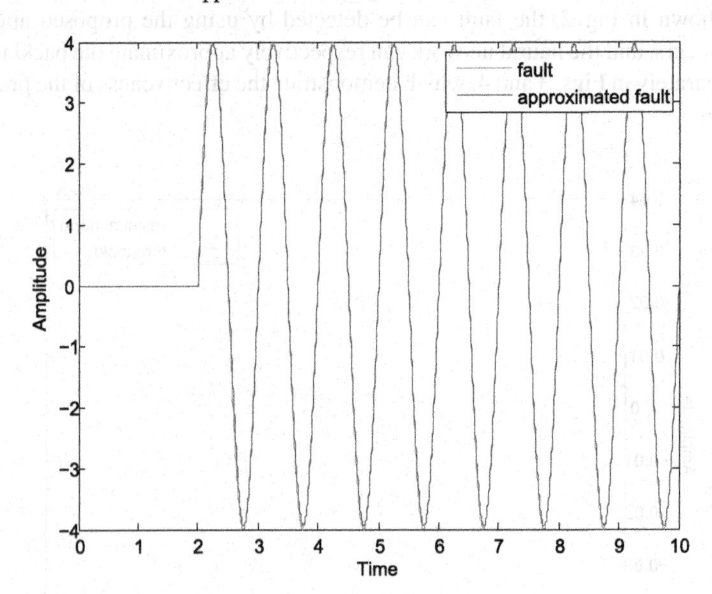

Fig. 4 The fault and the approximated fault

5 Conclusions

The fault detection and diagnosis problem for the single motor servo system with backlash nonlinear and actuator fault was investigated in this paper. Two RBF neural networks were used to identify the unknown backlash and the fault, and a threshold was computed to detect the occurrence of the fault. Simulation results showed the effectiveness and applicability of the proposed method.

Acknowledgements This work was supported by the National Natural Science Foundation of China under Grants 61433003, 61273150, 61321002.

References

1. Yoo SJ. Output-feedback fault detection and accommodation of uncertain interconnected systems with time-delayed nonlinear faults. IEEE Trans Syst Man Cybern Syst. 2017;47(5): 758–66.
2. Yi ZH, Etemadi AH. Fault detection for photovoltaic systems based on multi-resolution signal decomposition and fuzzy inference systems. IEEE Trans Smart Grid. 2017;8(3):1274–83.
3. Yang GH, Wang HM. Fault detection and isolation for a class of unvertain state-feedback fuzzy control systems. IEEE Trans Fuzzy Syst. 2015;23(1):139–51.
4. Davoodi M, Meskin N, Khorasani K. Event-triggered multiobjective control and fault diagnosis: a unified framework. IEEE Trans Ind Inform. 2017;13(1):298–311.
5. Wang YL, Shi P, Lim CC, Liu Y. Event-triggered fault detection filter design for a continuous-time networked control system. IEEE Trans Cybern. 2016;46(12):3414–26.
6. Han HG, Li Y, Qiao JF. A fuzzy neural network approach for online fault detection in waste water treatment process. Comput Electric Eng. 2014;40(7):2216–26.
7. Chai W, Qiao JF. Passive robust fault detection using RBF neural modeling based on set membership identification. Eng Appl Artif Intel. 2014;28(1):1–12.
8. Xiao YQ, He YG. A novel approach for analog fault diagnosis based on neural network and improved kernel PCA. Neurocomputing. 2011;74(7):1102–15.
9. Talebi HA, Khorasani K, Tafazoli S. A recurrent neural-network-based sensor and actuator fault detection and isolation for nonlinear systems with application to the satellite's attitude control subsystem. IEEE Trans Neural Netw. 2009;20(1):45–60.
10. Shi ZG, Zuo ZY. Backsetpping control for gear transmission servo systems with backlash nonlinearity. IEEE Trans Autom Sci Eng. 2015;12(2):752–7.
11. Merzouki R, Davila JA, Fridman L, Cadiou JC. Backlash phenomenon observation and identification in electromechanical system. Control Eng Pract. 2007;15(4):447–57.
12. Tan YH, Zhou ZP, Dong RL, He H. Fault detection of mechanical systems with inherent backlash. In: IEEE International Conference on Networking, Sensing and Control. 2013, p. 77–82.
13. Merzouki R, Medjaher K, Djeziri MA, Ould-Bouamama B. Backlash fault detection in mechatronic system. Mechatronics. 2007;17(6):299–310.
14. Zhu RJ, Chai TY, Shao C. Robust nonlinear adaptive observer design using dynamic recurrent neural networks. In: Proceedings of the American Conference. 1997, p. 1096–100.

Optimization of the Water Supply System Based on an Adaptive Particle Swarm Algorithm

Wanbiao Mao, Yupeng Zhao and Yongfang Mao

Abstract The optimization of water supply pipe networks based on traditional particle swarm algorithm is easy to trap into local optimum and slow to converge the optimum. In this paper, an adaptive particle swarm optimization algorithm with variation (VAPSO) is proposed. The inertia weight is dynamically adjusted by particle swarm complexity, and the variation threshold is dynamically adjusted by single particle complexity. The experiment of water supply system with VAPSO shows that the proposed algorithm has faster convergence speed and a significant advantage in optimizing performance. It can reduce the network diameter cost effectively.

Keywords Water supply pipe networks · VAPSO · Particle swarm complexity · Single particle complexity

1 Introduction

Urban water supply pipe network is the indispensable infrastructure of city construction and development. In order to meet the needs of urban development and urban residents, on the premise of meeting hydraulic condition, water supply pipe network delivery the water to the various nodes safely and reliably through the water transporting and distributing pipe. The investment of water supply pipe network generally accounts for 60–80% of the whole water supply system investment [1]. In order to meet the construction demands and hydraulic demands, and minimize the investment to the water supply pipe network efficiency, this paper

W. Mao
Key Laboratory of Reliability of Aerospace Launch Site, Haikou, Hainan Province 570100, People's Republic of China

W. Mao · Y. Zhao · Y. Mao (✉)
College of Automation, Chongqing University, Chongqing 400044, People's Republic of China
e-mail: yfm@cqu.edu.cn

Y. Jia et al. (eds.), *Proceedings of 2017 Chinese Intelligent Systems Conference*, Lecture Notes in Electrical Engineering 459,
https://doi.org/10.1007/978-981-10-6496-8_44

finds the lowest cost solution of the whole system, and finds the algorithm optimize the water supply pipe network design has a great practical significance [2].

Usually, the optimal design goal of water supply pipe network system is to satisfy the constraint conditions to find the lowest cost of pipe diameter collocation program [3]. Traditional optimization methods are mainly based on programming method, such as linear programming techniques, dynamic programming techniques and nonlinear programming techniques are widely used in the study of the optimization of water supply pipe network. Hopfield and Tank [4] etc used neural network to approximate nonlinear function, through the neural network to optimize the water supply pipe network, this method have high calculation accuracy and good optimization effect, but convergence speed of the algorithm is too slow, Kadu et al. [5] etc used genetic algorithm to optimize water supply pipe network, this method have high global optimization ability and fast convergence speed, but make the algorithm appear premature phenomenon easily. However, the optimization of the water supply pipe network is a high dispersed and binding combinatorial optimization problem, the traditional methods solve the problem with many shortcomings such as low precision and poor optimization effect.

In many optimization methods, particle swarm optimization (PSO) algorithm [6] has the advantages of global optimization and concise and understandability. PSO solve a variety of optimization problems by adjusting inertia weight [7]. Shi and Eberhart [8] etc put forward a parameter named inertia weight through the GA (Genetic Algorithm). This method improves the global search ability by linear decreasing inertia weight in the iteration process, but algorithm still has some shortcomings such as slow convergence speed and easily trap into local optimum in late iteration period. For such problems, this paper proposes a modified particle swarm optimization algorithm to achieve optimal design of the water supply system.

2 Model of Water Supply Pipe Network

The goal of water supply pipe network optimization is to find the minimum water supply pipe network construction cost on the premise of meeting the requirements of water quantity and water pressure and other conditions, the investment of water pipe is also included construction and maintenance expenses. The objective function of the water supply pipe network is simplified by objective function of pipe diameter [9].

$$\min C_n = \sum_{i=1}^{n} c_i L_i \tag{1}$$

Above formula, n is the total number of water supply pipe, C_n is the total cost of water pipe diameter, c_i is the unit cost of the pipe diameter in i pipe, L_i is the length of i pipe.

The objective function of network optimization model must satisfy the constraint conditions [10]. In this case, the hydraulic constraints are:

(1) The node flow continuity constraint

$$\sum(\pm q_{ij}) + Q_i = 0 \quad (2)$$

In formula above, Q_i is the flow of node i; q_{ij} is the flow of every pipe with connected to the node i. i is the start node number and j is the end one, each node in the network should satisfy the continuity equation.

(2) Energy balance constraint

$$\sum_{j=1}^{n_i} k_{ij} = 0 \quad (3)$$

Where k_{ij} is head loss of i to j, Energy balance constraint is also called the loop constraints, the head loss algebra sum is zero in the closed loop.

(3) Node pressure constraint

$$H_{\min} \leq H \leq H_{\max} \quad (4)$$

Where H is the water pressure.

(4) Pipe diameter constraint

$$d_i \geq d_{\min} \quad d_i \in D = \{D_1, D_2, \ldots, D_n\} \quad (5)$$

Where D is the standard pipe diameter set, $d_{\min}$ is the minimum pipe diameter. When design the pipe network optimization, the range of pipe diameter size within the standard discrete pipe diameter, and the chosen diameter size is bigger than the minimum diameter.

3 Modified Particle Swarm Optimization Algorithm

In optimization calculation, the basic idea of particle swarm optimization (PSO) algorithm is guiding the group to the possible solutions through sharing the information between the individuals and regarding an objective function extreme-value as a optimal solution space [11]. Iterative formula of particle swarm algorithm determined by the following several parts: cognition part, social part and inertia weight part, the iterative formula of velocity and position is as follows:

$$V_{t+1} = \omega_t V_t + rand()\left(X_{pt} - X_t\right) + rand()\left(X_{gt} - X_t\right) \quad (6)$$

$$X_{t+1} = X_t + V_{t+1} \tag{7}$$

In above formulas, V_t is the speed of the particle in the t iteration, X_t is the position of the particle in the t iteration, X_{pt} is the individual optimal position of the particles in the t iteration, X_{gt} is the global optimal position of the particles in the t iteration, ω_t is the inertia weight.

In the standard model of particle swarm optimization (PSO) algorithm, the cognition part enhances the diversity of population, the social part enhances the convergence speed of the algorithm, the inertia weight part balances the global and local search ability of the algorithm, but it still prone to premature convergence in the early stage of the search, and it has the slow convergence speed in the late period, and it's all because of the diversity of particles decreases in the iteration result in interactive information barriers. In order to describe the diversity degree of particle quantitatively in the iteration process, this paper proposed a new concept: the particle group complexity.

$$O^t = 1 + \sqrt{\frac{\sum_{i=1}^{N} \left(x_i^t - x_{ave}^t\right)^2}{N}} \tag{8}$$

$$x_{ave}^t = \frac{1}{N} \sum_{i=1}^{N} x_i^t \tag{9}$$

In formulas above, t is the number of iterations, O^t is the complexity of the particles in the t iteration, x_i^t is the distance of the i particle to the optimal particle in the t iteration, x_{ave}^t is the average distance of all particles to the optimal particle in the t iteration, N is the particle quantity.

In the iterative process, the particles are moving in the direction of the optimal particle, all particles will also closer to the optimal particle, the standard deviation of particle distance reflects the deviation degree of particle distance, when the deviation of particles distance is bigger, the diversity of particles is better and the complexity O^t is larger, on the other hand, the diversity of particles is worse and the complexity O^t is also smaller and close to 1, above shows that the range of complexity O^t is $(1, \infty)$.

According to the particle group complexity, the complexity of a single particle can be obtained. The definition formula is as follows:

$$O_i^t = 1 + \sqrt{\frac{\sum_{j=1}^{t} \left(x_i^j - x_i^{ave}\right)^2}{t}} \tag{10}$$

$$x_i^{ave} = \frac{1}{t} \sum_{j=1}^{t} x_i^j \tag{11}$$

Above formulas, O_i^t is the complexity of single particle i in the t iteration, x_i^j is the distance of the i particle to the optimal particle in the j iteration, x_i^{ave} is the average distance of i particle to the optimal particle in the all iterations.

3.1 The Strategy of Modified Inertia Weight

The most classical adjustment method [12] of modifying the Inertia weight w is to use an iterative linear descend way, the formula is:

$$\omega^t = \omega_{\max} - \frac{t(\omega_{\max} - \omega_{\min})}{T} \tag{12}$$

Above formula, $\omega_{\min}$ is minimum inertia coefficient, $\omega_{\max}$ is maximum inertia coefficient, T is the maximum number of iterations.

In the standard particle swarm optimization algorithm, when the inertia weight is a bigger value, it can enhance global optimization ability, avoiding algorithm falls into local optimum. And when the inertia weight is a smaller value, it can speed up the algorithm convergence, and enhance the local search ability. But the weight of all the particles are adjusted uniformly, the algorithm ignoring the diversity of particles. If have the particles find the global optimal point in early iterations, the inertia weight is too big so that the particle out of the optimal location, it will reduce the search algorithm efficiency greatly. Therefore, this paper proposed a strategy to dynamic adjust inertia weight by the complexity, the adjustment formula is:

$$\omega^t = \omega_{\max} - \frac{\omega_{\max} - \omega_{\min}}{O^t} \tag{13}$$

The larger complexity particles have bigger weight value, it speeds up the particles search for the whole solution space, on the contrary, the smaller weight value can make the particle to search detailed in the optimal solution space, and it speeds up the convergence.

3.2 The Strategy of Modified Particle Extreme-Value

In the iteration process, the population diversity of particles is descend, complexity is reduced, the particle is easy to fall into local optimum, this phenomenon leading particle swarm optimization (PSO) to premature convergence. In order to keep the population diversity, the particles can be mutated [13], and introduce a probability mutation mechanism, when the complexity of the particle is reduced, the mutation probability of particle fitness is reduced, on the contrary, the mutation probability of particle fitness is increase. This paper define a threshold of variation as follows:

$$R^t = O^t_{\min} + a(O^t_{\max} - O^t_{\min}) \tag{14}$$

Above formula, R^t is the threshold of variation, $O^t_{\max}$ is the maximum single particle complexity at the t iteration, $O^t_{\min}$ is the minimum single particle complexity at the t iteration, a is the coefficient of variation, the value is $(0, 1)$.

In this paper, the particle swarm is subjected to Gaussian variation. The particles with a single particle complexity lower than the threshold is called the First particle, and the variation low probability is P_1, the formula as follows:

$$x_i = x_{\min} + 0.5\gamma(x_{\max} - x_{\min}) \quad O^t_i < R^t \tag{15}$$

$$P_1 \in [0.01, 0.05] \tag{16}$$

The particles with a single particle complexity higher than the threshold is called the Second particle, and the variation high probability is P_2, the formula as follows:

$$x_i = x_{\max} - 0.618\gamma(x_{\max} - x_{\min}) \quad O^t_i > R^t \tag{17}$$

$$P_2 \in [0.08, 0.5] \tag{18}$$

The dual variation mechanism not only classifies the particles according to the different attributes of the particles, but also dynamically adjusts the variation thresholds with the change of the complexity of the particles.

3.3 The Particle Swarm Optimization Algorithm with Dynamic Adjustment

This paper defined a concept of complexity, in order to measure the diversity of the particle swarm quantitatively, and adjust the inertia weight of particle swarm algorithm through the complexity dynamically. In order to make the particle swarm optimization (PSO) is not easy premature convergence, a particle mutation probability was determined by the complexity, and the algorithm mutate the extreme-value of the particle, the algorithm described as follows:

- Step 1 Initialize the parameters: Determine the total number of particles N, the largest number of iterations T, maximum weight $\omega_{\max}$ and minimum weight $\omega_{\min}$.
- Step 2 Initialize the particle swarm, and calculate the complexity of the particles and the single particle i in the t iteration.
- Step 3 Initialize the global optimal value and the individual optimal value.
- Step 4 Calculate and update the inertia weight.
- Step 5 Determine the variation condition, the algorithm mutate the position of the particle.

Step 6 Update the global optimal position of the particle and the optimal position of the individual.

Step 7 If the number of iterations t reach the maximum number of iterations T, algorithm is terminated, if not reached, then $t = t + 1$, go to Step 4.

4 Application and Simulation Results

A water supply pipe network expansion project in Hunan Province is shown in Fig. 1, the minimum water pressure required for each node is 28 m, pipe network node flow and pipe length are shown in Figs. 2 and 3, node 1 and node 28 is the water plant node, and the different diameter prices are shown in Fig. 4.

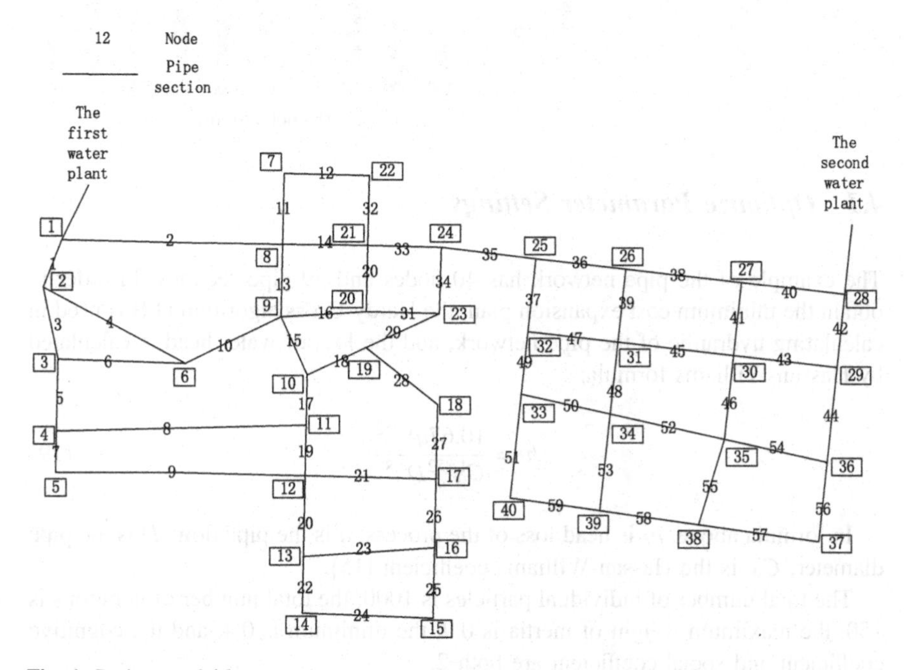

Fig. 1 Project model layout

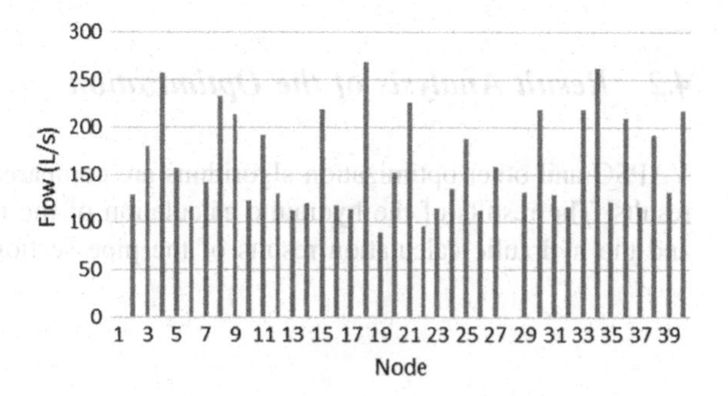

Fig. 2 Flow for each node of the pipe network

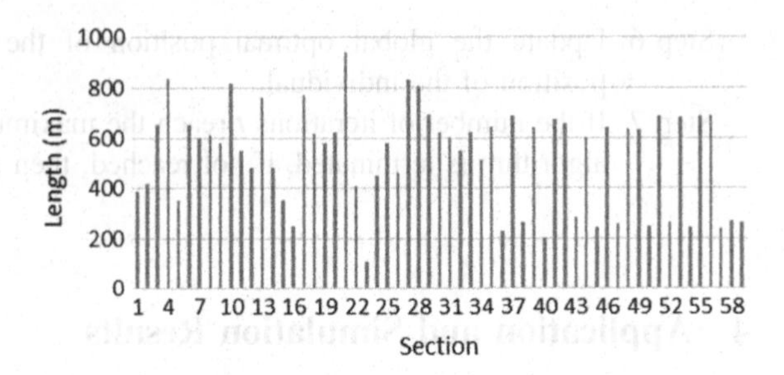

Fig. 3 Length of pipe section of pipe network

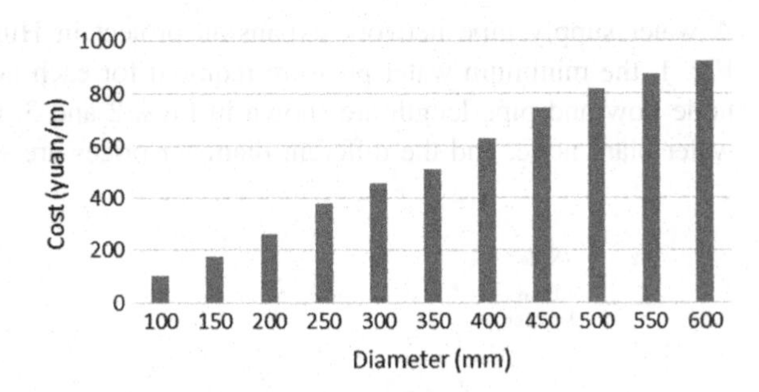

Fig. 4 Unit length pipe network pipe diameter cost

4.1 Optimize Parameter Settings

The example of the pipe network has 40 nodes and 59 pipe sections. In order to obtain the minimum cost expansion plan, the Hardy-Cross algorithm [14] is used in calculating hydraulic of the pipe network, and the loss of water head is calculated by Hassan-Williams formula:

$$h_f = \frac{10.67q^{1.852}}{C_W^{1.852} D^{4.87}} \tag{19}$$

In formula above, h_f is head loss of the process, q is the pipe flow, D is the pipe diameter, C_W is the Hassan-Williams coefficient [15].

The total number of individual particles is 1000, the total number of iterations is 150, the maximum weight of inertia is 0.9, the minimum is 0.4, and the cognitive coefficient and social coefficient are both 2.

4.2 Result Analysis of the Optimization

VAPSO and other optimization algorithms are compared by MATLAB simulation results. The results of the hydraulic calculation of the nodes are shown in Fig. 5, and the hydraulic calculation results of the pipe section are shown in Fig. 6, the

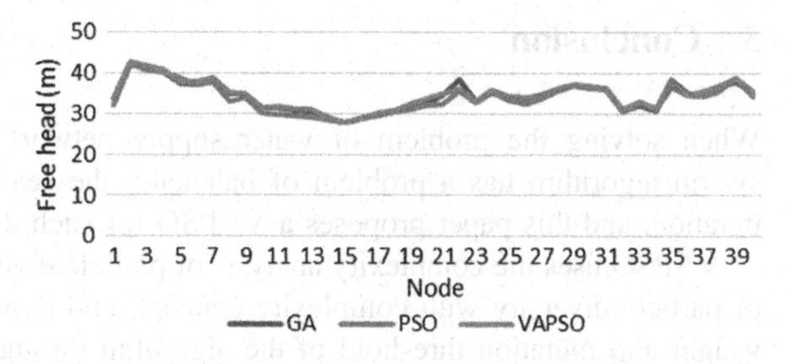

Fig. 5 Comparison of nodal hydraulic calculations

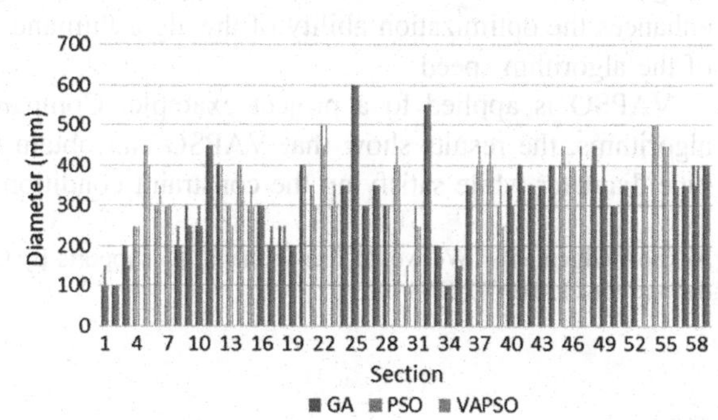

Fig. 6 Comparison of hydraulic calculations for pipe sections

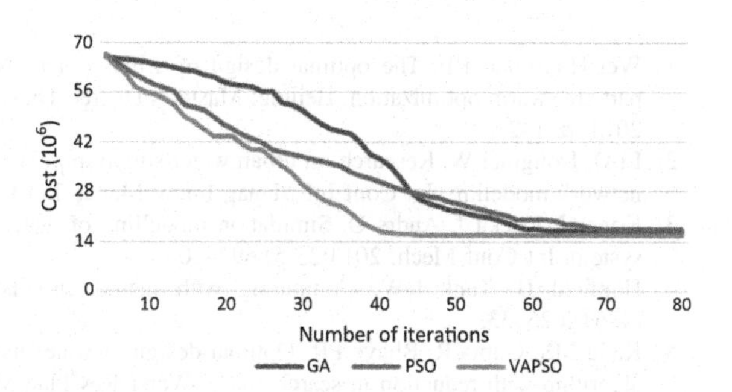

Fig. 7 The economic convergence curves

economic convergence curves are shown in Fig. 7. The economic cost of the pipe network is shown in Table 1.

As can be seen from the Table 1, VAPSO has a significant advantage in the economic cost of the pipe network. VAPSO saves 3.89% of the construction cost compared to GA, and VAPSO saves 6.37% of the construction cost compared to PSO.

Table 1 Comparison of economic cost of the pipe network

Method	GA	PSO	VAPSO
minC_n(yuan)	15504595	15915765	14900750

5 Conclusion

When solving the problem of water supply network optimization, the particle swarm algorithm has a problem of balancing the search ability before and after iteration, and this paper proposes a VAPSO for such defects.

VAPSO uses the complexity analysis of particle diversity to describe the degree of particle diversity with complexity concept, and dynamically adjusts the inertia weight and mutation threshold of the algorithm through the complexity. VAPSO enhances the optimization ability of the algorithm and accelerates the convergence of the algorithm speed.

VAPSO is applied to a project example. Compared with other optimization algorithms, the results show that VAPSO can obtain the optimal combination of pipe diameter while satisfying the constraint condition.

Acknowledgements We would like to thank the supports by China Central Universities Foundation (106112016CDJXY17003).

References

1. Wei HY, Qiao RF. The optimal design of urban water distribution systems via improved particle swarm optimization. Beijing: Master's Degree Thesis of Beijing Industry University; 2014. p. 1–2.
2. Li D, Dongmei W. Research on urban water supply pipe network accident locating based on network modeling. Int Conf Inf Manag Innov Manag Ind Eng. 2013;15(3):127–30.
3. Kovar J, Rucka J, Andrs O. Simulation modelling of water-supply network as mechatronic system. Int Conf Mech. 2014;25(5):697–700.
4. Hopfield JJ, Tank DW. Computing with neural circuits: a model. Science. 1986;233 (4764):625–33.
5. Kadu MS, Gupta R, Bhave PR. Optimal design of water networks using a modified genetic algorithm with reduction in search space. Water Res Plan Manage. 2008;134(2):147–60.
6. Kennedy J, Eberhart RC. Particle swarm optimization. Proc IEEE Int Conf Neural Netw. 1995;1(12):32–49.
7. Liu XM. Research on adaptive particle swarms algorithm for the optimization of urban water network. Chongqing: Master's Degree Thesis of Chongqing University; 2012. p. 24–25.
8. Shi Y, Eberhart RC. Fuzzy adaptive particle swarm optimization. Proc Cong Evolut Comput. 2001;24(8):1246–57.
9. Hong Wei L, Yan Jing L, Wu W, Mou L. Security assessment modeling for water supply network based on leakage analysis. World Automation Congres. 2012;37(4):1–4.
10. Puig V, Ocampo-Martinez C, Romera J, Rudy Q. Model predictive control of combined irrigation and water supply systems: application to the Guadiana river. IEEE Int Conf Netw Sens Control. 2012;23(4):85–90.
11. Sun R. A modified adaptive particle swarm optimization algorithm. Int Conf Comput Intel Security. 189(6):511–3.

12. Shilpa KC, Lakshminarayana C, Singh MK. Adaptive particle swarm optimization for best schedule in algorithmic-level synthesis. Int Conf Commun Electron Syst. 2013;30(3):1–5.
13. Pan G, Yuming X. Chaotic glowworm swarm optimization algorithm based on Gauss mutation. Int Conf Natural Comput Fuzzy Syst Knowl Discov. 2016;34(4):205–10.
14. Cunha EN, Dorea CET. Modeling, simulation and control of water supply system. IEEE First Int Smart Cities Conf. 2015;30(3):1–6.
15. Zhang Y, Jing W, Li N, Li S, Li K. Data-driven water supply systems modeling. Asian Control Conf. 2013;57(2):1–6.

A 3D Reconstruction Method Based on the Combination of the ICP and Artificial Potential Field Algorithm

Bo Fang and Chaoli Wang

Abstract For real-time and accurate three-dimensional (3D) reconstruction during autonomous mobile robot navigation, a method based on the combination of iterative closest points (ICP) and artificial potential field algorithm (APF) is proposed. In real-time path planning, the mobile robot uses the artificial potential field method to obtain the environment point-cloud image by Kinect. Then, the combination of the improved ICP method and the initial transformation matrix is applied to complete the 3D reconstruction. The experimental results show that the proposed algorithm is more efficient than normal distributions transform (NDT) and the traditional three-dimensional ICP method.

Keywords Artificial potential field (APF) · Mobile robot · Iterative closest points (ICP) · Three-dimensional (3D) reconstruction

1 Introduction

In the process of autonomous mobile robot navigation, the real-time accurate 3D reconstruction under complex environment is an important research topic. At present, a classic method for point-cloud matching is NDT [1–3], which uses the standard optimization technique to determine the optimal matching between point-cloud. Although NDT applies to processing a large amount of data, the calculation time of NDT is long. In addition, the ICP standard optimization techniques are used to determine the optimal matching between point-cloud. In this paper, a kind of algorithm (APF-ICP) which combines the artificial potential field (APF) [4–6] and iterative closest points (ICP) [7] is proposed. So when the amount

B. Fang · C. Wang (✉)
School of Optical-Electrical and Computer Engineering, University of Shanghai for Science and Technology, Shanghai 200093, China
e-mail: clclwang@126.com

B. Fang
e-mail: hfuu_fangbo@163.com

Y. Jia et al. (eds.), *Proceedings of 2017 Chinese Intelligent Systems Conference*, Lecture Notes in Electrical Engineering 459,
https://doi.org/10.1007/978-981-10-6496-8_45

of point-cloud data is large, ICP algorithm [8] compared with NDT algorithm has faster processing speed, where, ICP is used to convert the point set difference into a covariance matrix, and then the maximum eigenvector is obtained by the singular value SVD decomposition.

The maximum eigenvector corresponds to the quaternion, which is the European motion contrast value. Since the ICP needs to give the initial point for point-cloud, the direct determination of the initial point can affect the local optimal solution. In this paper, the path planning of the robot is carried out by the artificial potential field method. And according to the current pose transformation of the robot, we can obtain the initial transformation matrix of the point-cloud matching transformation. The combination of the matrix parameters and ICP algorithm for 3D reconstruction shows that APF-ICP algorithm compared with the traditional ICP algorithm [9, 10] has faster speed and higher accuracy.

2 Artificial Potential Field Method

Assume that the sizes of obstacles and robots have been reduced proportionately, which is shown as in Fig. 1. The FD is target for robot gravity, F1 and F2 are obstacles to the robot repulsion, and FI is the resultant force.

As shown in Fig. 1, we firstly synthesize repulsion, i.e., F1, 2 denoting the repulsion force is the synthesis of F1 and F2. The gravity of the robot and target are FD, the synthesis of the FD and F1 is the final resultant of the FI, robot will go toward the resultant force direction. According to the distance to the obstacles by the sonar [11] detection and the global coordinates $P_R(a_{rt}, y_R, a_R)$ of the robot, we can calculate the coordinates of the obstacle $P_O(x_O, y_O)$, where a_R is the current robot pose angle. Then, the convert relative coordinates can be transformed global coordinates.

$$x_O = x_R + d_O \cos(a_R + \alpha), \quad y_O = y_R + d_O \sin(a_R + \alpha) \tag{1}$$

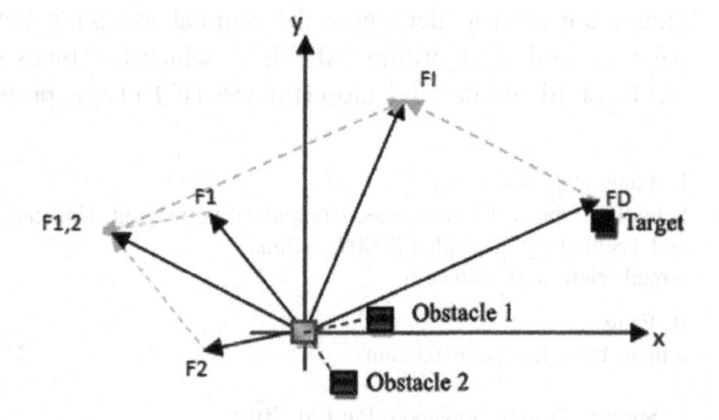

Fig. 1 Model of the robot

where d_O is distance between obstacle and robot, and$\alpha$is center angle between sonar and robot. The force between target point and robot is gravity ($P_I(x_I, y_I)$) [12], which can be resolved into the x, y, two directions component F_{IX}, F_{IY} under global coordinate.

$$F_{IX} = K_I \frac{1}{d_I^2} \frac{x_I - x_r}{d_I}, \quad F_{IY} = K_I \frac{1}{d_I^2} \frac{y_I - y_r}{d_I} \tag{2}$$

where d_I is the distance between the target and robots, K_I is a gravitational parameter. The gravity changes with the distance between robots and target, and K_I and d_I are function relationship. In addition, the repulsive force of robots can be computed through eight sonar sensors. The components of the F_{OX} and F_{OY} specific calculation formula are given as follows:

$$F_{OX} = K_O \frac{1}{d_O^2} \frac{x_O - x_R}{d_O}, \quad F_{OY} = K_O \frac{1}{d_O^2} \frac{y_O - y_R}{d_O} \tag{3}$$

To make the robot real-time obstacle information response, K_I and d_I are function relationship. Then, after getting robot repulsion and attraction, we can get the size of the force F of robots.

$$F = \sqrt{(F_{IX} + F_{OX})^2 + (F_{IY} + F_{OY})^2} \tag{4}$$

3 The Fusion of ICP and Robot Pose

3.1 ICP 3D Reconstruction

A. **Basic principle**

ICP algorithm proposed by Besl and Mckey is a point matching method based on contour features. General process of ICP algorithm is as follow:

The point-cloud data $P = \{P_i, i = 0, 1, \ldots, h\}$ and $U = \{U_i, i = 0, 1, \ldots, m\}$ are as shown in Fig. 2, which are respectively plotted by the red and blue dots. In addition, the elements number of U and P are not necessarily identical, here we suppose that h $\geq$ m. In order to make the distance between the point and the same point of the U and P point-cloud short, it is necessary to obtain the matrix of the point-cloud, which is the rotation and translation matrix.

B. **The objective function of iterative closest point**

In 3D space, the Euclidean distance between $\vec{p_i} = (x_i, y_i, z_i)$ and $\vec{q_i} = (x_i, y_i, z_i)$ can be expressed as:

Fig. 2 Point-cloud data

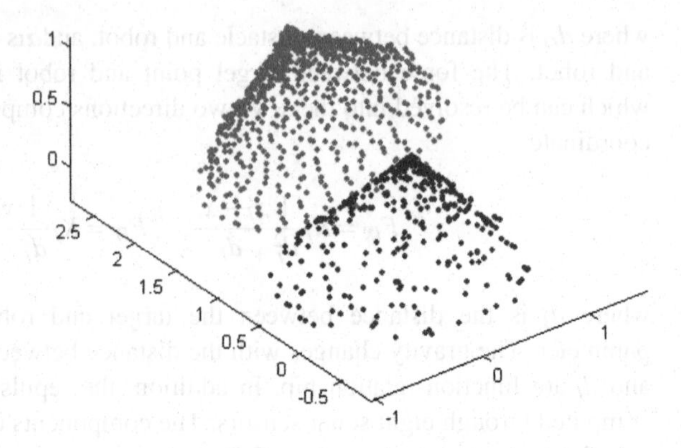

$$d(\vec{p_i}, \vec{q_i}) = \|\vec{p_i} - \vec{q_i}\| = \sqrt{(x_i - x_j)^2 + (y_i - y_j)^2 + (z_i - z_j)^2} \tag{5}$$

The purpose of 3D point-cloud matching problem is to find the transformation matrices H and T of the P and Q. For $\vec{q_i} = R\vec{p_i} + T + N_i, i = 1, \ldots, N$, the optimal solution is solved by least square method, then we can get H and T when E is the minimum value.

$$E = \sum_{i=1}^{N} \left|(R\vec{p_i} + T) - \vec{q_i}\right|^2 \tag{6}$$

C. **Parallel moving and rotating separation**

To estimate the initial value of the translation vector T, the specific method is to get the center of the point-cloud data P and Q:

$$\vec{p} = \frac{1}{n}\sum_{i=1}^{N} \vec{p_i}, \quad \vec{q} = \frac{1}{n}\sum_{i=1}^{N} \vec{q_i} \tag{7}$$

Point-cloud data P and Q are translated to the center point:

$$\overrightarrow{p_i} = \vec{p_i} - \vec{p}, \overrightarrow{q_i} = \vec{q_i} - \vec{q} \tag{8}$$

Thus, the optimization objective function can be transformed into:

$$E = \sum_{i=1}^{N} \left| \left(\overrightarrow{p_i} + \vec{p_i}\right) - \left[\mathrm{H}\left(\overrightarrow{q_i} + \vec{q_i}\right) + T \right] \right| \tag{9}$$

Then the optimization problem can be divided into the following questions:

(1) seek to minimize the H of the E,

(2) $T = \overrightarrow{q} - H\overrightarrow{p}$

D. **Using the control points to obtain the initial rotation matrix**

The method of control points is usually used to obtain the initial matrix, that is, two points ($pt_i^{'}$, $qt_i^{'}$, $i = 1, 2, 3$) are selected in the cloud. For the pair of the points, calculate the corresponding point matrix A_i:

$$A_i = \begin{bmatrix} 0 & (\overrightarrow{pt_i^{'}} - \overrightarrow{qt_i^{'}})^{\mathrm{T}} \\ (\overrightarrow{pt_i^{'}} - \overrightarrow{qt_i^{'}}) & D_i^M \end{bmatrix} \tag{10}$$

where $D_i^M = \overrightarrow{pt_i^{'}} + \overrightarrow{qt_i^{'}}$ denotes the transpose matrix of D_i.

For each pair of matrices A_i and B, we have

$$B = \sum_{i=1}^{m} A_i A_i^T \tag{11}$$

The eigenvalues of B are obtained by singular value decomposition (SVD), which is called H. However, in the process of 3D reconstruction of mobile robot, the control points cannot be accurately obtained. If the control points are given randomly, the calculation of the rotation matrix may lead to a local optimal problem. In this paper, the position and orientation of the robot are used to compute the rotation and translation matrix of the corresponding depth camera:

$$R^{'} = \begin{bmatrix} \cos a_R & \sin a_R & 0 & 0 \\ -\sin a_R & \cos a_R & 0 & 0 \\ 0 & 0 & 1 & 0 \\ 0 & 0 & 0 & 1 \end{bmatrix}, \quad T^{'} = \begin{bmatrix} 1 & 0 & 0 & 0 \\ 0 & 1 & 0 & 0 \\ 0 & 0 & 1 & 0 \\ dx & dy & 0 & 1 \end{bmatrix} \tag{12}$$

Here, dx and dy are respectively the translation distance of the robot in the XY plane of the two coordinates. Then we can get the rotation matrix by the following formula:

$$T = \overrightarrow{q} - H\overrightarrow{p} \tag{13}$$

The translation matrix is calculated by center point. In the process of mobile robot, the translation matrix can be obtained in real time, which is the translation matrix of the camera.

$$T = T^{*}\theta + T^{'} * (1 - \theta) \tag{14}$$

Remark 1 θ is the scale parameter, which represents the ratio of the translation matrix and the robot translation matrix from the center point. After the initial matching, all points in the point-cloud data $P^{'}$ take a matrix transformation.

The matching degree or iteration of comparing with the point-cloud data P' and Q' is the algorithm termination conditions.

E. **Iterative closest point (ICP)**

For each specific point in P, we can find the nearest point in Q as the corresponding point. At a given step, the following functions are minimized:

$$E=\sum_{i=1}^{N}\left|T^{k+1}\vec{p_i}-\vec{q_i}^{k}\right|^2,\quad \vec{q_i}^{k}=\vec{q}\left|\min_{q}\left|T^{k-1}\vec{p_i}-\vec{q_i}\right|\right| \tag{15}$$

Hence, the transformation matrices H and T based on the position and orientation are obtained.

3.2 *Pose Fusion*

In the process of autonomous navigation of mobile robot and real-time 3D reconstruction, robot transformation matrix is firstly obtained, which also denotes the transformation matrix of the depth camera in space. Specific algorithm flow chart is as follows: (Fig. 3)

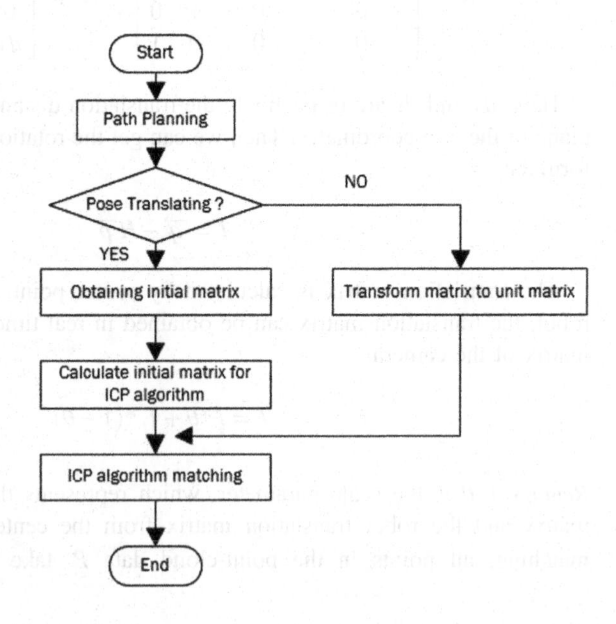

Fig. 3 Flow chart of position and ICP fusion algorithm

4 Experimental Results and Analysis

4.1 Path Planning

In order to verify the accuracy of path planning algorithm, experiments are carried out on the MobileSim platform. As show in Fig. 4, the red line represents the path of the robot:

4.2 3D Reconstruction

In order to verify the accuracy and real-time performance of the ICP algorithm and the robot pose fusion in 3D reconstruction, we do the experiment of point-cloud splicing. As show in Fig. 5, in the Fig. 5a, e of scenarios, we collect two point-cloud images at different locations, where Fig. 5b, f are left and right angle source image, Fig. 5c, g are the effect charts of Fig. 5a, b point-cloud through voxel filtering and only the spatial coordinates of the cloud point are retained.

It can be seen from Fig. 5 that the source point-cloud can make the source point-cloud become sparse by the voxel filter in the condition that the feature of the point-cloud is not lost.

As shown in Fig. 6, the average time for each iteration of point-cloud is 4, 3, 1.5 ms, and the average time decreases with the number of repeated superposition iteration. When the original point-cloud is matched and calculated for the first time, the difference between point-clouds is the largest. During the matching process, we need to calculate the deviation of all point-cloud distance and compare it with the matching degree, as shown in formula (15), so the number of point-cloud

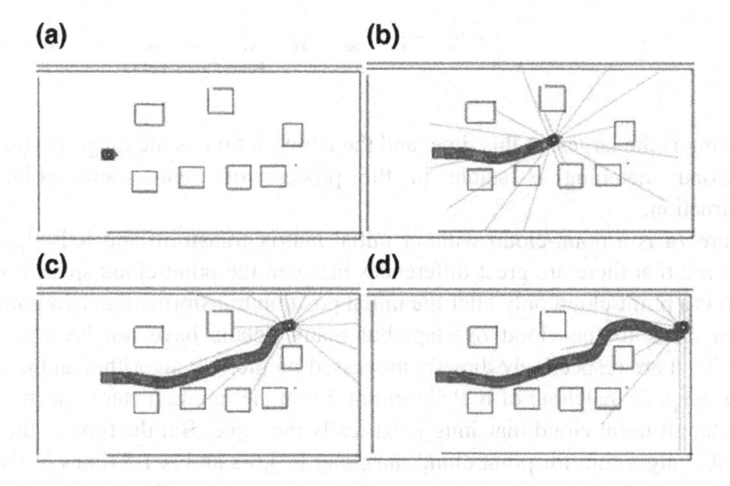

Fig. 4 Real time path planning of robot

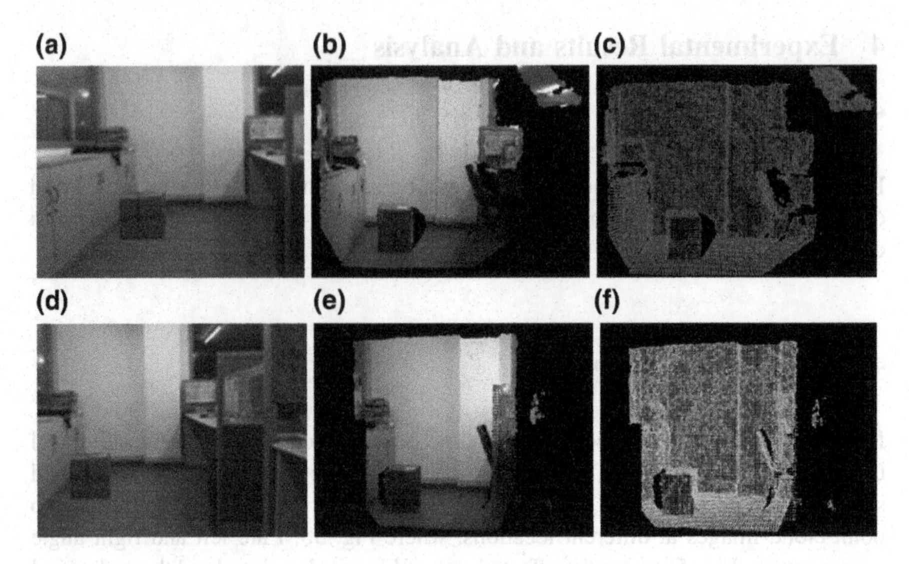

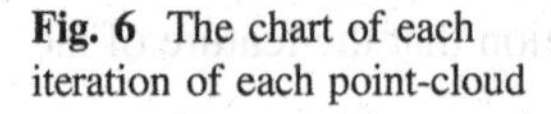
Fig. 5 Source point-cloud

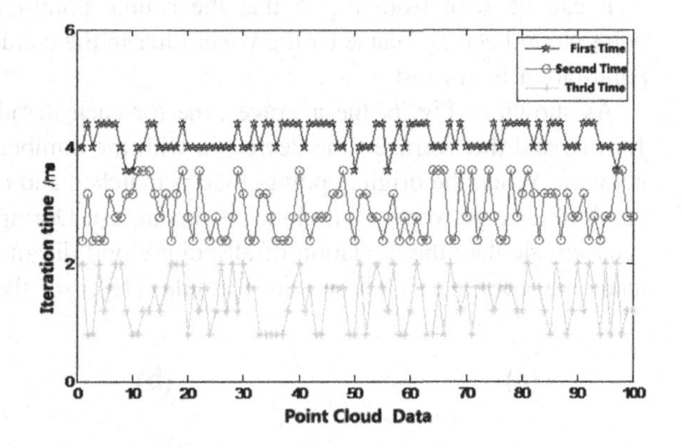

Fig. 6 The chart of each iteration of each point-cloud

processing is the largest at this time, and the iteration time is the longest. Therefore, point-cloud matching is stable in the process of multi frame point-cloud reconstruction.

Figure 7a is a point-cloud without initial matrix transform and ICP algorithm, we can see that there are great differences between the point-cloud space position, Fig. 7b is a point-cloud only after the initial position transformation, two points can be seen close to the cloud overlap, but many details have not been matched. Figure 7c, d are respectively directly processed by the ICP algorithm and the effect map of the pose matching of ICP algorithm. From the figure, it can be seen that the effect map of point-cloud matching is basically the same. But the time of the direct use of ICP algorithm for point-cloud matching is 3.6 s and is 1.5 times of the pose fusion, because in the matching process, if the reference coordinate system gap will

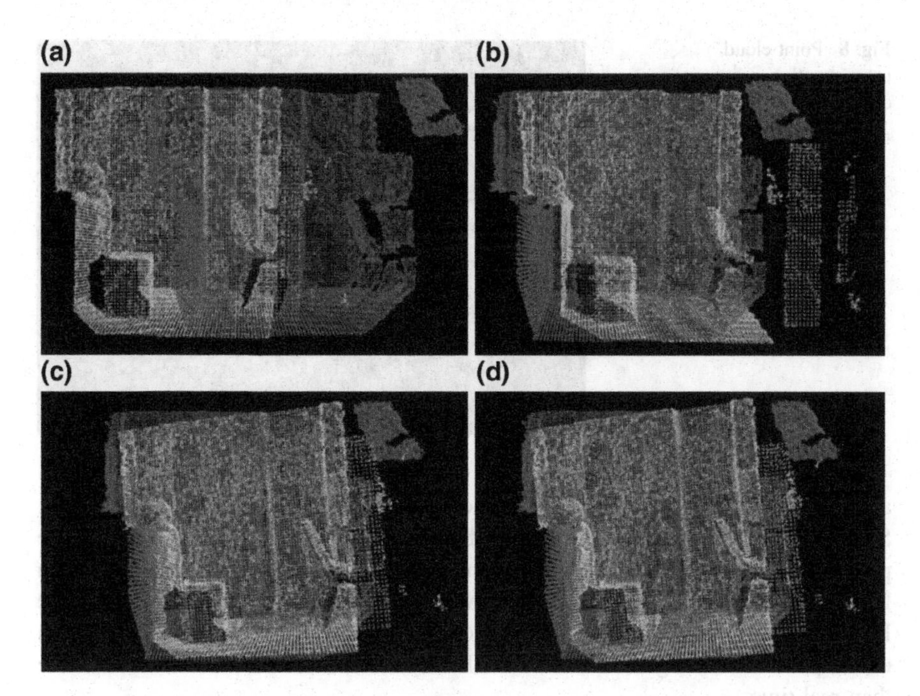

Fig. 7 Comparison of point-cloud reconstruction

increase the ICP algorithm iteration times, it will increase System running time. In order to verify the accuracy and real-time performance of the APF-ICP algorithm, we compare the classical ICP algorithm with the classical NDT algorithm. As shown in Table 1, we can register the 100-point-cloud frame and collect the time and Euclidean fitness mean: where, the European fitness is the output point to the nearest target point-cloud corresponding point iteration process distance square sum. It can be seen from Table 1 that the run time of NDT and ICP is 2.25 times and 10.69 times of APF-ICP respectively. Because NDT is directly on point-cloud matching, it will be affected by the point-cloud reference coordinate system. The APF-ICP algorithm uses the artificial potential field navigation to obtain the robot transformation matrix, and the matrix is the initial change matrix of the ICP algorithm. The APF-ICP algorithm uses the artificial potential field navigation to obtain the robot transformation matrix. So the number of iterations can be greatly reduced. So APF-ICP algorithm running time is better than the other two algorithms. At the same time, it can be seen from Table 1 that the NDT and ICP

Table 1 Algorithm comparison

Algorithm	Time (s)	European suitability (m)
Classic NDT	3.44	0.85
Classic ICP	16.35	0.59
APF-ICP	1.53	0.43

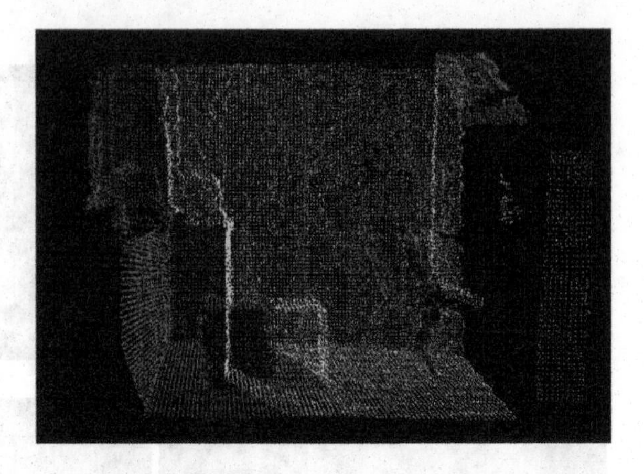

Fig. 8 Point-cloud reconstruction comparison chart

algorithms have 1.98 times and 1.37 times of the APF-ICP algorithm respectively. Since ICP can calculate the distance of between target point-cloud and the source point-cloud and find the nearest matching point, the degree of fitness is less than NDT. But because of the difference between the initial position of the cloud, the European suitability is greater than that of APF-ICP. However, the initial position of the point-cloud is greater than that of APF-ICP, so APF-ICP is more accurate than real-time.

As shown in Fig. 8, in the use of NDT algorithm for point-cloud matching process, the red dot cloud and the green point-cloud are mostly overlapped with the wall surface, but the side is largely failed to match. As the NDT algorithm make that a space (reference scan) is divided into the grid cell, each grid cell is used to load the point-cloud. And the lattice parameter size has a large effect. As shown in Fig. 8, the difference between the two image positions is large, and the probability distribution function of the corresponding lattice is large. Therefore, the reconstruction process will lose some features.

5 Conclusion

In this paper, a 3D reconstruction method of an effective mobile robot is proposed. To complete the 3D reconstruction, we need to use the APF method to plan the path of the robot in real time and obtain the initial transformation matrix of the robot. Then, the initial transformation matrix of the robot is combined with the ICP algorithm. Experiments show that the proposed algorithm is more suitable for 3D reconstruction of real-time path planning of mobile robots than traditional NDT and ICP algorithms. However, 3D reconstruction of point-cloud is only in the indoor environment, and the coordinate transformation of robots is very stable. And in outdoor environment, the accuracy of the algorithm may be affected by complex environments or sensors of mobile robots.

Acknowledgements This paper was partially supported by The National Natural Science Foundation (61374040, 61503205).

References

1. Magnusson M, Andreasson H, Nüchter A, et al. Appearance-based loop detection from 3D laser data using the normal distributions transform. IEEE Int Conf Robot Autom. IEEE Xplore 2009;23–28.
2. Saarinen J, Andreasson H, Stoyanov T, et al. Normal distributions transform occupancy maps: application to large-scale online 3D mapping. IEEE Int Conf Robot Autom. IEEE 2013;2233–8.
3. Ji WK, Lee BH. Robust and fast 3-D scan registration using normal distributions transform with supervoxel segmentation. Robotica. 2016;34(7):1630–58.
4. Chen T, Wen H, Hu H, et al. On-orbit assembly of a team of flexible spacecraft using potential field based method. Acta Astronaut. 2017;133:221–32.
5. Li C, Jiang X, Wang W, et al. A simplified car-following model based on the artificial potential field. Procedia Eng. 2016;137:13–20.
6. Zhou L, Li W. Adaptive artificial potential field approach for obstacle avoidance path planning. Seventh Int Symp Comput Intel Des. IEEE 2015;429–32.
7. Guo H, Wang G, Huang L, et al. A robust and accurate two-step auto-labeling conditional iterative closest points (TACICP) algorithm for three-dimensional multi-modal carotid image registration. PLoS One. 2016;11(2):e0148783.
8. Spinczyk D. Chapter 32—surface registration by markers guided nonrigid iterative closest points algorithm. Em Trends Image Process Comput Vis Pattern Recog. 2015;489–97.
9. Yang H, Jiang J, Zhao G, et al. An improved iterative closest points algorithm. World J Eng Technol. 2015;03(3):302–8.
10. Li C, Xue J, Du S, et al. A fast multi-resolution iterative closest point algorithm. Pattern Recog. IEEE 2010;1–5.
11. Bou-Maroun E, Rossignol J, Fonseca BD, et al. Feasibility of a microwave liquid sensor based on molecularly imprinted sol-gel polymer for the detection of iprodione fungicide. Sens Actuators B Chem. 2017;244:24–30.
12. Han C, Lee HM, Park M, et al. The diphoton resonance as a gravity mediator of dark matter. Phys Lett B. 2016;755:371–9.

A Method of Trajectory Prediction Based on Kalman Filtering Algorithm and Support Vector Machine Algorithm

Quan Cheng and Chaoli Wang

Abstract A trajectory prediction method based on kalman filter algorithm (KF) and support vector machine algorithm (SVM) is proposed to predict the trajectory prediction of fast flight ping-pong in the research of ping-pong robot. This method combines the real-time performance of KF and the stability of SVM. By comparing the correlation coefficient between the predicted value and the measured value, the method intelligently selects an appropriate algorithm for the trajectory prediction of ping-pong. Finally, the result of simulation experiment of fast flight ping-pong shows that the method has good stability to the trajectory prediction of ping-pong, and the prediction accuracy is obviously improved compared with the single algorithm.

Keywords Kalman filter · Support vector machine · Correlation coefficient · Trajectory prediction

1 Introduction

As a product of high electromechanical integration, ping-pong robot is an important platform for the study of robot vision and motion control by sensing, predicting, making decisions and coordinating the movement. Ping-pong robot can play ping-pong with human beings, and can perceive the routes and the trajectory of athletic objects, then, make reasonable judgment and return strategy for flexibly hit, which is a typical real-time intelligent robot [1]. The research of ping-pong robot system involves a wide range of fields, which combines the knowledge of computer

Q. Cheng · C. Wang (✉)
School of Optical-Electrical and Computer Engineering,
University of Shanghai for Science and Technology, Shanghai 200093, China
e-mail: clclwang@126.com

Q. Cheng
e-mail: cq951738624@163.com

Y. Jia et al. (eds.), *Proceedings of 2017 Chinese Intelligent Systems Conference*, Lecture Notes in Electrical Engineering 459,
https://doi.org/10.1007/978-981-10-6496-8_46

vision, AI, automatic control, robot kinematics and computer graphics, and has special research value and wide application prospect [2].

At present, both foreign and domestic, using high-speed vision to obtain the initial positioning information of the flight ball for ping-pong state estimation and future flight trajectory prediction is the current common practice of all ping-pong robots. For the motion state estimation and trajectory prediction task, many outstanding research results have emerged at home and abroad. For the scene that can realize the stable observation period through hardware synchronization, it is suggested to adopt the filtering method of motion modeling based on force analysis. Firstly, the filtering algorithm can use the more realistic motion model. Secondly, the filtering algorithm can better process the error of system and observation by a model, and obtain better calculation efficiency, good accuracy and robustness [3].

Kalman filter algorithm (KF) is the most common filtering algorithm used to solve the prediction problem. It is a mature predictive method based on linear regression analysis [4]. Support vector machine algorithm (SVM) is another algorithm that can be used as a prediction [5, 6]. SVM apply structural risk minimization principles to learn, and has a high generalization ability. On the basis of predecessors, this paper proposes a combinatorial trajectory prediction method based on KF and SVM for fast flight ping-pong trajectory prediction [7].

2 Prediction Algorithm and Modeling

2.1 KF Introduction and Modeling

KF is a recursive algorithm proposed by Kalman Re in 1960 [8]. It is a constant prediction and correction process. It has been successfully applied in many fields such as communication, navigation, guidance and control, and has high prediction accuracy before being applied to the prediction of flight object trajectory. KF is a filtering method for time-varying random signals. The mathematical model of the random signals and its measurement process is expressed respectively as:

$$x_k = Ax_{k-1} + Bu_{k-1} + w_{k-1}, \; z_k = Hx_k + v_k \tag{1}$$

where x_k is the state variable of the kth moment, A is the system matrix, that is, A is the state transition matrix from the $k-1$ th to kth moment, B is the input variable gain, u_{k-1} is the input variable, w_{k-1} is the system noise, z_k is the observation variable, H is the observation matrix, v_k is measurement noise. In this paper, we assume that the system noise is not correlated with the measurement noise, and the mean value is 0, and the covariance is the independent Gaussian white noise of Q and R respectively [9]. The statistical characteristics are as follows:

$$E[w_k]=0, E[w_k w_j^T]=Q_k\delta_{kj}, E[v_k]=0, E[v_k v_j^T]=R_k\delta_{kj}, E[w_k v_j^T]=0 \tag{2}$$

The minimum mean square error estimation $\hat{x}_k$ of the state variable x_k can be obtained by the following time updating equations and state updating equations of KF [10]. Time updating equations and state updating equations:

$$\begin{aligned} &\hat{x}_k^- = A\hat{x}_{k-1} + Bu_{k-1},\ P_k^- = AP_{k-1}A^T + Q \\ &K_k = P_k^- H^T(HP_k^- H^T + R)^{-1},\ \hat{x}_k = \hat{x}_k^- + K_k(z_k - H\hat{x}_k^-),\ P_k = (I - K_kH)P_k^- \end{aligned} \tag{3}$$

In this paper, KF model bases on motion modeling of force analysis. First of all, we establish a sports model for ping-pong which leaves the ping-pong bat. We view ball as the rigid body of quality uniform distribution. According to aerodynamic analysis, the main force of ping-pong is its own gravity, air friction, and Magnus force because of rotation. In order to facilitate the comparison of the advantages and disadvantages between a single algorithm and the combination algorithms, this paper only considers the gravity of ping-pong. At the same time, we assume that ping-pong conducts oblique throwing movement after leaving the bat, and θ is the angle on the horizontal direction.

As shown in the Fig. 1, the forward direction of the ping-pong is x-axis, and the descent direction of the ping-pong is y-axis. The direction which is perpendicular to the plane of x-axis and y-axis, and points to the observer is z-axis. These axises constitute the basic coordinate system of ping-pong motion.

Through modeling, we find some problems. If we directly describe the relationship between displacement and velocity by constructing the equation of displacement and velocity of ping-pong, it will emerge transcendental equations which are difficult to solve because of the addition of square terms. Accordingly, the components x_y and v_y of the displacement x and the velocity v on the y-axis are selected, and the state equation is easily constructed. Then, according to the angle relationship, the components x_y and v_y are converted into total displacement x_{total} and total velocity v_{total} again, and the problems can be resolved. According to the above figure and the above description, the kinematic equations of displacement and velocity of ping-pong are as follows:

$$x_t = x_{t-1} + \dot{x}_{t-1}\tilde{t} + \frac{1}{2}g\tilde{t}^2, \dot{x}_t = \dot{x}_{t-1} + g\tilde{t} \tag{4}$$

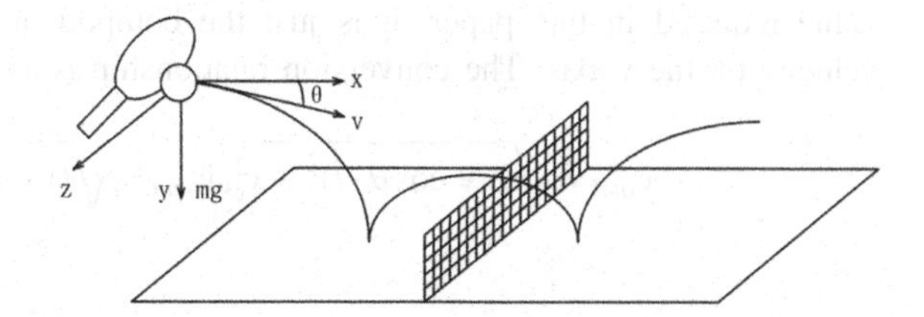

Fig. 1 Motion modeling of force analysis for fast flight ping-pong

where x_t is the displacement component of the ping-pong on the y-axis at time t, that is, $x_t = x_y$, $\dot{x}_t$ is the velocity component of the ping-pong on the y-axis at time t, at the same time, $\dot{x}_t$ is the derivative of x_t, that is, $\dot{x}_t = v_y = \dot{x}_y$, $\tilde{t}$ is the amount of time change, g is the gravitational constant.

In order to apply KF to predict state variable, the above kinematics equation is rewritten as a state space representation:

$$\begin{pmatrix} x_t \\ \dot{x}_t \end{pmatrix} = \begin{pmatrix} 1 & \tilde{t} \\ 0 & 1 \end{pmatrix} \begin{pmatrix} x_{t-1} \\ \dot{x}_{t-1} \end{pmatrix} + \begin{pmatrix} \frac{1}{2}\tilde{t}^2 \\ \tilde{t} \end{pmatrix} g + w_{t-1} \tag{5}$$

where w_{t-1} is the system noise. Corresponding to formula (1), that is, $x_k = \begin{pmatrix} x_t \\ \dot{x}_t \end{pmatrix}$,

$$A = \begin{pmatrix} 1 & \tilde{t} \\ 0 & 1 \end{pmatrix}, x_{k-1} = \begin{pmatrix} x_{t-1} \\ \dot{x}_{t-1} \end{pmatrix}, B = \begin{pmatrix} \frac{1}{2}\tilde{t}^2 \\ \tilde{t} \end{pmatrix}, u_{k-1} = g, w_k = w_{t-1}.$$

At the same time, in the experiment we adopt high-speed camera to shoot the fast flight ping-pong. After pretreating the video, we can get the measured value of displacement and velocity of ping-pong with noise. The measurement equation is as follows:

$$z_k = Hx_k + v_k \tag{6}$$

where z_k is the observation variable, H is the observation matrix, v_k is measurement noise. As mentioned above, we assume that the system noise is not correlated with the measurement noise, and the mean value is 0, and the covariance is the independent Gaussian white noise of Q and R respectively,that is, $p(w_k) \sim N(0, Q)$, $p(v_k) \sim N(0, R)$. Since there is no priori data, P and H are set to diagonal matrices and we set the corresponding initial values according to experience. Combining the Eqs. (1)–(7), then putting the initial value of the known A, B, Q, R, P and H to these equations, and we can obtain the predicted value of the next time, according to the iteration of the following equation, that is, $\hat{x}_k$.

$$\begin{cases} \hat{x}_k^- = A\hat{x}_{k-1} + Bu_{k-1}, P_k^- = AP_{k-1}A^T + Q \\ K_k = P_k^- H^T (HP_k^- H^T + R)^{-1}, P_k = (I - K_k H)P_k^- \end{cases} \tag{7}$$

where $\hat{x}_k = \hat{x}_k^- + K_k(z_k - H\hat{x}_k^-)$ is the formula which needs to be solved. $\hat{x}_k$ contains both displacement and velocity, but $\hat{x}_k = \begin{pmatrix} \hat{x}_y \\ \hat{\dot{x}}_y \end{pmatrix} = \begin{pmatrix} \hat{x}_y \\ \hat{v}_y \end{pmatrix}$ is not the final estimated value required in this paper, it is just the component of the displacement and velocity on the y-axis. The conversion relationship is as follows:

$$\hat{x}_{total} = \sqrt{(v \cdot \cos\theta \cdot \tilde{t})^2 + \hat{x}_y^2}, \hat{v}_{total} = \sqrt{(v \cdot \cos\theta)^2 + \hat{v}_y^2} \tag{8}$$

where $\hat{x}_{total}$ is the final displacement estimation, v is the initial velocity of the ping-pong, θ is the angle between the initial velocity and the horizontal direction, $\hat{x}_y$ is the component on the y-axis, $\hat{v}_{total}$ is the final velocity estimation, $\hat{v}_y$ is the component of the velocity $\hat{v}_{total}$ on the y-axis. Here, the establishment of the KF model is finished.

2.2 SVM Introduction and Modeling

SVM is a general learning machine developed by Vapnik and other scholars which is on the basis of statistical learning theory [11]. SVM applys structural risk minimization principle to learn, and has a high generalization ability. The simple derivation of the regression support vector machine algorithm is as follows: for the sample set $T=\{(x_i,y_i),i=1,2,\ldots,l\}$, where $x_i\in R^N$, $y_i\in R$. Firstly, we map the variable x to the high-dimensional feature space F through the non-linear mapping $\phi(\,)$, and then construct the linear regression function $f(x)=w^T\phi(x)+b$, where w and b are solutions of the following convex quadratic programming problem:

$$\begin{aligned}&\min_{w,b} P=\tfrac{1}{2}w^Tw+C\sum_{i=1}^{l}(\xi_i+\xi_i^*), s.t.\quad y_i-(w^Tx_i+b)\le\varepsilon+\xi_i\\&(w^Tx_i+b)-y_i\le\varepsilon+\xi_i^*,\xi_i,\xi_i^*\ge 0,i=1,2,\ldots,l\end{aligned}\tag{9}$$

where C is the penalty factor used to adjust the balance between empirical risk and expressive ability, ε is the insensitive loss function, ξ_i,ξ_i^* is the slack variable. Using the Lagrangian multiplier method, the original problem is transformed into the following dual problem:

$$\begin{aligned}&\min_{a,a^*} D=\tfrac{1}{2}\sum_{i=1}^{l}\sum_{j=1}^{l}Q_{ij}(a_i-a_i^*)(a_j-a_j^*)+\varepsilon\sum_{i=1}^{l}(a_i+a_i^*)-\sum_{i=1}^{l}y_i(a_i-a_i^*)\\&s.t.\quad 0\le a_i,a_i^*\le Cq_i,i=1,2,\ldots,l,\ \sum_{i=1}^{l}(a_i-a_i^*)=0\end{aligned}\tag{10}$$

where $Q_{ij}=\phi(x_i)^T\phi(x_j)=K(x_i,x_j)$, $K(,)$ is a kernel function, the introduction of a kernel function avoids the dimension disaster because of the use of non-linear mapping $\phi(\,)$, a_i,a_i^* is the solution of the dual problem (2), and the regression function that needs to be solved can be rewritten as:

$$f(x)=\sum_{i=1}^{l}(a_i-a_i^*)K(x_i,x)+b\tag{11}$$

Because the displacement and velocity of fast flight ping-pong vary with time, we can use the displacement and velocity of the few previous moments to predict the displacement and velocity of the next moment. Here, we only discuss the SVM modeling of the displacement here. We assume that x_i is a comprehensive factor which affects the trajectory prediction of ping-pong and y_i is the predicted value of the displacement of ping-pong at the next moment. The establishment of the fast flight ping-pong trajectory prediction model based on SVM is equal to seeking the establishment of the following expression:

$$f(x) = \sum_{i=1}^{k} (a_i - a_i^*) K(x, x_i) + b \tag{12}$$

where x is the comprehensive factor which affects the trajectory of ping-pong, x_i is the ith sample in k samples, $k(x, x_i)$ is the kernel function. There are four main categories of kernel functions such as linear, polynomial, radial basis function (RBF), sigmoid. In this paper, we use the radial basis function, as shown in the following formula.

$$K(x, y) = \exp\left| -\frac{\|x - y\|^2}{2\sigma^2} \right| \tag{13}$$

We assume that the current time is t, x_t is the displacement data of ping-pong at time t, x_{t-1} is the displacement data of ping-pong at time $t-1$. This paper adopts the displacement data of the ping-pong at the current time and the previous two moments to predict the displacement of the next moment. At time t, the input value of the sample is x_t, x_{t-1}, x_{t-2}; x_{t+1} is the output value of the sample, that is, $x_{t+1} = y_i$. The model can be expressed as:

$$x_{t+1} = \varphi_1 x_t + \varphi_2 x_{t-1} + \varphi_3 x_{t-2} \tag{14}$$

where $\varphi_1 + \varphi_2 + \varphi_3 = 1$, and $\varphi_1 > \varphi_2 > \varphi_3$. The impact of different historical moments on the predicted value is different, the closer to the forecasting time, the bigger the weight distribution. According to the established model, we apply the LIBSVM software package developed by Associate Professor Lin Ctlih-Jen of Taiwan University to predict the displacement of fast flight ping-pong [12].

3 Combination Method Based on KF and SVM

A standard assumption in the classical control theory is that the data transmission required by the control or state estimation algorithm can be performed with infinite precision [13]. In this paper, we combine the real-time performance of KF and the stability of SVM, and combine the idea of average linear combination forecasting and 0–1 combination forecasting. According to the prediction results of the previous three moments of the time to be predicted, we determine which prediction method will be adopted to conduct the prediction [14]. For the change law of the current ping-pong flight trajectory, the model flexibly takes the corresponding prediction algorithm. Here are two cases:

1. The simple average linear combination method is used to predict the state variable of the next moment, when the single prediction method has approximately the same forecasting error square sum for the previous three moments. Experience and experimental evidence show that this is reasonable. The expression is:

$$\hat{x}_{combine} = \frac{1}{2}(\hat{x}_{kalman} + \hat{x}_{svm}) \tag{15}$$

2. The 0-1 combination forecasting model is adopted, when there is a certain difference between the forecasting error square sums for the previous three moments of the single prediction method. According to the change rule of the current ping-pong flight path, we apply the appropriate model based on real-time decision-making to conduct the prediction with the correlation coefficient maximization criterion, so that different time takes different forecasting models to improve the forecasting accuracy. The mathematical expression of the model is:

$$\hat{x}_{combine} = \Theta(r_k - r_s)\hat{x}_{kalman} + \Theta(r_s - r_k)\hat{x}_{svm} \tag{16}$$

where r_k and r_s are respectively the correlation coefficient of the previous three moments of the predicted value and the measured value of KF and SVM, $\hat{x}_{combine}$, $\hat{x}_{kalman}$, $\hat{x}_{svm}$ are the predicted value of KF, SVM and the combined model at time t, respectively. The Θ $(\cdot)$ function is defined as: when $x \leq 0$, $\Theta(x) = 0$; when $x > 0$, $\Theta(x) = 1$. In order to make the predicted value close to the real value, we hope that the predicted value vector and the measured value vector are highly linearly related, the greater the correlation coefficient, the better the prediction effect, so the larger the r, the more effective the corresponding prediction method. The formula of correlation coefficient is as follows:

$$r_{XY} = \frac{\sum_{i=1}^{N}(X_i - \bar{X})(Y_i - \bar{Y})}{\sqrt{\sum_{i=1}^{N}(X_i - \bar{X})^2}\sqrt{\sum_{i=1}^{N}(Y_i - \bar{Y})^2}} \tag{17}$$

Assuming time is t, the predicted value of ping-pong displacement of the combined model is $x^t_{combine}$, the KF predicted value is x^t_{kalman}, the SVM predicted value is x^t_{svm}. The measured vector of previous three moments is $A = [A_{t-1}, A_{t-2}, A_{t-3}]$, The SVM predicted value of the previous three moments is $S = [S_{t-1}, S_{t-2}, S_{t-3}]$, The KF predicted value of the previous three moments is $K = [K_{t-1}, K_{t-2}, K_{t-3}]$.

4 Analysis for Simulation Experiment

In this paper, we respectively apply KF, SVM and combinatorial model algorithm to predict the displacement value of fast flight ping-pong. Due to experimental conditions, we use the simulation data to simulate the actual observation value in this paper. The sampling period is set to 0.002 s, the total iteration time is 0.12 s, the initial value of velocity is 5 m/s, and the angle with the horizontal direction is 20°. Substituting these initial data into kinematic equations, we can obtain 60 simulation datas of the actual observation value with noise.

As mentioned above, we assume that the system noise is not correlated with the measured noise, and the mean value is 0, and the covariance is the independent Gaussian white noise of Q and R respectively,that is, $p(w_k) \sim N(0, Q)$, $p(v_k) \sim N(0, R)$. Since there is no priori data, P and H are set to diagonal matrices and we set the corresponding initial values according to experience.

We adopt MATLAB software for simulation experiments in this paper.

From the simulation experiment, we obtain the prediction results of the three prediction methods of displacement and velocity respectively. The figures are follows:

Figures 2 and 3 illustrate the predicted results of KF, SVM and combined algorithm for the ping-pong displacement and velocity. We can clearly see that the results of combined algorithm for the ping-pong displacement prediction are significantly better than a single algorithm.

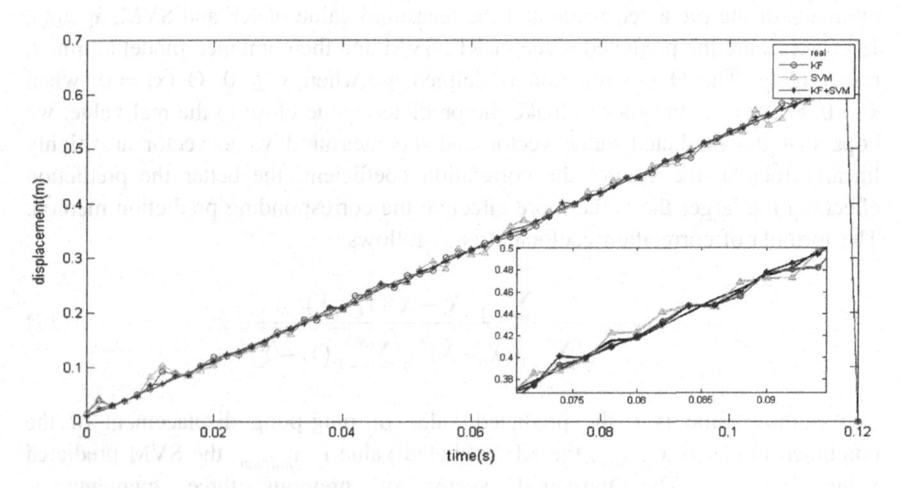

Fig. 2 The predicted value of the displacement of KF, SVM, KF + SVM

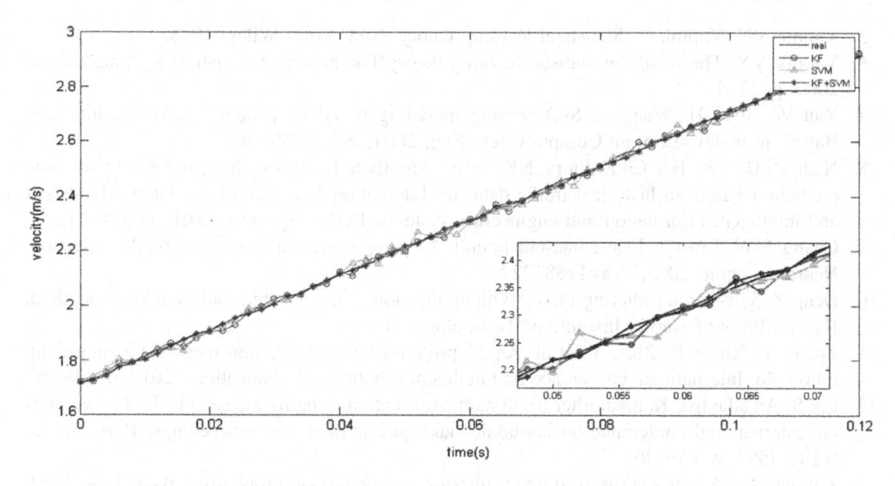

Fig. 3 The predicted value of the velocity of KF, SVM, KF + SVM

5 Conclusion

In this paper, a trajectory prediction method based on KF and SVM is proposed to predict the trajectory prediction of fast flight ping-pong in the research of ping-pong robot. This method combines the real-time performance of KF and the stability of SVM. By comparing the correlation coefficient between the predicted value and the measured value, the method intelligently selects an appropriate algorithm for the trajectory prediction of ping-pong. From the figures, we can see that the combined algorithm is better than the single algorithm. The result of simulation experiment of fast flight ping-pong shows that the method has good stability to the trajectory prediction of ping-pong, and the prediction accuracy is obviously improved compared with the single algorithm. The next job, we plan to try multi-step prediction for the trajectory of ping-pong.

Acknowledgements This paper was partially supported by the National Natural Science Foundation (61374040, 61503205).

References

1. Andersson RL. Dynamic sensing in a ping-pong playing robot. IEEE Trans Robot Autom. 1989;5(6):728–39.
2. Dan Yu, Wei Wei, Yuanhui Zhang. Study on dynamic target tracking algorithm based on Kalman prediction. Photoelectr Eng. 2009;36(1):52–6.
3. In: Proceedings of the 5th international conference on intelligent robotics and applications, ICIRA 2012, Montreal, Canada, 3–5 October 2012. Springer; 2012.
4. Zarchan P. Progress in astronautics and aeronautics: fundamentals of Kalman filtering: a practical approach. Aiaa; 2005.

5. Vapnik VN, Vapnik V. Statistical learning theory. New York: Wiley; 1998.
6. Vapnik VN. The nature of statistic learning theory/The nature of statistical learning theory. 2000. p. 17–4.
7. Yan W, Shao H, Wang X. Soft sensing modeling based on support vector machine and Bayesian model selection. Comput Chem Eng. 2004;28(8):1489–98.
8. Nath RPD, Lee HJ, Chowdhury NK, et al. Modified K-means clustering for travel time prediction based on historical traffic data. In: International conference on knowledge-based and intelligent information and engineering systems. Berlin: Springer; 2010. p. 511–21.
9. Chang MW, Lin CJ. Leave-one-out bounds for support vector regression model selection. Neural Comput. 2005;17(5):1188–222.
10. Deng Z. Self-Tuning filtering theory with applications—modern time series analysis method. Harbin: Press of Harbin Institute of Technology; 2003.
11. Zhang Y, Xiong R, Zhao Y, et al. An adaptive trajectory prediction method for ping-pong robots. In: International conference on intelligent robotics and applications. 2012. p. 448–59.
12. Liu S. An adaptive Kalman filter for dynamic estimation of harmonic signals. In: Proceedings 8th international conference on harmonics and quality of power proceedings, 1998, vol. 2. IEEE; 1998. p. 636–40.
13. Kalman RE. A new approach to linear filtering and prediction problems. J Basic Eng. 1960; 82(1):35–45.
14. Wessberg J, Stambaugh CR, Kralik JD, et al. Real-time prediction of hand trajectory by ensembles of cortical neurons in primates. Nature. 2000;408(6810):361–5.

Distributed Adaptive Control for Consensus of Unknown Nonlinear Multi-agent Systems

Shuai Chen and Chaoli Wang

Abstract This paper is concerned with consensus problem of multi-agent systems. Assuming that the communication among distinct agents in the entire group can be represented by a digraph structure, a distributed adaptive control protocol is presented to solve the leader-follower consensus problems for multi-agent systems having non-identical unknown nonlinear dynamics and uncertain disturbance. The proposed protocol is distributed because each agent's control protocol only utilizes the information of it's neighbor agents. In addition, the adaptive updating law for neural networks determined by projection method. It is shown that with our proposed protocol, all the consensus errors converge to zero asymptotically.

Keywords Multi-agent systems · Digraph · Distributed control · Projection method

1 Introduction

The study of multi-agent systems has received extensive attention in the past few decades, mainly because of its wide application such as spacecraft, mobile robots, etc. Some seminal works are [1–6], to name a few. Jadbabaie et al. provided a theoretical explanation for the Vicsek model in [1], it takes possible changes into account in nearest neighbors over times. Recently, three consensus problems were discussed in [2]. With further development, Moreau et al. studied multi-agents with nonlinear dynamics and switching communication signals in [3].

Most of the existing work takes into account the two control issues of the multi-agent systems, i.e., cooperative problem and distributed tracking problem. For

S. Chen · C. Wang (✉)
School of Optical-Electrical and Computer Engineering, University of Shanghai for Science and Technology, Shanghai 200093, China
e-mail: clclwang@126.com

S. Chen
e-mail: 245099262@qq.com

Y. Jia et al. (eds.), *Proceedings of 2017 Chinese Intelligent Systems Conference*, Lecture Notes in Electrical Engineering 459,
https://doi.org/10.1007/978-981-10-6496-8_47

consensus problem, the systems without a leader, design a controller for each agent and drive them to a common value, which depends on initial conditions in [7]. As for distributed tracking problem, there exists a leader as the instruction transmitter, which is also called leader-follower consensus problems. The leader only transmits information to a small part of the agents, and it does not get information from follower nodes. All followers are trying to reach the status of the leader. For these two consensus problems, some of existing works [8–10].

Recently, a series of papers has discussed multi-agent problems with nonlinear dynamics. In [11], Hou et al. studied the leaderless consensus problem, but the communication topology is an undigraph, and the proposed protocol is distributed. While Das and Lewis studied first-order integrator agents and the information communication graph among agents was directed, and the neural networks was used for nonlinear dynamics in [12]. In [13], Zhang and Lewis et al. further generalized the same problem to higher-order nonlinear systems. But a disadvantage in [12, 13] is that the control parameters of their controllers must be chosen above certain lower bounds, which are determined by the Laplacian matrix associated with the digraph. Wang et al. addressed higher-order nonlinear non-strict-feedback dynamics in [14]. Most of the present works focus attention on nonlinear problems without consider external disturbances, such as [15, 16], which frequently exist in real-world applications. Zhao et al. [16] investigated second-order nonlinear systems consensus tracking problems with distributed finite-time, but it does not take the unknown disturbance into consideration. In [17], Wang et al. used distributed adaptive consensus control the uncertain systems. Wang et al. designed a distribute protocol for second-order nonlinear system in [18].

In this paper, we design a distributed protocol to address the consensus problem of unknown nonlinear multi-agents system. The information transmits digraph among distinct nodes are general strongly connected. The proposed distributed algorithm using the neural network approximation and the projection technique for each agent is not dependent on the entry graph and only utilizes neighbor agents' information. Neural networks is utilized to approximate uncertain dynamics terms. In addition, the adaptive updating law of neural network is designed by projection method. Compared with Das and Lewis [12], the main contribution of this paper has two aspects. First, we proposed distributed protocol just using the neighbor agents information instead of utilizing entry graph. Second, projection method is used to design the adaptive updating law for neural networks. It is shown that with our proposed protocol, all the consensus errors converge to zero asymptotically.

The remainder of the paper is organized as follows. The basic definitions and graph theory are presented in Sect. 2. Then, we design distributed adaptive controllers and parameters update law in Sect. 3. Section 4 constructs a Lyapunov candidate and presents the main result. Section 5 presented some simulation results. Conclusions are given in Sect. 6.

2 Problem Statement and Preliminaries

Suppose the communication among the N agents can be represented by a digraph $G = (V, E)$, where $V = \{v_1, \ldots, v_N\}$ is a nonempty set of nodes and the edges is represent by $E \subseteq V \times V$. $(v_i, v_j) \in E$ means there exist a path from node i to node j. The set of neighbors of node i is denoted as $N_i = \{j | (v_j, v_i) \in E\}$. If node j is a neighbor of node i, then node i can get information from node j. A digraph is strongly connected if there exists a directed path between every pair of distinct nodes. The topology of a weighted graph is often represented by the adjacency matrix $A = [a_{ij}] \in R^{N \times N}$. The Laplacian matrix is define as $L = [l_{ij}]$, in which $l_{ii} = \sum_{j \in N} a_{ij}$ and $l_{ij} = -a_{ij}$, $i \neq j$.

The dynamics of the ith node is defined as

$$\dot{x}_i = f_i(x_i) + u_i + w_i(t) \tag{1}$$

where $x_i(t) \in R$ is the state of node i, $u_i \in R$ is the control input, and $w_i \in R$ is a disturbance. Note that each node may have it own distinct nonlinear dynamics, and $f_i(x_i)$ is local Lipschitz.

Assumption 1 (*Bounds on disturbance* $w_i(t)$) There exists an unknown positive bound $w_i{}^*$, such that

$$|w_i(t)| \leq w_i^*, i = 1, \ldots, N, \forall t \geq 0$$

Definition 1 The local consensus error for node i is define as

$$e_i = \sum_{j \in N_i} a_{ij}(x_i - x_j) + b_i(x_i - x_0) \tag{2}$$

$b_i > 0$ means leader transfer information to node i, exists at least one node i can get information from leader. We define the node i for which $b_i = 0$ as the node i can not get information from the leader. We define $B = diag(b_1, b_2, \ldots, b_N)$ and $e = [e_1, e_2, \ldots, e_N]^T$.

Note that (2) represents the error information for any node used for control purposes.

Assumption 2 The digraph G is strongly connected, every node i can transmit information to every other node j in digraph. The leader can transmit information to at least one node, i.e., $\sum_{i=1}^{N} b_i \geq 1$.

Definition 2 The tracking error for node is define as

$$\delta_i = x_i - x_0 \tag{3}$$

Let $\delta = [\delta_1, \delta_2, \ldots, \delta_N]^T$ be disagreement vector. We will use this error for analysis instead of distributed control designed for Lyapunov techniques, because it is a global error. And $\delta_i = 0$ for any node $i = (1, 2, \ldots, N)$ if and only if

$$x_i = x_0$$

we say all follower nodes can reach the leader's state.

The state of x_0 can be parameterized as

$$x_0(t) = {W_0}^T \theta_0(t) \tag{4}$$

where $\theta_0(t) \in R^{q_0}$ is a suitable basis function vector, and $W_0 \in R^{q_0}$ is a ideal constant NN weight vector.

Remark 1 [19] We assume the communication digraph is strongly connected. Therefore, if $b_i = 1$ for at least one node i, then $(L + B)$ is an irreducibly diagonally dominant M-matrix.

3 Distributed Control Design

The control objective in this paper is to design distributed controller u_i for each agent to achieve consensus state for the multi-agent system, i.e., design a suitable control protocol so that local consensus error $e(t)$ is converges to zero asymptotically. $\delta_i = 0$ means all follower nodes achieve consensus with the leader, so that $\|x_i - x_0\| = 0$ for $i, i = 1, \ldots, N$. The nonlinearities dynamic $f_i(x_i)$ and disturbances $w_i(t)$ are assumed to unknown.

According to the techniques in [20], we assume that the unknown nonlinearities in (1) are locally smooth and thus can be approximated on a compact set $\Omega \in R$ by

$$f_i(x_i) = {W_i}^T S_i(x_i) + \varepsilon_i \tag{5}$$

where $S_i(x_i) \in R^{l_i}$ is a suitable set of l_i basis functions and $W_i \in R^{l_i}$ a set of unknown coefficients, $\varepsilon_i \in R$ is the approximation error. According to the literature in [21], a variety of basis sets can be selected, including sigmoids, gaussians, etc.

Assumption 3 [18] On a compact set $\Omega \in R$,

$$|\varepsilon_i| \leq \varepsilon_i^*, i = 1, \ldots, N, \forall t \geq 0$$

where ε_i^* is an unknown positive constant.

Assumption 4 We assumption $\Delta_i = w_i + \varepsilon_i$, then

$$\left|\Delta_i\right| \le \Delta_i^*, i = 1, \ldots, N, \forall t \ge 0$$

where Δ_i^* is an unknown positive constant.

In general, how to determine the suitable ideal weight vector W_i is difficult, even with well knowledge of the consensus system. Therefore, for real applications, we use neural network to estimate the nonlinear function $f_i(x_i)$ as

$$\hat{f}_i(x_i) = \hat{W}_i^T S_i(x_i) \tag{6}$$

where $\hat{W}_i$ is the ideal weight vector estimated by neural network for W_i. Define the parameter estimation error is $\tilde{f}_i(x)$ and the function estimation error $\tilde{W}_i = W_i - \hat{W}_i$ is

$$\tilde{f}_i(x) = f_i(x) - \hat{f}_i(x_i) = \tilde{W}_i^T S_i(x_i) + \varepsilon_i \tag{7}$$

We introduce the following error variable

$$z_i = x_i - b_i x_0 - (1 - b_i)\hat{W}_{0i}^T \theta_0 \tag{8}$$

where z_i is the determined virtual control signals. $\hat{W}_{0i}$ is an estimate of W_0 when $b_i = 0$.

The distributed adaptive controllers are designed as

$$u_i = -k\hat{P}_i e_i - \hat{W}_i^T S_i(x_i) - \hat{\Delta}_i^* + b_i W_0^T \dot{\theta}_0 + (1 - b_i)(\dot{\hat{W}}_{0i}^T \theta_0 + \hat{W}_{0i}^T \dot{\theta}_0) \tag{9}$$

where k is a positive constant, $\hat{\Delta}_i^*$ is the estimate of Δ_i^*, $\hat{P}_i$ is the estimate of P_i.

By the projection algorithm with [22], the adaptive updating law for neural network weight matrix $\hat{W}_i$ is derived by (10)

$$\dot{\hat{W}}_i = \begin{cases} \lambda_i S_i(x_i) z_i, & \text{if } Tr(\hat{W}_i^T \hat{W}_i) < W_{\max i} \\ & \text{or } Tr(\hat{W}_i^T \hat{W}_i) = W_{\max i} \text{ and } \hat{W}_i^T S_i(x_i) z_i < 0 \\ \lambda_i S_i(x_i) z_i - \lambda_i \frac{\hat{W}_i^T S_i(x_i) z_i}{Tr(\hat{W}_i^T \hat{W}_i)} \hat{W}, & \text{if } Tr(\hat{W}_i^T \hat{W}_i) = W_{\max i} \text{ and } \hat{W}_i^T S_i(x_i) z_i \ge 0 \end{cases} \tag{10}$$

where $W_{\max i}$ is a given positive constant for limiting the neural network weight matrix $\hat{W}_i$, and $W_{\max i}$ is selected to satisfy $Tr(W_i^T W_i) \le W_{\max i}$, $\lambda_i > 0$ is adaption gain used to control the adaption rate of $\hat{W}_i$. The initial neural network weight matrix $\hat{W}_i(0)$ satisfy following

$$Tr(W_i^T(0) W_i(0)) \le W_{\max i}, \forall t \ge 0, i = 1, 2, \ldots, N \tag{11}$$

Lemma 1 *[22] Assuming (10) defined the updating law of neural network weight matrix and (11) represent the initial value of neural network weight matrix, then*

$$Tr(W_i^T(t)W_i(t)) \leq W_{\max i}, \forall t \geq 0, i = 1, 2, \ldots, N \tag{12}$$

The design parameter update rate are following

$$\dot{\hat{P}}_i = \mu_{P_i} z_i e_i \tag{13}$$

$$\dot{\hat{\Delta}}_i^* = \tau_i z_i \tag{14}$$

$$\dot{\hat{W}}_{0i} = -\Gamma_i e_i \theta_0 \tag{15}$$

where μ_{P_i} and τ_i are positive constants, Γ_i is positive define matrix.

Lemma 2 *[23] For a digraph satisfied Assumption 1, the L is irreducible M-matrix and B is positive matrix. Then $(L+B)$ is a nonsingular M-matrix. Define*

$$\begin{aligned} p &= [p_1, p_2, \ldots, p_N]^T = (L+B)^{-T}\underline{1}, \\ P &= diag(p_1, \ldots, p_N), Q = P(L+B) + (L+B)^T P, \end{aligned} \tag{16}$$

then $P > 0$ and $Q > 0$.

4 Stability Analysis

Theorem 1 *Under Assumptions 1–3, by ultiling distributed adaptive controllers in (10) and parameter update laws in (13)–(15), we have following results:*

(i) All signals in the closed-loop system are globally uniformly bounded.
(ii) All following agents achieved consensus asymptotically, i.e., $\lim_{t\to\infty}[x_i(t) - x_0(t)] = 0, \forall 1 \leq i \leq N$.

Proof Consider the Lyapunov candidate:

$$\begin{aligned} V = \sum_{i=1}^{N} \frac{1}{2}\Big[& z_i^2 + \frac{k}{\mu_{P_i}}\tilde{P}_i^2 + Tr(\frac{1}{\lambda_i}\tilde{W}_i^T\tilde{W}_i) + \frac{1}{\tau_i}\tilde{\Delta}_i^{*2} \\ & + Tr(k(1-b_i)P_i\tilde{W}_{0i}^T\Gamma_i^{-1}\tilde{W}_{0i})\Big] \end{aligned} \tag{17}$$

where $\tilde{P}_i = P_i - \hat{P}_i, \tilde{W}_i = W_i - \hat{W}_i, \tilde{W}_{0i} = W_0 - \hat{W}_{0i}$ are estimate error because μ_{P_i}, λ_i, τ_i are positive constant, and Γ_i is positive define matrices, then $V \geq 0$. The derivation of V is

$$\dot{V} = \sum_{i=1}^{N}\Big[z_i\dot{z}_i + \frac{k}{\mu_{P_i}}\tilde{P}_i(-\dot{\hat{P}}_i) + \frac{1}{\lambda_i}Tr(\tilde{W}_i^T(-\dot{\hat{W}}_i)) + \frac{1}{\tau_i}\tilde{\Delta}_i^*(-\dot{\hat{\Delta}}_i^*) + Tr(k(1-b_i)P_i\tilde{W}_{0i}^T\Gamma_i^{-1}(-\dot{\hat{W}}_{0i}))\Big] \tag{18}$$

From (8), the derivation of z_i is

$$\dot{z}_i = \dot{x}_i - b_i\dot{x}_0 - (1-b_i)(\dot{\hat{W}}_{0i}^T\theta_0 + \hat{W}_{0i}^T\dot{\theta}_0) \tag{19}$$

According to (1), (2), (4)–(9), the derivation of z_i is

$$\begin{aligned}\dot{z}_i &= f_i(x_i) + u_i + w_i(t) - b_iW_0{}^T\dot{\theta}_0 - (1-b_i)(\dot{\hat{W}}_{0i}^T\theta_0 + \hat{W}_{0i}^T\dot{\theta}_0)\\ &= -k\hat{P}_ie_i + \tilde{W}_iS_i(x_i) + \tilde{\Delta}_i^*\end{aligned} \tag{20}$$

Substituting (2), (3), (18) into (16), we have

$$\begin{aligned}\dot{V} &= \sum_{i=1}^{N} z_i[-k(P_i - \tilde{P}_i)e_i + \tilde{W}_iS_i(x_i) + \tilde{\Delta}_i^*] + \frac{k}{\mu_{P_i}}\tilde{P}_i(-\dot{\hat{P}}_i)\\ &\quad + Tr(\frac{1}{\lambda_i}\tilde{W}_i^T(-\dot{\hat{W}}_i)) + \frac{1}{\tau_i}\tilde{\Delta}_i^*(-\dot{\hat{\Delta}}_i^*) + Tr(k(1-b_i)P_i\tilde{W}_{0i}^T\Gamma_i^{-1}(-\dot{\hat{W}}_{0i}))\\ &= \sum_{i=1}^{N} -k(x_i - x_0)P_ie_i + \sum_{i=1}^{N} Tr(\frac{1}{\lambda_i}\tilde{W}_i^T(\lambda_iS_i(x_i)z_i - \dot{\hat{W}}_i))\end{aligned} \tag{21}$$

From (2), we have $e = (L + B)\delta$, then

$$\begin{aligned}\dot{V} &= -k\delta^TP(L+B)\delta + \sum_{i=1}^{N} Tr(\frac{1}{\lambda_i}\tilde{W}_i^T(\lambda_iS_i(x_i)z_i - \dot{\hat{W}}_i))\\ &= -\frac{k}{2}\delta^TQ\delta + \sum_{i=1}^{N} Tr(\frac{1}{\lambda_i}\tilde{W}_i^T(\lambda_iS_i(x_i)z_i - \dot{\hat{W}}_i))\end{aligned} \tag{22}$$

From (9), there exist two situations as follows:

(1) When $\dot{\hat{W}}_i = \lambda_iS_i(x_i)z_i$, hence

$$Tr(\frac{1}{\lambda_i}\tilde{W}_i^T(\lambda_iS_i(x_i)z_i - \dot{\hat{W}}_i)) = 0 \tag{23}$$

(2) When $\dot{\hat{W}}_i = \lambda_i S_i(x_i) z_i - \lambda_i \frac{\hat{W}_i^T S_i(x_i) z_i}{Tr(\hat{W}_i^T \hat{W}_i)} \hat{W}_i$, then $Tr(W_i{}^T W_i) = W_{\max i}$, $\hat{W}_i^T S_i(x_i) z_i \geq 0$. Hence

$$Tr(\frac{1}{\lambda_i}\tilde{W}_i^T(\lambda_i S_i(x_i)z_i - \dot{\hat{W}}_i)) = \frac{\hat{W}_i^T S_i(x_i)z_i}{Tr(\hat{W}_i^T \hat{W}_i)} Tr(\tilde{W}_i^T \hat{W}_i) \tag{24}$$

we have that

$$\begin{aligned} Tr(\tilde{W}_i^T \hat{W}_i) &= Tr(\tilde{W}_i^T W_i) - Tr(\tilde{W}_i^T \tilde{W}_i) \\ &= \frac{1}{2} Tr(W_i{}^T W_i) - \frac{1}{2} Tr(\tilde{W}_i^T \tilde{W}_i) - \frac{1}{2} Tr(\hat{W}_i^T \hat{W}_i) \\ &\leq 0 \end{aligned} \tag{25}$$

From this two case, we can obtain

$$Tr(\frac{1}{\lambda_i}\tilde{W}_i^T(\lambda_i S_i(x_i)z_i - \dot{\hat{W}}_i)) \leq 0 \tag{26}$$

So, we can obtain

$$\dot{V} = -\frac{k}{2}\delta^T Q \delta + \sum_{i=1}^{N} Tr(\frac{1}{\lambda_i}\tilde{W}_i^T(\lambda_i S_i(x_i)z_i - \dot{\tilde{W}}_i)) \leq 0 \tag{27}$$

From (17) and (27), we can obtain that z_i, $\hat{P}_i$, $\hat{\Delta}_i$,$\hat{W}_{0i}$, $\hat{W}_i$ and δ are bounded, so δ_i for $i = 1, \ldots, N$ are also bounded, we obtain that x_i are bounded for $i = 1, \ldots, N$. We can obtain e_i is bounded from (9) and $S_i(x_i)$ is bounded on an compact set $\Omega \subset R$, so we can conclude the control signal u_i is bounded. From (27), we also can obtain $\lim_{t\to\infty} \delta_i(t) = 0$ for $i = 1, \ldots, N$, i.e., $\lim_{t\to\infty}(x_i(t) - x_0(t)) = 0$, so follower node can reach consensus with the leader node.

5 Simulation Results

This section will show the effectiveness of the distributed adaptive protocol of Theorem 1. It is shown that the protocol effectively by using adaptive neural network at each node for unknown nonlinear dynamic and unknown disturbance.

For simplicity, consider the four node strongly connected digraph structure in Fig. 1 with a leader, we can see only node 1 get information from leader node.

Described the dynamics of the ith agent as the following form:

$$\dot{x}_i = u_i + f_i(x_i) + w_i(t), i = 1, \ldots, 4$$

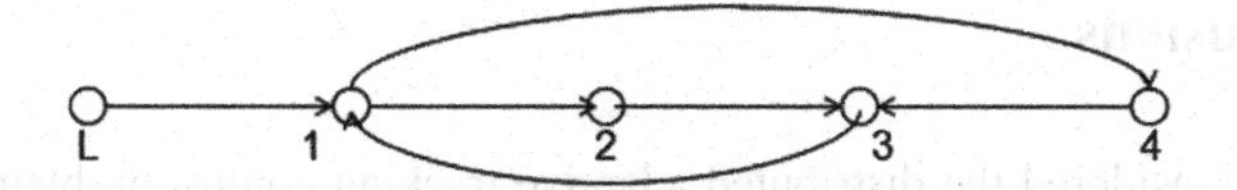

Fig. 1 Communication graph for 4 follower nodes and a leader

Fig. 2 Trajectory of x_i

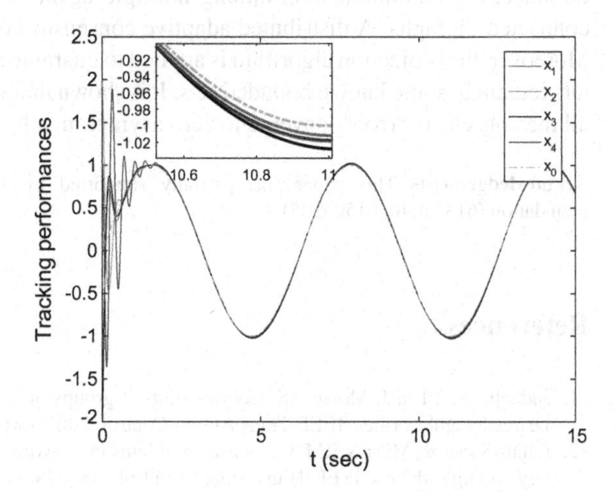

Fig. 3 Tracking error of $x_i(t) - x_0(t)$

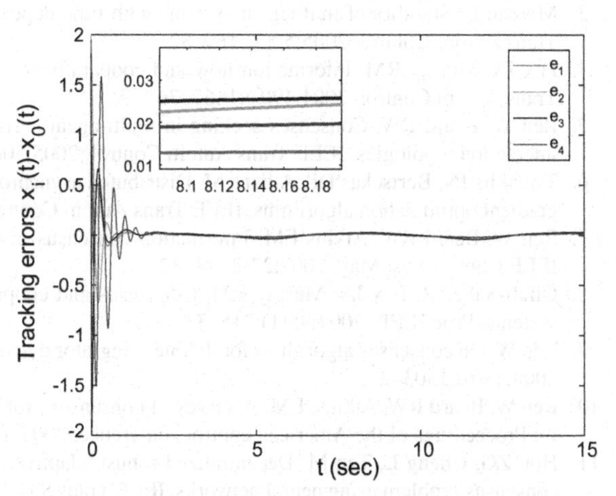

where $f_1 = 0.5\sin^2(x_1)$, $f_2 = e^{-x_2}$, $f_3 = \cos(x_3)$, and $f_4 = 0.8\mathrm{e}^{-0.2x_4}$ for $i = 1, \ldots, 4$. The leader's dynamic is described as $x_0(t) = W_0(t)^T\theta_0(t)$, where $W_0(t) = [0.5, 0.5]^T$, $\theta_0(t) = [\sin(t), \sin(t)]^T$. The initial states set as $x_1 = [1.0, 0.0]$, $x_2 = [2.0, 1.0]$, $x_3 = [-2.0, 0.0]$, $x_4 = [1.0, 1.0]$. The design parameters are choose as $k = \mu_{P_i} = \tau_i = 2$, $\Gamma_i = 2I$. The tracking states of $x_i(t)$ and associated tracking errors for $i = 1, \ldots, 4$, are show in Figs. 2 and 3 respectively.

6 Conclusions

This paper considered the distributed adaptive tracking control problem for multi-agent systems with non-identical unknown nonlinear dynamics and uncertain disturbance, the communication among multiple agents can be expressed by strong connected digraphs. A distributed adaptive consensus control law was constructed. Moreover, the projection algorithm is applied to guarantee that the estimated parameters remain in some known bounded sets. It is shown that with our proposed protocol, all the consensus errors converge to zero asymptotically.

Acknowledgements This paper was partially supported by The National Natural Science Foundation (61374040, 61503205).

References

1. Jadbabaie A, Lin J, Morse AS. Coordination of groups of mobile autonomous agents using nearest neighbor rules. IEEE Trans Autom Control. 2003;48(6):988–1001.
2. Olfati-Saber R, Murray RM. Consensus problems in networks of agents with switching topology and time-delays. IEEE Trans Autom Control. 2004;49(9):1520–33.
3. Moreau L. Stability of multiagent systems with time-dependent communication links. IEEE Trans Autom Control. 2005;50(2):169–82.
4. Fax JA, Murray RM. Information flow and cooperative control of vehicle formations. IEEE Trans Autom Control. 2004;49(9):1465–76.
5. Ren W, Beard RW. Consensus seeking in multi-agent systems under dynamically changing interaction topologies. IEEE Trans Autom Control. 2005;50(5):655–61.
6. Tsitsiklis JN, Bertsekas DP, Athans M. Distributed asynchronous deterministic and stochastic gradient optimization algorithms. IEEE Trans Autom Control. 1986;31(9):803–12.
7. Ren W, Beard RW, Atkins EM. Information consensus in multivehicle cooperative control. IEEE Control Syst Mag. 2007;27(2):71–82.
8. Olfati-Saber R, Fax JA, Murray RM. Consensus and cooperation in networked multi-agent systems. Proc IEEE. 2007;95(1):215–33.
9. Ren W. On consensus algorithms for double-integrator dynamics. IEEE Trans Autom Control. 2008;53(6):1503–9.
10. Ren W, Beard RW, Atkins, EM. A survey of consensus problems in multi-agent coordination. In: Proceedings of the American control conference. 2005. p. 1859–64.
11. Hou ZG, Cheng L, Tan M. Decentralized robust adaptive control for the multiagent system consensus problem using neural networks. IEEE Trans Syst Man Cybern Part B (Cybernetics). 2009;39(3): 636–47.
12. Das A, Lewis FL. Distributed adaptive control for synchronization of unknown nonlinear networked systems. Automatica. 2010;46(12):2014–21.
13. Zhang H, Lewis FL. Adaptive cooperative tracking control of higher-order nonlinear systems with unknown dynamics. Automatica. 2012;48(7):1432–9.
14. Wang G, Wang C, Yan Y, et al. Distributed adaptive output feedback tracking control for a class of uncertain nonlinear multi-agent systems. Int J Syst Sci. 2017;48(3):587–603.
15. Zhao-Xia W, Da-Jun DU, Min-Rui FEI. Average consensus in directed networks of multi-agents with uncertain time-varying delays. Acta Automatica Sinica. 2014;40(11):2602–8.
16. Zhao Y, Duan Z, Wen G, et al. Distributed finite-time tracking of multiple non-identical second-order nonlinear systems with settling time estimation. Automatica. 2016;64:86–93.

17. Wang W, Huang J, Fan H, et al. Decentralized adaptive consensus control of uncertain non-linear systems under directed topologies. In: 34th Chinese Control Conference (CCC). IEEE; 2015. p. 7090–95.
18. Wang G, Wang C, Li L, et al. Designing distributed consensus protocols for second-order nonlinear multi-agents with unknown control directions under directed graphs. J Franklin Inst. 2017;354(1):571–92.
19. Qu Z. Cooperative control of dynamical systems: applications to autonomous vehicles. Springer Science & Business Media; 2009.
20. Lewis FW, Jagannathan S, Yesildirak A. Neural network control of robot manipulators and non-linear systems. CRC Press; 1998.
21. Hornik K, Stinchcombe M, White H. Multilayer feedforward networks are universal approximators. Neural Netw. 1989;2(5):359–66.
22. Hou ZG, Cheng L, Tan M. Decentralized robust adaptive control for the multiagent system consensus problem using neural networks. IEEE Trans Syst Man Cybern Part B (Cybernetics). 2009;39(3):636-47.
23. Su S, Lin Z, Garcia A. Distributed synchronization control of multi-agent systems with unknown nonlinearities: The case of fixed directed communication topology. In: American control conference (ACC), 2014. IEEE; 2014. p. 5361–66.

A Protection Strategy of Microgrid Based on EtherCAT

Ziyi Fu, Liuyang Shen and Bing Cheng

Abstract This paper proposes a protection scheme of microgrid and it is based on EtherCAT communication protocol. There are two algorithms in this scheme. One is used to detect fault, and it is based on superposition theorem. The other one is used to locate and eliminate the fault. In order to verify the correctness of the algorithm which is used to detect fault, a simulation model in MATLAB is established. Use the topology's information of microgrid and fault's information to construct matrix, then complete the work about locating fault and eliminating fault by matrix operation. In this paper, the advantages of EtherCAT communication protocol are discussed, and the primary design of master station and slave station is carried out. Lay the foundation for detailed design of microgrid protection system.

Keywords Microgrid protection · Fault detection · Fault location · Matrix algorithm · EtherCAT

1 Introduction

Microgrid is a power system which consists of distributed power source, energy storage device, energy conversion device, load and monitoring protection device. Microgrid can achieve self control, protection and management [1]. Due to the complexity of fault in microgrid. Selectivity and sensitivity of the microgrid protection can not be realized [2]. Traditional protection methods may not be able to locate and isolate fault quickly and accurately. Therefore, it is urgent to study the protection method of microgrid for ensuring the stable operation of microgrid.

There are two main contents of microgrid protection: how to detect the fault and how to locate the fault. Literature [3] introduces a scheme of differential protection. It uses the direction and magnitude of the fault current to form differential signal.

Z. Fu (✉) · L. Shen · B. Cheng
School of Electrical Engineering and Automation, Henan Polytechnic University, Jiaozuo 454003, China
e-mail: fuzy@hpu.edu.cn

Y. Jia et al. (eds.), *Proceedings of 2017 Chinese Intelligent Systems Conference*, Lecture Notes in Electrical Engineering 459,
https://doi.org/10.1007/978-981-10-6496-8_48

Literature [4] connects the DG and the relay through the communication lines. Form a network to protect the whole microgrid.

Literature [5, 6] mainly introduces the protection scheme based on local information. It innovate the protection device and protection algorithm, but the disadvantage is the development of device needs long time.

Literature [7, 8] introduces the protection scheme based on communication. It can monitor the whole network, but the communication requires high reliability. Once the communication occurs fault, it means that the whole protection system will breaks down, therefore, the communication protocol of centralized protection scheme must guarantee safety and real time.

In order to realize fast fault identification, localization and removal, a protection scheme based on communication is proposed in this paper: microgrid protection unit collects the fault's information of circuit breaker, and use fault detection algorithms to identify the fault. Then protection unit uploads the fault's information to the central controller by EtherCAT. At last, the central controller uses the fault location algorithm to locate the fault, and eliminate it to complete the work.

2 Fault Detection Algorithm

2.1 The Principle of the Algorithm

Because there is a bidirectional power flow in microgrid, we should judge the fault's direction when we detect the fault. By the superposition theorem, we know that the fault can be regarded as the superposition of the non fault state and the fault state in the power network.

The voltage fault component can be expressed as:

$$\Delta u(t) = u_K(t) - u_{UNK}(t) \tag{1}$$

The current fault component can be expressed as:

$$\Delta i(t) = i_K(t) - i_{UNK}(t) \tag{2}$$

where $\Delta u_K(t)$ $\Delta i_K(t)$ are voltage and current that measured, and $u_{UNK}(t)$ $i_{UNK}(t)$ are voltage and current in non fault state.

Take the case of two terminal power network. As show in Fig. 1, there is an additional state network with forward direction fault. The protection device is installed on the M side, and assume the forward direction of the current from the bus to the line. When a short circuit fault occurs in F_1, fault component voltage $\Delta \dot{U} = -Z_M \cdot \Delta \dot{I}$, and additional power $\Delta \dot{E}_{F1}$ at the short-circuit point supply energy to the M side. At this time, the energy value be measured at the M should be negative.

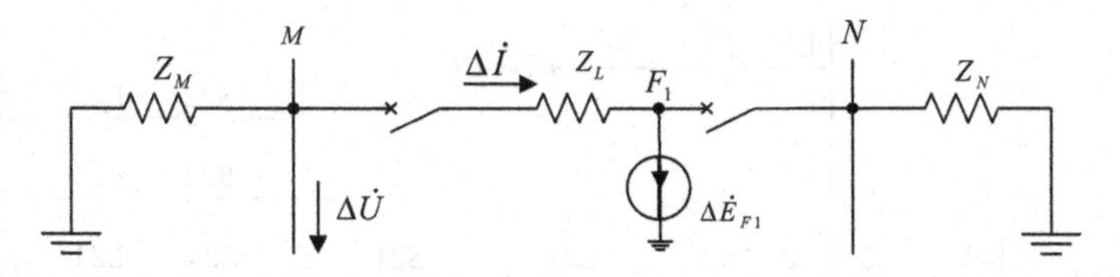

Fig. 1 Additional state network in forward direction fault

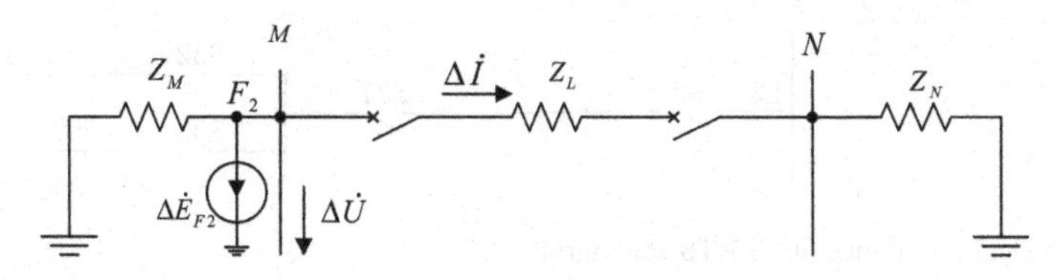

Fig. 2 Additional state network in backward direction fault

Figure 2 shows a additional state network in backward direction fault. If there is a fault in F_2, fault component voltage $\Delta\dot{U} = (Z_N + Z_L) \cdot \Delta\dot{I}$, then additional power $\Delta\dot{E}_{F2}$ supply energy to the N side. The energy value be measured at the M should be positive at the moment.

The formula for calculating energy value is:

$$E = \int_{t_1}^{t} (\Delta u_A(t) \times \Delta i_A(t) + \Delta u_B(t) \times \Delta i_B(t) + \Delta u_C(t) \times \Delta i_C(t))dt \tag{3}$$

where $\Delta u_A(t)$ $\Delta u_B(t)$ $\Delta u_C(t)$ are voltage fault components of A phase, B phase, C phase. $\Delta i_A(t)$ $\Delta i_B(t)$ $\Delta i_C(t)$ are current fault components of A phase, B phase, C phase. The fault occurs at t_1.

The criterion of fault's direction is defined as [9]:

$$E = \begin{cases} < 0, & \text{Forward direction fault} \\ = 0, & \text{No fault} \\ > 0, & \text{Backward direction fault} \end{cases} \tag{4}$$

2.2 Simulation and Analysis

The typical topology of the microgrid which is defined by CERTS is shown in Fig. 3:

In this paper, we build a model based on Fig. 3 in MATLAB. Source voltage level is 10 kV. Internal impedance is $(0.6 + j0.12)\Omega$. Line impedance is $(0.136 + j0.02)\Omega$. The power of DG is 20 kW. The power of Load1 is 3 MW, and

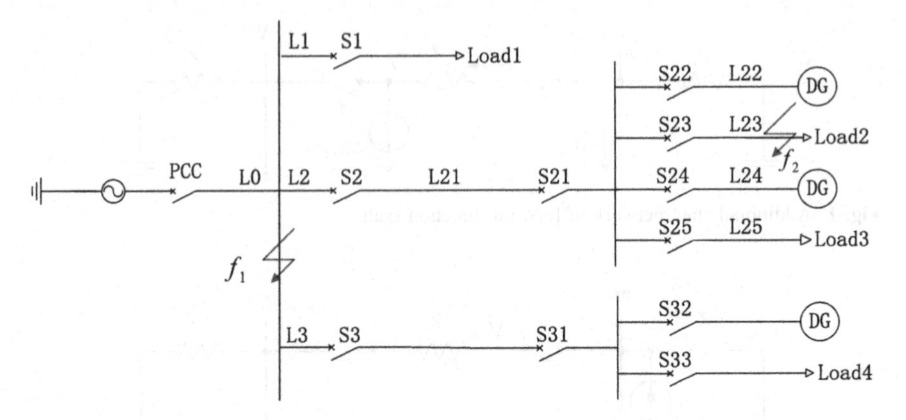

Fig. 3 Typical topologies of CERTS microgrid

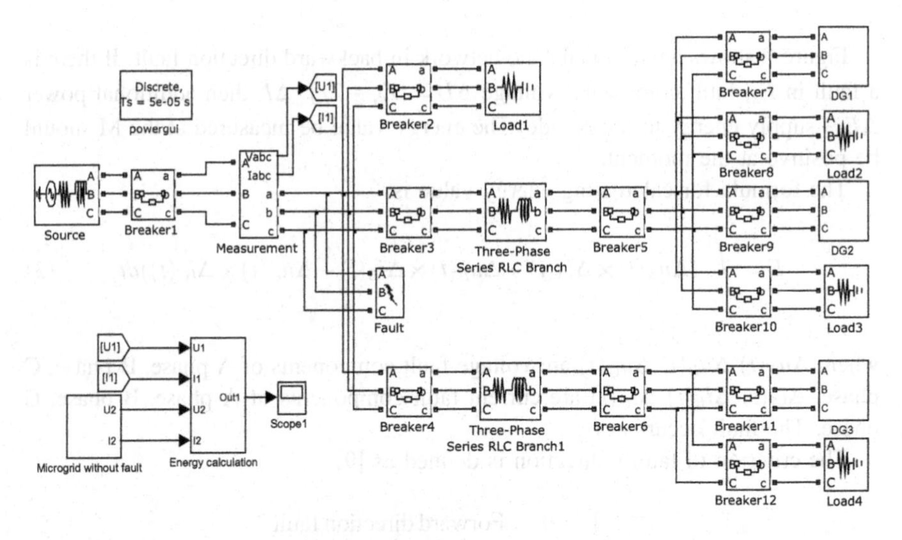

Fig. 4 Simulation diagram of CERTS microgrid

Load2, Load3, Load4 are 20 kW. Because the most common type of fault in the grid is the single phase to ground fault, set up a single phase short circuit fault at the bus and feeder individually. The simulation model is shown in Fig. 4.

Simulation example 1:
The single phase to ground fault occurs at 0.1 s in f_1. As can be seen from Fig. 5, the voltage and current of PCC point change suddenly. From Fig. 6 it can be seen that the fault's direction of PCC is forward direction, and the fault's direction of S1 S2 S3 is backward direction, so the fault occurred in the bus.

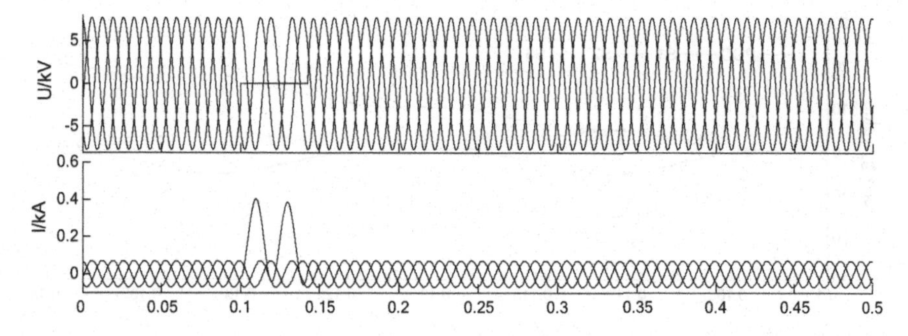

Fig. 5 The voltage and current waveform of PCC

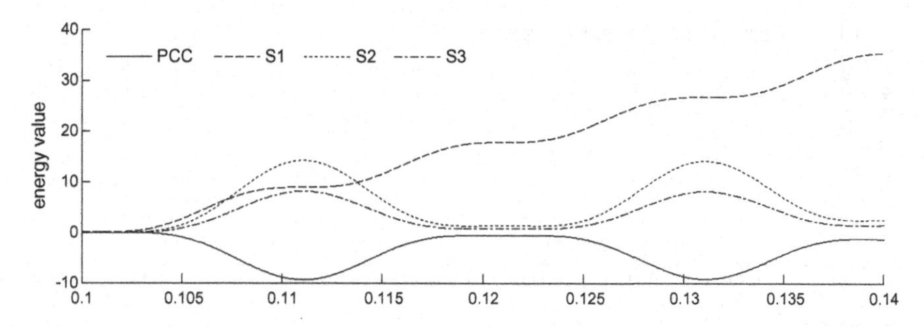

Fig. 6 The energy value waveform of circuit breaker

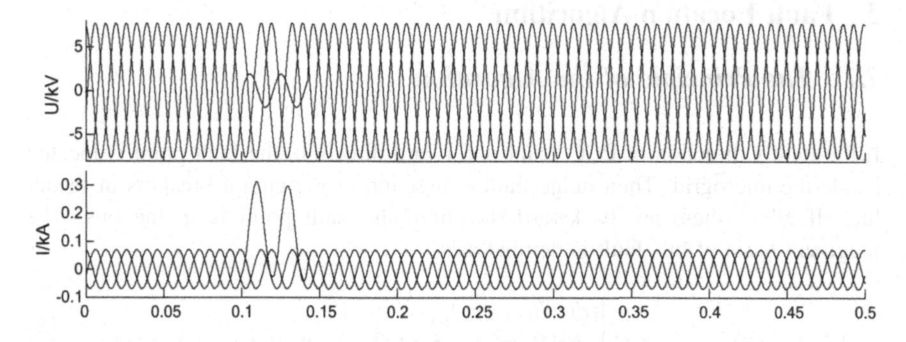

Fig. 7 The voltage and current waveform of PCC

Simulation example 2:

The single phase to ground fault occurs at 0.1 s in f_2. As can be seen from Fig. 7, the voltage and current of PCC point change suddenly. From Fig. 8 it can be seen that the fault's direction of PCC and S2 is forward direction, and the fault's direction of S1 S3 is backward direction, so the fault occurred in the feeder line L2. In Fig. 9, the fault's direction of S23 is forward direction and the fault's direction of S22 S24 S25 is backward direction. So the fault occurred in branch L23.

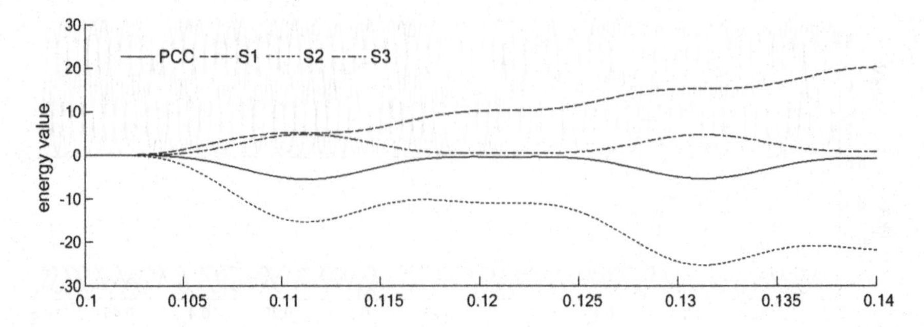

Fig. 8 The energy value waveform of circuit breaker

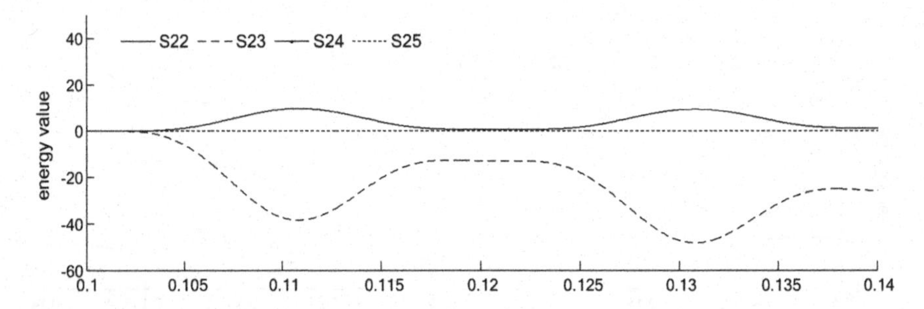

Fig. 9 The energy value waveform of circuit breaker

3 Fault Location Algorithm

3.1 The Principle of the Algorithm

If the fault's direction of PCC point is forward direction, the fault point is located inside the microgrid. Then judge fault's direction of the circuit breakers in feeder line. If all of them are backward direction, the fault point is in the bus. The judgment basis of bus fault is defined as:

$$\begin{cases} dir(B_{PCC}) = 1 \\ dir(B_i) = 0, \quad i = 1, 2, \ldots, n \end{cases} \tag{5}$$

where $dir(B) = 1$ indicates that the energy value of the breaker B is negative, in other words, there is a fault in the forward direction. $dir(B) = 0$ is an indication that the energy value of the breaker B is positive, so the fault occurs in the backward direction.

If the bus has no fault, the fault point is in feeder line whose fault's direction of the circuit breaker is forward direction. The judgment basis of feeder line fault is defined as [9, 10]:

$$\begin{cases} dir(B_{PCC}) = 1 \\ dir(B_k) = 1 \\ dir(B_i) = 0, \quad i = 1, 2, \ldots, k-1, k+1, n \end{cases} \tag{6}$$

The flow of fault location is shown in Fig. 10:

After determining the fault feeder, use the matrix algorithm to locate the fault branch quickly.

This paper defines the breaker-branch matrix $D_1 = \left[d_{1ij}\right]_{m \times n}$, and fault information matrix $F = [f_i]_{1 \times m}$. Where m is the number of circuit breakers, and n is the number of branches. In order to reduce the complexity of matrix, only take the circuit breakers which close to the bus side to construct the matrix, and the current direction from the bus to the feeder is used as the forward direction. The breaker-branch matrix is used to describe the connection relationship and direction information between the node and the branch. Matrix elements can be expressed as:

$$d_{1ij} = \begin{cases} 1, & \text{Branch } j \text{ is directly connected to the circuit breaker } i \text{ and } j \text{ is located in the forward direction of } i \\ 0, & \text{Branch } j \text{ is not directly connected to the circuit breaker } i \\ -1, & \text{Branch } j \text{ is directly connected to the circuit breaker } i \text{ and } j \text{ is located in the backward direction of } i \end{cases} \tag{7}$$

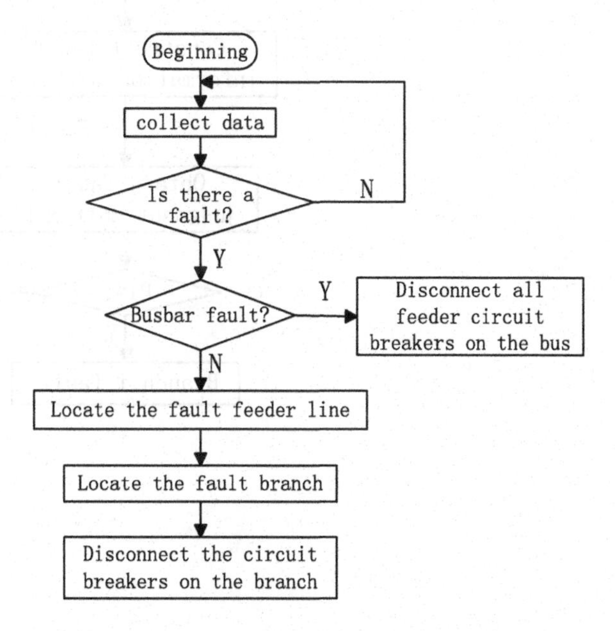

Fig. 10 Fault location flow chart

The fault information matrix describe the fault's information of circuit breaker. Matrix elements can be expressed as:

$$f_i = \begin{cases} 1, & \text{The fault's direction detected by the circuit breaker is in the forward direction} \\ 0, & \text{The fault's direction detected by the circuit breaker is in the backward direction or no fault} \end{cases} \quad (8)$$

The fault information matrix multiplied by the breaker-branch matrix to obtain the fault judgment matrix $P = [p_i]_{1 \times n}$. This matrix is used to describe the fault's information of the branch, and n is the number of branches. Matrix elements can be expressed as:

$$p_i = \begin{cases} 0, & \text{There is no fault in branch } i \\ 1, & \text{There is a fault in branch } i \end{cases} \quad (9)$$

The flow of fault branch location algorithm is shown in Fig. 11.

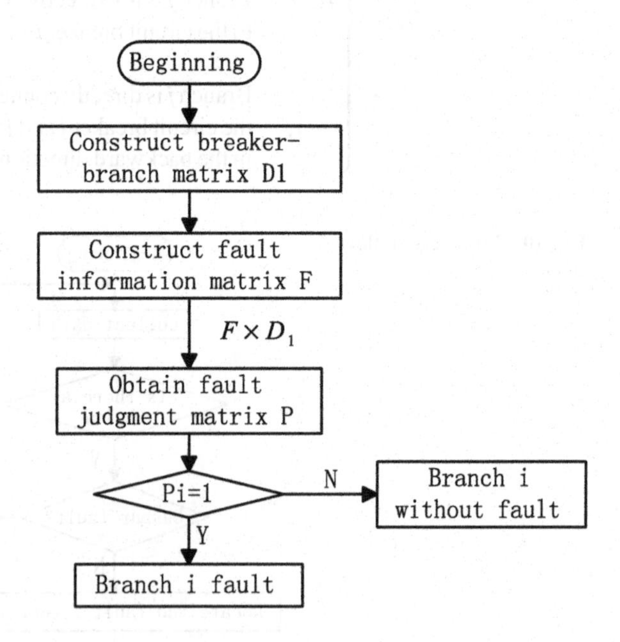

Fig. 11 fault branch location algorithmis flow chart

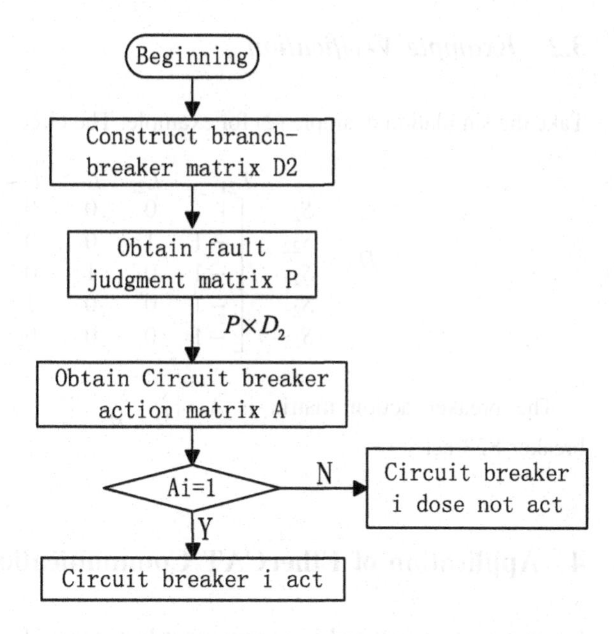

Fig. 12 fault isolation algorithm flow chart

After locating the fault branch by calculation, the corresponding circuit breaker is driven to trip, and the purpose of isolating faults is achieved. This paper defines branch-breaker matrix $D_2 = \left[d_{2ij}\right]_{n \times l}$. Where n is the number of branches, and l is the number of circuit breakers. The connection between the branch and the circuit breaker is described by this matrix. Matrix elements can be expressed as:

$$d_{2ij} = \begin{cases} 1, & \text{The branch } i \text{ is connected with the circuit breaker } j \\ 0, & \text{The branch } i \text{ is not connected with the circuit breaker } j \end{cases} \tag{10}$$

The fault judgment matrix multiplied by the branch-breaker matrix to obtain the breaker action matrix $A = [a_i]_{1 \times l}$, and l is the number of circuit breakers. The matrix is used to describe the action's information of the circuit breaker. Matrix elements can be expressed as:

$$a_i = \begin{cases} 1, & \text{Circuit breaker } i \text{ should act} \\ 0, & \text{Circuit breaker } i \text{ should not act} \end{cases} \tag{11}$$

The flow of fault isolation algorithm is shown in Fig. 12.

3.2 Example Verification

Take the simulation example (2) for example. The circuit breaker-branch matrix is

$$D_1 = \begin{array}{c} S_2 \\ S_{22} \\ S_{23} \\ S_{24} \\ S_{25} \end{array} \overset{\begin{array}{ccccc} L_{21} & L_{22} & L_{23} & L_{24} & L_{25} \end{array}}{\begin{bmatrix} 1 & 0 & 0 & 0 & 0 \\ -1 & 1 & 0 & 0 & 0 \\ -1 & 0 & 1 & 0 & 0 \\ -1 & 0 & 0 & 1 & 0 \\ -1 & 0 & 0 & 0 & 1 \end{bmatrix}}$$

The breaker action matrix is $A = \begin{bmatrix} \overset{S_2}{0} & \overset{S_{21}}{0} & \overset{S_{22}}{0} & \overset{S_{23}}{1} & \overset{S_{24}}{0} & \overset{S_{25}}{0} \end{bmatrix}$, so the circuit breaker S23 act.

4 Application of EtherCAT Communication Protocol

Communication protocol is an important factor that affects the effectiveness of the protection scheme. EtherCAT communication protocol is one of the fastest Ethernet protocols, and it is fully compatible with standard Ethernet. Because the data processing is based on hardware, it can ignore the impact of CPU performance and software factors. The delay is reduced and the real-time performance is improved. EtherCAT can process and transmit datas with a high speed. EtherCAT performance is shown in Table 1.

Synchronization of equipment is an important factor in the network. EtherCAT uses a precise array of distributed clocks to ensure the stable operation of the system [11]. In order to ensure the security of communication, EtherCAT uses a security protocol called safety over EtherCAT which based on application layer [12]. EtherCAT supports most of the topologies, such as linear, star and tree. So the flexibility of its connection mode has been improved, at the same time, the capacity of network is improved.

EtherCAT uses Master/Slave communication architecture. The hardware and software design of master station and slave station are very convenient. In this paper, each slave station (microgrid protection unit) completes the task of collecting

Table 1 EtherCAT performance profile

Process data	Update time
1000 distributed digital I/O	30 μs
200 analog I/O (16 bit)	50 μs ← → 20 kHz
100 servo axes, and each axis has 8 bytes of input and output data	100 μs
1 fieldbus master station-gateway (1486 byte input and 1486 byte output)	150 μs

and processing information, then the information is transmitted to the master station (centralized protection platform) through EtherCAT, and show it on human-computer interface in real time. Then master station determines the fault point and sends a command to slave station. Finally, slave station drives the breaker to act. The fault is eliminated in this way.

The master station is composed of industrial control computer C5102-0030 and control software TwinCAT. They are designed by the German Beckhoff company. This paper mainly uses TwinCAT system manager and TwinCAT PLC controller. The TwinCAT system manager configures the slave station, and completes the mutual mapping of I/O port between the master station and the slave station. The TwinCAT PLC controller provides the necessary programming environment for PLC control program to develop and debug.

The task of the master station is to receive information from the slave station, and to process the information and send instructions.

There are two main tasks about slave station:

(1) Collect the operation parameters of microgrid, then calculate and analyze the data, finally transmit it to the master station through EtherCAT.
(2) Receive instructions from the master station and execute the commands. According to the function of the slave station, the hardware structure is shown in Fig. 13.

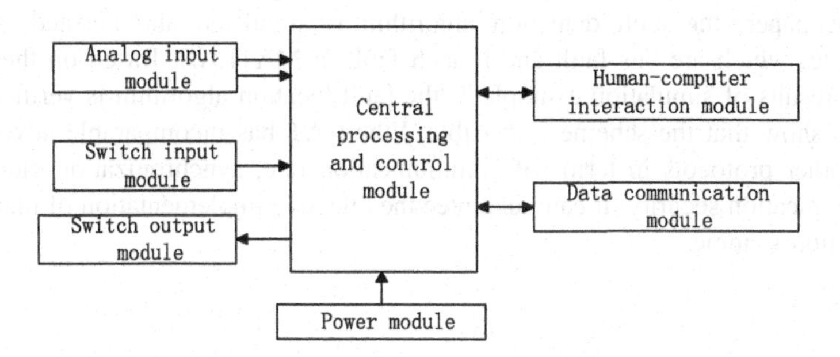

Fig. 13 Block diagram of slave station

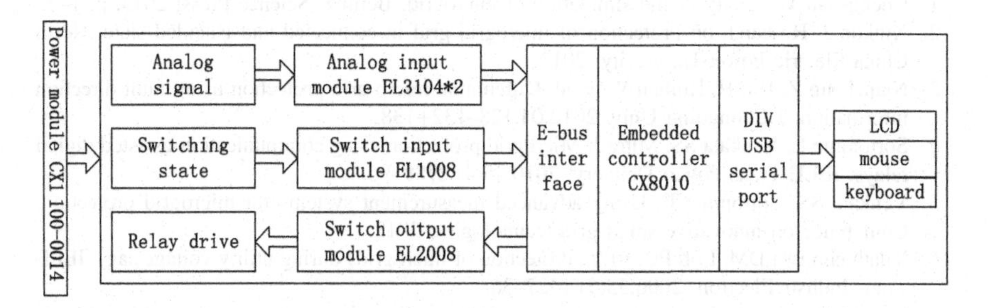

Fig. 14 Hardware structure diagram of slave station

Table 2 EtherCAT terminal module

Name of EtherCAT terminal module	Brief introduction
Analog input module EL3104	4 analog differential input channel. 16 bit sampling resolution. Signal range: $-10 \sim +10$ V
Switch input module EL1008	8 channel digital input port. Input voltage $-3 \sim 5$ V, output 0. Input voltage $11 \sim 30$ V, output 1
Switch output module EL2008	8 channel digital output port. Receives the control signal 1. Outputs 24 V DC voltage signal

As show in Fig. 14, slave microprocessor is embedded controller CX8010 which produced by Beckhoff company. EtherCAT terminal module uses analog input module EL3104, digital input module EL1008 and digital output module EL2008 [13].

Table 2 gives a brief introduction to the EtherCAT terminal module.

5 Conclusions

In this paper, the fault detection algorithm is simulated and verified by two examples which are bus fault and branch fault in MATLAB. Based on the simulation results of simulation example 2, the fault location algorithm is verified. The results show that the scheme is feasible. EtherCAT has incomparable advantages over other protocols in terms of communication rate, synchronization clock and communication security. It can guarantee the effective implementation of microgrid protection schemes.

References

1. Chengshan W. Analysis and simulation of microgrid. Beijing: Science Press; 2013. p. 1–2.
2. Xinkun J. Research on protection of microgrid grid in connected and islanded state. North China Electric Power University; 2012.
3. Niancheng Z, Rui H, Ruihan W, et al. Differential micro grid protection using fault direction information. J Chongqing Univ. 2012;04:128–132+138.
4. Sortomme E, Venkata SS, Mitra J. Microgrid protection using communication-assisted digital relays. IEEE Trans Power Delivery. 2010;25(4):2789–96.
5. Venkata SS, Sortomme E. Using advanced measurement systems for microgrid protection. Conference on innovative smart grid technologies; 2012.
6. Vilathgamuwa DM, Loh PC, Li Y. Protection of microgrids during utility voltage sags. IEEE Trans Industr Electron. 2006;53(5):1427–36.

7. Ustun TS, Ozansoy C, Zayegh A. Modeling of a centralized microgrid protection system and distributed energy resources according to IEC 61850-7-420. IEEE Trans Power Syst. 2012;27 (3):1560–7.
8. Sortomme E, Venkata SS, Mitra J. Microgrid protection using communication-assisted digital relays. IEEE Trans Power Delivery. 2010;25(4):2789–96.
9. Xiyuan M, Jinyong L, Aidong X, et al. Protection strategy of medium voltage photovoltaic microgrid based on fault direction information. South Power Syst Technol. 2015; 04:35–40.
10. Hongkun F. Research on coal mine integrated protector for preventing override trip based on Ethernet. Henan Polytechnic University; 2016.
11. EtherCAT-technology introduction and development overview. Int Mechatron 2006;06:17–22 +24.
12. Shengdong Y. Application of industrial Ethernet Fieldbus EtherCAT. InstrumTechnol. 2014;08:4–6.
13. Hongwei Z. Research on coal mine power monitoring system based on EtherCAT. Henan Polytechnic University; 2016.

The Admissions Big Data Mining Research Based on Real Data from a Normal University

Zhibin Tan, Jie Wang, Yan Peng and Fanghua Ma

Abstract In this paper, a Normal University's 2011–2016 real admissions data are analyzed by the Apriori, K-MEANS and KNN algorithm. The result shows that the university's normal students are more likely to choose other normal majors than to choose other non-normal majors related the normal majors and the overall situation of the Normal University's student enrollment is relatively stable. Liberal arts college is the most popular college. Chinese language and Literature (normal) and English (normal) are more popular in the Normal University. The result reveals the internal connection between the various majors and has a guiding role for specialties setup in the university.

Keywords Apriori algorithm · K-MEANS algorithm · KNN algorithm · Admissions data

1 Introduction

With the development of people's living standard and the progress of society, parents are more and more concerned about their children's education. With the further refinement of social division of labor, the role of students in the professional decision is also increasingly significant. At the same time, there is also a need for

Z. Tan · J. Wang (✉) · Y. Peng · F. Ma
School of Management, Capital Normal University, Beijing, China
e-mail: wangjie_cnu@126.com

Z. Tan
e-mail: 849135951@qq.com

Y. Peng
e-mail: pengyan@cnu.edu.cn

F. Ma
e-mail: mafanghua@cnu.edu.cn

Y. Jia et al. (eds.), *Proceedings of 2017 Chinese Intelligent Systems Conference*, Lecture Notes in Electrical Engineering 459,
https://doi.org/10.1007/978-981-10-6496-8_49

candidates to apply for professional analysis so that the school can adjust the professional settings to meet students' demand in the university.

College enrollment has become one of the hot spots in the community. There are Sina education channel, Baidu education to guide students to choose school. At the same time, some scholars have studied the college entrance examination. Li and Liu [1] explores the influencing factors of the college entrance examination of professional reporting decision. Ma et al. [2] does the structure study of professional college entrance examination decision. Shen and Sun [3] analyses decision based on the statistical model of the college entrance examination. Xu [4] establishes a voluntary recruitment decision support system based on data mining. Han and Liu [5] designs a set of college entrance examination prediction system based on Web mining.

These studies have analyzed the college entrance examination from the student's point of view and guide the candidates to fill their own volunteer instead of standing in the angle of college. So in this paper, we use some of the classic mature algorithm, to explore the internal linkage between candidates in the professional in the Normal University and provide help for specialties setup.

2 Data Analysis Based on Apriori Algorithm

This section introduces Apriori algorithm briefly and how to use Apriori algorithm to analysis data. The result from Apriori algorithm shows that which majors are applied together for a student and which major is more popular.

2.1 Introduction to Apriori Algorithm

Apriori [6] is an algorithm for frequent item set mining and association rule learning over transactional databases. It finds the frequent items by a iterative algorithm. The last iterative result is the next iterative base to find the large frequent items.

Apriori algorithm mainly includes the following three steps:

(a) First, Apriori algorithm needs to scan database D_S, gain frequent itemsets with an element and minimum support L_1 as the initial seed set.
(b) According to set L_(k-1)'s (k-1) frequent itemsets to generate k-candidate sequences and store it in the collection C_k. To find the sequence s_1 and s_2, which meets the elements of s_1 and s_2 only and just only have the last element is different.
(c) repeat (b). Until there is no new set of frequent sequences.

2.1.1 The Application of Apriori Algorithm in Admission Data Analysis

Because of the small amount of data in a single year, the author has analyzed the data of 6 years. The limit is 0.3 support and the frequent set is 2. The results are shown in the following figure.

Different colors mean support of different ranks. For example, in the upper left corner of the graph, that history (base class) points with an orange arrow at Chinese language mean 50–60% of students applying History (base class) also apply for the Chinese language and Literature. But Chinese language and Literature doesn't point at History (base class) mean that less 30% of students applying Chinese language and Literature also apply for the History (base class).

From the Fig. 1, most of these majors can be divided into 5 major groups: Language, Liberal arts (normal), Science (normal), Liberal arts (non-normal), Computer. In addition to language and liberal arts (normal), other groups are not connected. So these professional candidates select major with a certain purpose. Meanwhile, in the science (normal) group, the arrow between the any two professionals are two-way, and the color depth similar. It can be concluded that these majors are at the same level of attractiveness. But in other groups, the arrow between some specified major and other major are unidirectional or obvious differences in color depth. It can be say that the profession of being directed or directed by a colored arrow are more popular in the group. Obviously, Chinese language and Literature in Liberal arts (non-normal), English in Language, Electronic Information Engineering in Computer, Chinese language and Literature (normal) and English (normal) in Liberal arts (normal) are more popular in their group.

Furthermore, if we reduce the minimum support to 20%, all the groups on the picture will be linked together as a whole by this way that some majors in some group connects with Chinese language or Literature (normal) and English (normal). It can be concluded that these 2 majors may be more popular in all majors. And almost all of the majors including a major that is not present in the diagram have relevance related major except Social work, Remote sensing science and technology and Electronic Commerce. So Author thinks these three majors don't owe significant relevance.

3 Data Analysis of K-MEANS Algorithm

This section introduces K-means algorithm briefly and how to use K-means algorithm to analysis data including establishment of distance and coordinate system. The results include 2011–2013 result and 2014–2016 result. Then these two results are compared to find the changes which happen to students in 2011–2013 and 2014–2016.

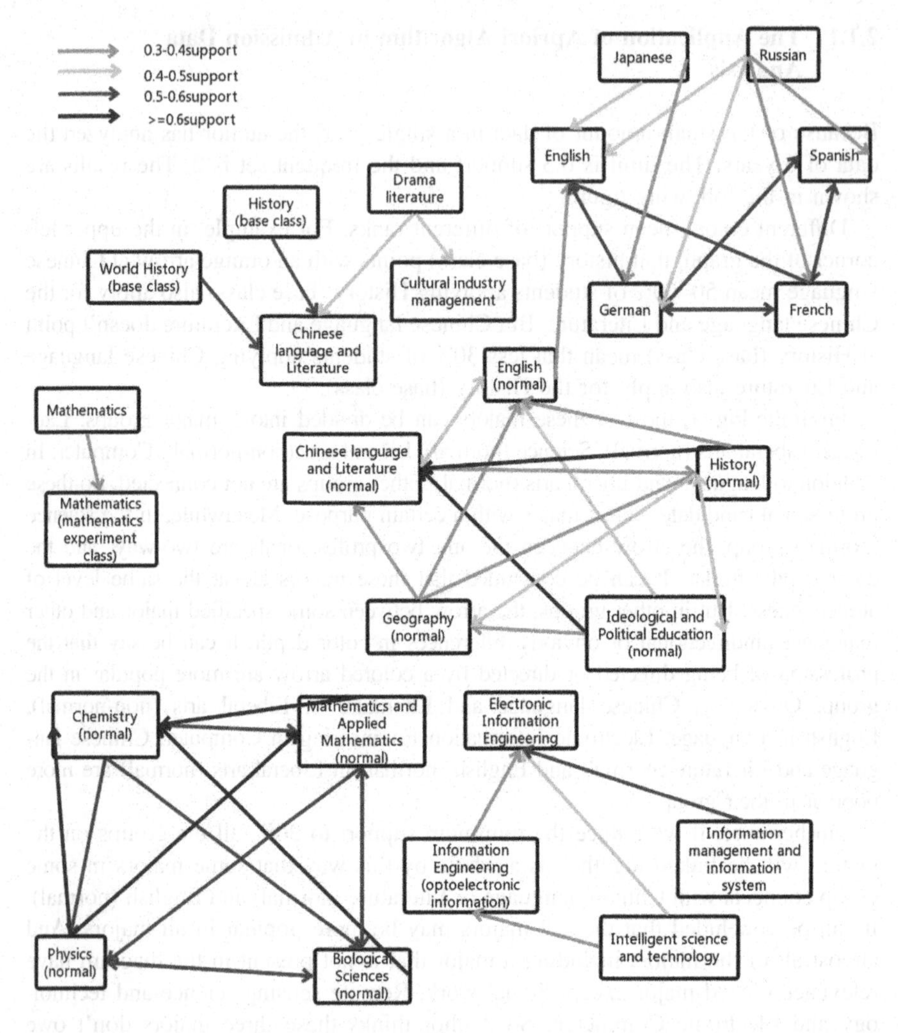

Fig. 1 Aprior result

3.1 Introduction to K-MEANS Algorithm

K-means [7] algorithm accepts input K; then put n data objects into K clusters in order to make the cluster available to meet the following conditions: high similarity within a cluster; and different clustering objects in lower similarity. Clustering similarity is calculated by using the average value of the objects in each cluster to obtain a "central object" (gravity center) (Fig. 2).

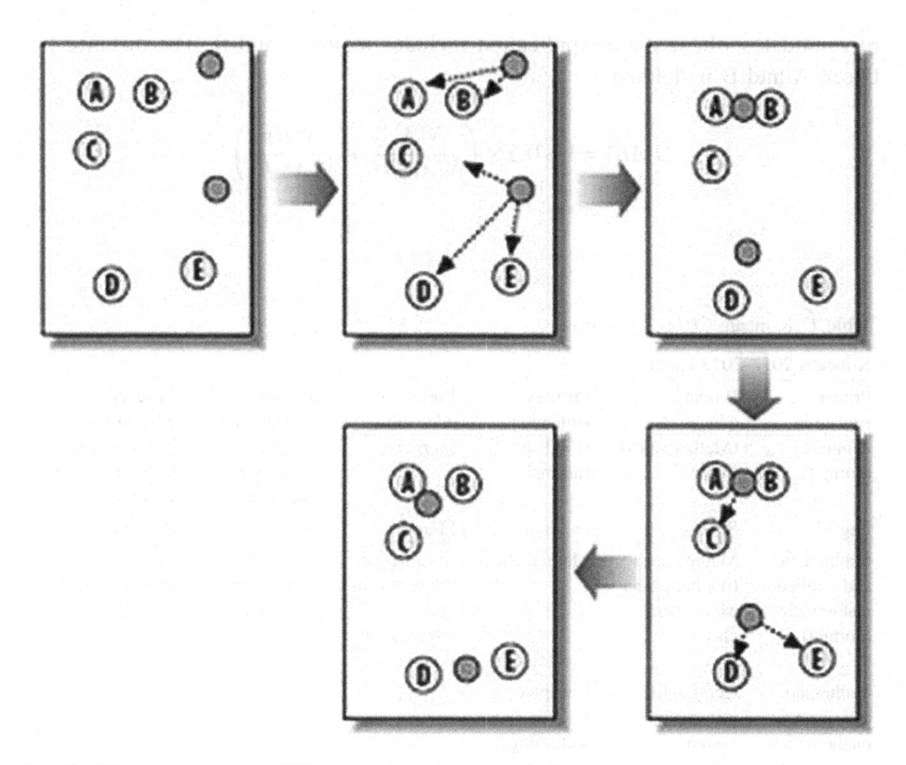

Fig. 2 Schematic diagram of K-means algorithm

K-means algorithm mainly includes the following four steps:

(a) Select k objects from N data objects as the initial cluster centers;
(b) Get the average value (central object) of each cluster, calculate the distance between each object and the central object. And according to the minimum distance to redistrict the object.
(c) Recalculate the average value of each (with change) clustering (center object);
(d) If all cluster don't change, terminate the program. If the condition is not satisfied, return to step (b).

3.2 *The Application of K-MEANS Algorithm in Admission Data Analysis*

In the application of K-MEANS algorithm, we must define the distance between the two major at first. The number of candidates whose major application includes major A is defined as $N(A)$. Similarly, $N(B)$ is defined. $N(A\&B)$ means the number of candidates whose major application includes major A and B. The distance between A and B is defined as

$$D(AB) = 1 - 0.5 \times \left(\frac{N(A)}{N(A\&B)} + \frac{N(B)}{N(A\&B)}\right) \quad (1)$$

Table 1 K-means 2011–2013 result

K-means 2011–2013 result					
Primary education (Chinese) (normal)	Primary education (Mathematics) (normal)	Primary education (English) (normal)	Preschool education (normal)	Primary education (science education) (normal)	Primary education (information technology) (normal)
English	Spanish	German	French	Japanese	Russian
Mathematics and applied mathematics (normal)	Mathematics (mathematical experiment class)	Mathematics	Information management and information system	Intelligent science and technology	Information engineering (optoelectronics)
Mathematics and applied mathematics	Geographic information system	Computer science and technology (normal)			
Chinese language and literature (normal)	English (normal)	Chinese language and literature	International economy and trade	Chinese language and literature (senior foreign secretary)	Chinese language and literature (senior foreign secretary)
Culture industry management	Drama film and television literature	Public management	Electronic commerce	Tourism management	
Chemical (normal)	Physics (normal)	Biological science (normal)	Geography science (normal)	Electronic information engineering	Educational technology
Education	Applied chemistry	Science and technology of remote sensing	Biological sciences	Biological sciences (base class)	
History (normal)	History (base class)	World history (base class)	Politics and administration	Archaeology	Ideological and political education (normal)
Social work	Philosophy	History (urban traditions and cultural management)			

Table 2 K-means 2014–2016 result

K-means 2014–2016 result					
Primary education (Chinese) (normal)	Primary education (Mathematics) (normal)	Primary education (English) (normal)	Preschool education (normal)	Primary education (science education) (normal)	Primary education (information technology) (normal)
English	Spanish	German	French	Japanese	Russian
Mathematics and applied mathematics (normal)	Mathematics (mathematical experiment class)	Mathematics	Information management and information system	Intelligent science and technology	Information engineering (optoelectronics)
mathematics and applied mathematics	geographic information system	Computer science and technology (normal)	Electronic commerce		
Chinese language and literature (normal)	English (normal)	Chinese language and literature	International economy and trade	Chinese language and literature (senior foreign secretary)	Chinese language and literature (senior foreign secretary)
Culture industry management	Drama film and television literature	Public management	Social work	Tourism management	
Chemical (normal)	Physics (normal)	Biological science (normal)	Geography science (normal)	Electronic information engineering	Educational technology
Education	Applied chemistry	Science and technology of remote sensing	Biological sciences	Biological sciences (base class)	
History (normal) philosophy	History (base class) History (urban traditions and cultural management)	World history (base class)	Politics and administration	Archaeology	Ideological and political education (normal)

For the K-MEANS algorithm, initial center points and K value will affect the experimental results. So the first step is to choose the appropriate initial center points and K values. According to the related literatures [8, 9] and the early test of our own, Author selects k = 6 and these initial center points are Primary education (Chinese) (normal), English, Mathematics and Applied Mathematics (normal), Chinese language and Literature (normal), Chemistry (normal) and History (normal). Then author establishes a six dimensional coordinate system and every initial center points is the standard of a dimension where a major's coordinate value is the distance mentioned in the previous paragraph between the major and the initial center points. Finally author uses 2011–2013 data and 2014–2016 data to operate K-MEANS algorithm. The results are shown in Tables 1 and 2.

According to the table, the results of 2011–2013 and 2014–2016 are similar. Only 2 majors (social work, E—commerce) transform their own clustering. These two majors don't owe significant relevance in 0.2 support Apriori algorithm. As can be seen from the results, the students' major application ideas of the Normal University did not change significantly in the past few years. It is interesting that English (normal) belongs to Chinese language and Literature (normal) cluster instead of English cluster.

4 Data Analysis of K-NN Algorithm

This section introduces K-NN algorithm briefly and how to use K-NN algorithm to analysis data. The result from K-NN algorithm shows for any major, the students in this major prefer to study in which college.

4.1 Introduction to K-NN Algorithm

The nearest neighbor [10], or K nearest neighbor algorithm (kNN, k-Nearest Neighbor) classification algorithm is one of the simplest methods of data mining classification technology. The so-called K nearest neighbor, is the meaning of the nearest neighbor K, saying that each sample can be represented by its nearest neighbor K (Fig. 3).

Fig. 3 Schematic diagram of K-NN algorithm

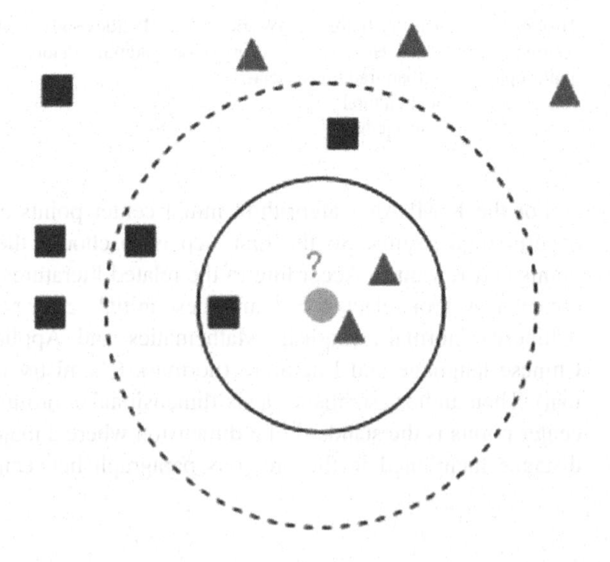

Table 3 K-NN result

	k-nn result					
Primary Education College	Primary education (Chinese) (normal)	Primary education (Mathematics) (normal)	Primary education (English) (normal)	Preschool education (normol)	Primary education (Science Education) (normal)	Primary education (Information Technology) (normal)
faculty of foreign languages	English	Spanish	German	French	Japanese	Russian
College of Arts	Mathematics and Applied Mathematics (normal)	Mathematics (mathematical experiment class)	Mathematics	Information management and information system	Information management and information system	Information Engineering (Optoelectronics)
Philosophy	mathematics and applied mathematics	geographic information system	Computer science and Technology (normal)	Electronic Commerce	Physics (normal)	Biological Science (normal)
Chemical (normal)	Chinese language and Literature (normal)	English (normal)	Chinese language and Literature	International economy and trade	Chinese language and Literature (senior foreign secretary)	Chinese language and Literature (senior foreign secretary)
Education	culture industry management	drama film and television literature	public management	Social work	Tourism Management	Geography Science (normal)
Electronic information engineering	educational technology	Applied chemistry	science and technology of Remote Sensing Science and technology	Biological Sciences	Biological Sciences (base class)	Ideological and Political Education (normal)
Politics and Administration						
college of history	History (normal)	History (base class)	World History (base class)	archaeology	History (Urban traditions and cultural management)	

For example, the green circle needs to be classified either to blue squares class or to red triangles class. If in the innermost circle (k = 3) it is regard as red triangles class because there are more triangles in the innermost circle. If in the outermost circle (k = 5) it is regard as blue squares class (3 squares compare to 2 triangles).

4.2 The Application of K-NN Algorithm in Admission Data Analysis

Here k-NN algorithm is used to determine the admission the students of one major whether to have enough strong desire to study in this college or not. For a particular major such as math, firstly, in the other majors which all students admitted by math also apply for, count first K major and make these major's college as their class. Second, add up this major's college, count which college appears at most in there. If one colleges is the same as another, and then compare the number of these two colleges' applications number. Finally, if the college which appears at most is same as the college which the major belongs to, it can be concluded that the students are willing to study in the College. Otherwise, put this major into college where students are most willing to go.

According to the test, the average number of top 7 majors compared to others have the obvious number of advantages, so the value of K is 7. Owing to the few admission number for each major, in order to ensure the reliability of the results, the data of 6 years as a whole is analyzed. The result is shown in the Table 3.

In Table 3, there are only four college exist on the figure (the normal university has 18 colleges): Primary Education College, Faculty of foreign languages, Liberal Arts Colleges, College of history. Liberal Arts Colleges has obvious advantages. Its number of major is far more than other college. This shows that the most popular college is Liberal arts.

5 Conclusion

By using these algorithms, the result shows that the overall situation of the Normal University student enrollment is relatively stable. There is a certain correlation between some specialties. Some of these specialties in accordance with the relevance can be divided into five groups: Language, Liberal arts (normal), Science (normal), Liberal arts (non-normal), Computer. In addition, for the major including normal or non-normal types, there are some differences between these two types of students' select. Normal students are more willing to choose other normal majors which are not similar to the major in curriculum setup. But non-normal students are more willing to choose majors which are similar to the major.

In k-NN algorithm's result, Liberal Arts Colleges is the most popular college. Its number of major is far more than other college. In Apriori algorithm's result, Chinese language and Literature (normal) and English (normal) in Liberal arts (normal) are more popular. Almost of the majors have their relevance major except Social work, Remote sensing science and technology and Electronic Commerce. In K-MEANS algorithm's result, the overall situation of Capital Normal University's student enrollment is relatively stable and most of correlations have not changed. The majors (Social work, Electronic Commerce) whose correlations has changed also don't have relevance related major in Apriori algorithm.

Among these 3 algorithm, Apriori algorithm can find clear relation between two specific majors. if we want to find more macroscopic result, we need k-means algorithm and k-nn algorithm. K-means algorithm reveals a new classification result which is different from the traditional college classification. But K-NN algorithm give us a result in the traditional classification.

So based on these result, the Normal University may raise the number of students enrolled in Liberal Arts Colleges and sets up more normal courses for normal students due to their desire that they want to become a teacher.

There are still some deficiencies in the study. Some of the data handle processes lack of full literature support. In the future, choose more algorithms for data analysis and further analysis of the data that has been excavated can become future research directions.

Acknowledgements This work was supported by the open research foundation of the machine intelligence and advanced computing key laboratory of education ministry (MSC-201707A), overlapping research project of Capital Normal University; science and technology innovation platform project of Capital Normal University. The study is approved for the school of management, Capital Normal University.

References

1. Li L, Liu Y. Analysis the factors influencing college entrance examinees filling the aspiration forms. China J Health Psychol. 2008.
2. Ma E, Kong X, Song G. Structure research on decision-making of filling the aspiration forms. China J Health Psychol. 2009.
3. Shen X, Sun S. Analysis of college entrance examination decision making based on statistical model. Stat Decis. 2014;21:57–9.
4. Xu G. Applying to college aided decision support system based on data mining. Comput Technol Autom. 2014;33(4):106–9.
5. Han XF, Liu XY. Design and implementation of a college entrance exam predict system based on web mining. Appl Res Comput. 2004;21(8):160–2.

6. Agrawal R, Srikant R. Fast algorithms for mining association rules in large databases. In: International conference on very large data bases. Morgan Kaufmann Publishers Inc.;1994. p. 487–99.
7. Hartigan JA, Wong MA. A K-means clustering algorithm. Appl Stat. 1979;28(1):100–8.
8. Yuan F, Zhou Z, Song X. K-means clustering algorithm with meliorated initial center. Comput Eng. 2007;33(3):65–6.
9. Yang SL, Yong-Sen LI, Xiao-Xuan HU, et al. Optimization study on k value of K-means algorithm. Syst Eng Theory Pract. 2006;26(2):97–101.
10. Altman NS. An introduction to Kernel and nearest-neighbor nonparametric regression. Am Stat. 1992;46(3):175–85.

The Immune System Model Based on B Method

Sheng-rong Zou, Si-ping Jiang, Xiao-min Jin and Chen Wang

Abstract Due to the lack of understanding of immunology, the immune theory is still controversial, which causes the difficulty in the immune system simulation. Based on the strict mathematics, the formal B method which is combined with the immune system as an example to transform the Unified Modeling Language class diagram into B formal specification in this paper. Through its strict verification technology, the model reliability is proved. Then the transformed class diagram will be transformed into the JAVA model of immune system, and some basic rules of immune system will be obtained.

Keywords B method · Immune model · Computer simulation · Class diagram

1 Introduction

It is difficult to find a unified specification to describe the immune system because of the diversity of the immune system, which also makes it difficult to establish a theoretical model of immune system. The traditional mathematical model can explain the local phenomenon of immunity, however it lacks the ability to describe the macroscopic properties of the immune system, so it has some limitations [1]. With the development of computer and electronic technology, it is easier for people to realize all kinds of methods on the technical level, so the computer simulation is used to simulate the immune system in this paper.

UML (Unified Modeling Language) is the industry standard of software analysis, design and visual modeling. It uses the object-oriented concept to support the whole process of software development [2]. However, UML uses the natural language description, it lacks semantic specification and rigorous reasoning mechanism. UML can not formalize and validate the consistency, completeness and

S. Zou (✉) · S. Jiang · X. Jin · C. Wang
College of Information Engineering, Yangzhou University, Yangzhou 225127, China
e-mail: srzou@qq.com

Y. Jia et al. (eds.), *Proceedings of 2017 Chinese Intelligent Systems Conference*, Lecture Notes in Electrical Engineering 459,
https://doi.org/10.1007/978-981-10-6496-8_50

correctness of the model, however the formal B method can make up for the disadvantages.

B method is one of the most important practical formal methods of software. It is built on the basis of Z language. It supports specification and all the refinement [3, 4], and also designs steps following the specification, and uses notation that generalized substitutions to express the transitions between states. From the software specification to the code both use AMN (Abstract Machine notation) specification language to describe the system specification and system design, to reduce the possibility of semantic errors. Each step of the validation requires the type checking and a series of proof obligations. Type checking was done before the formal verification, and all predicates including the set theory are checked by type checking. Its basic idea is that there is an upper bound on the set inclusion relation in the framework of a certain predicate which needs to be proved. This upper bound is called the superset of S (set) and the type of E (expression). Proof obligation is proved after the type checking. It proved that relevant specifications maintain the invariant as long as we can prove a series of proof obligations.

Compared with the traditional software development modeling technology, this paper proposes the way of formal B method combined with UML, which not only reduces the difficulty of formalized description of the system, but also solves some practical problems [5]. For one thing, people pay more attention on code design in previous software development process while ignoring the demand of establishing a model, this reason leads to the inconsistency and ambiguity of each model in the whole software development process; for another, establishing the requirements model is the first stage of software development and it is also a very important stage, whether the requirements model is good or not will directly affects the accuracy and correctness of the later period of the model, then affects the efficiency and quality of software development.

2 Immune System Description

The immune system is an important system for performing immune function [1]. It is the material basis of the immune response and is also a natural barrier that fight against foreign virus. To ensure the health of the body, it is able to identify and eliminate the invading microbes, allogeneic cells or macromolecules (antigens, virus-infected cells). The immune system is a complex system of life, but many urgent problems related to diseases need to be solved by scientists [6], so computer simulation is a useful way [7].

The natural immune system is a complex distributed information processing learning system, which functions in immune defense, immune tolerance, immune memory, immune surveillance. The immune system is a multi-layer structure, which is mainly divided into three layers:

First layer: it is made up of skin and mucous and can prevent the invasion of pathogens and use secretions to kill germs;

Second layer: the innate immune system, which is also known as non-specific immune system, is a series of defense against pathogens and formed in the long process of evolution. The innate immunity does not have specific pathogens. It has the ability to destroy them in the initial contact with the pathogens. Phagocytes and dendritic cells play major roles in this layer;

Third layer: adaptive immune system. This layer is mainly composed of T cells and B cells. One of the important reasons why immune system is adaptable is that they can produce a variety of specific antibodies. Adaptive immunity can identify all microorganisms, even "invaders" had never encountered before, through this, the innate immune can complete the immune function that can not complete before and effects on certain pathogens directly. And the system will remember the "invaders" features in order to response quickly to the second invasion. The main research of this paper is focused on this layer.

Since the immune system is made up of immune cells, the key to design the immune system is to design immune cells. The process of removing foreign substances is called cellular immune [8], the cells which take part in the process called effector cells. Cellular immunity is dominated by T cells, T cells will proliferate, differentiate into effector T cells after stimulated by antigens. When a same antigen enters the body again, the effector T cells against the antigen directly and the cytokines which released by effector T cells will also have a synergistic killing effect. Memory cells can be produced when the antigens were wiped out first time. Memory cells can not perform effector function directly, only when the same or similar antigen invading again, it can proliferate and differentiate into effector cells rapidly. A handful of memory cells can divide into memory cells again, executing the function of specific immunity. Influenza virus is used as a example in this paper. Influenza virus invade normal cells, so the immune response is mainly the cellular immune response of T cells. Attributes of the participants are described below:

Antigen (Ag): it is a kind of substance that can stimulate the body to make immune response and can react with the corresponding antibodies or sensitized lymphocytes in vivo or in vitro. The Ag used in this paper is influenza virus. Influenza, in human, is caused by a virus that attacks mainly the upper respiratory tract, the nose, throat and bronchi and rarely also the lungs [9, 10].

T cell: it is also known as T lymphocytes. T cells play an important role in cellular immune response by directly executing the immune function after differentiating into effector T cells. It recycled by lymphatics, tissue fluid and contacts with Ag. It can strengthen the immune response maintain the the immune memory and attack In. It also plays the function of cellular immunity and immune regulation.

Antigen presenting cell (APC): APC are cells that can be infected by Ag, usually located on the surface of skin or lumen. For example, the upper respiratory cells. APC may be infected by Ag when they contact with the Ag. When an epithelial cell exposures to Ag regardless of whether it is infected or not, it will have a function of antigen presentation, that is it can submit the information to the T cells. We choose the influenza virus as Ag, so the corresponding APC are located on the surface of the skin or lumen (In this paper, the antigen presenting cells known as epithelial cells).

Infected cell (In): Epithelial cells infected by Ag become In, and have antigen presentation function. Ag replicate in infected cells and released by budding.

3 Formalization of Immune System Class Diagram

Through the establishment of UML, class diagram, we convert the class diagram into the B formal specification. On this basis, the formal description of the target system is realized, which avoids the direct use of the B method to model the system, reduces the difficulty of formal development, and enhances the practicability of the B method [11]. Meanwhile, formal specification also provides a reference for the precise semantics of UML. With the method of converting UML class diagram into B method formal specification, the UML class diagram can be used as the starting point of the formal development of B method.

UML class diagram is the core of object modeling and processing. It can be used to explain the associations between the classes in the target system, and describe the types of the objects and their relationships in the system. It not only explains the structure of information, but also explains the actions of the system. Class diagram contains all the functions and data of the system operation. It expresses the static structure of object model. For one thing, it describes the composition of each class, that is the attributes, operations and constraints to be specific, for another, it describes the static associations between classes in the system, the main static associations types are generalization, abstraction, permission, and binding. It describes the static associations between classes in the system, the main static associations types are generalization, abstraction, permission, and binding. It can not only improve the efficiency of software development and reduce the cost, but also improve the quality of software engineering.

Since the class diagram and the B abstract machine have the characteristics of encapsulation, which lays the theoretical foundation for the formalization of the

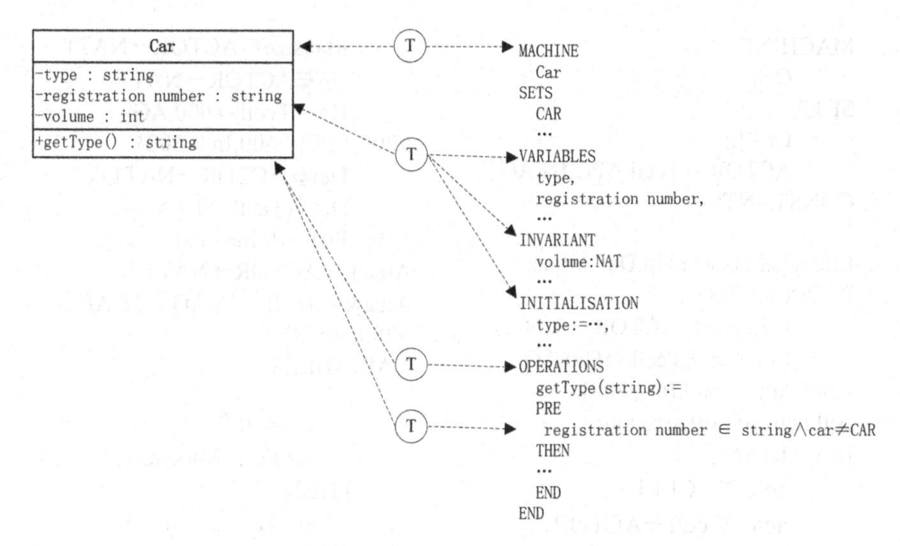

Fig. 1 Conversion frameswork of the basic class diagram

class diagram. Next, we should establish the semantic mapping between them. Specific rules are as follows:

Rule 1: Class name is translated into an abstract machine name corresponding to the class, attribute name is converted to the variable name, and defined according to number one by one;
Rule 2: Restrictions and constraints in the class are given in the invariant clause;
Rule 3: The initial value of the attribute variables are given in the initialization clause;
Rule 4: The operations in the class correspond to the operations in the abstract machine.

According to the rules above, a basic class framework diagram of the B formal specification as shown in Fig. 1.

The B machine of the immune system is described as follows:

The first machine packages a class of cells, which defines two sets, the first set is CELL. It is an extension set, representing the present or future cells. Second sets is ACTOR, which represents the main actors in the immune system. There are five constants in the system: lifecycle, maxsize, hp, def, attack, which represent the corresponding life cycle, max cell size, cell life, cell defense values and cell attack values. There are six variables: cell, actor, dx, dy, size, age, which represent the corresponding cell, cell type, cell horizontal coordinate, cell ordinate, cell size, cell age. And the operation of the machine is described to create a new cell (Creatcell), cell growth (Growth) cell division (Division), cell death (Death).

```
MACHINE
      Cell
SETS
      CELL;
      ACTOR={Tcell,APC,IN,AG}
CONSTANTS

Lifecyle,Maxsize,Hp,Def,Attack
PROPERTIES
      Lifestyle∈ACTOR→0..600∧
      Lifecyle={Tcell|->600,AG|-
>300,APC|->600,In|->200}
cell,actor,dx,dy,size,age
INVARIANT
      cell ⊆ CELL∧
      actor∈cell→ACTOR∧
      dx∈NAT∧
      dy∈NAT ∧
      size∈ACTOR→0..Maxsize
      age∈ACTOR→NAT ∧
INITIALIZATION
      cell, actor: =Ø, Ø||
      dx:∈NAT,dy:∈NAT||
      size:=0...Maxsize||
      sge:=0..lifecycle
OPERATION
i ← Creatcell=
   PRE
      CELL-cell≠Ø
   THEN
      ANY j WHERE
      j∈CELL-cell
      THEN
      cell:=cell∪{j}||
      i:=j
      END
   END
END;
Growth(c)=
   PRE
      Maxsize∈ACTOR→NAT1
      Hp∈ACTOR→NAT∧
      Hp={Tcell|->600,AG|-
>200,APC|->600,In|->300}
      Def∈ACTOR→NAT1∧
      Def={Tcell|->10,AG|-
>10,APC|->10,In|->10}
Attack∈ACTOR→NAT1∧
Attack={Tcell|->18,AG|->18,APC|-
>28,In|->28}
VARIABLES

         c∈cell∧
         size(c )<Maxsize(c)
      THEN
         size(c):=size(c)+1
      END
   END;
   Division(c)
      PRE
         c∈cell∧
         size(c)=Maxsze(c)
      THEN
         size(c):=maxsize(c)/2
      END
   END;
   Death(c)
      PRE
         c∈cell
      THEN
      VAR age IN
        age←getage(c);
         IF age(c)=Max Age(c)
      THEN
         cell◁dx=Ø;
         cell◁dy=Ø
         END
      END;
   END
```

As shown in Fig. 2, the relationship between TCELL and CELL is generalization which can be understood as the inheritance operation in the program design, that is, through inheritance, subclass not only can use their own attributes and operations, but also use the parent class attributes and operations. Due to the limit length, TCELL, APC, AG, IN are inherited from the parent class CELL, the following is only TCELL machine as an example to illustrate.

```
MACHINE
    TCELL
EXTENDS
    CELL
SET
    TCELL;
    STATUS={unactive,active}
VARIABLES
    tcell,status
    tcell≠TCELL
ANY j WHERE
 j∈TCELL→tcell
THEN
 Tcell:=tcell∪{j}
END
  END;
Growth(c);
Recognition(c)=
  PRE
INVARIANT
        tcell⊆TCELL∧
        status∈CELL→STATUS
INITIALIZATION
        status:=unactive
OPERATION
i←Creatcell(c)=
  PRE
        c∈cell
        c∈cell
  THEN
WHILE   status(c)=active ∧ (c=tcell)
DO
 Produce(c)
  END
END;
Death(c)
END
```

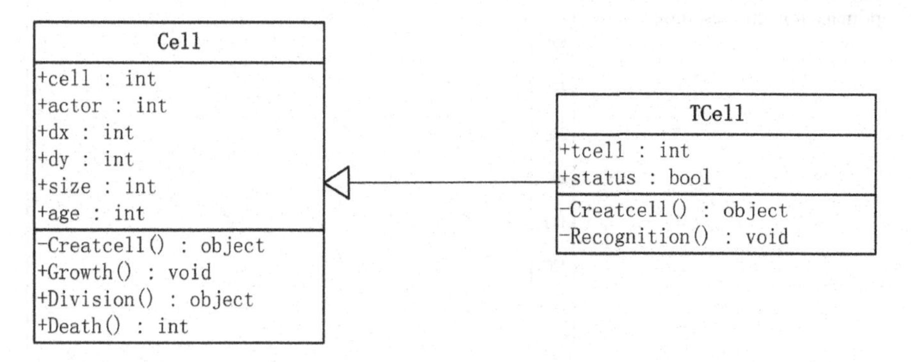

Fig. 2 Classes of Cell and TCell

4 JAVA Model of Immune System

A simulation of the Java program has been design based on the formal specifications of B method. The key of immune system modeling is to simulate the immune process [12]. When the program runs, all the cells in the system act under the rule of given actions, and at the same time a cell number graph can be generated, which reflects the real-time number of different types of cells. The horizontal axis represents the time scale, and each time point represents 4.5 h. The vertical axis represents the number of cells.

Figure 3 shows the development curve of immune response. When the time reached about 5 time points, free influenza virus in the system reached a number of lower values. Then with the influenza virus replicate in the APC and be released, the number of free influenza virus was increasing, which also led to the virus and T cells contact frequently. Through touching T cells, immune system began to study and identify the influenza virus, and then a large number of T cells were activated and proliferated rapidly. T cells in the system were also grown rapidly. The influenza virus completely cleared out when final time point 35 (about 6–7 days). This result is consistent with the results obtained from experiments in biology [13, 14], which shows that our simulation model is effective.

As shown in Fig. 4, it is clear that the number of T cells at the beginning of the simulation (the virus has just invaded). After the initial response, there are many memory cells corresponding to the antigens has been stored in the system, so the T cell does not need to spend time to learn, it can quickly activate and proliferate. This also leads to rapid shortening immune cycle, it only needs 18.5 time points (about 3–4 days) to remove the virus completely. It shows that the secondary immune response is significantly faster than the initial immune response, and memory cells can make the immune response more rapidly and powerfully.

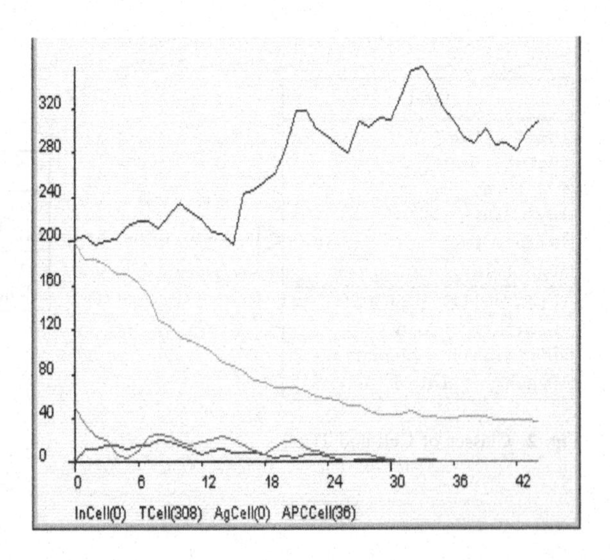

Fig. 3 Cell number curve of primary immune response

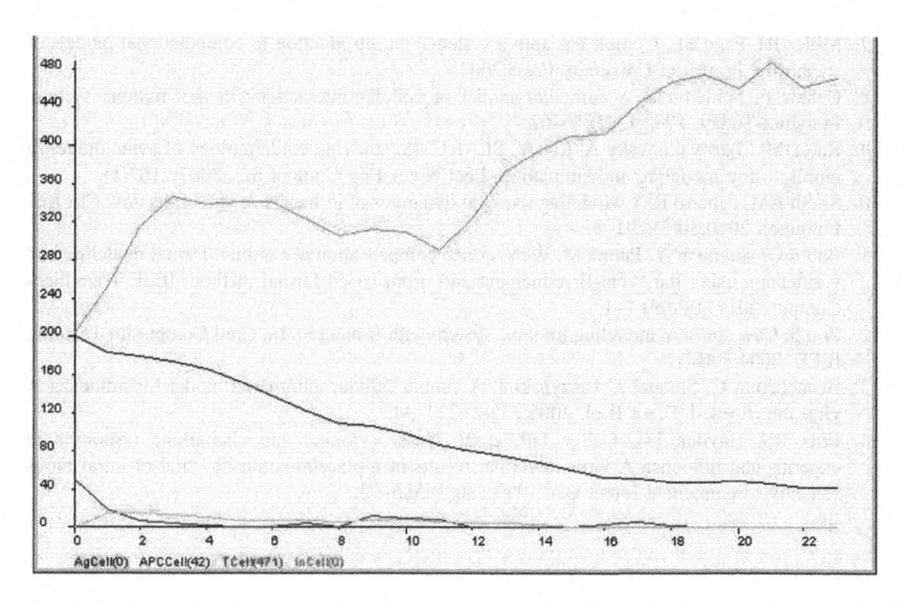

Fig. 4 Cell number curve of secondary immune response

5 Conclusion

In this paper, the immune system which designed by B method combined with UML has the advantages of visualization, with no ambiguity and high reliability. Through the analysis which based on the simulation results of JAVA model, it proves that the dynamic behavior of the model is consistent with the basic laws of biology, and the simulation model is effective. Compared with the experimental results in biological tissue and test tubes, the computer model requires shorter time, lower cost and easier analysis.

References

1. Hongwei M, Xingquan Z. Artificial immune system. Beijing: Science Press; 2009.
2. Selic B, Bock C, Cook S, et al. OMG unified modeling language (Version 2.5). 2015.
3. Zou S. Modeling distributed algorithm using B. In: Proceeding of the international grid and cooperative computing conference. Lecture notes in computer science. 2004. p. 683–9.
4. Lano K. The B Language and method: a guide to practical formal development. New York: Springer Inc.; 1996.
5. Fachada N. SimulIm: an application for the modelling and simulation of complex systems, using the immune system as an example. Graduation project report, Higher Technical Institute, Technical University of Lisbon; 2005.
6. Dasgupta D. An artificial immune system as a multi-agent decision support system. IEEE Int Conf Syst Man Cybern. IEEE; 1998; 4:3816–20.

7. Miller JH, Page SE. Complex adaptive systems: an introduction to computational models of social life. Princeton University Press;2007.
8. Celada F, Seiden PE. A computer model of cellular interactions in the immune system. Immunol Today. 1992;13(2):56–62.
9. Rao DM, Chernyakhovsky A, Rao V. SEARUMS: studying epidemiology of avian influenza rapidly using modeling and simulation. Lect Notes Eng Comput Sci. 2007;2167(1).
10. Smith AM, Ribeiro RM. Modeling the viral dynamics of influenza A virus infection. Crit Rev Immunol. 2010;30(3):291–8.
11. Babin G, Aitameur Y, Pantel M. Web service compensation at runtime: formal modeling and verification using the event-B refinement and proof based formal method. IEEE Trans Serv Comput. 2017; PP(99):1–1.
12. Zou S. Case study of modeling immune system with B method. Int Conf Comput Inf Technol. IEEE; 2004:890–5.
13. Beauchemin C, Samuel J, Tuszynski J. A simple cellular automaton model for influenza A viral infections. J Theor Biol. 2005;232(2):223–34.
14. Fritz RS, Hayden FG, Calfee DP, et al. Nasal cytokine and chemokine responses in experimental influenza A virus infection: results of a placebo-controlled trial of intravenous zanamivir treatment. J Infect Dis. 1999;180(3):586–93.

A Shallow-Dense Network Approach to Synchronization Pattern Classification of Multivariate Epileptic EEG

Hengjin Ke, Siwei Chen, Hao Zhang, Yunbo Tang, Yongyan Liu, Dan Chen and Xiaoli Li

Abstract A long-standing issue in the field of neuroscience is identifying evolving patterns from multivariate electroencephalography (EEG) signals superimposed with intensive noise. With insufficient prior knowledge, it becomes even more important to (1) accurately detect synchronization dynamics among data channels and (2) adaptively classify evolving patterns to better characterize the intrinsic nature of brain activities represented by the EEG. This study uses a shadow-dense network approach to solve these problems. The maximal information coefficient (MIC) method is extended to enable global synchronization measurement of all data channels embedded in the EEG. The global MIC measures are organized in time sequence to represent the evolving synchronization patterns. A shallow-dense neural network is designed to adaptively characterize the nonstationary patterns and then classify them. Experiments are performed to evaluate this approach over an epileptic EEG dataset. It is found that this approach can classify seizure states with accuracy, sensitivity, and specificity of 97.292%, 98.696%, and 96.116%, respectively; these results are superior to those of most existing methods. The proposed approach achieves this performance without denoising the EEG; in contrast, denoising is essential in existing methods. Furthermore, the proposed approach requires only one hyperparameter, which avoids the potential errors caused by excessive parameter settings in existing methods.

Keywords Maximal information coefficient · Global synchronization pattern · Shallow-dense neural network · Classification

---

H. Ke · S. Chen · H. Zhang · Y. Tang · Y. Liu · D. Chen (✉)
School of Computer Science, Wuhan University, Wuhan 430072, China
e-mail: dan.chen@whu.edu.cn

X. Li (✉)
Beijing Normal University, Beijing 100875, China
e-mail: XiaoLi@bnu.edu.cn

Y. Jia et al. (eds.), *Proceedings of 2017 Chinese Intelligent Systems Conference*, Lecture Notes in Electrical Engineering 459,
https://doi.org/10.1007/978-981-10-6496-8_51

1 Introduction

Current research on brain functioning and brain pathologies often assesses the synchronization pattern between EEG signals to characterize interactions in different brain regions. Neurons are believed to communicate with and share information among each other through synchronous firing [1]. Reliable methods have long been pursued in various disciplines to measure the synchronization coupling strength of multidimensional data superimposed with intensive noise and interference. The ability to find synchronization patterns in multidimensional data has become increasingly important in current scientific and engineering problems such as feature extraction [2], complex oscillator networks, neural computing [3], and brain pathology [4].

The role of synchronization in brain activity is not fully understood, and many experimental and theoretical studies are needed to establish the communication between different brain regions [4]. Early studies focused on bivariate synchronization analyses, such as mutual information (MI), Perason product moment correlation, and Spearman rank order correlation. Among these methods, MI is one of the most significant information theoretic interdependence measures [3], and it achieved superior performance on discrimination and robustness to noise [5]. After comparing MIC with the correlation coefficient and MI in terms of linear, nonlinear, antinoise, robustness, and finding nonfunction relation properties, Reshef concluded that MIC is the best strength measurement for bivariate analyses [6]. In recent years, multivariate synchronous analysis methods have been developed, such as phase synchronization cluster analysis (PSCA), S-estimator [3], and correlation matrix analysis (CMA). An S-estimator can effectively measure global synchronization; however, it cannot obtain synchronization details among variables. PSCA can obtain topological details about different variables, but it cannot obtain global synchronization information. CMA affords both advantages. This study extends MIC to enable the global synchronization measurement of all data channels embedded in the EEG. Global MIC measures are organized in time sequence to represent evolving synchronization patterns.

Synchronization patterns are often diverse and uncertain. To address this problem, nonlinear adaptive pattern classification based on a deep neural network (NN) is essential in clinical application because it affords (1) nonlinearity, (2) adaptivity, and (3) fault tolerance. The classifier performance is not dependent on the NN layer. Forrest suggested a smaller CNN architecture called SqueenzeNet that achieved the same accuracy level because it afforded three advantages: (1) more efficient distributed training, (2) less overhead, and (3) easy deployability on embedded platforms with limited resources [7]. However, an extremely deep NN suffers from issues such as (1) vanishing/exploding gradients and (2) degradation problem.

Numerous previous works on EEG multidimensional analysis have been based on possessing sufficient prior knowledge. However, the suffer from the following disadvantages: (1) needing extra preprocessing for denoising and removing artifacts based on expert knowledge, (2) needing long latency to track frequency features [2], and (3) poor understanding of brain activity across different brain regions and across multiple subjects with a patient-specific strategy, the main reason for which is only special channels related strongly to disease are deeply examined, but they lack synergy between brain regions. All classifiers need a series of preprocessing filters to estimate the influence of intense noise and interference, and therefore, classifiers may limit noise relatively weakly.

To achieve these objectives, global synchronization features that can effectively limit high-level noise and interference extracted without prior knowledge, and offline model training and an online predictor are proposed for detection across all brain regions using multiple subject records.

2 Related Work

Cui proposed a new method called S-estimator based normalized weighted permutation mutual information (SNWPMI) for analyzing multichannel EEG synchronization strength [3]. It combines bivariate signal synchronization analysis of normalized weighted-permutation mutual information (NWPMI) with an S-estimator measure. The proposed NWPMI and SNWPMI can be suggested as an effective index to estimate the synchronization strength.

Li investigated permutation entropy as a tool to predict absence seizures [8]. It can not only track dynamical changes in EEG records but also successfully detect a preseizure state with 53.8% accuracy in 4.9 s.

Cai proposed a paralleled GAD-PDSA approach that measured the synchronization strength of bivariate nonstationary nonlinear data against phase differences and extended it to global synchronization measurement with a parallelized GPGPU [9]. It is based on adaptive decomposition phase difference synchronization analysis. GAD-PDSA can largely improve the scalability of data processing; the synchronization analysis served as an effective indicator for epileptic focus localization.

Chen proposed a parallelized multiple-channel synchronization analysis method called G-NLI for measuring the synchronization direction and strength of a global neural signal [4]. It performs synchronization measurement in a massively parallel manner. G-NLI can largely improve the runtime performance by more than 1000 times, and it can support real-time global synchronization measurement for successful epileptic focus localization.

Piotr proposed patient-specific classification of patterns based on synchronization to predict epilepsy seizure onset [2]. It is computed using EEG synchronization, including cross-correlation, nonlinear interdependence, dynamical entrainment or wavelet synchrony, and extracted spatiotemporal patterns. It achieved 71% sensitivity and 0 false positives.

To the best of our knowledge, this paper is the first to apply only the MIC to detect the synchronization of multiple channels and to predict seizure onset and offset points.

3 Shallow-Dense Net with Correlation Matrix Based on Maximal Information Coefficient

This section describes the proposed approach. First, an overview of the process flow is presented. Then, the theoretical principle is discussed. Finally, the design strategies for the classification model are described.

3.1 Overview

This approach involves the following phases: (1) feature extractor, (2) building model to predict seizure, and (3) evaluation. In this study, preprocessing is not used, unlike in existing studies, because of (1) the difficulty of online analysis and (2) the need for sufficient prior knowledge to set extreme hyperparameters manually. Figure 1 shows the detailed process flow.

3.2 Correlation Matrix Based on Maximal Information Coefficient

Maximal Information Coefficient MIC is a useful measurement for identifying a subset of the strongest relationships in a dataset. It is a positive real value that shows boundness, symmetry, and invariance. Details can be referred in [6].
Correlation Matrix based on Maximal Information Coefficient MIC can only evaluate the synchronization strength of bivariate coupling time sequence. To analyze the global synchronization strength of multichannel signals, MIC and a correlation matrix are combined into a correlation matrix based on MIC (CMMIC). Every element in the matrix indicates the synchronization strength between each channel. Formally, we define CMMIC as

$$CMMIC = \begin{bmatrix} MIC_{11} & MIC_{12} & \cdots & MIC_{13} \\ MIC_{21} & MIC_{22} & \cdots & MIC_{23} \\ \vdots & \vdots & \ddots & \vdots \\ MIC_{n1} & MIC_{n2} & \cdots & MIC_{nn} \end{bmatrix} \tag{1}$$

where $I_{ij}(i, j = 1, ..., n)$ denotes the synchronization strength between channels i and j. According to the properties of MIC, CMMIC is a positive definite matrix because $MIC_{ij} \geq 0$ && $MIC_{ii} = 1$. Its eigenvalues have properties according to linear algebra theory; obviously, the trace value of CMMIC is equal to the number of channels. With a small probability event, one ZERO matrix is obtained if and only if all channels are independent of each other.

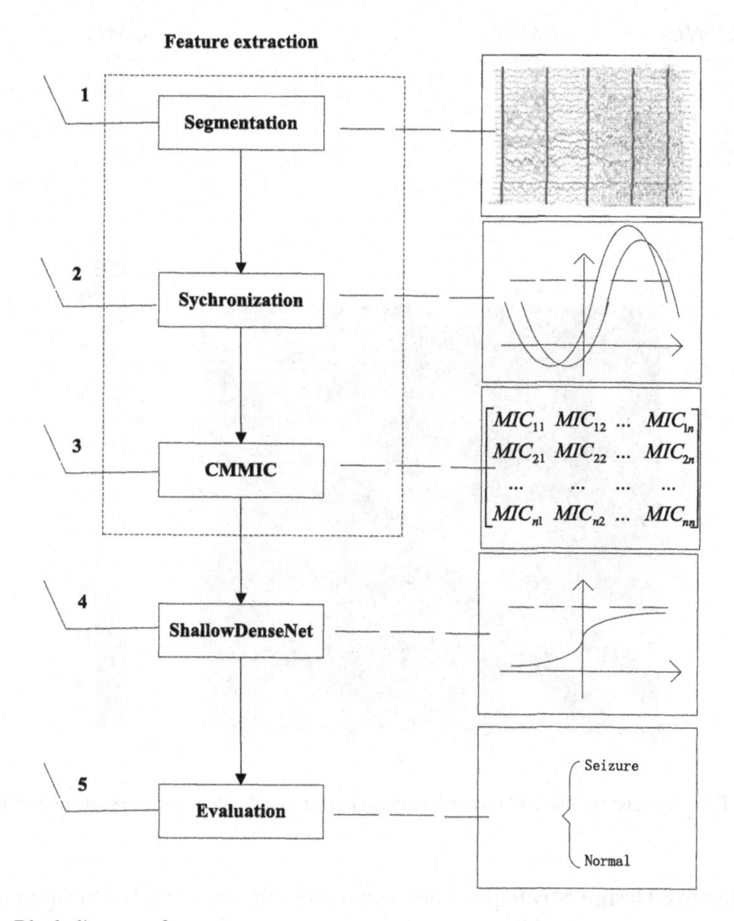

Fig. 1 Block diagram of process

1. $\lambda \geq 0$
2. $p = \sum_{i=1}^{N} \lambda_i = tr(CMMIC) = \sum_{i=1}^{N} MIC_{ii} = \#Channels$

EEG data is organized as a time sequence. Each pattern in the sequence is one $N \times N$ (*N is the channel count*) matrix whose elements express the synchronization strength between each channel, which is measured as MIC in Sect. 3.2. Figure 2 shows a time sequence in which each pattern consists of $N \times N$ gray images. The gray image is symmetric because of the symmetry property of MIC.

3.3 Shallow-Dense Neural Network

This section outlines our design strategies for *ShallowDenseNet* with a few parameters and uses these strategies to construct *ShallowDenseNet*, which mainly comprises a dense layer.

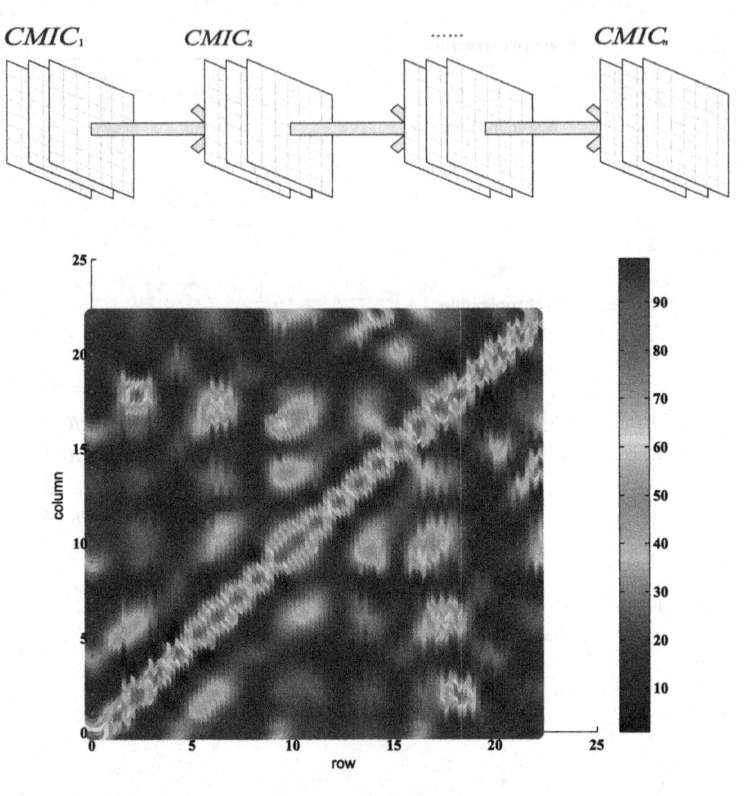

Fig. 2 Time sequence of one EEG record, with each pattern indicated as a grayscale image (0–255)

Architecture Design Strategies The overarching objective of classification in this paper is to identify an architecture with few layers that shows competitive accuracy. To achieve this, the following three main strategies are used:

Strategy (1) **Dropout**. Dropout is a technique for addressing overfitting in deep neural nets by randomly dropping units from the neural network during training. Srivastava suggested optimal strategies for dropping out 20% of the input units and 50% of the hidden units [10].

Strategy (2) **No convolution and pooling layer**. To reduce model parameters, convolution and pooling have been investigated in detail and have shown great success; however, if a pattern shows diversity and nonstationarity, information loss and reduced performance may occur.

Strategy (3) **Dense Net**. In DenseNet, each layer has $\frac{L(L+1)}{2}$ direct connections in a feedforward fashion [11].

Shallow-Dense Net Architecture This section describes a strategy for shallow net that achieves the same accuracy level and a dense net with reduced redundant feature maps [11]. *ShallowDenseNet* begins with a standalone Dropout layer with rate of 20%, followed by three Dense layers, each of which has direct connections from previous layers, and ending with a final sigmoid activation function to classify the synchronization pattern.

Other Shallow-Dense Net Details In this section, we briefly provide design choices that have been omitted from the above architecture. The intuition behind these choices is described below.

- Optimizers. Stochastic gradient descent (SGD) was used with a mini-batch size of 50.
- Learning rate. The learning rate is set to 0.01.
- Decay and Momentum. Decay coefficient of learning rate decreases at each epoch as $1e-6$, and the Nesterov parameter is applied to update the momentum to 0.9.
- Objectives. Minimize the mean squared error in *ShallowDenseNet*.
- Activations. A sigmoid function was used.

4 Experimental Results and Discussion

4.1 Data Description

Unlike studies that used the BONN dataset that contains only one channel or the FRE dataset that contains six channels, the CHB-MIT database used in this study contains 23 channels. In this study, the CHB-MIT scalp EEG database was used; it is available for free at http://physionet.org/physiobank/database/chbmit [12].

4.2 Training Samples Build

To address the problem of biased performance caused by imbalanced samples, Markov Chain Monte Carlo (MCMC) sampling is used to balance the seizure state and normal state samples.

4.3 Performance Measures

To evaluate our proposed *ShallowDenseNet* classifier, a K-fold cross-validation algorithm is employed. With a 10-fold cross-validation method, the data is divided into ten folds by shuffling, and 10 iterations are performed. In each iteration, eight folds are trained, and the remaining two folds are used for testing the classifier performance. The final performance result is the average of all folds. Figure 3 shows the accuracy & loss metrics for the training and validation processes with a window size of 2048. Here, *acc* and *loss* indicate the accuracy and error in the training step,

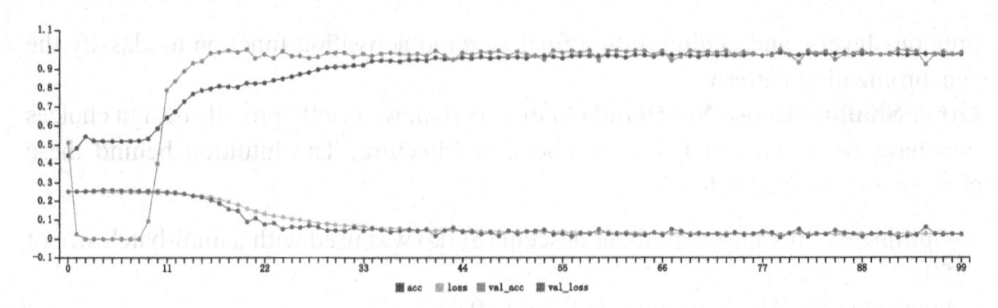

Fig. 3 Accuracy & Loss in training and testing processes

respectively, and *val_acc* and *val_loss* indicate the accuracy and error in the validation, respectively. Obviously, the model does not overfit because (1) *acc* and *val_acc* are really high, and (2) there is no large difference between the *acc* and *val_acc* curves.

The performance of the *ShallowDenseNet* classifier is measured by SEN ($TP/(TP + FN)$), SPE ($TN/(TN + FP)$), ACC ($(TN + TP)/(TP + FN + TN + FP)$), Precision ($TP/(TP + FP)$), and Recall ($TP/(TP + FN)$). *SEN* and *SPE* describe the number of correctly detected epileptic seizures and normal performances, respectively. *ACC* denotes the average performance of the classifier. *Precision* calculates the proportion of all correctly detected seizure onsets from all that were actually classified. *Recall* calculates the proportion of all correctly detected seizure onsets from all correctly detected seizures and negative normals. Fine-tuning the segmentation size results in better performance. Figure 4 shows a box chart of the classification performance with respective to segmentation. As the size increases, *SEN*, *SEP*, and *ACC* increase almost linearly when the window size is smaller than 1556 and then increase slowly. The box height indicates the amount of variance, which shows the stationarity of the classification performance. With a window size of 2048 (8 s), we achieve accuracy, sensitivity, and specificity of [97.29% ± 0.04%], [98.70% ± 0.13%], and [96.12% ± 0.12%], respectively. The smaller standard deviation illustrates the stationarity of our *ShallowDenseNet* classifier.

Accuracy and loss cannot express the model performance for unbalanced data effectively because high accuracy will be achieved even if all class labels are predicted as the majority classification. To determine the performance of imbalanced classification in a biclass problem, $Gmean(\sqrt{(Sensitivity \times Specificity)})$ and F_1-$Score(2 \times \frac{Precision \times Recall}{Precision+Recall})$ are measured for imbalanced performance [13].

Except for achieving high performance indexes including *SEN*, *SPE*, and *ACC*, several estimates of the balance of performance indexes including *GMEAN* and F_1-*Score* show that our approach can not only effectively detect epileptic seizure onset but also effectively find negative classifications to decrease the false alarm rate. Table 1 shows the details.

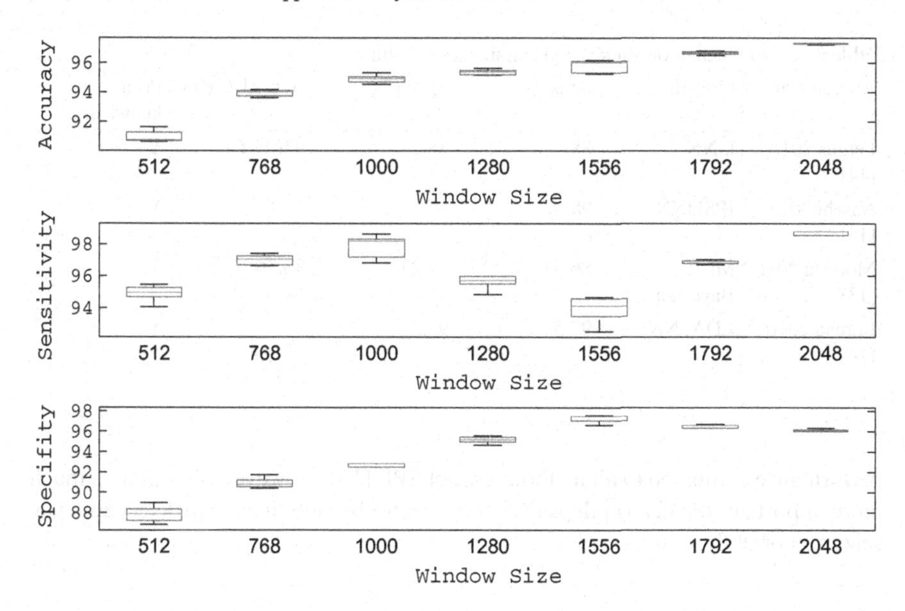

Fig. 4 Relationship between performance (accuracy, sensitivity, and specificity) and window size

Table 1 F_1*-Score* and *GMEAN* performance related to different window sizes

Window size	GMEAN	F_1-Score
512	90.10869461	90.37097186
768	94.68170377	94.75146386
1000	94.88503541	94.94009876
1280	96.93832055	96.94115133
1556	96.33352522	96.3467291
1792	97.09728934	97.09848491
2048	97.994	97.994

4.4 Classification Performance Comparison

We now focus on evaluating *ShallowDenseNet* on CMMIC. The results found in this paper can be compared in more detail with the papers listed in Table 2, which are all based on the CHB-MIT dataset.

The column "Prior Knowledge" indicates whether the classifier is dependent on prior knowledge, which is (1) those which need hyperparameters to be set based on expert knowledge, (2) those focusing on patient-specific classifiers, and (3) phases such as filtering noise and removing artifacts based on sufficient prior knowledge. "Y" and "N" indicates that prior knowledge was and was not used, respectively. To the best of our knowledge, all existing classifiers need sufficient prior knowledge; however, this is not necessary with the proposed approach, and it achieves superior

Table 2 Seizure detection studies and classification results

Author year	Classifier	Sens (%)	Spec (%)	Acc/AUC (%)	Prior knowledge
Fergus 2016 [14]	k-NN	88	88	93(*AUC*)	Y
Nasehi 2013 [15]	IPSONN	98	–	–	Y
Morteza 2016 [13]	MLP, Bayesian	86.53	97.27	86.56	Y
Lorena 2016 [16]	LDA, NN	97.5	99.9	–	Y

performance compared to all methods except SPE [16]. However, SEN plays a much more important role in an epilepsy seizure detector because it aims to detect epilepsy seizures correctly.

5 Conclusions

It is a important issue to find synchronization patterns in multivariate *EEG* superimposed with intensive noise and accurately to classify them on this basis under the circumstance of insufficient prior knowledge. Such capability can significantly benefit brain dysfunction research and practices, e.g., epilepsy.

This study extended the MIC method to measure global synchronization of multivariate *EEG*. The global MIC measures (CMMICs) have been organized in time sequence to represent the evolving synchronization patterns. CMMICs maintain abundant useful information to differentiate the seizure states from the rest.

A shallow-dense neural network is designed to adaptively characterize the nonstationary patterns related to seizures and then classify them. The *ShallowDenseNet* feeds the features of the earlier layer into the later layer to make the dense blockers arranged as a forward dense net. The design alleviates the vanishing gradient problem and strengthens feature propagation, which lead to a substantial reduction of parameters. The *ShallowDenseNet* adaptively classifies the CMMICs and captures the intrinsic nature of seizure activities represented by the EEG.

Experiments have been performed to evaluate the proposed approach over the CHB-MIT scalp EEG dataset. The results show an improvement relative to existing methods, with accuracy, sensitivity, and specificity of [97.29% ± 0.04%], [98.70% ± 0.13%], and [96.12% ± 0.12%], respectively. The standard deviation of most results are small, which indicate that the performance of *ShallowDenseNet* is relatively stable.

The proposed approach achieves this performance without the need for denoising the EEG; in contrast. Furthermore, the approach requires only one hyperparameter, which avoids the potential errors caused by excessive parameter settings. The overall work in this study held the potentials to classify complicated synchronization patterns of raw *EEG* with insufficient prior knowledge.

References

1. Gysels E. Phase synchronization for classification of spontaneous EEG signals in brain-computer interfaces. Lausanne EPFL;2006.
2. Mirowski P, Madhavan D, LeCun Y, Kuzniecky R. Classification of patterns of *EEG* synchronization for seizure prediction. Clin Neurophysiol. 2009;120:149–71.
3. Cui D, Pu W, Liu J, Bian Z, Li Q, Wang L, Gu G. A new *EEG* synchronization strength analysis method: S-estimator based normalized weighted-permutation mutual information. Neural Netw. 2016;82:30–8.
4. Chen D, Li X, Cui D, Wang L, Lu D. Global synchronization measurement of multivariate neural signals with massively parallel nonlinear interdependence analysis. IEEE Trans Neural Syst Rehabil Eng. 2014;22(1):33–43.
5. Bonita JD, Ambolode LCC II, Rosenberg BM, Cellucci CJ, Watanabe TAA, Rapp PE, Albano AM. Time domain measures of inter-channel EEG correlations: a comparison of linear, nonparametric and nonlinear measures. Cogn Neurodyn. 2014;8:1–15.
6. Reshef DN, Reshef YA, Finucane HK, Grossman SR, McVean G, Turnbough PJ, Lander ES, Mitzenmacher M, Sabeti PC. Detecting novel associations in large datasets. Science. 2011;334(6062):1518–24.
7. Iandola FN, Han S, Moskewicz MW, Ashraf K, Dally WJ, Keutzer K. Squeezenet: Alexnet-level accuracy with 50x fewer parameters and <0.5 MB model size. In: International conference on learning representations, Feb 2017.
8. Li X, Ouyang G, Richards DA. Predictability analysis of absence seizures with permutation entropy. Epilepsy Res. 2007;77(1):70–4.
9. Cai C, Zeng K, Tang L, Chen D, Peng W, Yan J, Li X. Towards adaptive synchronization measurement of large-scale non-stationary non-linear data. Futur Gener Comput Syst. 2015;43(44):110–9.
10. Srivastava N, Hinton G, Krizhevsky A, Sutskever I, Salakhutdinov R. Dropout: a simple way to prevent neural networks from overfitting. J Mach Learn Res. 2014;15(1):1929–58.
11. Huang G, Liu Z, Weinberger KQ. Densely connected convolutional networks. In: IEEE conference on computer vision and pattern recognition, vol 00; 2016.
12. Goldberger AL, Amaral LAN, Glass L, Hausdorff JM, Ivanov PCh, Mark RG, et al. PhysioBank, PhysioToolkit, and PhysioNet: components of a newresearch resource for complex physiologic signals, Circulation 101 (23)e215Ce220[EB/OL]. http://physionet.org/physiobank/database/chbmit.
13. Behnam M, Pourghassem H. Real-time seizure prediction using RLS filtering, interpolated histogram feature based on hybrid optimization algorithm of Bayesian classifier, Hunting search. Comput Methods Program Sin Biomed. 2016;136:115–36.
14. Fergus P, Hussain A, Hignett D, Al-Jumeily D, Abdel-Aziz K, Hamdan H. A machine learning system for automated whole-brain seizure detection. Appl Comput Inf. 2016;12(1):70–89.
15. Nasehi S, Pourghassem H. Patient specific epileptic seizure onset detection algorithm based on spectral features and IPSONN classifier. In: International conference on communication systems, network technologies. Gwalior: IEEE;2013. p. 186–90.
16. Orosco L, Correa AG, Diez P, Laciar E. Patient non-specific algorithm for seizures detection in scalp *EEG*. Comput Biol Med. 2016;71:128–34.

Reliability Analysis of Distribution Network with Microgrid Based on Hybrid Energy Storage

Yumei Wang and Zhaochu Song

Abstract The reliability evaluation of distribution network with microgrid is an important aspect of reliability research of power system. Firstly, the stochastic models of output power of micro power source are established, and the charge/discharge strategy of hybrid energy storage is established. The microgrid is formed in a certain way. Secondly, through dividing the area of system and reducing load when the output power is insufficient, the outage time of each load point is determined. The system reliability indexes are calculated by sequential Monte Carlo simulation. The effectiveness of the evaluation method is verified by simulation results of IEEE RBTS BUS6, to which the microgrid is connected.

Keywords Distribution network · Microgrid · Monte carlo · Hybrid energy storage

1 Introduction

With the access of the microgrid (MG), the traditional distribution network becomes a new network containing distributed generators (DG), and its operation and control made an enormous change [1]. The reliability evaluation of distribution network also appears new problems. So it is necessary to analyze the reliability of the distribution network containing MG.

At present, many researches on the reliability evaluation of distribution network containing MG have been done. On the basis of the models of wind turbines (WT), photovoltaics (PV) and battery energy storage (BES), a sequential Monte Carlo simulation method (sMC) is proposed to quantitatively evaluate the influence of MG on reliability of distribution network in [2]. Reference [3] adopts the minimal path method to evaluate the reliability, with the goal of maximizing the equivalent

Y. Wang (✉) · Z. Song
School of Electric Engineering and Automation, Henan Polytechnic University,
Jiaozuo 454000, China
e-mail: wangym@hpu.edu.cn

Y. Jia et al. (eds.), *Proceedings of 2017 Chinese Intelligent Systems Conference*, Lecture Notes in Electrical Engineering 459,
https://doi.org/10.1007/978-981-10-6496-8_52

payload, but it does not involve the load reduction strategy of island operation. Reference [4] establishes the reliability model of the combined operation of DG and energy storage (ES), and uses Monte Carlo method to analyze the reliability of the distribution network containing MG, but does not consider the storage's capacity and different forms of ES.

The paper establishes a hybrid energy storage (HES) model of ultracapacitor (UC) and battery and formulates the charge/discharge strategy. The sMC based on region partition is proposed to analyze the reliability of system. The paper analyzes the effect of MG containing HES on the reliability of distribution network by simulation.

2 Component Reliability Model

The paper uses the components including WT, PV, HES and loads.

According to statistics, the state and output power of WT is mainly determined by the wind speed, and the relation is approximated by a piecewise function as follows:

$$P_w = \begin{cases} P_r(a+bv) & v_{ci} < v_t \le v_r \\ P_r & v_r < v_t \le v_{co} \\ 0 & 0 < v_t \le v_{ci} \text{ or } v_t > v_{co} \end{cases} \quad (1)$$

where: P_w, P_r are the output and rated power of WT; v_{ci}, v_r, v_{co} are the cutting in, rated and cutting out wind speed of WT; a, b are the functional expression of v_{ci}, v_r, v_{co} [5].

According to statistics, Beta distribution can approximate the probability distribution of the actual light intensity, and its probability density function is as follows [6]:

$$f(x) = \frac{\Gamma(\alpha+\beta)}{\Gamma(\alpha)\Gamma(\beta)}\left(\frac{x}{x_{\max}}\right)^{\alpha-1}\left(1-\frac{x}{x_{\max}}\right)^{\beta-1} \quad (2)$$

where: x and $x_{\max}$ are actual and maximum intensity; α and β are shape parameters of Beta distribution and determined by the mean and variance of light intensity [7].

For solar PV cells, the corresponding output power is as follows:

$$P_p = \eta x S \quad (3)$$

where: S is the area of PV; η is the photoelectric conversion efficiency of PV.

The HES can quickly smooth power fluctuation, frequently charge/discharge, fully replenish insufficient energy and absorb excess energy [8].

HES must ensure sufficient output power, and its maximum output energy must meet the load demand. The expressions of UC and battery are as follows:

$$\begin{cases} W_{uc,i+1} = W_{uc,i} + \Delta W_{uc,i} = W_{uc,i} + P_{uc,i}\Delta t \\ S_{oc,i+1} = S_{oc,i} + \Delta S_{oc,i} = S_{oc,i} + \frac{P_{bat,i}\Delta t}{n} \end{cases} \tag{4}$$

$$\Delta W_{uc,i} + \Delta S_{oc,i} = \Delta W_i \tag{5}$$

where: $P_{uc,i}$ and $P_{bat,i}$ are output power of UC and battery; Δt is control cycle; n is matching coefficient; $\Delta W_{uc,i}$ and $\Delta S_{oc,i}$ are transform energy of UC and battery; ΔW_i is total energy shortage.

The power of UC and battery can not exceed their maximum power. If the power is loss suddenly, HES must quickly stabilize system, namely the total power of HES P_i must not be less than the shortage of the maximum missing power $\Delta P_{\max}$:

$$P_i = P_{uc,i} + P_{bat,i} \geq \Delta P_{\max} \tag{6}$$

The time series model of load is given in [9], and it can accurately simulate the time-varying state of load.

3 Operation Strategy of Distribution Network with MG

The coordinated control strategy of MG mainly depends on the operation mode of MG and the relationship between the output of micro power and the loads' demand.

When the MG operates in grid-connected mode, the load is supplied by DG firstly. If the output power of DG is greater than the load demand, the extra energy will charge for ES. If it exceeds the storage capacity limit, the output of DG should be reduced. When the output power of DG can not meet the load demand, the distribution network will output power to maintain the power balance of system. Until the power supply of the system is insufficient, ES will be discharged within the capacity constraints.

When MG operates in island mode, the load is firstly supply by DG [10]. If the output power of DG is greater than the load demand, ES will be charged until it reaches the max capacity. If the total output of DG and ES can not meet the load demand, partial loads must be removed. If the DG fails, ES can also be a separate power supply. But if ES system fails, DG can not be a separate normal power supply to loads [11].

If the micro sources' output is not enough in island mode of the MG, it is necessary to remove some of the load in order to maintain the system power balance. The objective function of load reduction is the sum of the power supply load and the load weight is the largest, and its the constraint is that the total power of the load supply is not greater than the maximum power of ES providing. The functions are as follows:

$$f = \max \sum_{i \in D} \omega_i L_i \quad s.t. \quad \sum_{i \in D} L_i \leq P_{DG} + P_t \tag{7}$$

where: D is the largest possible connectivity area in island mode; ω_i and L_i are the weight coefficient and the size of load i; P_{DG} is the sum of the output power of DG; P_t is the output power provided by ES in island mode [12].

4 Reliability Evaluation of Distribution Network with MG

The system reliability indexes of the traditional distribution network, such as system average interruption frequency index (SAIFI), system average interruption duration index (SAIDI), average service availability index (ASAI), energy not served (ENS), and so on, can still be equipped with micro-network distribution network to describe the power supply capacity [13].

Fault mode effect analysis (FMEA) is the basis of reliability evaluation of distribution network. It is complex to analyze the influence of each component directly. So the distribution network must be partitioned to simplify. All components with common switches are divided into the same area, and the effect of all components in the same area is the same [14]. On the basis of the feeder partition, the reliability index is calculated by sMC. And the specific steps are as follows:

(1) The system network is simplified. The failure condition of all regions is traversed, and FMEA library is established. Enter the original data, the initial simulation time is 0. And all components are in the normal state.
(2) All components are sampled by the pseudo-random number generated by the computer. The time to repair TTF of each component is obtained by the formula (8), and the component with minimum time to repair $TTF_{\min}$ is found, it's faulty components. At this time, system time is $TTF_{\min}$, and simulation time starts calculating. By generating a new random number, failure repair time of component is determined by the formula (9). At the same time, the isolation time of failure is generated.

$$TTF = -\left(\frac{1}{\lambda_i}\right) \ln U_i \tag{8}$$

$$TTR = -\left(\frac{1}{\mu_i}\right) \ln V_i \tag{9}$$

where: λ_i and μ_i are the outage and repair rate of the component i; U_i and V_i are random numbers obeying (0, 1) uniform distribution.

(3) Judge the faulty component. If faulty component is the smallest path component of the system, then $T = T + TTR$. If the faulty component is not the smallest path component of the external distribution network, then $T = T + RT$ and turn to step (4). If the faulty component is non-powered component of the MG, turn to step (5).
(4) Generate a random number u obeyed (0, 1) uniform distribution, and compare the u and probability p of MG switch success. If $u \leq p$, the island mode is successful. The charge/discharge power of ES and output power of MG are calculated. The output power of MG and load value are compared to judge whether the load is cut. If $u > p$, the island mode fail.
(5) According to FMEA library, the load's power failure type is found to form the reliability index of load point.
(6) The simulation time required to judge whether the accuracy requirements are met. If the time is not reached, turn to step (2). If the time is reached, simulation ends. The reliability index of load point is obtained by counting the cumulative blackout time and the number of blackouts to calculate the reliability evaluation index of system.

5 Example Analysis

The improved IEEE RBTS Bus6 system is used as an example to simulate. Figure 1 shows the systems' wiring diagram. The example includes 30 lines, 23 fuses (load points), and a plurality of switches. Buses, transformers and switches of the system are 100% reliable. Other components' parameters refer to Ref. [15]. The p is 0.9. MG contains WT, PV and ES. In WT, $V_{ci} = 3\,\text{m/s}$, $V_r = 12\,\text{m/s}$, $V_{co} = 24.8\,\text{m/s}$. The single PV area is 2.16 m^2, and its efficiency is 15.47%.

The MG 1 is constructed at load 21, 22. And ES are simulated respectively by HES and BES. In HES, the maximum power of UC and battery are 50 and 25 kW. The charge/discharge efficiency of ES is 90%. The state of charge of battery and UC is 0.2–1.0, and initial state of charge is 0.8. The capacity of UC and battery are 3 and 10 MW h. In BES, the charging state and range of the battery are the same as above, and capacity is 13 MW h. According to the simulation results, the reliability evaluation indexes of system is shown in Table 1.

After MG access, the reliability of the system is improved obviously. The capacity of HES is less than BES, Table 1 shows that HES compared to BES on the improvement of the system reliability indexes is more obvious. This is due to the complementary effect of UC and battery, and HES can smooth the fluctuation of renewable energy to achieve better matching of DG and loads, thereby reducing the probability of load reduction. And the reliability level of the system is improved.

The MG 2 is constructed at load 8, 9, 10. The system contains the same WT, PV and ES. According to the simulation results, we can get the evaluation index of the system as shown in Table 2.

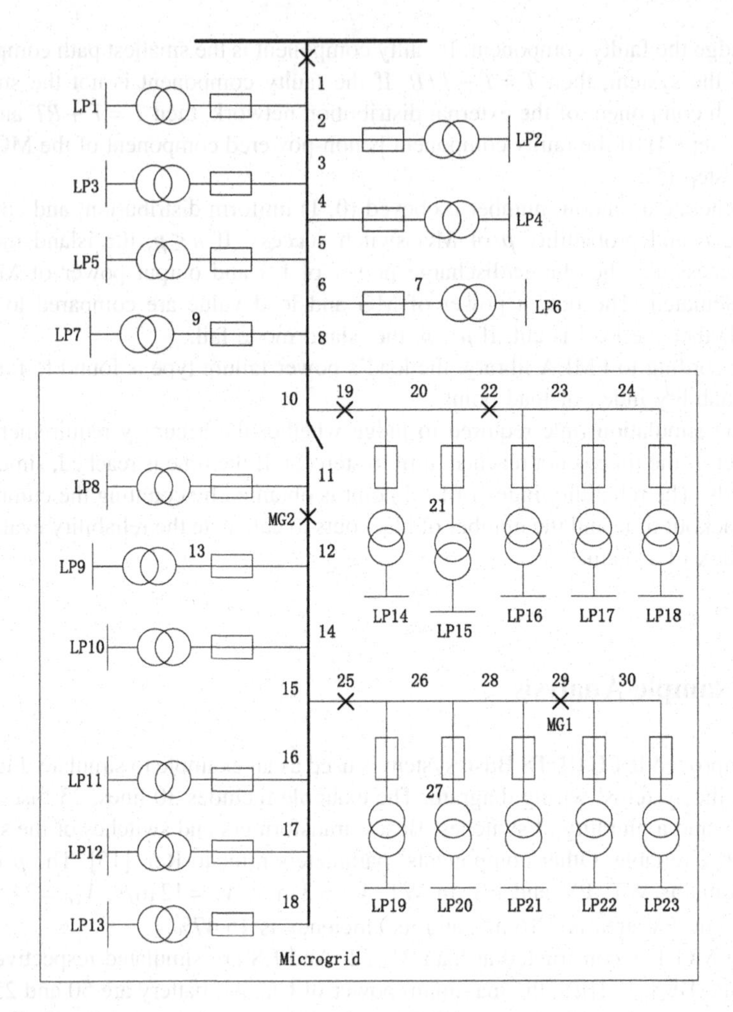

Fig. 1 Testing system

Table 1 Reliability indexes of distribution network

Distribution network	SAIFI	SAIDI	ASAI	EENS
Without any MG	1.5258	7.8812	99.915	39.938
With BES MG	1.2869	6.6457	99.929	33.221
With HES MG	1.0876	5.6189	99.963	30.175

From the data in Table 2, it can be seen that the system reliability of MG 2 is lower than that of MG 1. The results show that accessing MG the poor level reliability of loads in the original system can improve the system reliability better. It has reference value for selecting locations of important load in practical application.

Table 2 Reliability indexes of two schemes

Reliability index	Original system	System with MG 1	System with MG 2
SAIFI	1.5253	0.8876	0.9785
ASAI	99.917	99.963	99.951
EENS	40.236	30.172	33.981

6 Conclusion

The paper studies the output power characteristics of WT, PV and the limit of HES operation. Considering the coordinated operation strategy of ES and system, the reliability model of WT, PV and HES is established and connected to distribution network. After the component failure, the system is divided into several areas. According to results of the division, the load shedding is considered when the power is insufficient in island mode of the MG. Thus the outage time of each load point affected by the fault is determined. On this basis, the sMC is given to quantitatively analyze the influence of MG access on the system reliability.

The results show that the MG can effectively improve the system reliability, and the contribution of HES to the system reliability level is greater than BES. And MG accessed at loads of poor reliability can improve the system reliability better.

References

1. Bie Z, Li G, Wang X. Review on reliability evaluation of new distribution system with micro-grid. Electr Power Autom Equip. 2011;31(1):1–6.
2. Ge S, Wang H, Wang Y, et al. Reliability evaluation of distribution system including distributed wind turbines, photovoltaic arrays and batteries. Autom Electr Power Syst. 2012;36(5):16–23.
3. Liu C, Zhang Y. Distribution network reliability considering distribution generation. Autom Electr Power Syst. 2007;31(22):46–9.
4. Liang H, Cheng L, Liu S. Monte Carlo simulation based reliability evaluation of distribution system containing microgrids. Power Syst Technol. 2011;35(10):76–81.
5. Karki R, Hu P, Billinton R. A simplified wind power generation model for reliability evaluation. IEEE Trans Energy Convers. 2006;21(2):533–40.
6. Abouzahr I, Ramakumar R. An approach to assess the performance of utility-interactive photovoltaic systems. IEEE Trans Energy Convers. 1993;8(2):145–53.
7. Karaki SH, Chedid RB, Ramadan R. Probabilistic performance assessment of autonomous solar-wind energy conversion systems. IEEE Trans Energy Convers. 1999;14(3):766–72.
8. Wei C, Jing S, Li R, et al. Composite usage of muti-type energy storage technologies in microgrid. Autom Electr Power Syst. 2010;34(1):112–5.
9. Wang P, Billinton R. Time sequential distribution system reliability worth analysis considering time varying load and cost models. IEEE Trans Power Deliv. 1999;14 (3):1046–51.
10. Xu Y, Wu Y. Reliability evaluation for distribution system connected with wind-turbine generators. Electr Power Autom Equip. 2011;4:154–8.

11. Bahramirad S, Reder W, Khodaei A. Reliability-constrained optimal sizing of energy storage system in a microgrid. IEEE Trans Smart Grid. 2012;3(4):2056–62.
12. Savier JS, Das D. Impact of network reconfiguration on loss allocation of radial distribution systems. IEEE Trans Power Deliv. 2007;22(4):2473–80.
13. Wang Y, Wu Z. Reliability analysis of grid-connected microgrid using sequential simulation. Electr Power Autom Equip. 2015;13(4):101–8.
14. Yongji G. Reliability analysis of power system. Tsinghua University Press;2003.
15. Billinton R, Jonnavithula S. A test system for teaching overall power system reliability assessment. IEEE Trans Power Syst. 1996;11(4):1670–6.

Coordinated Control of Multi-satellites Formation Flying for Pulsar Observation

Qiang Chen, Xiaomin Bei, Hengbin Zhang and Jianjun Zhang

Abstract X-ray pulsar navigation is a critical and frontier technology for spacecraft's autonomous navigation in future. Now an effective database of pulsars angle position with high precision has not been developed. The precision of pulsar angle position measurement could be greatly improved by observing a pulsar jointly at the same time with several satellites. While observing the same pulsar the satellites have to set their attitude and orbit coordinately. There should be a constraint of tasks, spatial graphs, agreements and time in the observing process. In paper a plan of observing a pulsar at the same time by several satellites is put forward, a method of multi-satellites coordinated control is presented, and a digital experiment is carried out for simulating the proposed scheme. As results, the plan is feasible, the controller of the satellites orbit position and attitude is stable, the plan and control method is especially suitable for X-ray pulsar interferometry so that the accuracy of pulsar angle position can be improved by observation with multi-satellites formation.

Keywords Formation flying · Pulsar navigation · Coordinated control · Attitude and orbit control

1 Introduction

The pulsar is known as the most accurate natural astronomical clock for stable signal cycle so that it can provide time and space reference with high precision.

Q. Chen (✉)
School of Automation Science and Electrical Engineering, Beihang University (BUAA), Beijing 100191, China
e-mail: cq_01@sina.com

X. Bei · H. Zhang · J. Zhang
Qian Xuesen's Lab of Space Technology, Chinese Academy of Space Technology, Beijing 100094, China

Y. Jia et al. (eds.), *Proceedings of 2017 Chinese Intelligent Systems Conference*, Lecture Notes in Electrical Engineering 459,
https://doi.org/10.1007/978-981-10-6496-8_53

Space navigation using X-ray pulsar is now considered to be next-generation method which has attracted considerable attention of scientists all over the world. The most important requirement of pulsar navigation is to establish a database of the pulsar's position in sky. It is of great practical value to research the control method of multi-satellites formation flying for establishing the pulsars navigation database.

The use of multi-satellites formation flying to replace the traditional single spacecraft is now the hot spot in the development of space technology, such as the current applications: ESA's Darwin program (infrared space interferometer), NASA's TPF program (life exploration), France's CNES program (SAR satellite imaging), and NASA and ESA's LISA program (detecting gravitational waves), GRACE plan (gravitational gradient measurement). There are many people who research on the control method of formation flying of satellites [1, 2]. In 1985, Vassar proposed a formation flying control method for circular orbits based on decoupling of the C-W equation's in-plane motion and normal-direction motion of the orbital plane [3]. In 1995, C.R. Mcinnes applied the behavioral approach to the spacecraft formation control, based on the distributed ring structure of the spacecraft formation information flow map, and presented a coordinated algorithm to avoid the inter-satellites collision [4]. In 1998, Terui used a sliding mode control algorithm to design position and attitude control algorithm of a single satellite [5]. In 2005, Gaulocher put forward a linear fractional transformation model of the varied and coupled translational and rotational relative dynamics for formation flying control [6]. In 2007, Wei Ren addressed the method of formation keeping and attitude alignment for multiple spacecraft with local neighbor-to-neighbor information exchange [7]. In 2010, Y.H Wu designed a control algorithm of coupling orbit and attitude for satellites' formation flying [8]. Recently, Y Liu solved the formation control problem of discrete-time multi-agent systems with unknown nonlinear dynamics by means of the iterative learning approach [9]. L Zhao designed an adaptive multi-spacecraft controller by nonsingular fast terminal sliding mode considering inertia uncertainties and external disturbances with unknown bounds [10]. Y Huang applied Robust $H\infty$ control method to solve the problem of spacecraft formation flying with coupled translational and rotation dynamics [11]. For the need of pulsars observation application, the multi-satellites coordinated control method is studied in this paper, which focus on the design of finite time orbit and attitude coordinated controller from the theoretical method based on terminal sliding mode control to build high-precision pulsar angle position database for the construction of next generation navigation.

2 Scheme of Pulsar Observation

At present, there have been more than 2000 pulsars, according to the existing observation records, select 4 of them as candidates for the observation of the pulsars, as Table 1.

Table 1 The parameters of final selected pulsars for observation

No.	Name	Cycle (s)	p_f	W (s)	Fx (ph/cm^2/s)	RA (°)	Dec (°)
1	B0531+21	0.033085	0.7	0.00167	1.54E+00	83.63322	22.01446
2	B1937+21	0.001558	0.86	0.000021	4.99E−05	294.91066	21.58309
3	B1821−24	0.003054	0.98	0.000055	1.93E−04	276.13337	−24.86975
4	B0540−69	0.050499	0.67	0.0025	5.15E−03	85.0465	−69.33164

It is supposed that the detector effective area is 0.1 m^2, timing residual error is less than 0.1 μs, the pulsars radiation X-ray signal is weak, and the pulsars parameters in above list are introduced, according to the formula

$$\sigma_{\mathrm{TOA}} = \frac{W\sqrt{([R_b + R_s(1-p_f)](A_d W f_0) + R_s A_d p_f) t_{obs}}}{2R_s A_d p_f t_{obs}} \tag{1}$$

The minimum observation time that the profile of observed pulsars could be calculated as 400 s, 4000 s, 3700 s and 300000 s respectively. Under the condition that the same pulsar can be observed on the same time by the satellites in different orbit, the orbit sections of multi-satellites synchronous observation and the corresponding observation time will be designed, and using STK software the observation constellation parameters of multi-satellites formation flying can be figured out, including the number of satellites, orbital parameters of satellites, constellation configuration and attitude coordination time. In paper the simultaneous observation model with the most simplified master-slave binary satellites is used to research coordinated control method, as Fig. 1.

3 Dynamics and Kinematics Coordinated Control Model of Multi-satellites Attitude and Orbit

3.1 Dynamics and Kinematics Model of Coordinated Control of Multi-satellites Attitude

The dynamics equation of rigid body satellite attitude is generally expressed as:

$$J\dot{\omega} = -\omega \times J\omega + u + d \tag{2}$$

J is the inertia moment of the spacecraft in body coordinate.

While the attitude kinematics equation is described by the modified Rodrigue parameters, it is shown that:

$$\dot{\sigma} = G(\sigma) \cdot \omega \tag{3}$$

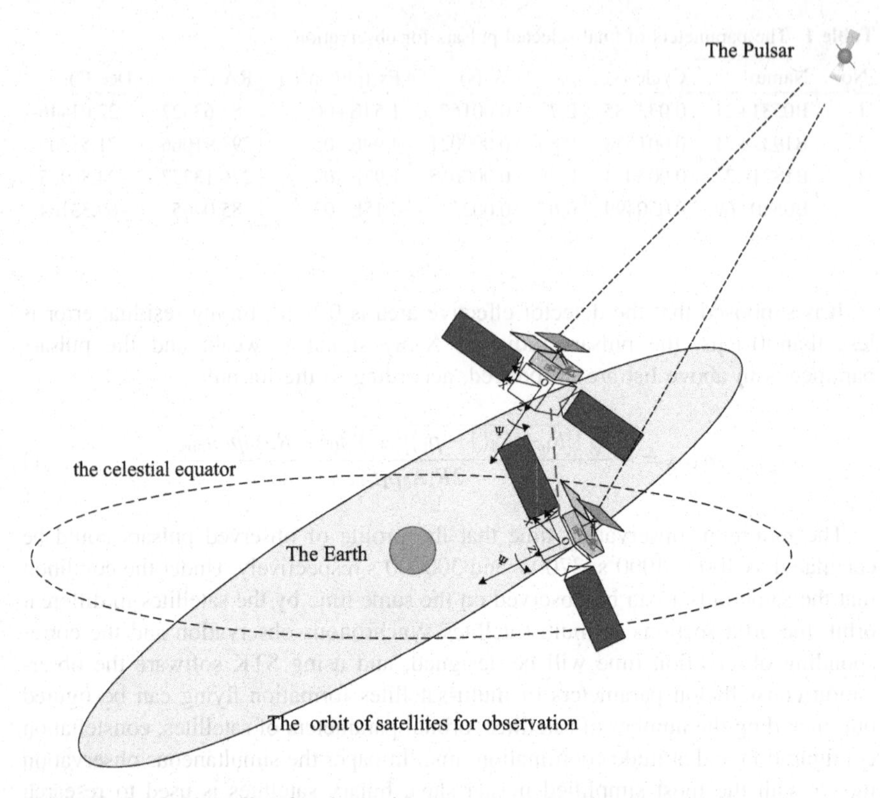

Fig. 1 The orbit of satellites for observation

$$\sigma = [\sigma_1 \quad \sigma_2 \quad \sigma_3] = \mathrm{e} \cdot \tan\left(\frac{\theta}{4}\right) \tag{4}$$

$$G(\sigma) = \frac{1}{2}\left(\frac{1-\sigma^T\sigma}{2} I + \sigma^{\times} + \sigma\sigma^T\right) \tag{5}$$

$$\sigma^{\times} = -(1/\sigma^2)\sigma \tag{6}$$

For the attitude control of formation flying satellites, relative attitude error is expressed as

$$\dot{\sigma}_{lf} = G(\sigma_{lf}) \cdot \omega_{lf} \tag{7}$$

By formula (3) $\omega = G^{-1}(\sigma) \cdot \sigma$ can be obtained.

And then it is replaced into (2)

$$H(\sigma)\ddot{\sigma} + Q(\sigma, \dot{\sigma})\dot{\sigma} = G^{-T}(\sigma)d_\tau + G^{-T}(\sigma)u_\tau \tag{8}$$

$$H(\sigma) = G^{-T}(\sigma)JG^{-1}(\sigma) \tag{9}$$

$$Q(\sigma,\dot{\sigma}) = -G^{-T}(\sigma)JG^{-1}(\sigma)\dot{G}(\sigma)G^{-1}(\sigma) + G^{-T}(\sigma)(JG^{-1}(\sigma)\dot{\sigma})^{\times}G^{-1}(\sigma) \tag{10}$$

3.2 Dynamics and Kinematics Model of Coordinated Control of Multi-satellites Orbit

The reference coordinate system is defined as the orbital coordinate system of the leader satellite. The orbit position of the follower satellite in the reference coordinate system can be expressed as $\rho = [x\ y\ z]'$.

Assuming that leader satellite is not subject to active control, the dynamic equation of the leader satellite and follower satellite in the geocentric equatorial inertial coordinate system is:

$$\ddot{\vec{r}}_l + \frac{\mu}{r_l^3}\vec{r}_l = \frac{f_l}{m_l} \tag{11}$$

$$\ddot{\vec{r}}_f + \frac{\mu}{r_f^3}\vec{r}_f = \frac{f_f}{m_f} + \frac{d_f}{m_f} \tag{12}$$

Take (11) and (12) into formula $\bar{\rho} = \vec{r}_l - \vec{r}_f$

$$\ddot{\bar{\rho}} = \frac{\mu}{r_l^3}\vec{r}_l - \frac{\mu}{r_f^3}\vec{r}_f - \frac{f_l}{m_l} + \frac{f_f}{m_f} + \frac{d_f}{m_f} \tag{13}$$

The part coordinates of $\bar{\rho}$ in the leader satellite orbit coordinate system are expressed as $\rho = [x \quad y \quad z]'$. The coordinates of $\vec{r}_l$ in the leader satellite orbit coordinate system are expressed as $\vec{r}_l = [r_l \quad 0 \quad 0]'$. Thus, the Eq. (13) is expanded in the leader satellite orbit coordinate system as that:

$$\ddot{\rho} + 2\omega^{\times}\dot{\rho} + \dot{\omega}^{\times}\rho + \omega^{\times}(\omega^{\times}\rho) = -\frac{\mu}{r_f^3}(\rho + r_l) - \frac{\mu}{r_l^3}r_l - \frac{f_l}{m_l} + \frac{f_f}{m_f} + \frac{d_f}{m_f} \tag{14}$$

ω is the angular velocity of the leader satellite orbit.

$$\omega = \frac{n(1 + e\cos(f))}{(1 - e^2)^{\frac{3}{2}}} \tag{15}$$

Among the formula (15)

$$n=\sqrt{\frac{\mu}{a^3}} \tag{16}$$

ax, ay, az are the acceleration component of all the driving force and control force distributed at three axes in the leader star orbit coordinate system, which is in addition to the gravity of the earth center.

The formula (16) can be written as Lagrange-Euler form:

$$M\ddot{\rho}+C(\omega)\dot{\rho}+D(\omega,\dot{\omega})\rho+n(\rho)=d_f+u_f \tag{17}$$

Among the formula (17):

$$M=m_f\times I_{3\times 3} \tag{18}$$

$$C(\omega)=2m_f\begin{bmatrix} 0 & \omega & 0 \\ -\omega & 0 & 0 \\ 0 & 0 & 0 \end{bmatrix}) \tag{19}$$

$$D(\dot{\omega},\omega)=2m_f(\frac{\mu}{r_f^3}I_{3\times 3}+\begin{bmatrix} \omega^2 & \dot{\omega} & 0 \\ -\dot{\omega} & \omega^2 & 0 \\ 0 & 0 & 0 \end{bmatrix}) \tag{20}$$

$$n(\rho)=\mu m_f\begin{bmatrix} -\frac{r_l}{r_f^3}+\frac{1}{r_l^2} & 0 & 0 \end{bmatrix} \tag{21}$$

3.3 Orbit and Attitude Coupling Dynamics Model of Formation Flying

The formula expressions of orbit and attitude dynamics are all Lagrange-Euler form so that they can be combined, as formula (25).

It can be noted that

$$x=\begin{bmatrix} \Delta\sigma \\ \Delta\rho \end{bmatrix}=[\Delta\sigma_1 \quad \Delta\sigma_2 \quad \Delta\sigma_3 \quad \Delta r_x \quad \Delta r_y \quad \Delta r_z]^T \tag{22}$$

$$u=\begin{bmatrix} \tau \\ f \end{bmatrix} \tag{23}$$

$$d=\begin{bmatrix} d_\tau \\ d_f \end{bmatrix} \tag{24}$$

According to formulas (7) and (16), the coupling dynamic model of orbit and attitude of satellites for the coordinated formation flying is:

$$M^{*}\ddot{x}+C^{*}(x,\dot{x})\dot{x}+g^{*}(x)x=d+u \tag{25}$$

$$M^{*}=\begin{bmatrix} H & 0 \\ 0 & m_f I_{3\times 3} \end{bmatrix} \tag{26}$$

$$C^{*}=\begin{bmatrix} Q(\sigma,\dot{\sigma}) & 0 \\ 0 & C(\omega) \end{bmatrix} \tag{27}$$

$$g^{*}=\begin{bmatrix} 0 \\ -D(\omega,\dot{\omega})\rho-n(\rho) \end{bmatrix} \tag{28}$$

4 Orbit and Attitude Coordinated Control Algorithm of Multi-satellite Formation Flying for Pulsar Observation

The orbit and attitude control system of multi-satellites formation flying satellites is a nonlinear time-varying system. The problems of how to control well under the conditions of uncertain model parameters and system disturbance in multi-satellites formation flying satellites can be solved with the application of sliding mode control method.

Note that:

$$\mathrm{e}=\begin{bmatrix} \Delta\sigma_{1d}-\Delta\sigma_{1} \\ \Delta\dot{\sigma}_{1d}-\Delta\dot{\sigma}_{1} \\ \Delta\sigma_{2d}-\Delta\sigma_{2} \\ \Delta\dot{\sigma}_{2d}-\Delta\dot{\sigma}_{2} \\ \Delta\sigma_{3d}-\Delta\sigma_{3} \\ \Delta\dot{\sigma}_{3d}-\Delta\dot{\sigma}_{3} \\ \Delta r_{xd}-\Delta r_{x} \\ \Delta\dot{r}_{xd}-\Delta\dot{r}_{x} \\ \Delta r_{yd}-\Delta r_{y} \\ \Delta\dot{r}_{yd}-\Delta\dot{r}_{y} \\ \Delta r_{zd}-\Delta r_{z} \\ \Delta\dot{r}_{zd}-\Delta\dot{r}_{z} \end{bmatrix}=\begin{bmatrix} e_1 \\ \dot{e}_1 \\ e_2 \\ \dot{e}_2 \\ e_3 \\ \dot{e}_3 \\ e_4 \\ \dot{e}_4 \\ e_5 \\ \dot{e}_5 \\ e_6 \\ \dot{e}_6 \end{bmatrix}, \tag{29}$$

so

$$\dot{e}=\begin{bmatrix}\Delta\dot{\sigma}_{1d}-\Delta\dot{\sigma}_1\\ \Delta\ddot{\sigma}_{1d}-\Delta\ddot{\sigma}_1\\ \Delta\dot{\sigma}_{2d}-\Delta\dot{\sigma}_2\\ \Delta\ddot{\sigma}_{2d}-\Delta\ddot{\sigma}_2\\ \Delta\dot{\sigma}_{3d}-\Delta\dot{\sigma}_3\\ \Delta\ddot{\sigma}_{3d}-\Delta\ddot{\sigma}_3\\ \Delta\dot{r}_{xd}-\Delta\dot{r}_x\\ \Delta\ddot{r}_{xd}-\Delta\ddot{r}_x\\ \Delta\dot{r}_{yd}-\Delta\dot{r}_y\\ \Delta\ddot{r}_{yd}-\Delta\ddot{r}_y\\ \Delta\dot{r}_{zd}-\Delta\dot{r}_z\\ \Delta\ddot{r}_{zd}-\Delta\ddot{r}_z\end{bmatrix} \tag{30}$$

$$s(x)=Ce=\begin{bmatrix}c_1e_1+\dot{e}_1 & c_2e_2+\dot{e}_2 & c_3e_3+\dot{e}_3\\ c_4e_4+\dot{e}_4 & c_5e_5+\dot{e}_5 & c_6e_6+\dot{e}_6\end{bmatrix}^T=\begin{bmatrix}s_1 & s_2 & s_3 & s_4 & s_5 & s_6\end{bmatrix}^T \tag{31}$$

$$\begin{aligned}\dot{s}(x)&=\begin{bmatrix}c_1\dot{e}_1+\ddot{e}_1 & c_2\dot{e}_2+\ddot{e}_2 & c_3\dot{e}_3+\ddot{e}_3 & c_4\dot{e}_4+\ddot{e}_4 & c_5\dot{e}_5+\ddot{e}_5 & c_6\dot{e}_6+\ddot{e}_6\end{bmatrix}^T\\ &=\begin{bmatrix}c_1e_1 & c_2\dot{e}_2 & c_3\dot{e}_3 & c_4\dot{e}_4 & c_5\dot{e}_5 & c_6\dot{e}_6\end{bmatrix}^T+\begin{bmatrix}\ddot{e}_1 & \ddot{e}_2 & \ddot{e}_3 & \ddot{e}_4 & \ddot{e}_5 & \ddot{e}_6\end{bmatrix}\\ &=\begin{bmatrix}c_1e_1 & c_2\dot{e}_2 & c_3\dot{e}_3 & c_4\dot{e}_4 & c_5\dot{e}_5 & c_6\dot{e}_6\end{bmatrix}^T\\ &\quad+\begin{bmatrix}\Delta\ddot{\sigma}_{1d}-\Delta\ddot{\sigma}_1 & \Delta\ddot{\sigma}_{2d}-\Delta\ddot{\sigma}_2 & \Delta\ddot{\sigma}_{3d}-\Delta\ddot{\sigma}_3 & \Delta\ddot{r}_{xd}-\Delta\ddot{r}_x & \Delta\ddot{r}_{yd}-\Delta\ddot{r}_y & \Delta\ddot{r}_{zd}-\Delta\ddot{r}_z\end{bmatrix}^T\\ &=\begin{bmatrix}c_1e_1 & c_2\dot{e}_2 & c_3\dot{e}_3 & c_4\dot{e}_4 & c_5\dot{e}_5 & c_6\dot{e}_6\end{bmatrix}^T\\ &\quad+\begin{bmatrix}\Delta\ddot{\sigma}_{1d} & \Delta\ddot{\sigma}_{2d} & \Delta\ddot{\sigma}_{3d} & \Delta\ddot{r}_{xd} & \Delta\ddot{r}_{yd} & \Delta\ddot{r}_{zd}\end{bmatrix}^T\\ &\quad-\begin{bmatrix}\Delta\ddot{\sigma}_1 & \Delta\ddot{\sigma}_2 & \Delta\ddot{\sigma}_3 & \Delta\ddot{r}_x & \Delta\ddot{r}_y & \Delta\ddot{r}_z\end{bmatrix}^T\\ &=\begin{bmatrix}c_1e_1 & c_2\dot{e}_2 & c_3\dot{e}_3 & c_4\dot{e}_4 & c_5\dot{e}_5 & c_6\dot{e}_6\end{bmatrix}^T\\ &\quad+\begin{bmatrix}\Delta\ddot{\sigma}_{1d} & \Delta\ddot{\sigma}_{2d} & \Delta\ddot{\sigma}_{3d} & \Delta\ddot{r}_{xd} & \Delta\ddot{r}_{yd} & \Delta\ddot{r}_{zd}\end{bmatrix}^T\\ &\quad-inv(M^*)*\left(u+d-C^*(x,\dot{x})\dot{x}-g^*(x)x\right)\end{aligned} \tag{32}$$

Let

$$\dot{s}(x) = -\varepsilon\,\mathrm{sgn}(s) - ks = \begin{bmatrix} -\varepsilon_1 \mathrm{sgn}(s_1) - ks_1 \\ -\varepsilon_2 \mathrm{sgn}(s_2) - ks_2 \\ -\varepsilon_3 \mathrm{sgn}(s_3) - ks_3 \\ -\varepsilon_4 \mathrm{sgn}(s_4) - ks_4 \\ -\varepsilon_5 \mathrm{sgn}(s_5) - ks_5 \\ -\varepsilon_6 \mathrm{sgn}(s_6) - ks_6 \end{bmatrix} \tag{33}$$

So

$$u = M^* \Big[[c_1 e_1 \quad c_2 \dot{e}_2 \quad c_3 \dot{e}_3 \quad c_4 \dot{e}_4 \quad c_5 \dot{e}_5 \quad c_6 \dot{e}_6]^T + [\Delta\ddot{\sigma}_{1d} \quad \Delta\ddot{\sigma}_{2d} \quad \Delta\ddot{\sigma}_{3d} \quad \Delta\ddot{r}_{xd} \quad \Delta\ddot{r}_{yd} \quad \Delta\ddot{r}_{zd}]^T$$

$$+ \varepsilon\,\mathrm{sgn}(s) + ks] \quad + C^*(x, \dot{x})\dot{x} + g^*(x)x - d \tag{34}$$

And $V = \frac{1}{2} s^T s$, from that we can get $\dot{V} = s^T \dot{s} < 0$, so the control is stable.

5 Mathematical Simulation Analysis

Simulation conditions: leader satellite is designed as the US RXTE satellite.

Semi-major axis a = 7371 km; eccentricity e = 0.0; inclination i = 28.5°; right ascension of ascending node Ω = 0°; argument of perigee ω = 150° true anomaly f = 0°;

The initial distance between master satellite and follower satellite is 2 km.

The initial orbital position offset is set to [100 10 20], and the speed offset is 0.

The initial attitude offset is set to [0.06585 0.06585 0.0931119], the attitude angular velocity offset is set to 0.

The total time for the satellites to perform a joint observation is 3000 s.

The inertia moment matrix of the follower satellite is set to [25577.32 0 0; 0 4446.18 0; 0 0 23848.43] × 10^{-2}. The mass of the follower satellite is 100 kg. The gravity gradient torque which affects attitude of the satellites is set to 5 × 10^{-5} [3cos(wt) + 2sin(wt) 5cos(wt) + 3 4cos(wt) − 3sin(wt) − 1]

Considering the main factors of the orbit perturbation, the orbital perturbation force matrix is set to: 10^{-2} × sin(wt) × [1.2 2.0 1.6]′.

The simulation results are shown in the following figure that follower satellite can track and maintain with master satellite. Follower satellite spent 50 s in an orbit correction of 100 m distance relative to master satellite. (Black for the target value, red for the tracking value, as Figs. 2, 3, 4, 5).

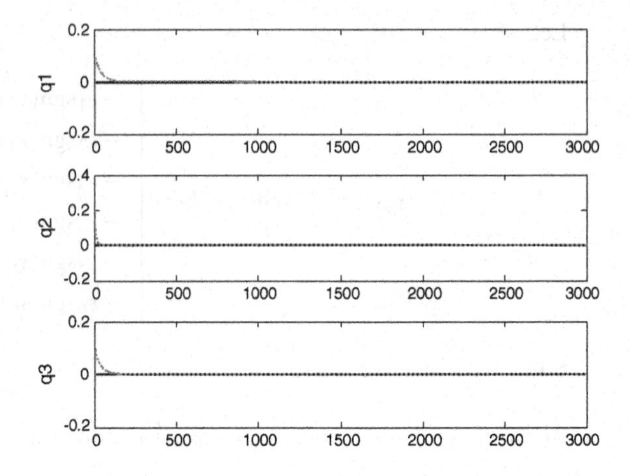

Fig. 2 Attitude angular

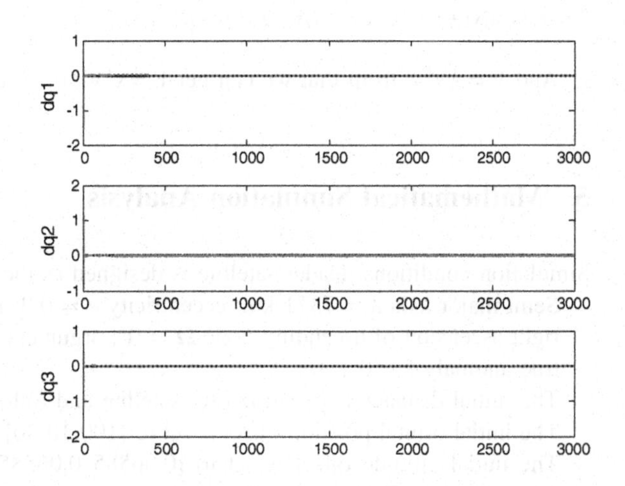

Fig. 3 Attitude angular velocity

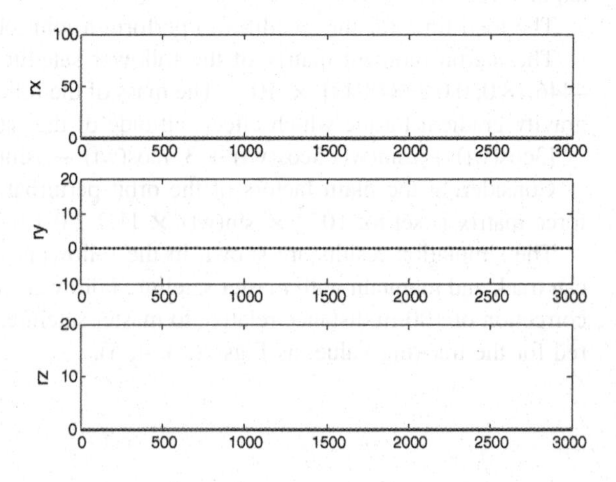

Fig. 4 Relative orbital position

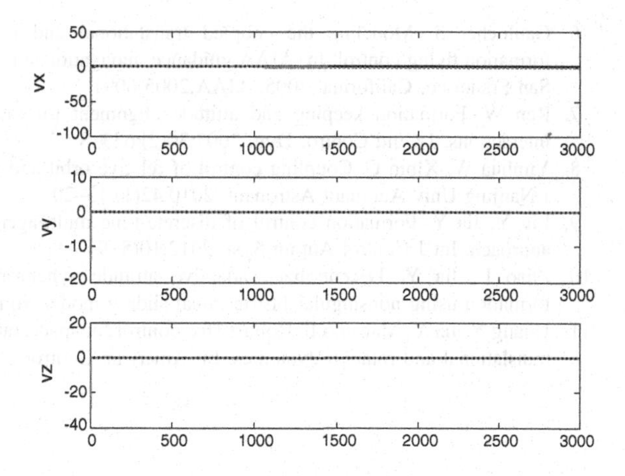

Fig. 5 Relative velocity

6 Conclusion

Pulsar navigation is an important direction of next-generation space navigation. In the paper, referred to multi-agents coordinated control technology, the dynamic model and control algorithm of satellite formation flying was studied for the application of constructing the database of pulsars angle position. The coupling relationship between orbit and attitude of the master and follower spacecrafts was considered to establish the coupling kinematic equation. The sliding mode variable structure control method was used in the coordinated control of spacecraft formation flying. The results of mathematical simulation show that the method can be used to maintain the synchronization of orbit and attitude of the spacecrafts for pulsar observation.

Acknowledgements Supported by the National Key Research and Development Program of China (Grant No. 2017YFB0503300, 2017YFB0503304).

References

1. Ren W, Beard RW. Distributed consensus in multi-vehicle coordinated control: theory and application. Publishing House of Electronics Industry;2014. p. 223–45.
2. Liu J. Simulation of sliding mode variable structure control. Beijing: The Press of Tsinghua University;2014. p. 495–539.
3. Vassar RH. Formation keeping for a pair of satellite in a circular orbit. J Guid Control Dyn. 1985;8(2):235–42.
4. Mcinnes CR et al. Autonomous ring formation for a planar constellation of satellites. J Guid Control Dyn. 1995;18(5):1215–7.
5. Terui F. Position and attitude control of a spacecraft by sliding mode control. Trans Jpn Soc Mech Eng C. 1998;64(621):1691–8.

6. Gaulocher S. Modeling the coupled translational and rotational relative dynamics for formation flying control. In: AIAA guidance, navigation, and control conference and exhibit. San Francisco, California, 2005. AIAA;2005-6091.
7. Ren W. Formation keeping and attitude alignment for multiple spacecraft through local interactions. J Guid Control Dyn. 2007;30(2):633–8.
8. Yunhua W, Xibin C. Coupling control of relative orbit and attitude for formation satellites. J Nanjing Univ Aeronaut Astronaut. 2010;42(1):13–20.
9. Liu Y, Jia Y. Formation control of discrete-time multi-agent systems by iterative learning approach. Int J Control Autom Syst. 2012;10(5):913–9.
10. Zhao L, Jia Y. Decentralized adaptive attitude synchronization control for spacecraft formation using nonsingular fast terminal sliding mode. Nonlinear Dyn. 2014;78:2779.
11. Huang Y, Jia Y, Matsuno F. Robust H∞ control for spacecraft formation flying with coupled translational and rotation dynamics. In: American Control Conference;2016. p. 4059–64.

An Adaptive Finite Time Control for the Electrical Drive with Elastic Coupling

Shubo Wang, Xuemei Ren and Siqi Li

Abstract In this paper, a novel adaptive nonsingular terminal sliding mode control (ANTSMC) based parameter estimation is developed for servo system with unknown system parameters. An auxiliary filter variable is employed to derive parameter estimation error information without measuring the acceleration. The estimation error is used as new leakage term in parameter update law. The key idea of the proposed parameter estimation scheme is that a sliding mode technique is introduced to ensure the finite time convergence of the estimation error in the presence of persistent excitation (PE). Moreover, an adaptive observer is designed to estimate the unmeasured state variables. Then, an adaptive nonsingular terminal sliding mode controller is designed for the servo system to achieve high performance tracking control. Finally, simulation results are used to illustrate the effectiveness of the proposed control method.

Keywords Adaptive observer · Finite-time parameter estimation · Servo system · Nonsingular terminal sliding mode control (NTSMC)

1 Introduction

In recently years, the permanent-magnet synchronous motors (PMSM) have been widely applied in industry applications, due to their compact structure, high power density, high torque to inertia ratio and efficiency [12]. However, the presence of potential nonsmooth dynamics such as friction [9], dead-zone and backlash [1] introduced by the transmission devices (e.g. gears, lead screws) may diminish the motion control precision.

S. Wang (✉)
College of Automation and Electrical Engineering, Qingdao University,
Qingdao 266071, China
e-mail: wangshubo1130@126.com

X. Ren · S. Li
School of Automation, Beijing Institute of Technology, Beijing 100081, China

Y. Jia et al. (eds.), *Proceedings of 2017 Chinese Intelligent Systems Conference*, Lecture Notes in Electrical Engineering 459,
https://doi.org/10.1007/978-981-10-6496-8_54

Various control methods have been proposed to reduce the effect of friction in servo mechanisms. For example, the conventional proportional-integral-derivative (PID) control method is adopted to compensate for the effect of friction from the control loop by using a feedback or feedforward configuration [6]. However, the PID control method may limit applicability due to the amplification of sensor noise and the occurrence of limit cycles. Thus, the nonlinear control techniques have been utilized to control the PMSM. Many researchers aimed to develop nonlinear control schemes to handle the nonlinearity exists in PMSM, and various nonlinear control methods have been presented, such as adaptive control (AC) [23], robust control (RC) [20], backstepping control [21] and dynamic surface control (DSC) [22], artificial intelligence [14] and so on. These control methods can improved the control performance of the servo system from different views.

However, there are some disturbances and parameter uncertainties which may reduce the robustness of the control system. To enhance the robustness of the control system, sliding mode control (SMC), which is a robust control method, is utilized to handle the bounded disturbances. Nevertheless, a drawback of the TSMC is that there exists singular problem. To overcome singular problem, a novel sliding technique, named non-singular TSMC (NTSMC) [8], is developed to achieve finite-time fast convergence and eliminate singularity problem in TSMC. Among these approaches, the controller design relies on the knowledge of the system parameter (e.g. system inertia, stiffness coefficients). However, the mechanical components are not unknown and the system parameters are difficult to calculate in practice. Thus, the parameter estimation methods are used to resolve this problem.

In the past decades, there were many parameter estimation/system identification methods are developed and widely applied in many fields. In particular, the gradient algorithms (GA) [18], least squares (LS) or recursive least-squares (RLS) algorithms [19], and projection methods [10] have been successfully in industry applications, where the adaptive law is updated by the tracking error or observer error, and regressor vector or matrix should be satisfied persistently excited (PE) [18]. However, when the noise or disturbance exists in the control system, the estimation values can not guarantee the robustness of the parameter estimation. To overcome this drawback, some modified parameter estimation methods, such as e-modification and σ-modification have been proposed [10], where the estimation values can not converge to actual values, only converges to a small set around their true values. It is noted that the update laws of aforementioned estimation schemes depend on observer error or tracking error [11]. In this case, the adaptive controller need a large adaptive gains to achieve fast adaptation, which will result in high-frequency oscillations when the noise or disturbance exists in control system. To improve the control and estimate performance, the adaptive laws should preferably include some information of the parameter estimation error [3]. Motivated by this fact, in our recently work [16], a novel parameter estimation method based on the filter operations is developed to remedy the aforementioned phenomena, which can achieve exponential and/or finite-time error convergence without using the derivative of the system states. The main feature of this method is that we can not calculate derivative of the system states and avoids online test for the invertibility of a regressor matrix and the computation of its inverse. This idea combined with model reference adaptive control to estimate the system parameters of a class of nonlinear systems [13].

In this paper, we will propose an adaptive non-singular terminal sliding mode controller (ANTSMC) based on parameter estimation for servo system. This control strategy is achieved based on a novel adaptive nonlinear observer design. A novel parameter estimation method is developed to guarantee the estimation values which can converge to actual true in finite time. An auxiliary filtered variable is firstly designed to derive the information of parameter estimation error which can be used as a new leakage term to drive the parameter update law. Hence, the exponential convergence of the control error and estimation error is achieved simultaneously. The proposed ideas are further improved via a sliding mode technique to achieve the finite-time convergence. An ANTSMC based on parameter estimation is proposed to control the servo system for robust, fast and high-precision tracking performance. Finally, the effectiveness of the proposed methods is validated by simulations.

2 Dynamic Model and Problem Formulation

A schematic of classical servo drive system is shown in Fig. 1, which is composed of a drive motor connected to a load machine through a low stiffness shaft and flexible. The actual control input of the servo drive system is input voltage. As compared to the motor inertia and load inertia, the value of the moment of inertia of elastic shaft is smaller than motor inertia and load inertia. This assumption involves the neglecting of the moment of inertia of the elastic shaft. The gearbox and nonlinear friction are take into in this paper. The dynamic mathematical model of such system can be described as

$$J_m \ddot{\theta}_m = T_m - B_m \dot{\theta}_m - T_s \tag{1}$$

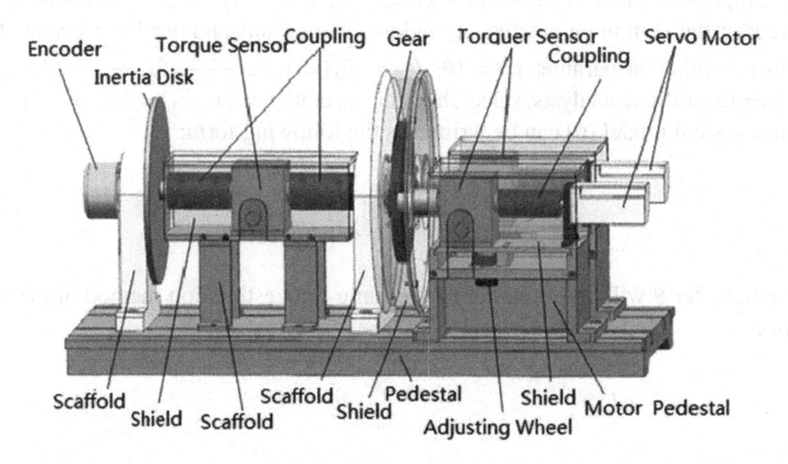

Fig. 1 Schematic of the proposed closed-loop control system

$$T_m = K_t \frac{u - K_e \dot{\theta}_m}{R} \tag{2}$$

$$J_l \ddot{\theta}_l = T_l - B_l \dot{\theta}_l - T_L - T_f \tag{3}$$

where θ_m and θ_l denote the motor angular speed and load angular velocity; J_m and J_l are the motor inertia and load inertia, respectively. T_m is the motor torque, T_f is the friction force, T_s and T_l are the input and output of the gearbox, B_m and B_l denote the damping constants of the motor and load, T_L is the disturbance torque of the load. u is the input voltage, K_t is the motor torque constant, K_e is the electromotive force constant, R is the stator resistance.

In this paper, the friction force is described by the following model

$$T_f = F_c \text{sgn}(\dot{\theta}_l) \tag{4}$$

where F_c is the Coulomb friction coefficient, and the backlash model can be written as

$$T_L = nT_s + d(T_s) \tag{5}$$

where n is the backlash rate, and $d(\cdot)$ is a nonlinear function and bounded, i.e. $|d| \leq \rho$, where ρ is a constant.

By analyzing, the control system (1)–(5) can be simplified as the following form

$$\begin{cases} \ddot{\theta}_l = \omega \\ J\dot{\omega} = -(B + \dfrac{n^2 K_t K_e}{R})\omega + \dfrac{nK_t}{R}u - F_c \text{sgn}(\omega) + \tau_d \end{cases} \tag{6}$$

where ω is the load speed. $J = J_l + n^2 J_m$, and $B = B_l + n^2 B_m$ are the lumped inertia and damping constant of the control system, $\tau_d = d - T_L$ is the total disturbance. Moreover, the system parameters are unknown constants and need to be estimated. Further defining the parameter $\theta = [\theta_1, \theta_2, \theta_3, \theta_4] = [(B+\frac{n^2 K_t K_e}{R})/J, \frac{nK_t}{JR}, F_c/J, \tau_d/J]$. In order to facilitate analysis, select the state variable $x = [x_1, x_2] = [\theta_l, \omega]$, thus, the control system model (6) can be written as the following form:

$$\begin{cases} \dot{x}_1 = x_2 \\ \dot{x}_2 = -\theta_1 x_2 + \theta_2 u - \theta_3 \text{sgn}(\omega) + \theta_4 \end{cases} \tag{7}$$

The parameter θ will be estimated by the finite-time estimation method in the next section.

3 Adaptive Observer Design

In practice, the system state vectors (e.g. position θ_m and ω_m) are not measured by sensors fixed on equipment due to the space limitation and high cost. In order to design controller for the servo system, it is necessary to employ the state observer to estimation state variables.

For convenience, we first consider the following nonlinear system:

$$\begin{cases} \dot{x} = Ax + \psi(y, u) + \varphi(x, \Theta, u) + \zeta \\ \varphi(x, \Theta, u) = B_1\varphi_1(x, u) + B_2\varphi(x, u)\Theta \\ y = Cx \end{cases} \tag{8}$$

where $\psi(y, u) = 0$, $B_1\varphi_1(x, u) = 0$, $A = \begin{bmatrix} 0 & 1 \\ 0 & 0 \end{bmatrix}$, $B_2 = \begin{bmatrix} 0 \\ 1 \end{bmatrix}$, $C = \begin{bmatrix} 1 & 0 \\ 0 & 1 \end{bmatrix}$, $\varphi(x, u) = [-x_2, u, -\text{sgn}(x_2)]$, $\Theta = [\theta_1, \theta_2, \theta_3]^T$.

According to [7], the adaptive state observer for the nonlinear servo mechanical control system (6) can be design as

$$\dot{\hat{x}} = A\hat{x} + \psi(y, u) + \varphi(\hat{x}, \hat{\Theta}, u) + (L - C\hat{x}) \tag{9}$$

where $\hat{x}$ is the estimation of state vector x, and $\hat{\Theta}$ is the estimated parameter vector. L is the observer matrix, which should satisfy $A_c = A - LC$. Then, the estimation error dynamics can be described by

$$\dot{\tilde{x}} = (A - LC)\tilde{x} + B_2\varphi\tilde{\Theta} + \zeta \tag{10}$$

where $\tilde{\Theta} = \Theta - \hat{\Theta}$ is the parameter estimation error.

We can obtain the following Lemma.

Lemma 1 *([7]) Since the parameter Θ in (8) is bounded, the nonlinear function $\varphi(x, \Theta, u)$ satisfies the Lipschitz condition such that $\|\varphi(x, \Theta, u) - \varphi(\hat{x}, \Theta, u)\| \leq \alpha\|x - \hat{x}\|$. There exists positive definite matrices, P and Q such that*

$$A_c^T P + PA_c = -Q \tag{11}$$

ensures that the error dynamics system (10) is asymptotical stability.

4 Parameter Estimation

Through the parameter adaptation, the traditional robust adaptive control can reduce the adverseness of the servo mechanisms. However, the adaptive laws may result in a long time to achieve the control performance. According to [2], if the parameter converge to true value in a finite time, the control performance of the servo mechanisms can be improved. Therefore, a novel robust adaptive finite-time parameter estimation and control approach is proposed in this paper.

4.1 Filter Design

To design adaptive law for system (8), we define the filtered variables x_{2f} and Ψ_f of the dynamics x_2 and φ as

$$\begin{cases} k\dot{x}_{2f} + x_{2f} = x_2, x_{2f}(0) = 0 \\ k\dot{\varphi}_f + \varphi_f = \varphi, \varphi_f(0) = 0 \end{cases} \tag{12}$$

where $k > 0$ is the filter time constant. Moreover, we introduce another auxiliary filter variable for the bounded disturbance (only used for analysis)

$$k\dot{\zeta}_f + \zeta_f = \zeta, \zeta_f(0) = 0 \tag{13}$$

From (12) to (13), one can obtain that

$$\dot{x}_{2f} = \frac{x_2 - x_{2f}}{k} = \varphi\Theta + \zeta_f \tag{14}$$

Finally, we define new filtered regressor matrix $P(t)$ and vector $Q(t)$ as follows

$$\begin{cases} \dot{P} = -lP + \varphi_f\varphi_f^T, P(0) = 0 \\ \dot{Q} = -lQ + \varphi_f\left[(x_2 - x_{2f}/k)\right], Q(0) = 0 \end{cases} \tag{15}$$

where $l > 0$ is a design parameter. The solution of (15) can be derived as

$$\begin{cases} P(t) = \int_0^t e^{-l(t-r)}\varphi_f^T(r)\varphi_f(r)dr \\ Q(t) = \int_0^t e^{-l(t-r)}\varphi_f^T(r)\left[(x_2 - x_{2f}/k)\right]dr \end{cases} \tag{16}$$

where

$$\begin{aligned} Q(t) &= P(t)\Theta + \int_0^t e^{-l(t-r)} l\varphi_f^T(r)\zeta_f dr \\ &= P(t)\Theta + \zeta_N \end{aligned} \tag{17}$$

where $\zeta_N = \int_0^t e^{-l(t-r)} l\varphi_f^T(r)\zeta_f dr$.

Define another auxiliary vector H can be obtained based on PQ in (16) as

$$H = P\hat{\Theta} - Q \tag{18}$$

where $\hat{\Theta}$ is the estimation of unknown parameter.

By substituting (15) into (16) or (17), we have

$$Q = P\Theta \tag{19}$$

Then, (18) can be rewritten as

$$H = P\hat{\Theta} - P\Theta = -P\tilde{\Theta} + \zeta_N \tag{20}$$

where $\tilde{\Theta} = \Theta - \hat{\Theta}$ is the parameter estimation error. ζ_N is bounded since (x_1, x_2) is assumed to be bounded, implying the φ and φ_f is bounded. Thus, $\|\zeta_N\| \leq \varepsilon_{\zeta_N}$ for a constant $\varepsilon_{\zeta_N} > 0$.

Then, one has the following Lemma:

Lemma 2 *([15, 17]) The matrix P in* (17) *is positive definite, there exists the minimum eigenvalue* $\lambda_{min}(P) > \sigma > 0$ *for a positive constant* $\sigma > 0$. *The detailed proofs of Lemma 2 is given in [15, 17].*

4.2 Finite-Time Parameter Estimation

This section improves the adaptive control and adaptive law in terms of sliding mode techniques [4] to achieve finite-time convergence of both the tracking error and estimation error. Then the adaptive law for updating $\hat{\Theta}$ is given by

$$\dot{\hat{\Theta}} = \Gamma\left(\varphi^T F(y - C\hat{x}) - \kappa\frac{P^T H}{\|H\|}\right) \tag{21}$$

where Γ is a diagonal matrix, F is a constant matrix, and κ is a design parameter.

Remark 1 In Eq. (20), a sliding mode term $P^T H/\|H\|$ is introduced, thus, finite-time error convergence to a small region can be proved. In particular, the convergence performance depends on the adaptive gain Γ, and ensure the error $\tilde{\Theta}$ is bounded, that is $\lim_{t\to\infty} \tilde{\Theta} = P_1^{-1}\zeta$.

5 Controller Design

5.1 Adaptive Controller Design

Consider the control system (8), we will utilize the NTSMC technique to design the adaptive controller, which can guarantee the reference can reach the origin in finite time. The control structure is shown in Fig. 2. Define tracking error

$$e = x_1 - x_d \tag{22}$$

To avoid the singularity problem in linear SMC and terminal sliding mode (TSM), a modified nonsingular terminal sliding mode developed in [5], which can be defined as

$$s = \dot{e} + \lambda_1 e + \lambda_2 \beta(e) \tag{23}$$

where

$$\beta(e) = \begin{cases} |e|^{\gamma}\mathrm{sgn(e)}\; \mathrm{s} = 0 \;\mathrm{or}\; \mathrm{s} \neq 0, |\mathrm{e}| > \mu \\ \beta_1 e + +\beta_2 |e|^2 \mathrm{sgn(e)}\; \mathrm{s} \neq 0, |\mathrm{e}| \leq \mu \end{cases} \tag{24}$$

where $\beta_1 = (2-\gamma)\mu^{\gamma-1}$, $\beta_2 = (\gamma-1)\mu^{\gamma-2}$, and $\mu > 0$. λ_1 and λ_2 are positive constants.

The derivative of s is

$$\dot{s} = \ddot{x}_1 - \dot{x}_r \tag{25}$$

where

$$\dot{x}_r = \begin{cases} \ddot{x}_d + \lambda_1 \dot{e} + \lambda_2 \gamma |e|^{\gamma-1} \dot{e}\; s = 0 \;\mathrm{or}\; \mathrm{s} \neq 0, |\mathrm{e}| > \mu \\ \ddot{x}_d + \lambda_1 \dot{e} + \lambda_2 \beta_1 \dot{e} + 2\lambda_2 \beta_2 |e| \dot{e}\; s \neq 0, |\mathrm{e}| \leq \mu \end{cases} \tag{26}$$

Finally, the adaptive nonsingular terminal sliding mode (ANTSMC) is designed as

$$u = \frac{1}{\hat{\theta}_2}\left[- k_1 s_1 - k_2 |s|^{\gamma} \mathrm{sgn(s)} + \hat{\theta}_1 x_2 + \hat{\theta}_3 + \sigma \mathrm{sgn(s)} \right] \tag{27}$$

where k_1, k_2 denote the design parameters, $-k_1 s_1 - k_2 |s|^{\gamma}\mathrm{sgn(s)}$ is a feedback control to guarantee the finite time convergence of the sliding mode s, $\sigma\mathrm{sgn(s)}$ is a robust term.

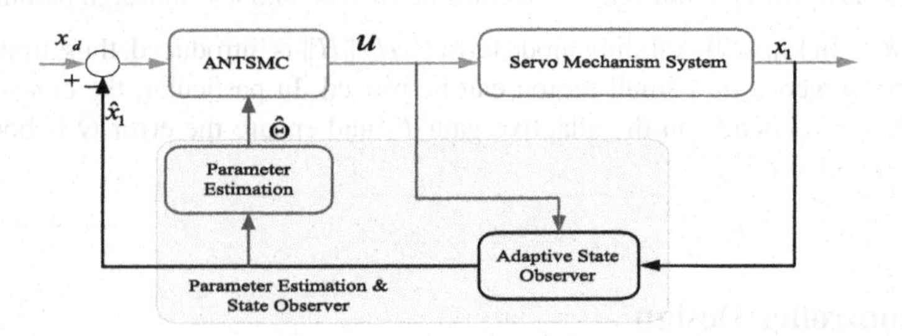

Fig. 2 Schematic of the proposed closed-loop control system

5.2 *Stability Analysis*

In order to certify stability of closed-loop servo control system with the proposed adaptive controller (27) and parameter update law (21), we have the following result.

Theorem 1 *Consider the control system (8) with adaptive state observer (9) with adaptive update law (21), and the adaptive controller (27), then,*

1. *all the signal of the closed-loop system are semiglobally uniformly ultimate bounded (SGUUB).*
2. *For $\zeta = 0$, the estimated parameter is finite time convergence.*
3. *The nonsingular terminal sliding mode surface s can converge to the equilibrium point within finite time.*

Proof (1) Selected a Lyapunov candidate as follows:

$$
\begin{aligned}
V(t) &= \frac{1}{2}\tilde{x}^T P\tilde{x} + \frac{1}{2}s^2 + \frac{1}{2}H^T P^{-1}\Gamma^{-1}P^{-1}H \\
&= \frac{1}{2}[\tilde{x}_1\tilde{x}_2]^T \begin{bmatrix} P_1 & 0 \\ 0 & P_2 \end{bmatrix} \begin{bmatrix} \tilde{x}_1 \\ \tilde{x}_2 \end{bmatrix} + \frac{1}{2}s^2 \\
&+ \frac{1}{2}H^T P^{-1}\Gamma^{-1}P^{-1}H \\
&= \frac{1}{2}\tilde{x}_1^T P_1\tilde{x}_1 + \frac{1}{2}\tilde{x}_2^T P_2\tilde{x}_2 + \frac{1}{2}s^2 + \frac{1}{2}H^T P^{-1}\Gamma^{-1}P^{-1}H \\
&= V_1 + V_2 + V_3 + V_4
\end{aligned}
\tag{28}
$$

where $V_1 = (1/2)\tilde{x}_1^T P_1\tilde{x}_1$, $\tilde{V} = V_2 + V_3 + V_4 = (1/2)\tilde{x}_2^T P_2\tilde{x}_2 + (1/2)s^2 + (1/2)H^T P^{-1}\Gamma^{-1}P^{-1}H$.

The derivative of $\tilde{V}$ is given as

$$
\begin{aligned}
\dot{\tilde{V}} &= \frac{1}{2}(\tilde{x}_2^T P_2\dot{\tilde{x}}_2 + \dot{\tilde{x}}_2^T P_2\tilde{x}_2) + s\dot{s} + \frac{d}{dt}\left[\frac{1}{2}H^T P^{-1}\Gamma^{-1}P^{-1}H\right] \\
&= -\frac{1}{2}\tilde{x}_2^T Q\dot{\tilde{x}}_2 + \tilde{x}_2^T P_2 B_2\varphi\tilde{\Theta} - k_1 s^2 - k_2|s|^{\gamma+1} + \varphi\tilde{\Theta}s \\
&+ H^T P^{-1}\Gamma^{-1}(\dot{\hat{\Theta}} + P^{-1}PP^{-1}\zeta + P^{-1}\dot{\zeta}) + \tilde{x}_2^T P_2\zeta
\end{aligned}
\tag{29}
$$

Define

$$
\zeta_B = P^{-1}PP^{-1}\zeta + P^{-1}\dot{\zeta}
\tag{30}
$$

Then,

$$
\begin{aligned}
\dot{\tilde{V}} &= -\frac{1}{2}\tilde{x}_2^T Q\dot{\tilde{x}}_2 + \tilde{x}_2^T P_2 B_2\varphi\tilde{\Theta} - k_1 s^2 - k_2|s|^{\gamma+1} + \varphi\tilde{\Theta}s \\
&+ H^T P^{-1}\Gamma^{-1}(\dot{\hat{\Theta}} + \zeta_B) + \tilde{x}_2^T P_2\zeta
\end{aligned}
\tag{31}
$$

Substituting parameter update law (21) into (31), one has

$$\begin{aligned}\dot{\hat{V}} &= -\frac{1}{2}\tilde{x}_2^T Q\dot{\tilde{x}}_2 + \tilde{x}_2^T P_2 B_2 \varphi\tilde{\Theta} - k_1 s^2 - k_2|s|^{\gamma+1} + \varphi\tilde{\Theta}s \\ &+ H^T P^{-1}\Gamma^{-1}\left[\Gamma\varphi^T(FC\tilde{x} + s) - k\frac{P_1^T H_1}{\|H_1\|} + \zeta_B\right] + \tilde{x}_2^T P_2\zeta \\ &= -\frac{1}{2}\tilde{x}_2^T Q\dot{\tilde{x}}_2 - k_1 s^2 - k_2|s|^{\gamma+1} - k\frac{H^T P^{-1} P_1^T H_1}{\|H_1\|} \\ &+ H^T P^{-1}\Gamma^{-1}\zeta_B + \tilde{x}_2^T P_2\zeta \\ &\le -\lambda_{\min}(Q)\|\tilde{x}_2\|^2 - \left(k - \|P^{-1}\Gamma^{-1}\zeta_B\|\right)\|H\| - k_1 s^2 \\ &- k_2|s|^{\gamma+1}\end{aligned} \tag{32}$$

By using the Young's inequality, we can obtain the following inequality

$$\tilde{x}_2^T P_2\zeta \le 2\|\tilde{x}_2\|^2 + \frac{1}{4}\|P\|^2\|\zeta\|^2 \tag{33}$$

Substituting (33) into (32), one has

$$\begin{aligned}\dot{\hat{V}} &\le -(\lambda_{\min}(Q) - 2)\|\tilde{x}_2\|^2 - \left(k - \|P^{-1}\Gamma^{-1}\zeta_B\|\right)\|H\| \\ &- k_1 s^2 - k_2|s|^{\gamma+1} + \frac{1}{4}\|P\|^2\|\zeta\|^2 \\ &\le -\mu\sqrt{V} + \ell\end{aligned} \tag{34}$$

where $\ell = \frac{1}{4}\|P\|^2\|\zeta\|^2$, $\mu_1 = \min\{(\lambda_{\min}(\mathrm{Q}) - 2), -\mathrm{k}_1, -\mathrm{k}_2, (\mathrm{k} - \|\mathrm{P}^{-1}\Gamma^{-1}\zeta_{\mathrm{B}}\|)\sigma \sqrt{2/\lambda_{\max}(\Gamma)}\}$ is a positive constant. Hence, $\lim_{t\to\infty} V = 0$ and $\lim_{t\to\infty} H = 0$ holds in finite time $t_a \le 2\sqrt{V(0)}/\mu_2$. This together with the fact $H = P\tilde{\Theta} + \zeta$ further guarantees that the estimation error converges to $\lim_{t\to\infty} P\tilde{\Theta} = \zeta$ in finite time.

From (34), one can see that the $(\tilde{x}_2, \tilde{\Theta})$ will converge to a set of ultimate boundedness.

(2) when $\zeta = 0$, (34) can be reduced as

$$\begin{aligned}\dot{\hat{V}} &\le -(\lambda_{\min}(Q) - 2)\|\tilde{x}_2\|^2 - \left(k - \|P^{-1}\Gamma^{-1}\zeta_B\|\right)\|H\| \\ &- k_1 s^2 - k_2|s|^{\gamma+1} + \frac{1}{4}\|P\|^2\|\zeta\|^2 \\ &\le -\mu\sqrt{V}\end{aligned} \tag{35}$$

Thus, observer error $\tilde{x}_2$ and parameter estimation error $\tilde{\Theta}$ will decay exponentially to zero. According to, it follows from (35) that $\lim_{t\to\infty} V = 0$ holds in finite time $t_a \le 2\sqrt{V(0)}/\mu_2$. Hence, the control error s and estimation error $\tilde{\Theta}$ all converge to zero in finite time, which implies $\tilde{\Theta} \to 0$ in finite time because $H = P\tilde{\Theta}$ for $\zeta = 0$.

6 Simulation

In this section, the extensive simulations are used to illustrate the effectiveness of the proposed control schemes. The servo system parameters (6) are selected as $J = 2 \times 10^{-4}$, $K_t = 0.185$, $K_e = 0.3$, $R = 1.5\,\Omega$, $n = 10$. The friction torque is described by $T_f = T_c \mathrm{sgn}(x_2)$, where friction coefficient is $T_c = 0.07$ which denotes the Coulomb friction. Hence, the unknown parameter to be estimated can be given as $\Theta = [\theta_1, \theta_2, \theta_3, \theta_4] = [18.9, 1, 0.35, 6.16]$.

The proposed adaptive NTSMC (27) and update law (21) are simulated with the parameters $k_1 = 25$, $k_2 = 15$, $\lambda_1 = 15$, $\lambda_2 = 10$, $\gamma = 17/12$. The update law parameters are selected as $l = 1, k = 0.1, \Gamma = 5.12 * \mathrm{diag}\,([1.1\ 0.03\ 0.035\ 0.105])$, the parameter condition is $\hat{\Theta} = [0\,0\,0.1\,0.05]$, $k_1 = 4$. The observer gain matrix in (10) are given as $A - LC = \begin{bmatrix} 2 & 0 \\ 0 & 0.5 \end{bmatrix}$. It is noted that if we set $\kappa = 0$, the proposed adaptive law (21) will become the classical gradient method. To illustrate the necessity for using the novel modification terms $\kappa P^{TH}/\|H\|$, the parameter estimation performance of (21) is compared with the gradient method in this simulation.

The simulation results are depicted in Figs. 3, 4 and 5. Figure 3 shows the tracking performance of proposed adaptive NTSMC (27) and observer performance of the designed adaptive state observer (9). From Fig. 3, one can see that the output position x_1 and speed signal can precisely track the reference signal x_d, $\dot{x}_d$, and the state estimation values $\hat{x}_1$ and $\hat{x}_2$ can achieve the desired trajectory x_d and $\dot{x}_d$. These results show the effectiveness of the our design adaptive state observer (9). The parameters estimation are shown in Figs. 4 and 5. Figure 4 depicts the parameter estimation values of our proposed adaptive finite time estimation method, the gradient method is given in Fig. 5. From these figures, it can be observed that tracking performance and parameter estimation are all retained with the proposed method. Specifically, Fig. 4 illustrates that the parameter estimates with the suggested adaptive law (21)

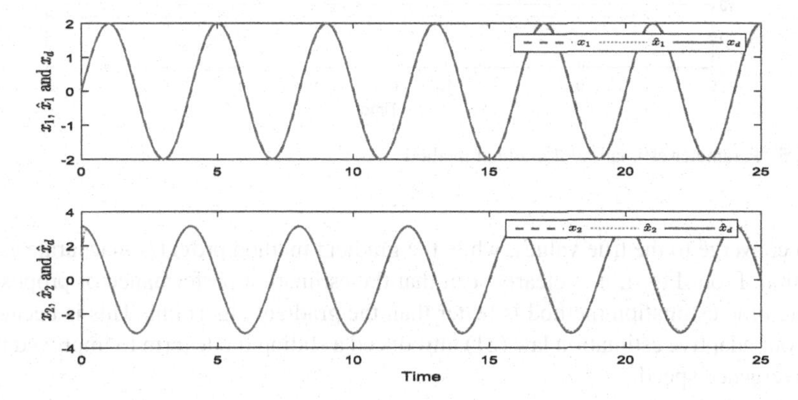

Fig. 3 Tracking performance and observer performance

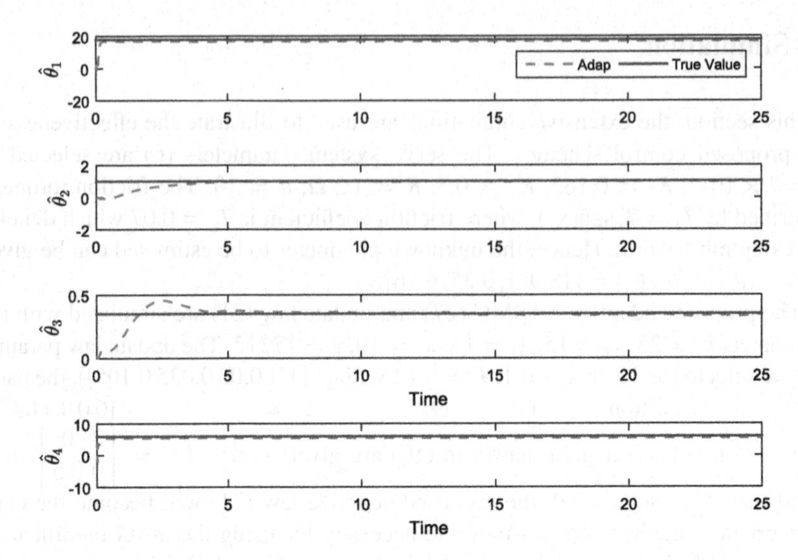

Fig. 4 Parameter estimation of proposed method

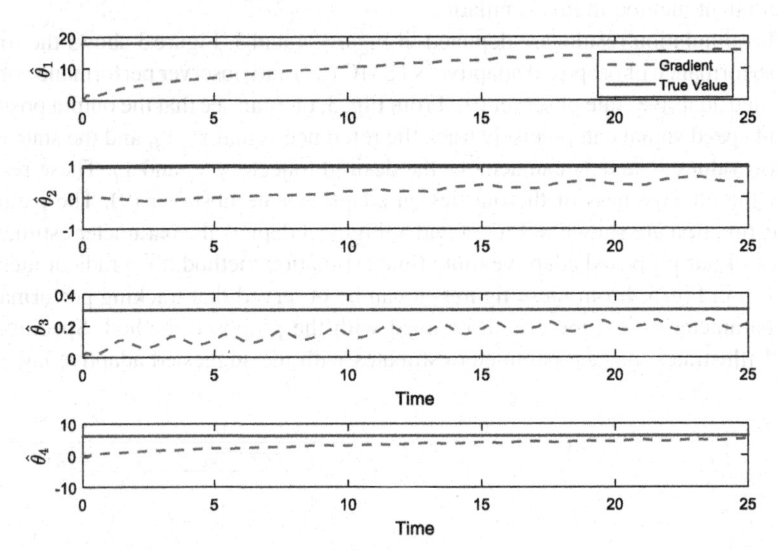

Fig. 5 Parameter estimation of gradient method

can converge to the true values, while the gradient method provides inaccurate estimation. From Fig. 4, it is clearly seen that the estimation performance of proposed finite time estimation method is better than the gradient algorithm. This is because the our adaptive estimation law (21) introduced a sliding mode term to improved the convergence speed.

All the aforementioned simulation results clearly show that the proposed ANTSMC can achieve better control performance. Compared with gradient method, the proposed parameter estimation method can converge true values in finite time. Moreover, our design adaptive state observer can precisely estimate state variables.

7 Conclusion

This paper proposed a novel ANTSMC method based on parameter estimation for the servo system with unknown system parameters. An auxiliary filter variable is exploited to derive parameter estimation error information without measuring the differential of state. Moreover, a sliding mode term is introduced in parameter update law to achieve finite time convergence. Simulation results are used to illustrate the effectiveness of the proposed method.

Acknowledgements This work is supported by National Natural Science Foundation of China (No. 61433003 and No. 61273150).

References

1. Armstrong-Hélouvry B, Dupont P, De Wit CC. A survey of models, analysis tools and compensation methods for the control of machines with friction. Automatica. 1994;30(7):1083–1138.
2. Adetola V, Guay M. Finite-time parameter estimation in adaptive control of nonlinear systems. IEEE Trans Autom Control. 2008;53(3):807–11.
3. Adetola V, Guay M. Performance improvement in adaptive control of linearly parameterized nonlinear systems. IEEE Trans Autom Control. 2010;55(9):2182–6.
4. Bhat SP, Bernstein DS. Continuous finite-time stabilization of the translational and rotational double integrators. IEEE Trans Autom Control. 1998;43(5):678–82.
5. Chen Q, Ren X, Na J, Zheng D. Adaptive robust finite–time neural control of uncertain PMSM servo system with nonlinear dead zone. Neural Comput Appl. 2016;1–12.
6. Dupont P. Avoiding stick-slip through PD control. IEEE Trans Autom Control. 1994;39(5):1094–7.
7. Ekramian M, Sheikholeslam F, Hosseinnia S, Yazdanpanah M. Adaptive state observer for lipschitz nonlinear systems. Syst Control Lett. 2013;62(4):319–23.
8. Feng Y, Yu X, Man Z. Non-singular terminal sliding mode control of rigid manipulators. Automatica. 2002;38(12):2159–67.
9. Hu C, Yao B, Wang Q. Adaptive robust precision motion control of systems with unknown input dead-zones: A case study with comparative experiments. IEEE Trans Ind Electron. 2011;58(6):2454–64.
10. Ioannou P.A, Sun J. Robust adaptive control. Courier Corporation. 2012.
11. Joensen A, Madsen H, Nielsen HA, Nielsen TS. Tracking time-varying parameters with local regression. Automatica. 2000;36(8):1199–204.
12. Na J, Chen Q, Ren X, Guo Y. Adaptive prescribed performance motion control of servo mechanisms with friction compensation. IEEE Trans Ind Electron. 2014;61(1):486–94.
13. Na J, Herrmann G, Ren X, Mahyuddin M.N., Barber, P. Robust adaptive finite-time parameter estimation and control of nonlinear systems. In: IEEE International Symposium on Intelligent Control. 2011. p. 1014–19

14. Na J, Ren X, Zheng D. Adaptive control for nonlinear pure-feedback systems with high-order sliding mode observer. IEEE Trans Neural Netw Learn Syst. 2013;24(3):370–82.
15. Na J, Mahyuddin, M.N, Herrmann G, Ren X, Barber P. Robust adaptive finite-time parameter estimation and control for robotic systems. Int J Robust Nonlinear Control. 2014.
16. Na J, Mahyuddin MN, Herrmann G, Ren X, Barber P. Robust adaptive finite-time parameter estimation and control for robotic systems. Int J Robust Nonlinear Control. 2015;25(16):3045–71.
17. Na J, Ren X, Xia Y. Adaptive parameter identification of linear siso systems with unknown time-delay. Syst Control Lett. 2014;66:43–50.
18. Narendra K.S., Annaswamy A.M. Stable adaptive systems. Courier Corporation. 2012.
19. Sastry S, Bodson M. Adaptive control: Stability, convergence and robustness. Courier Corporation. 2011.
20. Umeno T, Hori Y. Robust speed control of dc servomotors using modern two degrees-of-freedom controller design. IEEE Trans Ind Electron. 1991;38(5):363–8.
21. Wang S, Ren X, Na J, Zeng T. Extended-state-observer-based funnel control for nonlinear servomechanisms with prescribed tracking performance. IEEE Trans Autom Sci Eng. 2016;99:1–11.
22. Wang S, Ren X, Na J, Gao X. Robust tracking and vibration suppression for nonlinear two-inertia system via modified dynamic surface control with error constraint. Neurocomputing. 2016;203:73–85.
23. Xu L, Yao B. Adaptive robust precision motion control of linear motors with negligible electrical dynamics: theory and experiments. IEEE/ASME Trans Mechatron. 2001;6(4):444–52.

On Line Partial Discharge Localization in Cable Based on Time Varying Kurtosis and Time Window Energy Ratio

Kang Sun, Jingdie Guo, Ziqiang Li and Lei Wei

Abstract On the basis of using traveling wave method to carry out location of PD of high voltage cable, the time varying kurtosis method which used in the field of Seismic signal detection is introduced to solve the problem that the positioning accuracy is dependent on the accuracy of time delay estimation. The time window energy ratio method is introduced to solve the problem that the time-varying kurtosis which can not identify the PD pulse is difficult to obtain the first break time on line. Firstly, the time window of the partial discharge is detected by the energy ratio of the time window, and then the local time-varying kurtosis curve is obtained. Finally, based on the time varying kurtosis, the online picking up time is achieved. The simulation results show that this method not only has high precision and strong anti-interference ability, but also can realize that relative error in −4 dB noise environment is 0.23%.

Keywords Traveling wave method · Kurtosis · Time window energy ratio · Partial discharge

1 Introduction

In recent years, the high voltage cable which has the advantages of high reliability and simple wiring has been widely used. But the insulation of high voltage cable is becoming more and more serious [1]. Partial discharge (PD) is an effective method to detect the insulation condition of high voltage cable [2], and One of the key points is how to quickly locate the local sources. The traveling wave method is the most common method of location for high voltage cables, and its positioning

K. Sun · J. Guo (✉) · Z. Li · L. Wei
School of Electrical Engineering and Automation, Henan Polytechnic University, Jiaozuo 454000, China
e-mail: 709100780@qq.com

Z. Li · L. Wei
XJ ELECTRIC CO., LTD, Xuchang 461000, China

Y. Jia et al. (eds.), *Proceedings of 2017 Chinese Intelligent Systems Conference*, Lecture Notes in Electrical Engineering 459,
https://doi.org/10.1007/978-981-10-6496-8_55

accuracy is highly dependent on the time difference between PD pulses reaching both ends of the sensor [3]. Based on wavelet, Lv Qinghua put forward the signal arrival time analysis method to estimate the signal delay, but this method can only analyze the signal with sine function [4]. At present, t are all based on the generalized correlation method. All of these methods require that the noise and signal are independent of each other and the prior knowledge of noise and signal, which is not consistent with the actual situation [5].

The higher order statistics has become a common signal processing technique in recent years. It can describe the signal, improve the self correlation and provide more signal information. Kurtosis is a classical statistic which can be used to characterize the steepness of signal probability distribution in higher order statistics [6]. The time varying kurtosis curve can be used to pick up the arriving-time in seismic source location. In this paper, it is introduced to estimate the time delay of PD pulse. But the time varying Kurtosis Algorithm is only applicable to the determined time window, it is difficult to detect the PD events effectively for the on-line monitoring of the time series. In seismic wave detection, the energy characteristics of the time window can be used to realize the on-line identification of seismic wave and the initial pick-up of the wave arrival time [7]. Therefore, combining the advantages of two methods, time delay estimation based on kurtosis and time window energy ratio is proposed in this paper, and applied in fault location of cable insulation on line. The results of simulation showed that this method not only has high precision and strong anti-interference ability, but also can realize that relative error in −4 dB noise environment is 0.23%, and has high application value in Engineering.

2 The First Break Time Picking up Based on Kurtosis

Compared with the two order statistics, higher order statistics have more comprehensive information. So, higher order statistics is more widely used in the field of signal processing. If x is a random variable, its mathematical expectation is:

$$E[X] = \int_{-\infty}^{+\infty} xp(x)dx \tag{1}$$

$p(x)$ is the probability density of x. The k order statistics of X can be expressed as:

$$m_k = E[(X - E[X])^k], \quad k > 1 \tag{2}$$

In the actual data processing of cable insulation fault detection, the computation data is massive, and the data calculation of higher order statistics itself is also complex. If higher order statistics is used to analyze the detected data, the computational efficiency is very low. Therefore, in order to improve the efficiency of data processing, the special points or slices of higher order statistics are selected to reflect the characteristics of the data [8].

Kurtosis is an important parameter to measure the time series of non symmetric and non Gauss distribution, which reflects the degree of concentration of signal distribution, and its value indicates the steepness of probability distribution of signal [9].

According to formulas (1) (2), the expression of kurtosis can be described as:

$$K = \frac{m_4}{(m_2)^2} \quad (3)$$

The noise and the PD pulse in the time series which can be obtained by on-line detection have different characteristics of waveform. At the arrival time of PD signal, the signal asymmetry is the strongest, and the wave steepness is the most obvious.

The time-varying kurtosis curve can well characterize the asymmetry of signal. $[x_n]$ is set as a fixed length window, and then a small time window is designed. The core of the small window is i, and the length of the window is a.

The kurtosis $K(i)$ is expressed by formulas (2) and (3). K indicates the kurtosis of the window $[x_n]$. The time varying kurtosis of i can be expressed as follows:

$$K_t(i) = \frac{K(i) - K}{K} \quad (4)$$

$K(i)$ represents the kurtosis of the small time window in the formula.

By the above analysis, it can be seen that the maximum point of the time varying kurtosis curve is the first break point of the PD pulse.

Figure 1 is a low noise window that a single exponential function pulse is added at 630th sampling points.

Figure 2 is the time varying kurtosis variation curve of the time window. The maximum value of the curve is 631, which is consistent with the first arrival time of the function pulse.

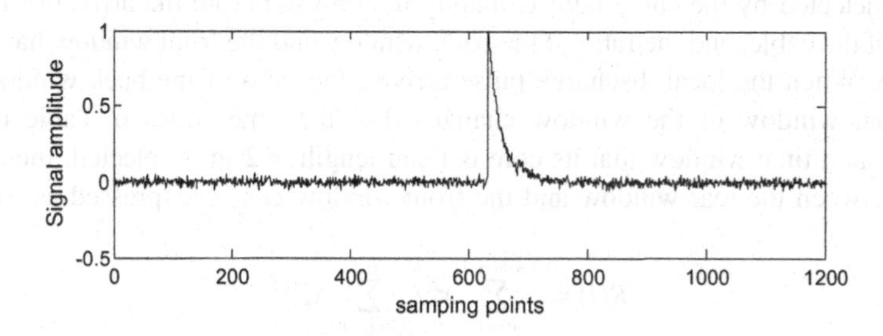

Fig. 1 Low noise time window with function pulse

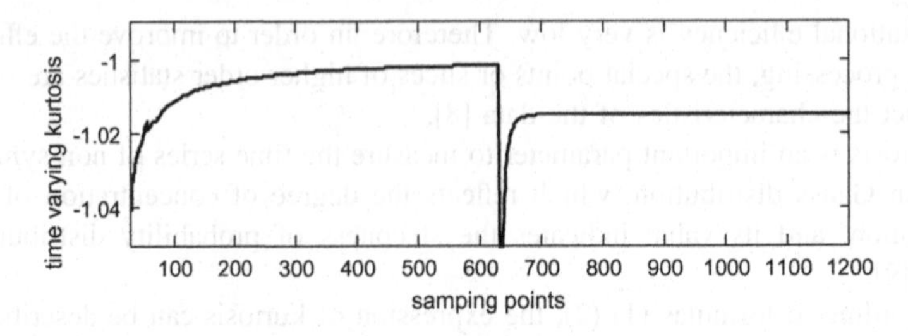

Fig. 2 The corresponding time varying kurtosis curve in Fig. 1

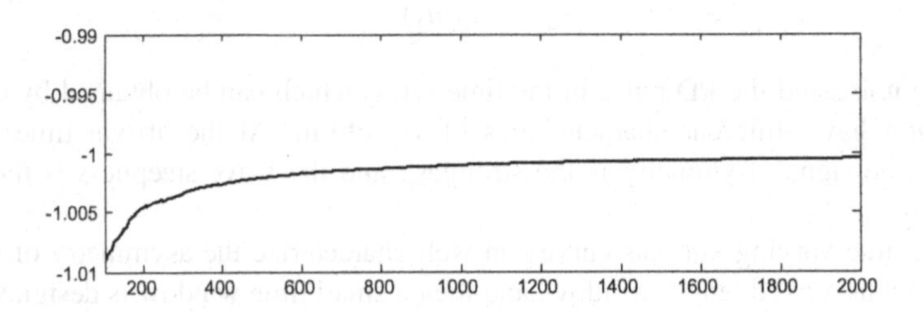

Fig. 3 Time-varying kurtosis curve of pure noise signal

3 Detection of PD Based on Time Window Energy Ratio

In the time series including the PD event, the algorithm of time kurtosis can achieve the high precision pickup at the first arrival time, but it can not realize the on-line identification of PD events. As is shown in Fig. 3, even in the pure noise time series without PD pulses, there is always a time-varying kurtosis maximum point, but this value is meaningless. Therefore, the primary task of online pickup is to identify the local release events and to determine the time windows when the release occurs.

Time window energy ratio method is often used in seismic wave identification. In this paper, it is introduced to the identification of PD events and the determination of the time window. For a time series, when a fixed length sliding window is selected, the maximum value of the energy ratio of the rear window and the front window in the window is a sign to detect the first arrival time of the pulse. The signal detected by the cable fault is mainly field noise before the arrival of the PD pulse of the cable, and the ratio of the back window and the front window have little change. When the local discharge pulse arrives, the ratio of the back window and the front window of the window changes. For the time series of cable on-line detection, a time window that its core is i and length is 2 m is selected, the energy ratio between the rear window and the front window can be expressed as follows:

$$R(i) = [\sum_{k=i}^{i+m-1} x_k^2 / \sum_{k=i-m}^{i-1} x_k^2]^{1/2} \tag{5}$$

In order to increase the stability of formula (5), a stability factor can be added. So the ratio of energy can be defined as follows:

$$R(i) = [(\sum_{k=i}^{i+m-1} x_k^2)^{1/2} + \lambda]/[(\sum_{k=i-m}^{i-1} x_k^2)^{1/2} + \lambda] \quad (6)$$

λ is the stability factor. The value of λ should be more less than the energy of the front window $(\sum_{k=i-m}^{i-1} x_k^2)^{1/2}$.

The significance of the time window energy ratio is to determine whether there is a PD event in the time window by judging whether the value is larger than the given threshold value. The maximum value of the energy ratio of the front window and the back window can also be used to achieve the initial pick-up time of the PD pulse. But its accuracy is not high, usually only used as a preliminary estimate of the time to wave [10].

4 On Line Localization of Cable Local Discharge Source Based on Two Step Method

The time window energy ratio method can realize the on-line detection of partial discharge events, and determine the outgoing window, and can determine the first arrival time by detecting the maximum value of R, but the picking precision of this method is not high. Kurtosis algorithm has higher picking accuracy, but it can not detect and recognize on-line events, and can not achieve online positioning. Therefore, a new two step method, which combines the energy ratio of the time window and the time-varying kurtosis, is proposed to obtain the high accuracy online pickup. According to the time window energy ratio to determine whether there is partial discharge events and determine the out window of the discharge, then Then, the time-varying kurtosis curve is obtained by using the time-varying Kurtosis Algorithm, and the initial arrival time is picked up by finding the maximum value. The result of time window energy ratio and the two step method are shown in Fig. 4. For the signal with PD pulse, the first break time is 632, and the extraction result of time window energy ratio is about 624. The result of the two step method is as follows: 631. It can be seen that the picking accuracy of the two step method is much higher than the time window energy ratio method.

The method which combines time window energy ratio with time varying kurtosis can pick up the first break time by high precision and accurately estimate the transmission delay of the PD pulse. At present, the method of cable fault location based on time delay estimation is mainly traveling wave method which includes single ended traveling wave method and double terminal traveling wave

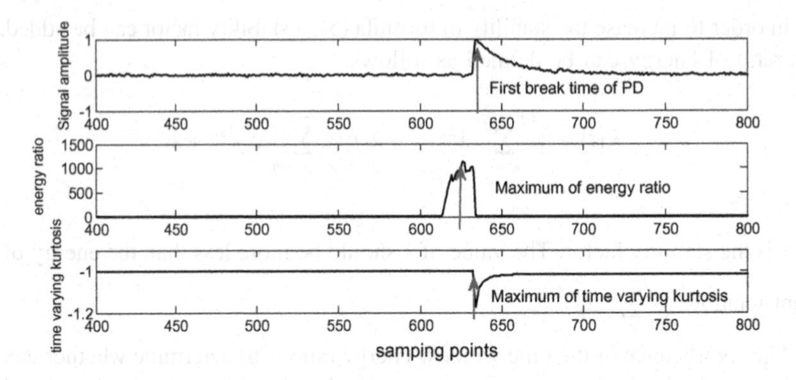

Fig. 4 Comparison of detection results of PD events

Fig. 5 Cable configuration

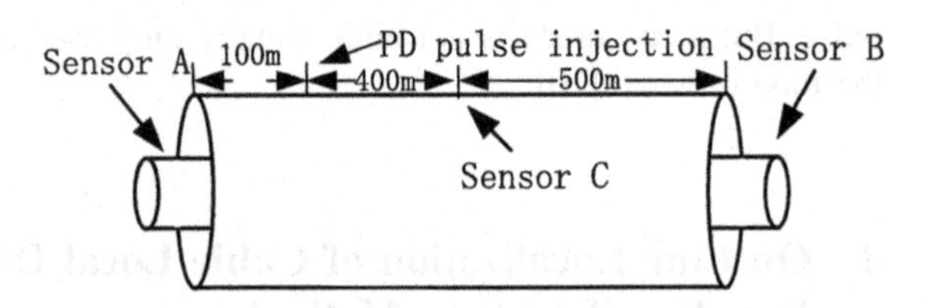

method. With installing the sensor on one end of the cable, the single terminal traveling wave which is based on the reflection principle of traveling wave mainly solves the fault location of short cable [11]. Two sensors are installed at the two ends of the cable, which can effectively solve the problem of the attenuation of the PD pulse. The formula (7) is location algorithm of double terminal traveling wave method.

$$X = 0.5[(t_1 - t_2)v + L] \quad (7)$$

t1 and t2 are the time that the PD pulse reaches the sensor at both ends, and V is the propagation velocity of the PD pulse in the cable. The formula (7) shows that the premise of the double terminal traveling wave method is the velocity is known, but the propagation velocity of PD pulse in the cable is uncertainty because of the type, aging conditions and operating environment of cable. Even for the same cable, the propagation velocity of the different frequencies of the PD pulse is also different. Therefore, the prior wave velocity will increase the positioning error. Based on this, this paper puts forward the method of traveling wave location with three sensors. The configuration of cable is shown in Fig. 5.

The time of PD pulse based on the energy ratio of the time window and the time-varying kurtosis to reach the A, B, C three sensors is t_A, t_B and t_C respectively.

If $|t_A - t_C| < |t_B - t_C|$, then the source is located in the AC segment, the spread of the signal is:

$$v = \frac{L}{2(t_B - t_C)} \tag{8}$$

If $|t_A - t_C| > |t_B - t_C|$, then the source is located in the BC segment, the spread of the signal is:

$$v = \frac{L}{2(t_C - t_A)} \tag{9}$$

If $|t_A - t_C| = |t_B - t_C|$, then the source is located outside the AB cable ends, he spread of the signal is:

$$v = \frac{L}{2|(t_B - t_C)|} = \frac{L}{2|(t_A - t_C)|} \tag{10}$$

When the calculated wave velocity is substituted (1), if the source is located in the AC segment, the distance to the A end is:

$$X = 0.5[(t_A - t_C)\frac{L}{2(t_B - t_C)} + L] \tag{11}$$

If the source is located in the BC segment, the distance to the B end is:

$$X = 0.5[(t_B - t_C)\frac{L}{2(t_C - t_A)} + L] \tag{12}$$

5 Simulation Experiment and Result Analysis

5.1 The Model and Method of Simulation

By using the Bergero Model cable model in PSCAD to simulate the cable in the PD system, the feasibility of the time window energy ratio and the time varying kurtosis method to estimate the time delay of the cable PD is verified. As shown in Fig. 6, a 1 km 10 kV cable is configured and simplified into 3 parts: conductor layer, shield layer and skin layer.

The waveform of PD pulse is very steep, which can be equivalent to the function model of Exponential:

$$f(t) = Ae^{-(t - t_0)/\tau} \tag{13}$$

In the formula, A is Signal amplitude, t_0 is Signal generation time and τ is attenuation coefficient. When A = 2 mV, τ = 1 μs. The discharge model is shown in Fig. 7.

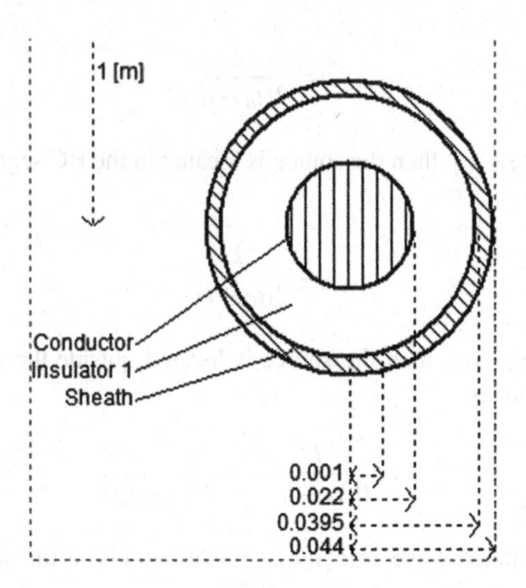

Fig. 6 The simulation model of cable

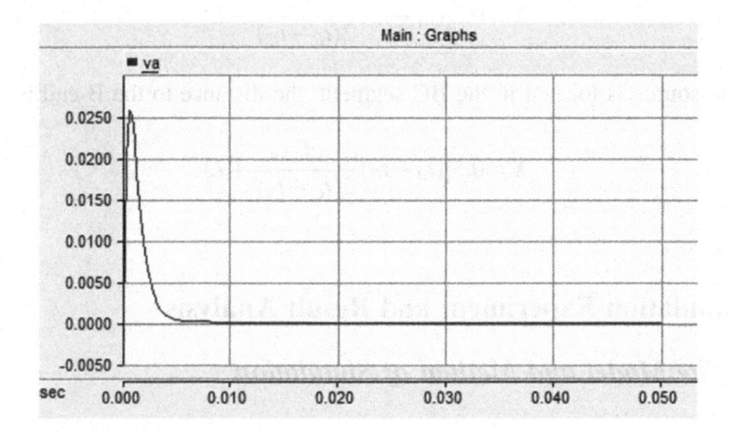

Fig. 7 The discharge curve of exponential function

The exponential function pulse is added at one end of the distance 100 m of the 1 km 10 kV cable in the PSCAD. The simulation model is shown in Fig. 8, and the measuring position is consistent with the Fig. 5.

5.2 *The Influence of the Length of Time Window on the Pickup Accuracy*

It is known that the length of the selected window is closely related to the precision of time delay estimation when the time window is used to estimate the time delay.

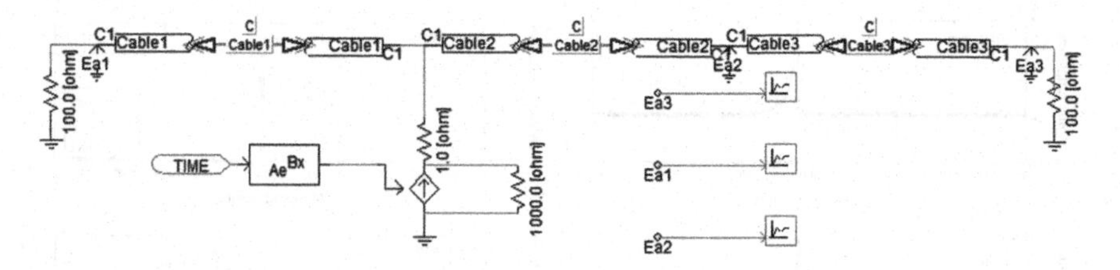

Fig. 8 Pscad simulation model of PD

Table 1 The effect of window length on the result of wave moment of PD

Length of time window	Pickup time by different methods		
	Energy ratio	Two-step method	Manual picking
50	726	730	731
100	704	730	731
150	697	730	731
200	679	730	731
300	678	730	731

In order to explore the relationship between the picking results based on time window energy ratio and time varying kurtosis time delay estimation and the time window length, the different time windows are used to pick up the first time of the same time series in this paper, and the results are shown in Table 1.

As can be seen from Table 1, the results of the two step extraction which is not affected by the length of time window in this paper are almost the same as manual.

5.3 *Experimental Results and Analysis*

Gauss white noise is the main interference in the on-line detection of the cable. In order to verify the practicability of the method mentioned in this paper, the Gauss white noise is added to the input pulse, and the signal to noise ratio is 20, 4, −4 dB. By using the method described in this paper, the time varying kurtosis curve is shown in Fig. 9.

The first arrival time of the PD pulse by three sensors is 425, 503 and 631 respectively. The first break time of the signal, the results of cable online positioning and its error in the case of different signal to noise ratio, which can be obtained by the method described in this paper, are shown in Table 2.

In Table 2, the relative error is defined as the percentage of the total length of the line, the formula can be described as:

$$\text{relative error} = \frac{|\text{Experimental value} - \text{True value}|}{\text{Line length}} \times 100\% \tag{14}$$

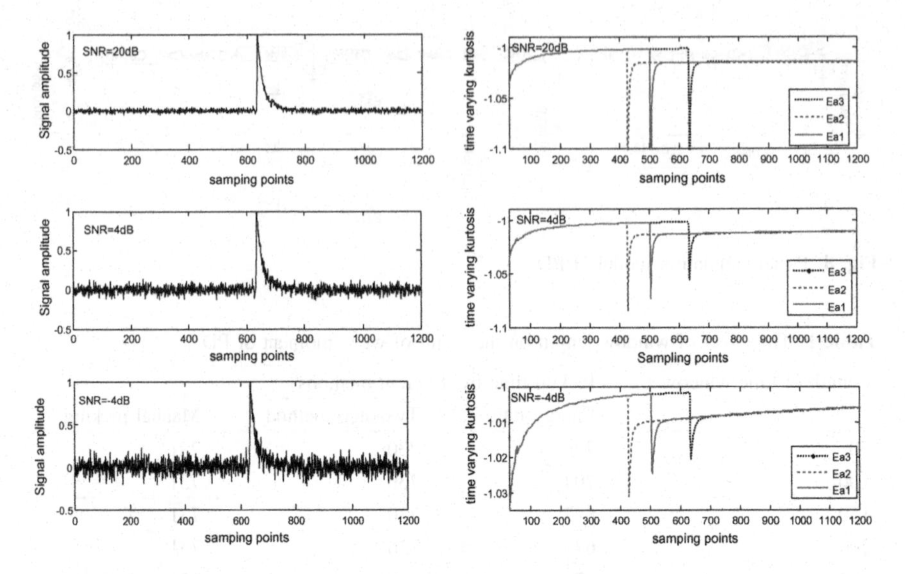

Fig. 9 The curves of different noises and corresponding characteristic

Table 2 The picking and location results of partial discharge in different noise environments

SNR (dB)	Two-step method				
	Ea1	Ea2	Ea3	Location (m)	Error/relative error
20	426	503	632	100.8	0.8 m/0.08%
4	426	503	632	100.8	0.8 m/0.08%
−4	426	504	632	97.7	2.3 m/0.23%

It can be seen from Table 2 that the time delay estimation method based on the energy ratio of time window and time-varying kurtosis is of high precision, and the positioning error is 2.3 m when the signal to noise ratio is −4 dB.

6 Conclusion

1. Combined the advantages of time window energy ratio and time varying index effectively, two step method proposed in this paper can identify the PD events on line and realize the high accuracy of the first break time of the signal, which can effectively improve the online positioning accuracy.
2. Based on three sensor, the traveling wave location method not only realizes the on-line detection of partial discharge of the long cable, but also solves the problem of the uncertainty of the wave speed, and improves the accuracy of the on-line positioning.

3. The two step method proposed in this paper is more adaptive to noise, and it has important theoretical significance and engineering application value.

References

1. Guo C, Li Z, Qian Y, et al. Current status of partial techniques discharge detection and location in XLPE power cable. High Volt Appar. 2009;45(3):56–60.
2. Bin W, Wei W, Chengrong L, et al. Partial discharges on-line detection of 110 kV XLPE cables using VHF clamp current transducer. High Volt Technol. 2004;30(7):37–9.
3. Tan J, Ge W, Qiu J. The contrast of single-terminal traveling wave fault location method and two-terminal traveling wave fault location method. Autom Electr Power Syst. 2006;30(6):92–5.
4. Lv Q, Tang H. Signal arrival time location method based on wavelet. J Wuhan University of Technology (Transportation Science and Engineering). 2006;30(5):873–6.
5. Du Y, Liao R, Zhou Q, et al. Study on time delay estimation algorithm in cable partial discharge location based on wavelet transform. High Volt Appar. 2007;43(5):389–93.
6. Hu Y, Yi C, Pan S, et al. Improved kurtosis time-varying gradient method of micro seismic signal recognition technology. Geophys Prospect Pet. 2012;51(6):625–32.
7. Zhang H, Zhu G, Wang Y, et al. Automatical picking up of seismic first arrival based on time window energy ratio and AIC two-step. Geophys Geochem Explor. 2013;37(2):269–73.
8. Liu J, Wang Y, Yao Z, et al. On micro-seismic first arrival identification: a case study. Chin J Gerophys. 2013;56(5):1660–6.
9. Jie G, Xiangxian C, Hai H. Application of kurtosis in on line detection of transformer iron core looseness. Chin J Sci Instrum. 2010;31(11):2401–7.
10. Qin H, Song W. Method of picking up microseismic first arrivals based on time—window energy ratio and mutual information. Geophys Geochem Explor. 2016;40(2):374–9.
11. Song J. Some key techniques of cable lengths measurement based on time domain reflectometry. Harbin Institute of Technology;2010.

Synchronous Control of Multi-motor Driving Servo Systems

Shuangyi Hu, Xuemei Ren and Wei Zhao

Abstract A nonlinear continuous predictive control method based on sliding mode is proposed to realize the synchronous control of multi-motor driving servo systems. The continuous-time recursive least-squares algorithm with forgetting factor is developed to estimate the disturbance and the unknown parameters, which compensates the influence of noise and guarantees that the parameter estimation converges to the true values. The continuous prediction control law is improved by using the sliding mode variable structure scheme, which ensures the rapid synchronization of the motors and deals with the problem of model uncertainty. The simulation results are presented to demonstrate the effectiveness of the method.

Keywords Multi-motor driving servo system · Least-squares algorithm · Nonlinear continuous predictive control · Sliding mode control

1 Introduction

The multi-motor driving servo systems play an irreplaceable role in driving the large power or large inertia system [1, 2]. The synchronous control is an important problem to be addressed. Therefore, the synchronous control strategy is needed and it is widely used in industrial and military fields.

The control strategy of servo system is mainly divided into classical control theory, modern control theory, intelligent control theory and compound control theory [3]. Although a variety of control strategies have obtained some results, there are still some problems. For example, the control accuracy is not high enough, dynamic tracking performance is poor, and anti-jamming performance is not good. Model predictive control is a form of control in which the current control action is obtained to solve a finite horizon open-loop optimal control problem. An important advantage of this type of control is its ability to cope with hard constraints on

S. Hu · X. Ren (✉) · W. Zhao
School of Automation, Beijing Institute of Technology, Beijing, China
e-mail: xmren@bit.edu.cn

Y. Jia et al. (eds.), *Proceedings of 2017 Chinese Intelligent Systems Conference*, Lecture Notes in Electrical Engineering 459,
https://doi.org/10.1007/978-981-10-6496-8_56

controls and states [4]. Sliding mode variable structure control was proposed in the early 1960s. It is a class of nonlinear control [5]. Since the design of the sliding surface is independent of the system parameters and external disturbances, the algorithm has strong robustness and fast response to parameter variation and perturbation. Therefore, the sliding mode algorithm is widely concerned and applied to the field of multi-motor synchronization.

For unknown parameters and model-based control algorithms, parameter estimation strategies are indispensable and have drawn much more attention. In order to obtain true parameter values, Ding improved the classical gradient algorithm, and proposed a hierarchical stochastic gradient algorithm [6] and a gradient-based iterative algorithm [7], which can effectively avoid redundant estimation and provide better estimation accuracy. Although the gradient estimation algorithm can achieve parameter estimation at lower computational cost, the parameter convergence rate is too slow. In order to improve the estimation rate of the parameters, Li et al. [8] designed a maximum likelihood least squares method for nonlinear systems with external noise, which can accurately estimate system parameters and external noise. Although the least squares method improves the accuracy and speed of parameter estimation, the larger computational complexity can not be ignored. In order to solve this problem, the literature [9] which combined with the data filtering technology, proposed an adaptive parameter estimation strategy based on the parameter error, which was applied to the adaptive observer to achieve the simultaneous estimation of the system state and parameters.

In this paper, to obtain the unknown parameters of the servo system, the recursive least-squares algorithm based on filters is designed. Then a nonlinear continuous predictive controller is provided with a variable structure controller. Furthermore, based on Lyapunov stability theory, the stability of the multi-motor driving servo system is discussed.

The rest of this paper is organized as follows. In Sect. 2, the dynamic model of the multi-motor driving servo system is given. The recursive least-squares algorithm with forgetting factor is designed in Sect. 3. And the nonlinear continuous predictive controller is developed in Sect. 4. The simulation results are then provided in Sect. 5. Finally, Sect. 6 draws the conclusion.

2 System Modeling

According to [10], the dynamic equations of the multi-motor driving servo system are given as follows:

$$J_{mi}\ddot{\theta}_{mi}(t) + b_i\dot{\theta}_{mi}(t) = u_i(t) - \tau_i(t), i = 1, 2, 3, 4, \tag{1}$$

where θ_{mi} is the position of motor i, $\dot{\theta}_{mi}$ represents the angular velocity of motor i, J_{mi} denotes the inertia of moment, b_i is the viscous friction coefficient, u_i denotes the control signal, τ_i is the transmission torque, which is represented by the following model [11]:

$$\tau_i = k_i\left[\psi_i - \alpha\left(\frac{2}{1+e^{-\nu\psi_i}} - 1\right)\right], \tag{2}$$

where $\psi_i = \theta_{mi} - \theta_l$ (θ_l is the position of load) and ν is a constant which satisfies $\nu = 8$, the unknown parameters k_i and α denote torsional coefficient and backlash width. Define the state $[x_{1i}, x_{2i}] = [\theta_{mi}, \dot{\theta}_{mi}]$. Then the system (1) is rewritten as follows:

$$\begin{cases} \dot{x}_{1i} = x_{2i} \\ \dot{x}_{2i} = \theta_i^T \varphi_i \end{cases} \tag{3}$$

where

$$\theta_i = \left[\frac{1}{J_{mi}}, \frac{k}{J_{mi}}, \frac{k\alpha}{J_{mi}}, \frac{b_i}{J_{mi}}\right]^T, k = k_i, i = 1,2,3,4,$$
$$\varphi_i = \left[u_i, -\psi_i, \frac{e^{-\nu\psi_i} - 1}{1+e^{-\nu\psi_i}}, -x_{2i}\right]^T, i = 1,2,3,4.$$

3 Parameter Estimation of System

Inspired by the literatures [12, 13], a new continuous-time recursive least-squares algorithm with forgetting factor is presented to estimate the unknown parameters of the systems. Considering the driven system (3), the following filters are proposed:

$$\begin{aligned} k_{fi}\dot{x}_{fi} + x_{fi} &= x_{2i}, x_{2i}(0) = 0 \\ k_{fi}\dot{\phi}_{fi} + \phi_{fi} &= \phi_i, \phi_i(0) = 0, \end{aligned} \tag{4}$$

where x_{fi} and ϕ_{fi} are treated as the filter value of x_{2i} and ϕ_i, k_{fi} is a positive constant.

Define the filter matrix $P_i \in \Re^{4\times4}, Q_i \in \Re^{1\times4}$ as follows:

$$\begin{aligned} \dot{P}_i &= -\rho P_i + \phi_{fi}\phi_{fi}^T, P_i(0) = 0 \\ \dot{Q}_i &= -\rho Q_i + \left(\tfrac{x_{2i} - x_{fi}}{k_{fi}}\right)\phi_{fi}^T, Q_i(0) = 0. \end{aligned} \tag{5}$$

where ρ is a constant which satisfies $\rho > 0$. It is easy to know that P_i is a symmetric matrix. Then the solution of matrix differential Eq. (5) is obtained as follows:

$$P_i(t) = \int_0^t e^{-\rho(t-\tau)} \phi_{fi}(\tau)\phi_{fi}^T(\tau)d\tau$$
$$Q_i(t) = \int_0^t e^{-\rho(t-\tau)} (\frac{x_{2i}(\tau) - x_{fi}(\tau)}{k_{fi}})\phi_{fi}^T(\tau)d\tau. \quad (6)$$

From (4), it is easy to know that

$$\dot{x}_{fi} = \frac{x_{2i} - x_{fi}}{k_{fi}} = \theta_i^T \phi_{fi}.$$

Then an equation is obtained as follows:

$$Q_i = \theta_i^T P_i. \quad (7)$$

Define the parameter estimation error $\theta_i = \theta_i - \theta_i$, where θ is the estimated value of θ_i. Thus the auxiliary variable R_i is given by

$$R_i = Q_i - \hat{\theta}_i P_i = \tilde{\theta}_i P_i. \quad (8)$$

Consider the following cost function

$$J(\theta_i) = \frac{1}{2}\int_0^t e^{-\beta(t-\tau)}\left[Q_i(\tau) - \hat{\theta}_i^T(t)P_i(\tau)\right]\left[Q_i(\tau) - \hat{\theta}_i^T(t)P_i(\tau)\right]^T d\tau + \frac{1}{2}e^{-\beta t}(\hat{\theta}_i - \hat{\theta}_{i0})^T M_0(\hat{\theta}_i - \hat{\theta}_{i0}), \quad (9)$$

where $M_0 = M_0^T > 0, \beta \geq 0, \hat{\theta}_{i0} = \hat{\theta}_i(0)$. Hence, any local minimum satisfies

$$\nabla J(\theta(t)) = 0, \forall t \geq 0$$

i.e.,

$$\nabla J(\theta_i) = e^{-\beta t} M_0(\hat{\theta}_i - \hat{\theta}_{i0}) - \int_0^t e^{-\beta(t-\tau)} P_i(\tau)\left[Q_i^T(\tau) - P_i^T(\tau)\hat{\theta}_i(t)\right]d\tau = 0. \quad (10)$$

From (10), the estimated value $\hat{\theta}_i$ is given by

$$\hat{\theta}_i(t) = K_i(t)\left[e^{-\beta t} M_0 \hat{\theta}_{i0} + \int_0^t e^{-\beta(t-\tau)} P_i(\tau) Q_i^T(\tau) d\tau\right], \quad (11)$$

where

$$K_i(t) = \left[e^{-\beta t} M_0 + \int_0^t e^{-\beta(t-\tau)} P_i(\tau) P_i^T(\tau) d\tau \right]^{-1},$$

which is a symmetric matrix.

Because $M_0 = M_0^T > 0$ and $P_i(\tau)P_i^T(\tau)$ is positive definite, $P_i(t)$ exists at each time t. Using the identity

$$\frac{d}{dt} K_i K_i^{-1} = \dot{K}_i K_i^{-1} + K_i \frac{d}{dt} K_i^{-1} = 0,$$

it is found that K_i satisfies the following differential equation

$$\dot{K}_i = -K_i \left(\frac{d}{dt} K_i^{-1}\right) K_i = \beta K_i - K_i P_i(t) P_i^T(t) K_i, K_i(0) = M_0^{-1}. \tag{12}$$

Similarly, differentiating $\theta_i(t)$ and using (12) and $R_i = Q_i - \hat{\theta}_i P_i$, the adaptive law is obtained as

$$\dot{\hat{\theta}}_i = K_i P_i R_i^T. \tag{13}$$

Theorem 1 *For system* (3) *with the adaptive law* (13), *if the regressor vector* ϕ_i *is persistent excitation, there's no approximation error, then the estimate error* $\tilde{\theta}_i$ *converges to zero exponentially.*

Proof Consider the Lyapunov function as

$$V = \frac{1}{2} \tilde{\theta}_i^T K_i^{-1} \tilde{\theta}_i,. \tag{14}$$

then its derivative of V can be calculated as

$$\begin{aligned} \dot{V} &= -\tilde{\theta}_i^T P_i R_i^T + \frac{\tilde{\theta}_i^T P_i P_i^T \tilde{\theta}_i}{2} - \frac{\tilde{\theta}_i^T \beta K_i^T \tilde{\theta}_i}{2} \\ &= \tilde{\theta}_i^T P_i P_i^T \tilde{\theta}_i + \frac{\tilde{\theta}_i^T P_i P_i^T \tilde{\theta}_i}{2} - \frac{\tilde{\theta}_i^T \beta K_i^T \tilde{\theta}_i}{2} \\ &= -\frac{\tilde{\theta}_i^T P_i P_i^T \tilde{\theta}_i}{2} - \frac{\tilde{\theta}_i^T \beta K_i^T \tilde{\theta}_i}{2} \\ &< -\frac{\tilde{\theta}_i^T P_i P_i^T \tilde{\theta}_i}{2} < -\frac{\left\| P_i^T \tilde{\theta}_i \right\|^2}{2}. \end{aligned} \tag{15}$$

According to (15), it is concluded that the estimation error $\tilde{\theta}_i$ converges to zero exponentially.

4 Design of Synchronous Controller

In order to achieve the synchronization control of the multi-motor system, from (1), the nonlinear equations are established as follows

$$\ddot{\theta}_{mi} = h_i(\theta_{mi}, t) + \gamma_i u_i(t), i = 1, 2, 3, 4, \tag{16}$$

where

$$h_i(\theta_{mi}, t) = -\frac{f_i(\dot{\theta}_{mi})}{J_{mi}} - \frac{\tau_i}{J_{mi}}, \gamma_i = \frac{1}{J_{mi}}$$

If the system (16) is the nominal system, define the following function

$$s_i = \dot{e}_i + k_i e_i, \tag{17}$$

where $e_i = \theta_{mi} - \theta_0, k_i > 0$ and θ_0 is the desired location. Thus, the following equation is obtained

$$\dot{s}_i = \ddot{e}_i - k_i \dot{e}_i = \ddot{\theta}_{mi} - \ddot{\theta}_0 - k_i \dot{e}_i. \tag{18}$$

Define $\delta > 0$, then the predictive value of $s(t + \delta)$ is given by

$$s_i(t + \delta) \approx s_i(t) + \delta \dot{s}_i(t). \tag{19}$$

Then consider the following cost function

$$J(\theta_{mi}, u_i, t) = \frac{1}{2} Q s_i^2(t + \delta), \tag{20}$$

where Q is treated as a matrix which satisfies $Q > 0$. Then solve the following predictive control problem

$$\min_{u_i(t)} J(\theta_{mi}, u_i, t).$$

From $\partial J / \partial u_i = 0$, the single-step continuous prediction control law is given by

$$u_{inpc} = -\delta^{-1} \gamma_i^{-1} \left[s + \delta(h_i + k_i \dot{e}_i - \ddot{\theta}_0) \right], i = 1, 2, 3, 4. \tag{21}$$

However, the parameters of the system are uncertain. It is difficult to ensure the robust stability of the system when there is model uncertainty. Therefore, it is necessary to apply a variable structure controller which guarantees the synchronization of motors. Assume that the uncertainty of the system parameters is described as

$$h_i = h_i^{'} + \Delta h_i, \gamma_i = \gamma_i^{'} + \Delta \gamma_i,$$

where $h_i^{'}, \gamma_i^{'}$ are nominal parameters, and $\Delta h_i, \Delta \gamma_i$ denote the uncertainty of the system parameters. An equation is given by

$$\dot{s}_i = h_i^{'} + \Delta h_i + (\gamma_i^{'} + \Delta \gamma_i)(u_{inpc} + u_{ivsc}) - \ddot{\theta}_0 - k_i \dot{e}_i, \tag{22}$$

where u_{ivsc} is variable structure control law.

So the nonlinear continuous predictive control method based on sliding mode is described as follows

$$u_i = u_{inpc} + u_{ivsc} = -\delta^{-1} \gamma_i^{-1} [s_i + \delta(h_i + k_i \dot{e}_i - \dot{\theta}_0)] - (\gamma_i^{'})^{-1} \varphi_i sign(s_i) \tag{23}$$

Theorem 2 *The nonlinear continuous predictive control method based on sliding mode can achieve the synchronization of system* (16), *and the tracking error satisfies* $\lim_{t \to \infty} e_i(t) = 0$.

Proof Consider the Lyapunov function as $V = \frac{1}{2} s_i^2$. Substituting Eq. (23) into Eq. (22), the following equation is obtained

$$\dot{s}_i = -\delta^{-1}(1 + \Delta \gamma_i \gamma_i^{-1}) s_i - (1 + \Delta \gamma_i \gamma_i^{-1}) \varphi_i sign(s_i) + \eta_i,$$

where $\eta_i = \Delta h_i - \Delta \gamma_i \gamma_i^{-1}(h_i + k_i \dot{e}_i - \ddot{\theta}_0)$.

From Theorem 1, it is found that $\dot{\tilde{\theta}}_i < 0$. Then $|\tilde{\theta}_i| < |\tilde{\theta}_i(0)| = |\theta_i - \hat{\theta}_i(0)| = |\theta_i|$. So, $|\Delta \gamma_i| < |\gamma_i|$. The following inequality is obtained

$$|\Delta \gamma_i \gamma_i^{-1}| < 1.$$

If the following inequality is obtained

$$\varphi_i > (1 + \gamma_i \gamma_i^{-1})^{-1} \eta_i \cdot sign(s_i),$$

$$\begin{aligned} \dot{V} = s_i \dot{s}_i &= -\delta^{-1}(1 + \Delta \gamma_i \gamma_i^{-1}) s_i^2 - (1 + \Delta \gamma_i \gamma_i^{-1}) \varphi_i |s_i| + \eta_i s_i \\ &\leq -\delta^{-1}(1 + \Delta \gamma_i \gamma_i^{-1}) s_i^2 - \eta \cdot s_i + \eta_i s_i \\ &= -\delta^{-1}(1 + \Delta \gamma_i \gamma_i^{-1}) s_i^2 < 0 \end{aligned}$$

From (18), $\lim_{t \to \infty} e_i(t) = 0$ is obtained if selecting appropriate parameter k_i. This completes the proof.

5 Simulation Results

In this section, the control algorithm proposed in this paper is simulated and verified. Determine the system initial state $\theta_{m1}(0) = \theta_{m2}(0) = 0.2, \theta_{m3}(0) = \theta_{m4}(0) = -0.2$, $\dot{\theta}_{mi}(0) = 2.51, i = 1, 2, 3, 4$. The parameters of system (1) are taken as $J_m = 0.028, b = 1.2, k = 56, \alpha = 0.2, \delta = 0.1, R = 0, Q = 100$ (Figs. 1, 2, 3 and 4).

The results shown in figures above indicate that this predictive control algorithm achieves the rapid position and speed synchronization of the motors with a small steady-state synchronization error.

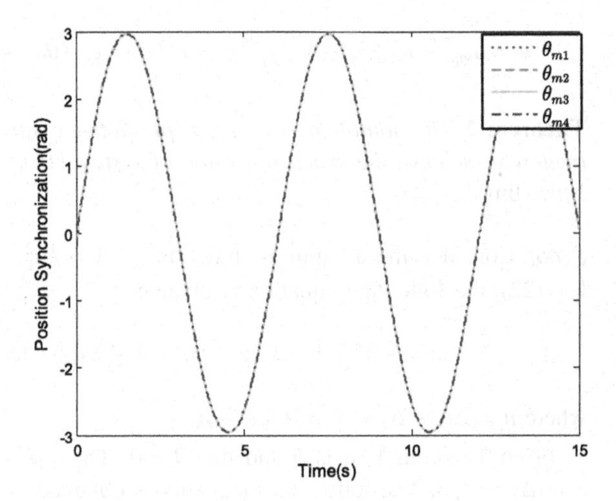

Fig. 1 Position synchronization

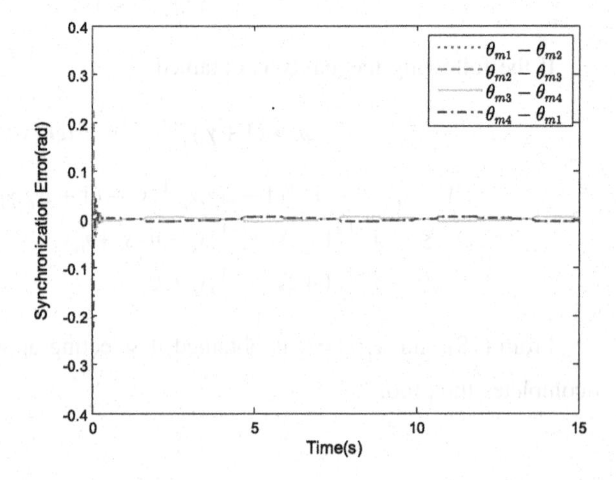

Fig. 2 Position synchronization error

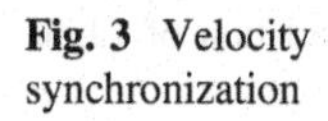

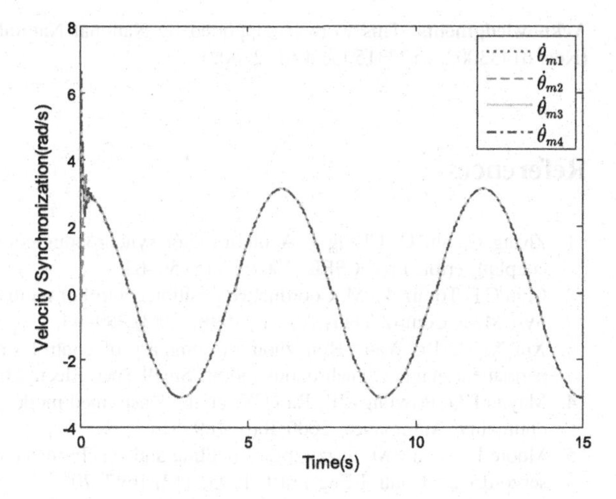

Fig. 3 Velocity synchronization

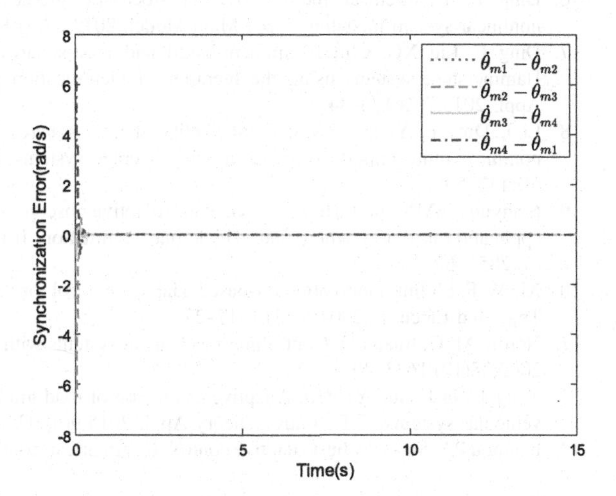

Fig. 4 Velocity synchronization error

6 Conclusion

In this paper, a predictive control method based on sliding mode is proposed to achieve the synchronous control of multi-motor driving servo systems with uncertainty. The estimation algorithm combined with filter and auxiliary variable is proposed to eliminate the effect the external disturbance effectively and guarantees that the estimation precision is improved. The continuous robust predictive controller is proposed based on the sliding mode variable structure scheme, which ensures the rapid synchronization of motors. The simulation results verify the effectiveness of the algorithm.

Acknowledgments This work is supported by National Natural Science Foundation of China (Nos. 61433003, 61273150 and 61321002).

References

1. Zhang C, Shi Q, Cheng J. A multi-motor synchronous control strategy based on adjacent coupling error. Proc CSEE. 2007;27(15):59–63.
2. Chiu GT, Tomizuka M. Coordinated position control of multi-axis mechanical systems. J Dyn Syst Meas Control-Trans ASME. 1998;120(3):389–93.
3. Xin X, He Li, Wang Hongzhou. A summary of control strategy for AC servo system of permanent magnet synchronous motor. Small Spec Electr Mach. 2010;2:67–70.
4. Mayne DQ, Rawlings JB, Rao CV, et al. Constrained model predictive control: stability and optimality. Automatica. 2000;36(6):789–814.
5. Moore P, Chen CM. Fuzzy logic coupling and synchronized control of multiple independent servo-drives. Control Eng Pract. 1995;3(12):1697–708.
6. Ding F. Hierarchical multi-innovation stochastic gradient algorithm for Hammerstein nonlinear system modeling. Appl Math Model. 2013;37(4):1694–704.
7. Ding F, Liu XG, Chu J. Gradient-based and least-squares-based iterative algorithms for Hammerstein systems using the hierarchical identification principle. IET Control Theory Appl. 2013;7(2):176–84.
8. Li J, Ding F, Yang G. Maximum likelihood least squares identification method for input nonlinear finite impulse response moving average systems. Math Comput Model. 2012;55 (3):442–50.
9. Mahyuddin MN, Na J, Herrmann G, et al. Adaptive observer-based parameter estimation with application to road gradient and vehicle mass estimation. IEEE Trans Ind Electron. 2014;61 (6):2851–63.
10. Xie W-F. Sliding-mode-observer-based adaptive control for servo actuator with friction. IEEE Trans Ind Electron. 2007;54(3):1517–27.
11. Nordin M, Gutman PO. Controlling mechanical systems with backlashła survey. Automatica. 2002;38(10):1633–49.
12. Yang J, Na J, Guo Y, et al. Adaptive estimation of road gradient and vehicle parameters for vehicular systems. IET Control Theory Appl. 2015;9(6):935–43.
13. Ioannou PA, Sun J. Robust adaptive control. In: American control conference; 1996. p. 1574–8.

Solar Power Based Wireless Charging System Design

Chenxi Zhang, Zetao Li, Yingzhao Zhang and Zhongbin Zhao

Abstract This paper designs a solar charging system which can convert solar energy into electrical energy and wirelessly charge devices such as mobile phones. First, we research the related documents to get the information of the features of solar energy wireless charging system; then we select components which are suitable for this system and use PROTEL software to draw the schematic diagram and PCB diagram. Then we process and weld the PCB to obtain the hardware circuit of solar wireless charging system. At last, we test and process the system data to obtain the electrical circuit parameters.

Keywords Solar energy · Wireless charging · PROTEL · Test1 introduction

1 Introduction

1.1 Significance of Solar Energy

Currently, fossil fuels account for a large proportion in the total use of global energy resource. However, as the fossil energy is non-renewable energy, there exists many problems.

Energy shortage: Energy supply is insufficient in most of the world due to the limitation of conventional energy sources and the heterogeneity of its distribution,

C. Zhang (✉) · Z. Li · Y. Zhang · Z. Zhao
The Electrical Engineering College, Guizhou University, Guiyang 550025, Guizhou, China
e-mail: 87376464@qq.com

Z. Li
e-mail: gzgylzt@163.com

Y. Zhang
e-mail: 847302325@qq.com

Z. Zhao
e-mail: 1090663695@qq.com

Y. Jia et al. (eds.), *Proceedings of 2017 Chinese Intelligent Systems Conference*, Lecture Notes in Electrical Engineering 459,
https://doi.org/10.1007/978-981-10-6496-8_57

which lead to dissatisfaction of the requirements of the economic development. In the long term, the world's proven oil reserves can only be used till 2020 while natural gas can only be used till 2040, even the coal resources with abundant reserves can only be maintained for two or three hundred years. Thus, if human-beings do not try to find the fossil fuels alternatives as soon as possible, human being will face the crisis of fossil fuel depletion sooner or later.

Environmental pollution: Due to the burning of coal, oil and other fossil fuels, hundreds of thousands of tons Sulphur dioxide and other harmful substances discharged to the atmosphere, causing the serious pollution to the atmospheric environment and affecting the health and quality of life of residents directly as well as causing acid rain in some areas which brings severe pollution to soil and water. These problems will eventually force people to adjust the energy structure and take advantage of solar energy as well as other renewable clean energy.

Greenhouse effect: The use of fossil fuels not only causes environmental pollution, but also produces a greenhouse effect due to the emission of greenhouse gases which finally will cause global climate change. The issue of greenhouse effect has been listed in the global agenda because its impact has even exceeded the environmental pollution. Relevant international organizations have held several meetings trying to find a way to set limit on emissions of carbon dioxide and other greenhouse gases worldwide [1].

Solar energy is the most important energy in all kinds of basic renewable energy which can produce biomass energy, wind energy, ocean energy, water power and so on directly or indirectly. Broadly speaking, solar energy contains all kinds of renewable energy listed above. As a kind of renewable energy, solar energy refers to the direct conversion and utilization of solar energy. The technology that convert solar radiant energy by conversion device into heat energy is called Solar thermal utilization technology; technology that use heat energy to generate electricity is called solar thermal power generation These two technology are belonging to the same technical field. The technology that convert solar radiant energy into electrical energy by conversion device is called Solar photovoltaic technology. The photoelectric conversion device usually uses the photovoltaic effect principle of the semiconductor device to carry out the photoelectric conversion. Therefore, it is called Solar photovoltaic technology [1].

In 1950s, there were two major technical breakthroughs in solar energy utilization: the first one is the practical monocrystalline silicon battery developed by Bell Laboratory in 1954; the second one is the theory of selective absorbing surface and the selective solar absorbing coating developed by an Israelite named Tabor in 1955. These two technical breakthroughs laid the technical foundation for promoting the use of solar energy into the modern development period [2].

Since 1970s, many countries set off an upsurge in the development and utilization of solar energy and renewable energy in view of the limitation of conventional energy supply and the growth of environmental protection pressure. In 1973, the United States draft a government level solar power generation program, and then officially list the photovoltaic power generation into public power planning in 1980. The cumulative investment for these two project is more than 800 million US dollars. In 1992, the U.S. government issued a new photovoltaic power generation plan and set a grand development goals. Japan formulated the "sunshine plan" in 1970s, and integrated the "Moonlight plan" (energy conservation plan), "Environmental planning", "sunshine plan" into the new "sunshine plan" in 1993. Germany and other European countries as well as some developing countries have also developed a corresponding development plan. Since 1990s, the United Nations held a series of summit meetings attended by national leaders of various countries, discussing and formulating the world solar energy strategic planning as well as international solar energy convention, establishing the international solar energy fund, promoting the development and utilization of solar energy and renewable energy. It has become an international consensus that utilize renewable energy plays important role in a country's sustainable development strategy [2].

1.2 Meaning of Wireless Charging System

Wireless charger refers to a charger which do not need the traditional charging power supply line when recharging a terminal equipment. It uses the latest wireless charging technology and won 20 patents in 2007. Mobile phones, MP3 players, power tools and other power adapters do not need charging wire any more by using one charging base station. Inductive coupling technology transmits electrical power by using a magnetic field between coils, which play a role as a bridge between charging base stations and equipment. Most of the current chargers are all charging the battery inside the equipment through the metal wire. Wireless charging technology has the advantage of convenience and versatility while the disadvantage is low efficiency and can only provide electricity. While the Apple's Dock connector not only provides power, but can also synchronize audio and video files into devices through the USB socket. However, wireless charging technology does bring progress to Wi-Fi and battery technology. For devices that do not require data transmission, this new technology will greatly reduce the number of chargers required. Additionally, public mobile device charging stations may finally become a reality with wireless charging technology.

2 Design of Solar Wireless Charger General Circuit

2.1 General Design Requirements of the Circuit

The purpose of this design is to produce a solar wireless charger. Therefore, it is necessary to carry out the research and design of solar regulator and wireless charging circuit. After the research and design, we need to design and assemble the circuit board based on the designed circuit in order to get a set of circuit board with complete function. For this purpose, the project shall first carry out the design of overall circuit structure.

2.2 Overall Design of Circuit Structure

The solar wireless charging circuit is mainly composed of the solar panels, wireless transmitting circuits, wireless receiving circuits, charging socket circuits, 5 V step-down circuits, and singlechip circuits, etc. Among them, the singlechip circuit obtains the voltage of the solar panel and the buck regulator circuit through the multiplex voltage acquisition chip. If the voltages of the solar panel and the buck regulator circuit are both functional, then the liquid crystal display (LCD) will show that users can charge. If the voltage is abnormal, then the LCD will show that users should not to charge the phone and should carry out the circuit troubleshooting. This circuit is designed as shown in Fig. 1.

2.3 Components Selection

2.3.1 Model Selection of Solar Panel

Currently, there are two types of the solar panels, one is folding type while the other is plate type. Considering the actual demand and cost control, we use the plate type solar panels. At present, there are three kinds of solar cells, such as monocrystalline silicon, polycrystalline silicon and amorphous silicon, the first two types of silicon are easiest to buy on the market while the last one is hard to purchase because of low production. Monocrystalline battery board has high price because of its high efficiency and good quality while polycrystalline silicon has the lowest price compared with the other two silicon because of its ordinary effect. We are using the monocrystalline battery board to achieve the desired effect. The panel includes output voltage and short circuit current, power and so on.

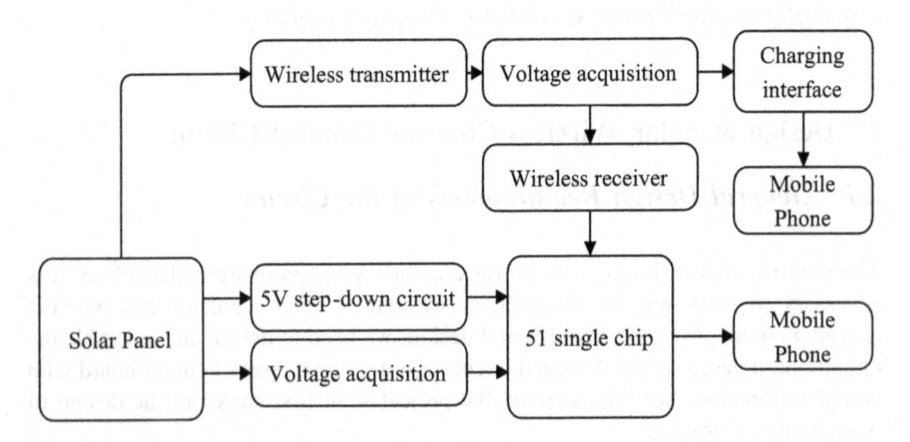

Fig. 1 Overall design of circuit diagram

Open circuit voltage (Voc): Open circuit voltage is the voltage shows in the open circuit condition between positive and negative poles. It means that the output voltage value is tested under the condition which the Photovoltaic cell is placed under the 100 mW/cm^2 light irradiation meanwhile the output of the photovoltaic cell is open at both poles. The open circuit voltage of the photovoltaic cell is proportional to the logarithm of the incident light irradiance and inversely proportional to the ambient temperature and independent of the battery size.

Short circuit current (Isc): Short circuit current is the current shows in Short circuit condition between positive and negative poles. It refers to the current flowing through both poles of photovoltaic cells when the output is short-circuited under the irradiation of standard light source. The general method to measure the short circuit current is using an ammeter whose internal resistance is less than 1 Ω.

Output power (Pm): maximum output voltage (Voc) × maximum output current (Isc).

In this design, we use the solar panel of which the nominal value of open circuit voltage is 6 V and the nominal value of short-circuit current is 550 mA. It's power rating is 3.3 W. In the actual test, the value of open circuit voltage and short circuit current is 5.98 V and 535 mA respectively and the value of output power is 3.20 W under these circumstances.

2.3.2 Selection of Wireless Power Transmitting Chip

The wireless power transmitting circuit can be designed with a special wireless power transmitting chip or a discrete component. If it's designed by using discrete component, it is likely to cause accident because of the negligence or misoperation. This design uses a special wireless power transmitting chip to play the role as wireless power transmitting system taking account of the requirements of security and stability. The power transmission circuit is designed by combining XKT-408 chip with T5336 chip.

XKT-408 integrated circuits have high precision and good stability by using CMOS production process. Therefore, it is especially suitable for the system of wireless inductive intelligent charging and power supply management. XKT-408 is responsible for processing the radio energy transfer of the Wireless inductive intelligent charging and power supply management system. It prompts energy conversion, implements real-time monitoring of the circuit by using electromagnetic energy principle and manages the intelligent control of battery charging. XKT-408 can be used to achieve reliable and fast wireless charger or wireless power supply with only a few external components.

T5336 integrated circuit, which is especially used in wireless intelligent charging, has also adopt the CMOS production process. It can form a good control circuit by cooperating with XTK-408A, which can automatically control the voltage and frequency of the electromagnetic wave emitted by the transmitting coil. Because the internal resistance of the coil cannot be ignored as well as the energy loss during energy transfer, the oscillation current in the LC oscillation circuit will be greatly attenuated. In that case, the electromagnetic wave emitted by the transmitting coil can be remain stable by controlling the voltage of No. 7 and No. 8 output port of T5336, adjusting the voltage at both ends of the LC oscillation circuit, compensating loss voltage causing for the impedance in circuit and energy transfer (Table 1).

2.3.3 Selection of Wireless Power Receiving Chip

Wireless power receiving chips adopt T3168 chip which has a simple peripheral circuit meanwhile it also has the function of receiving electromagnetic energy DC-DC voltage reducing and DC-DC voltage stabilizing. This is the reason why it is widely used. The circuit of the chip is completely in a shutdown state when standby. The value of no-load current is about 8 mA. When a magnet is close to 452L1, the circuit starts to work and the working indicator light turns on, so that it can avoid being triggered by common metal objects.

2.3.4 SCM Selection

It is necessary to use an intelligent microprocessor to monitor the voltage of the solar panel and the wireless receiving circuit, and to analyze whether the requirements of charging are achieved. In this part 89C51 SCM is used as the system microprocessor.

It is a low-voltage and high-performance CMOS 8-bit microprocessor with a 4 K bytes FPEROM (FPEROM—Flash and Erasable Read Only Memory). It's commonly known as SCM. SCM EEPROM can be erased 100 times. The device is manufactured with ATMEL's high-density nonvolatile memory technology and is compatible with the order set and efferent under the MCS-51 industrial standard. ATMEL 89C51 is a highly efficient micro controller because it combines the 8-bit CPU and flash memory in a single chip. 89C2051 is a streamlined version of it. 89C SCM provides a high flexibility and inexpensive solution for many embedded control system [3].

2.3.5 Selection of A/D Conversion Chip

PCF8591 is a monolithic integrated, separately supplied, low power, 8-bit CMOS data acquisition device. It has 4 analog inputs, 1 analog output and 1 I^{2}C serial bus

Table 1 Characteristics of T5336 chip

Symbol	Working parameters	Working conditions	Minimum value	Typical value	Maximum value	Unit
VDD	Working voltage	25 °C	3	12	15	V
AOUT	Output current	VDD = 12	100	500	800	mA
IA	No-load power consumption	VDD = 12	1	3	10	mA
M	Sensing distance		1	5	300	mm

socket. 3 address pins, A0, A1 and A2 of PCF8591, can be used in hardware address programming, which allows 8 PCF8591 devices exist in the same I^2C bus access without additional hardware. In PCF8591 device, input and output address, control signal and data signal are transmitted through double bidirectional I^2C bus by serial mode [4].

2.3.6 Selection of 5 V Voltage Reduction Chip

This design adopted 51 singlechip microcomputer for voltage acquisition and display. So, it is necessary to provide 5 V voltage source for SCM. LM2575 chip is adopted to design 5 V voltage circuit. LM2575 series switching power integrated circuit is a 1 A integrated circuit voltage produced by National Semiconductor Corporation Ns. It is integrated with a fixed oscillator. Only a few peripheral devices are needed to form a high-efficiency and stable voltage reduction circuit. This design adopts the chip as the SCM 5 V buck chip.

3 The Hardware Circuit Design of the Wireless Charging System

3.1 The Electrical Schematic Diagram Design

3.1.1 The Design of Wireless Power Transmitting Circuit

The core components of the wireless power transmitting circuit are XKT-408A and T5336. The circuit diagram is shown in the Fig. 2.

The DC input voltage of the wireless power transmission circuit is within the range of 5.5–5.98 V. The T5336 can output a controllable low voltage under the control of XKT-408A. The voltage difference between the DC voltage and the output voltage of the T5336 controls the copper coil and the LC oscillation circuit of the C11, thus emits a stable high frequency electromagnetic wave. The circuit is stable and reliable with simple peripheral device.

3.1.2 The Design of Wireless Power Receiving Circuit

The core chip of the wireless receiving circuit is T3168. The circuit diagram is shown in Fig. 3.

The energy transmitted from the transmitting terminal is high frequency oscillatory wave. If we charge to the phone directly, very adverse effects will happen to the phone. So, we need to convert the high frequency oscillation wave into a DC voltage first, and then get a step-down regulator voltage. After receiving the power

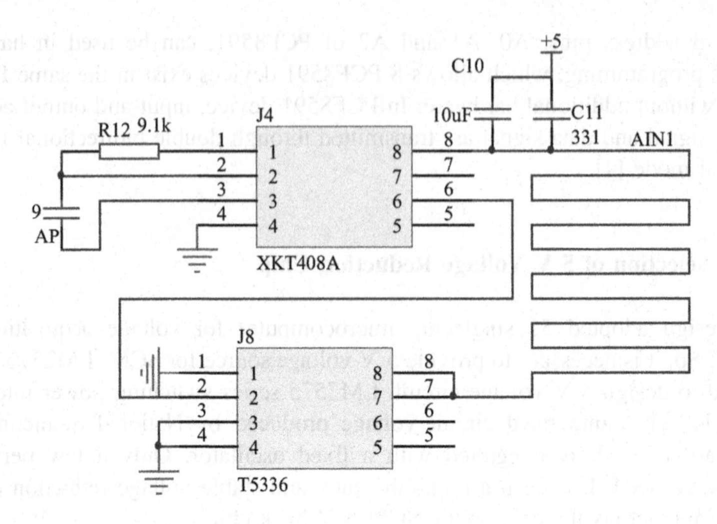

Fig. 2 Wireless transmission module

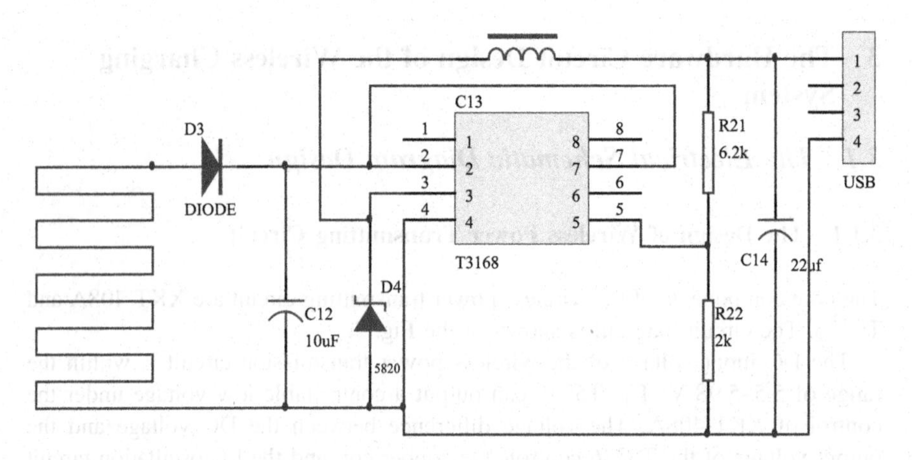

Fig. 3 Wireless receiving module

of the wireless terminal, the circuit can be rectified by a diode firstly, and the DC voltage can be obtained. Then through the step-down regulator circuit composed of the T3168 chip, the circuit gets 5 V output voltage. T3168 chip can adjust the output voltage through the R21 and the R22. To achieve output of 5 V, this example uses the combination of 6.2 and 2 k.

3.1.3 The Design of 5 V Step-Down Circuit

5 V step-down chip adopts LM2575. The circuit diagram is shown in Fig. 4.

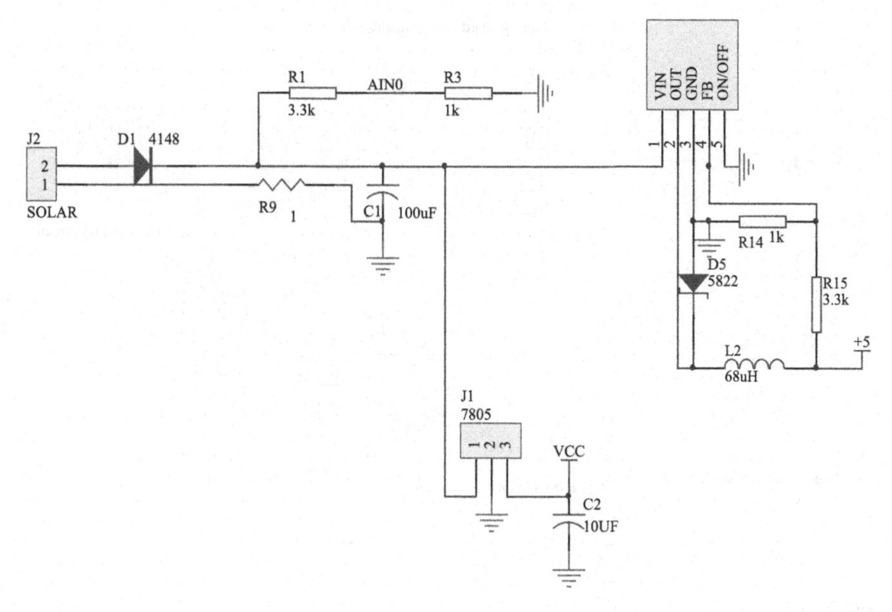

Fig. 4 5 V Step-down circuit

LM2575 series switching power integrated regulator is a 1A integrated regulator produced by National Semiconductor Corporation Ns. It is integrated with a fixed oscillator. Only with a few peripheral devices, it can form efficient voltage stabilizing circuit to achieve reduction of voltage. When the value of R14 and R15 are 10 k and 3.3 k respectively, the chip output the standard 5 V voltage, supporting the work of SCM.

3.1.4 The Circuit Design of Single Chip Microcomputer System

The circuit diagram of 51 SCM is shown in Fig. 5. The introduction of the core circuit of 51 SCM.

Power supply circuit: supply power to SCM. This design adopts LM2575 chip to provide 5 V power supply.

Clock circuit: the time reference of SCM determine its operating speed. This design adopts 11.0592M crystal oscillator to supply external clock.

Reset circuit: confirm the starting state of SCM and complete the starting procedure of it. This design adopts capacitance and resistance to achieve the reset function of power.

The circuit of LCD (Liquid Crystal Display) is shown in Fig. 6. This design adopts 1602 LCD screen which is also known as the 1602 Hollerith type liquid crystal display. It is a kind of dot matrix LCD module which is used to display letters, numbers and symbols. It consists of many dot matrix character bits such as

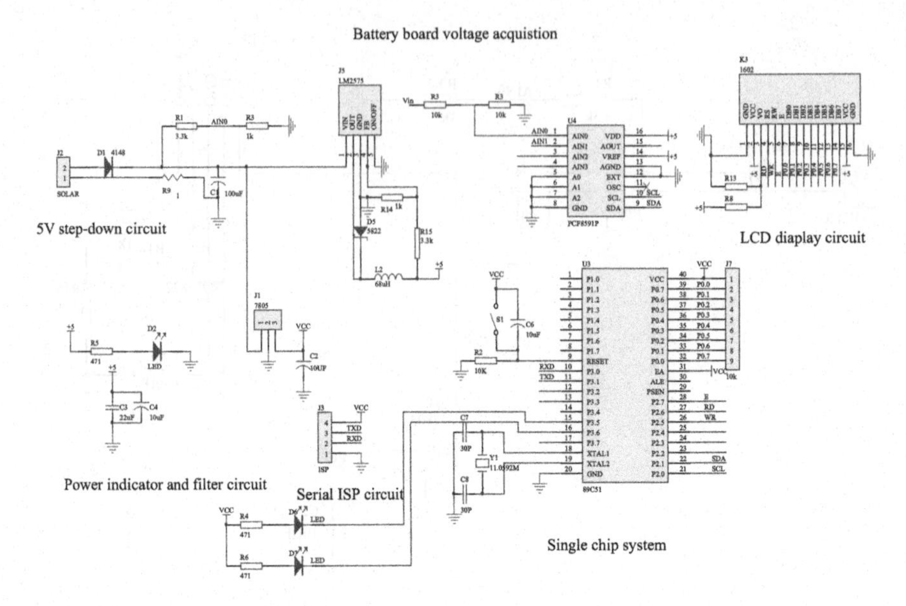

Fig. 5 SCM system circuit diagram

5 × 7 or 5 × 11, etc. Each dot character bit can display a character. There is a blanking of dot pitch between each interval. There are intervals between each line so there are also character spacing and vertical spacing. This design adopts 51 SCM to drive the LCD screen directly to display the voltage of the solar panel and the supply voltage of the SCM.

The voltage collecting circuit of solar panel (shown in Fig. 7). PCF8591 chip can only measure voltage within 0–5 V. The output voltage of the battery board is commonly more than 5 V so we need to measure partial voltage. This design adopts two 10 k resistors to halve the voltage, and then double the actual collecting voltage in the SCM operation. After a lot of experiments, we found that the deviation is no more than 5% when reading the battery voltage in practice.

Fig. 6 The circuit of LCD diagram

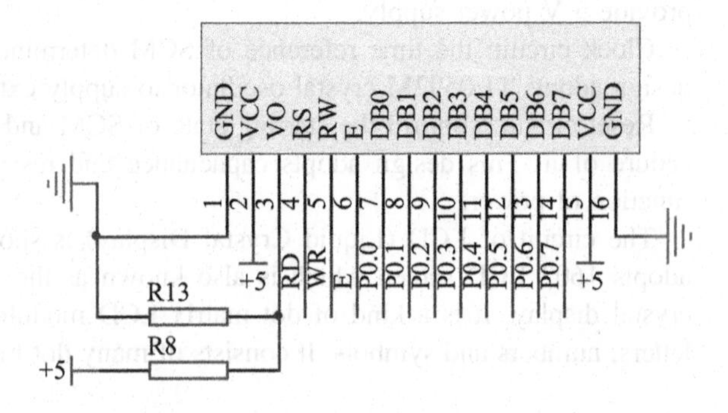

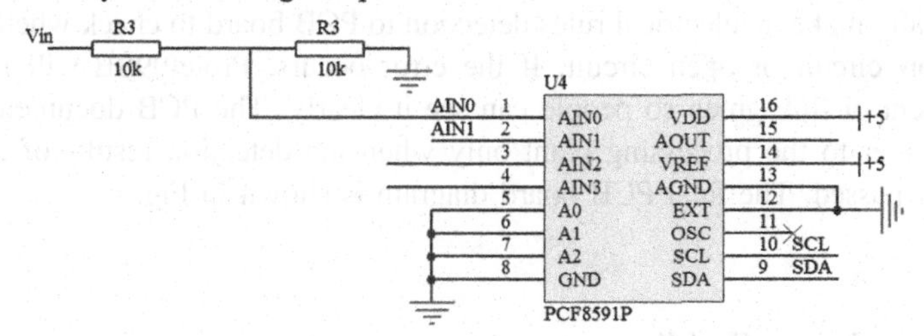

Fig. 7 Voltage collecting circuit of solar panel diagram

3.2 The Design of PCB Circuit

After the selection and design of the hardware lectotype of the system, we must use PCB board to test the rationality of the design. We use Protel99SE to design the system board. PCB board is an important part of the whole system design. Mainly divided into several steps.

The overall design of the circuit board. Before drawing the PCB, there are preliminary designs for circuit board firstly such as the number of layers of circuit boards, the package of component and their location of the installment, the board size, etc. These aspects are important parts because they determine the design framework of circuit board.

The establishment of components' package. In general, the package of some components has been built in Protel99SE and users can directly get them. But for special components, users should establish their own schematic diagram and PCB component package and then they can invoke those in schematic diagram and PCB diagram. Users can also load their own components' package into the internal components library of the Protel99SE and invoke them directly in the future.

Drawing the circuit diagram. Firstly, we should load the components into the schematic and place them neatly to make the circuit line easy as far as possible. And then connect the circuit wiring based on their own design to establish a network label. All the pin together will share unified network label. Drawing circuit schematic diagram should be checked with ERC. We can make a one-to-one mapping between PCB package and schematic package when there is no electrical error.

Create a network table and import it into the PCB diagram. After drawing the circuit schematics, we should create a network table firstly, and then import the network table and component package together into the PCB diagram. In accordance with the layout of the pre-planning, the components will be placed on the PCB board reasonably.

The wiring can be either manual or automated in Protel99SE. There is often a lot of unsatisfactory details of Protel99SE in automatic routing. So, we use manual wiring to complete the PCB board design.

Electrical rules detection of PCB. After the completion of the manual wiring, we will usually make an electrical rules detection to PCB board to check whether there is a short circuit or open circuit. If the error occurs, Protel99SE will mark the component or link green so people can see it clearly. The PCB document can be handed over to the processing plant only when the detection results of electrical rules are passed. The final PCB board diagram is shown in Fig. 8.

3.3 Hardware Welding

After the circuit board has been processed, firstly we need check whether there is short circuit, open circuit, the loss of the components and so on. Then we can take a hard welding in good condition. The system uses the SMT chip. It is necessary to use the knife iron, flux, 0.5 mm solder wire, tweezers and other tools.

We should weld from low place to high place in accordance with the height of components onboard followed by the core chip, SMD resistor capacitor, crystal, single row pin, power plug, power module, etc. Welded circuit board is shown in Fig. 9.

3.4 Hardware Debugging

The instruments will be used in debugging contains on-off regulated power supply, digital multimeter and so on. The debugging process is as follows.

Fig. 8 System hardware PCB board diagram

Fig. 9 Welded circuit board diagram

Firstly, we supply 6 V power directly to the circuit board as system input rather than to connect it with solar panel. Then we use multimeter to measure SCM supply circuit voltage which transmitted by LM2575. Secondly, we check the voltage of wireless radiating circuit and wireless receiving circuit. If the voltage is normal, we can connect the circuit board with solar panel and take an outdoor testing. If the voltage is abnormal, we need to take a troubleshooting process.

3.5 Performance Testing

To get the performance parameters of the system, we will have a targeted testing for the hardware circuit.

4 The Software Design of Solar Energy Wireless Charging System

4.1 Process Design of Software System

This design uses 89C51 microcontroller as the microprocessor, which adopt KEIL software to compile C language program. After programming, it downloads the program through the ISP to the microcontroller. The software process of the design is shown in Fig. 10.

First, the power on reset, the initial configuration of the microcontroller, and then initialize the PCF8591 microcontroller and LCD screen. Second, the system starts the A/D conversion program to detect the output voltage of the solar panel and the input voltage of the wireless power transmission circuit. If the voltage is not within

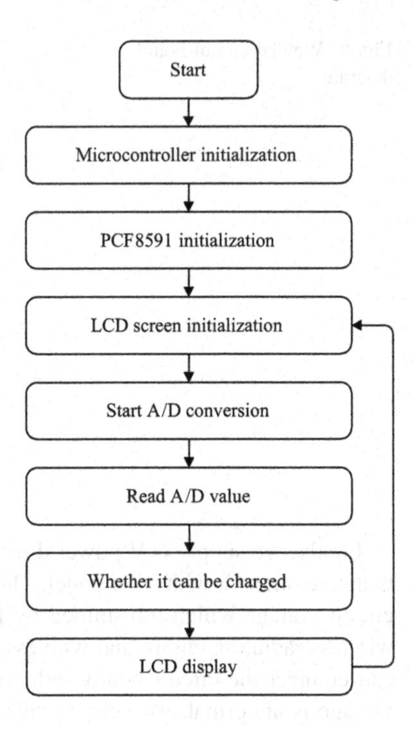

Fig. 10 Software flow chart

the range of safe charging, the user is noticed not to charge. If the voltage is normal, the user can charge.

4.2 Programming PCF8591

PCF8591 chip, which is the chip of internal reference voltage source, is adapted the use of IIC communications to achieve data transceiver. In the use of PCF8591 chip, SCM had to activate the chip. The way of activation is through the IIC communication, sending effective address information to PCF8591 chip. PCF8591 is activated when it is available. However, PCF8591 has not yet officially started the A/D conversion. One needs to send A/D conversion control word via IIC to officially start the conversion. The control word generally includes setup of conversion channel, conversion mode and so on. After setting, the PCF8591 begin to A/D conversion, the 51 microcontroller can access the converted data in real-time. The converted data is not a true voltage value. The actual voltage value can be calculated by dividing the collected A/D value with 1024(value of A/D full range), then 5 times the quotient.

4.3 LCD Screen Design

1602 LCD screen uses parallel data transmission mode for communication. Before using, it needed to be initialized by SCM. The initial configuration usually needs to set screen content and display mode. After configuration, microcontroller can send data to 1602 LCD screen though 8 data pins. One can use microcontroller control the RS register to realize the data read, and set the EN register to realize data display.

5 Summarizes and Prospects

5.1 Summarizes

In the fourth chapter, according to the technical requirements of solar charging, chooses the appropriate electronic components and accessories. On this basis, consulting the relevant technical manuals, it designed a set of solar wireless charging circuit schematic. In order to check the design drawings are correct, it would be drawing by PCB circuit board, and will be sent to the manufacturer of the drawings for processing. Finally, the circuit function is realized by the circuit board welding and software programming. Through data testing, it acquires the circuit parameters and other information.

5.2 Prospects

Through effort of several months, as well as the instructor's tireless teaching, It is completed the production of material object and the writing of thesis. Relevant data also need to continue to improve in future.

References

1. Jiang L. A new dissertation of solar energy refrigeration and heating. North China Electric Power University; 2008.
2. Wang H. Study on heat conduction of solar focus heat pipe. Tianjin University; 2008.
3. Deng Y. Research on the pipe inspection robots system based on ethernet and its control and image transmission. Shanghai Jiao Tong University; 2015.
4. Yang W, Zhang L, Gong H, Luo W. Design and implementation of data collection system for unattended station. School of Physical and Electronic Information Engineering, Qinghai University for Nationalities. Modern Electron Tech. 2014;37:50–53.

Modeling and Dynamic Characteristic Analysis of Flexible Manipulator

Yuzhen Zhang, Qing Li and Weicun Zhang

Abstract The dynamic model of the flexible link manipulators (FLMs) is established and dynamic characteristic is analyzed in the paper. First, in order to improve the accuracy of the system, the model of the FLMs is constructed based on the assumed mode method with boundary conditions; second, the natural frequency and vibration mode functions are analyzed in detail. The residual vibration of the loaded flexible arm tip is derived. A simple controller is adopted to inhibit vibration. This research has provided a foundation for refinement of the FLMs model and for the active control of the vibration analysis, which possess a high practical value in engineering.

Keywords Flexible link manipulators · Modal analysis · Elastic deformation · Dynamic model · Vibration analysis

1 Introduction

With the rapid development of aerospace industry and the deepening of space exploration, space technology and robot technology have a great development. Flexible link manipulators exhibit many advantages, such as lightweight, high-speed operation, lower energy consumption, and better payload carrying capacity, in applications requiring large workspace where rigid ones may not be suitable [1]. Research on FLMs is motivating in recent years, due to the fact that industrial requirements for quicker response times and lower power consumption, especially when the weight of the robots is a concern to prevent unnecessary energy consumption and to achieve higher payload-to-mass ratio [2].

Tip position control of a FLM is challenging due to occurrence of vibration owing to distributed link flexible, which makes the system non-minimum phase, under actuated and infinite dimensional. Therefore, how to accurately describe the

Y. Zhang · Q. Li · W. Zhang (✉)
School of Automation and Electrical Engineering, University of Science and Technology Beijing, Beijing 100083, China
e-mail: weicunzhang@ustb.edu.cn

Y. Jia et al. (eds.), *Proceedings of 2017 Chinese Intelligent Systems Conference*, Lecture Notes in Electrical Engineering 459,
https://doi.org/10.1007/978-981-10-6496-8_58

vibration of the flexible structure is the vital content. Recently, some researchers have made many studies on the FLMs. M. Benosman and G. Le Vey give a survey of the control of flexible manipulators in 2004 [3]. Santosha Kumar Dwivedy and Peter Eberhard provide a literature review of dynamic analysis of flexible manipulators [4]. Flexible modeling, dynamic analysis and control of mechanical arm are covered. Reference [5] shows the dynamic characteristic analysis for the humanoid rigid-flexible coupling robotic arm with three rotating joints. The vibration suppression of the system has been studied, including PD control [6], adaptive sliding mode control [7], neural network control [8], boundary control [9], etc. As the description of the FLMs model is the basis of the active vibration control, especially there is tip payload, it is very important to analysis modal and dynamic characteristics. Although there are some description in literatures, dynamic characteristics and tip payload influence on the system are not detailed comparative analysis, which is the important factor to control strategy selection.

In this paper, the elasticity deformation problem of flexible manipulator is analysis. And the dynamic model is established, obtaining its natural frequency and vibration mode. Meanwhile, the influences of modal truncation method and tip payload are proposed and analyzed, which lay a foundation for the research of control methods.

2 Preliminaries and Problem Formulation

Flexible manipulator is a highly nonlinear and strong coupling dynamic system. Therefore, a first step towards designing an efficient control strategy for these manipulators must be aimed at developing accurate dynamic models that can characterize the above flexibilities along with the rigid dynamics [10]. Especially in order to guarantee accurately the tip position of FLM, the flexible link deformation and vibration should be determined precisely.

Considering the complicated flexible manipulator mechanism, it's very difficult to build the suitable precise mathematical model. There are three aspects need to be considered: the selection of flexible robotic arm link model; the establishment of flexible manipulator boundary condition; the description of the flexible manipulator link deformation. So, the rest of the paper is organized as follows. According to the Lagrange dynamics and assumed mode method, the mathematical model of the flexible manipulator with tip mass and inertia rotation is derived. Then, the vibration changing regularity is researched considering different mode orders, different structure parameters.

The flexible link is assumed to be an Euler-Bernoulli beam. The following assumptions are made [10–12].

(1) The motion of each link is assumed to be in the horizontal plane, so that the influence of gravity is ignored.
(2) Each link is assumed to be long and slender, therefore, it can have deformations in the horizontal direction only.

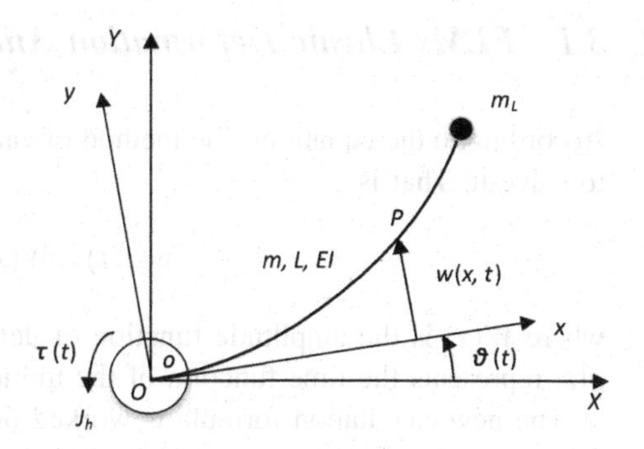

Fig. 1 Structural model of flexible manipulator with tip payload

(3) Links are considered to have constant cross-sectional area and uniform material properties.
(4) Motor shaft and clamping device are regarded as rigid body.
(5) The control torque can just be applied at the joint.
(6) The number of sensors that can be used, is limited, and they are at restricted locations along the manipulator.

The flexible manipulator system is described in Fig. 1. The frame *XOY* is the inertia frame and the frame *xoy* is the local rotating reference frame with the hub. *EI* is the uniform flexural rigidity of the beam. *L* is the length of the beam. m and m_L represent respectively the mass of the beam and the tip payload. $\tau(t)$ and J_h are the control torque of the manipulator and the hub inertia. *P* is one point on the beam and $w(x,t)$ is horizontal deformation. θ represents the angle of the hub.

3 Dynamic Model of the FLMs

Due to the detailed modeling of the FLM has been shown in reference [13] and [14], some portion of the results will be directly given as follow. The FLM's corresponding equations and boundary conditions are shown

$$EI\frac{\partial^4 w(x,t)}{\partial x^4}+\rho A\frac{\partial^2 w(x,t)}{\partial t^2}=0 \tag{1}$$

$$\begin{cases} w(0,t)=0 \\ \frac{\partial w(0,t)}{\partial x}=0 \\ EI\frac{\partial w^2(L,t)}{\partial x^2}=0 \\ EI\frac{\partial w^3(L,t)}{\partial x^3}=m\frac{\partial w^2(L,t)}{\partial t^2} \end{cases} \tag{2}$$

where ρ is density of the link and A is cross-sectional area.

3.1 FLMs Elastic Deformation Analysis

According to the equations, the method of variable separation [15, 16] can be adopt to solve it. That is

$$w(x,t) = W(x)q(t) \tag{3}$$

where $W(x)$ is the amplitude function of deformations in the horizontal direction. $q(t)$ represents the time function of the motion law.

The new calculation formula is worked out

$$-\frac{\rho A}{EIq(t)}\frac{d^2q(t)}{dt^2} = \frac{1}{W(x)}\frac{d^4W(x)}{dx^4} \tag{4}$$

It can be defined that β^4 is the constant for the right and left sides of the equation, and $\beta^4 = (\rho A/EI)\omega^2$, where ω is the natural frequency actually. The partial differential equation can be divided to two differential equations [17]. Base on the differential equation numerical analytical method, the general solution can be obtained. Then, combined with the formula of boundary conditions, vibration mode function can be given

$$W_i(x) = \sin(\beta_i x) - \sinh(\beta_i x) + \frac{\sin(\beta_i L) + \sinh(\beta_i L)}{\cos(\beta_i L) + \cosh(\beta_i L)}(\cosh(\beta_i x) - \cos(\beta_i x)) \tag{5}$$

where $i = 1, 2, \ldots, \infty$.

According to the above analysis, we know that elastic deformations is infinite order. In practical application, it's impossible to solve the infinite order. The front modal can be used only, and the precision is satisfied with the demands. So, a finite dimensional expression can be represented by an assume mode method [18].

$$w(x,t) = \sum_{i=1}^{N} W_i(x)q_i(t) \tag{6}$$

where N is the number of assume modes; $W_i(x)$ is the mode shapes, which is spatial coordinate, for the i-th order; $q_i(t)$ is the modal coordinates, which is time coordinate, for the i-th order.

3.2 Dynamic Modeling

The procedures of the flexible link modeling based on the Lagrange equation includes the following content. First, according to the spatial and time coordinate, finite dimensional model can be established. Then, kinetic energy and potential energy of the flexible manipulator system are shown. Finally, the dynamic equation can be present by using the Lagrange equation [19].

From Fig. 1, the position coordinates of the point $P(X, Y)$ can be given

$$\begin{cases} X = x\cos\theta - w(x,t)\sin\theta \\ Y = x\sin\theta + w(x,t)\cos\theta \end{cases} \tag{7}$$

Further, the time derivative of the position is the velocity for the point. And, the total kinetic energy of the flexible manipulator can be given, which includes the electric motor rotor kinetic energy, the flexible link kinetic energy and the kinetic energy of tip payload.

According to the kinetic energy formulas, the three parts of the kinetic energy for the flexible link system can be calculated respectively. So, there is no detailed account for this section.

As the motion of each link is assumed to be in the horizontal plane, the potential energy is generated by the elastic deformation. Therefore, the elastic potential energy of flexible link manipulator is shown

$$V = \frac{1}{2}EI\int_0^L \left(\frac{\partial^2 w(x,t)}{\partial x^2}\right)^2 dx \tag{8}$$

Then, kinetic energy and potential energy can be put into Lagrange equation, and the dynamic equation is obtained.

Actually, it's inevitable that there is damping factor in the flexible manipulator. The joint friction, structure damping of the mechanical arm material, even the air damping when moving, can affect the vibration of flexible manipulator. Therefore, the damping should be considered. Reference [16] gives the detailed discussion about the damping, the results are used directly for simplicity.

Above all, the dynamic equation of the flexible link manipulator can be given

$$\begin{bmatrix} M_r(q) & M_{rf} \\ M_{fr} & M_N \end{bmatrix}\begin{bmatrix} \ddot{\theta} \\ \ddot{q} \end{bmatrix} + \begin{bmatrix} C_H & 0 \\ 0 & C_{fi} \end{bmatrix}\begin{bmatrix} \dot{\theta} \\ \dot{q} \end{bmatrix} + \begin{bmatrix} 0 & 0 \\ 0 & K_N \end{bmatrix}\begin{bmatrix} \theta \\ q \end{bmatrix} + \begin{bmatrix} h_r(\dot{\theta}, q, \dot{q}) \\ h_f(\theta, q) \end{bmatrix} = \begin{bmatrix} \tau \\ 0 \end{bmatrix} \tag{9}$$

where $M_r(q) = J_h + \rho A\int_0^L x^2 dx + m_L L^2$, $M_{rf} = M_{fr}^{\mathrm{T}} = [a_1\ a_2 \cdots a_N]$, $a_i = m\int_0^L xW_i(x)dx + m_L LW_i(L)$, $M_N = \mathrm{I}_{N\times N}$, $K_N = EI\int_0^L \left[\frac{d^2W_i(x)}{dx^2}\right]^2 dx$, $h_r(\theta, q, \dot{q}) = \sum_{i=1}^{N} 2\theta\dot{q}_i q_i$, $h_f(\theta, q) = \left[-\theta^2 q_1\ \ -\theta^2 q_2 \cdots -\theta^2 q_N\right]^{\mathrm{T}}$, $i = 1, 2, \ldots, N$, and C_H is mechanical arm joints damping coefficient, C_{fi} is material structure damping coefficient.

4 Modal Analysis

For a flexible link manipulator system, vibration suppression is the crucial content. How flexible link vibration is discussed in this section.

Base on the elastic deformation analysis, we can give the frequency equation of the flexible link with tip payload. That is

$$1 + \cosh(\beta L)\cos(\beta L) + \frac{m_L}{m}\beta L(\cos(\beta L)\sinh(\beta L) - \sin(\beta L)\cosh(\beta L)) = 0 \quad (10)$$

As above shown, when m_L tends to zero, the frequency equation is exactly the case that flexible link without tip payload.

According to the vibration mode function and nature frequency in the section three, the numerical analytical resolutions can be obtained. Then, the modals can be solved. Thus, the vibration mode shape is obtained. Now, some parameters of the flexible link manipulator are listed in Table 1.

The numerical simulation of time domain with MATLAB software can be done. Thus, the vibration curves of flexible manipulator for the six lowest order modal can be obtained, which is present in Fig. 2.

Indeed, while the dynamic model of FLM is theoretically infinite, which includes stiff movements and infinite elastic modal, the modal truncation should be done for engineering application. In general, with the increase of elastic modal, the numerical results is closed to the distributed parameter systems. However, the control system is very complex in its computing process. Therefore, under the premise of ensuring the accuracy, combined with the analysis of dynamic characteristics, the modal truncation will be illustrated.

Table 1 Parameters of the flexible link manipulator

Material and geometric parameter	Value
Density of the FLM (ρ)	2.7×10^3 kg/m^3
Uniform flexural rigidity (EI)	300 N m^2
Cross-sectional area (A)	1.05×10^{-4} m^2
Length of the FLM (L)	1 m
Tip payload (m_L)	0.1 kg
Hub inertia (J_h)	0.8 kg m^2

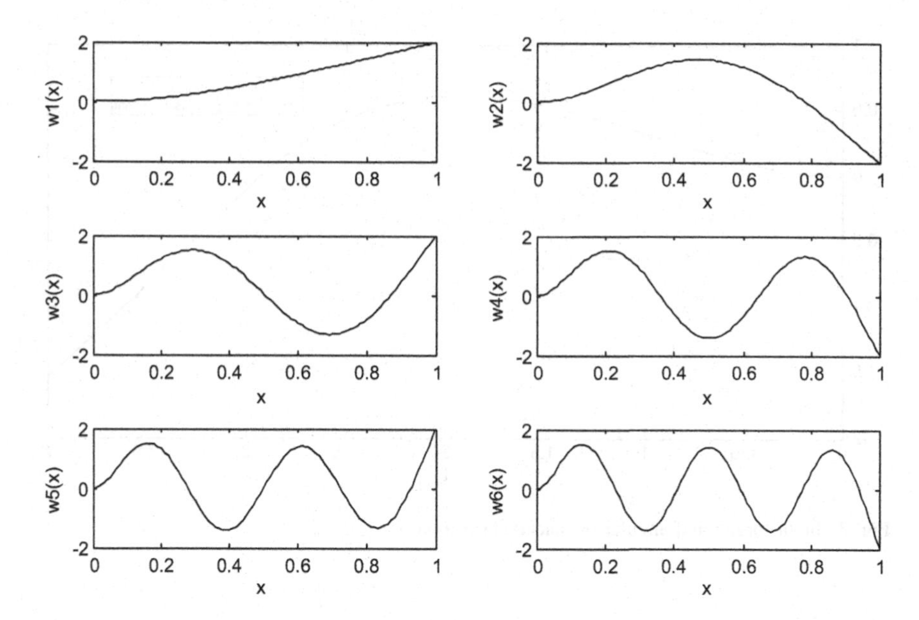

Fig. 2 Vibration curves of flexible manipulator

5 Dynamic Characteristic Analysis

The flexible manipulator is a highly nonlinear and strong coupling dynamic system. Base on the model above, dynamic characteristic is analyzed when different driven torques and tip additional mass.

The simulation is done in the MATLAB/Simulink according to the dynamic model. The force applied on the flexible link manipulator by the joint is

$$\tau = 1 - t \quad (0 \leq t \leq 3) \tag{11}$$

In Fig. 3, the input torque acted on flexible manipulator joint and angular displacement response of flexible manipulator are shown. In order to analyze the influence of modal truncation on the flexible manipulator dynamic characteristics, the first order modal and the two lowest order modal are compared. It's shown in Fig. 4.

Considering the flexible manipulator is used to grab objects, the influence of tip payload on the model is necessary to analyze. The Fig. 5 gives the contrast curve with different tip payloads.

As show in Fig. 4, the different between the two modal truncation methods is exist but tiny. Because of the existence of damping, vibration will gradually become smaller. So, in different occasions the modal truncation is according to the actual situation. From Fig. 5, with the increase of tip payload, the amplitude gets bigger. Thus, the controller must have the adaptability and robustness.

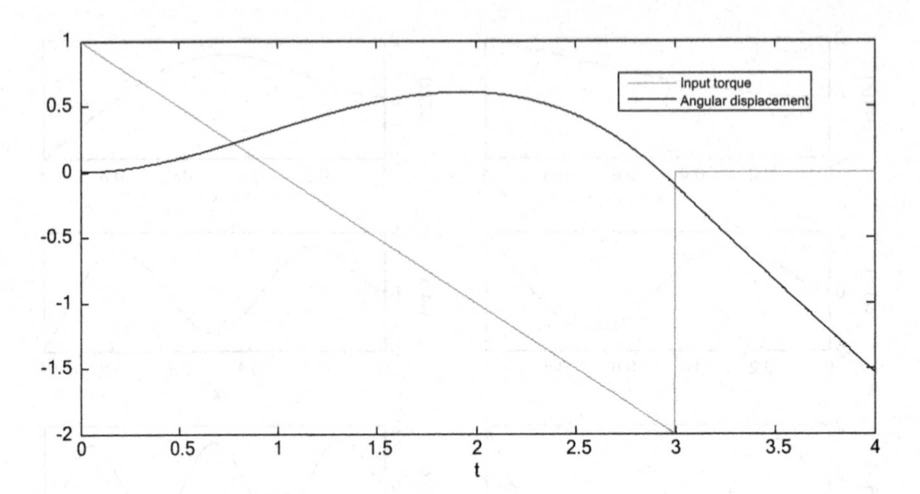

Fig. 3 Input torque and angular displacement response

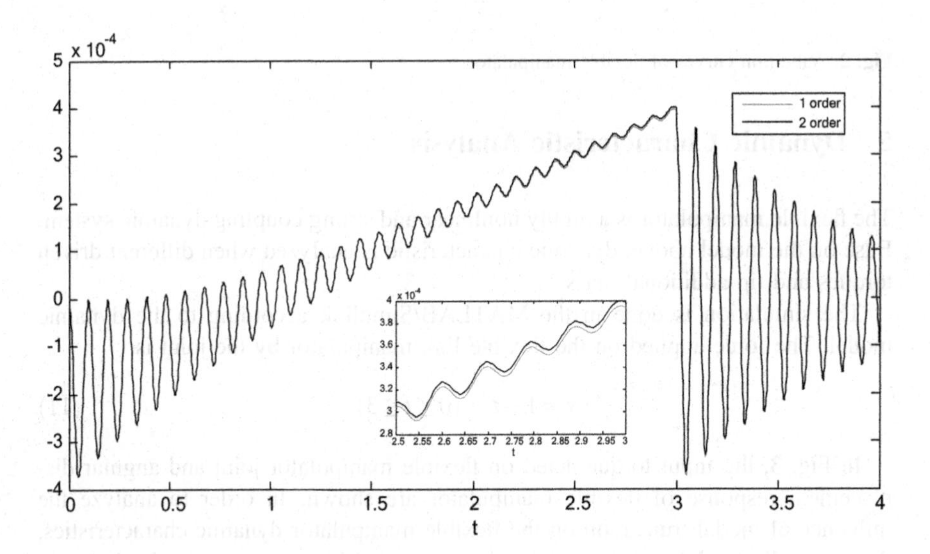

Fig. 4 Response curve in different modal truncation

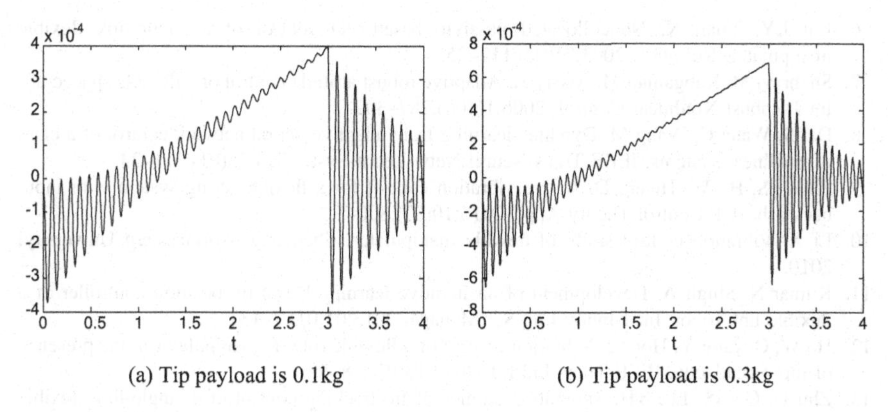

(a) Tip payload is 0.1kg (b) Tip payload is 0.3kg

Fig. 5 Vibration curve with different tip payloads

6 Conclusions and Future Work

With the help of the method of variable separation and modal analysis, the dynamic modeling issue on the flexible link manipulator has been addressed with the simple methods. The results show that the model is correct and effective, which provides a foundation for refinement of the FLMs model and for the active vibration control.

However, the discretization would result in more inaccuracies. The dynamic model could be more appropriate and accurate. To a certain extent, assume mode method possesses a high practical value in engineering, it can bring some problems for system simultaneously, especially in vibration suppression. Thus, the future research will be focused on the active vibration suppression method.

Acknowledgements The author would like to thank the anonymous reviewers for their constructive and insightful comments for further improving the quality of this work. This work was supported by National Natural Science Foundation of China (No. 61520106010), National Key Technologies R&D Program (No. 2013BAB02B07) and National Natural Science Foundation of China (No. 61603362).

References

1. Wu L, Yang G, Kuai X, Sun F. Flexible-link manipulator modeling, analysis and control. Beijing: Higher Education Press; 2012.
2. Jiang D. Nonlinear vibration research of the flexible manipulator in out-space environment. Shenyang: Northeastern University; 2010.
3. Benosman M, Vey GL. Control of flexible manipulators: a survey. Robotica. 2004;22(05):533–45.
4. Dwivedy SK, Eberhard P. Dynamic analysis of flexible manipulators, a literature review. Mech Mach Theory. 2006;41(07):749–77.
5. Liu X, Huang Y, Cui P, Xu Z. Modeling and dynamic characteristic analysis of flexible robotic arm. Noise Vib Control. 2014;34(06):7–11.

6. Liu LY, Yuan K. Noncollocated passivity-based PD control of a single-link flexible manipulator. Robotica. 2003;21(02):117–35.
7. Shahravi M, Kabganian M, Alasty A. Adaptive robust attitude control of a flexible spacecraft. Int J Robust Nonlinear Control. 2006;16(06):287–302.
8. Dai S, Wang C, Wang M. Dynamic learning from adaptive neural network control of a class of nonlinear systems. IEEE Trans Neural Netw Learn Syst. 2014;25(01):111–23.
9. Zhang S, He W, Huang D. Active vibration control for a flexible string system with input backlash. IET Control Theory Appl. 2016;10(07):800–5.
10. Li X. Vibration control study of flexible manipulator. Shenyang: Northeastern University; 2010.
11. Kumar N, Singh A. Development of an iterative learning based tip position controller of a flexible link robot. Int J Innov Eng Sci Manage. 2013;01(01):7–13.
12. He W, Ouyang Y, Hong J. Vibration control of a flexible robotic manipulator in the presence of input deadzone. IEEE Trans Ind Inf. 2017;13(01):48–59.
13. Zhu G, Ge SS, Lee TH. Simulation studies of tip tracking control of a single-link flexible robot based on a lumped model. Robotica. 1999;17(01):71–8.
14. Sakawa Y, Matsuno F, Fukushima S. Modeling and feedback control of a flexible arm. J Robotic Syst. 1985;02(04):453–72.
15. Lou J. Research on integrated control of trajectory tracking and vibration suppression of a space flexible manipulator system using piezoelectric actuators. Hangzhou: Zhejiang University; 2013.
16. Yang H. Study on dynamic modeling theory and experiments for rigid-flexible coupling systems. Shanghai: Shanghai Jiao Tong University; 2002.
17. Ding X. Research on robot control. Hangzhou: Zhejiang University Press; 2006.
18. Cui L, Zhang J, Gao L, Xiao Z. Research on dynamic modeling of flexible manipulator system. J Syst Simul. 2007;19(06):1205–8.
19. An K, Bi Y, Ma J. Vibration modal analysis of the single-link manipulator with end-effector payload. Opto-Electron Eng. 2016;43(07):22–7.

Graphic Modelling Approach as a Support for Event-B Modelling

Xiaolong Li, Jun Liu, Keming Wang and Yang Xu

Abstract Event-B method, as an evolution of B-method, is a formal method for system-level modelling and analysis based on extended first order logic and set theory, which provides flexible approaches of refinement and decomposition to construct large systems gradually with some successful applications in their formal verifications. However, it is hard to manipulate and grasp this method for many researchers because of its highly abstraction. In order to reduce the burden of developers' work and the complexity of Event-B model, a graphic modeling approach, called Event-B graph, is introduced in this paper, which is used as an alternative way to clearly describe the state flow of the model and provide a graphic way of system-level modelling to bridge the real problem and the Event-B model construction. After introducing some new concepts and structure of Event-B graph, the transformation algorithm from Event-B graph to Event-B model is provided and the equivalence of the transformation is then proved. An example is finally provided to illustrate the procedure of Event-B graph construction.

Keywords Formal method · Event-B method · Event-B graph · Refinement

1 Introduction

With the development of the software formalization, formal methods gained more and more attentions in the area of software security. Event-B method [1], as a formal language based on predicate calculus and theorem proving, plays an important role in modelling and verification of software systems. Solving the

X. Li (✉) · J. Liu · K. Wang · Y. Xu
National Local Joint Engineering Laboratory of System Credibility
Automatic Verification, School of Mathematics, Southwest Jiaotong
University, Chengdu 610031, Sichuan, China
e-mail: 1845253023@qq.com

J. Liu
School of Computing and Mathematics, Ulster University, Belfast,
Northern Ireland, UK

Y. Jia et al. (eds.), *Proceedings of 2017 Chinese Intelligent Systems Conference*, Lecture Notes in Electrical Engineering 459,
https://doi.org/10.1007/978-981-10-6496-8_59

complexity and maneuverability for formal method are always the key research areas in the last few years. Because of the highly abstract of Event-B method, it's hard for engineers to learn it and use it for system modeling, and also difficult for nonprofessional formal developers to read and understand. In order to enhance readability and reduce complexity of Event-B model, it is of urgent need to introduce efficient ways to Event-B model analysis and specification.

Generally, Event-B model consists of context and machine, which describes the static and dynamic properties of system separately. Various techniques and tools have been proposed to analysis requirement document and construct Event-B model. One of the most popular ways is refinement [2]. By using the refinement, engineers can develop models in a high efficiency, and the complexity of systems is controlled in a certain degree [3]. At the same time, many theories and tools are developed, which tried to enhance the efficiency of Event-B model development. However, most of them have failed to put the heavy workload of modelling and revision into consideration, and give strong readability, or support component reusability explicitly [4]. Besides, the states flow [5] are not clearly represented between different events in those methods. Graphic method and theory is an interesting and important way to visualize the structure and system in a more clear way, so is regarded as one important modeling method. Among different methods, graphic methods to support Event-B modeling have not yet been investigated extensively.

Motivated by this situation, a graph-based modeling method, called Event-B graph, is introduced in this paper. It is an extensible development method based on Event-B language [6] and timed automata [7], also a refinement supported method designed with an attempt at modelling large and complex systems. More specifically, this new method is designed to (1) represent Event-B model into an equivalent Event-B graph which structure includes context, locations and edges; (2) depart the variable set and invariants into the context and defines the events of the Event-B model using the graph structure; (3) use the combination of edges and locations, the idea of states flow, and variables filter to define the guards and actions of the event; (4) gain guards from edges and variable filter, which only treat the needed variable state as guards in the state flow from source location to target location; (5) add the strategy of refinement into graph modelling in order to support stepwise model specification; (6) abstract the same mechanism from Event-B method, which keeps the consistence among a set of refinement graph; (7) provide the equivalent transformation rule between Event-B graph and Event-B model, which ensures the consistency between two modeling approaches. The procedure of Event-B graph construction is provided and the equivalence of transformation is proved. An example is given to illustrate the procedure and effectiveness of the method.

The remaining of the paper is organized as follows. Some background information about Event-B is overviewed in Sect. 2, along with some related work about Event-B modeling methods and toolset. The key concepts and features of Event-B graph including definitions, the structure, the transformation rule, and refinement scheme are detailed in Sect. 3. In addition, a press controller example is provided to illustrate the extensibility and flexibility of Event-B graph method in Sect. 4. The paper is concluded in Sect. 5.

2 Background

Event-B method, as a formal method, is inherited from B method [8] by using typed set theory and mathematical languages for modelling concurrent systems. Comparing to the original B method, Event-B method has more flexible refinement skill for modelling systems gradually [9].

Rodin platform [10] is an open toolset for Event-B model construction from requirement to verification and it supports a top-down development process [11]. Thus, developers may define new function in the Rodin platform by developing a plugin package. To extend the basic functionality of the Rodin platform, developers concern several functions of plug-ins, such as AnimB, ProR, Theory plug-in, and UML-B and so on. AnimB is an animator for Rodin platform and ProR supports integration of natural language requirements and Event-B models. Theory Plug-in provides capabilities to extend the Event-B mathematical languages and the Rodin proving infrastructure. UML-B state machine animation provides an animation of UML-B state-machine [12].

Event-B model makes a clear distinction between the static and dynamic part of a system with context and machine separately. To precisely define the static properties of a system, context contains four elements of an Event-B model: carrier sets, constants, axioms, and theorems. To precisely define dynamic behavior of a system, machine contains four elements of an Event-B model: variables, invariants, theories and events. Variables are used to describe state of a system and they are constrained by invariants.

In order to capture the possible state change of a system, machine use the events to define guards and actions to describe the state transition. Two parts are contained in an event are (1) using guards to justify whether the variable substitutions are satisfied or not, collectively denoted by $G(s, c, t, v)$; (2) using actions to implement variables substitution, collectively denoted by $A(s, c, t, v, v')$, where s represents carrier sets, c represents constants, t represents parameters and v is a subset of variables. Each machine may see multiple contexts and each context can be extended by other contexts.

Refinement, as a key modelling technique in Event-B method, is able to help developers to do rigorous modeling while reducing the burden of modelling and verification [13]. Event-B method uses strategy refinement to support stepwise refinement of both context and machine. If a machine N refines another machine M, then M is called the abstract machine and N is a concrete machine. Event-B uses two principal types of refinement: superposition refinement and data-refinement. Superposition refinement corresponds to a spatial and temporal extension of a model, while data refinement is used in order to modify the state of the machine [14].

3 Event-B Graph

3.1 Concept and Structure of Event-B Graph

The graphic modeling framework, called Event-B graph, introduced in this section is a formalism for specifying Event-Based systems, especially to support for Event-B modeling. The basic concepts and the structure of Event-B graph is illustrated in Fig. 1. A generation method will be given to build an equivalent transformation from Event-B graph to Event-B model.

In general, Event-B method precisely define the relation between previous guards and the subsequent actions since the state flow is hided in events. Obviously, it fails to represent model into a graph form which is more convenient to write and read. Event-B graph is trying to transform events into edges and locations. The benefit of doing this is to simplify the guards design process by hiding the inter-communication of locations in edges. This change connects a relation between different locations and gives an equivalent transformation rule between Event-B model and Event-B graph.

According to Fig. 1, Event-B graph makes explicit distinction between the context and graph. Compared to the original Event-B model, variables and invariants are classified into context. Graph, using locations and edges, generates events under the specific rule. It uses the variable filter and additive predicates which are transformed between locations and defined in edges respectively to

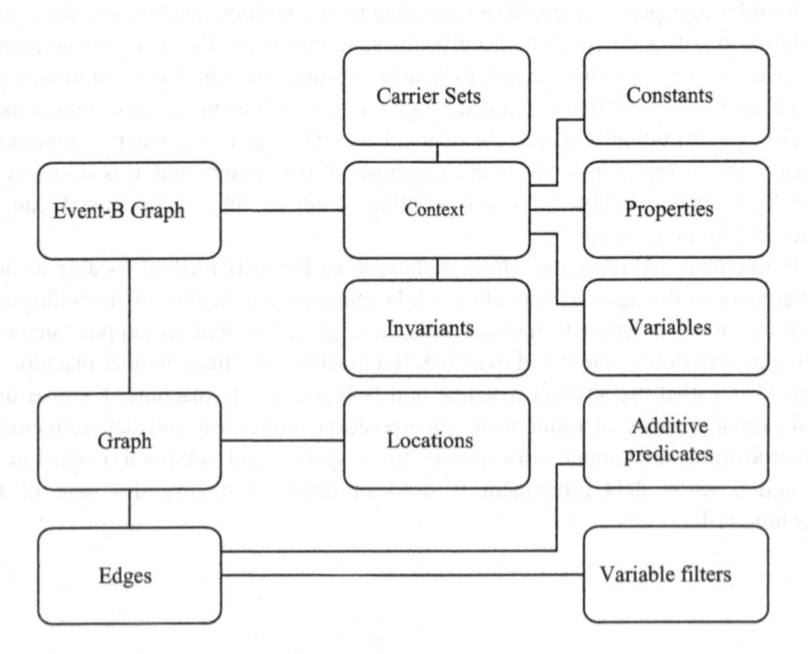

Fig. 1 Structure of Event-B graph

specify guards of machine. Two elements are involved into a graph: (1) the locations for storing actions of Event-B model, collectively denoted by *L*; (2) the edges for specifying the connection among locations and generating guards for target locations, collectively denoted by *E*.

3.2 Edges and Locations of Event-B Graph

Event-B graph uses the locations and edges to support the definitions of guards and actions separately. Using locations and edges to build Event-B graph is based on the idea of graph modelling and it provides a visual method to construct an Event-B model. In general, guards of Event-B model are generated from edges, and guards may be derived from source locations or added by hand. Actions are stored in locations, and actions are all added by hand.

Edge, as a basic element of Event-B graph, is responsible for states flow between locations. The edges transform all of the variable states from its target locations to source locations. The main functions of an edge are defined in variables filter and additional guards. Three elements are involved into edges: (1) the variables filter for choosing necessary variables states required by guards of next locations and keeping the predicates of these variables as guards for target location; (2) the parameter for adding parameter variables for guards; (3) the additional guards for writing additive predicates about variables and parameters. Combining these elements in edges, a complete guards of Event-B is generated without any loss of functions.

The locations help define the name and the actions set of the Event. Two elements are involved into locations: (1) the name for marking event; (2) the actions set for describing the substitution of variables. Obviously, there is no difference between the actions of Event-B graph and the actions of Event-B model.

In order to verify the feasibility and invariance of Event-B graph, an equivalent transformation rule from Event-B graph to Event-B model is necessary. A generation rule of guards is given in Fig. 2, as a key relation between Event-B graph and Event-B model, is responsible for guards generation.

The generation rule is showed in a graph form in Fig. 2. Each edge is a directed line segment point from source location to target location. Parameters are denoted

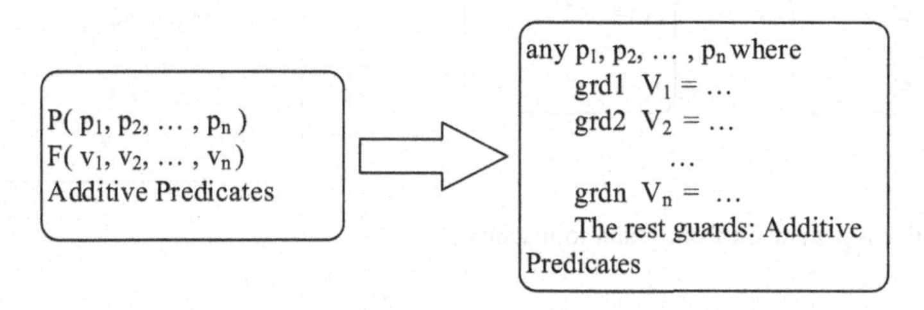

Fig. 2 Guards generation from edge to event

by an n vector $P(p_1, p_2, \ldots, p_n)$ and variables filter is denoted by an n vector $F(v_1, v_2, \ldots, v_n)$, the rest guards are supplemented by additive predicates. One of essential features of edge is supporting the states transition and the guard generation by adding specific rule. Each location connects source location with target location. The variable filter vector captures necessary variable states which will be used to construct the guards of target location, the rest variables keep changeless in the state flow until them are substituted.

The event helps describe dynamic properties of Event-B model which could be generated by Event-B graph. In most cases, except for the initial location, every location has one or more source locations. Event-B graph uses the combination of edges and location to generate events. The combination of edge and location can generate an equal event. Edge and location, as the basic elements in Event-B graph, is responsible for generating guards and substitution. The combination of location and its source edges under a specific rule can generate an equivalent event.

A complete event generation process from Event-B graph to Event-B model is shown in Fig. 3. The source location_SR connects to the target location_TA by the connection of an edge. The whole predicates generated by edges are the guards of the target location, collectively denoted by $F(v1, v2, \ldots, vn)$ and $Add(s, c, v)$. The whole predicates generated by the target locations, collectively denoted by Add(s, c, v, v'). The guards of event are defined into $G(s, c, v) = F(v_1, v_2, \ldots, v_n) \wedge Add(s, c, v)$ in an edge. And the combination of them generates an event location_TA = array t where $G(s, c, v)$ then $A(s, c, v')$ end. Above all, a complete generation from Event-B graph to Event-B model is satisfied without any loss of information. Certainly, the inverse process is established too.

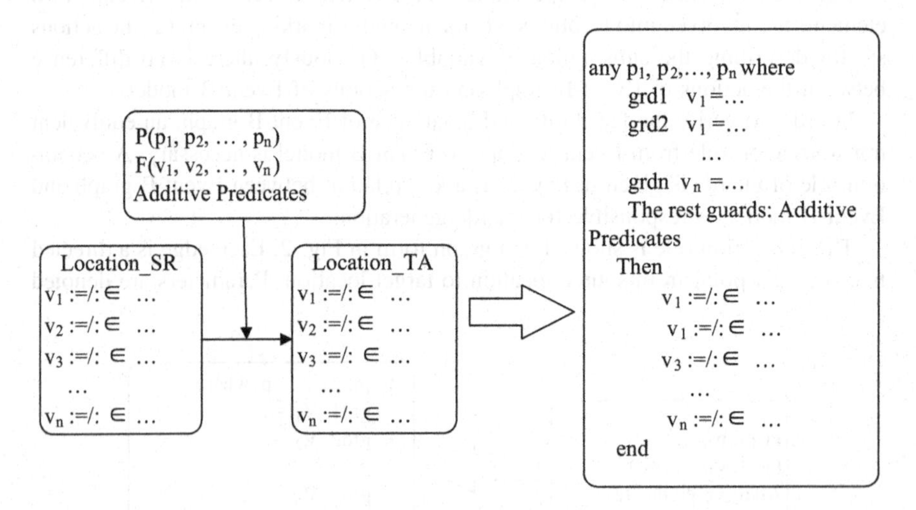

Fig. 3 Event generation from graph to machine

3.3 Event-B Graph Refinement

Event-B graph maintains a clear separation between the context and the graph. Generally, most software systems may contain complex function requirements which may work independently and interact frequently. In most cases, stepwise model construction method is an effective way to solve the complexity of large systems. To master the complexity of large system, Event-B method uses the refinement mechanism to support stepwise modelling. That means Event-B method enables developers to build a model by making it more and more precise, and every aspect of requirement can be added into a model gradually.

To achieve modeling in a sequence of graphs and keep the consistence of these graphs, Event-B graph abstracts the same refinement mechanism from Event-B method. As a result, an ordered sequence of models is formed where each model is supposed to be a refinement of one proceeding model in the sequence.

From a given graph G, a new graph H can be built and asserted to be a refinement of G in Fig. 4. Graph G is said to be an abstraction of H, and H is said to be a refinement of G or a concrete version of it. Likewise, context C, seen by a model G, can be refined to a context D. Like the context extension of Event-B model, the sets and constants of abstract context are kept in the concrete context. In addition, the carrier sets are defined in new properties in the concrete context. Unlike the context extension of Event-B model, the adding of variables which are constrained by new invariants of concrete context also cause the extension of context. The refinement of graph enables developers to refine or merge existing events, and add new events to abstract graph.

Event-B, using locations and edges, maintains merging and refining existing graphs. It also introduces new events trough the adding of locations and edges. Unlike the refinement of Event-B method, the graph refinements are defined in edges and locations. During the existing graph refinement, the adding of guards and

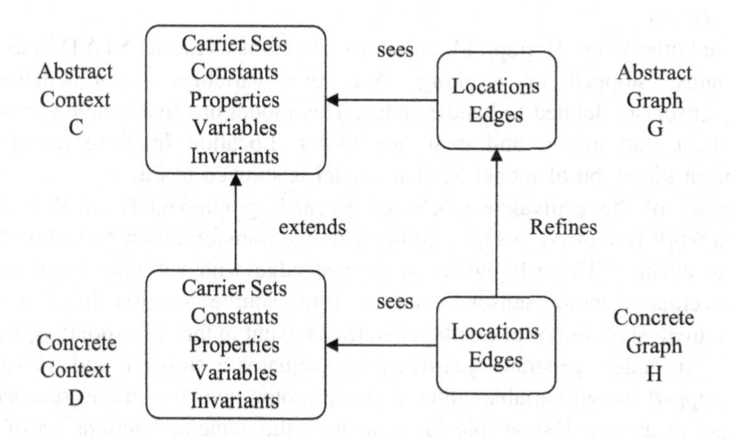

Fig. 4 Event-B graph refinement

P(s, c)∧I(s, c, v)∧H(s, c, w)∧J(s, c, v, w) ⊢ ∃ w′. S(s, c, w, w′)	FIS_REF
P(s, c)∧I(s, c, v)∧J(s, c, v, w)∧H(s, c, w) ⊢ G(s, c, v)	GRD_REF
P(s, c)∧I(s, c, v)∧J(s, c, v, w) ∧H(s, c, w) ∧S(s, c, w, w′) ⊢ ∃ v′. (R(s, c, v, v′)∧J(s, c, v′, w′))	INV_REF

Fig. 5 Refinement laws

actions can be realized by adding of variables in variables filter and additive guards. To introduce new events during refinement, Event-B graph extends the graph according to the adding of locations and edges.

The refinement laws of Event-B graph are given in Fig. 5. Suppose we have an abstract event with guard *G(s, c, v)* and before-after predicate *R(s, c, v, v′)* and a refining concrete event with guard *H(s, c, w)* and before-after predicate *S(s, c, w, w′)*. And *J(s, c, v, w)* are the invariants in concrete model.

4 A Press Controller Example of Event-B Graph

A small Event-B graph and its refinement process is outlined in this section. For expressing the transformation process from Event-B graph to Event-B model, a detail generation is illustrated in Fig. 4. Its aim is to illustrate the transformation between Event-B graph and Event-B model. During the Event-B graph construction procedure, refinement skill is involved too.

A short example is proposed here to show the process how to transform Event-B graph into Event-B model. This is a mechanical press controller example from the book 'Modelling in Event-B: system and software engineering' which is written by Abrial [15]. A transformation from Event-B graph to Event-B model is also showed in this section.

In the initial Event-B graph M, a defined context uses the set STATUS to define two status: stopped or working. And two variables motor_actuator and motor_sensor are defined to handle status. This model has five locations named in Ini_0, treat_start_motor, and treat_stop_motor. Location Ini_0 represent initial location in the graph of model M. The model is showed in Fig. 6.

Because of the equivalence between Event-B graph and Event-B model. A detailed work is showed in Fig. 7 to describe the transformation procedure from a graph to event of Event-B model. A defined edge with variables filter vector F (motor_actuator, motor_sensor) connects from source location Ini_0 to target location treat_start_motor of graph generates an event in the corresponding Event-B model. The edge generate guards motor_actuator = stopped and motor_sensor = stopped trough variables filter vector F(motor_actuator, motor_sensor). And the target location is responsible for generating the name and actions set of Event

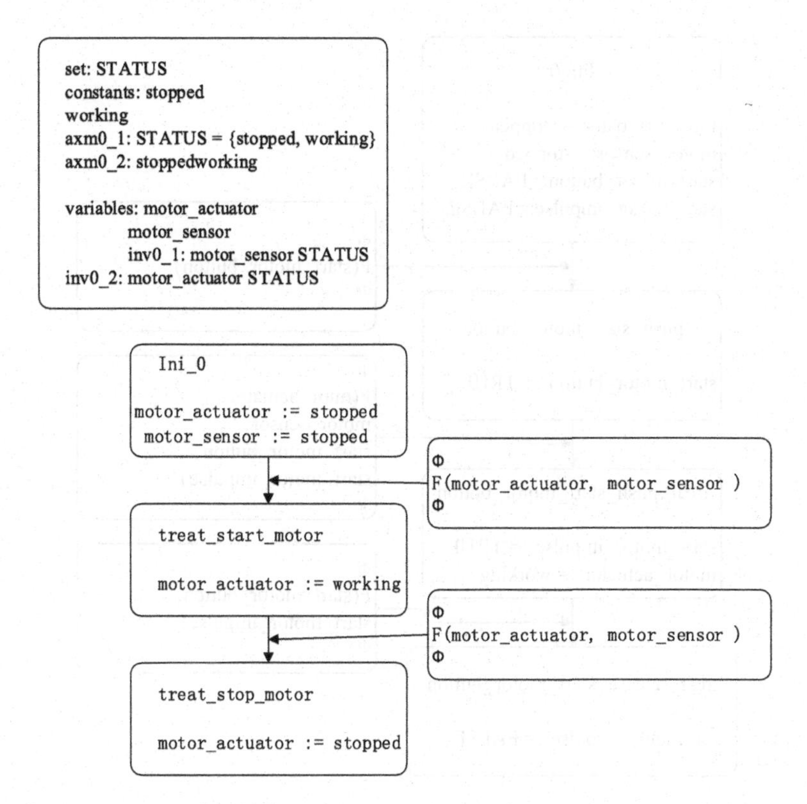

Fig. 6 An Event-B graph model M

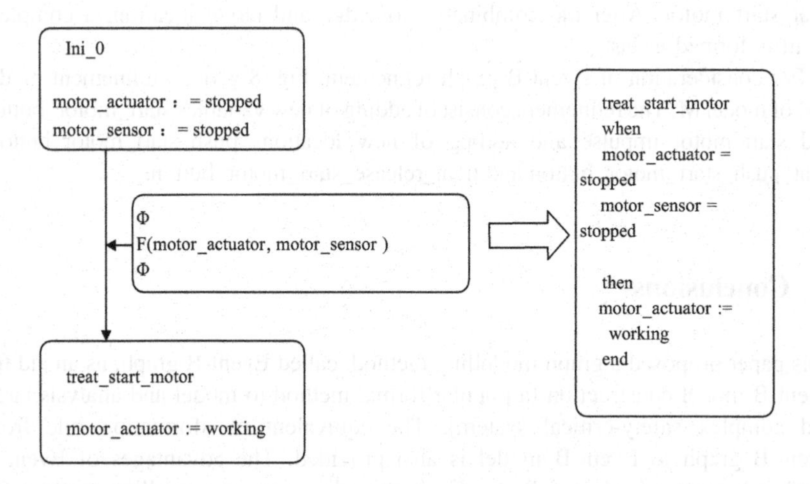

Fig. 7 A generation of event treat_start_motor from graph to machine

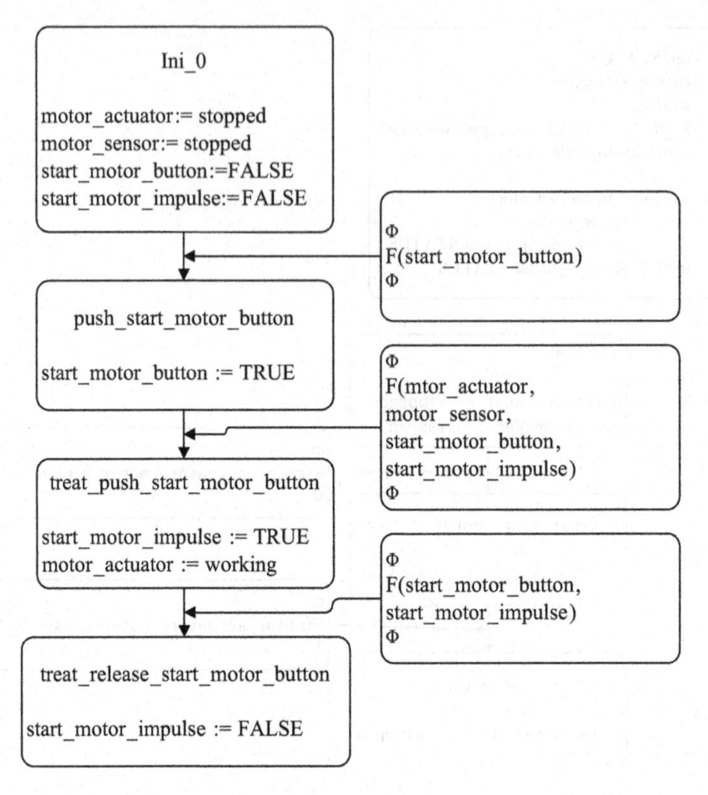

Fig. 8 A concrete graph RM of refined graph M

treat_start_motor. After the combination of edge and target location, a complete event is formed at last.

For consideration of Event-B graph refinement, Fig. 8 gives a refinement model RM of model M. The refinement consist of adding of new variables start_motor_button and start_motor_impulse and adding of new locations push_start_motor_button, treat_push_start_motor_button and treat_release_start_motor_button.

5 Conclusions

This paper proposed a graph modelling method, called Event-B graph, as an aid for Event-B model construction (a popular formal method to model and analysis large and complex safety-critical system). The equivalent transformation rule from Event-B graph to Event-B model is also provided. The advantages of Event-B graph are summarized as follows: (1) it provides a graph modelling method for Event-B modeling based on Event-B language therefor reduce the burden of developers' work and the complexity of Event-B model; (2) it enhances the flexibility and extensibility of Event-B method; (3) it clearly shows the data flow among locations; (4) it eventually enables developers to construct model with refinement strategy.

Further research work is focused on developing a Event-B graph plug-in on Rodin platform. Event-B graph will be further enhanced to define decomposition and include more functions to enhance its modeling capability and range, therefore, to enhance the capability of Event-B method.

References

1. Cansell Dominique, Méry Dominique. The Event-B method—concept and case studies. Monographs in Theoretical Computer Science, Berlin: Springer; 2008. p. 47–152.
2. Abrial J -R, Hallerstede S. Refinement, decomposition, and instantiation of discrete models: application to Event-B. In: International workshop on abstract state machine, vol. 77, no. 1–2; 2007; p. 17–40.
3. Kobayashi T, Ishikawa F, Honiden S. Refactoring refinement structure of Event-B machine. Springer International Publishing AG;2016. p. 444–59.
4. A Edmunds, C Snook, M Walden. On Component-Based Reuse for Event-B [C]. International Conference on Abstract State Machines;2016: 151–166.
5. Dghaym D, Trindade MG, Bulter M, Fathabali AS. In: A graphical tool for Event-B refinement structure. Abstract state machines, alloy, B, Tla, VDM, and Z: international conference;2016.
6. C. Métayer (ClearSy), J.-R. Abrial, L. Voisin (ETH Zürich). Event-B language [EB/OL]. http://rodin.cs.ncl.ac.uk/;2005.
7. Alur R, Dill DL. A Theory of Timed Automata [J]. Theoret Comput Sci. 1994;126(2):183–235.
8. Abrial JR, Lee MKO, Neilson D, Scharbach PN, Sorensen IH. The B-mehod. In: Vdm 91-formal software development, international symposium of Vdm Europe, vol. 552, no. 1;1991. p. 398–405
9. Dghaym D, Trindade MG, Bulter M, Fathabali AS. A graphical tool for Event-B refinement structure. In: abstract state machines, Alloy, B, Tla, VDM, and Z: international conference;2016. p. 269–74.
10. Abrial J -R, Butler MJ, Hallerstede S, Voisin L. A Roadmap for the rodin toolset. German: Springe;2008. p. 347–47.
11. Thai Son Hoang. Steve Schneider, Helen Treharne and David M. Williams. Reasoning about Action System Using the B-Method [J]. Formal Aspects Comput. 2016;28(6):1–27.
12. Said MY, Butler MJ, Snook CF. Language and tool support for class and state machine refinement in UML-B. In: World Congress on Formal Methods;2009. p. 579–95.
13. Iliasov A. Use case scenarios as verification conditions: Event-B/Flow approach; 2011, 8968: 9–23.
14. Schneider S, Treharne H, Wehrheim H. The behavioural semantics of Event-B refinement. Formal Aspects Comput. 2014; 26(2):251–80.
15. Abrial J-R. Modeling in Event-B: system and software engineering. New York: Cambridge University Press; 2010. p. 100–48.

Optimization of Mobility Pattern for Underwater Wireless Sensor Networks

Liping Liu and Meng Chen

Abstract Due to the fluidity of water and limitations of nodes, it is challenging to update node's location and movement at all times. The paper proposes optimization of mobility pattern for underwater wireless sensor networks. Based on the ocean current model, Gauss radial basis function is utilized as spatial function to construct the mobility pattern for underwater nodes. Considering centers and coefficients changing with node's location and movement, the cost function of average dissimilarity is selected to choose centers to increase accuracy. The extended Kalman algorithm is used to update coefficients when movements changing. According to the real-time mobility pattern, nodes can estimate future location. Results show that the optimal mobility pattern is more accurate and suitable for underwater wireless sensor networks in the seashore environment.

Keywords Underwater wireless sensor networks · Mobility pattern · Optimization · Extended kalman

1 Introduction

Last several years have overseen a rapidly growing interest in underwater wireless sensor networks (UWSNs) due to its wide spectrum of applications in aquatic environment, such as oceanographic data collection, pollution monitoring, disaster prevention and assisted navigation [1–3].

The localization of UWSNs is indispensable in the real application environment, which means the data information collected by underwater sensor nodes needs to be combined with their position to make it useful. However, compared with the terrestrial environment, the aquatic features make the localization process difficult. For example, underwater sensor nodes move with water currents, making the location updated periodically. So many proposed algorithms suitable for terrestrial envi-

L. Liu (✉) · M. Chen
School of Electrical and Information Engineering, Tianjin University, Tianjin 300072, China
e-mail: lipingliu@tju.edu.cn

Y. Jia et al. (eds.), *Proceedings of 2017 Chinese Intelligent Systems Conference*, Lecture Notes in Electrical Engineering 459,
https://doi.org/10.1007/978-981-10-6496-8_60

ronment cannot be applied directly. On the other hand, the acoustic signal is utilized to achieve communication among underwater sensor nodes instead of electromagnetic wave. The features of acoustic signal such as high error rate, propagation delay and limited bandwidth [4, 5] make underwater sensor nodes more difficult to get location. In recent years, some new schemes have been proposed for aquatic environment to solve the problem of localization coverage and localization error.

Erol et al. [6] proposed AAL which utilizes the Autonomous Underwater Vehicle (AUV) to localize nodes. AUV can get location through GPS and then move underwater following the planned path. When receiving information from at least three non-collinear AUVs, nodes can get location through trilateration method. Due to the accurate location of AUV, the localization error of un-known nodes is small. [7] proposed UDB scheme which also utilizes AUV to get located. The advantage for UDB is that it is more energy-efficient due to the silent state of nodes. In order to localize the large-scale UWSNs, [8] proposed LSLS which increases a complementary phase. In this process, un-known nodes send localization requests to other anchor nodes and select a different set of reference nodes to achieve localization. However, the scheme is suitable for static UWSNs, which means nodes of the network should be fixed to a certain position. [9] proposed DNRL scheme which can solve the mobile UWSNs localization with the mobile located DNR nodes. However, since the number of DNR nodes is limited, the localization coverage is finite. In order to solve the problem, [10] used the well-known nodes to localize other un-known nodes, increasing the localization coverage. However the localization error increases due to the less accurate reference nodes. Zhou et al. [11] proposed SLMP algorithm which is a predictable localization method. In SLMP, AR model is utilized to simulate the movement of underwater nodes and the Durbin algorithm is exploited to update the mobility pattern. According to the node mobility pattern and the past location, nodes can achieve self-localization. The advantage of the method is that it can decrease communication frequency so as to reduce communication cost. However, the localization error increases because AR model is an approximate mobility pattern of underwater nodes.

In the paper, optimization of UWSN mobility pattern is proposed. The node mobility pattern is constructed based on the ocean current model so that it can be closer to the movement of nodes. Considering centers and coefficients of mobility pattern influencing its accuracy, the cost function of average dissimilarity is adopted as basis to choose centers and the extended Kalman algorithm is utilized to optimize coefficients. As a result, underwater nodes can get real-time and accurate mobility pattern and estimate the future position according to it and the past location.

2 Construction of Node Mobility Pattern

2.1 UWSN Architecture

The UWSN is deployed in the seashore environment where water current is relatively flat. There are three types of nodes in the UWSN, which are beacon nodes, anchor nodes and ordinary nodes. Beacon nodes flow on the water and obtain locations through GPS. They can help anchor nodes estimate positions. Anchor nodes spread underwater and have large energy and strong computing ability. They can communicate with beacon nodes as well as ordinary nodes. Ordinary nodes have simple structure and limited energy. They can only communicate with anchor nodes within the communication radius. In order to save cost, the number of beacon nodes and anchor nodes is small while ordinary nodes are the largest.

2.2 Construction of Mobility Pattern

Considering that the mobility of objects in the seashore environment is not totally random process and movements of underwater sensor nodes have time and spatial correlations [12], we construct the node mobility pattern based on the ocean current model. The movement of ocean current is mainly caused by the tide. Therefore the ocean current model comprising of temporal and spatial basis functions of tidal frequency [13] is chosen as the basis model. It is described as Eq. (1).

$$y(x,t) = y_0(x) + \sum_{i=1}^{N} [g_i(x)\cos(\omega_i t)] + \sum_{i=1}^{N} [h_i(x)\sin(\omega_i t)] \tag{1}$$

where x is the position to compute velocity at time t, y is the velocity at position x and time t, N is the number of tides, $\sin(\omega_i t)$ and $\cos(\omega_i t)$ are the temporal basis functions with tidal frequency ω_i, respectively, $y_0(x)$ is the average speed of ocean current during observation time and $y_0(x)$, $g_i(x)$ and $h_i(x)$ are functions of the position x.

The ocean current model has low temporal and spatial solution, making it more suitable for the large geographical scope, such as global climate monitoring, assisted remote navigation. However, the ocean current model needs to be more precise so that it can be used as mobility pattern for UWSN. Considering that Gauss radial basis function has small approximation error and high smoothness and accuracy, we choose it as spatial basis function to insert to the formula (1) to improve temporal and spatial solution. Then we can get the equations as follows:

$$\varphi_j(x) = \exp\left(\frac{-\|x - c_j\|^2}{2\sigma^2}\right) \tag{2}$$

$$y_0(x) = \sum_{j=1}^{M} \theta_{1,j}\varphi_j(x) \tag{3}$$

$$g_i(x) = \sum_{j=1}^{M} \theta_{2i,j}\varphi_j(x) \tag{4}$$

$$h_i(x) = \sum_{j=1}^{M} \theta_{2i+1,j}\varphi_j(x) \tag{5}$$

where M is the number of Gauss radial basis functions, c_j is the jth center of Gauss radial basis function, σ is the width of Gauss radial basis function and $\theta_{i,j}$ is the coefficient of Gauss radial basis function. Then according to the above equations, the mobility pattern for underwater sensor nodes is as follows:

$$y(x,t) = \sum_{j=1}^{M} \lambda_j(t)\varphi_j(x) \tag{6}$$

$$\lambda_j(t) = \theta_{1,j} + \sum_{i=1}^{N} \theta_{2i,j}\cos(\omega_i t) + \sum_{i=1}^{N} \theta_{2i+1,j}\sin(\omega_i t) \tag{7}$$

3 Optimization of Node Mobility Pattern

With time and position changing, the mobility pattern needs to be updated to simulate movements of the node in time. Considering centers and coefficients are the main factors to determine accuracy of the mobility pattern, they need to be optimized according to node's different location to keep high accuracy.

3.1 Selection of Center

The center of Gauss radial basis function is affected by node's location. That is to say, when the node moving from one place to another, the center needs to be selected a new one. However, if the update frequency of center is high, the energy consumption of computation will increase greatly. Therefore, the center is chosen to make it simulate movement of the node when it moves in a certain region so as to decrease update frequency. In the paper, we choose three Gauss radial basis functions to construct the mobility pattern. So three centers need to be chosen in the region where the node moves. As shown in Fig. 1, three centers are chosen in Region 1 to simulate movement of the node. The centers won't change as long as

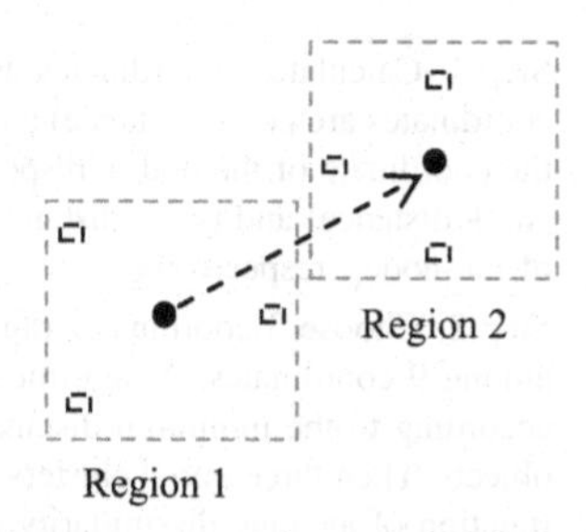

Fig. 1 Node movement diagram

the node is in the Region 1. While it moves to Region 2, another three centers need to be selected.

In order to spread centers to different positions as far as possible, the cost function of average dissimilarity is chosen as the basis to select centers. In this way, the node can have at least one center around it when it moves in one region, increasing accuracy of estimated velocity. The cost function of average dissimilarity is described as follows:

$$E = \sum_{j=1}^{3} \sum_{p \in cluster_j} \left(\|p - o_j\|^2 \right) \tag{8}$$

where $cluster_j$ is the jth cluster, o_j is the represent object in the jth cluster, namely the center of Gauss Radial basis function, and p are the remaining objects except o_j in the jth cluster.

The algorithm of selecting centers in the mobility pattern is described in Table 1.

3.2 Update of Coefficients

When the node's movements change, the coefficients also need to be updated so that the mobility pattern can track the node in time. Considering the coefficient Eq. (7) is a nonlinear one and the extended Kalman algorithm has less historical data, fast computation speed and small estimation error, the extended Kalman algorithm is utilized to update them.

First, we discretize the Eq. (7) at $t = t_k$ with localization period T_1 and $t_k = k \times T_1$. The discrete equation is as follows:

$$\lambda_j(\mathrm{k}T_1) = \theta_{1,j} + \sum_{i=1}^{N} \theta_{2i,j} cos(\omega_i kT_1) + \sum_{i=1}^{N} \theta_{2i+1,j} \sin(\omega_i kT_1) \tag{9}$$

Then, we adopt Taylor formula to expand $\sin(\omega kT_1)$ and $\cos(\omega kT_1)$ as follows:

Table 1 The algorithm to update the center

Step 1: Calculate 9 coordinates around the node. The x coordinates are (x_0 + distance) and (x_0 − distance) and x_0 (x_0 is the coordinate of the node), respectively. The y coordinates are (y_0 + distance) and (y_0 − distance) and y_0 (y_0 is the coordinate of the node), respectively
Step 2: Choose 3 coordinates randomly as represent objects among 9 coordinates. Assign the remaining coordinates according to the minimum distance from them to the represent objects. Then three initial clusters are formed. Calculate the cost function of average dissimilarity E_0 with Eq. (8)
Step 3: Choose another coordinate as new represent object substituting the previous one in each cluster to calculate the cost function of average dissimilarity E_1 with Eq. (8)
Step 4: If $E_1 < E_0$, the new represent object replaces the previous one and E_1 substitutes for E_0. Otherwise, the present object and E_0 don't change
Step 5: Repeat step 3 and 4 until the cost function of average dissimilarity E_0 doesn't change
Step 6: The last represent object in each cluster will serve as centers of Gauss radial basis functions

$$\sin(\omega k T_1) = \sin(\omega(k-1)T_1) + \omega T_1 \cos(\omega(k-1)T_1) \tag{10}$$

$$\cos(\omega k T_1) = \cos(\omega(k-1)T_1) - \omega T_1 \sin(\omega(k-1)T_1) \tag{11}$$

With Eqs. (10) and (11), we can get the state Eq. (12).

$$\lambda(\mathrm{k}) = \lambda(k-1) + \sum_{i=1}^{N} \omega_i T_1 \begin{bmatrix} \theta_{2i,1} & \theta_{2i+1,1} \\ \vdots & \vdots \\ \theta_{2i,M} & \theta_{2i+1,M} \end{bmatrix} \begin{bmatrix} -\sin(\omega_i(k-1)T_1) \\ \cos(\omega_i(k-1)T_1) \end{bmatrix} + w(k-1) \tag{12}$$

where $\lambda(k)$ is $\lambda(kT_1) = [\lambda_1(kT_1), \lambda_2(kT_1), \ldots, \lambda_j(kT_1), \ldots, \lambda_M(kT_1)]^{\mathrm{T}}$,$w(k-1)$ is process noise and $w(k-1) = [w_1(k-1), w_2(k-1), \ldots, w_M(k-1)]^{\mathrm{T}}$.

The observation equation is described as Eq. (13).

$$y(k-1) = H(k-1)\lambda(k-1) + v(k-1) \tag{13}$$

where $\mathrm{y}(k-1)$ is the real velocity of node, $v(k-1)$ is the observation noise and$v(k-1) = [v_1(k-1), v_2(k-1), \ldots, v_M(k-1)]^{\mathrm{T}}$, and $H(k-1)$ is the observation matrix, $H(k-1) = [\varphi_1(k-1), \varphi_2(k-1), \ldots, \varphi_M(k-1)]$.

According to the state Eq. (12) and the observation Eq. (13), we utilize the extended Kalman algorithm to get the optimal $\hat{\lambda}(k)$, namely $\hat{\lambda}_j(k)(j = 1, 2, \ldots, M)$. Then, the optimal coefficient $\theta_{i,j}$ can be updated with Eq. (14).

$$\hat{\theta}_j(k) = \left[\alpha^{\mathrm{T}}(k)\alpha(k)\right]^{-1}\alpha^{\mathrm{T}}(k)\hat{\lambda}_j(k) \tag{14}$$

where $\alpha(\mathrm{k}) = [1, \cos(\omega_1 kT_1), \sin(\omega_1 kT_1), \ldots, \cos(\omega_N kT_N), \sin(\omega_N kT_N)]$.

3.3 *Optimization of Mobility Pattern*

Considering that the three types of nodes are different, optimization process for the mobility pattern is different. Beacon nodes needn't estimate their mobility pattern because they can get positions through GPS. Anchor nodes choose centers with the cost function of average dissimilarity as basis and update coefficients through the extended Kalman algorithm to optimize the mobility pattern. For ordinary nodes, which cannot utilize the same method to update coefficients as anchor nodes do due to their limited energy, they calculate coefficients with Eqs. (15) and (16).

$$\gamma_{sn} = \frac{\frac{1}{d_{sn}}}{\sum_{s=1}^{sum} \frac{1}{d_{sn}}} \tag{15}$$

$$\theta_n(j) = \sum_{s=1}^{sum} \gamma_{sn}\theta_s(j) \tag{16}$$

where d_{sn} is distance from anchor node s to ordinary node n, and sum is the total number of anchor nodes in the communication radius of ordinary node n.

After getting the real-time mobility pattern, nodes can calculate their locations with Eq. (17).

$$Loc(t+1) = Loc(t) + \hat{v}(t) \times T_1 \tag{17}$$

4 Analysis of Experimental Results

We use MATLABR2014b to simulate the UWSN in the region from 117.25 to 132.2°E and from 24 to 43.45°N. Some parameters are listed in Table 2.

4.1 *Influences on Node's Velocity*

Node density influences the number of anchor nodes within communication radius of ordinary nodes. Prediction window is another factor to affect frequency of

Table 2 Simulation parameters

Parameters	Values
The number of nodes	500
Beacon node proportion	5%
Anchor node proportion	10%
Ordinary node proportion	85%
Prediction window size (T_w)	30 s
Localization period (T_1)	1 s
Communication radius (R)	20 m
Error threshold of updating anchor node	0.05R

updating coefficients. Therefore we analyze the average velocity error varying with them and compare it with SLMP algorithm.

In Fig. 2a, the average velocity error in the two algorithms increases monotonically with node density, while the velocity error of the proposed algorithm is lower than that in SLMP algorithm. Besides, with the increase of anchor nodes, the velocity error decreases and it is also lower in our algorithm. It is reasonable since the node mobility pattern constructed in the paper is suitable for complex underwater environment, while AR model in SLMP is a linear one. So when movements of underwater sensor nodes are complex, the error is smaller in our algorithm.

Figure 2b shows that the average velocity error in the proposed algorithm shows a slight growing trend. In SLMP, it decreases at beginning, then it increases since the length of prediction window is about 50 s. With the length of prediction window growing, the mobility pattern of anchor nodes in the proposed algorithm won't update for a long time, thus increasing the speed error. However it doesn't influence the speed error too much. In SLMP, if the prediction window is too small, there isn't enough data information to construct AR model. Therefore with increasing of prediction window at beginning, it can describe movements clearly with more data information. However, if the prediction window continues to increase, movements

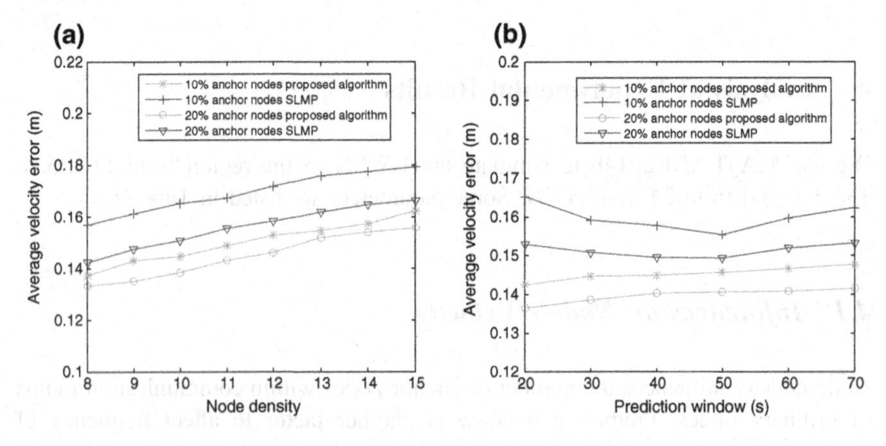

Fig. 2 The velocity changing with **a** node density **b** prediction window

in one prediction window become complex and it is difficult to estimate the future mobility pattern with the formal one. Therefore, the velocity error drops first and increases when the prediction window becomes larger.

4.2 Influences on Localization Coverage

The localization coverage is defined as the rate of located nodes whose localization error is smaller than half of communication radius among all nodes. Results of localization coverage varying with node density and prediction window are shown in Fig. 3a, b.

In Fig. 3a, the localization coverage increases with node density. The localization coverage is larger in the proposed algorithm than that in SLMP algorithm. Besides, it also increases with the proportion of anchor nodes. In SLMP algorithm, to ensure the accuracy of ordinary nodes' mobility pattern, un-known nodes need to get at least four anchor nodes to achieve localization. In the proposed algorithm, due to the accurate mobility pattern, it doesn't have the requirement of the number of anchor nodes, which increasing the localization coverage.

Figure 3b shows the localization coverage in the proposed algorithm decreases slowly with the prediction window growing. In SLMP algorithm, it increases when the prediction window is about from 20 s to 50 s, and then it decreases when the predict window stills rises. In the proposed algorithm, with the prediction window growing, average speed error becomes larger. So the number of nodes satisfying position requirement reduces, thus decreasing localization coverage. In SLMP, average speed error decreases at beginning and then increases with larger prediction window. Therefore, localization coverage has similar trend as speed error.

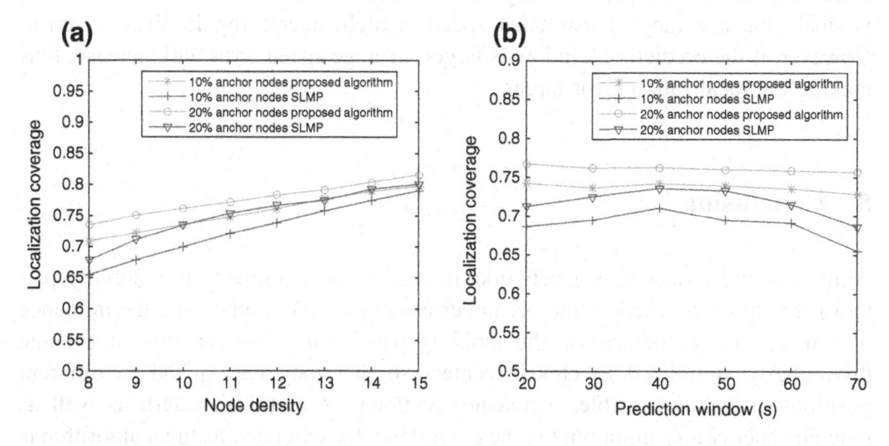

Fig. 3 The localization coverage changing with **a** node density **b** prediction window

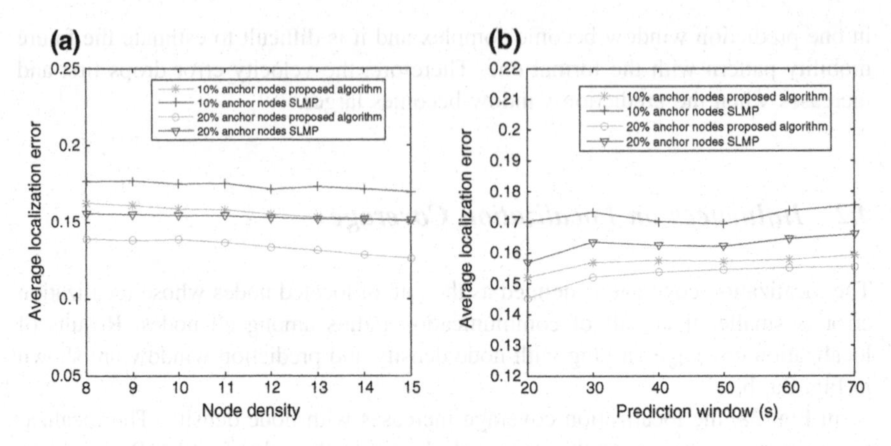

Fig. 4 The average localization error changing with **a** node density **b** prediction window

4.3 *Influence on Average Localization Error*

The localization error is used to measure the accuracy of located nodes. We compare it with SLMP algorithm and they are shown in Fig. 4a, b.

Figure 4a shows the average localization error in the two algorithms has a declining trend with the increasing node density and it is lower in the proposed algorithm. As we know, the average localization error is influenced by localization coverage, which makes it decrease. In Fig. 4b, the localization error increases with the prediction window growing in the proposed algorithm. While in SLMP, the average localization error increases at first, then it decreases slowly from 30 s to 50 s. When the prediction window still increases, it rises again. There is accumulative error for ordinary nodes. The longer the prediction window, the larger the accumulative error is. Therefore the average localization error shows a slight increasing trend in the proposed algorithm. In SLMP, when the prediction window is small, the accuracy of estimated speed is high, decreasing localization error. However, if the prediction window is larger, average speed error will increase, thus making the localization error larger.

5 Conclusion

Aiming at underwater sensor networks in seashore environment, the mobility pattern is constructed based on the ocean current model. After analyzing the influence of centers and coefficients on the mobility pattern, the cost function of average dissimilarity is utilized to choose centers so that they can spread to different positions as far as possible, increasing accuracy of mobility pattern as well as reducing energy consumption. On the other hand, the extended Kalman algorithm is selected to update coefficients when node's movement changing, improving real time of mobility pattern. The proposed mobility pattern is compared with SLMP algorithm in average speed error, localization coverage and average localization error. Results show that the mobility pattern proposed in the paper has better performances and is suitable for complex environment.

References

1. Silva AR, Vuran MC. Development of a test bed for wireless underground sensor networks. EURASIP J Wirel Commun Netw. 2010;1:1–14.
2. Lin M, Wu Y, Wassell I. Wireless sensor network: water distribution monitoring system. In: 2008 IEEE radio and wireless symposium. 2008. p. 775–8.
3. Yoon S-U, Cheng L, Ghazanfari E, Pamukcu S, Suleiman MT. A radio propagation model for wireless underground sensor networks. Glob Telecommun Conf. 2011;57(4):1–5.
4. Akyildiz IF, Sun Z, Vuran MC. Signal propagation techniques for wireless underground communication networks. Elsevier J Phys Commun. 2009;2(3):167–83.
5. Vuran MC, Akyildiz IF. Channel model and analysis for wireless underground sensor networks in soil medium. Elsevier J Phys Commun. 2010;3(4):245–54.
6. Erol M, Vieiral LFM, Gerla M. AUV-aided localization for underwater sensor networks. In: 2007 international conference on wireless algorithms, systems and applications. 2007. p. 44–54.
7. Luo H, Guo Z, Wei D, et al. LDB: Localization with directional beacons for sparse 3D underwater acoustic sensor networks. J Netw. 2010;5(1):28–38.
8. Cheng W, Thaeler A, Cheng X, Liu F, et al. Time-synchronization free localization in large scale underwater acoustic sensor networks. In: 29th IEEE international conference on distributed computing systems workshops, vol. 9, no.11, 2009; p. 80–7.
9. Yao X. Survey of underwater wireless sensor networks localization technology. Mod Electron Tech. 2013;36(7):11–5.
10. Erol M, Vieira LFM, Caruso A, et al. Multi stage underwater sensor localization using mobile beacons. In: The second international conference on sensor technologies and applications. 2008. p. 710–14.
11. Zhou Z, Zheng P, Cui J, Shi Z, et al. Scalable localization with mobility prediction for underwater sensor networks. IEEE Trans Mob Comput. 2011;10(3):335–48.
12. Beerens SP, Ridderinkhof H, Zimmerman JTF. An analytical study of chaotic stirring in tidal areas. Chaos Solut Fractals. 1994;4(6):1011–29.
13. Pawlowicz R, Beardsley B, Lentz S. Classical tidal harmonic analysis including error estimates in MATLAB using T-TIDE. Comput Geosci. 2002;28(8):929–37.

Continuous Prediction of Joint Angle of Lower Limbs from sEMG Signals

Yihao Du, Hao Wang, Shi Qiu, Jinming Zhang and Ping Xie

Abstract In order to realize the rehabilitation training of mirror movement in stroke patients, a new motion analysis method of EMG signal is proposed. First, surface electromyography (sEMG), hip joint and knee joint angles of 6 lower limb muscles are collected synchronously. Then, by introducing the coherence analysis and calculating the significant area index, the coupling relationship between the sEMG and the joint angle is quantitatively described, and the muscles of the most coupling relationship are set to the input channels of the model. Next, we introduce the least squares extreme learning machine algorithm based on golden section (GS-LSELM), and establish a nonlinear prediction model between sEMG and joint angle. Finally, the experimental results show that the proposed method can quickly build the model under different motion periods, and it could be used in the tracking control of the rehabilitation robot.

Keywords sEMG · Coherence analysis · GS-LSELM · Angle prediction · Motion analysis

---

Y. Du · H. Wang · S. Qiu · J. Zhang · P. Xie (✉)
School of Electrical Engineering, Yanshan University, Qinhuangdao 066004, China
e-mail: pingx@ysu.edu.cn

Y. Du
e-mail: 154060324@qq.com

H. Wang
e-mail: 1214817535@qq.com

S. Qiu
e-mail: 1548962898@qq.com

J. Zhang
e-mail: 1363789634@qq.com

Y. Jia et al. (eds.), *Proceedings of 2017 Chinese Intelligent Systems Conference*, Lecture Notes in Electrical Engineering 459,
https://doi.org/10.1007/978-981-10-6496-8_61

1 Introduction

According to WHO, stroke has become the leading cause of premature death after coronary heart disease and lower respiratory tract infection, and 75% of the patients show varying degrees of limb motor dysfunction, which seriously affected the quality of patients' life. In recent years, rehabilitation robot technology has developed rapidly, and has been widely used in clinical rehabilitation [1]. Studies have shown that the active training mode based on human-computer interaction can improve the participation of patients and accelerate the recovery of motor function [2].

One of the major sequelae of stroke patients is unilateral motor dysfunction. Therefore, according to the motion analysis of sEMG, injured limb can achieve active rehabilitation exercises by mirroring healthy limb [3]. However, there are still some problems in the motion analysis of sEMG, such as poor real-time and low accuracy, which restricts the development of human-computer interaction technology on rehabilitation exercises.

In this paper, we select the muscle channel based on the analysis of the coupling relationship between the joint angle of the lower limb and the EMG signal, which can reduce the time-consuming and instability of model caused by data redundancy. We propose the least squares extreme learning machine algorithm based on golden section(GS-LSELM), and establish a prediction model between the sEMG and joint angle, which is used to predict the hip angle and knee angle, and achieve the motion analysis finally. The experimental results show that, compared with the BP neural network, the model establishment time is reduced by 99.85% and the prediction error is also reduced, which satisfy real-time and accuracy requirements of the sEMG motion analysis and can be used to active rehabilitation robot tracking control.

2 Joint Motion Analysis Algorithm Based on GS-LSELM

2.1 *Algorithm Principle*

The principle of sEMG motion analysis algorithm is shown in Fig. 1. The whole process is divided into three parts: synchronizing data acquisition, data processing and analysis, model training and angle prediction.

2.2 *Data Acquisition*

In this paper, we make the lower limb flexion and extension as experimental mode of operation, and simultaneously collect sEMG, hip and knee angle of seven

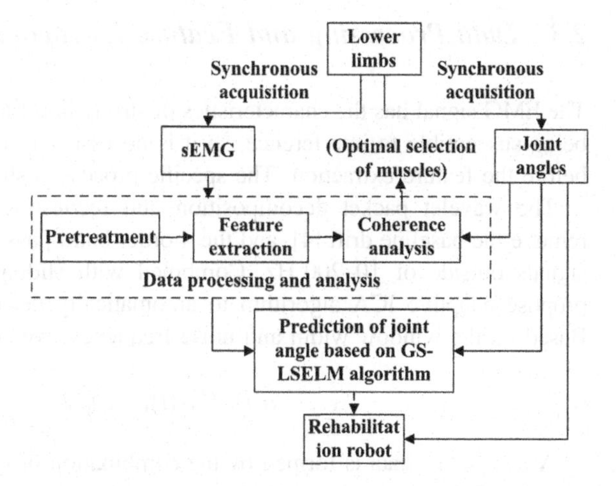

Fig. 1 The principle of joint motion analysis algorithm

healthy subjects (5 boys, 2 girls, (25 ± 2) years old). Before the experiment, the subjects were asked to have no muscle fatigue and had a good mental state, and were familiar with the experimental process. With the EMG acquisition equipment of the US Delsys company, we recorded vastus rectus (VR), vastus lateralis (VL), vastus medialis (VM), semitendinosus muscle (SM), biceps muscle(BM), and tibialisa-nterior (TA) of the subjects.

The experimental procedure is as follows: Before experiment, a subject lies on the experimental platform with the feet fixed on the slideway and pedal, as shown in Fig. 2, and does continuous motion in three cycles (5 s, 3.5 s, 2 s). To avoid muscle fatigue, the subject rest 3–5 min before the start of each experiment. The data of the seven subjects are recorded and repeated three times according to the above procedure.

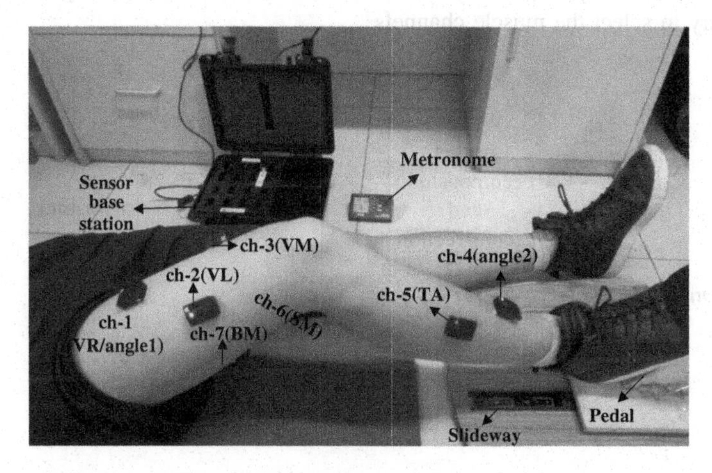

Fig. 2 Data acquisition equipment and mark point location

2.3 Data Processing and Feature Extraction

The EMG signal has the characteristics of strong non-linearity, non-stationarity and being susceptible to interference, so it is necessary to preprocess the EMG signal before the feature extraction. The specific process is shown in Fig. 3:

The wavelet packet decomposition and reconstruction technique is used to remove the baseline drift [4], and the 4 order band-pass filter is used to remove the signals outside of 10–200 Hz. Combined with sliding window technology, we propose adaptive ICA algorithm to automatically detect power frequency noise. Based on the window width and noise frequency, we construct signals:

$$a(t) = [a_1(t), \cdots, a_6(t)]^{\mathrm{T}} \quad (1)$$

A new set of data is formed by the combination of the structural signal and the original sEMG, and its power frequency noise and harmonic signal is separated by traditional ICA method. After preprocessing, the EMG is expressed by ξ, and its characteristics *WL*(wave length) is abstracted, which represents the accumulated wavelength of ξ over a period.

$$WL = \sum_{i=1}^{N-1} |\xi_{i+1} - \xi_i| \quad (2)$$

where, N is the sampling number within a period of time, and 13 sampling points are selected as a data segment in this paper.

As shown in Fig. 4, with the change of joint angle, sEMG of vastus rectus muscle (VR) shows a strong periodicity, and the biceps muscle (BM) has poor periodicity and robustness, which has a small change in 0 ~ 5 s and a big change in 10 ~ 15 s. At the same time, too many input channels will increase the complexity of the model, reduce the stability of the model and increase the training time, so it is necessary to select the muscle channels.

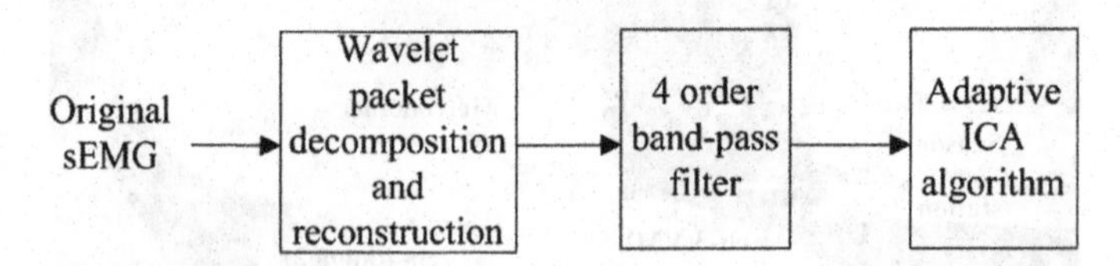

Fig. 3 Original sEMG preprocessing

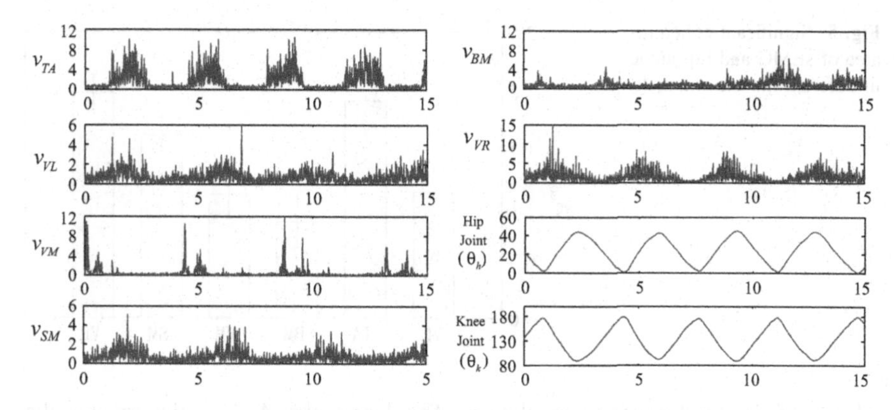

Fig. 4 Characteristic signal and joint angle signal of sEMG

2.4 Muscle Optimal Selection

In this paper, the local frequency coherence analysis method [5] is used to calculate the coherence value between the EMG feature signal v and the joint angle signal θ. Taking hip for example, the hip angle frequency spectrum is mainly concentrated in the 0~2 Hz. We record the frequency channel as ω, and have a coherence analysis of v and θ_h in the frequency channel ω by the following formula.

$$c_{v_\omega\theta_h}(f) = \frac{|<S_{v_\omega\theta_h}(f)>|^2}{|<S_{v_\omega v_\omega}(f)>|*|<S_{\theta_h\theta_h}(f)>|} \tag{3}$$

where $S_{\theta_h\theta_h}(f)$ and $S_{\theta_h\theta_h}(f)$ are separately the self spectral density function of v and θ_h in ω, and $S_{v_\omega\theta}(f)$ is the cross spectral density function of v and θ_h in ω. $c_{v_\omega\theta_h}(f)$ is used to describe the linear correlation of the two signals in ω, whose span is from 0 (no correlation) to 1(perfect correlation).

The significant coherence threshold CL is used to describe the coherence degree between v and θ_h.

$$CL(\alpha) = 1 - (1-\alpha)^{\frac{1}{n-1}} \tag{4}$$

where n is the number of data segments involved in spectral estimation, α is the confidence level.

The significant coherence area index $A_{coh(\omega)}$ is used to describe the coherence of v and θ_h in ω.

$$A_{coh(\omega)} = \sum_f \Delta f \cdot (C_{v_\omega\theta_h}(f) - CL) \tag{5}$$

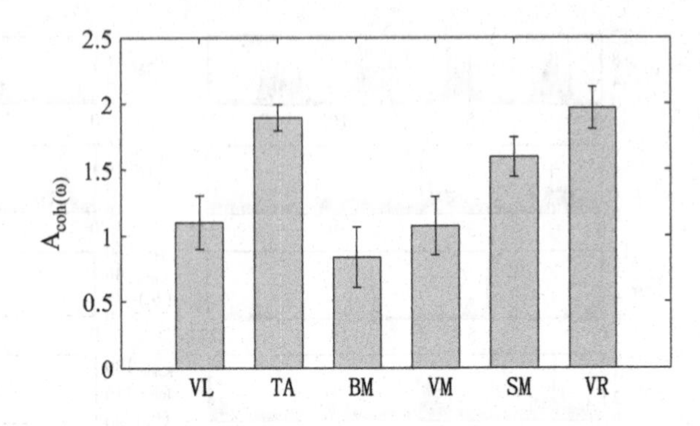

Fig. 5 Significant coherence area of sEMG and hip angle in different muscle channels

where Δf is the frequency resolution. The larger the $A_{coh(\omega)}$, the greater the coherence of v and θ_h is in ω.

As shown in Fig. 5, in the frequency channel ω, the coherence between the anterior tibial muscle and the hip joint is the largest, followed by the vastus rectus muscle and semitendinosus muscle. The same conclusion is found in the analysis of the knee joint. So the anterior tibial muscle and vastus rectus are used as the input channel to predict the hip and knee angles.

2.5 Joint Angle Prediction Based on GS-LSELM

In this paper, we use the least squares method [6] to optimize the input weight and bias of the limit learning machine, and combine the golden segmentation algorithm to optimize the number of hidden layer nodes and simplify the network structure to obtain the optimal prediction accuracy.

Network input of the predicted model are the tibial anterior muscle signal u_{AT} and vastus rectus muscle signal u_{VR}, and the difference signals $\Delta u_{AT} = u_{AT}(i+1) - u_{AT}(i)$, $\Delta u_{VR} = u_{VR}(i+1) - u_{VR}(i)$, and network output are hip and knee joint angles. Taking hip angle prediction for example, $U = \{u_{j,1}, u_{j,2}, \ldots, u_{j,n}\}$(j = 1,2,3,4) and $\theta_h = \{\theta_1, \theta_2, \ldots, \theta_n\}$ are the input and output of the network respectively, the number of samples and hidden layer nodes are n and L respectively, and the implicit layer excitation function is $G(\cdot)$. We chooses the sigmode function as an excitation function:

$$G(z) = \frac{1}{1 + e^{-z}} \tag{6}$$

Expected mathematical model is

$$\theta_h = \sum_{i=1}^{L} \beta_i G_i(\alpha_i \times u_i + b_i) \tag{7}$$

where $\alpha_i = [\alpha_{i1}, \alpha_{i1}, \ldots, \alpha_{in}]^T$ is the weight of the i-th hidden layer node and the input node, b_i is the i-th hidden layer node threshold, $\beta_i = [\beta_{i1}, \beta_{i2}, \ldots, \beta_{iL},]^T$ is the connection weight of the output layer node and the i-th hidden layer node. Formula (8) can be simplified as: $\theta_h = H \cdot \beta$.

$$H = G\begin{bmatrix} \alpha_1 \cdot u_1 + b_1 & \cdots & \alpha_L \cdot u_1 + b_L \\ \vdots & \ddots & \vdots \\ \alpha_1 \cdot u_n + b_1 & \cdots & \alpha_L \cdot u_n + b_L \end{bmatrix} = G(u \cdot \alpha) \tag{8}$$

where H is the hidden layer output matrix,

$$u = \begin{bmatrix} u_1 & u_2 & \cdots & u_n \\ 1 & 1 & \cdots & 1 \end{bmatrix}^T \tag{9}$$

$$\alpha = \begin{bmatrix} \alpha_1 & \alpha_2 & \cdots & \alpha_L \\ b_1 & b_2 & \cdots & b_L \end{bmatrix} \tag{10}$$

According to the Mohr-Penrose inverse matrix theory $u \cdot \alpha = \theta_h \theta_h^+ G^{-1}(\theta_h \beta^+)$, assuming $Z = \theta_h^+ G^{-1}(\theta_h \beta^+)$, then we can obtain the linear relationship between the input weight and the output value: $u \cdot \alpha = \theta_h Z$. By the principle of least squares solution, when Z is randomly generated, the input weights u and the bias α can be obtained and can be substituted into the formula (7) and (8) to calculate the hidden layer output matrix H and output weight β, so as to obtain the parameters of the model.

3 Results

According to the description of Sect. 2.2, 7 subjects (S1 ~ S7) carried out lower limb flexion and extension with three cycles (5 s, 3.5 s, 2 s). sEMG, hip and knee angle were collected synchronously, which were used to calculate the characteristics u_{AT}, u_{VR} and the difference signal Δu_{AT}, Δu_{VR} of tibialis anterior muscle and vastus rectus muscle. The root mean square error *RMSE* and training time *T* are selected as the performance verification indexes. The smaller the *RMSE*, the higher the prediction accuracy of the model; and the less the training time *T*, the faster the model is established and the better the real-time.

Taking S2 as an example, when the motion cycle is 5 s, the RMSE of the hip joint angle is 8°, and it is 2.9° when the motion cycle is about 2 s, which shows that motion velocity has a great influence on prediction error. So, it is necessary to choose the right speed in the rehabilitation training of mirror image motion to avoid

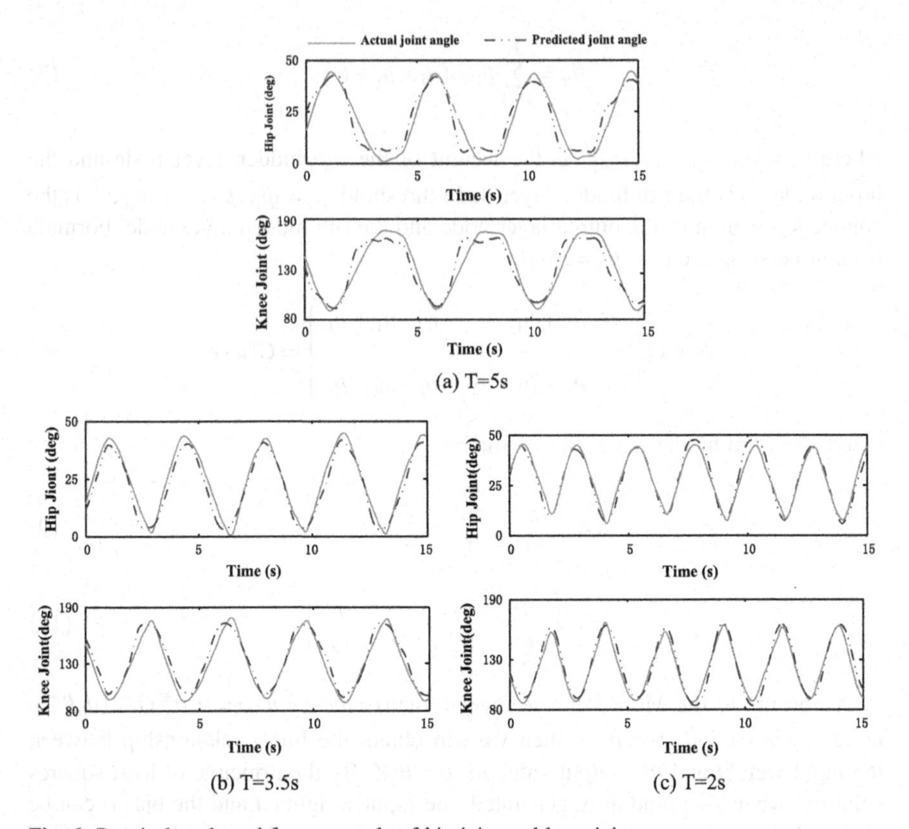

Fig. 6 Practical angle and forecast angle of hip joint and knee joint

Table 1 Comparisons of real-time between BP and GS-LSELM

	Training sample number	Estimated number of groups	Forecast sample number	Average training time (s)	Average prediction time (s)
BP	1200	7	2960	0.863	0.0895
GS-ELM	1200	7	2960	0.0013	0.0012

large prediction error and patient fatigue. The prediction results of the three kinds of motion cycle are shown in Fig. 6. It can be seen that the deviation between the predicted value and the actual value is larger at low speed, and is mainly concentrated at the inflection point.

Table 1 shows the comparison results of the two algorithms on the training time and the verification time. From Table 1, we can know that, on the same training set, the training time of GS-LSELM is 0.0013 s, which is only 0.15% of BP, and its prediction time is only 1.2 ms, which shows that the GS-LSELM algorithm is better than traditional BP neural network in real-time.

4 Conclusion

In this paper, we studied the coupling relationship between the EMG signals of lower limbs and the joint angles by the method of coherence analysis, and filtered the model input channel and used a first order recursive filter to realize the data synchronization. On this basis, we proposed GS-LSELM algorithm and established the prediction model of joint angles based on the EMG signals, and the performance of the model was verified by experiment. In future studies, the proposed method can be used in the rehabilitation training system, making the rehabilitation robot tracking the joint angle so as to achieve the active movement of patients.

References

1. Lo AC, Guarino PD, Richards LG, et al. Robot-assisted therapy for long-term upper-limb impairment after stroke. N Engl J Med. 2010;362(19):1772–83.
2. Costandi M. Machine recovery. Nature. 2014.
3. Hanghua W, Ni C. Characteristics and prevention of motor control in hemiplegic stroke. Chin Tissue Eng Res. 2005;9(29):140–1.
4. Zhong L, Wei G. ECG signal processing based on wavelet decomposition and reconstruction based on Mallat algorithm. Electron Des Eng. 2012;20(2):57–9.
5. Ma P, Chen Y, Yihao D. Analysis of EEG EMG coherence in stroke rehabilitation exercise. J Biomed Eng. 2014;5:971–7.
6. Huynh HT, Won Y, Kim JJ. An improvement of extreme learning machine for compact single-hidden-layer feed forward neural networks. Int J Neural Syst. 2011;18(5):433–41.

Research on the Smooth Switching and Coordinated Control System of Microgrid Based on Master-Slave Control

Ziyi Fu, Bing Cheng and Liuyang Shen

Abstract There is a problem of smooth switching between grid-connected mode and the island mode under the master-slave control structure of microgrid. This paper uses the simulation software MATLAB to build a simulation model of dual power supply low voltage microgrid. In grid-connected mode,the main power is disconnected, the slave power supply uses PQ control method. Once the fault occurs on the side of the large power grid or microgrid, microgrid must switch to the island mode. At this time, the main power supply is controlled by V/f to support the stability of voltage and frequency. The slave power supply uses PQ control method for constant power output. If there is no communication channel between master and slave controller, The control strategy may misoperate or operation failed. In this paper, EtherCAT communication protocol is used to realize the communication between the master controller and the slave controller, and the TwinCAT software is used as the main controller to study the coordinated control system based on EtherCAT.

Keywords Microgrid · Master-slave control · PQ control · V/f control · EtherCAT

1 Introduction

With the depletion of energy and the deterioration of environment, the application of new energy has shown great prospects for development. It has the advantages of strong environmental protection, good flexibility and high efficiency. The emergence of microgrid to solve the problem of distributed power supply access to the grid.

Z. Fu (✉) · B. Cheng · L. Shen
School of Electrical Engineering and Automation, Henan Polytechnic University, Jiaozuo 454003, China
e-mail: fuzy@hpu.edu.cn

Y. Jia et al. (eds.), *Proceedings of 2017 Chinese Intelligent Systems Conference*, Lecture Notes in Electrical Engineering 459,
https://doi.org/10.1007/978-981-10-6496-8_62

There are two kinds of steady-state operation modes of microgrid, which are grid-connected mode and the island mode. Of course, there are two transient operation modes which are grid-connected switches to islanding and islanding switches to grid-connected. Therefore, the microgrid needs a good control system to ensure stable and flexible operation. In the process of operation control, the microgrid needs to make a quick and independent response to the emergency in the grid based on the real-time monitoring of the local information. When the fault occurs in the microgrid or the large power grid, the microgrid should be transformed into the island operation mode quickly and automatically, in order to improve the reliability and the power quality of microgrid [1].

According to the different functions of distributed power supply, the control mode can be divided into master-slave control mode, hierarchical control mode and peer-to-peer control mode. Peer-to-peer control mode requires a high level of system parameters, So it is still in the stage of laboratory research(such as the United States Wisconsin microgrid experimental system, Spain's Catalunya microgrid laboratory system) [2]. At present, there are a lot of scholars research on the master-slave control. There are three control strategies of micro power supply: PQ control, V/f control and droop control.

Literature [3, 4] design the switching strategy of PQ control and V/f control based on current loop control. It can realize the smooth switching between the two modes of grid-connected and islanding. However, this method is not applicable when there is no current control.

Literature [5] describes the microgrid system with DC source, and adopts the PQ control and droop control according to different modes. However, It only considers the voltage synchronization, and the problem of state matching in the switching process is not studied deeply.

Literature [6] proposes the control strategy of multi loop master-slave switching. The master power supply adopts a multi loop control structure with a built-in parameter loop and the slave power supply adopts a constant power control structure. The switching process is stable and the control structure is reliable.

Literature [7] presents an integrated control strategy. Before and after the switching mode of microgrid, the internal power of microgrid is kept in equilibrium, the voltage and frequency fluctuate within the allowable range. However, the suppression of transient oscillation is not obvious.

This paper uses the control strategy which is combined by PQ control method and V/f control method. Research on coordinated control system of microgrid, and it uses EtherCAT communication protocol to transmit data. It can improve the real-time performance of the control system, and realize the smooth switching of microgrid operation mode. Finally, this paper builds the simulation of the microgrid to verify the effectiveness and feasibility of the control strategy.

2 Master-Slave Control Mode

2.1 The Principle of the Algorithm

Master-slave control selects a master power supply in the island mode, which provides the voltage and frequency support for the power system. At this time, the master power supply adopts V/f control strategy, and the slave power supply adopts PQ control strategy. The structure of the master-slave control system is composed of power supply, inverter, PQ controller, V/f controller, voltage and current sampling module, AC bus, static switch and load. The power supplys connect to the AC bus of microgrid through the power electronic conversion devices, and the microgrid connects to the large power grid through the point of common coupling (PCC). The structure of master-slave control is shown in Fig. 1 [8].

3 Distributed Power Supply Control Mode

3.1 The Constant Power Control Mode (PQ)

When the microgrid connects to the power grid, the output of the active power and reactive power are coupled, and the essence of the PQ control is to decouple the output power of the inverter, then it realizes the independent control of active power and reactive power.

The distributed power supply adopts PQ control, the three-phase fundamental voltage output of the grid-connected inverter is $u.U_m$ is the amplitude of the phase voltage. The dq transform is defined as:

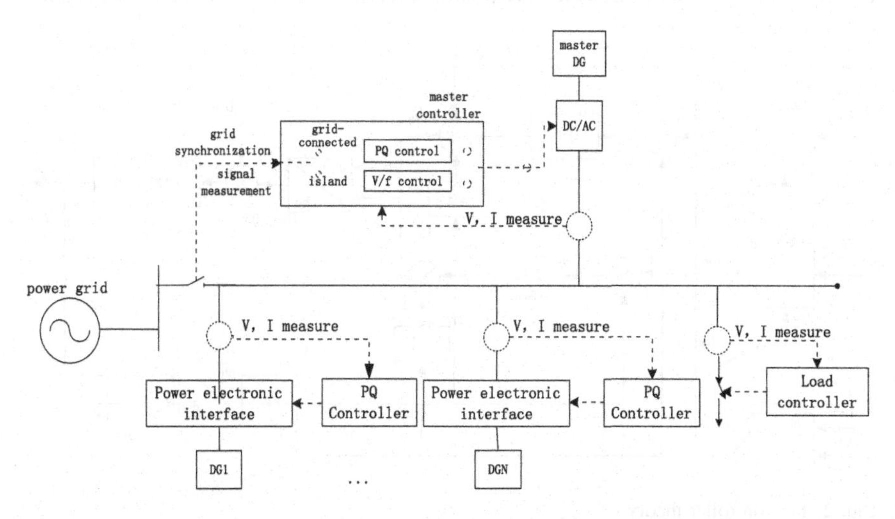

Fig. 1 Master-slave control architecture of microgrid

$$\begin{bmatrix} u_a \\ u_b \\ u_c \end{bmatrix} = \begin{bmatrix} U_m \cos(\omega t) \\ U_m \cos(\omega t - \frac{2\pi}{3}) \\ U_m \cos(\omega t + \frac{2\pi}{3}) \end{bmatrix} \quad (1)$$

$$\begin{bmatrix} u_d \\ u_q \end{bmatrix} = \begin{bmatrix} \cos \omega t & \cos(\omega t - \frac{2\pi}{3}) & \cos(\omega t + \frac{2\pi}{3}) \\ -\sin \omega t & \sin(\omega t - \frac{2\pi}{3}) & -\sin(\omega t + \frac{2\pi}{3}) \end{bmatrix} \begin{bmatrix} u_a \\ u_b \\ u_c \end{bmatrix} = \begin{bmatrix} U_m \\ 0 \end{bmatrix} \quad (2)$$

After the transformation from the three-phase stationary abc coordinate system to the dq synchronous rotating coordinate. The voltage and current in the abc coordinates are decoupled from dq. The expressions of active power and reactive power are:

$$\begin{cases} P = u_d i_d + u_q i_q \\ Q = u_d i_q + u_q i_d \end{cases} \quad (3)$$

The Preset reference active power P_{ref} and reactive power Q_{ref}, then calculate the dq axis reference current I_{dref} and I_{qref}. Because $U_q = 0$, the expressions of I_{dref} and I_{qref} are:

$$\begin{cases} I_{dref} = \frac{P_{ref}}{u_d} \\ I_{qref} = -\frac{Q_{ref}}{u_d} \end{cases} \quad (4)$$

Therefore, the change of the current controls the output power of the inverter. i_d determines the active power and i_q determines the reactive power. PQ control schematic is shown in Fig. 2:

Double loop control with outer loop power control and inner loop current control. The outer loop provides the power reference for the output of the inverter.

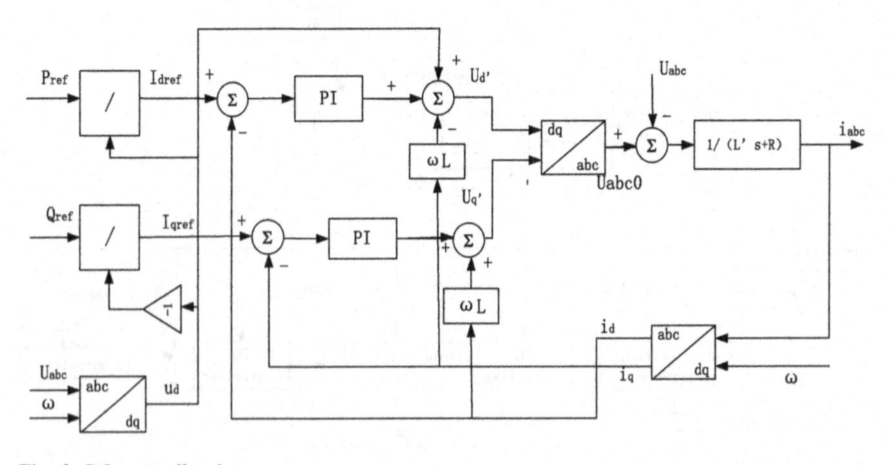

Fig. 2 PQ controller theory

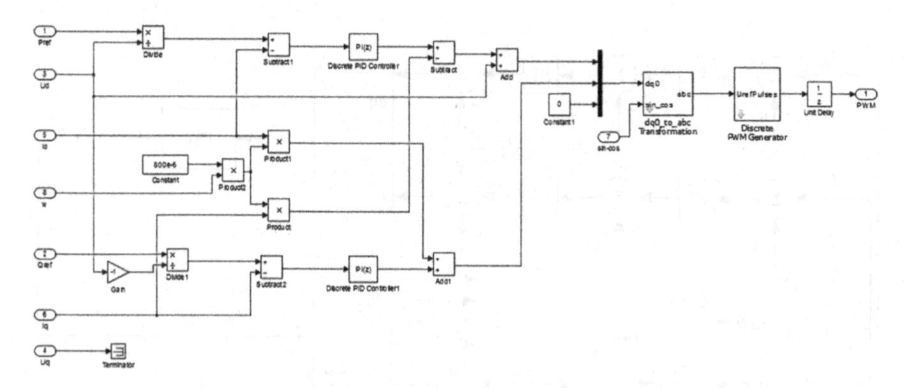

Fig. 3 Power module simulation diagram of PQ control

The inner loop plays a role of fine tuning and tracks the reference current, in order to provide the correct reference voltage to the inverter. Compare the power reference value and the actual measured value, then adjust the difference with PI. The reference signals I_{dref} and I_{qref} for the inner loop current control are obtained. Next, compare the reference signals I_{dref} and I_{qref} with the measured signals i_d and i_q. The error is adjusted by PI, and the coupling effect of the circuit inductance is taken into account, then obtain the dq axis reference signals $U_d^{'}$ and $U_q^{'}$. Use parker inverse transform for $U_d^{'}$ and $U_q^{'}$ to obtain the three-phase control signal U_{abc0}, which is used to drive the PWM control, finally, the output power of the inverter is constant. ω is angular frequency, L is filter inductance. PQ control simulation model is shown in Fig. 3:

3.2 *Constant Voltage and Constant Frequency Control Mode (V/f)*

The V/f control is suitable for islanding operation. Its main function is to provide stable voltage and frequency for other micro power supply, In order to ensure the continuous power supply of sensitive load in microgrid. V/f control mode uses the double loop control mode which contains voltage outer loop and current inner loop. The amplitude of the output voltage is regulated by the external voltage loop, which guarantees the steady-state accuracy of the output voltage. The current inner loop improves the dynamic response speed and reduces the harmonic content of the output voltage. The output value of the voltage loop is the input reference value of the current loop, then the PWM control signal is generated by the current inner loop. Finally, output voltage and frequency Stably. U_{dref} and U_{qref} are dq axis reference component of voltage.The V/f control block diagram is shown in Fig. 4:

The simulation module is divided into conversion module, voltage and current double loop control module.In the simulation, the power frequency 50 Hz is set as

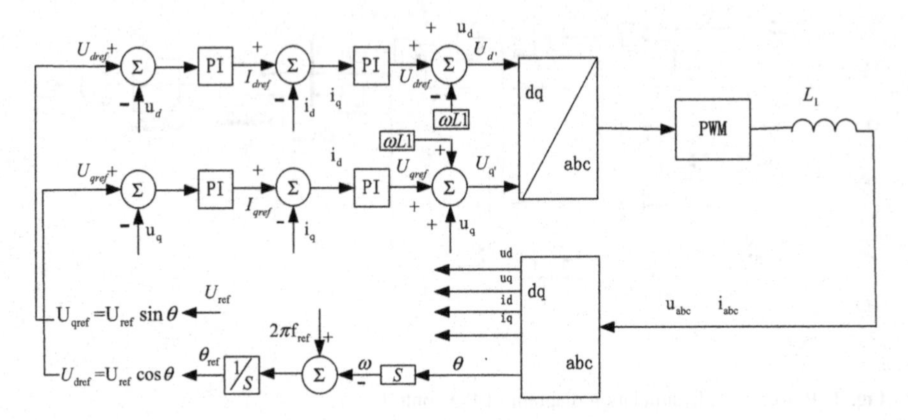

Fig. 4 V/f controller theory

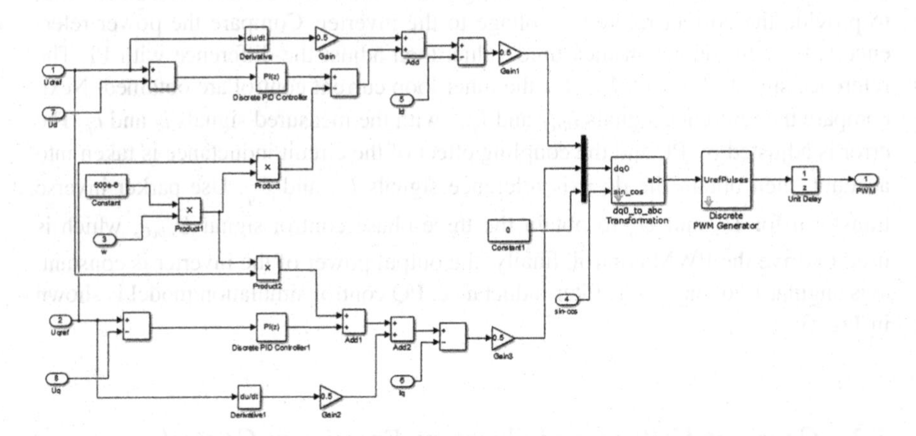

Fig. 5 Power module simulation diagram of V/f control

the reference frequency, and a unit delay is added to prevent the error in the simulation solution. Some simulation diagrams are shown as follows (Fig. 5).

4 Simulation of Master-Slave Control in Microgrid

The main power supply is controlled by V/f. In order to improve the power, the output current must be increased, because the load changes need to adjust the main control power supply.Therefore, the main power supply capacity should have a margin.The overall structure drawing is shown in the following figure (Fig. 6).

The simulation figure is shown in Fig. 7.

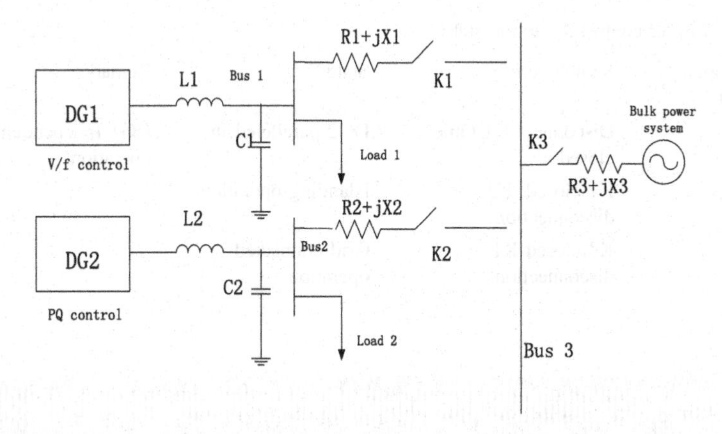

Fig. 6 Example of master-slave control in microgrid

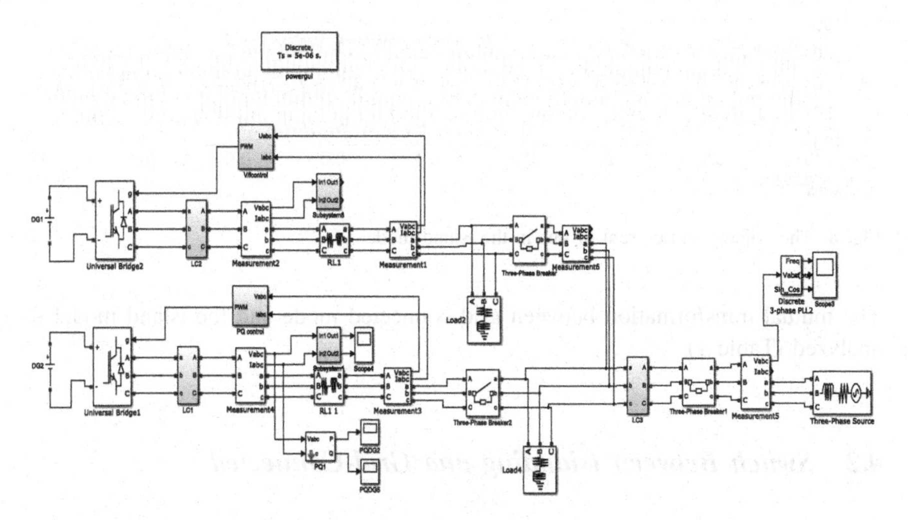

Fig. 7 Simulation model of master-slave structure in microgrid

4.1 Parameter Setting of Master-Slave Control in Microgrid

The LC filter is used in the simulation, It can effectively suppress the high frequency harmonics of the inverter output voltage. DC sources instead of DG units, Load L1: 40 kW, 20 kVar, Load L2: 20 kW, 20 kVar, Filter capacitor: 0.0015 F. Filter inductance: 0.0006 H, PLL coefficients in PQ control module: 40,1600. Current control loop PI coefficients: 0.5, 20. In the V/f control module: Reference voltage: 380 V, 0 V. In the PQ control module: Reference power: 20 kW, 0 kW. In order to verify the universality and rationality of the constructed simulation model,

Table 1 Master control set action state

Simulation time (s)	Switch action	State	Remarks
0–0.3	Disconnect K1 Other closed	DG2 paralleled in	DG1 Independent operation
0.3–0.6	K1 closed, K3 disconnection	Islanding operation	
0.6–0.8	K3 closed K1 disconnection	Grid connected operation	

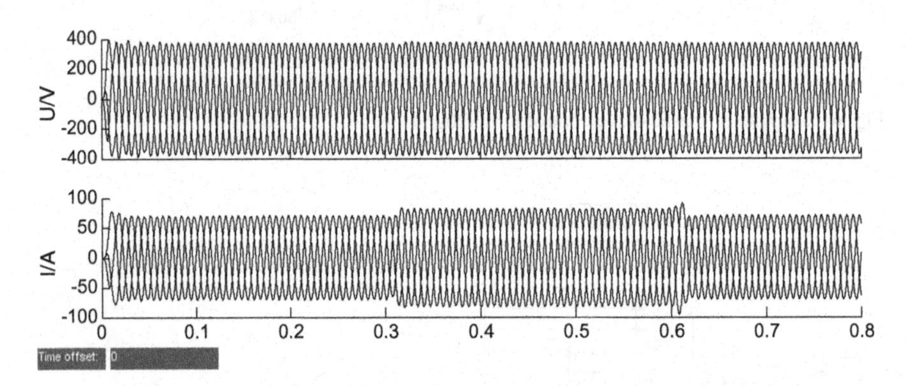

Fig. 8 The voltage and current of bus in the island mode

The mutual transformation between grid-connected mode and the island model is analyzed (Table 1).

4.2 Switch Between Islanding and Grid-Connected

For the wave form shown in Fig. 8, Within 0–0.3 s, DG2 constants power output in grid-connected mode, The large power grid provides voltage and frequency support, the voltage and current on the power grid side remain stable. When the microgrid is disconnected from the power grid at 0.3 s, DG1 and DG2 are in the island operation mode, The main power supply is immediately switched to V/f operation mode,in order to maintain the stability of the internal voltage and frequency of the microgrid. Within 0.6–0.8 s, microgrid exits island operation mode, once again connects the large power grid. Voltage and frequency of microgrid again supported by the large power grid, The output current remains stable. In the switching process of microgrid mode, there is harmonic interference in the PQ control current loop, so current disturbances exist on the microgrid side.

For the frequency curve shown in Fig. 9: Grid-connected and island operation mode are switching at 0.3 and 0.6 s, there exist certain oscillation of the system

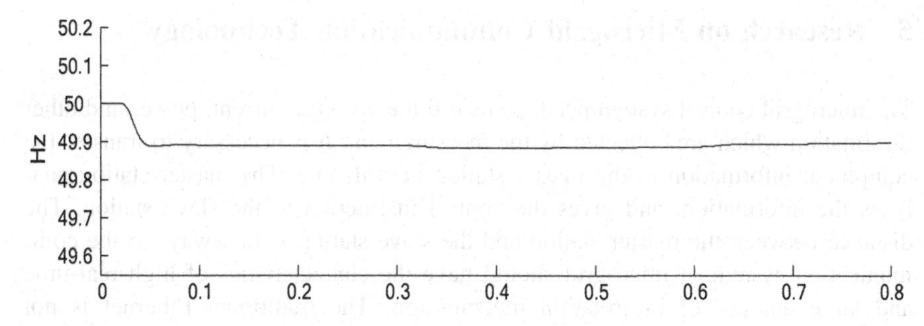

Fig. 9 The frequency of microgrid system

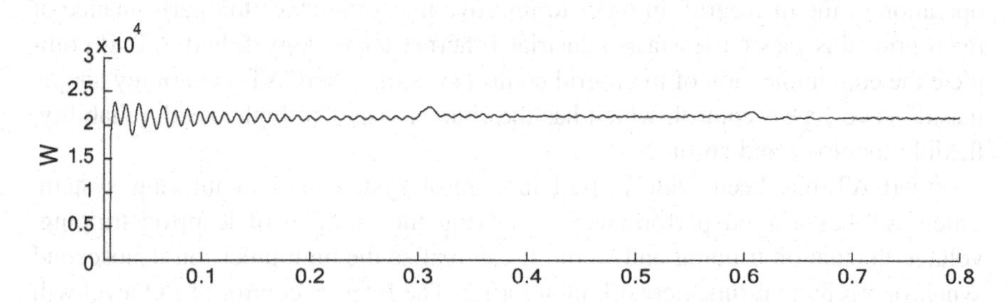

Fig. 10 DG2 output active power

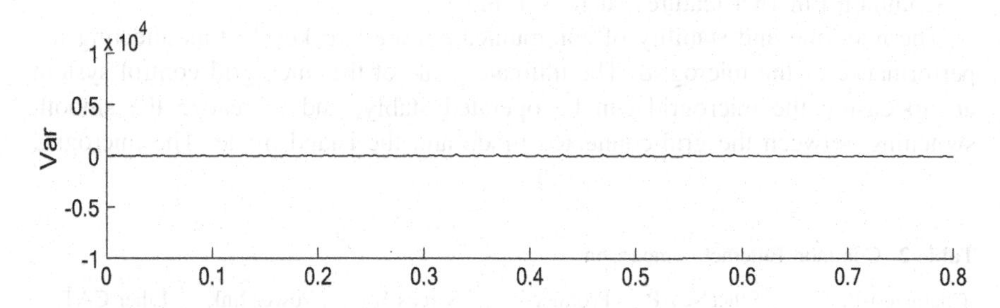

Fig. 11 DG2 output reactive power

frequency, But the overall frequency is always stable between 49.87 and 50.02 Hz, Within ±0.2 Hz,This meets the system requirements.

Figures 10 and 11 show the DG2 output active power and reactive power, Because DG2 is running in PQ mode, It performs constant power output, In grid-connected mode, the power shortage is provided by the large power grid, After switching to the island mode, the power shortage is provided by the main power supply. At this time, switching system will have a greater oscillation,It may bring the entire system of microgrid a certain degree of disorder. So we need Coordinated control system for real time monitoring of microgrid operation.

5 Research on Microgrid Communication Technology

The microgrid control system needs to record the voltage, current, power and other information which are collected by the measurement. It is necessary to transfer the equipment information to the master station in real time. The master station analyzes the information, and gives the control instruction to the slave station. The distance between the master station and the slave station is far away, so the communication system of microgrid should have the characteristics of high real-time and large amount of information transmission. The traditional Ethernet is not real-time, so in the control process, there may be conflicts between the superior control strategy and the underlying control strategy, which will affect the normal operation of the microgrid. In order to improve the communication performance of microgrid, this paper uses fast industrial Ethernet technology EtherCAT to complete the communication of microgrid control system. EtherCAT technology adopts master-slave access control, which has the characteristics of high speed, reliability, flexible topology and so on.

EtherCAT has been widely used in control system and monitoring system. EtherCAT has a good performance in solving the problem of leapfrog tripping, voltage fluctuation tripping and so on. It can realize the high precision nanosecond synchronous by real-time network monitoring. The Internet control of I/O level will be extended to the mine site by Ethernet communication, and it can realize the real-time monitoring from the local area network to the field level [9–12].

Common Ethernet features such as Table 2:

The real-time and stability of communication are two key factors affecting the performance of the microgrid. The ultimate goals of the microgrid control system are to ensure the microgrid can be operated stably, and to realize the smooth switching between the grid-connected mode and the island mode. The microgrid

Table 2 Common Ethernet comparison

Characteristic	EtherNet/IP	Profinet	Sercos-II	Power link	Ether CAT
Topological flexibility	Poor performance	General performance	General performance	General performance	High performance
Construction cost	Lower cost	Higher cost	General cost	General cost	Lower cost
Redundant cable	Unsupported	Unsupported	Supportive	Supportive	Supportive
The number of organizations that support it	General quantity	General quantity	General quantity	General quantity	More quantity
Special hardware	CIP Sync	Master station, slave station	Master station, slave station	Slave station	Slave station
Synchronization performance	Poor performance	General performance	General performance	General performance	High performance
Cycle time	Longer time	General time	General time	General time	Shorter time

control system collects the information data of the distributed power supply and the load circuit. Then it monitors the running state and controls the field equipment and measurement point remotely. Microgrid determines the operation mode by monitoring and collecting the information of the running state with the bottom equipment in real time. At the same time, according to the monitoring information to determine the control strategy of microgrid, and the control method of inverter.

The microgrid control system includes upper monitor, EtherCAT master station and slave station, and all kinds of implementing agencies. The upper monitor uses the SCADA monitoring system. The master station is an industrial control computer, and selects the TwinCAT software. TwinCAT integrates EtherCAT protocol stack and real-time kernel. The embedded PLC controller CX8010 is selected as the main controller of the EtherCAT slave station. The slave station includes the analog acquisition terminal, the data acquisition terminal, the digital output terminal, the communication converter, the human-computer interaction module and so on. Action routine is written in the TwinCAT PLC of the master station.

The coordinated control system of microgrid is in Fig. 12:

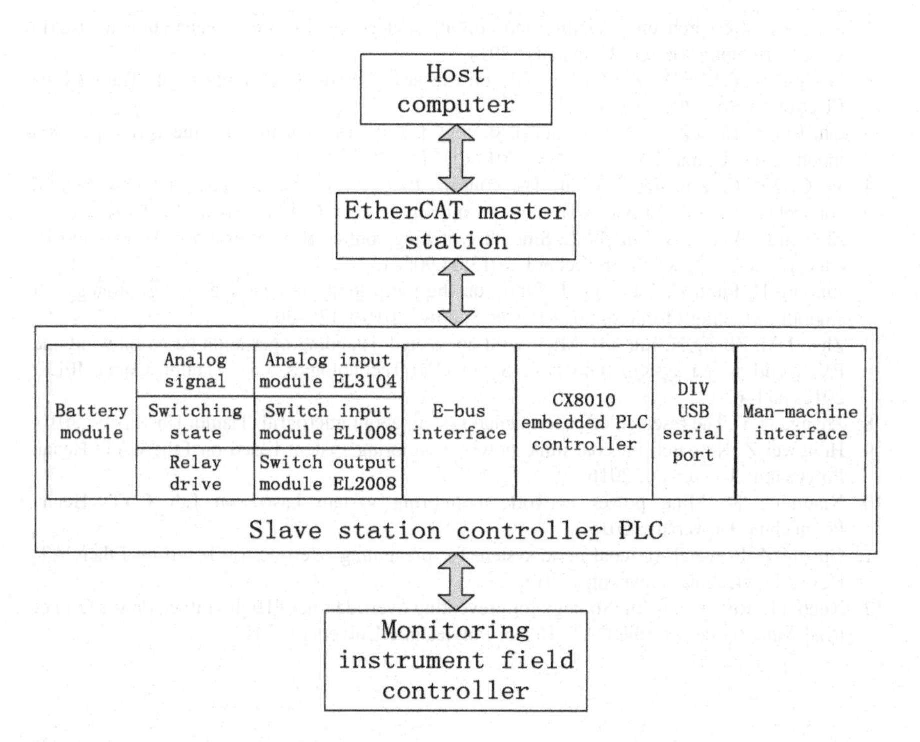

Fig. 12 Hardware structure of control system

6 Conclusions

In this paper, the simulation of the master-slave structure of the dual power supply microgrid is established theoretically. The slave power supply uses PQ control strategy in grid-connected mode, and the master power uses V/f control strategy in the island mode. According to the results of simulation, the master-slave control strategy can effectively realize the smooth switching of the microgrid and maintain the voltage and frequency stability. In order to improve the communication speed between master and slave controllers, this paper uses the EtherCAT communication protocol in the microgrid system, and achieves the fast switching from PQ control to V/f control. At last, the designs of the hardware structure about the master station and slave station are carried out.

References

1. Shouyi Z. Research on modeling and control strategy of distributed generation microgrid system. Beijing: Jiaotong University; 2014.
2. Wang Chengshan W, Zhen LP. Study on the key technology of micro grid. Trans China Electrotech Soc. 2014;2:1–12.
3. Qiu Lin X, Lie ZZ, et al. The control strategy for smooth switching of microgrid operation mode. Trans China Electrotech Soc. 2014;2:171–6.
4. Jie C, Xin C, Zhiyang F, et al. The control strategy for smooth switching between grid connected and isolated operation mode in microgrid. Proc CSEE. 2014;19:3089–97.
5. Zhiwen L, Wenbo X, Mingbo L. Smooth switching control of microgrid based on composite energy storage. Power Syst Technol. 2013;04:906–13.
6. Junping H, Pupu C, Yaodong T. Study on the multi loop master-slave control strategy for smooth switching of microgrid. J Power Supply. 2016;2:128–36.
7. Zhao DM, Zhang N, Liu YH. Micro-grid connected/ islanding oper-ation based on wind and PV hybrid power system Innovative Smart Grid Technologies-Asia. Tianjin, China: IEEE; 2012. p. 1–6.
8. Zhangang Y. The research of experimental system about microgrid. Tianjin University; 2010.
9. Hongwei Z. Research on coal mine power monitoring system based on EtherCAT. Henan Polytechnic University; 2016.
10. Xiaoming R. Mine power network monitoring system based on EtherCAT. Henan Polytechnic University; 2015.
11. Guozhi Z. Research on coal mine system for preventing override trip based on EtherCAT. Henan Polytechnic University; 2015.
12. Xuebin L. Research on the strategy for preventing override trip of High-voltage Power Grid of Coal Mine based on EtherCAT. Henan Polytechnic University; 2013.

Effect of Parametric Variation of Center Frequency and Bandwidth of Morlet Wavelet Transform on Time-Frequency Analysis of Event-Related Potentials

Guanghui Zhang, Lili Tian, Huaming Chen, Peng Li, Tapani Ristaniemi, Huili Wang, Hong Li, Hongjun Chen and Fengyu Cong

Abstract Time-frequency (TF) analysis of event-related potentials (ERPs) using Complex Morlet Wavelet Transform has been widely applied in cognitive neuroscience research. It has been widely suggested that the center frequency (fc) and bandwidth (σ) should be considered in defining the mother wavelet. However, the issue how parametric variation of fc and σ of Morlet wavelet transform exerts influence on ERPs time-frequency results has not been extensively discussed in previous research. The current study, through adopting the method of Complex Morlet Continuous Wavelet Transform (CMCWT), aims to investigate whether time-frequency results vary with different parametric settings of fc and σ. Besides, the nonnegative canonical polyadic decomposition (NCPD) is used to further confirm the differences manifested in time-frequency results. Results showed that different parametric settings may result in divergent time-frequency results, including the corresponding time-frequency representation (TFR) and topographical distribution. Furthermore, no similar components of interest were obtained from different TFR results by NCPD. The current research, through highlighting the importance of parametric setting in time-frequency analysis of ERP data, suggests that different parameters should be attempted in order to get optimal time-frequency results.

G. Zhang · H. Chen · F. Cong (✉)
Department of Biomedical Engineering, Faculty of Electronic Information and Electrical Engineering, Dalian University of Technology, Dalian 116024, China
e-mail: cong@dlut.edu.cn

G. Zhang · T. Ristaniemi · F. Cong
Department of Mathematical Information Technology, University of Jyväskylä, 40014 Jyväskylä, Finland

L. Tian · H. Wang · H. Chen (✉)
School of Foreign Languages, Dalian University of Technology, Dalian 116024, China
e-mail: chenhj@dlut.edu.cn

P. Li · H. Li
College of Psychology and Sociology, Shenzhen University, Shenzhen 518060, China

Y. Jia et al. (eds.), *Proceedings of 2017 Chinese Intelligent Systems Conference*, Lecture Notes in Electrical Engineering 459,
https://doi.org/10.1007/978-981-10-6496-8_63

Keywords Complex Morlet Wavelet Transform • Event-related potentials • Center frequency • Bandwidth • Time-frequency representation

1 Introduction

Electroencephalogram (EEG) has been extensively applied in cognitive neuroscience research. EEG, according to different experimental paradigms and external stimuli, can be divided into three categories: spontaneous EEG [1], event-related potentials (ERP) [2], and ongoing EEG [3]. The main methods employed in ERP data processing are as the following: (1) Time-domain analysis, (2) Frequency-domain analysis and (3) Time-frequency analysis [4–8]. As ERP signals are nonstationary and time-varying, neither the time-domain nor the frequency-domain analysis can be used to effectively reveal the time-frequency information of ERP data. Time-frequency analysis, by focusing on the time-varying features of ERP components, is conducted to transform a one-dimensional time signal into a two-dimensional time-frequency density function, which aims to reveal the number of frequency components and how each component varies over time.

In 1996, Tallon-Baudry et al. introduced the Morlet wavelet for time-frequency analysis of ERP data [9]. Since then, the Morlet wavelet has been widely applied by researchers in conducting time-frequency analysis, with its citations over 1100 times (From the Google scholar). However, a synthesis of previous research showed that in most cases the value of K is fixed (e.g., $K = 7$) [9–12], therefore leaving the issue whether parametric variation of fc and σ has an impact on time-frequency results unresolved. This study is devoted to investigation of the issue.

2 Method

2.1 Data Description

The data was collected to investigate whether a short delay in presenting an outcome affects brain activity. For the detailed information of experimental procedure, readers can refer to Wang et al. research [13]. Twenty-two undergraduates and graduate students participated in the experiment as volunteers. All the participants, aged from 18 to 24, were right-handed with normal or corrected-to-normal vision and no one was reported to have neurological or psychological disorders. EEG was recorded using a 64-channel system (Brian Products GmbH, Gilching, Germany) with reference on the left mastoid. The vertical and horizontal electrooculogram (EOG) was recorded from electrodes placed above and below the right eye and on the outer canthi of the left and right eyes respectively. Electrode impedance was maintained below 10 k Ohm. The EEG and EOG were sampled continuously at 500 Hz with 0.01–100 Hz bandpass filtering.

2.2 Complex Morlet Wavelet Transform

The CMCWT method, based on the Complex Morlet Wavelets, was adopted for time-frequency analysis in the present study.

If $x(t)$ is a discrete sequence of length T, the definition of the Continuous Wavelet Transform (CWT) can be expressed as follows:

$$X(a,b) = \frac{1}{\sqrt{|a|}} \sum_{t=0}^{T-1} x(t)\Phi\left(\frac{t-b}{a}\right) \tag{1}$$

In the above formula, $x(t)$ represents the signal to be transformed; a refers to the scaling and b the time location or shifting parameters; $\Phi(t)$ stands for the mother wavelet. In this study, the Complex Morlet Wavelets is defined as the mother wavelet [9–12]:

$$\Phi(t,fc) = \frac{1}{\sqrt{\pi\sigma^2}}\, e^{i2\pi tf} e^{\frac{-t^2}{2\sigma^2}} \tag{2}$$

According to the above formula, a Gaussian shape respectively in the time and frequency domain around its fc can be obtained.

A wavelet family is characterized by a constant ratio:

$$K = fc/\sigma f = 2\pi\sigma fc \tag{3}$$

In this formula, $\sigma_f = 1/2\pi\sigma$, K should be greater than 5 [9].

Taken together, this method (CMCWT) can be described as below:

$$\mathrm{CMCWT(t,f)} = |\Phi(t,f)^{*}x(t)|^2 \tag{4}$$

In the above formula, '*' refers to convolution.

2.3 Nonnegative Canonical Polyadic Decomposition

Nonnegative Canonical Polyadic Decomposition (NCPD) has been widely applied to study time-frequency representation (TFR) of EEG [14, 15]. For example, given a third-order tensor including the modes of time, frequency and space, $\underline{\mathrm{X}} \in \mathcal{R}^{\mathrm{I}_1 \times \mathrm{I}_2 \times \mathrm{I}_3}$, the NCPD can be defined:

$$\underline{\mathrm{X}} = \sum_{t=1}^{R} t_r \circ f_r \circ s_r + \underline{E} = \sum_{r=1}^{R} \underline{X}_r + E = \hat{\underline{\mathrm{X}}} + \underline{E} \approx \hat{\underline{\mathrm{X}}} \tag{5}$$

In this formula, the symbol '∘' denotes the outer product of vectors. The t_r,f_r, and s_r correspond to the temporal component #r, the spectral component #r, and the spatial component #r, and the three components reveal the properties of the multi-domain properties of an ERP in the time, frequency and space domains [14].

For the same multi-channel EEG data, different parameters of CMCWT may produce different TFR (indeed, third-order tensors in this study) in terms of visual inspection. Then, the application of NCPD on those tensors can assist to investigate whether the similar components of interest can be extracted from different tensors resulting different TFR parameters of the same EEG data. For the detailed information of the number of extracted components for each mode, and the criteria of selecting multi-domain features, readers can refer to Cong et al. research [15].

3 Data Processing and Analysis

The ERP data were pre-processed in MATLAB and EEGLAB [16], including the following steps: a 50 Hz notch filter to remove line noise, a low-pass filtering of 100 Hz, segmentation of the filtered continuous EEG into single trials (each trial was extracted offline from 200 ms pre-stimulus onset to 1000 ms post-stimulus onset), baseline correction, artifact rejection and averaging.

In CMCWT analysis, the frequency range was set from 1 to 30 Hz, respectively in 0.1 Hz step (fc = 9, 10, respectively), in 0.2 Hz step (fc = 5, 6, 7, 8, 9, 10, respectively), in 0.3 Hz step (fc = 3, 4, 5, 6, 7, 8, 9, 10, respectively), in 0.4 Hz step (fc = 3, 4, 5, 6, 7, 8, 9, 10, respectively), in 0.5 Hz step (fc = 2, 3, 4, 5, 6, 7, 8, 9, 10, respectively), in 0.6 Hz step (fc = 2, 3, 4, 5, 6, 7, 8, 9, 10, respectively), in 0.7 Hz (fc = 2, 3, 4, 5, 6, 7, 8, 9, 10, respectively), in 0.8 Hz step (fc = 2, 3, 4, 5, 6, 7, 8, 9, 10, respectively), in 0.9 Hz step (fc = 1, 2, 3, 4, 5, 6, 7, 8, 9, 10, respectively) and in 1 Hz step (fc = 1, 2, 3, 4, 5, 6, 7, 8, 9, 10, respectively). All the above parametric settings met the requirement of constant ratio (greater than 5).

To further investigate whether parametric variation of fc and σ has an impact on time-frequency results, four steps are carried out in the following sequence:

(1) Select a typical topographical distribution of TFR results as the template. When $\sigma_0 = 1$, the value of fc can be respectively set as 1, 2, 3, 4, 5, 6, 7, 8, 9 and 10. The topographical distribution of $fc_4 = 4$ is finally chosen as the template $\mathrm{T}_{\mathrm{template}}(\sigma_0, fc_4)$ in terms of the prior knowledge of the ERP of interest.
(2) Define a fc_n, calculate the Correlation Coefficients (CCs) between the template $\left(\mathrm{Y}_{\mathrm{template}}\right)$ and each spatial component $s_r(\sigma_0, fc_n)$ obtained by NCPD (R components were extracted in each mode), which can be described as:

$$\mathrm{Y}(\sigma_0, fc_n, \mathrm{r}) = \rho\left(s_r(\sigma_0, fc_n), T_{template}(\sigma_0, fc_4)\right) \tag{6}$$

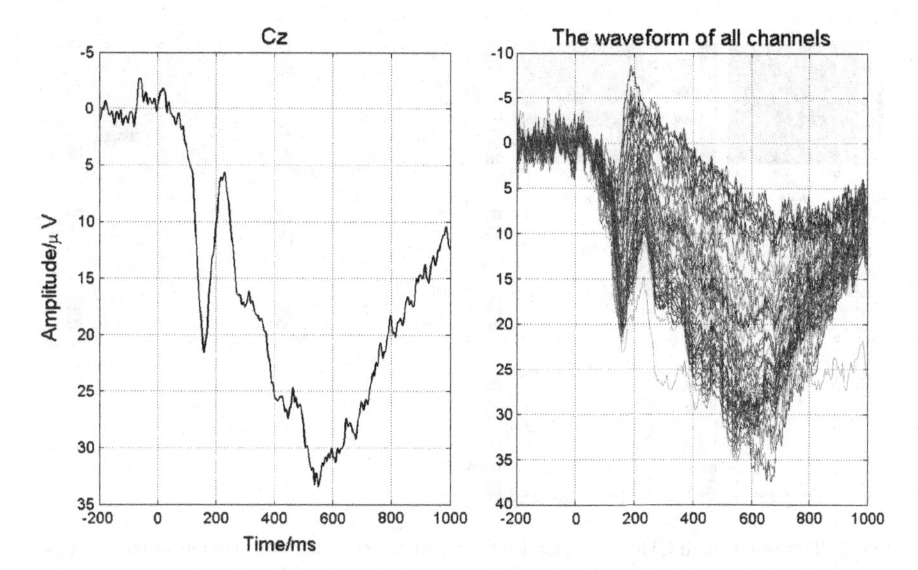

Fig. 1 Waveforms at Cz and all channels

In the above formula, $r = 1, 2, \cdots, 21, n = 1, 2, \cdots, 10$. Subsequently, the maximal CC is chosen as:

$$q(\sigma_0, fc_n) = \max(Y(\sigma_0, fc_n, 1), Y(\sigma_0, fc_n, 2), \cdots, Y(\sigma_0, fc_n, R)) \quad (7)$$

Then, the corresponding *r*th components with the maximum CC were obtained.

(3) Based on the obtained components of each dimension and their corresponding TFR results, we need to judge whether the TFR results of different parameters are similar or not. The TFR results are different when the *fc* is respectively set as 1 and 9. Besides, the corresponding 16th components of $fc_1 = 1$ and 1th components of $fc_9 = 9$ are similar in the spacial dimension, but not the temporal and spectral dimension (as shown in the first and third row of Fig. 4).
(4) With the same procedure mentioned above, we can analyze the results of other σ and *fc* parameters to explore potential differences in the time-frequency results.

4 Results

Results showed that parametric variations of σ and *fc* lead to different time-frequency representation and topographical distribution. The data results shown in Figs. 1, 2, 3, 4 and 5 are all from one subject in one condition (short-gain

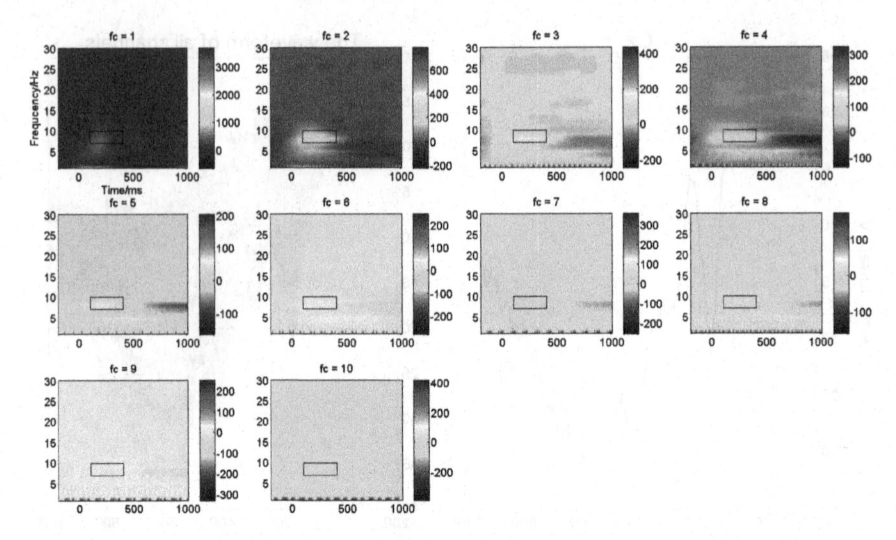

Fig. 2 TFR results with CMCWT method at Cz electrode. $\sigma = 1$, the time window of the rectangle area was from 100 to 400 ms and its frequency range from 7 to 10 Hz

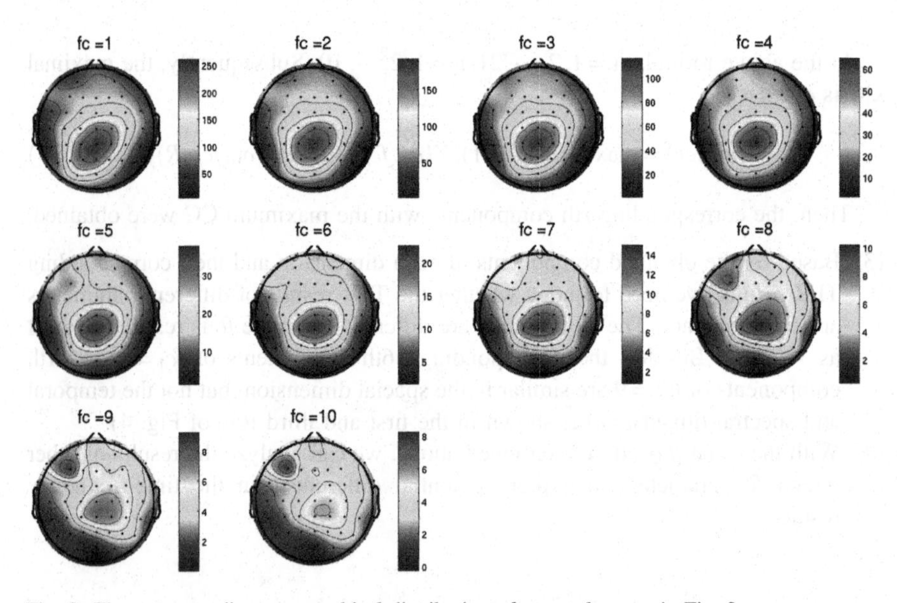

Fig. 3 The corresponding topographical distribution of rectangle areas in Fig. 2

condition), and the data used for statistical analysis (Fig. 6) are from 22 subjects in four conditions (waiting time (short, long) × feedback valence (loss, gain)). Due to the space limitation, only the results of $\sigma = 1$, fc = 1, 2, 3, 4, 5, 6, 7, 8, 9, 10 are presented here.

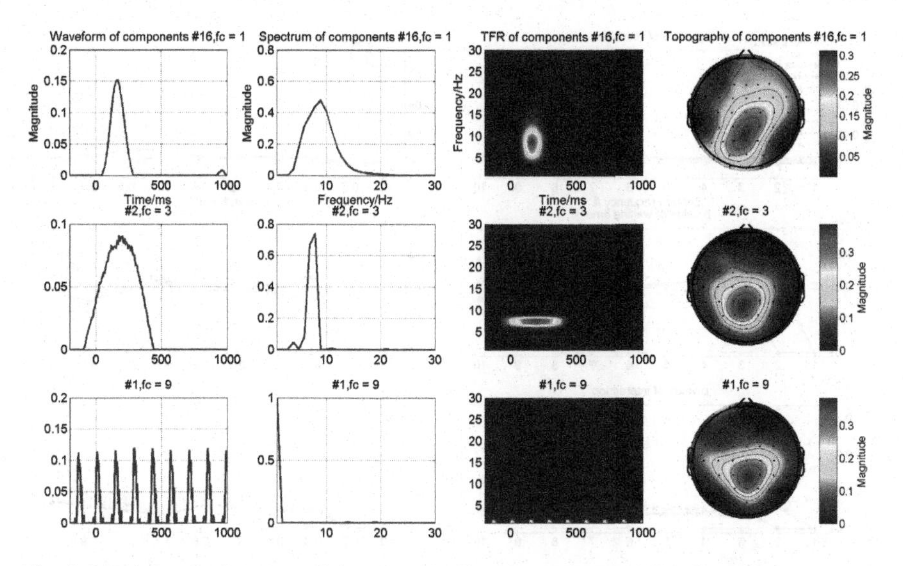

Fig. 4 Multi-domain features and the corresponding temporal, spectral, and spatial components. The third-order ERP tensor of the TFR for NCPD includes frequency (30 frequency bins), time (600 samples), and feature modes (58 channels). 21 components were extracted from each mode for NCPD, the order and variance of each component for NCPD are not determined. The TFR is based on the outer product of the temporal and spectral components

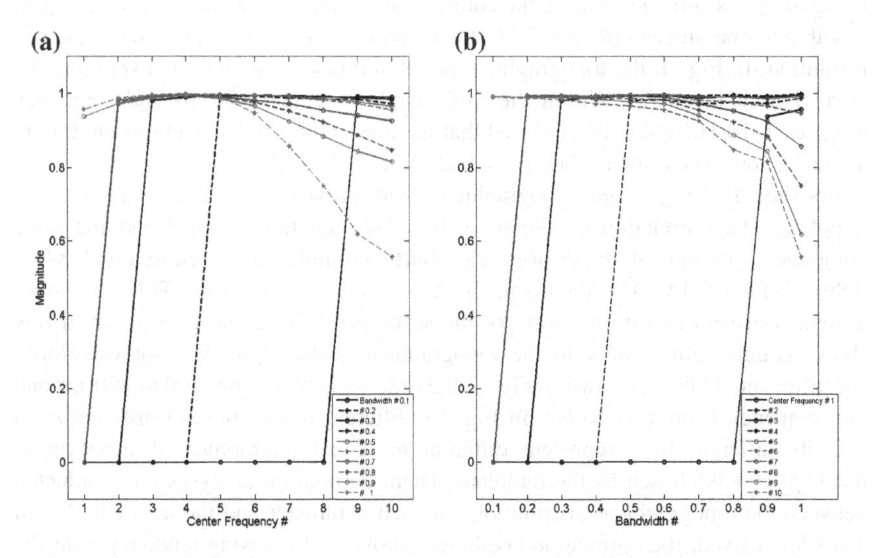

Fig. 5 The Correlation coefficients between topographical distributions of each parameter and the template ($\sigma = 1$, $fc = 4$)

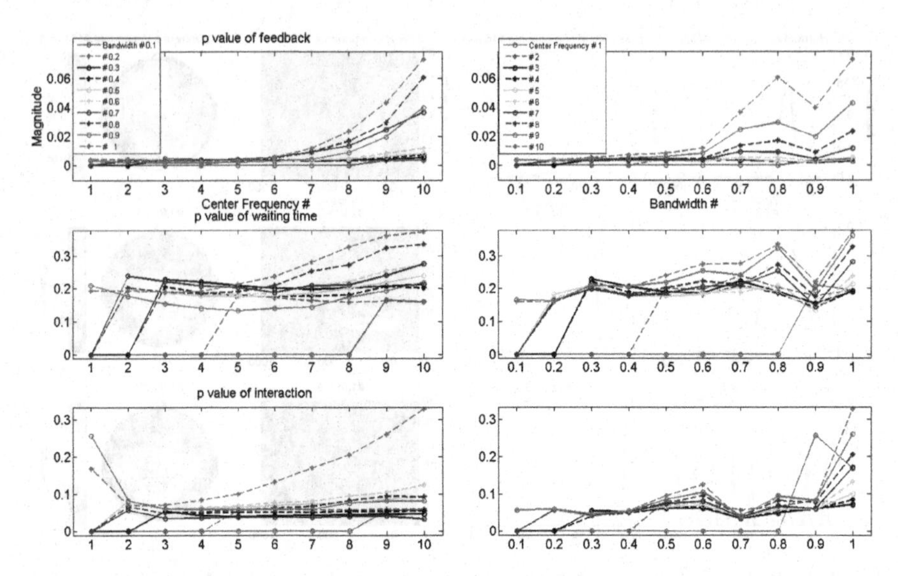

Fig. 6 Two-way repeated measurements ANOVA results of different *fc* and σ parameters. Two factors refer to waiting time (short, long) and feedback valence (loss, gain)

Figure 1 shows the waveform at Cz electrode in the left panel and the waveform of all channels in the right. From the waveform, certain ERP components can be recognized. As shown in Fig. 2, the comparison of the TFR results of *fc* = 1, 2, 3, 4, 5 with other parameters (*fc* = 6, 7, 8, 9, 10) shows differences in the time frequency resolution. In Fig. 3, the topographical distributions are obtained by averaging the area of each rectangle. Within the time window of 100–400 ms and frequency range of 7–10 Hz, it can be observed that the topographical distributions are highly similar among each other when *fc* = 1, 2, 3, 4, 5, 6, 7, 8.

As shown in Fig. 4, the topographical distribution of *fc* = 4 is selected as the template. The correlation coefficient values between the template and the three components of spatial dimension (the fourth column) are respectively 0.8495, 0.8947 and 0.8915. A comparison of the waveform, spectrum, TFR and topographical distribution in first row (or the second row) with those of the third row shows commonalities only in the topographical distribution, but not waveform, spectrum and TFR. As shown in Fig. 5, the time window is from 100 to 400 ms and the frequency from 7 to 10 Hz. In Fig. 5a, different lines represent different σ; in Fig. 5b, different lines represent different *fc*. The topographical distribution of $\sigma = 1$, *fc* = 4 is chosen as the template. Then, correlation analyses are conducted between the topographical distributions of each parameter and the template. When σ (or *fc*) is fixed, the correlation coefficient shows a decreasing tendency with the increasement of *fc* (or σ).

In Fig. 6, the power of the region of interest (time: 100–400 ms, frequency: 7–10 Hz) of Cz with different parametric settings of *fc* and σ is analyzed. The first/second/third row respectively shows the *p* value of the waiting time condition/feedback condition/interaction. The figure shows how the p value changes with fc (or σ), when σ (or fc) is a constant. When σ (or fc) is fixed, the p value shows an increasing tendency with the increasement of fc (or σ).The corresponding statistical analysis results may also differ with different settings of fc and σ.

5 Conclusion

The current study, through employing the methods of CMCWT, explored the influence of fc and σ variation on the time-frequency and topographical results of ERP data. Besides, NCPD was used to further confirm the differences manifested in time-frequency results. Results showed that parametric variation of σ and fc had an effect on time-frequency results. Moreover, it was found that different components would be obtained from different TFR results by NCPD. The current study therefore suggests that different parameters should be examined in order to get optimal time-frequency results. Meanwhile, the NCPD method is highly encouraged to be applied for the further confirmation of differences in time-frequency results.

Acknowledgements This work was supported by National Natural Science Foundation of China (Grant No. 81471742) and the Fundamental Research Funds for the Central Universities [DUT16JJ(G)03] in Dalian University of Technology in China.

References

1. Niedermeyer E, da Silva FL. Electroencephalography: basic principles, clinical applications, and related fields. Lippincott Williams & Wilkins; 2005.
2. Luck SJ. An introduction to the event-related potential technique 2005;66.
3. Cong F, et al. Linking brain responses to naturalistic music through analysis of ongoing EEG and stimulus features. IEEE Trans Multimed. 2013;15(5):1060–9.
4. Herrmann CS, et al. Time–frequency analysis of event-related potentials: a brief tutorial. Brain Topogr. 2014;27(4):438–50.
5. Sáncheznàcher N, et al. Event-related brain responses as correlates of changes in predictive and affective values of conditioned stimuli. Brain Res. 2011;1414(1):77–84.
6. Sanchezalavez M, Ehlers CL. Event-related oscillations (ERO) during an active discrimination task: Effects of lesions of the nucleus basalis magnocellularis. Int J Psychophysiol Off J Int Organ Psychophysiol. 2016;103:53.
7. Mathes B, et al. Maturation of the P3 and concurrent oscillatory processes during adolescence. Clin Neurophysiol Off J Int Fed Clin Neurophysiol. 2016;127(7):2599.
8. Ergen M, et al. Time-frequency analysis of the event-related potentials associated with the Stroop test. Int J Psychophysiol Off J Int Organ Psychophysiol. 2014;94(3):463–72.
9. Tallon-Baudry C, et al. Stimulus specificity of phase-locked and non-phase-locked 40 Hz visual responses in human. J Neurosci. 1996;16(13):4240–9.
10. Tallon-Baudry C, et al. Induced γ-band activity during the delay of a visual short-term memory task in humans. J Neurosci. 1998;18(11):4244–54.
11. Tallon-Baudry C, et al. Oscillatory γ-band (30–70 Hz) activity induced by a visual search task in humans. J Neurosci. 1997;17(2):722–34.

12. Tallon-Baudry C, Bertrand O. Oscillatory gamma activity in humans and its role in object representation. Trends Cogn Sci. 1999;3(4):151–62.
13. Wang J, et al. P300, not feedback error-related negativity, manifests the waiting cost of receiving reward information. NeuroReport. 2014;25(13):1044–8.
14. Cong F, et al. Multi-domain feature extraction for small event-related potentials through nonnegative multi-way array decomposition from low dense array EEG. Int J Neural Syst. 2013;23(02):1350006.
15. Cong F, et al. Tensor decomposition of EEG signals: a brief review. J Neurosci Methods. 2015;248:59–69.
16. Delorme A, Makeig S. EEGLAB: an open source toolbox for analysis of single-trial EEG dynamics including independent component analysis. J Neurosci Methods. 2004;134(1):9–21.

Clustering Personalized 3D Printing Models with Multiple Modal CNN

Jianwei Chen, Lin Zhang and Xinyu Dong

Abstract Clustering personalized 3D printing models is very useful for a cloud manufacturing management system, but it is difficult to cluster directly because of the complexity and abstraction of the 3D print model input. In this paper we use the convolution neural networks (CNNs) to learn the similarities of 3D print model pairs in different input modes and integrate these similarities by multi-channel spectral clustering. The three-dimensional CNN and the view-based CNN are used for different input modes. Our experiments show that the accuracy of the clustering can be improved by merging training results of different input modes.

Keywords Spectral cluster • Personalized 3D print model • CNNs • Similarity classifier

1 Introduction

3D printing technology plays a very important role in the modern manufacturing system, and is gradually becoming commonly used means of manufacturing in emerging manufacturing systems such as cloud manufacturing systems [1–3]. Personalized 3D printing model are 3D printing models designed by ordinary users without professional design experience and generally depict common things in everyday life. Unlike other ordinary 3D models, the personalized 3D printing models are characterized by simple structure and have more specific semantic

J. Chen · L. Zhang (✉)
School of Automation Science and Electrical Engineering, Beihang University, Beijing 100191, China
e-mail: johnlin9999@163.com

J. Chen · L. Zhang
Engineering Research Center of Complex Product Advanced Manufacturing Systems, Ministry of Education, Bejing 100191, China

X. Dong
Department of Computer Science, Stony Brook University, Stony Brook, NY, USA

Y. Jia et al. (eds.), *Proceedings of 2017 Chinese Intelligent Systems Conference*, Lecture Notes in Electrical Engineering 459,
https://doi.org/10.1007/978-981-10-6496-8_64

content. The Internet and cloud processing centers produce a large number of 3D printing model without high-quality semantic label every day. For unlabeled personalized 3D printing models, as shown in Fig. 1, effective aggregation can reduce the difficulty of management and improve the retrieval speed [4, 5].

However, the features extracted from the personalized 3D model are abstract features, and this feature form does not directly form an effective clustering standard, which makes the goal of semantic clustering become blurred [6]. Therefore, for this task, we provide a basic assumption that the morphological features of the personalized 3D model determine its semantic features [7].

The 3D printing model is typically stored as a STL file, which records the vertex position and the normal vector of the triangular slice of the the model. Different modal input form can be extracted from the 3D printing model and we can train the corresponding similarity classifier respectively. Common input modes are space voxels, optical view, skeleton view, etc. [8–11].

In fact, there are many related work on these input modalities. Some classification works have been completed on special 3D models similar with the personalized 3D printing model with volumetric and multi-view input modalities [8]. real-time object recognition with a deep network called Voxnet use the voxels representation as input mode, this can provide good clustering effect but the extraction of the model is not complete [9, 10]. Rotation-invariant 3D local features is essentially and providing a more advantageous speed performance, but the same, this is not perfect for the handling of personalized 3D model details [11].

Due to the complexity and abstraction of the personalized 3D printing model input feature form, traditional cluster methods don't work well [12, 13]. Spectral clustering is a solution to this kind of problems, especially in the field of machine vision [14–16], it can transform the clustering task into an easier task, to learn the similarity between two 3D printing models. In the industry, through the migration

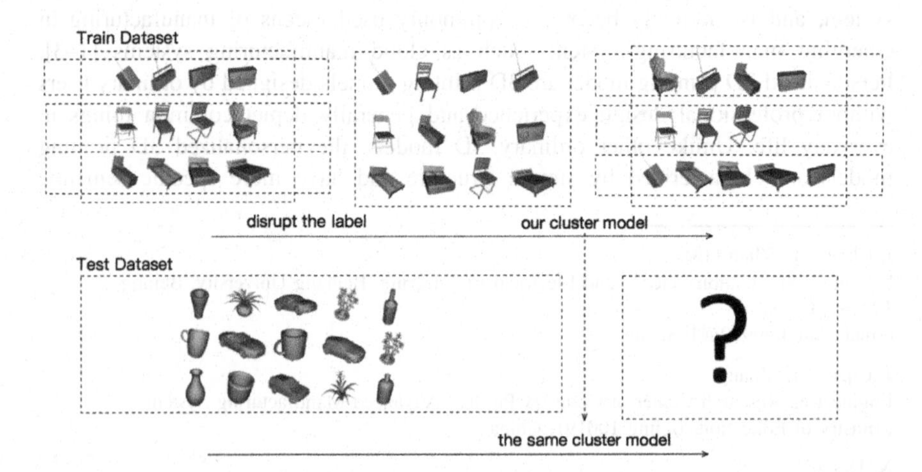

Fig. 1 We train our semi-cluster model with the cluster result and the original label. Test results on test sets without any labels reflect the performance of our clustering model

of existing results in the field of computer vision [17, 18], the complex personalized 3D printing model of the vectorization method becomes more mature. At the same time, the transformation can also greatly increase the training sample size, for example, the training sample size of a dataset with 30 semantic classes, each with 200 3D printing models, increases from 6000 to 36 million, and 2.4 million even if the positive and negative cases are balanced.

In this paper, we design a clustering method for 3D printing model as shown in Fig. 2. We extract four input modes from the original format (STL file) of 3D printing model: the optical view, the sketch view, the original voxels and the sketch voxels. We use the three-dimensional CNN and the view-based CNN to learn the similarity between two 3D printing models. The similarity is used for spectral clustering, and the accuracy of the training set is optimized by adjusting the threshold. The similarity results for different input modalities will be integrated when performing spectral clustering.

Section 2 specifies the input mode extraction method, the network structure and loss function of our similarity classifiers and the spectral cluster method. The experimental results are demonstrated in Sect. 3. Finally, in Sect. 4, we present our conclusions and improvement work directions in future.

2 Multi-modality Clustering Model

In this section, we discuss the three key parts of the clustering model shown in Fig. 2 in detail. The basic research object is model pair which consist of two 3D printing models. First, we extract feature from the original format of a 3D printing model pair in four input modes. Next, based on these input modes, we train four independent similarity classifiers. We then merge four similarity scores from one 3D printing model pair with a spectral cluster.

2.1 Input Modality

The 3D printing model is typically stored in STL file format. In this format, the 3D print model is stored as a set of triangular patches, and the vertex three-dimensional coordinates and normal vectors of each triangular element are recorded. In general, these triangular pieces will enclose a closed space.

Unfortunately, the similarity classifier implemented with neural networks can not directly use the original form of the 3D model as input. We need vector form input features, familiar with image or word bag. Given the abstraction and complexity of the original format, we have a number of ways to characterized the actual characteristics of the model, which are called input mode.

One basic assumption is that different input modes reflect the actual characteristics of the 3D printing model from different angles. In order to improve the effect

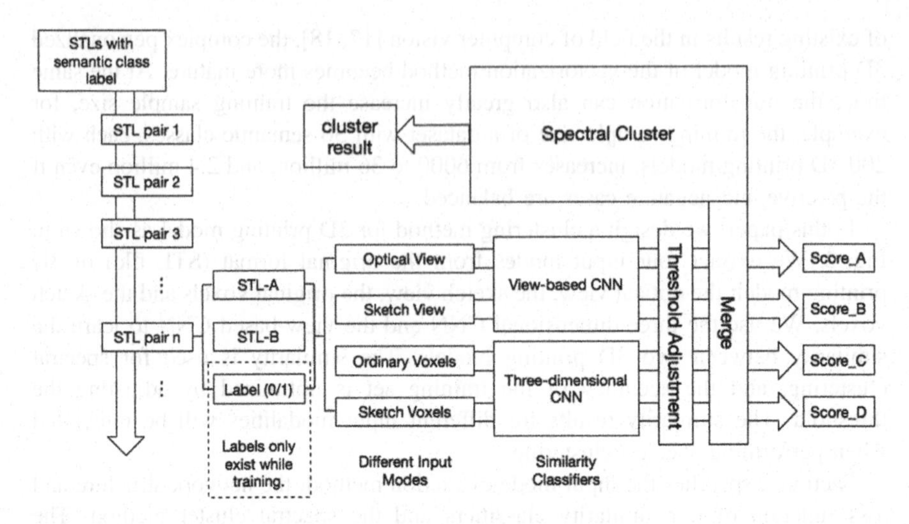

Fig. 2 Basic structure of our clustering systems for personalized 3D printing models. Similarity scores from different input modes are merged with a spectral cluster. There will be no "label" for test sets

of the clustering model in general, we trained separately on different input modes. Here we use four input modes.

The first modality is sketch view. If we show all the edges of the triangular pieces in the projection, we get the sketch view. As shown in Fig. 3, the sketch view can distinct between the parts that are described in detail and the parts that are briefly described. For example, on a keyboard 3D printing model, compared to its other parts, the button part has more display details, no matter from which direction to do projection. We draw the projection in 64 directions. An interesting fact is that we can not divide the sphere into 64 identical parts, because there are only five regular polyhedrons and the number of faces can only be up to 20. So we use the approximate average segmentation algorithm.

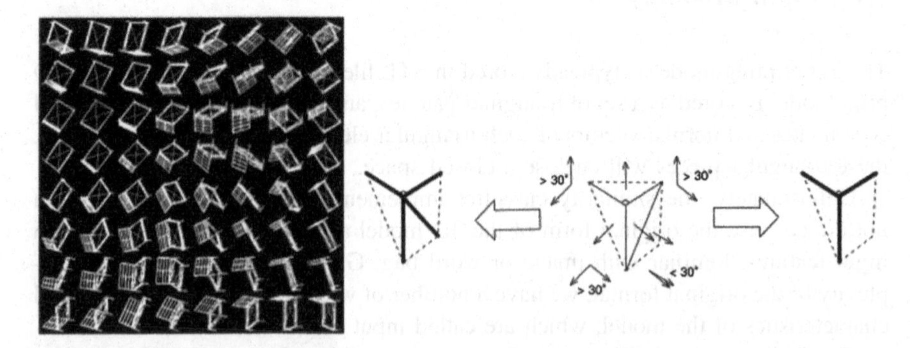

Fig. 3 The left part is a sketch view. The right part explains which lines are displayed for sketch view and optical view. The dotted line represents what we do not care about

The second modality is optical view. Unlike the sketch view, the optical view only shows some of the edges of the triangle pieces. As shown in Fig. 3, we can get the normal vector of each triangular piece. Given every edge is shared by two triangular pieces, if the angle between the two normal vectors is more than 30 degrees, we show this edge. This input mode is named after that the lines on most curved surfaces will be hidden and the projection is closer to the optical projection.

Voxel is an extension of the two-dimensional image matrix to the three-dimensional space. Divide the three-dimensional space into a number of small lattices by a custom unit length and count the number of vertexes falling into each lattice. The input form in this modality is a three-dimensional matrix. It is the third modality, called original voxel.

The fourth modality is sketch voxels. Taking into account the part of the personalized 3D printing model of the triangular pieces is very large, we will miss a lot of important information if only consider the vertex of triangular piece. We take such measures: on the edge of the triangular element, take a point every unit length, and treat these points and the original vertexes equally. It is the fourth modality.

2.2 *Similarity Classifiers*

Based on the improved convolution neural network, we construct a classifier to predict whether two personalized 3D printing models are similar. For each set of 3D models, the similarity classifier produces a similarity matrix.

The four modalities we have proposed can be divided into two groups: lattice structure and view group structure. So we propose two improved CNN from two different angles: the three-dimensional CNN and the view-based CNN.

2.2.1 Three-Dimensional CNN

The three-dimensional CNN is mainly for two voxel modes mentioned above. There is a great similarity between the voxel modal description and the pixel matrix description for image, but the voxel modes are extended to the three-dimensional space. In the adjacent range, the information carried in the lattice pixels is often continuous, which is similar to the natural image, so the convolution of the two-dimensional space can be extended to three dimensions. This constitutes our basic connection for the voxel modality, as shown in the formula below.

$$z_{p,q,r}^{c_s} = f\left(\sum_{(i,j,k)\in V} \omega_{i,j,k}^{c_s} x_{p+i,q+j,r+k}\right)$$

$$z_{p,q,r} = \left(z_{p,q,r}^{c_1}, z_{p,q,r}^{c_2}, \ldots, z_{p,q,r}^{c_n}\right)$$

The x represents a lattice pixel. The z represents the convolutional result and it has multiple channels. The ω represents the convolution parameter to be trained. The f represents the activate function. The V represents the convolution range.

The three-dimensional convolution units transform the complex voxel modal input into a high-dimensional vector and further transforms it into a similarity value through the fully connected layer.

2.2.2 View-Based CNN

For the view-type modalities, we focus not only on the projection binary image itself, but also on the relationship between the projections. Unlike the convolution of three-dimensional lattice pixels, the convolution between adjacent projection images shown in Fig. 4 helps to solve this problem.

As shown in the right part of Fig. 4, the view-based convolution helps to synthesize the information of adjacent projection images. With the full connection layer, the similarity of the two personalized 3D printing models can be predicted by the projection group.

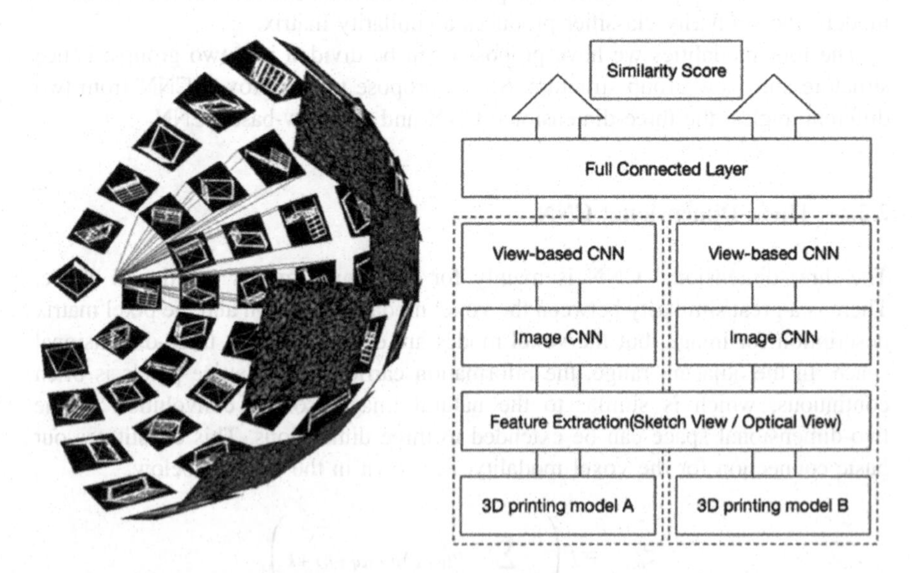

Fig. 4 Convolution of adjacent projection image is essentially a concatenation of all projection images within a convolution range in a different channel form into a multichannel image. Resulting in a higher dimension of a plurality of projection images that are adjacent to the centre one

2.3 Spectral Cluster

Based on the above processing, we get the similarity matrix on the set of personalized 3D printing model without labels. However, this similarity matrix generally does not translate into a valid clustering result. We use the minimum flow segmentation to deal with the similarity matrix and do spectrally cluster on the result of the minimum flow segmentation.

Before the minimum flow segmentation algorithm, the similarity needs to be transformed into a Boolean type. The traditional spectral clustering achieves better results by adjusting the thresholds of similarity judgements. When the similarity score of any modal output is less than T_{min}, we determine that it is dissimilar. when the output fraction of any modality is greater than T_{max}, we determine that it s similar. In other cases, the average similarity of each modality is compared to T_{ave}. (Under normal circumstances, $T_{min} = 0.2$, $T_{max} = 0.8$, $T_{ave} = 0.5$)

We define the clustering error function as:

$$E_{cluster} = \frac{1}{N^2} \sum_{1 \leq i \leq N, 1 \leq j \leq N} \left(S_{i,j} - T_{i,j}\right)^2$$

The S represents the similarity matrix from our CNN similarity model. The T represents the ground truth similarity matrix.

3 Experiments

Based on the clustering model we described above, this section describes clustering experiments on the ModelNet40 dataset (Table 1).

3.1 ModelNet40 Dataset

Modelnet40 is a generic dataset provided by Princeton University for 3D model classification [19]. It consists of 40 categories and there are hundreds of models in each category. Each Category describes a concrete class of things, such as beds,

Table 1 Neural network structure configuration

View-type modality			Voxel-type modality	
Image	View-based	Full connected	3D CNN	Full connected
64 × 64 × 1	64 × 256	2 × 2048	16 × 16×16 × 1	2 × 1024
16 × 16 × 16	8 × 1024	1024	4 × 4 × 4 × 256	256
1 × 1 × 256	1 × 2048	1	1 × 1024	1

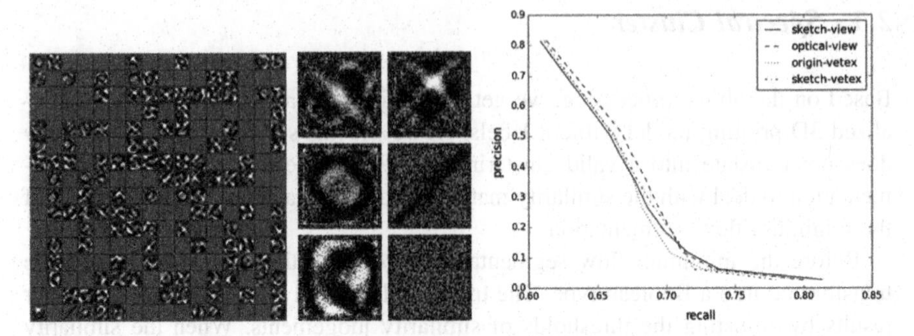

Fig. 5 Left: The visualization effect for convolutional parameters of the first layer of view-based CNN, after 30 rounds of training on 100,000 3D printing model pairs. Right: The precision-recall curve of our similarity classifiers of four modalities

plants, aircraft and so on. In view of the high semantic components, the 3D model in Modelnet40 dataset and the personalized 3D printing model is almost exactly the same. We use 20 categories to train our clustering model and 10 of the remaining categories to test.

3.2 *Result Analysis*

As shown in the left part of Fig. 5, some patterns have been extracted from 3D model projections. This layer compresses the 64 × 64 view image into 256 channels vector. Each thumbnail represents a group of parameters which corresponds to the corresponding output neuron and the image of the thumbnail represents the pattern extracted from the original projection. As shown in the right part of Fig. 6, our similarity classifier has achieved good results.

Based on the results of the similarity classifier, we use the spectral clustering strategy described above to process each model separately and merge them as the "merged group". Five categories are randomly selected from train set, and the 10 categories used for the test are divided into two groups. The results of the clustering experiments are shown in the Table 2. Multimodal clustering reduces the clustering error by 2.8%. We can conclude that the analysis of multiple modalities helps to improve the performance of clustering models.

Table 2 clustering errors of four modalities

	s-view	o-view	o-voxel	s-voxel	Merged
Train	0.349	0.372	0.294	0.277	0.242
dev1	0.525	0.562	0.552	0.527	0.501
dev2	0.573	0.532	0.581	0.528	0.513

4 Conclusion

In this paper, we propose to train a neural network model for clustering the personalized 3D printing model without labels. First, we extract the feature of the 3D printing model for different modalities. Then, we construct classifiers based on CNN that can classify similarities for two individuals. With combining the similarity classification results of these modalities, we use the spectral clustering algorithm to give a clustering result. We have experimented on the ModelNet40 dataset and achieved good results which verified that the combination of classifications in different modalities can effectively improve the clustering effect.

In the future, we will focus on extending this clustering algorithm from personalized 3D printing models with high semantic component to common 3D printing models with low semantic component, such as mechanical parts.

Acknowledgements The research is supported by the National High-Tech Research and Development Plan of China under grant No. 2015AA042101.

References

1. Gibson I, Rosen D, Stucker B. Additive manufacturing technologies: 3D printing, rapid prototyping, and direct digital manufacturing. Springer; 2014.
2. Mai J, Zhang L, Tao F, et al. Customized production based on distributed 3D printing services in cloud manufacturing. Int J Adv Manuf Technol. 2016;84(1–4):71–83.
3. Tao F, Zhang L, Liu Y, et al. Manufacturing service management in cloud manufacturing: overview and future research directions. J Manuf Sci Eng. 2015;137(4):040912.
4. Gao Z, Wang D, Zhang H, et al. A fast 3D retrieval algorithm via class-statistic and pair-constraint model. In: Proceedings of the 2016 ACM on multimedia conference. ACM; 2016. p. 117–21.
5. Li B, Lu Y, Godil A, et al. A comparison of methods for sketch-based 3D shape retrieval. Comput Vis Image Underst. 2014;119:57–80.
6. Chang AX, Funkhouser T, Guibas L, et al. Shapenet: An information-rich 3d model repository 2015. arXiv:1512.03012.
7. Leng B, Zhang X, Yao M, et al. A 3D model recognition mechanism based on deep Boltzmann machines. Neurocomputing. 2015;151:593–602.
8. Qi CR, Su H, Nießner M, et al. Volumetric and multi-view cnns for object classification on 3D data. In: Proceedings of the IEEE conference on computer vision and pattern recognition. 2016. p. 5648–56.
9. Maturana D, Voxnet SS: A 3D convolutional neural network for real-time object recognition. In: 2015 IEEE/RSJ international conference on intelligent robots and systems (IROS), IEEE; 2015. p. 922–8.
10. Brock A, Lim T, Ritchie JM, et al. Generative and discriminative voxel modeling with convolutional neural networks 2016. arXiv:1608.04236.
11. Furuya T, Ohbuchi R. Deep Aggregation of Local 3D geometric features for 3D model retrieval. 2016.
12. Banfield JD, Raftery AE. Model-based gaussian and non-gaussian clustering. Biometrics; 1993. p. 803–21.

13. Coretto P, Hennig C. Robust improper maximum likelihood: tuning, computation, and a comparison with other methods for robust Gaussian clustering. J Am Stat Assoc. 2016;111 (516):1648–59.
14. Ding L, Gonzalez-Longatt FM, Wall P, et al. Two-step spectral clustering controlled islanding algorithm. IEEE Trans Power Syst. 2013;28(1):75–84.
15. Galasso F, Keuper M, Brox T, et al. Spectral graph reduction for efficient image and streaming video segmentation. In: Proceedings of the IEEE conference on computer vision and pattern recognition. 2014. p. 49–56.
16. Ohn-Bar E, Trivedi MM. Learning to detect vehicles by clustering appearance patterns. IEEE Trans Intell Trans Syst. 2015;16(5):2511–21.
17. Simo-Serra E, Trulls E, Ferraz L, et al. Discriminative learning of deep convolutional feature point descriptors. In: Proceedings of the IEEE international conference on computer vision. 2015. p. 118–26.
18. Liu Z, Li Z, Zhang J, et al. Euclidean and Hamming Embedding for image patch description with convolutional networks. In: Proceedings of the IEEE conference on computer vision and pattern recognition workshops. 2016. p. 72–78.
19. Wu Z, Song S, Khosla A, et al. 3d shapenets: A deep representation for volumetric shapes. In: Proceedings of the IEEE conference on computer vision and pattern recognition. 2015. p. 1912–20.

State Feedback Stabilization of Stochastic Non-holonomic Mobile Robots Under Arbitrary Switchings

Dongkai Zhang, Hongmei Zhang, Huining Wu, Ye Yue and Qinghui Du

Abstract This work is aimed at the stabilization problem of stochastic nonholonmic mobile robots under arbitrary switchings. The model of stochastic nonholonomic mobile robots under arbitrary switching is given. Based on this model, state feedback controllers and switching control strategy are given. Furthermore, the system states are asymptotically stabilized at the zero equilibrium point in probability. Finally, the efficiency of controllers is demonstrated by a numerical example.

Keywords Stochastic nonholonomic mobile robots · State feedback stabilization · Arbitrary switching

1 Introduction

On the one hand, nonholonomic mobile robots, which were classified into four types [1], play an important role in the control of nonholonomic systems. The problems of stabilization and tracking were discussed with parametric uncertainties [2–4] and under visual servoing model [5–8]. In recent years, with the development of stochastic control, the control of stochastic nonholonomic systems was considered. The problem of stabilization was discussed in [9–14]. The problem of trajectory tracking of stochastic nonholonomic dynamic systems was considered [15]. The stabilization controller of a class of stochastic high-order nonholonomic systems with Markovian switching was designed [16]. Based on the models in [2], the stabilization of nonholonomic mobile robots was studied [10, 17–22].

D. Zhang (✉) · H. Zhang · H. Wu · Y. Yue
School of Science, Shijiazhuang University, Shijiazhuang 050035, China
e-mail: zdkmailhot@126.com

Q. Du
School of Mathematics, Luoyang Normal University, Luoyang 471934, China

Y. Jia et al. (eds.), *Proceedings of 2017 Chinese Intelligent Systems Conference*, Lecture Notes in Electrical Engineering 459,
https://doi.org/10.1007/978-981-10-6496-8_65

On the other hand, the control of switched systems under arbitrary switchings is a very important problem. The control of a class of nonlinear switched systems was given: stabilization [23–25] and global finite-time stabilization [26]. The stabilization in probability of stochastic nonlinear systems under arbitrary switchings was considered: state feedback stabilization [27] and output feedback stabilization [28, 29]. The controller of stabilisation problem of random non-linear systems with arbitrary switching was given [30]. The adaptive tracking controllers for stochastic nonlinear systems in nonstrict-feedback under arbitrary switching was discussed [31].

It is well known that a mobile robots unavoidably experiences switching. However, to authors' knowledge, there are few available results for the stabilization of stochastic nonholonomic mobile robots under arbitrary switching. So, there exists a problem which is how to find the models of stochastic nonholonomic mobile robots under arbitrary switching and deal with its problem of stabilization.

The main idea of this paper can be characterised as follows. (i) We give the model of stochastic nonholonomic mobile robots under arbitrary switching. State feedback stabilization is discussed by back stepping technique. (ii) A switching method is presented, which guarantees that the system is asymptotically stabilized in probability.

2 Problem Formulation

The nonholonomic wheeled mobile robot can be described by following differential equations [1]:

$$\begin{cases} \dot{x}_c = v\cos\theta, \\ \dot{y}_c = v\sin\theta, \\ \theta = w, \end{cases} \tag{1}$$

where (x_c, y_c) is the states, v is the forward velocity while w is the angular velocity of the robot.

The stochastic model with arbitrary switchings are given as follows

$$\begin{cases} d\theta = w_1 dt + w_{[\sigma(t),\,2]} dB, \\ dx_c = v_1 \cos\theta dt + v_{[\sigma(t),\,2]} cos\theta dB, \\ dy_c = v_1 \sin\theta dt + v_{[\sigma(t),\,2]} \sin\theta dB, \end{cases} \tag{2}$$

where $B(t)$ is independent standard Wiener processes, $w_{[\sigma(t),\,2]} = w_{[\sigma(t),\,2]}(\theta)$ and $v_{[\sigma(t),\,2]} = v_{[\sigma(t),\,2]}(\theta, x_c, y_c)$, $\sigma(t)$: $[0, +\infty) \to M = \{1, 2, \ldots, m\}$ is a piecewise constant switching signal.

For system (2), by the following transformation

$$\begin{cases} x_0 = \theta, u_0 = w_1, u = v_1, \\ x_1 = x_c \sin\theta - y_c \cos\theta, \\ x_2 = x_c \cos\theta - y_c \sin\theta, \end{cases} \tag{3}$$

one can obtain

$$dx_0 = u_0 dt + w_{[\sigma(t),\,2]} dB, \tag{4.1}$$

$$\left.\begin{aligned} dx_1 &= x_2 u_0 dt - \left\{ \frac{1}{2} x_1 w^2_{[\sigma(t),\,2]} - v_{[\sigma(t),2]} w_{[\sigma(t),\,2]} \right\} dt + x_2 w_{[\sigma(t),\,2]} dB \\ dx_2 &= u dt - \left\{ x_1 u_0 + \frac{1}{2} x_2 w^2_{[\sigma(t),\,2]} \right\} dt + \left\{ v_{[\sigma(t),\,2]} - x_1 w_{[\sigma(t),\,2]} \right\} dB \end{aligned}\right\}. \tag{4.2}$$

Remark 1 The main difference between this paper and [17, 21, 22] is the arbitrary switching signal $\sigma(t)$ existed.

3 Controllers Design

Assumption 1 There exist positive functions

$$p_{[\sigma(t),\,2]} \triangleq p_{[\sigma(t),\,2]}(x_0), q_{[\sigma(t),\,2]} \triangleq q_{[\sigma(t),\,2]}(x_0, x_1, x_2),$$

and positive constants τ_1 and τ_2, such that

$$\begin{cases} w_{[\sigma(t),\,2]} = p_{[\sigma(t),\,2]} x_0, \quad p_{[\sigma(t),\,2]} \le \tau_1, \\ \left| v_{[\sigma(t),\,2]} \right| = (|x_1| + |x_2|) q_{[\sigma(t),\,2]}, \quad q_{[\sigma(t),\,2]} \le \tau_2. \end{cases}$$

In the following, the stabilization of system (4) will be considered with $x_0(t_0) \neq 0$. If $x_0(t_0) = 0$, it will be discussed in the following section.

Firstly, we will design the controller u_0. One can take u_0 as follows:

$$u_0 = -\eta_0 x_0, \eta_0 = \lambda + \frac{3}{2}\tau_1^2, \tag{5}$$

where $\lambda > 0$.

Choosing a Lyapunov function

$$V_0(x_0) = \frac{1}{4} x_0^4, \tag{6}$$

from (5), (6), (7) and Assumption 1, one can obtain

$$\mathcal{L}V_0 \le x_0^3 u_0 + \frac{3}{2} x_0^2 w_{[\sigma(t),\,2]}^2 \le -\lambda x_0^2. \tag{7}$$

Theorem 1 *There exist positive constant* λ, *controller* u_0 *as* (5) *such that the closed-loop subsystem* (4.1) *and* (5) *is asymptotically stable in probability.*

Remark 2 From above Theorem, there exists a constant $m_1 > 0$, such that one has x_0 is bounded in probability, i.e.,

$$\lim_{t\to\infty} \sup_{t>t_0} P\{|x_0(t_0)| > m_1\} = 0. \tag{8}$$

Remark 3 Substituting (6) into the subsystem (4.1), one gets

$$\mathrm{d}x_0 = -\eta_0 x_0 \mathrm{d}t + p_{[\sigma(t),\,2]} x_0 \mathrm{d}B. \tag{9}$$

So, for $x_0(t_0) \ne 0$, one has $x_0 \ne 0$.

In fact, from Lemma 2.3 in [33], its solution can be expressed as

$$x_0(t) = x_0(t_0)\exp\left\{\int_{t_0}^{t}\left(-\eta_0 - \frac{1}{2} p_{[\sigma(t),\,2]}^2\right)\mathrm{d}s + \int_{t_0}^{t} p_{[\sigma(t),\,2]}\mathrm{d}B\right\} \tag{10}$$

Secondly, the controller u will be designed and the following transformation is needed.

$$z_1 = \frac{x_1}{x_0},\ z_2 = x_2. \tag{11}$$

By (11) and (4.2), one has

$$\begin{cases} \mathrm{d}z_1 = -\eta_0 z_2 \mathrm{d}t + \phi_1 \mathrm{d}t + \psi_1 \mathrm{d}B, \\ \mathrm{d}z_2 = u\mathrm{d}t + \phi_2 \mathrm{d}t + \psi_2 \mathrm{d}B, \end{cases} \tag{12}$$

where

$$\phi_1 = \eta_0 z_1 + z_1 \frac{w_{[\sigma(t),\,2]}^2}{x_0^2} - \frac{1}{2} z_1 w_{[\sigma(t),\,2]}^2 + \frac{v_{[\sigma(t),\,2]} w_{[\sigma(t),\,2]}}{x_0} - \frac{x_2 w_{[\sigma(t),\,2]}^2}{x_0^2},$$

$$\psi_1 = \frac{x_2 w_{[\sigma(t),\,2]}}{x_0} - \frac{z_1 w_{[\sigma(t),\,2]}}{x_0},\ \phi_2 = -\left\{x_1 u_0 + \frac{1}{2} x_2 w_{[\sigma(t),\,2]}^2\right\}_2,$$

$$\psi_2 = v_{[\sigma(t),\,2]} - x_1 w_{[\sigma(t),\,2]}.$$

The error variables ε_1 and ε_2 can be defined by

$$\varepsilon_1 = z_1, \varepsilon_2 = z_2 - \alpha_1(z_1), \tag{13}$$

Step 1 Defining the 1st Lyapunov function

$$V_1 = \frac{1}{4}\varepsilon_1^4, \tag{14}$$

by (12)–(14) and Itô formula, one has

$$\mathcal{L}V_1 = \varepsilon_1^3\{-\eta_0 z_2 + \phi_1\} + \frac{3}{2}\varepsilon_1^2\psi_1^2. \tag{15}$$

We choose α_1 as

$$\alpha_1 = c_1\varepsilon_1, \tag{16}$$

where constant $c_1 > 0$. From (15) and Lemma 2.2 in [32], the following inequalities hold.

$$\begin{aligned}
-\eta_0 z_2\varepsilon_1^3 &\le \eta_0\left\{\frac{3d}{4}\varepsilon_1^4 + \frac{1}{4d^3}\varepsilon_2^4\right\} - \eta_0 c_1\varepsilon_1^4,\\
-\frac{1}{2}\varepsilon_1^3 z_1 w_{[\sigma(t),2]}^2 &\le \frac{1}{2}\tau_1^2 m_1^2\varepsilon_1^4,\\
\varepsilon_1^3\frac{v_{[\sigma(t),2]}w_{[\sigma(t),2]}}{x_0} &\le \tau_1\tau_2\left\{\frac{3}{4} + m_1 + c_1\right\}\varepsilon_1^4 + \frac{1}{4}\tau_1\tau_2\varepsilon_2^4,\\
\varepsilon_1^3 z_1\frac{w_{[\sigma(t),2]}^2}{x_0^2} &\le \tau_1^2\varepsilon_1^4, 3\varepsilon_1^2 z_1^2\frac{w_{[\sigma(t),2]}^2}{x_0^2} \le 3\tau_1^2\varepsilon_1^4,\\
-\varepsilon_1^3 x_2\frac{w_{[\sigma(t),2]}^2}{x_0^2} &\le \tau_1^2\left\{c_1 + \frac{3}{4}\right\}\varepsilon_1^4 + \frac{1}{4}\tau_1^2\varepsilon_2^4,\\
3\varepsilon_1^2 x_2^2\frac{w_{[\sigma(t),2]}^2}{x_0^2} &\le 3\tau_1^2\{1 + 2c_1^2\}\varepsilon_1^4 + 3\tau_1^2\varepsilon_2^4,
\end{aligned}$$

where constant $d > 0$. By above inequalities and (15), we have

$$\begin{aligned}
\mathcal{L}V_1 \le &\left\{-\eta_0 c_1 e + \left(1 + \frac{3d}{4}\right)\eta_0 - c_1\eta_0(1-e) + \frac{\tau_1\tau_2(4m_1 + 4c_1 + 3)}{4}\right.\\
&\left.+ \frac{\tau_1^2(2m_1^2 + 31 + 4c_1 + 24c_1^2)}{4}\right\}\varepsilon_1^4 + \left\{\frac{\eta_0}{4d^3} + \frac{1}{4}\tau_1\tau_2 + \frac{13}{4}\tau_1^2\right\}\varepsilon_2^4,
\end{aligned}$$

where constant $0 < e < 1$. Letting

$$\begin{cases} c_1 \geq \frac{4+3d}{4e}, \\ \eta_0 \geq \frac{\tau_1\tau_2(4m_1+4c_1+3)+\tau_1^2\left(2m_1^2+31+4c_1+24c_1^2\right)}{2c_1(1-e)}, \end{cases} \tag{17}$$

one has

$$\mathcal{L}V_1 \leq -\frac{\eta_0 c_1(1-e)}{2}\varepsilon_1^4 + \left\{\frac{\eta_0}{4d^3} + \frac{1}{4}\tau_1\tau_2 + \frac{13}{4}\tau_1^2\right\}\varepsilon_2^4. \tag{18}$$

Step 2 By (12) and (13), one gets

$$\begin{aligned} \mathrm{d}\varepsilon_2 = u\mathrm{d}t - \left\{-\eta_0 z_1 x_0^2 - \frac{1}{2}z_2 w_{[\sigma(t),\,2]}^2\right\}\mathrm{d}t + c_1\{\eta_0 z_2 - \eta_0 z_1 \\ - z_1\frac{w_{[\sigma(t),\,2]}^2}{x_0^2} + \frac{1}{2}z_1 w_{[\sigma(t),\,2]} - \frac{v_{[\sigma(t),\,2]}w_{[\sigma(t),\,2]}}{x_0} + \frac{x_2 w_{[\sigma(t),\,2]}^2}{x_0^2}\right\}\mathrm{d}t \\ + \left\{v_{[\sigma(t),\,2]} - x_1 w_{[\sigma(t),\,2]} - c_1 z_2\frac{w_{[\sigma(t),\,2]}}{x_0} + c_1 z_1\frac{w_{[\sigma(t),\,2]}}{x_0}\right\}\mathrm{d}B. \end{aligned} \tag{19}$$

Define the 2nd Lyapunov candidate function

$$V_2 = V_1 + \frac{1}{4}\varepsilon_2^4. \tag{20}$$

So, one can obtain

$$\begin{aligned} \mathcal{L}V_2 \leq -\frac{c_1\eta_0(1-e)}{2}\varepsilon_1^4 + \left\{\frac{\eta_0}{4d^3} + \frac{1}{4}\tau_1\tau_2 + \frac{13}{4}\tau_1^2\right\}\varepsilon_2^4 \\ + \varepsilon_2^3\{u + \phi_2 + c_1\eta_0 z_2 - c_1\phi_1\} + 3\varepsilon_2^2\{\psi_2^2 + c_1^2\psi_1^2\}. \end{aligned} \tag{21}$$

By (21) and Lemma 2.2 in [32], one has

$$\begin{aligned} \varepsilon_2^3\eta_0 z_1 x_0^2 &\leq \frac{1}{4}\varepsilon_1^4 + \frac{3}{4}\left(\eta_0 m_1^2\right)^{\frac{4}{3}}\varepsilon_2^4, \\ -\frac{1}{2}\varepsilon_2^3 z_2 w_{[\sigma(t),\,2]}^2 &\leq \frac{1}{8}\varepsilon_1^4 + \frac{1}{2}\tau_1^2 m_1^2\varepsilon_2^4 + \frac{3}{8}\left(c_1\tau_1^2 m_1^2\right)^{\frac{4}{3}}\varepsilon_2^4, \\ \varepsilon_2^3 c_1\eta_0 z_2 &\leq \frac{1}{4}\varepsilon_1^4 + \left(c_1\eta_0 + \frac{3}{4}\left(\eta_0 c_1^2\right)^{\frac{4}{3}}\right)\varepsilon_2^4, \\ \varepsilon_2^3 c_1\eta_0 z_1 &\leq \frac{1}{4}\varepsilon_1^4 + \frac{3}{4}\left(\eta_0\tau_1^2\right)^{\frac{4}{3}}\varepsilon_2^4, \end{aligned}$$

$$-\varepsilon_2^3 c_1 z_1 \frac{w^2_{[\sigma(t),\,2]}}{x_0^2} \le \frac{1}{4}\varepsilon_1^4 + \frac{3}{4}\left(c_1\tau_1^2\right)^{\frac{4}{3}}\varepsilon_2^4,$$

$$\frac{1}{2}\varepsilon_2^3 c_1 z_1 w^2_{[\sigma(t),\,2]} \le \frac{1}{8}\varepsilon_1^4 + \frac{3}{8}\left(c_1\tau_1^2 m_1^2\right)^{\frac{4}{3}}\varepsilon_2^4,$$

$$-c_1\varepsilon_2^3 \frac{v_{[\sigma(t),2]} w_{[\sigma(t),\,2]}}{x_0} \le \frac{1}{2}\varepsilon_1^4 + \left\{\frac{3}{4}\left(c_1\tau_1\tau_2 m_1\right)^{\frac{4}{3}} + c_1\tau_1\tau_2 + \frac{3}{4}\left(c_1^2\tau_1\tau_2\right)^{\frac{4}{3}}\right\}\varepsilon_2^4,$$

$$c_1\varepsilon_2^3 \frac{x_2 w^2_{[\sigma(t),\,2]}}{x_0^2} \le \frac{1}{4}\varepsilon_1^4 + \left\{c_1\tau_1^2 + \frac{3}{4}\left(c_1^2\tau_1^2\right)^{\frac{4}{3}}\right\}\varepsilon_2^4,$$

$$\begin{aligned} &6\varepsilon_2^2\left\{v^2_{[\sigma(t),\,2]} + z_1^2 x_0^2 w^2_{[\sigma(t),\,2]} + c_1^2\left(z_1^2+z_2^2\right)\frac{w^2_{[\sigma(t),\,2]}}{x_0^2}\right\} \\ &\le 2.5\varepsilon_1^4 + \left\{72\tau_2^4 m_1^4 + 24\tau_2^2 + 288c_1^4\tau_2^4 + 18\tau_1^4 m_1^8 + 12c_1^2\tau_1^2 \right. \\ &\left. + 72c_1^8\tau_1^4 + 18c_1^4\tau_1^4\right\}\varepsilon_2^4. \end{aligned}$$

Adding $c_2\varepsilon_2^4$ while subtracting it into (21) together with above inequalities, one gets

$$\mathcal{L}V_2 \le -\left\{\frac{c_1\eta_0(1-e)}{2} - 4.5\right\}\varepsilon_1^4 - c_2\varepsilon_2^4 + \varepsilon_2^3 u + H_1\varepsilon_2^4, \tag{22}$$

where

$$\begin{aligned} H_1 = {} & c_2 + \frac{3}{4}\left(\eta_0 m_1^2\right)^{\frac{4}{3}} + \frac{1}{2}\tau_1^2 m_1^2 + \frac{3}{8}\left(c_1\tau_1^2 m_1^2\right)^{\frac{4}{3}} + c_1\eta_0 + \frac{3}{4}\left(\eta_0 c_1^2\right)^{\frac{4}{3}} + \frac{3}{4}\left(c_1\eta_0\right)^{\frac{4}{3}} \\ & + \frac{3}{4}\left(c_1\tau_1^2\right)^{\frac{4}{3}} + \frac{3}{8}\left(c_1\tau_1^2 m_1^2\right)^{\frac{4}{3}} + \frac{3}{4}\left(c_1\tau_1\tau_2 m_1\right)^{\frac{4}{3}} + c_1\tau_1\tau_2 + \frac{3}{4}\left(c_1^2\tau_1\tau_2\right)^{\frac{4}{3}} \\ & + 72\tau_2^4 m_1^4 + 24\tau_2^4 + 288c_1^4\tau_2^4 + 18\tau_1^4 m_1^8 + 12c_1^2\tau_1^2 + 72c_1^8\tau_1^2 + 18c_1^4\tau_1^4 \end{aligned}$$

Letting

$$u = -H_1\varepsilon_2, \tag{23}$$

and substituting it into (22), one gets

$$\mathcal{L}V_2 \le -\bar{c}_1\varepsilon_1^4 - c_2\varepsilon_2^4, \tag{24}$$

where $\bar{c}_1 = \frac{c_1\eta_0(1-e)}{2} - 4.5$.

Choosing the following Lyapunov function

$$V = V_0 + V_2,$$

which together with (8) and (25), it is easy to see

$$\mathcal{L}V \leq -\lambda x_0^2 - \bar{c}_1\varepsilon_1^4 - c_2\varepsilon_2^4. \tag{25}$$

Theorem 2 *There exist positive constants* $c_1, c_2, \lambda, m_1, d, \tau_1, \tau_2$ *and* $0 < e < 1$ *to satisfy* (17) *and* (25), *controllers* u_0 *and* u *as* (5) *and* (23) *such that the closed-loop system* (12) *and* (23) *is asymptotically stable in probability.*

4 Switching Control Stability

For $x_0(t_0) \neq 0$, an open loop control $u_0 = -u_0^* \neq 0$ is chosen to satisfy $x_0(t_s^*) \neq 0$ in a limited time. So, one has the following theorem.

Theorem 3 *For* (2), *applying the following switching method,*

(i). If

$$\{(\theta(t_0), x_c(t_0), y_c(t_0)) \in \mathbb{R}^3 | \theta(t_0) \neq 0\},$$

u_0 *and* u *are chosen as* (5) *and* (23);

(ii). If

$$\{(\theta(t_0), x_c(t_0), y_c(t_0)) \in \mathbb{R}^3 | \theta(t_0) = 0\},$$

we apply $u_0 = -u_0^* \neq 0$ *and* $u = u^*$ *on* $[t_0, t_s^*)$, u_0 *and* u *as* (5) and (23) on $[t_s^*, \infty)$.

Then, the states of (2) are asymptotically regulated to zero in probability.

5 A Simulation Example

For initial conditions $\theta(0) = 2, x_c(0) = -2.4412$ and $y_c(0) = -3.3167$, consider the system (2) with $\sigma(t)$: $[0, +\infty) \to M = \{1, 2\}$. When $s = 1$, the systems are as follows

$$\begin{cases} d\theta = w_1 dt + w_{[1,\,2]} dB, \\ dx_c = v_1 \cos\theta dt + v_{[1,\,2]} cos\theta dB, \\ dy_c = v_1 \sin\theta dt + v_{[1,\,2]} \sin\theta dB, \end{cases}$$

where $w_{[1,2]} = 0.01\theta \cos\theta$ and

$$v_{[1,2]} = 0.01(x_c \cos\theta + y_c \sin\theta)\sin(x_c \sin\theta - y_c \cos\theta).$$

When $s = 2$, the systems are as follows

$$\begin{cases} d\theta = w_1 dt + w_{[2,\,2]} dB, \\ dx_c = v_1 \cos\theta dt + v_{[2,\,2]} cos\theta dB, \\ dy_c = v_1 \sin\theta dt + v_{[2,\,2]} \sin\theta dB, \end{cases}$$

where $w_{[2,2]} = 0.01\theta \sin\theta$ and

$$v_{[2,2]} = 0.01(x_c \cos\theta + y_c \sin\theta)\cos(x_c \sin\theta - y_c \cos\theta).$$

By Assumption 1, we have $p_{[1,2]} = 0.01\cos(x_0)$, $p_{[2,2]} = 0.01\sin(x_0)$, $q_{[1,2]} = 0.01\sin(x_1)$, $q_{[2,2]} = 0.01\cos(x_1)$, i.e., $\tau_1 = 0.01$ and $\tau_2 = 0.01$. From Remark 2, m_1 equals to 2. Letting $d = 1$ and $e = 0.5$, by (17) and (25), one has $c_1 = 1.58$, $\eta_0 = 11.5$ and $c_2 = 0.1$. Figure 1 gives the responses of states. Figure 2 show the responses of controllers u_0 and u. Figure 3 demonstrates the response of switching signals $\sigma(t)$.

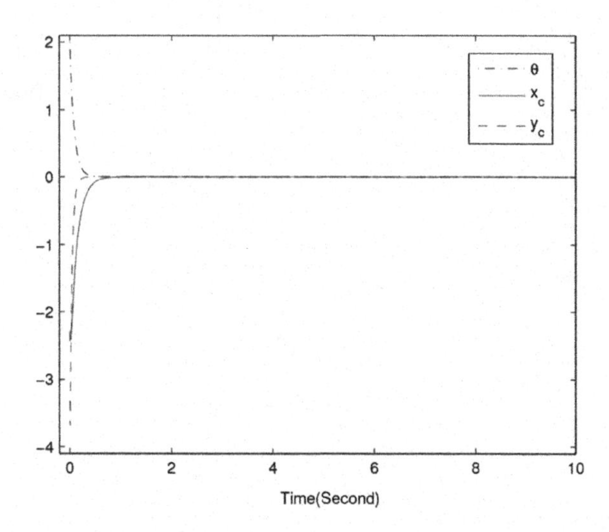

Fig. 1 The responses of states θ, x_c and y_c

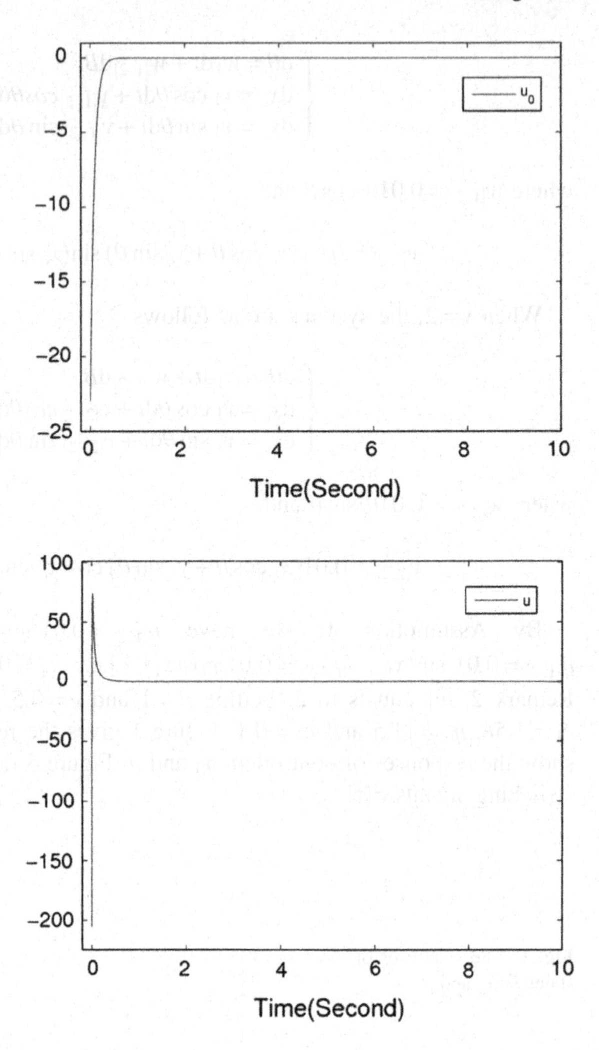

Fig. 2 The responses of controllers u_0 and u

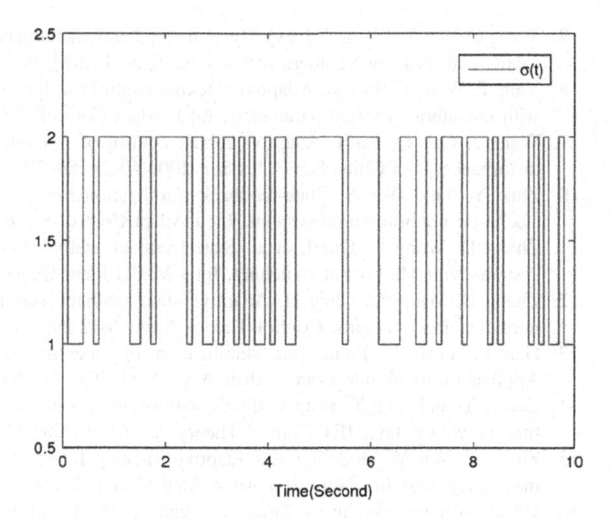

Fig. 3 The responses of switching signals $\sigma(t)$

6 Conclusions

We give a model of stochastic nonholonomic mobile robots under arbitrary switching. State feedback controllers are designed by backstepping technique. To guarantee system states asymptotically stabilized at the zero equilibrium point in probability, a switching control strategy is given.

Acknowledgements This work is supported by the National Natural Science Foundation of China (no. 61503262 and 61374040), Natural Science Foundation of Hebei Province (no. A2014106035), Foundation for High-level Talents of Hebei Province (no. A2016001144), Natural Science Foundation of Henan Educational Committee (no. 17A110027).

References

1. Campion G, Bastin G, D'Andréa-Novel B. Structural properties and classification of kinematic and dynamic models of wheeled mobile robots. IEEE Trans Rob Autom. 1996;12 (1):47–62.
2. Hespanha J, Liberzon S, Morse A. Towards the supervisory control of uncertain nonholonomic systems. In: Proceedings of the American control conference, San Diego, U.S.A., 1999. p. 3520–3524.
3. Jiang Z. Robust exponential regulation of nonholonomic systems with uncertainties. Automatica. 2000;36(2):189–209.
4. Xi Z, Feng G, Jiang Z, Cheng D. A Switching algorithm for global exponential stabilization of uncertain chained systems. IEEE Trans Autom Control. 2003;48(10):1793–8.
5. Liang Z, Wang C. Robust stabilization of nonholonomic chained form systems with uncertainties. Acta Automat Sinica. 2011;37(2):129–42.
6. Yang F, Wang C. Adaptive stabilization for uncertain nonholonomic dynamic mobile robots based on visual servoing feedback. Acta Automat Sinica. 2011;37(7):857–64.

7. Wang C, Mei Y, Liang Z, Jia Q. Dynamic feedback tracking control of non-holonomic mobile robots with unknown camera parameters. Trans Inst Meas Control. 2010;32(2):155–69.
8. Yang F, Wang C, Jing G. Adaptive tracking control for dynamic nonholonomic mobile robots with uncalibrated visual parameters. Int J Adapt Control Sig Process. 2013;67(8):688–700.
9. Wang J, Gao Q, Li H. Adaptive robust control of nonholonomic systems with stochastic disturbances. Sci China Ser F: Inf Sci. 2006;49(2):189–207.
10. Zhao Y, Yu J, Wu Y. State-feedback stabilization for a class of more general high order stochastic nonholonomic systems. Int J Adapt Control Sig Process. 2011;25(8):687–706.
11. Zhang D, Wang C, Qiu J, et al. State-feedback stabilisation for stochastic non-holonomic systems with Markovian switching. Int J Model Ident Control. 2012;16(3):221–8.
12. Zhang D, Wang C, Chen H. Adaptive state-feedback stabilization for stochastic nonholonomic chained systems. Control Theory Appl. 2012;29(11):1479–87.
13. Gao F, Yuan F. Finite-time stabilization of stochastic nonholonomic systems and its Application to Mobile Robot. Abstr Appl Anal. 2012;2012:1–18.
14. Gao F, Yuan F, Wu Y. State-feedback stabilisation for stochastic non-holonomic systems with time-varying delays. IET Control Theory A. 2012;6(17):2593–600.
15. Zhang Z, Wu Y. Modeling and adaptive tracking for stochastic nonholonomic constrained mechanical systems. Nonlinear Anal: Anal Model. 2016;21(2):166–84.
16. Du Q, Wang C, Wang G, Zhang D. State-feedback stabilization for stochastic high-order nonholonomic systems with Markovian switching. Nonlinear Anal: Hybrid Syst. 2015;18:1–14.
17. Shang Y, Meng H. Exponential stabilization of nonholonomic mobile robots subject to stochastic disturbance. J Inf Comput Sci. 2012;9(9):2635–42.
18. Liu Y, Wu Y. Output feedback control for stochastic nonholonomic systems with growth rate restriction. Asian J Control. 2011;13(1):177–85.
19. Zhang D, Wang C, Chen H, et al. Adaptive stabilization of stochastic non-holonomic systems with nonhomogeneous uncertainties. Trans Inst Meas Control. 2013;35(5):648–63.
20. Zhang D, Wang C, Chen H. Adaptive state-feedback stabilization of stochastic uncertain nonholonomic systems. In: the 31th chinese control conference, Hefei, China, 2012. p. 1517–1522.
21. Wu Z, Liu Y. Stochastic stabilization of nonholonomic mobile robot with heading-angle-dependent disturbance. Math Probl Eng. 2012;2012:1–17.
22. Feng W, Sun Q, Cao Z, et al. Adaptive state-feedback stabilization for stochastic nonholonomic mobile robots with unknown parameters. Discrete Dyn Nat Soc. 2013;2013:1–9.
23. Wang C, Jiao X. Adaptive control under arbitrary switching for a class of switched nonlinear systems with nonlinear parameterization. Int J Control. 2015;88(10):2044–54.
24. Wu L, Yang R, Shi P, Sue X. Stability analysis and stabilization of 2-D switched systems under arbitrary and restricted switchings. Automatica. 2015;59:206–15.
25. Ma R, Zhao J. Backstepping design for global stabilization of switched nonlinear systems in lower triangular form under arbitrary switchings. Automatica. 2010;46(11):1819–23.
26. Liang Y, Ma R, Wang M, et al. Global finite-time stabilisation of a class of switched nonlinear systems. Int J Syst Sci. 2015;46(16):2897–904.
27. Hou M, Fu F, Duan G. Global stabilization of switched stochastic nonlinear systems in strict-feedback form under arbitrary switchings. Automatica. 2013;49(8):2571–5.
28. Liu L, Zhao X, Niu B, et al. Global output-feedback stabilisation of switched stochastic non-linear time-delay systems under arbitrary switchings. IET Control Theory A. 2015;9(2):283–92.
29. Liang X, Hou M, Duan G. Output feedback stabilization of switched stochastic nonlinear systems under arbitrary switchings. Int J Auto Comput. 2013;10(6):571–7.
30. Jiao T, Xu S, Li Y, et al. Adaptive stabilisation of random systems with arbitrary switchings. IET Control Theory A. 2015;9(18):2634–40.
31. Zhao X, Shi P, Zheng X, Zhang L. Adaptive tracking control for switched stochastic nonlinear systems with unknown actuator dead-zone. Automatica. 2015;60:193–200.

32. Lin W, Qian C. Adaptive control of nonlinearly parameterized systems: a nonsmooth feedback framework. IEEE Trans Autom Control. 2002;47(5):757–74.
33. Mao X. Stochastic differential equations and their applications. Chichester: Horwood Publishing; 1997.

A Review of Retinal Vessel Segmentation and Artery/Vein Classification

Dongmei Fu, Yang Liu and Zhicheng Huang

Abstract The Fundus blood vessel is the only vascular system that can be observed noninvasively in the human body. Through the fundus photograph, we can get arterial and venous structures. Changes in the shape and size of blood vessels are important features for the diagnosis of diabetes, hypertension and other diseases. The segmentation of the blood vessels and the classification of arteries and veins are the basis for obtaining the characteristics and quantitative indicators. This paper discusses the research progress of retinal vessel segmentation and arteriovenous classification on the fundus images, and summarizes the research background, various methods along with advantages and disadvantages. It aims to guide the researchers to understand the research content and progress in this field, and to provide a comprehensive foundation for the follow-up research work.

Keywords Fundus images · Retinal vessels segmentation · Artery/vein classification

1 Introduction

The Fundus blood vessel is the only vascular system that can be observed noninvasively in the human body. Through fundus photograph obtained from the fundus camera we can observe the distribution and morphology of fundus arteriovenous vessels and other fundus tissues (Fig. 1). Medical studies have shown that the curvature of retinal vessels, the change of its diameter and color depth are important clinical features, and are important references for the diagnosis of diseases, such as diabetes, hypertension and so on.

Most changes of the morphological features and vessels diameter can be obtained by accurately segmenting blood vessels and correctly distinguishing

D. Fu (✉) · Y. Liu · Z. Huang
School of Automation and Electrical Engineering, University of Science and Technology Beijing, 100083 Beijing, China
e-mail: fdm_ustb@ustb.edu.cn

Y. Jia et al. (eds.), *Proceedings of 2017 Chinese Intelligent Systems Conference*, Lecture Notes in Electrical Engineering 459,
https://doi.org/10.1007/978-981-10-6496-8_66

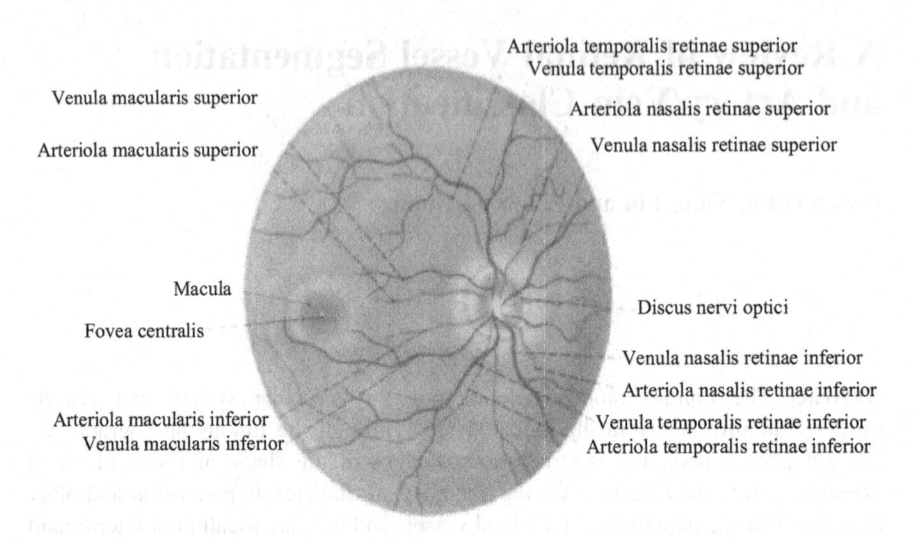

Fig. 1 Fundus retinal image and its structure

arteries and veins. But the retinal images have the following characteristics: Illumination is not uniform, the blood vessels stagger with each other complexly, and the difference between arteries and veins is small. It is difficult to improve the accuracy and robustness of vascular segmentation and classification, which makes it become a hot topic in the field of ophthalmic medical image processing.

At present, many scholars at home and abroad have paid attention to the important role of retinal images in clinical diagnosis, and have carried out a lot of research work, which are very important in two aspects: the first is to achieve accurate segmentation of retinal blood vessels automatically; the second is to achieve automatic classification of artery and vein. This article will sum up and analyze their researches about these two aspects.

2 The Methods of Retinal Vascular Segmentation

From the perspective of image processing [1], the existing methods of retinal vessel segmentation can be broadly divided into six categories: (1) Methods based on vascular tracking, (2) Methods based on matched filtering, (3) Methods based on morphological operations, (4) Methods based on deformation models, (5) Methods based on traditional machine learning, (6) Methods based on deep learning.

(2.1) **Methods based on vascular tracking**

The earliest vascular tracking method was proposed by Liu [2] in 1993. The method is based on the continuous structure of the blood vessels. Firstly, the initial seed points are set, and then track along the blood vessels until the end condition is satisfied. The main difficulties of this method lie in the selection of the initial seed points and the design of the tracking algorithm. Many scholars have already done a lot of researches, for example: Can [3] proposed an adaptively automatic vascular tracking algorithm in 1999, using a pixel-wide grid line to find the local gray-level minimum points on the blood vessels as the initial seed points for cyclic tracking; Vlachos [4] proposed a method based on multi-scale line-tracking, and post-processing with morphological operations; Nayebifar [5] presented a new approach based on particle filtering to determine and to track locally the vessel paths in retina; Mouloud Adel [6] proposed a method that used the Maximum a posteriori probability criterion to achieve vessel segmentation based on the vascular diameter; Yin [7] made some improvements on the basis of [6], firstly, the initial blood vessel boundary points and directions were given manually, during the tracking process, the local grey levels statistics information could be obtained by using a semi-ellipse as the dynamic search window; Zhang [8] proposed a method based on Bayesian theory and multi-scale linear detection, and designed respectively three models for the non-interference, cross and branch, then the Bayesian maximum posteriori probability criterion was used to determine the vascular boundary points.

In general, the methods based on vascular tracking are simple and intuitive, and easy to understand, and we can obtain some local information such as changes in blood vessel diameter. But there are also deficiencies, for example, the result of segmentation is sensitive to the initial seed points and is easily affected by branch or cross points, and the fault tolerance is low.

(2.2) **Methods based on matched filtering**

Matched filtering is an effective method for retinal vessels segmentation. The method based on the matched filtering was firstly proposed by Chaudhuri [9]. Because the gray distribution of the retinal vascular profile is consistent with the Gaussian properties, the image can be convoluted with a filter to extract the target object.

Hoover [10] used a threshold descent search algorithm to extract the blood vessels after the matched filtering, and selected a reasonable threshold by analyzing the region's features. The algorithm considers the local features of retinal vessels and the regional characteristics of vascular network distribution simultaneously. It can greatly reduce the error rate, but the overall calculation is complicated.

Researchers have improved the method of matched filtering. Jiang [11] proposed an adaptive local threshold method based on multi-threshold detection. Zhang [12] applied a local vessel cross section analysis that used double sided thresholding to improve a matched filter. This method reduced the rate of blood vessel omission in abnormal fundus images. After that, Zhang [13] proposed a MF-FDOG (matched filtering, first-order derivative of the Gaussian) method, which used the local response of the MF and the FDOG to identify the vascular boundaries and improved the accuracy of vascular segmentation. Cinsdikici [14] proposed a hybrid model of matched filter and ant colony algorithm to accurately extract retinal vessels. In order to better extract fine blood vessels and calculate the width of vessels, Li [15] proposed a MPMF (multi-scale production of the matched filter) method for vessel extraction. In the image preprocessing stage, multi-scale matched filter was used to enhance the contrast of images while suppressing noise, and then a double thresholding method was used to detect vessels.

The effect of matched filter algorithm depends on the degree of matching between the template and the blood vessel, which is affected by many factors, such as central light reflection, radius change, noise interference, lesion interference and so on.

(2.3) Methods based on morphological operations

Mathematical morphology is very important in the field of image processing, and is a powerful tool in the edge detection and skeleton extraction. The image is corroded, expanded and carried out other operations by using structural elements, and then subtracted from the original image to get edges of the image.

According to the characteristics of retinal vessels with connectivity and local linear distribution, Zana [16] used a method of mathematical morphology to extract the retinal vascular structure. Since Zana's approach was too dependent on structural elements, Ayala [17] proposed a new method and its main idea was to define the averages of a given fuzzy set by using different definitions of the mean of a random compact set. Mendonca [18] enhanced the retinal vessels through using multi-scale top hat transformation and then combined with the extracted vascular centerline to obtain retinal vessel segmentation results. Fraz [19] improved on the basis of [18]. The vessel centerlines detection was combined with the morphological bit plane slicing to achieve the vascular segmentation. Yang [20] presented a novel hybrid approach that combined mathematical morphology with a fuzzy clustering method. First, the morphological operation was used to enhance vessels and to suppress noise in different directions. And then FCM (fuzzy c-means) was used for vascular extraction. Miri [21] used the curvelet transform and multi-structural element morphological reconstruction to analyze and process the fundus images. Experiments showed that the method had a high accuracy of 96%. Wang [22], respectively, used morphological reconstruction and regional growth method to extract blood vessels after Gabor wavelet transformation.

Methods based on mathematical morphology are generally fast and efficient, and they have the advantage of suppressing noise. However, these methods do not make full use of vascular characteristics such as vascular profile. The effect of segmentation is seriously dependent on the selection of structural elements, so the method is often used for the extraction of blood vessel centerline, and is combined with other methods for vascular segmentation and arteriovenous classification.

(2.4) Methods based on deformation models

The basic principle of extracting retinal blood vessels based on the deformation model is to describe the target boundary with a continuous curve. The boundary curve is defined by the energy function, which is deformed under the action of external force and internal force guidance. Therefore, the problem of extracting retinal vessels is converted to finding the minimum value of the energy function. The methods based on deformation models can be divided into two types: parameter deformation models and geometric deformation models.

The method based on the parameter deformation model is also called the active contour model, initially called "snake" model. Espona [23] used the classic snake model to segment the retinal blood vessels in the fundus images, and introduced morphological operations to establish a deformation model based on the vascular centerline. Al-Diri [24] proposed an efficient model called ROT (Ribbon of twins) to extract retinal blood vessels, and combined the concept of ribbon snake and twin snakes to find the vascular contours. After that, on the basis of this, [25] proposed a novel algorithm for segmenting and measuring retinal blood vessels, which can accurately locate the edge of blood vessels under difficult conditions, including noisy blurred edges, light reflex phenomena, closely parallel vessels and others.

The geometric deformation model is based on the evolutionary theory of deformation curves with implicit parameters, and the level set method is used to track the shape of target. Based on the Active Contours Without Edges, Vese [26] proposed a multiphase level set architecture using the Mumford-Shah model for image segmentation. Sum [27] presented a level set method based on the Active Contour that can extract blood vessels under nonuniform illumination. Zhao [28] proposed a new retinal vessel segmentation method based on level set and region growing. Liang [29] used the background variance of the retinal vessel to set the weight coefficients such as shape constraints and area constraints, and used the connectivity information to remove artifacts and lesions to further improve the accuracy of vascular segmentation.

The level set methods are used to describe the active contour and solve the curve evolution equation, which can deal with the boundary topology change, and detect the edge of multiple objects in the image. But the vessel segmentation methods based on the deformation model are more sensitive to the position of the initial curve, and the calculation is more complicated.

(2.5) Methods based on traditional machine learning

Machine learning can be divided into two categories: supervised learning and unsupervised learning. The supervised learning methods need to be manually marked in advance, of which the main idea is using various methods to extract the features in the image, construct the feature vectors, and then we use the feature vectors of the sample set to train a classifier to get the classification model.

Soares [30] applied 2D Gabor wavelet transform function to deal with the retinal image and constructed the feature space, then used the Bayesian classifier to classify each image pixel as vessel or non-vessel. Ricci [31] proposed a method of using line operators and support vector classification. Marin [32] first extracted a 7D feature vector, which was composed of a gray level and features based on invariant moments, and then used a neural network to identify the retinal vessels. Fraz has done a lot of work in the retinal vessel extraction using supervised learning methods, and in [33] proposed a method of using morphological linear operators, line strengths and oriented Gabor filters at multiple scales, and using the Gaussian mixture model for retinal vessels classification. Zhu [34], on the basis of method [35], reduced the number of feature vectors by selecting the most effective features, and then combined CART (classification and regression tree) with AdaBoost to train a strong classifier for retinal vessel segmentation.

Different with the supervised learning method, unsupervised learning methods do not require manually labeled training samples, they attempt to use the unsupervised learning method to find the inherent characteristics of retinal vessels. Lupascu [36] proposed an algorithm for minimal path tracking in the retinal vascular skeleton based on the graph structure of the retinal vessels. Later, Xie [37] presented a segmentation method that combined the genetic algorithm and FCM. This method makes full use of the strong local search ability of FCM and the global convergence of genetic algorithm.

The methods based on machine learning are mostly simple and easy to understand, but the data set requirement is higher, and during the training process, it may over-fit. The methods based on machine learning are the currently main research direction of retinal vascular segmentation, and their average accuracy is higher than other methods.

(2.6) **Methods based on deep learning**

In recent years, methods based on deep learning, which is an important branch of machine learning, have achieved remarkable results in target classification and segmentation tasks in the field of computer vision. They have great potentials in medical image processing and are aroused great attention at home and abroad academia.

In the field of fundus retinal vessel segmentation, Wang [38] proposed a new hierarchical retinal vascular segmentation method. In this method, CNN (Convolutional Neural Network) was trained as a feature extractor, and RF (Random Forest) worked as an integrated classifier. Fu [39] combined HED (holistically-nested edge detection) and CRFs (fully-connected Conditional Random Fields) for retinal vessel segmentation, which took full account of the high order constraints between the vessel point and other surrounding pixels. Liskowski [40] first used GCN and ZCA to preprocess the fundus images, and then utilized a deep neural network trained on a large dataset to achieve the retinal vessel segmentation. Deep neural network can learn the image features directly from the unprocessed data set, which is an end-to-end learning method, and avoids the complexity of manually designing features. However, this method is based on having a large number of training data set. Currently, the number of labeled fundus images is small, and marking the sample data is difficult. These factors limit the development of deep learning in the field of retinal vessel segmentation.

(2.7) **Summary**

In short, the researchers have developed a number of efficient and rapid methods of vessel segmentation (Table 1). But in the computer-aided diagnosis process, the goal is to automatically extract effective diagnostic information, such as changes in arteriovenous diameter and so on. Therefore, only achieving the retinal vessel segmentation is far from enough. It is necessary to be able to distinguish the arteries and veins, and to achieve the measurement of vessel diameter and the calculation of related parameters.

3 Methods of Artery/Vein Classification

In recent years, there are many research achievements on retinal vessel segmentation at home and abroad. However, there are just few researches focusing on automatic classification of arteries and veins, and most studies of fundus retinal images remain on the extraction of vascular trees or on the detection of other lesions. In this section, we will review the existing methods of retinal arteriovenous vascular classification and then make a conclusion.

Table 1 The accuracy comparison of some vessel segmentation algorithm

Method categories	Years	Method in some literatures	Accuracy	DataBase
2.1	2010	Method in [4]	0.9290	DRIVE
2.2	2010	Method in [13]	0.9382	DRIVE
			0.9484	STARE
	2009	Method in [14]	0.9293	DRIVE
2.3	2012	Method in [19]	0.9430	DRIVE
			0.9442	STARE
	2015	Method in [22]	0.9457	DRIVE
			0.9451	STARE
2.4	2014	Method in [28]	0.9477	DRIVE
			0.9509	STARE
	2016	Method in [29]	0.9535	DRIVE
			0.9503	STARE
2.5	2011	Method in [33]	0.9476	DRIVE
			0.9578	STARE
	2014	Method in [34]	0.9607	DRIVE
2.6	2015	Method in [38]	0.9767	DRIVE
			0.9813	STARE
	2016	Method in [39]	0.9470	DRIVE
			0.9545	STARE

(3.1) **Methods based on color features**

Retinal vessels have following color characteristic: the arteries are brighter and lighter than the veins. Vázquezz [41] used "snake" model to extract the feature points from vessels near the optic disc, and selected multiple sets of feature vectors in the two color spaces of RGB and HSL, and then proposed a localized arteriovenous classification method based on K-means clustering algorithm. After that, he used the minimal path approach to revise the results of classification mentioned above [42]. Relan [43] automatically classified retinal vessels as arteries or veins based on color features using GMM-EM (Gaussian Mixture Model, Expectation-Maximization) unsupervised classifier and a quadrant-pairwise approach. Niemeijier [44] proposed a supervised automatic method based on intensity and derivative information for distinguishing arteries from veins in retinal vessels. In [45, 46], the arteries and veins were classified by analyzing the extracted graph from the retinal vasculature. The final classification of entire vascular tree was decided on the type of each intersection point and the label of each vessel segment. Vijayakumar [47] proposed a classification method based on Random Forest and SVM (Support Vector Machines). Methods mentioned above highly depend on the color information, for the fundus retinal images where the background is complex and the brightness is not uniform, leading to the accuracy of classification is limited.

(3.2) **Methods with structural characteristics**

The color difference between arteries and veins is very small in retinal images, and the brightness is not uniform, so the artery/vein classification of simply relying on color information is not accurate enough. Medical experience shows that the blood vessels have connectivity, and have the following structural characteristic: vein accompanies artery and vice versa. Many researchers fully consider the structural characteristics of blood vessels and effectively improve the accuracy of artery/vein classification. Mirsharif [48] divided the vessel tree into several subsets, and then integrated vascular tracking techniques and color information to classify the global vessels. According to the topological structure of fundus vessels, Estrada [49] proposed a global arteriovenous classification method based on graph theory. The above two methods take the vascular structural properties into account, but the use of vascular color information is not enough. [50] proposed a vessel segmentation method based on the B-COSFIRE filter, and then extracted the color characteristics of blood vessels, respectively, completing the artery/vein classification based on SVM and CNN. Finally, the classification results of the two methods were modified by introducing the vascular structure information. At present, there are few studies on retinal artery/vein classification, and the accuracy of classification still need to be improved.

4 Summary and Outlook

In this paper, we have discussed the research progress of retinal vessel segmentation in the fundus images and the artery/vein classification. The retinal vessel is the only vascular system that can be observed noninvasively in the human body. The study of retinal vessels is of great significance for the diagnosis of common diseases, such as diabetes, hypertension and so on. Although many researchers have made a lot of contributions in this field, and have developed a variety of efficient and rapid methods for retinal vessel segmentation and artery/vein classification, there are still many problems to be solved, such as:

(1) The contrast between fine blood vessels and background is weak, the accuracy of fine blood vessel segmentation remains to be improved in the future.
(2) For a class of the fundus images from people who have diseases, the accuracy of segmentation is affected by bleeding, hard infiltration, micro-aneurysm and other lesions.
(3) It is important to segment the retinal vessels precisely and to classify the artery and vein accurately in specific areas like the region of 1−1.5DD (disc diameter) from the optic disc, for the calculation of AVR (Arteriole to Venule Ratio) and the auxiliary diagnosis.
(4) The detection of key points, such as branch points, cross points and so on.
(5) The identification of key morphological features, such as silver-like, copper-like and other vascular lesions features.

The problems mentioned above are the focuses and difficulties in the retinal image processing. In the near future, these problems are worthy further exploration and research.

References

1. Zhu Ch, Zou B, Xiang Y, et al. A survey of retinal vessel segmentation in fundus images [J]. J Comput-Aided Design & Comput. 2015;27(11):2046–57.
2. Liu I, Sun Y. Recursive tracking of vascular networks in angiograms based on the detection-deletion scheme [J]. IEEE Trans Med Imaging. 1993;12(2):334–41.
3. Can A, Shen H, Turner JN, et al. Rapid automated tracing and feature extraction from retinal fundus images using direct exploratory algorithms [J]. IEEE Trans Inf Technol Biomed. 1999;3(2):125–38.
4. Vlachos M, Dermatas E. Multi-scale retinal vessel segmentation using line tracking [J]. Comput Med Imaging Graph. 2010;34(3):213–27.
5. Nayebifar B, Moghaddam HA. A novel method for retinal vessel tracking using particle filters [J]. Comput Biol Med. 2013;43(5):541–8.
6. Adel M, Moussaoui A, Rasigni M, et al. Statistical-based tracking technique for linear structures detection: application to vessel segmentation in medical images [J]. IEEE Signal Process Lett. 2010;17(6):555–8.
7. Yin Y, Adel M, Bourennane S. Retinal vessel segmentation using a probabilistic tracking method [J]. Pattern Recogn. 2012;45(4):1235–44.

8. Zhang J, Li H, Nie Q, et al. A retinal vessel boundary tracking method based on Bayesian theory and multi-scale line detection [J]. Comput Med Imaging Graph. 2014;38(6):517–25.
9. Chaudhuri S, Chatterjee S, Katz N, et al. Detection of blood vessels in retinal images using two-dimensional matched filters [J]. IEEE Trans Med Imaging. 1989;8(3):263–9.
10. Hoover AD, Kouznetsova V, Goldbaum M. Locating blood vessels in retinal images by piecewise threshold probing of a matched filter response [J]. IEEE Trans Med Imaging. 2000;19(3):203–10.
11. Jiang X, Mojon D. Adaptive local thresholding by verification-based multithreshold probing with application to vessel detection in retinal images [J]. IEEE Trans Pattern Anal Mach Intell. 2003;25(1):131–7.
12. Zhang L, Li Q, You J, et al. A modified matched filter with double-sided thresholding for screening proliferative diabetic retinopathy [J]. IEEE Trans Inf Technol Biomed. 2009;13 (4):528–34.
13. Zhang B, Zhang L, Zhang L, et al. Retinal vessel extraction by matched filter with first-order derivative of Gaussian [J]. Comput Biol Med. 2010;40(4):438–45.
14. Cinsdikici MG, Aydın D. Detection of blood vessels in ophthalmoscope images using MF/ant (matched filter/ant colony) algorithm [J]. Comput Methods Programs Biomed. 2009;96 (2):85–95.
15. Li Q, You J, Zhang D. Vessel segmentation and width estimation in retinal images using multiscale production of matched filter responses [J]. Expert Syst Appl. 2012;39(9):7600–10.
16. Zana F, Klein JC. Segmentation of vessel-like patterns using mathematical morphology and curvature evaluation [J]. IEEE Trans Image Process. 2001;10(7):1010–9.
17. Ayala G, León T, Zapater V. Different averages of a fuzzy set with an application to vessel segmentation [J]. IEEE Trans Fuzzy Syst. 2005;13(3):384–93.
18. Mendonca AM, Campilho A. Segmentation of retinal blood vessels by combining the detection of centerlines and morphological reconstruction [J]. IEEE Trans Med Imaging. 2006;25(9):1200–13.
19. Fraz MM, Barman SA, Remagnino P, et al. An approach to localize the retinal blood vessels using bit planes and centerline detection [J]. Comput Methods Programs Biomed. 2012;108 (2):600–16.
20. Yang Y, Huang S, Rao N. An automatic hybrid method for retinal blood vessel extraction [J]. Int J Appl Math Comput Sci. 2008;18(3):399–407.
21. Miri MS, Mahloojifar A. Retinal image analysis using curvelet transform and multistructure elements morphology by reconstruction [J]. IEEE Trans Biomed Eng. 2011;58(5):1183–92.
22. Wang XH, Zhao YQ, Liao M. Automatic segmentation for retinal vessel based on multi-scale 2D Gabor wavelet [J]. Acta Automatica Sinica. 2015;41(5):970–80.
23. Espona L, Carreira M, Ortega M, et al. A snake for retinal vessel segmentation [J]. Pattern Recogn Image Anal. 2007;178–85.
24. Aldiri B, Hunter A. A ribbon of twins for extracting vessel boundaries [C]. In: Proceedings of the 3rd European medical and biological engineering conference. 2005. p. 1–6.
25. Aldiri B, Hunter A, Steel D. An active contour model for segmenting and measuring retinal vessels [J]. IEEE Trans Med Imaging. 2009;28(9):1488–97.
26. Vese LA, Chan TF. A multiphase level set framework for image segmentation using the Mumford and Shah model [J]. Int J Comput Vision. 2002;50(3):271–93.
27. Sum KW, Cheung PYS. Vessel extraction under non-uniform illumination: a level set approach [J]. IEEE Trans Biomed Eng. 2008;55(1):358–60.
28. Zhao YQ, Wang XH, Wang XF, et al. Retinal vessels segmentation based on level set and region growing [J]. Pattern Recogn. 2014;47(7):2437–46.
29. Liang LM, Huang ChL, ShiF, et al. Retinal vessel segmentation using level set combined with shape priori [J]. Chin J Comput.2016;39. Online Publishing No.173.
30. Soares JVB, Leandro JJG, Cesar RM, et al. Retinal vessel segmentation using the 2D Gabor wavelet and supervised classification [J]. IEEE Trans Med Imaging. 2006;25(9):1214–22.
31. Ricci E, Perfetti R. Retinal blood vessel segmentation using line operators and support vector classification [J]. IEEE Trans Med Imaging. 2007;26(10):1357–65.

32. Marín D, Aquino A, Gegúndez-Arias ME, et al. A new supervised method for blood vessel segmentation in retinal images by using gray-level and moment invariants-based features [J]. IEEE Trans Med Imaging. 2011;30(1):146–58.
33. Fraz MM, Remagnino P, Hoppe A, et al. A supervised method for retinal blood vessel segmentation using line strength, multiscale Gabor and morphological features [C]. In: IEEE international conference on signal and image processing applications, 2011:pp. 410–5.
34. Zhu C, Yao X, Zou B, et al. Retinal Vessel Segmentation in Fundus Images Using CART and AdaBoost [J]. J Comput-Aided Des Comput Graph. 2014;26(3):445–51.
35. Lupascu CA, Tegolo D, Trucco E. FABC: retinal vessel segmentation using AdaBoost [J]. IEEE Trans Inf Technol Biomed. 2010;14(5):1267–74.
36. Lupascu CA, Tegolo D. Graph-based minimal path tracking in the skeleton of the retinal vascular network [C]. In: IEEE international symposium on computer-based medical system, 2012. pp. 1–6.
37. Xie S, Nie H. Retinal vascular image segmentation using genetic algorithm plus FCM clustering [C]. In: 3rd international conference on intelligent system design and engineering applications, 2013. pp. 1225–8.
38. Wang S, Yin Y, Cao G, et al. Hierarchical retinal blood vessel segmentation based on feature and ensemble learning [J]. Neurocomputing. 2015;149:708–17.
39. Fu H, Xu Y, Wong D WK, et al. Retinal vessel segmentation via deep learning network and fully-connected conditional random fields [C]. In: 2016 IEEE 13th international symposium on biomedical imaging, 2016. pp. 698–701.
40. Liskowski P, Krawiec K. Segmenting retinal blood vessels with deep neural networks [J]. IEEE Trans Med Imaging. 2016;35(11):2369–80.
41. Vázquez SG, Barreira N, Penedo MG, et al. Improvements in retinal vessel clustering techniques: towards the automatic computation of the arterio venous ratio [J]. Computing. 2010;90(3–4):197–217.
42. Vázquez SG, Cancela B, Barreira N, et al. Improving retinal artery and vein classification by means of a minimal path approach [J]. Mach Vis Appl. 2013;24(5):919–30.
43. Relan D, MacGillivray T, Ballerini L, et al. Retinal vessel classification: sorting arteries and veins [C]. In: 2013 35th annual international conference of the IEEE, engineering in medicine and biology society (EMBC), 2013. pp. 7396–9.
44. Niemeijer M, Ginneken BV, Abràmoff MD. Automatic classification of retinal vessels into arteries and veins [C]. In: International society for optics and photonics, 2009. pp. 72601F-72601F-8.
45. Dashtbozorg B, Mendonça AM, Campilho A. An automatic graph-based approach for artery/vein classification in retinal images [J]. IEEE Trans Image Process. 2014;23(3): 1073–83.
46. Joshi VS, Reinhardt JM, Garvin MK, et al. Automated method for identification and artery-venous classification of vessel trees in retinal vessel networks [J]. PLoS ONE. 2014;9 (2):e88061.
47. Vijayakumar V, Koozekanani DD, White R, et al. Artery/vein classification of retinal blood vessels using feature selection [C]. In: 2016 38th annual international conference of the IEEE, engineering in medicine and biology society (EMBC), 2016. pp. 1320–3.
48. Mirsharif Q, Tajeripour F, Pourreza H. Automated characterization of blood vessels as arteries and veins in retinal images [J]. Comput Med Imaging Graph. 2013;37(7):607–17.
49. Estrada R, Allingham MJ, Mettu PS, et al. Retinal artery-vein classification via topology estimation [J]. IEEE Trans Med Imaging. 2015;34(12):2518–34.
50. Yang Y. Study on classification of retinal vascular segmentation and arteriovenous [D]. Harbin Institute of Technology, 2016.

New Methods for Utilization of Predictive Information in Neural Network PID Controller

Shukun Jia and Liang Wang

Abstract In this paper, we analyze the utilization of predictive information in neural network PID controller (NN-PID). Based on the accuracy of predictive model, two novel methods are proposed to improve control performance. When predictive model is high-accuracy, two-step ahead predictive information is incorporated into loss function to adjust the weight of NN. When the predictive model is low-accuracy, only one-step ahead information is used and learning rate is adjusted based on the prediction error. Consequential simulation are conducted with each method.

Keywords Neural network PID control · Predictive information · Loss function · Variable learning rate

1 Introduction

Since the introduction of NN-PID, there are many papers dedicated to improving its performance. In [1], particle swarm optimization (PSO) algorithm is introduced to improve the convergent speed and prevent the search for global-optimal from trapping into local optima. Paper [2, 3] come up with an effective way to initialize both parameters and topology of neural network. Paper [4] proposes a smart method to combine margin stability of system with the momentum rate. A majority of papers [4, 5, 6] incorporate the predictive information in loss function, directly and simply, to motivate backward propagation (BP), but few paper study how to use predictive information more appropriately and effectively.

S. Jia (✉) · L. Wang
School of Automation Science and Electrical Engineering, Beihang University, Beijing Shi, China
e-mail: jsk0011@163.com

L. Wang
e-mail: wangliang@buaa.edu.cn

Y. Jia et al. (eds.), *Proceedings of 2017 Chinese Intelligent Systems Conference*, Lecture Notes in Electrical Engineering 459,
https://doi.org/10.1007/978-981-10-6496-8_67

In this paper, we propose two novel methods which use predictive information in a different way based on two kinds of situations. The first method is used when prediction is high-accuracy, in which situation we involve more predictive information to train NN than general. The second method is used in a converse situation. Without satisfied accuracy, we just use one-step ahead information and adjust learning rate by evaluating the accuracy of prediction. Simulations show the effectiveness of both methods.

This paper is organized as following: In Sect. 2, the topology of NN-PID with its conventional forward propagation(FP) formula is given. In Sect. 3, the method used in high-accuracy situation is introduced and corresponding BP formulas are given. Relative simulation results is presented. Section 4 proposes the solution for low-accuracy predictive model and consequential simulation results show the effectiveness of proposed methods. Section 5 provides a conclusion of this paper.

2 Fundamental Structure and Forward Propagation

In order to utilize sufficient information of the system state, and avoid redundant information as well, we give $error(k)$, $error(k-1)$ and $y(k)$ as $input(k)$.

The structure of NN-PID is presented as (Fig. 1):

Generally we choose sigmoid function or its derivation as active function. There are $\sigma_1(\cdot)$ and $\sigma_2(\cdot)$:

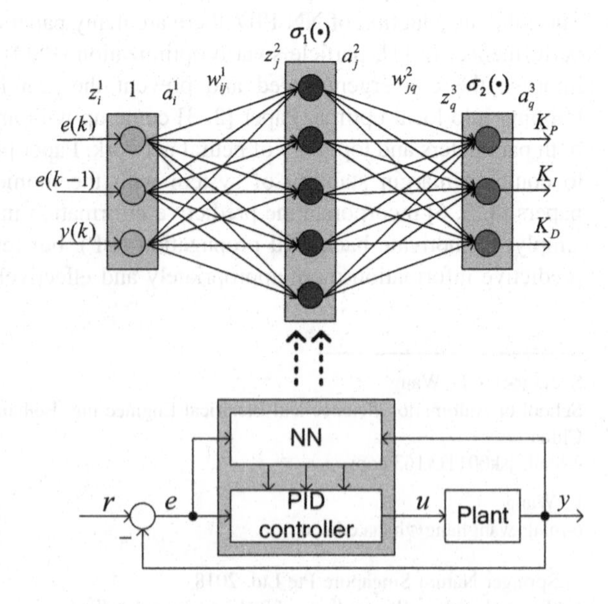

Fig. 1 Proposed NN-PID architecture and NN topology

$$\sigma_1(x) = 1 - \frac{2}{1+e^{-x}} \quad \sigma_1(x) \in (-1, 1) \tag{1}$$

$$\sigma_2(x) = \frac{1}{1+e^{-x}} \quad \sigma_2(x) \in (0, 1) \tag{2}$$

Conventional FP formula is:

$$\begin{cases} a_1^1 = z_1^1 = e(k) \\ a_2^1 = z_2^1 = e(k-1) \\ a_3^1 = z_3^1 = y(k) \end{cases} \tag{3}$$

$$\begin{cases} z_j^2 = \sum_{i=1}^{3} w_{ij}^1 a_i^1 \\ a_j^2 = \sigma_1(z_j^2) \end{cases} \quad j = 1, 2 \ldots 5 \tag{4}$$

$$\begin{cases} z_q^3 = \sum_{j=1}^{5} w_{jq}^2 a_j^2 \\ a_q^3 = \sigma_2(z_q^3) \end{cases} \quad q = 1, 2, 3 \tag{5}$$

$$\begin{cases} a_1^3 = K_P \\ a_2^3 = K_I \\ a_3^3 = K_D \end{cases} \tag{6}$$

Through typical discrete-time PID algorithm, we can get the final output of PID controller:

$$u(k) = u(k-1) + K_p[e(k) - e(k-1)] + K_I e(k) + K_D[e(k) - 2e(k-1) + e(k-2)] \tag{7}$$

3 Backward Propagation with High-Accuracy Predictive Model

3.1 Method Introduction

In conventional NN-PID algorithm, one-step ahead information is used to update NN's weights. So every BP crosses with one FP (FP-BP-FP-BP, FB). In this paper, we take two-step ahead information and conduct each BP after two FPs (FP-FP-BP-FP-FP-BP, FFB). The output of predictive model at time k + 1, i.e. $\hat{y}(k+1)$, is used as transition between two FPs.

When the prediction is high-accuracy, this advantage should be fully used. Firstly, FFB takes two FP operations with one group of $K_P K_I K_D$, by which the

NN's weight act more than usual and its performance gets fully represented. Secondly, FFB doesn't only take the information at next moment into consideration, but also the information following that moment. So it incorporates more future information, through which its pre-judgment performance get improved. Both of these points help to adjust NN's weight more faster and effectively.

New process of the parameter tuning is presented as (Fig. 2):

Inspired by paper [7], we set the loss function as following:

$$J(k) = \frac{1}{2}[\hat{E}^2(k+1) + \hat{E}^2(k+2)] \tag{8}$$

where $\hat{E}(k+1) = r(k+1) - \hat{y}(k+1)$.

Gradient descent method is used to adjust the value of NN's weight [8]:

$$\Delta w_{jq}^2(k+1) = -\lambda \frac{\partial J(k)}{\partial w_{jq}^2(k)} + \eta \Delta w_{jq}^2(k) \tag{9}$$

where λ represents learning rate and η represents momentum coefficient.

The factorization of $\frac{\partial J(k)}{\partial w_{jq}^2(k)}$ is as following:

$$\frac{\partial J(k)}{\partial w_{jq}^2(k)} = \frac{\partial(\hat{E}(k+1) + \hat{E}(k+2))}{\partial w_{jq}^2(k)} = \frac{\partial[\hat{E}(k+1)]}{\partial w_{jq}^2(k)} + \frac{\partial[\hat{E}(k+2)]}{\partial w_{jq}^2(k)} \tag{10}$$

where

$$\frac{\partial[\hat{E}(k+1)]}{\partial w_{jq}^2(k)} = \frac{\partial[\hat{E}(k+1)]}{\partial \hat{y}(k+1)} \cdot \frac{\partial \hat{y}(k+1)}{\partial u(k)} \cdot \frac{\partial u(k)}{\partial a_q^3(k)} \cdot \frac{\partial a_q^3(k)}{\partial z_q^3(k)} \cdot \frac{\partial z_q^3(k)}{\partial w_{jq}^2(k)} \tag{11}$$

$$\frac{\partial[\hat{E}(k+2)]}{\partial w_{jq}^2(k)} = \frac{\partial[\hat{E}(k+2)]}{\partial \hat{y}(k+2)} \cdot \frac{\partial \hat{y}(k+2)}{\partial u(k+1)} \cdot \frac{\partial u(k+1)}{\partial a_q^3(k+1)} \cdot \frac{\partial a_q^3(k+1)}{\partial z_q^3(k+1)} \cdot \frac{\partial z_q^3(k+1)}{\partial w_{jq}^2(k)} \tag{12}$$

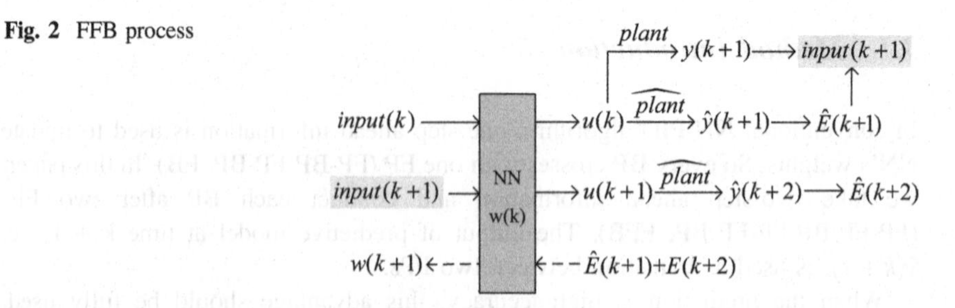

Fig. 2 FFB process

In equations above, $a_q^3(k)$ and $z_q^3(k)$ get from the NN's FP when we give *input*(k); $a_q^3(k+1)$, $z_q^3(k+1)$ get from the given *input*$(k+1)$.

Similarly we can get the adjustment value $\Delta w_{ij}^1(k+1)$.

3.2 Simulation and Analysis

Plant model and predictive model:

$$y(k+1) = \frac{y^2(k)}{1+2y(k)} + u(k) \tag{13}$$

Input signal:

$$r(k) = \begin{cases} 0.5 & k < 50 \\ 0.577\sin(k\pi/150) & 50 \le k < 150 \\ 1.0 & 150 \le k < 250 \\ 0.5 & 250 \le k < 300 \end{cases} \tag{14}$$

The predictive model in this simulation is exactly the plant model itself, which means high-accuracy and the prediction outcome is high reliable. The simulation results are presented as following. Figure 3 results from one-step ahead information and Fig. 4 is from two-step. Figure 5 shows the error comparison:

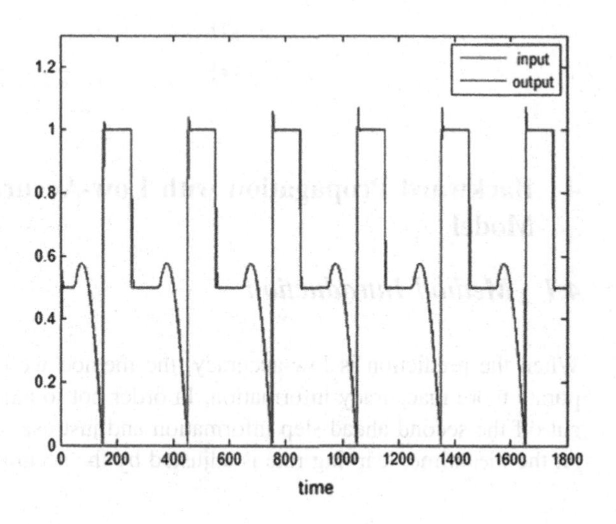

Fig. 3 One-step

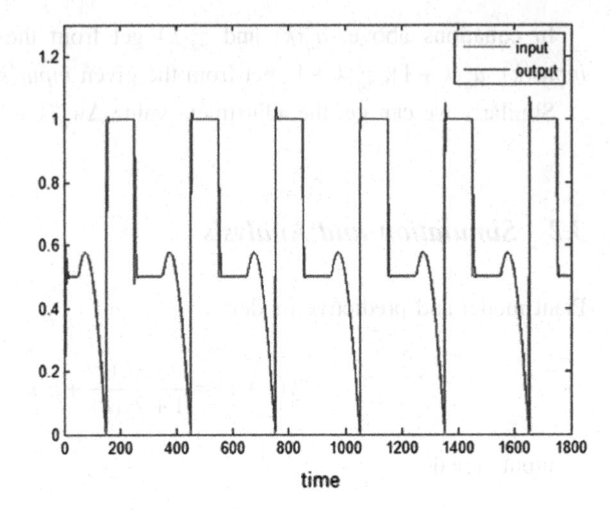

Fig. 4 Two-step

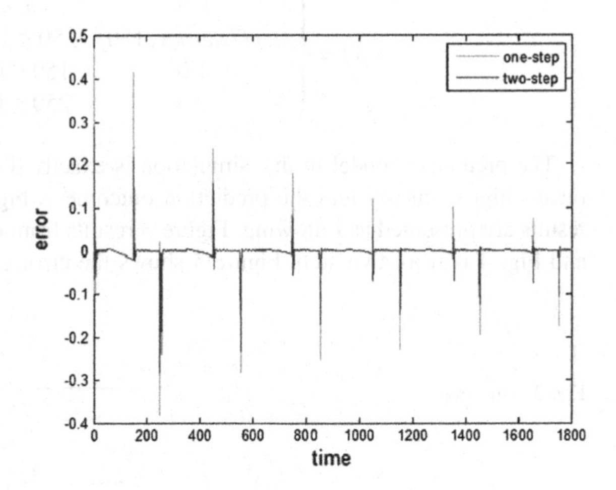

Fig. 5 Error comparison

4 Backward Propagation with Low-Accuracy Predictive Model

4.1 Method Introduction

When the prediction is low-accuracy, the method we proposed above will incorporate more inaccuracy information. In order not to amplify this disadvantage, we cut off the second ahead step information and just use one-step ahead information. At the meantime, learning rate is adjusted by the accuracy evaluation of prediction near the moment. If prediction is high accuracy at the former moment, its information should be trusted. So the learning rate is increased and adjustment speed should be accelerated. But if not, learning rate is decreased and adjustment speed should slow down. This action is conducted through a coefficient μ.

$$\mu(x) = \begin{cases} 1 - 4x^2 & x < 0.5 \\ 0 & x \geq 0.5 \end{cases} \tag{15}$$

where $x = |y(k) - \hat{y}(k)|$

And the learning rate $\lambda = \mu \cdot \lambda_0$

4.2 Simulation and Analysis

Plant model:

$$y(k+1) = 0.5y(k) + u(k) \tag{16}$$

Prediction model:

$$\hat{y}(k+1) = 0.5y(k) + u(k) + \omega \tag{17}$$

where ω is white noise.

Input signal:

$$r(k) = \begin{cases} 0.5\sin(k\pi/25) & k < 250 \\ 0.5 & 250 \leq k < 500 \\ 0.3\sin(k\pi/25) + 0.4\sin(k\pi/32) + 0.3\sin(k\pi/40) & 500 \leq k < 750 \end{cases} \tag{18}$$

In this simulation we give a linear plant. The prediction model is constructed by adding white noise to the original plant. It means that the outcome of this prediction is doubtful. By adjusting learning rate based on the accuracy of prediction, the simulation result is presented as following. Figure 6 results from constant-learning rate and Fig. 7 results from variable-learning rate. Figure 8 shows the error comparison:

The key of the process is variable learning rate, its dynamic value is presented as Fig. 9:

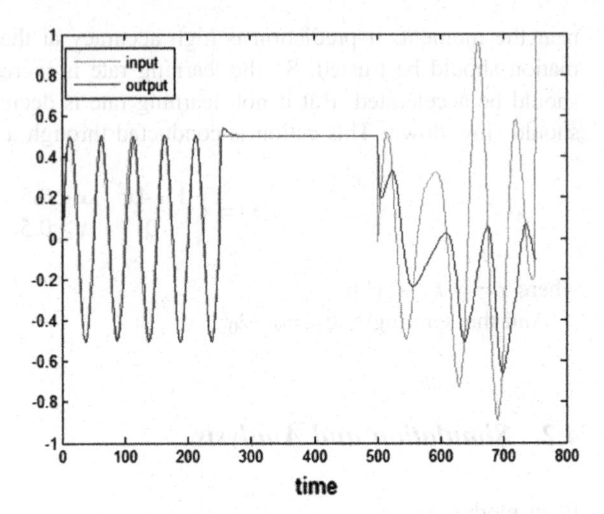

Fig. 6 Constant-learning rate

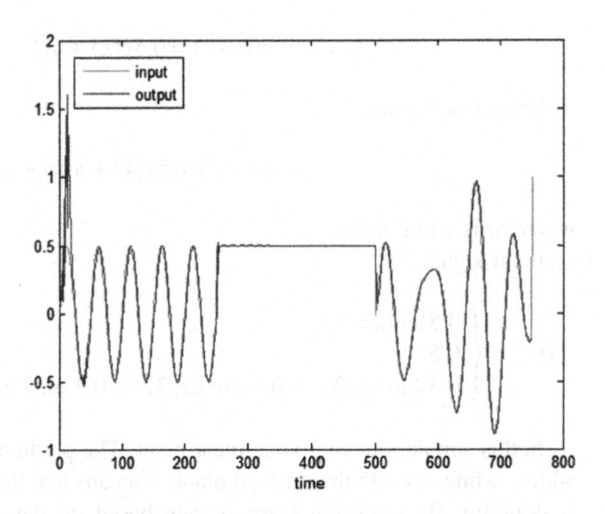

Fig. 7 Variable-learning rate

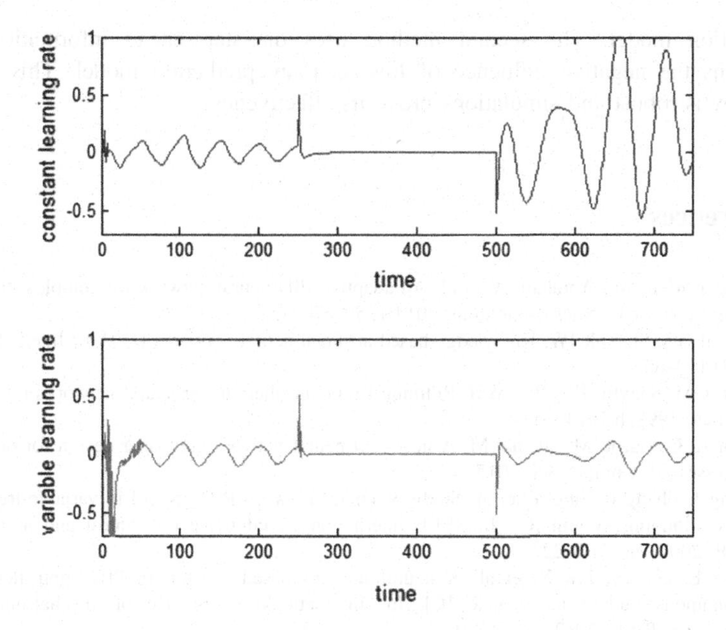

Fig. 8 Error comparison

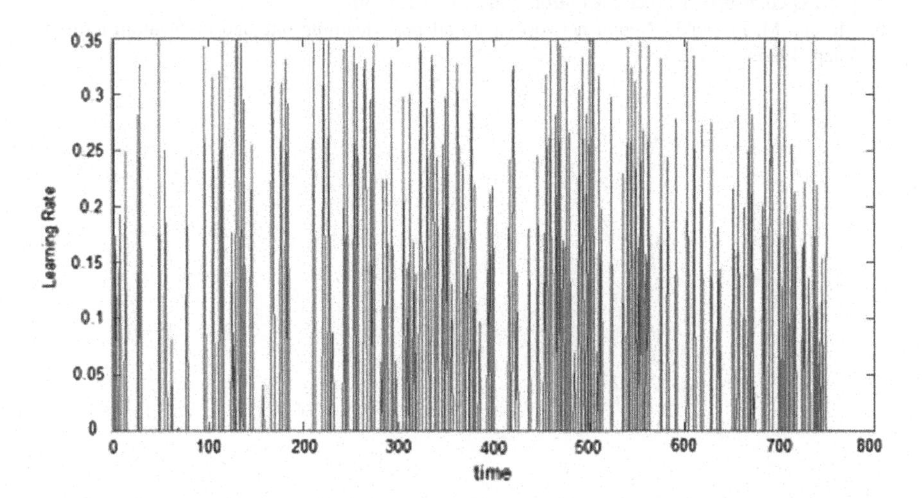

Fig. 9 Dynamic learning rate

5 Conclusion

We have proposed two new methods to use predictive information based on the accuracy of prediction. The first method bases on more future information to adjust NN's weight. Therefore it takes full use of the advantage of high-accuracy prediction model. The second method uses one-step ahead information and weakens the negative influence of low-accuracy prediction model. This whole strategy is robust and simulations prove its effectiveness.

References

1. Kang J, Meng W, Abraham A, et al. An adaptive PID neural network for complex nonlinear system control[J]. Neurocomputing. 2014;135(8):79–85.
2. Towell GG, Shavlik JW. Knowledge-based artificial neural networks [J]. Artif Intell. 1994;70 (1–2):119–65.
3. Scott GM, Shavlik JW, Ray WH. Refining PID controllers using neural networks [J]. Neural Comput. 1992;4(5):746–57.
4. Zribi A, Chtourou M, Djemel M. A new PID neural network controller design for nonlinear processes [J]. Comput Sci. 2015.
5. Wang J, Zhang C, Jing Y, et al. Study of neural network PID control in variable-frequency air-conditioning system [C]. In: IEEE international conference on control and automation IEEE. 2007. pp. 317–22.
6. Dong E, Guo S, Lin X, et al. A neural network-based self-tuning PID controller of an autonomous underwater vehicle [C]. In: International conference on mechatronics and automation IEEE. 2012. pp. 898–903.
7. Chen J, Huang TC. Applying neural networks to on-line updated PID controllers for nonlinear process control[J]. J Process Control. 2004;14(2):211–30.
8. Moreira M, Fiesler E. Neural networks with adaptive learning rate and momentum terms[J]. Idiap, 1995.

Hand-Eye Calibration of IMU and Camera Without External Equipments

Yacong Wang and Long Zhao

Abstract To relate measurements made by visual and inertial sensors, we must solve the hand-eye calibration (HEC) equation $AX = XB$, where X denotes an unknown rotation and translation between camera and inertial measurement unit (IMU), and A and B respectively represent the calculated movement transformations related with the camera and IMU. This paper introduces a systematic framework to jointly calibrate IMU and relative transformations X with a linear decomposition algorithm which is composed by Kronecker product and singular value decomposition. Without the requirements of external equipments including Robot Operating System or specific hardware and of the A and B featured the identical rotation angle, it enables the extension to the arbitrary set-up and the noise in rotation. The details of our framework are given, together with a validity of A and B movements followed by results of real experiments, showing that the enough precision and more robustness can be achieved.

Keywords Visual and inertial sensor · Hand-eye calibration · Linear decomposition algorithm · Kronecker product

1 Introduction

To make an ideal choice for accurate Visual-Inertial Odometry (VIO) or Simultaneous Localization and Mapping (SLAM), it is popular to combine visual and inertial measurements in robotics and mobile phones due to the complementary

Y. Wang · L. Zhao (✉)
School of Automation Science and Electrical Engineering, Beihang University,
Beijing 100191, China
e-mail: buaa_dnc@buaa.edu.cn

Y. Wang · L. Zhao
Digital Navigation Center, Beihang University, Beijing 100191, China

L. Zhao
Science and Technology on Aircraft Control Laboratory, Beihang University,
Beijing 100191, China

Y. Jia et al. (eds.), *Proceedings of 2017 Chinese Intelligent Systems Conference*, Lecture Notes in Electrical Engineering 459,
https://doi.org/10.1007/978-981-10-6496-8_68

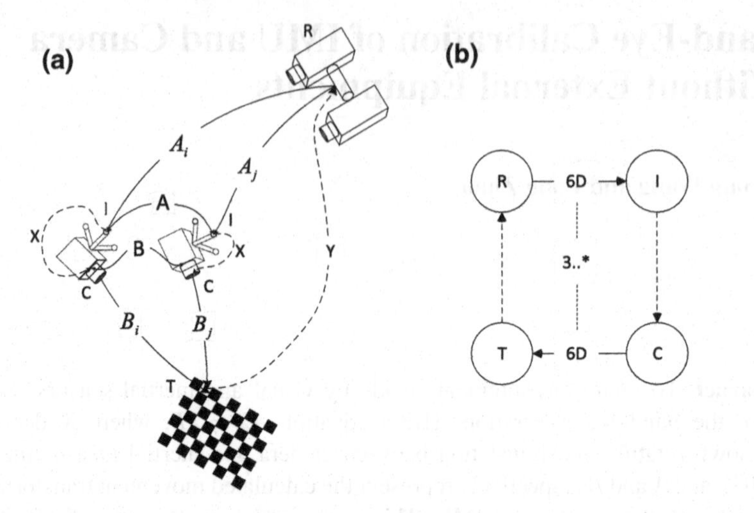

Fig. 1 **a** The standard hand-eye calibration; **b** hand-eye-calibration spatial transformation pattern

characteristics of the two sensing modalities. For the accuracy and robustness of state estimation for VIO, these different sensors must be temporally and spatially registered with respect to each other. Using a camera to estimate the 3D transformation of the object relative to the inertial sensors (Inertial Measurement Unit, IMU) within the work volume, the relative rotation and translation between the IMU and the camera, between the target and the camera, between the IMU and the reference frame is necessary to required. Hand-eye calibration (HEC) is the process used in the field of VIO for determining the relative position and orientation between the camera and the IMU. Commonly, hand-eye calibration problem got its name for that the camera (called as eye) was mounted on the gripper (called as hand) in the robotics community. Through a calibration pattern and the control commands, the camera motion and the gripper motion is acquired respectively, meaning that the camera is rigidly connected to the robot gripper.

The hand-eye calibration is usually described by homogeneous transformation matrices. As seen in the Fig. 1, the transformation from IMU to camera is denoted by X, and A_i, B_i, respectively, indicate the transformation matrix from the reference to the IMU coordinate system and from the camera to the target coordinate system (world coordinate system) at the ith pose. Figure 1 illustrates the standard hand-eye calibration process: Estimate the camera-sensor position and orientation X with several measurements A_i, B_i (that gives you A, B). Using the camera extrinsic calibration, the camera-target transformation B_i is obtained. The reference-IMU transformation A_i is given by the Attitude and Heading Reference System (AHRS). For one pose of IMU and camera, these two transformations, the target-reference and the IMU-camera, are unknown. One motion that has two poses, the reference-IMU or the camera-target transformation, should be required to yield the hand-eye equation, $AX = XB$, first formulated by Tsai and Shiu [1, 2]. Early solutions regard the rotation part of X decoupled from the translational one, the latter following the former, since rotational estimation errors can propagate to the translational part. Several approaches have been proposed to calculate the rotation, such as Chou et al. [3] gave closed-form solution with two-step approach in which singular value decomposition (SVD) is to be used in addition to the quaternion-based rotational presentation. Liang et al. [4] gave a new SVD-based linear decomposition algorithm to solve the equation, where the error caused by the noise is minimized, as well as A and B do not need to be rigid transformations with an identical rotation angle. For the robustness and accuracy, the hand-eye calibration equation in this paper employ that new linear decomposition operations, and the screw vector is not to be calculated.

It is obvious that if the exact 3D positons of points on the calibration target in the world coordinate system is known, as well as the 3D poses of IMU with respect to reference coordinate system, i.e. initial frame coordinate, can be determined by AHRS algorithm, the camera can utilize Zhang [5] to calibrate extrinsic parameters, then it is a trivial matter to compute the 3D homogeneous transformation between the camera and the IMU.

The remainder of this paper is organized as follows. The next section describes how the IMU is calibrated to correct the readings. Then, the detailed exposition on hand-eye calibration equation and the solutions via linear decomposition algorithm are given in Sect. 3. Our experimental results of matching of Rodrigues angles and HEC calibration are presented in Sect. 4. Finally, Sect. 5 contains a conclusion and pointers to future developments.

2 IMU Calibration

In order to calibrate the low-cost IMUs, multi-positon scheme that calibrates scales and misalignment factors, as well as estimates sensor biases, for both the accelerometers and gyroscopes triads [6] is proposed in this paper. We move the sensor by hand and place it at various static positions and attitudes to reliably detect the static intervals with parameterless static filter within the sensor measurements. Taking samples measured in the intervals, the accelerometers triad can be first calibrated. Then, these results are to be exploited to calibrate the gyroscopes with numerical integration algorithm.

2.1 Sensor Error Model

In real IMU, the three axes of the accelerometers frame (AF) are usually non-orthogonal to each other, as well as the gyroscopes frame (GF). In David [6], small angles transformed from the non-orthogonal frame to the orthogonal body frame (BF, AOF, BOF) are denoted as S^S.

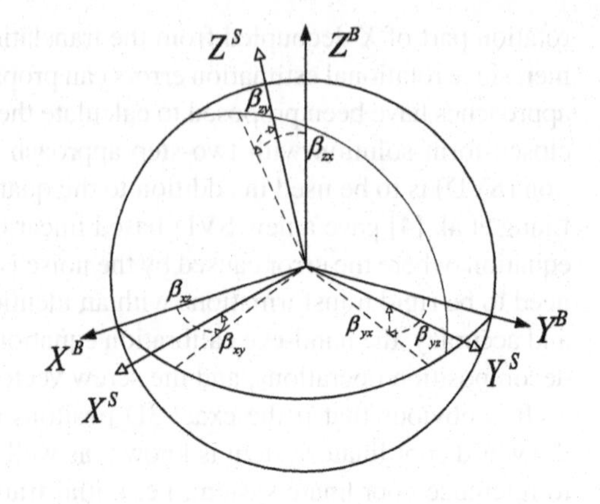

Fig. 2 The rotation of accelerometers or gyroscopes axes (X^S, Y^S, Z^S) around body frame axes (X^B, Y^B, Z^B)

$$S^B = TS^S, \quad T = \begin{bmatrix} 1 & -\beta_{yz} & \beta_{zy} \\ \beta_{xz} & 1 & -\beta_{zx} \\ -\beta_{xy} & \beta_{yx} & 1 \end{bmatrix}. \tag{1}$$

As seen in the Fig. 2, β_{ij} denotes the Euler angle of ith AF or GF axis rotates around the jth BF axis.

To simplify the calibration, the angles $\beta_{xz}, \beta_{xy}, \beta_{yx}$ become zero assuming that the BF coincides with the AOF. For the accelerometers (a^O, α_{ij}) and gyroscopes (ω^O, γ_{ij}), Eq. 1 becomes:

$$a^O = T^a a^S, T^a = \begin{bmatrix} 1 & -\alpha_{yz} & \alpha_{zy} \\ 0 & 1 & -\alpha_{zx} \\ 0 & 0 & 1 \end{bmatrix}; \; \omega^O = T^g \omega^S, T^g = \begin{bmatrix} 1 & -\gamma_{yz} & \gamma_{zy} \\ \gamma_{xz} & 1 & -\gamma_{zx} \\ -\gamma_{xy} & \gamma_{yx} & 1 \end{bmatrix} \tag{2}$$

where a^O and a^S are the specific physical quantities in the orthogonal body frame and accelerometers frame respectively, and the same as the ω^O and ω^S.

Finally, the accelerometer a^O and gyroscope ω^O sensor error model are:

$$a^O = T^a K^a (a^S + b^a + v^a), \quad \omega^O = T^g K^g (\omega^S + b^g + v^g) \tag{3}$$

where v^a, v^g are the measurement noise and K^a, K^g are scaling matrixes, b^a, b^g are two bias vectors of accelerometer and gyroscope respectively, as follows:

$$K^a = \begin{bmatrix} S_x^a & 0 & 0 \\ 0 & S_y^a & 0 \\ 0 & 0 & S_z^a \end{bmatrix}, K^g = \begin{bmatrix} S_x^g & 0 & 0 \\ 0 & S_y^g & 0 \\ 0 & 0 & S_z^g \end{bmatrix}; \; b^a = \begin{bmatrix} b_x^a \\ b_y^a \\ b_z^a \end{bmatrix}, b^g = \begin{bmatrix} b_x^g \\ b_y^g \\ b_z^g \end{bmatrix}. \tag{4}$$

2.2 Calibration Procedure

The length of static interval determines the accuracy of the IMU calibration since the static intervals is used to calibrate the accelerometer, and of the motion intervals located between two consecutive static intervals for the gyroscopes calibration. Static detector operator that uses the variance of accelerometer signals determines the length t_{wait} seconds. Computing the variance magnitude as follows:

$$\varsigma(t) = \sqrt{[var_{t_w}(a_x^t)]^2 + [var_{t_w}(a_y^t)]^2 + [var_{t_w}(a_z^t)]^2} \tag{5}$$

where (a_x^t, a_y^t, a_z^t) describes each accelerometer sample in the t_w seconds static interval centered at t.

The measurements noise ν^a, ν^g can be neglected through averaging signal measured in each static interval. Then, accelerometers triad calibration can be done with the unknown parameter vector:

$$\theta^{acc} = [\alpha_{yz}, \alpha_{zy}, \alpha_{zx}, S_x^a, S_y^a, S_z^a, b_x^a, b_y^a, b_z^a] \tag{6}$$

As in the conventional IMU calibration scheme, the cost function used to estimate θ^{acc} is:

$$L(\theta^{acc}) = \sum_{i=1}^{N} (||g||^2 - ||h(a_k^S, \theta^{acc})||^2)^2 \tag{7}$$

where $||g||$ is the actual local gravity magnitude (i.e., the latitude and altitude of the calibration location), and $h(a_k^S, \theta^{acc}) = T^a K^a (a_k^S + b^a)$ with N acceleration vectors a_k^S measured in the AF. In order to obtain optimal θ^{acc}, Levenberg-Marquardt (LM) is employed to minimize Eq. 7.

Using Allan variance to denote the random gyroscope bias drifts, as follows:

$$\sigma_a^2 = \frac{1}{2}\langle (x(\tilde{t}, k) - x(\tilde{t}, k-1))^2 \rangle = \frac{1}{2K} \sum_{i=1}^{N} (x(\tilde{t}, k) - x(\tilde{t}, k-1))^2 \tag{8}$$

where $x(\tilde{t}, k)$ is the average gyroscope signals in the kth time interval which spans $\tilde{t}$ seconds, and $k, k-1$ is the two consecutive intervals. A good static initial period T_{init} is that the Allan variances of the three axes converge to an expected small value.

With the bias-free gyroscope signals obtained over T_init period and the corrected accelerometer readings based on Eq. 7, the cost function for gyroscope calibration is:

$$L(\theta^{gyro}) = \sum_{k=2}^{M} ||u_{a,k} - u_{g,k}||^2 \tag{9}$$

where $\theta^{gyro} = [\gamma_{yz}, \gamma_{zy}, \gamma_{xz}, \gamma_{zx}, \gamma_{xy}, \gamma_{yx}, S_x^g, S_y^g, S_z^g]$ and $u_{a,k}$ denotes gravity versor (i.e., acceleration versor) given by the calibrated accelerometers. Acceleration versor $u_{g,k} = \psi[\omega_i^S, u_{a,k-1}]$, where ψ defines the Runge-Kutta integration operator that computes the final orientation through the consecutive gyroscope readings ω_i^S. Then, LM algorithm is exploited to minimize Eq. 9 to obtain optimal θ^{gyro}.

3 Hand-Eye Calibration

A general hand-eye system comprises of a sensor (or a camera) and a robotic arm (or a hand, a IMU as the substitute in our work), and the sensor is always mounted on one joint of the hand. Then, hand-eye calibration (HEC) translates to the solving solution of the equation $AX = XB$, where X is a homogeneous matrix representing an unknown 3D rotation and translation from the camera to the IMU, and A and B are the matrices describing the 3D transformation of camera and IMU, respectively.

3.1 HEC Equation

Figure 3 describes the HEC coordinate transformation that involved in the HEC process: the planar calibration target is fixed in a place and the IMU-stereo-camera is moved from location ith to location jth in front of the target by the hand. Two closed loops formed separately by the two coordinate transformations $t \longrightarrow r \longrightarrow i^{(i)} \longrightarrow c^{(i)} \longrightarrow t$ and $t \longrightarrow r \longrightarrow i^{(j)} \longrightarrow c^{(j)} \longrightarrow t$ are the key realization, and thus the following equations are described:

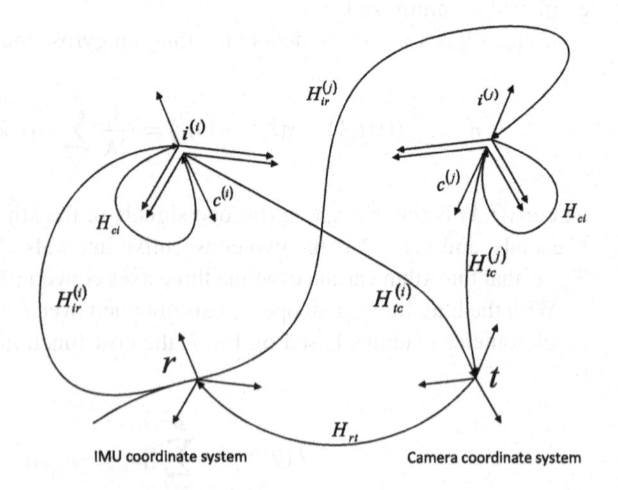

Fig. 3 Coordinate transformations involved in the process (The IMU and reference, camera and target coordinate systems are denoted by i, r, c, and t, respectively. H_{xy} transforms from the x to the y coordinate system)

$$(H_{ir}^{(i)})^{-1}H_{ir}^{(j)}H_{ci} = H_{ci}H_{tc}^{(i)}(H_{tc}^{(j)})^{-1} \tag{10}$$

From the Eq. 10 we have

$$C_{ij}H_{ci} = H_{ci}D_{ij} \tag{11}$$

where H_{ci} is the Euclidean transformation matrix representing rotation and translation from the camera to the IMU coordinate, i.e., the hand-eye calibration. The HEC depends on (i) a known calibration target and the traceable corner on the target to obtain corresponding origin and the points for camera motion, (ii) a reliable algorithm to give the consecutive poses expressed with quaternions for IMU motion.

To solve for the HEC, a set of coordinate systems are required to define here:

$F_r : (O_r, X_r, Y_r, Z_r)$ is the reference world coordinate system. It locates in the initial IMU coordinate frame where IMU is power-on at that moment. All other IMU frames (i.e., F_i) are situated with respect to it.

$F_t : (O_t, X_t, Y_t, Z_t)$ is the referable target coordinate system. All camera poses are relative to this frame.

$F_i : (O_i, X_i, Y_i, Z_i)$ is the IMU coordinate system. It describes each IMU pose when the IMU-stereo-camera is moved from one position to another, that is the frame denotes the IMU motion given by the position and orientation w.r.t. F_r. The pose of the IMU, i.e., the transformation of F_i relative to the F_r, is constantly provided by the AHRS technique.

$F_c : (O_c, X_c, Y_c, Z_c)$ is the camera coordinate system. The origin is situated at the optical center of the vision system (i.e., also known as the principal point on the image plane). Its Z axis is coincident with the optical axis of the camera and the X axis is parallel to the image rows, while Y axis is decided by the right-hand rule. The intrinsic camera parameters calibrated in advance allows expressing the extrinsic camera parameters (i.e., also known as the matrix about camera motion) relative to the F_t.

Due to camera calibration beforehand, the transformation matrices D_{ij} are expressed as the product of extrinsic matrices. And the transformation matrices C_{ij} denotes the transformations between the two consecutive positions of IMU, which could be computed from the AHRS technique.

3.2 HEC Solution

Let us concentrate on the solution of the HEC equation $AX = XB$, where A, B and X express the 4×4 homogeneous transformation matrices:

$$A = \begin{bmatrix} R_A & t_A \\ 0^T & 1 \end{bmatrix}, \quad B = \begin{bmatrix} R_B & t_B \\ 0^T & 1 \end{bmatrix}, \quad X = \begin{bmatrix} R_X & t_X \\ 0^T & 1 \end{bmatrix} \tag{12}$$

where $A = (H_{ir}^{(i)})^{-1} H_{ir}^{(j)}$ defines coordinate transformation from the *j*th pose to the *i*th pose of the IMU, and $B = H_{tc}^{(i)} (H_{tc}^{(j)})^{-1}$ expresses homogeneous transformation matrix from the *j*th to the *i*th position and orientation of the camera. Matrix B must be performed by extrinsic calibration with Zhang [5]. However, in most cases the extrinsic calibration is greatly dependent on intrinsic camera parameters and image corner detections. So, the estimation of the camera motion B might sometimes be quite inaccurate.

Therefore, solving the H_{ci} is equivalent to solving the X in equation $AX = XB$. With Eq. 12 the HEC equation $AX = XB$ can transform to one 3×3 rotation matrix and one 3×1 vector equation, as follows:

$$R_A R_X = R_X R_B \tag{13}$$

$$(R_A - I_3) t_X = R_X t_B - t_A \tag{14}$$

where I_3 is a 3×3 identity matrix. Here, R_X and t_X are the solution of the equation, but only R_X is the desired one in our work. For t_X solved either from classic quaternion algorithms or from new matrix screw theory is not accurate, we measure the distance on each axis manually or use the factory parameters directly. Therefore, the key realization is to apply the Kronecker product to the rotation matrices R_X, that is to the orientational component of the HEC [4].

There are two main algorithms to solve the equation $AX = XB$, one is separate estimation of rotational and translational parts with independence, as well as error propagation from the first stage to the second stage, and another is simultaneous estimation with the complementary solution to each other. Our HEC solution employs the former method to merely estimate the rotation matrix. Since A and B are real matrices, subsequent operations for rotation matrices are performed in a real closed field. We propose the following formulations to solve the R_X.

Using Kronecher product, Eq. 13 can be represented as:

$$(R_A \otimes I_3 - I_3 \otimes R_B^T) vec(R_X^T) = 0 \tag{15}$$

Then, we have $R_A R_X I_3 - I_3 R_X R_B = 0$, which is equivalent to the Eq. 13.
Representing Eq. 15, as:

$$\begin{bmatrix} R_{A_1} \otimes I_3 - I_3 \otimes R_{B_1}^T \\ \vdots \\ R_{A_n} \otimes I_3 - I_3 \otimes R_{B_n}^T \end{bmatrix} vec(R_X^T) = 0 \tag{16}$$

where *vec* is a linear operator vectorizing the matrix R_X column-wise and $\otimes$ is a Kronecher product operator. The $vec(R_x)$ is orthogonal according to the HEC physical model.

And $F(i) = (R_A \otimes I_3 - I_3 \otimes R_B^T) vec(R_X)$, a 9×1 vector, represents the mean error of the rotation matrix R_X. We can obtain the optimal rotation matrix to minimize the mean error item.

Then, we define $L = (R_A \otimes I_3 - I_3 \otimes R_B^T)$ and Eq. 15 is equivalent to $LX = 0$ which can be solved by the least squares solution of $||LX||$. We apply SVD on L, having $L = USV^T$, and acquire some equations.

The eigenvectors y corresponding to the minimal eigenvalue of L can reshape to a 3×3 matrix, that is matrix M^{pre}, (i.e., $M^{pre} = (vec^{-1}(y))^T$). Then, according to [7] we have:

$$M = \frac{sign(det(M^{pre}))}{\sqrt[3]{|det(M^{pre})|}} M^{pre}, \quad sign(x) = \begin{cases} 1 & \text{if } x > 0 \\ 0 & \text{if } x = 0 \\ -1 & \text{if } x < 0 \end{cases}. \tag{17}$$

From Eq. 17, we ensure $det(M) = 1$, that is the matrix M is orthogonal, and the rotation matrix can be M.

In practice, non-accurate solutions of rotation matrices may be given by the above equations due to noise. Therefore, re-orthogonalizing the matrices computed above may be beneficial to guarantee the indeed rotations.

If $det(M) \neq 0$, the SVD of M is $M = USV^T$, then the most approximate orthogonal matrix of M is $R = |UV^T|$, that is the optimal rotation matrix is $R_x = |UV^T|$.

Here,

$$|A| = \begin{cases} A & \text{if } det(A) \geq 0 \\ -A & \text{if } det(A) \leq 0 \end{cases}.$$

From the above equations, we could always obtain the optimal solution of the orientation of the HEC, whether the IMU and camera have the same rotation angle or not; that is, the solution of the HEC can also be applied in non-rigid transformations.

4 Experiments and Discussions

A real HEC experiment has been performed using the Loitor camera, which is a stereo camera with CMOS MT9V034 and the MPU-6050 as the IMU sensor, while the left mono camera is only employed for the IMU-camera calibration. This equipment has 6 degrees of freedom (DOF) and moves before the calibration target of the checkerboard with 8×6 and 28×28 cm. The calibration target is visible at each camera station. Raw sensor data is logged to a PC at 3.30 GHz and then AHRS and OpenCV camera calibration algorithm are used to provide IMU motion and camera motion respectively. For the best performance in HEC, the two sensors must be temporally registered with respect to each other, that is the raw data is aligned at the timestamps of IMU and camera, which is the most critical preprocessing stage before the computation of R_A and R_B.

The IMU calibration in Sect. 2 provides the correct accelerometers and gyroscopes data without misalignment and scale deviation for AHRS technique used to

Fig. 4 Hardware configuration for hand-eye calibration

calculate the IMU transformations. It is an indispensable part of the systematic work as the data pre-processing stage, and the same to intrinsic calibration. More detailed results of IMU calibration can be found in [6].

In the experiment, intrinsic camera parameters calibration is performed beforehand followed by correcting radial distortion and neglecting all other optical deformations for the all images. And the extrinsic rotation parameters are calculated by the PnP algorithm using OpenCV library. It should be stressed that more than two transformations with nonparallel rotation axes and no pure translation are required to solve the HEC problem.

The priori numerical values of R_X are known, and the accuracy of HEC is gauged by placing the camera-IMU equipment at various position. Each parameter of R_X can be independently estimated by this experiment which has two stages and the hardware is described in Fig. 4.

4.1 Matching of Rodrigues

Since the Rodrigues angle and the rotation matrix are one-to-one correspondence, we compute the Rodrigues angles of the IMU and the camera rotation matrix to describe the match of R_A and R_B to obtain correct results. In reality, IMU and camera are fixed on board resulting rigid transformations for R_A and R_B, that is the two Rodrigues angles should be equal roughly, but our method can tolerate the noise in rotations. After the valid matching of Rodrigues angles, we can obtain the desired solution of HEC equation.

From the Fig. 5, we can see that the undesirable matching of Rodrigues angles, which has outliers in the zoomed boxes, is always occurred in the camera extrinsic parameters calibration, while the ideal matching is the two equal angles with

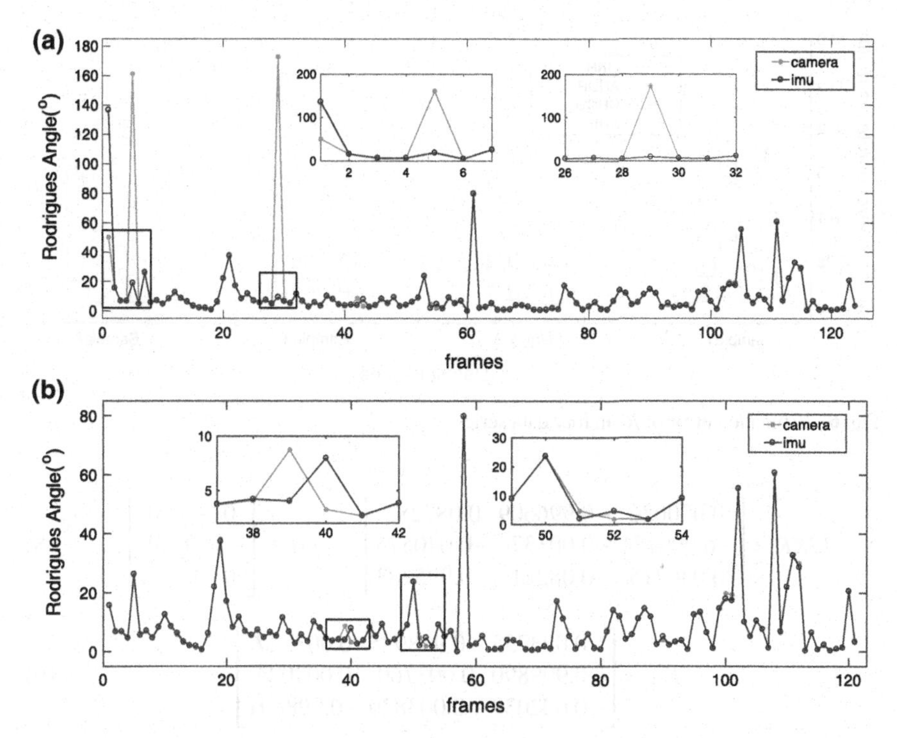

Fig. 5 The Rodrigues angles of IMU and camera: **a** the camera extrinsic parameters include the two error estimation, seen in the box; **b** the correct match of IMU and camera transformations with noise, seen in the box (the *small box* is zoomed in the *large box*)

allowable distinction (the difference is 10 degree in our work). In addition to, the outliers in Fig. 5a are obviously error match due to the large difference between the two. So the results of HEC are sensitive to the camera calibration. Here we believe that the IMU calibration and AHRS algorithm are accurate and this is also in line with the reality.

4.2 Calibration Results

It is common to quantify orientation error between the computed rotation matrices and the ideal theory matrix of HEC. From the Fig. 6 we can observe that the difference (i.e., the estimation error) between the estimated values and the theory value which is known as 1 or 0. The Eq. 18 formulate the mean estimated matrix of our work labeled as *LIKR* and the ideal matrix labeled as *TH* to show the difference of accuracy. And the Eq. 19 formulate the Kalibr estimation result. We can see that the estimation value is the more closer to 1 (or the 0), the more precise it is.

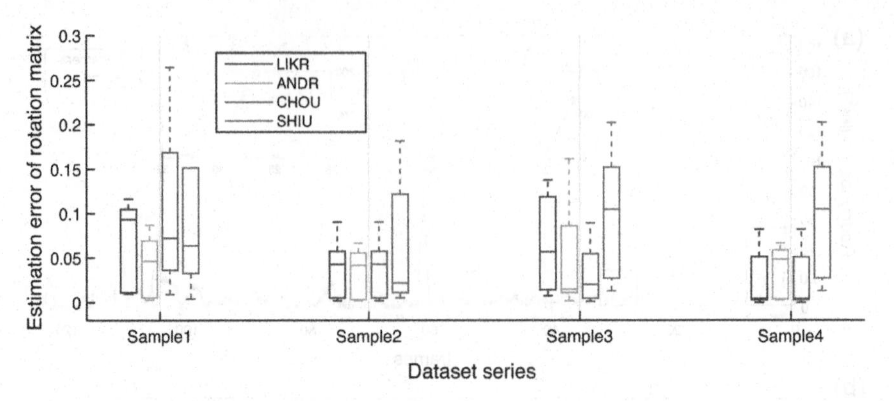

Fig. 6 Estimation error of R_x in four data series

$$LIKR = \begin{bmatrix} 0.000026 & -0.996609 & 0.082283 \\ -0.999171 & -0.003376 & -0.040575 \\ 0.040715 & -0.082214 & -0.995783 \end{bmatrix}, \quad TH = \begin{bmatrix} 0 & -1 & 0 \\ -1 & 0 & 0 \\ 0 & 0 & -1 \end{bmatrix}, \tag{18}$$

$$KA = \begin{bmatrix} -0.003753 & -0.998985 & 0.003827 \\ -0.999890 & 0.003760 & 0.002024 \\ -0.002039 & -0.003820 & -0.998891 \end{bmatrix}. \tag{19}$$

In the Fig. 6, the abscissa indicates the experimental data series where the Sample 1, 2 include outliers (as known in the Fig. 5) and the Sample 3, 4 are the ideal matching. Comparison on the four samples with LIKR (i.e., our work), ANDR [8], CHOU [3] and SHIU [2] has been performed to display the difference of accuracy directly.

The statistics of the obtained estimation error of each value in the HEC matrix depicts values $1/4$–$3/4$ quantile as a box with horizontal line at median. From the Fig. 6, we can obtain that LIKR is stable in the outliers even though the error match with the reasonable accuracy while other methods are either instable or error estimation. The simultaneous solution ANDR needs the translation to solve the rotation so we assume the identical translation of the IMU and the camera. The first three methods have similar desirable accuracy within the ideal match, while the SHIU presents the worse performance since it is the most primitive algorithms with no optimization. Above all, our method can obtain the relative ideal results in the presence of noise, and the common performance without outliers.

Due to the implementation code is either C++ (OpenCV) or MATALB without other environment dependence (e.g., Robot Operating System, ROS), our method can be employed to arbitrary hardware and many industrial applications. For example, the tool box called Kalibr can calibrate multiple cameras, multiple IMUs and Camera-IMU using the continuous-time batch estimation and maximum likelihood estimation theory [9]. The results are more accurate than HEC within the ROS and without the shake for equipment, seen in the Eq. 19.

5 Conclusions

In this work, we present a systematic method to jointly calibrate IMU and spatial transformations between IMU and camera. Using the linear decomposition operations and Kronecker product to the rotation component of the Hand-eye calibration $AX = XB$ allows us to treat the orientational component as the linear system to obtain the optimal and robust solution without the identical angle for the IMU and the camera motion. Experimental results show that our method is precise enough and the accuracy of error items is close to the Kalibr when A and B are rigid transformations. But in the presence of noise, our HEC is more applicable than the Kalibr.

However, one of the main limitation is that the translation estimation of HEC is sensitive to the rotation error. In addition, the extrinsic calibration of camera greatly depends on intrinsic parameters and corner detection, as well as the magnitudes of rotation angle is little so that the calibration target remains in the view of cameras, which all limit the accuracy of the estimation.

Furthermore, we plan to integrate the temporal calibration of sensor fusion in this framework and future work will focus on solving the rotational and translational parts of the matrix X at the same time to further reduce errors.

Acknowledgements Project supported by the National Science and Technology Major Project of the Ministry of Science and Technology of China (Grants No.2016YFB0502102, 2016YFB050 2004), the National Natural Science Foundation of China (Grant No.41274038,41574024), the Beijing Natural Science Foundation (Grant No.4162035), and the Aeronautical Science Foundation of China (Grant No.2016ZC51024).

References

1. Tsai YR, Lenz KR. A new technique for fully autonomous and efficient 3D robotics hand/eye calibration. IEEE Trans Robot Autom. 1989;5(3):345–58.
2. Shiu YC, Ahmad S. Calibration of wrist-mounted robotic sensors by solving homogeneous transform equations of the form AX=XB. IEEE Trans Robot Autom. 1989;5(1):16–29.
3. Chou JC, Kamel M. Finding the position and orientation of a sensor on a robot manipulator using quaternions. Int J Robot Res. 1991;10(3):240–54.
4. Liang R, Mao J. Hand-eye calibration with a new linear decomposition algorithm. J Zhejiang Univ-Sci A. 2008;9(10):1363–8.
5. Zhang Z. A flexible new technique for camera calibration. IEEE Trans pattern Anal Mach Intell. 2000;22(11):1330–4.
6. Tedaldi D, Pretto A, Menegatti E. A robust and easy to implement method for IMU calibration without external equipments. In: IEEE international conference on robotics and automation. 2014. p. 3042–9.
7. Shah M. Solving the robot-world/hand-eye calibration problem using the Kronecker product. J Mech Robot. 2013;5(3):031007.
8. Andreff N, Horaudand R, Espiau B. On-line hand-eye calibration. In: Second international conference on 3D digital imaging and modeling. 1999. p. 430–6.
9. Furgale P, Barfoot TD, Sibley G. Continuous-time batch estimation using temporal basis functions. In: IEEE international conference on robotics and automation. 2012. p. 2088–95.

Feature Extraction for Target Spacecraft in the Final Approaching Phase of Rendezvous and Docking

Wenjing Pei and Yingmin Jia

Abstract The rendezvous and docking technology is the key issue to accomplish the spacecraft maintenance in-orbit, space station supply, and astronauts visiting. The technology of the vision measurement for spacecraft directly determines that its results are successful or fail in the final approaching phase of rendezvous and docking. Feature extraction for target spacecraft includes edges, lines, circles, and especially motion information. In this paper, image smoothing based on 2-Dimensional Adaptive Wiener Filtering is introduced, then Canny Edge Detector is used for edge detection, finally using Standard Hough Transform to extract lines and the characteristic circle. Experimental results show that the computation is obviously improved, meanwhile, the precision of detection is also improved. What' s more, lines and circles detection as the fundamental step of extracting motion information, even the vision measurement for spacecraft position and attitude determination, is absolutely necessary.

Keywords 2-D adaptive wiener filtering · Lines and the characteristic circle detection · Standard hough transform · Rendezvous and docking

1 Introduction

In spatial rendezvous and docking, images are obtained by CCD Optical Sensor. However, light is so weak in space that image acquisition is prone to poor exposure. Meanwhile, spatial images are inevitably mixed with different levels of noise and distortion. The major purpose to smooth image is to remove partial noise, while preserving the images' features.

W. Pei · Y. Jia (✉)
The Seventh Research Division and the Center for Information and Control,
School of Automation Science and Electrical Engineering,
Beihang University (BUAA), Beijing 100191, China
e-mail: ymjia@buaa.edu.cn

Y. Jia et al. (eds.), *Proceedings of 2017 Chinese Intelligent Systems Conference*, Lecture Notes in Electrical Engineering 459,
https://doi.org/10.1007/978-981-10-6496-8_69

Numerous smoothing methods have been proposed. The classical methods can be divided into linear filtering and nonlinear filtering. Mean filtering is a common linear filtering, which is usually thought of a convolution operation as the mask is successively moved across the image until every pixel has been covered [1, 2]. Similarly, Gaussian filtering is an another linear filtering, and has been intensively studied in image processing [3, 4]. It is considered as the optimal filter in a sense, but there are still some outstanding problems. Using Gaussian filter for noise suppression, but the signal will be also distorted at the same time. Median filtering is a nonlinear smoothing method to reduce edge blurring, in which the current pixel is replaced by the median value of pixels in neighborhood. Although median filtering can smooth images, it usually need a high overhead, because of the requirement to order the pixel in neighborhood [5, 6]. Wiener Filtering is the optimal estimator in the sense of mean squared error (MSE) for stationary Gaussian process [7, 8].

Edges contain much more important information of an image. Comparing with other approaches, Canny Edge Detector have better performance for edge detection [9, 10]. Most people consider it as the best edge detection algorithm. Standard Hough Tranform is widely used to identify geometry, such as lines, circles and so on [11]. Not only is it relatively accurate, but also having high running speed.

In this paper, we describe an algorithm for detection of lines and the characteristic circle in spacial images. 2-Dimensional Adaptive Wiener Filtering is used to smooth images, since it can apply lower-order rectangular window to calculate higher order, so that it save more time for whole program. According to the local variance, the output values of the filtering are adjusted, thereby most of noise is eliminated. Edge of the target spacecraft can be extracted by Canny Edge Detector. Finally, utilizing Standard Hough Transform extract lines and circles of the target spacecraft, accurately and rapidly. It lays the foundation for further extracting motion information.

The rest of this paper is organized as follows. Section 2 presents specific methodology, including images preprocessing and Standard Hough Transform. Section 3 shows the experimental results and compares to the previous methods. Section 4 concludes this paper and proposes the future work of this research.

2 Methodology

The feature of target aircraft model, as shown in Fig. 1, contain two solar panels (lines) and the front part of the cabin (characteristic circle). Therefore, in order to extract motion information in the final approaching phase of spacecraft rendezvous and docking, we need first to detect the aircraft's edge, lines and the characteristic circle from acquired images by CCD Optical Sensor, accurately and rapidly. Vision measurement system is consist of a camera on the spacecraft and a target marker on the target spacecraft.

Figure 2 shows the overall process of the proposed method for lines and the characteristic circles detection. After acquiring images, the detection process starts with preprocessing. The images are grayed first and smoothed by using 2-D

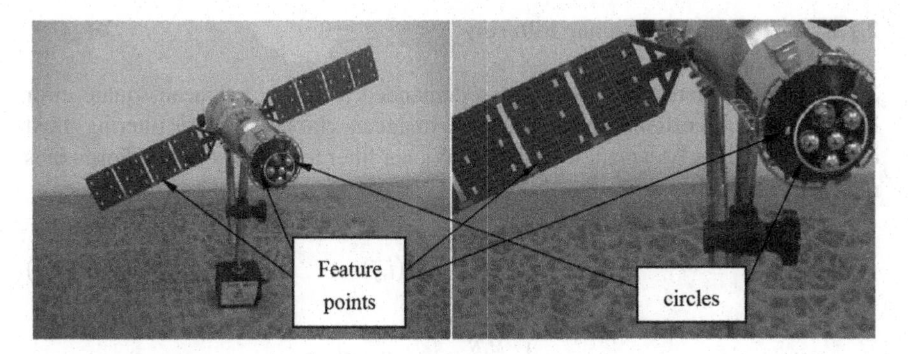

Fig. 1 A diagram of the surface of target spacecraft

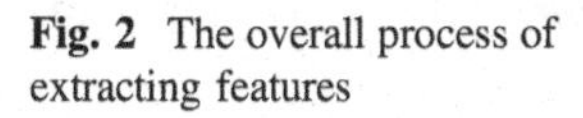

Fig. 2 The overall process of extracting features

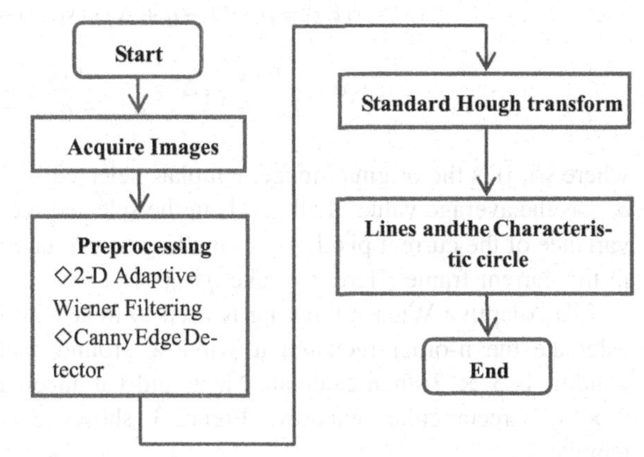

Adaptive Wiener Filtering, since the obtained images mixed with some noise. Then Canny edge detector is applied to extract all possible edges. We set upper and lower threshold to select contours. Standard Hough transform is utilised to filter out non-linear and non-circular shaped edges. Finally, the correct lines and characteristic circle are selected. The detailed procedures are presented in the following subsections.

2.1 *Image Preprocessing*

Image pre-processing is divided into two parts: image smoothing and Canny Edge Detector. Image smoothing is necessary for the acquired image to eliminate noise. 2-D Adaptive Wiener Filtering can smooth the necessary noise points. For Canny Edge Detector, we can select the spacecraft's edge by setting upper and lower threshold.

2.1.1 2-D Adaptive Wiener Filtering

2-D Adaptive Wiener Filtering is also demanded to make the mean square error least between the original image and gray image as classical wiener filtering. First of all, we calculate Mean and variance of Wiener filter template selected. Formula is represented as follows:

$$\mu = \frac{1}{MN} \sum_{i,j}^{M,N} s(i,j) \tag{1}$$

$$\sigma^2 = \frac{1}{MN} \sum_{i,j}^{M,N} s^2(i,j) - \left(\frac{\mu}{MN}\right)^2 \tag{2}$$

$$r(i,j) = \mu + (1 - q + \Delta) \cdot (s(i,j) - \mu) \tag{3}$$

$$q = \frac{\sigma_{avg}}{\sigma_{\mathrm{var}} + 1}, \Delta = \frac{\sigma_{\mathrm{var}}}{\sigma_{avg} + \sigma_{\max} + 1} \tag{4}$$

where s(i, j) is the original image, template selected is M × N, i, j = 0, 1,…, n−1. σ_{avg} is the average value of all pixels in the selected rectangular window. σ_{var} is the variance of the current pixel. σ_{var} is the largest variance in the variance of all pixels in the current frame. Here, we take $q \in [0, 1]$.

2-D Adaptive Wiener Filtering is used to treat with individual pixels point. We calculate the n-order rectangular window around each pixel. Initial rectangular window is 3 × 3, then calculate Mean and variance each pixel in 5 × 5, 7 × 7, 9 × 9… rectangular window. Figure 3 shows 2-D Adaptive Wiener Filter templates.

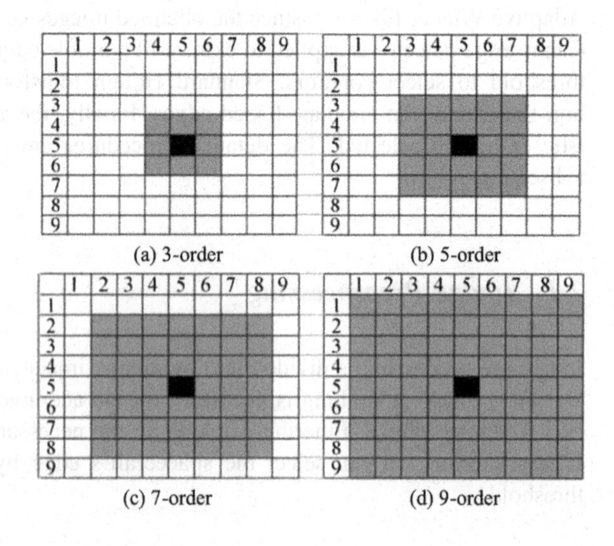

Fig. 3 2-D adaptive wiener filter templates

The detailed procedure is presented as followed.

① Calculate the mean and variance, according to formulas 1, 2, get average variance and mean of each pixel.

② Calculate the average variance of all pixels, the maximum variance of pixels is saved separately.

③ Set threshold T1.

④ By comparing the average variance of each pixel in each-order rectangular window, the minimum variance value is chosen as the final processing window.

⑤ Using formula 3 can calculate the gray value of each pixel, output results.

Here, calculation result of the 3-order rectangular window is applied directly to other orders. For example, apply 3-order rectangular window to 5-order, 5-order to 7-order, and so on. In this way, we can avoid repeating operations, thus saving the computing time.

2.1.2 Canny Edge Detector

We extract all possible edge segments via Canny Edge Detection and setting proper threshold, contained the upper T2 and lower threshold T3, respectively.

2.2 Standard Hough Transform

Here, Standard Hough Transform is applied to detect the lines located in two solar panels of target aircraft and the characteristic circle in the front part of the cabin. It is essentially a voting process where each point belonging to the patterns votes for all the possible patterns passing through that point. Then we accumulate these votes in an accumulator array, the pattern receiving the maximum votes is recognized as the desired pattern.

2.2.1 Detection of Lines

Parametric equations of lines are defined by formula 5,

$$\rho = x\cos\theta + y\sin\theta \tag{5}$$

where (x, y) is the points on the line in x-y coordinates, $\theta(\theta \in [0, \pi])$ denotes the angle the normal line makes with x-axis, and is ρ the normal distance from the origin to the line. As shown in Fig. 4, (ρ, θ) is defined as the parameter space in the pixel space.

Following is the detail procedure of Standard Hough Transform.

① Use images after Canny Edge Detection.

② Calculate the gradient of (x, y) by Sobel Operator.

③ Using obtained gradient Sets up accumulate array H(ρ, θ) which represents the number of curves intersecting at one point in the pixel space, and it is initialized to zero.

④ Calculate the parameter values (θ, ρ) in the pixel space. For a point (x_0, y_0), a cluster of straight lines through this point is defined by formula 6.

$$\rho_\theta = x_0 \cdot \cos\theta + y_0 \cdot \sin\theta \tag{6}$$

⑤ Draw all the lines through the point (x_0, y_0) in polar coordinates, and we will get a sine curve. If the curves intersect, after doing this for all points in the image, it means that they are in the same straight line. At this time, update accumulated array H(ρ, θ) = H(ρ, θ) + 1.

⑥ Set the threshold T4. When H(ρ, θ) is larger than T4, this intersection (θ, ρ_θ) represents a line in the x-y coordinates.

⑦ Draw all of the detected lines in the x-y coordinates.

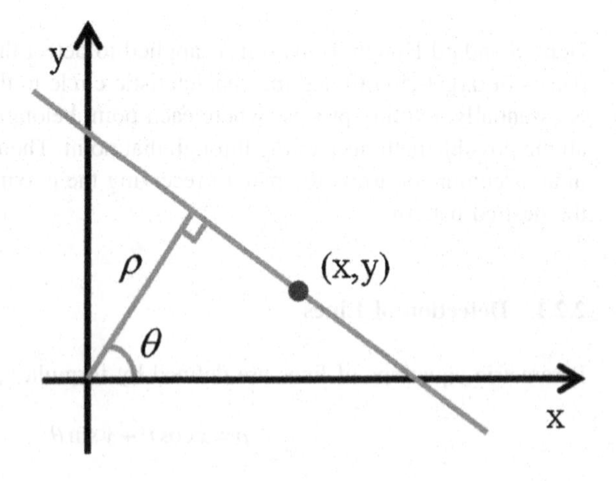

Fig. 4 Relationship (x, y) and (θ, ρ)

2.2.2 Circle Hough Transform

Detection of the characteristic circle is the same as lines detection. It's just increase parameters to 3, which are coordinates of the center and radius of the circle, respectively. Meanwhile, accumulate array is H(r, a, b, θ). Others are similar to detection of lines in the Sect. 2.2.1.

3 Analysis of Experimental Results

In this experiment, we use model of Shenzhou-7 spacecraft as the target spacecraft. Binocular vision system is consist of two CCD cameras and a computer, which is regarded as tracking spacecraft. Due to the limitation of experimental conditions, the ratio of the target spacecraft model to the original size is 1:40. Figure 5 displays the original image.

Figure 5 shows the results of detection of lines by using the Standard Hough transform. We preproccess the original image by 2-D Adaptive Wiener Filtering and Canny Edge Detection (see Fig. 5a). Figure 5b is the results of lines detection of two solar panels.

Figure 6 illustrates the progress of Circle Hough Transform. The characteristic circle is represented by the yellow circle, where red plus is the center of it in Fig. 6c. In addition, accumulation array from Circle Hough transform and 3-d view the accumulation array are in Fig. 6d, e respectively.

Both Tables 1 and 2 show that lines and circles detection based on 2-D Adaptive Wiener Filtering together with Standard Hough Transform can improve the running speed, while the accuracy is almost not affected.

Fig. 5 The original image

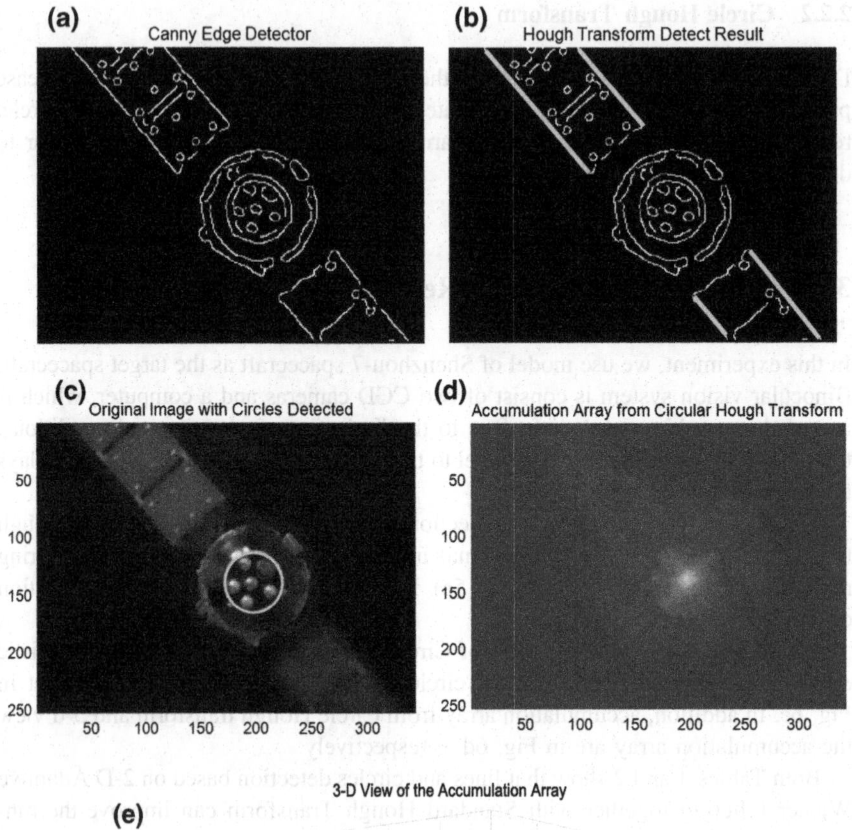

Fig. 6 **a,b** Detection of lines **c, d, e** Circle hough transform

Table 1 Lines detection

Results	No filtering	2-D adaptive wiener filtering
Time(s)	0.423146	0.367304

Table 2 The circle detection

Results	No filtering	2-D adaptive wiener filtering
Time(s)	0.315195	0.268808
Rad	26	26
Center	(196.7,140.3)	(197.1,139.9)

4 Conclusion

In the final approaching phase of rendezvous and docking, the accurate image feature extraction have directly influence the success of spacecrafts' rendezvous and docking. As a first step, it's necessary to detect lines and the characteristic circles. In this paper, smoothing images by using 2-D Adaptive Wiener Filtering, Canny Edge Detector and Standard Hough Transform are applied to detect edge, lines and the characteristic circle. Comparing with the previous algorithms, we improve the running speed without affecting the accuracy. In future work, we will focus on further extracting motion information about the target spacecraft.

Acknowledgements This work was supported by the NSFC (61327807, 61521091, 61520106010, 61134005), and the National Basic Research Program of China (973 Program: 2012CB821200, 2012CB821201).

References

1. Gonzalez RC, Woods RE. Digital image processing, second ed Pearson Education, 2002.
2. Rakshit S. Fast mean filtering technique (FMFT). Pattern Recogn. 2007; 890–7.
3. Bergholm F. Edge focusing. IEEE Trans Putt Anal Mach Intell. 1987;PAMI-9:726–41.
4. Williams DJ, Shah M. Normalized edge detector. In: Proceedings of 1990 IEEE Intelligence 5th Conference on Pattern Recognition. 1990. p. 942–6.
5. Jayant NS. Average and median-based smoothing techniques for improving digital speech quality in the presence of transmission errors. IEEE Trans Commun. 1976;COM-24:1043–5.
6. Shrestha S. Mage denoising using new adaptive based median filter. Signal & image processing: An international journal(SIPIJ). 2014;5(4).
7. Lei F, Iton F, Yatagai T. Adaptive binary transform correlator for image recognition. APPLIDE OPTICS. 2002;41(35):7416–21.
8. KinTak U, Xiaoyu He. A novel image denoising algorithm based on non-uniform rectangular partition and interpolation. In: International conference on multimedia communications. 2010; p. 26–7.
9. Wang W, Wang L. Edge detection of the Canny algorithm based on maximum between-class posterior probability. Comput Appl. 2009;29A: p. 962–1027.
10. Xue L, Li T, Wang Z. Adaptive Canny edge detection algorithm. Comput Appl. 2010;27(9): pp. 3588–90.
11. Ye H, Shang G. A new method based on hough transform for quick line and circle detection. In: International conference on biomedical engineering and Informatics, 2015. p. 52–6.

Research and Application of Extracting Data Sampling Point from Time Series Database

Nana Shen, Xinjian Lu and Dewen Miao

Abstract In order to solve the problem of extracting data points and displaying time curve of industrial process, the linear difference algorithm, median algorithm, maximum algorithm and minimum algorithm are proposed. The result shows that proposed algorithms can solve those problems and meet requirements of real-time data extraction in industry process. As a result, it has played a catalytic role for the application of time series database and the development of industrial process.

Keywords Time series database · Linear difference algorithm · Median algorithm · Maximum algorithm minimum algorithm

1 Introduction

Real-time database technology is a combination procduct of real-time systems and database technology, and it is suitable to deal with rapidly changing data and time-limited transaction processing. Domestic popular real-time databases are Wonderware company's SQL, OSIsoft company's PI, AspenTech company's IP21. Take 5000 points and 20 client as an example, the price of database are: InfoPlus. 21—$110000, each interface is $10000, OPC is free charge, Industrial SQL Server —$65000, each IDAS is $1200, OPC Link free charge. It can be seen that the cost of the real-time database is high, for some small and medium-sized enterprise is unbearable. At present a feasible scheme is to use time series database to instead of real-time database for data collection and storage [1].

Time series database is mainly applied to deal with data which has time tag, namely time data, the data having time tag also is referred to as time series data. As

N. Shen (✉) · X. Lu
Nanjing ChemCyber Technology Company Ltd.,
JiangDong North Road No. 388, Gulou District, Nanjing 210000, China
e-mail: nana.shen@chemcyber.com

D. Miao
School of Software College, NanJing University of Technology, Nanjing 210000, China

Y. Jia et al. (eds.), *Proceedings of 2017 Chinese Intelligent Systems Conference*, Lecture Notes in Electrical Engineering 459,
https://doi.org/10.1007/978-981-10-6496-8_70

for function, time series database and real-time database are basically same. Real-time database has been the basic data platform of enterprise information, it can collect a variety of data from working process and those data can be converted into effective information for all kinds of business. As for real-time database meeting needs of production management, enterprise process monitoring and security information sharing, it plays an important role on improving industrial process, reducing material, increasing production.

The time series database provides fast and efficient industrial information for users [2]. Factory real-time data is stored in real-time database, the factory owner can see and analyze the information. The client application makes it easy for the user to manage the plant level, such as improving industrial process, quality control, and fault prevention maintenance. Through real-time database, product planning, maintenance management, expert systems, laboratory information systems, simulation and optimization applications can be integrated, and it plays a bridge role between business management and real-time production (Fig. 1).

Customer software and reports are two commonly used methods,they are used to display and analysis data which will be extracted from time series database. There are two problems when data being extracted from time series database, Such as some point data does not exist in the time series database or some information of displaying time curve of industrial process are lost. According to needs, some point data or a certain period data of industrial will be displayed in report form, As the

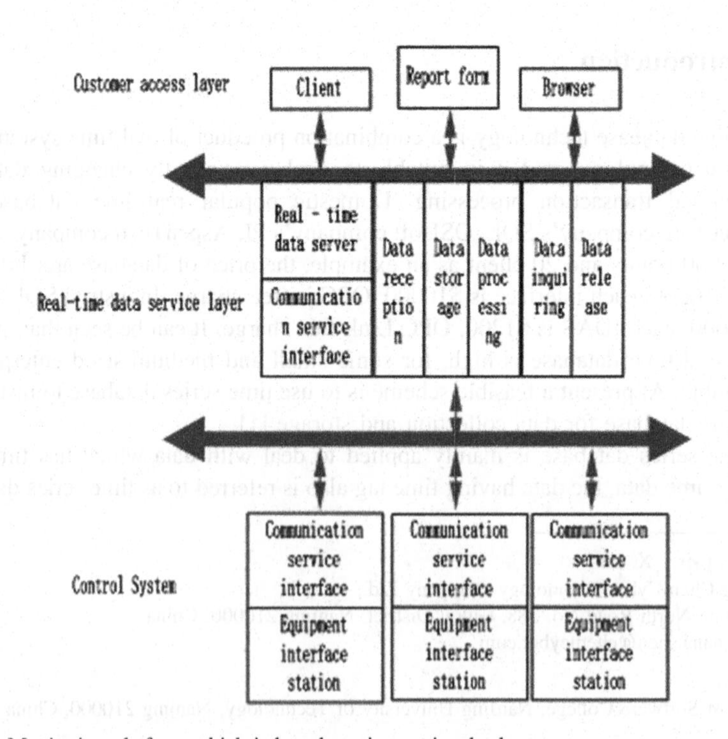

Fig. 1 Monitoring platform which is based on time series database

time series database sampling frequency is a fixed value, some time points data does not exist in the time series database. When user view the warning information curve from customer software, the larger the selected time range is, the more the number of points being described is. Due to the display interface being certain, it is not easy to describe each point clearly which will lead to the curve be distortion. These problems are contrary to the idea which is that time series database acquisition and storage process information can be used to improve the industrial process, reduce the material, increase production and then decision-makers could make more flexible business decision quickly at any time. As a result, those problems are hinder the construction of enterprise information.

Based on application background of the time series database and the problems of data acquisition in the time series database, the linear difference algorithm, the median algorithm and the maximum and minimum algorithm [3, 4] are proposed in this paper. The results show that this method solves the problem of data points and data curve accumulation.

2 Extracting Data Point of Time Series Database Data

2.1 Check the Data of the Specified Time Point

Based on needs of industry applications, customers view industrial data information through the excel report. Time series database has a fixed sampling frequency, sampling frequency is generally 15 s or other cycles. As the sampling frequency, some data can not to be extracted in the time series database, When according to needs to extracting the data. It will produce a null value in excel report.

2.2 Drawing Data Point Curve of the Specified Time Period

For better management, business managers inquires a period of time warning information through the mobile phone, pc and other terminal application. This warning information is presented through curve in the terminal interface. The width of display interface is a fixed value, a curve of a reasonable period time can be described clearly and warning information of device is presented accurately. As the width of display interface is a fixed value, if the time range of view becomes larger and even it becomes into the original several times, the curve will become distorted. In this situation, it is difficult for user to obtain the alarm information intuitively and accurately from the distortion curve. For example, assume that setting the current time as a starting point, drawing liquid level information curves of 4, 8, 24 h, time series database sampling frequency is $p = 15$ s and then there are 460 data points of 4 h, 1920 data points of 8 h, 5760 data points of 24 h. In short, the greater scope of the query period, the more points will be described. In case of the value of front-end display interface width unchanged, the more points are expressed, the deeper degree of stacking is, the greater distortion of curve is.

3 Algorithm of Data Point Acquisition

According to the background of industrial application, the information which is presented in excel report has statistical properties, that is, the user can get basic situation information of one point in time or a certain period time through excel report; and the information of warning curve which is in front-end display interface has personalized properties, the user can get detail and meticulous information of one point in time or a certain period time through viewing the alarm curve, such as the time of occurrence of the alarm and the duration of the alarm information. As the properties of excel report and warning curve of front-end display interface, the data of alarm curve can be blurred, but the data of excel report should to be maintained real-time nature. According to the properties of two different application situation and high sampling frequency of time series database, the following algorithms are used to solve the problem of data acquisition.

3.1 Algorithm of Excel Report Data Describing

3.1.1 Algorithm of Linear Interpolation

Function y = f (x) is used to described many practical problems and a considerable number of functions are obtained through experiments or observations. Although f (x) is present on [a, b], function value of a series of points xi on [a, b] can be got only, this is just a function table. Some functions have analytical expressions, but because of the complexity calculation and inconvenient using, it is also to be made a function table, such the trigonometric function table, logarithmic tables and so on.

Linear interpolation is an interpolation method which is widely used in mathematics, computer graphics and other fields [5–7]. In order to study the changing rule of function, it is necessary to find the function value which is not on the table. Thus, a function p(x) can be made according to the given function table, p(x) can reflect the properties of the function f(x) and is easy to calculate. In a way, p(x) is stands of f(x). Usually selecting a simple function as p(x) and $p(x_i) = f(x_i)$ is true when i = 1, 2,…, n, and then p(x) is the interpolation function we want, this is the linear interpolation method (Fig. 2).

The required time points will be converted into seconds, that is, xxx seconds. If the required time point is not found in time series database, linear interpolation method will be used to calculate two time points which are on both sides of the required time point: Assuming coordinates of the two time points are: (x_0, y_0),

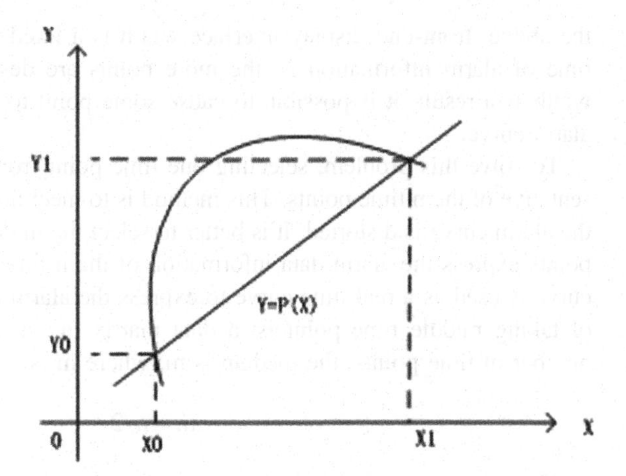

Fig. 2 Diagram of linear interpolation function

(x_1, y_1) the coordinate of the required time point is (x_2, y_2), Where x_2 is known, y_2 is unknown, then:

$$p(x) = \frac{x_0 - x_1}{y_0 - y_1}(x)$$

$$y_2 = \frac{x_0 - x_1}{y_0 - y_1}(x_2)$$

y_2 is the value of the required time point.

3.1.2 Mean Algorithm

Based on linear interpolation algorithm, y_2 is calculate according to on both sides of it, as a result the time range is limited. As describing above, information of excel report reflects the generality of production data. Due to the high sampling frequency of the time series database, y_2 is not enough to describe generality of data. In order to fully express the generality, taking the average value of a certain time range as the value of the required time point. Namely the time range should to be extended, the way to extend time range is that taking required time point as the center and extending its left and right range. Such as taking the time point as the midpoint, and its left and right range were extended to the 3 min or any other time. Namely the required time point value is the average of the time period.

3.2 Algorithm of Drawing Alarm Curve

Through the alarm curve, user can get the information of alarm history. Data of curve should have real-time nature properties which can not to be processed. From the above, front-end display interface width is a fixed value, the bigger inquiring time of alarm information is, the more points are described in display interface width.as a result, it is possible to cause some point to be lost or distortion of the alarm curve.

To solve this problem, selecting one time point from n time points as a representative of the n time points. This method is to meet the demand requirements and the alarm curve is distorted. It is better to select the middle time point of the n time points express the alarm data information of the n time points, that is, the median curve is used as a real-time curve to express the alarm information. The algorithm of taking middle time point is: a data matrix [a_0, a_1,..., a_n] is made up with a number of time points, the median is m, where m is:

$$m = n/2$$

The rule of five into will be used when n can not be divisible by two. For example, when the time range is 24 h, there are 5760 points being obtained from the time series database, time series database sampling frequency is p = 15 s, time of six points is 90 s, take the midpoint value as the level value of this time period. The data matrix which is made up with by this 6 time points is: [a_0, a_1, a_2, a_3, a_4, a_5], and then the middle of the subscript is 5/2, about 3, that is a_3;

To improve this algorithm further, Hawkeye function is added to this algorithm. Hawkeye function is that selecting some part of real-time curve and the selected part will be enlarged,and then the remaining n−1 points are displayed.it is better to for user to observed and understand alarm information in detail.

In order to enrich the alarm information, besides the above median algorithm, the maximum and minimum algorithm can be used to describe maximum curve and minimum curve [8]. That is, the upper and lower ends of the real-time curve are added with the maximum curve and minimum curve, the algorithm is as follows:

Select the minimum point from n points, to form a data matrix [a_0, a_1,..., a_n] is made up with n points,supposing a_0 is the minimum value and comparing a_0 with the remaining n−1 point, the calculation is as follows:

```
mixDevation = 0.0;
min = a0;
for(int i = 1; i < n + 1; i + +){
double temp = a(i);
if (min > temp){
min = temp;
}
}
```

mixDevation = min;

By comparing, mixDevation is the minimum value of the n points, maximum value can also be obtained through this method [9]. Taking 114 points from time series database whose sampling frequency is 15 s, each group is constituted by six

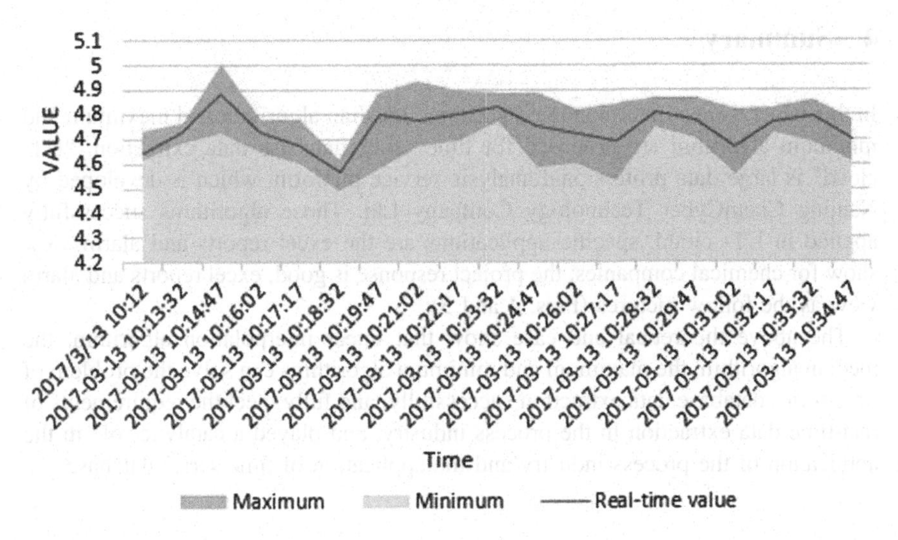

Fig. 3 Alarm curves graph

points and then there are 19 groups. And through the algorithm of drawing alarm curve, the final complete alarm curves are as shown (Fig. 3):

Data table		
Serial number	Time	Value(UT_LI2001.PV)
1	2017-3-13 10:12:17	4.636838913
2	2017-03-13 10:12:32	4.73251152
3	2017-03-13 10:12:47	4.629489422
4	2017-03-13 10:13:02	4.620019436
5	2017-03-13 10:13:17	4.616096973
6	2017-03-13 10:13:32	4.731715679
…	…	…
109	2017-03-13 10:39:17	4.698540211
110	2017-03-13 10:39:32	4.710143089
111	2017-03-13 10:39:47	4.728112221
112	2017-03-13 10:40:02	4.610611439
113	2017-03-13 10:40:17	4.650400162
114	2017-03-13 10:40:32	4.691971779

4 Summary

In this paper, Linear interpolation algorithm, amedian algorithm and maximum and minimum algorithm are proposed for time series database data extraction. "UT-cloud" is large data professional analysis service platform, which is developed by Nanjing ChemCyber Technology Company Ltd. Those algorithms successfully applied in UT- cloud, specific applications are the excel reports and alarm view show for chemical companies, the project response is good, excel reports and alarm view as the follow pictures (Figs. 4 and 5):

The above theoretical and case show that linear interpolation algorithm, the median algorithm, the maximum and minimum algorithms can solve the problem of time series database data extraction successfully, and fully meet the requirements of real-time data extraction in the process industry, and played a catalytic role in the application of the process industry and the application of time series database.

	A	B	C	D	E	F	G	H	I	J	K
1	Serial number	Department	Unit	Name	Type	Instructions	Enable	Instrument item	09/09 17:00	09/09 16:00	08/31 08:00
2	A-001	Logistics department	Oil working	G-901A	Liquid level		√	A.WL.MT.LT901A	3.84	3.62	2.96
3	A-002	Logistics department	Oil working	G-901B	Liquid level		√	A.WL.MT.LT901B	10.78	10.67	9.86
4	A-003	Logistics department	Oil working	G-901C	Liquid level		√	A.WL.MT.LT901C	13.41	13.87	13.44
5	A-004	Logistics department	Oil working	G-901D	Liquid level		√	A.WL.MT.LT901D	0.68	0.66	0.57
6	A-005	Logistics department	Dock working	G-1110A	Liquid level		√	A.WL.MT.LT-1110A	1.02	1.22	0.98
7	A-006	Logistics department	Dock working	G-1110B	Liquid level		√	A.WL.MT.LT-1110B	1489.01	1475.23	1487.89
8	A-007	Logistics department	Dock working	G-1110C	Liquid level		√	A.WL.MT.LT-1110C	314.00	315.03	315.76
9	A-008	Logistics department	Dock working	G-1110D	Liquid level		√	A.WL.MT.LT_1110D	10.78	10.54	9.78
10	A-009	Logistics department	Dock working	G-1110E	Liquid level		√	A.WL.MT.LT_1110E	121.89	123.00	123.98
11	A-010	Logistics department	Oil working	G-1110E	Instantaneous flow rate		√	A.WL.MT.FT-1021	77.98	76.90	77.65
12	A-011	Logistics department	Oil working	G-901B	Instantaneous flow rate		√	A.WL.MT.FT-1025	23.00	21.98	22.78

Fig. 4 Excel report

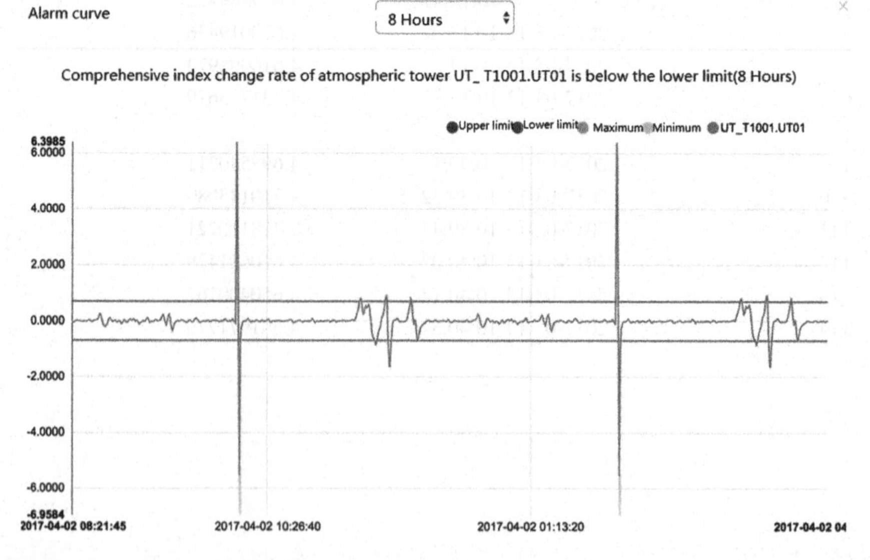

Fig. 5 Alarm graph of Nanjing ChemCyber Technology Company Ltd

References

1. Hamilton JD. Time series analysis. Princeton: Princeton University Press; 1994.
2. Keogh E, Lonardi S, Chiu BY. Finding surprising patterns in a time series database in linear time and space. In: Proceedings of the 8th ACM SIGKDD international conference on Knowledge discovery and data mining. ACM, 2002. p. 550–6.
3. Wang X, Mueen A, Ding H, et al. Experimental comparison of representation methods and distance measures for time series data. Data Min Knowl Disc. 2013; 1–35.
4. Tsai HH, Tseng HC, Lai YS. Robust lossless image watermarking based on α-trimmed mean algorithm and support vector machine. J Syst Softw. 2010;83(6):1015–28.
5. Malvar HS, He L, Cutler R. High-quality linear interpolation for demosaicing of Bayer-patterned color images. Acoustics, speech, and signal processing. In: Proceedings. (ICASSP'04). IEEE international conference on. IEEE. 2004; 3: iii–485.
6. Blu T, Thévenaz P, Unser M. Linear interpolation revitalized. IEEE Trans Image Process. 2004;13(5):710–9.
7. Rippa S. Long and thin triangles can be good for linear interpolation. SIAM J Numer Anal. 1992;29(1):257–70.
8. .Remenyi D, White T. Sherwood-Smith M. Achieving maximum value from information systems: a process approach[M]. Wiley, Inc., 1997.
9. Pinçe Ç, Ferguson M, Toktay B. Extracting maximum value from consumer returns: allocating between remarketing and refurbishing for warranty claims. Manuf Serv Oper Manag. 2016; 18(4):475–92.

Printed in the United States
By Bookmasters